Oxygen and Oxy-Radicals in Chemistry and Biology

Academic Press Rapid Manuscript Reproduction

Proceedings of the International Conference on Oxygen and Oxy-radicals held at the University of Texas at Austin, in May 1980.

OXYGEN AND OXY-RADICALS IN CHEMISTRY AND BIOLOGY

Edited by

M. A. J. Rodgers

Center for Fast Kinetics Research
University of Texas at Austin
Austin, Texas

E. L. Powers

Laboratory of Radiation Biology
Center for Fast Kinetics Research
University of Texas at Austin
Austin, Texas

ACADEMIC PRESS 1981

A Subsidiary of Harcourt Brace Jovanovich, Publishers

New York London Toronto Sydney San Francisco

ACADEMIC PRESS, INC.
111 Fifth Avenue, New York, New York 10003

United Kingdom Edition published by
ACADEMIC PRESS, INC. (LONDON) LTD.
24/28 Oval Road, London NW1 7DX

Library of Congress Cataloging in Publication Data
Main entry under title:

Oxygen and oxy-radicals in chemistry and biology.

Proceedings of the Conference on Oxygen and Oxy-Radicals in Chemistry and Biology, held May 25-29, 1980, at the University of Texas at Austin, and sponsored by the Center for Fast Kinetics Research of the University.
Includes index.
1. Oxygen--Physiological effect--Congresses. 2. Oxidation, Physiological--Congresses. 3. Oxidation --Congresses. 4. Radicals (Chemistry)--Congresses. I. Rodgers, M. A. J. II. Powers, E. L. (Edward Lawrence), Date. III. Conference on Oxygen and Oxy-Radicals in Chemistry and Biology (1980 : University of Texas at Austin) IV. University of Texas at Austin. Center for Fast Kinetics Research.

QP535.Q1Q934   574.19'214   81-10846
ISBN 0-12-592050-4   AACR2

PRINTED IN THE UNITED STATES OF AMERICA

81 82 83 84   9 8 7 6 5 4 3 2 1

CONTENTS

ROLE AND TOXICITY OF SUPEROXIDE IN CELLULAR SYSTEMS

G. Czapski, Organizer, D. C. Borg, Chairman

O_2 AND CELLULAR RADIATION SENSITIVITY

E. L. Powers, Chairman

OXYGEN-LIKE RADIATION SENSITIZERS: MECHANISMS OF ACTION

Robin Willson, Chairman

EXCITED STATE CHEMISTRY OF O_2

Christopher S. Foote, Chairman

SENSITIZED PHOTOOXIDATION OF BIOMOLECULES
J. D. Spikes, Chairman

REVERSIBLE O_2 BINDING
A. Riggs, Chairman

OXYGENASES
D. T. Gibson, Chairman

GAS PHASE OXIDATION AND ENVIRONMENTAL IMPLICATIONS

H. Niki, Chairman

POSTER SESSION

M. A. J. Rodgers, Chairman

Participants in the Conference

Numbers in parentheses indicate the pages on which the authors' papers begin.

A. H. Ahmed **(593)**, School of Pharmacy and Pharmacology, University of Bath, Avon, United Kingdom

K. A. Almqvist **(397)**, Department of Organic Chemistry, University of Geneva, Switzerland

Robert F. Anderson **(597)**, CRC Gray Laboratory, Mount Vernon Hospital, Northwood, Middlesex, HA6 2RN United Kingdom

Rudolf Anker **(61)**, Institut für Strahlenchemie im Max Planck Institut für Kohlenforschung, Mülheim, Federal Republic of Germany

Abraham Antebi **(636)**, Department of Chemistry, Bar-llan University, Ramat-Gan, Israel

K.-E. Arfors **(615, 757)**, Department of Experimental Medicine, Uppsala, Sweden

E. Arro **(757)**, Experimental Medicine, University of Uppsala, Uppsala, Sweden

K.-D. Asmus **(69,666)**, Bereich Strahlenchemie, Hahn-Meitner Institut Berlin, Glienicker Str. 100, 1000 Berlin 39, Federal Republic of Germany

Roger Atkinson **(559)**, Environmental Research and Technology, Inc., 2625 Townsgate Road, Suite 360, Westlake Village, California 91316

Steven D. Aust **(147)**, Department of Biochemistry, Michigan State University, East Lansing, Michigan

J. E. Ayling **(601)**, University of Texas at San Antonio, San Antonio, Texas 78282

D. Bahnemann **(69)**, Berich Strahlenchemie, Hahn-Meitner Institut Berlin, Berlin, Federal Republic of Germany

S. W. Bailey **(601)**, University of Texas at San Antonio, San Antonio, Texas

Alan C. Baldwin **(543)**, Department of Chemical Kinetics, SRI International, 333 Ravenswood Avenue, Menlo Park, California 94025

A. Balland **(610)**, Laboratoire de Radiobiochimie, Centre d'Etudes Nucléaires de Grenoble, Grenoble, France

G. Barchietto **(397)**, Department of Organic Chemistry, University of Geneva, Geneva, Switzerland

A. J. Bard University of Texas at Austin, Austin, Texas

G. M. Bartoli **(641)**, Institute of General Pathology, Catholic University of Rome, Rome, Italy

S. Bartoli **(641)**, Institute of General Pathology, Catholic University of Rome, Rome, Italy

Mary-Jane Bell **(461)**, Medical Biophysics Branch, Whiteshell Nuclear Research Establishment, Atomic Energy of Canada Limited Research Company, Pinawa, Manitoba, Canada

Jesse H. Bennett **(604)**, Agricultural Research Center, SEA, USDA, Beltsville, Maryland 20705

John E. Biaglow **(349, 766)**, Division of Radiation Biology, Case Western Reserve University, 2064 Abington Road, Cleveland, Ohio 44106

E. A. Blakely **(716)**, Division of Biology and Medicine, Lawrence Berkeley Laboratory, Berkeley, California

Joseph Bonaventura, Duke University Marine Laboratory, Beaufort, North Carolina

M. Bonifačić **(69)**, Hahn-Meitner Institut Berlin, Berlin, Federal Republic of Germany

Donald C. Borg **(177, 197, 734)**, Medical Research Center, Brookhaven National Laboratory, Upton, New York 11973

Wolf Bors **(75)**, Institut für Biologie, Abteilung für Strahlenbiologie 8042, München, Federal Republic of Germany

Eberhard Bothe **(61)**, Institut für Strahlenchemie im Max-Planck-Institut für Kohlenforschung, Mülheim, Federal Republic of Germany

John C. Brown **(604)**, Agricultural Research Center, SEA, USDA, Beltsville, Maryland

U. T. Brunk **(757)**, University of Uppsala, Uppsala, Sweden

Garry R. Buettner **(606, 679)**, Department of Chemistry, Wabash College, Crawfordsville, Indiana 47933

E. P. Burrows **(475)**, Division of Environmental Chemistry, The Johns Hopkins University, Baltimore, Maryland

J. Cadet **(610)**, Laboratoire de Radiobiochimie, Centre d'Etudes Nucléaires de Grenoble, Grenoble, France

S. Cannistraro **(730)**, Istituto di Fisica Università di Perugia, Perugia, Italy

Josephine A. Carew **(628)**, Department of Biology, Tufts University, Medford, Massachusetts

Mary P. Carpenter **(743)**, Oklahoma Medical Research Foundation, Oklahoma City, Oklahoma

Winslow S. Caughey **(787)**, Department of Biochemistry, Colorado State University, Fort Collins, Colorado

Arthur I. Cederbaum (**612**), Department of Biochemistry, Mount Sinai School of Medicine, New York, New York 10029

A. Chatterjee (**716**), Division of Biology and Medicine, Lawrence Berkeley Laboratory, Berkeley, California

Michael A. Chirigos (**622**), National Cancer Institute, Bethesda, Maryland

C. N. Christy (**746**), Department of Chemistry, University of South Florida, Tampa, Florida

L. Ciborowski (**649**), Department of Therapeutic Radiology, Tufts-New England Medical Center, Boston, Massachusetts

Gerald Cohen (**45, 612**), Department of Neurology, Mount Sinai School of Medicine, New York, New York 10029

Robert B. Cundall (**739**), Department of Biochemistry, University of Salford, Salford, United Kingdom

G. Czapski , Department of Physical Chemistry, The Hebrew University, Jerusalem, Israel

D. J. G. Davies (**593**), School of Pharmacy and Pharmacology, University of Bath, Bath, Avon, United Kingdom

R. F. Del Maestro (**615, 757**), Department of Neurosurgery, University of Western Ontario, London, Ontario, Canada

Miral Dizdaroglu (**619**), Chemistry Department, University of Maryland, Baltimore, Maryland 21228

Carroll R. Dobbs (**622**), Biochemistry Department, Armed Forces Radiobiology, Research Institute, Bethesda, Maryland 20014

N. J. F. Dodd (**161**), Paterson Laboratories, Christie Hospital, Manchester, England

J. Draper (**663**), University of Kentucky, Lexington, Kentucky

H. Brian Dunford (**625**), Department of Chemistry, University of Alberta, Edmonton, Alberta, Canada T6G 2G2

John J. Elmore, Jr. (**177**), Medical Research Center, Brookhaven National Laboratory, Upton, New York

Edward R. Epp (**335**), Department of Radiation Medicine, Harvard Medical School, Boston, Massachusetts

M. Jiménez Estrada (**397**), Department of Organic Chemistry, University of Geneva, Geneva, Switzerland

David Ewing (**269**), Department of Radiation Therapy and Nuclear Medicine, Hahnemann Medical College and Hospital, 230 N. Broad Street, Philadelphia, Pennsylvania 19102

James A. Fee (**205**), Department of Biological Chemistry, The University of Michigan, Ann Arbor, Michigan 48109

Ross S. Feldberg (**628**), Department of Biology, Tufts University, Medford, Massachusetts 02155

E. Martin Fielden (**301, 329**), Institute of Cancer Research, Sutton, Surrey, United Kingdom

L. J. Fischer (**631**), Department of Pharmacology, University of Iowa, Iowa City, Iowa 52242

Robert A. Floyd (**743**), Oklahoma Medical Research Foundation, Oklahoma City, Oklahoma

Christopher S. Foote (**425**), Department of Chemistry, University of California, Los Angeles, California 90024

J. Lawrence Fox (**771**), Department of Zoology, University of Texas at Austin, Austin, Texas

Hans Frauenfelder, Department of Physics, University of Illinois, Urbana-Champaign, Illinois

Albert W. Frenkel (**634**), Department of Botany, University of Minnesota, St. Paul, Minnesota 55108

Irwin Fridovich (**197**), Department of Biochemistry, Duke University Medical Center, Durham, North Carolina 27710

Aryeh A. Frimer (**636, 639, 676**), Department of Chemistry, Bar-llan University, Ramat-Gan, Israel

G. Gabor (**55**), Israel Institute for Biological Research, P.O. Box 19, Ness-Ziona, Israel

T. Galeotti (**641**), Institute of General Pathology, Catholic University of Rome, Via Pineta Sacchetti 644, Roma, Italy

J. E. Garst (**507**), Department of Animal Science, University of Illinois, Urbana, Illinois 61801

D. T. Gibson (**495**), Department of Microbiology, University of Texas at Austin, Austin, Texas 78712

Pessiah Gilinsky (**639**), Department of Chemistry, Bar-llan University, Ramat-Gan, Israel

J. L. Glenn (**747**), Venom Research Laboratory, VA Medical Center, Salt Lake City, Utah

M. Göbl (**666**), Bereich Strahlenchemie, Hahn-Meitner-Institut für Kernforschung, Berlin, Germany

W. A. Goddard, Chemistry Department, California Institute of Technology, Pasadena, California

L. Golder (**753**), Université Louis Pasteur, Strasbourg, France

Klaus Gollnick (**379**), Institut für Organische Chemie der Universität München, Karlstrasse 23, D-8000, München, Germany

Nigel T. Goodchild (**645, 649**), Department of Therapeutic Radiology, Tufts-New England Medical Center, Boston, Massachusetts 02111

H. Graf (**675**), Institute of Physiological Chemistry, The University of The Saarland, Homburg, Federal Republic of Germany

Brian H. Gray (**652**), Experimental Hematology Department, Armed Forces Radiobiology Research Institute, Bethesda, Maryland 20014

Clive L. Greenstock (**119**), Atomic Energy of Canada Limited, Whiteshell Nuclear Research Establishment, Pinawa, Manitoba, R0E 1L0 Canada

Brenda Walker Griffin (**141**), Biochemistry Department, The University of Texas Health Science Center, Dallas, Texas 75235

S. Grossman **(676)**, Life Science Department, Bar-llan University, Ramat-Gan, Israel

L. I. Grossweiner **(479)**, Illinois Institute of Technology, Chicago, Illinois 60616

B. S. Hahn **(610)**, Division of Environmental Chemistry, The Johns Hopkins University, Baltimore, Maryland

Michael L. Hair **(655)**, Xerox Research Centre of Canada, Mississagua, Ontario, Canada

M. Halmann **(55)**, Israel Institute for Biological Research, Ness-Ziona, Israel

John R. Harbour **(655)**, Xerox Research Centre of Canada, Mississagua, Ontario, L5L 1J9 Canada

C. R. Hardy **(657)**, Institute of Cancer Research, Department of Physics, Sutton, Surrey, United Kingdom

H. A. Harrop **(285, 660)**, Cancer Research Campaign Gray Laboratory, Mount Vernon Hospital, Northwood, Middlesex, England

Herta Hartmann **(379)**, Institut für Organische Chemie der Universität München, München, Germany

K. Hefter **(683)**, Tufts-New England Medical Center, Department of Therapeutic Radiology, Boston, Massachusetts

Howard E. Heggestad **(604)**, Agricultural Research Center, SEA, USDA, Beltsville, Maryland

K. D. Held **(285, 660)**, Cancer Research Campaign Gray Laboratory, Mount Vernon Hospital, Northwood, Middlesex, HA6 2RN England

C. L. Hicks **(663)**, University of Kentucky, Lexington, Kentucky 40506

K. O. Hiller **(666)**, Bereich Strahlenchemie, Hahn-Meitner-Institut für Kernforschung, Berlin, D-1000 Germany

Morton Z. Hoffman **(769)**, Department of Chemistry, Boston University, Boston, Massachusetts

Carleton J. Howard **(535)**, Aeronomy Laboratory, National Oceanic and Atmospheric Administration, U. S. Department of Commerce, Boulder, Colorado 80303

Rolf D. Issels **(667)**, Ludwig-Maximilians-University of Munich, Klinikum Grosshadern, Medizinische Klinik III, Marchion Instrasse 15, 8000 München 70, West Germany

Sandra L. Issler **(655)**, Xerox Research Centre of Canada, Mississagua, Ontario, Canada

Leland S. Jahnke **(634)**, Department of Botany and Plant Pathology, University of New Hampshire, Durham, New Hampshire

C. W. Jefford **(397)**, Department of Organic Chemistry, University of Geneva, 1211 Geneva 4, Switzerland

T. C. Jenkins **(329)**, Radiobiology Unit, Physics Department, Institute for Cancer Research, Sutton, Surrey, United Kingdom

Giulio Jori **(441, 730)**, Istituto di Biologia Animale, Università di Padova, 35100 Padova, Italy

M. Kartha (**725**), Department of Radiotherapy, The Ohio State University, Columbus, Ohio

J. Kiefer (**293**), Strahlenzentrum der Justus-Liebig-Universität, Giessen, Germany

Ganesh M. Kishore (**521**), Department of Chemistry, The University of Texas at Austin, Austin, Texas 78712

W. Knöpfel (**397**), Department of Organic Chemistry, University of Geneva, Geneva, Switzerland

S. Kohmoto (**397**), Department of Organic Chemistry, University of Geneva, Geneva, Switzerland

W. H. Koppenol (**671**), Department of Biochemistry and Molecular Biology, Northwestern University, Evanston, Illinois 60201

Grant W. Koroll (**461, 736, 739**), Whiteshell Nuclear Research Establishment, Atomic Energy of Canada Limited Research Company, Pinawa, Manitoba, Canada

Walter Kremers (**461**), Whiteshell Nuclear Research Establishment, Atomic Energy of Canada Limited Research Company, Pinawa, Manitoba, Canada

Norman I. Krinsky (**153**), Department of Biochemistry and Pharmacology, Tufts University School of Medicine, Boston, Massachusetts 02111

K. Sree Kumar (**622**), Biochemistry Department, Armed Forces Radiobiology Research Institute, Bethesda, Maryland

W. E. Kurtin (**791**), Chemistry Department, Trinity University, San Antonio, Texas

Hartmut Kuthan (**497, 675**), Department of Physiological Chemistry, Homburg-Saar, Federal Republic of Germany

Lester Kwock (**645, 649, 683**), Department of Therapeutic Radiology, Tufts-New England Medical Center, Boston, Massachusetts

C. Larroque (**764**), INSERM, Montpellier, France

Edward H. Lee (**604**), Agricultural Research Center, SEA, USDA, Beltsville, Maryland

Y. Y. Leshem (**676**), Life Science Department, Bar-llan University, Ramat-Gan, Israel

Susan W. C. Leuthauser (**679**), Radiation Research Laboratory, University of Iowa, Iowa City, Iowa 52242

Peter H. Levine (**153**), The Memorial Hospital and the Department of Medicine, The University of Massachusetts School of Medicine, Worcester, Massachusetts

Y. Liftman (**676**), Department of Chemistry, Bar-llan University, Ramat-Gan, Israel

Peck-Sun Lin (**645, 649, 683**), Department of Therapeutic Radiology, Tufts-New England Medical Center, Boston, Massachusetts 02111

B. A. Lindig (**687**), Center for Fast Kinetics Research, The University of Texas at Austin, Austin, Texas 78712

*C. Clifton Ling** (**335**), Massachusetts General Hospital[1], Harvard Medical School, Boston, Massachusetts

Alan C. Lloyd (**559**), Environmental Research and Technology, Inc., Westlake Village, California

Dean P. Loven (**690**), Radiation Research Laboratory, University of Iowa, Iowa City, Iowa 52242

J. Luggen-Hölscher (**293**), Strahlenzentrum der Justus-Liebig-Universität, Giessen, Germany

H. Ma (**657**), Department of Physics, Institute of Cancer Research, Sutton, Surrey, United Kingdom

Stefan Marklund (**693**), Department of Clinical Chemistry, University Hospital of Umeå, Umeå, Sweden

Jerzy E. Maruchin (**698**), Medical Center of Postgraduate Education, 01-813 Warszawa, Poland

B. Masloch (**666**), Bereich Strahlenchemie, Hahn-Meitner-Institut für Kernforschung, Berlin, Germany

L. Masotti (**641**), Institute of Biological Chemistry, University of Parma, Parma, Italy

B. D. Michael (**285, 660**), Cancer Research Campaign Gray Laboratory, Mount Vernon Hospital, Northwood, Middlesex, HA6 2RN England

Howard B. Michaels (**335**), Department of Radiation Medicine, Massachusetts General Hospital, Harvard Medical School, Boston, Massachusetts 02114

Christa Michel (**75**), Institut für Biologie, Gesellschaft für Strahlen-und Umweltforschung, München, Federal Republic of Germany

Theodore Mill (**779**), SRI International, 333 Ravenswood Avenue, Menlo Park, California 94025

Barbara C. Millar (**301, 329**), Physics Division, Institute of Cancer Research, Sutton, Surrey, United Kingdom

Hara P. Misra (**700**), Laboratory of Energy-Related Health Research, University of California, Davis, California 95616

G. Misslbeck (**683**), Department of Therapeutic Radiology, Tufts-New England Medical Center, Boston, Massachusetts

R. E. J. Mitchel (**703**), Radiation Biology Branch, Atomic Energy of Canada Limited, Chalk River, Ontario, K0J 1J0 Canada

Keith Moffat (**489**), Section of Biochemistry, Molecular and Cell Biology, Cornell University, Ithaca, New York 14853

A. Momzikoff (**731**), Institut Océanographique, Paris, France

D. P. Morrison (**703**), Radiation Biology Branch, Atomic Energy of Canada Limited, Chalk River, Ontario, Canada

Mary Treinen Moslen (**169**), Department of Pathology, The University of Texas Medical Branch, Galveston, Texas

*Present address: George Washington University Medical Center, Washington, D. C.

S. H. Moss **(593)**, Department of Radiology, Stanford University Medical School, Stanford, California

Alexander D. Nadezhdin **(625)**, Department of Chemistry, University of Alberta, Edmonton, Alberta, Canada

T. Nagamatsu **(475)**, Division of Environmental Chemistry, The Johns Hopkins University, C. O. E. H. - U. S. P. H. S. Hospital, Baltimore, Maryland

Edward J. Nanni, Jr. **(15)**, Department of Chemistry, University of California, Riverside, California

G. Neri **(641)**, Institute of Human Genetics, Catholic University of Rome, Rome, Italy

P. Neta **(83)**, Radiation Laboratory, University of Notre Dame, Notre Dame, Indiana 46556

H. Niki , Research Staff, Ford Motor Company, Dearborn, Michigan

Larry W. Oberley **(606, 679, 690)**, Radiation Research Laboratory, University of Iowa, Iowa City, Iowa

Terry D. Oberley **(679)**, Department of Pathology, University of Wisconsin, Madison, Wisconsin

Peggy L. Olive **(766)**, Radiobiology Section, Johns Hopkins Oncology Center, Baltimore, Maryland

Ralph A. Olsen **(604)**, Agricultural Research Center, SEA, USDA, Beltsville, Maryland

P. O'Neill **(657)**, Institute of Cancer Research, Department of Physics, Sutton, Surrey, United Kingdom

J. A. Ors **(746)**, IBM Watson Research Center, Yorktown Heights, New York

L. K. Patterson **(89)**, Radiation Laboratory and the Department of Chemistry, University of Notre Dame, Notre Dame, Indiana 46556

Hanns Paur **(379)**, Institut für Organische Chemie der Universität München, München, Germany

G. Peters **(705)**, Center for Fast Kinetics Research, The University of Texas at Austin, Austin, Texas 78712

J. A. Peterson, Department of Biochemistry, University of Texas, Health Science Center, Dallas, Texas

J. Petryka **(634)**, University of Minnesota, Medical Research Unit, Abbott-Northwestern Hospital, Minneapolis, Minnesota

M. Planker **(615)**, Department of Medicine A, University of Dusseldorf, West Germany

E. L. Powers **(241)**, Laboratory of Radiation Biology, Center for Fast Kinetics Research, The University of Texas at Austin, Austin, Texas 78712

William A. Pryor **(133, 761)**, Department of Chemistry, Louisiana State University, Baton Rouge, Louisiana 70803

J. A. Raleigh **(355)**, Department of Radiation Oncology, Cross Cancer Institute, 11560 University Avenue, Edmonton, Alberta, Canada

Elmer J. Rauckman (**97**), Department of Surgery, Duke University Medical Center, Durham, North Carolina

Edward S. Reynolds (**169**), Department of Pathology, The University of Texas Medical Branch, Galveston, Texas 77550

R. C. Richmond (**708**), Norris Cotton Cancer Center, Dartmouth-Hitchcock Medical Center, Hanover, New Hampshire 03755

Helen W. Richter (**711**), Department of Chemistry, Carnegie-Mellon University, Pittsburgh, Pennsylvania 15213

Austen Riggs , Department of Zoology, The University of Texas at Austin, Austin, Texas

W. Roberts (**725**), Department of Radiotherapy, The Ohio State University, Columbus, Ohio

M. A. J. Rodgers (**687, 705, 714**), Center for Fast Kinetics Research, The University of Texas at Austin, Austin, Texas 78712

R. Roots (**716**), Division of Biology and Medicine, Lawrence Berkeley Laboratory, Berkeley, California 94720

Gerald M. Rosen (**97**), Department of Pharmacology, Duke University Medical Center, Durham, North Carolina 27710

H. S. Ryang (**475**), Division of Environmental Chemistry, The Johns Hopkins University, C. O. E .H. - U. S. P. H. S. Hospital, Baltimore, Maryland

Arthur L. Sagone, Jr. (**719, 725, 742**), Department of Medicine, The Ohio State University, Columbus, Ohio 43210

C. Salet (**731**), Laboratoire de Biophysique, Muséum National d'Histoire Naturelle, Paris, France

B. Salvato (**730**), Istituto di Biologia Animale, Centro CNR Emocianine, Università di Padova, Padua, Italy

S. Santini (**641**), Institute of General Pathology, Catholic University of Rome, Rome, Italy

A. Santomasso (**263**), Memorial Sloan-Kettering Cancer Center, New York, New York

R. Santus (**731**), Laboratoire de Physico-Chimie de l'Adaptation Biologique, Muséum National d'Histoire Naturelle, Paris, France

Manfred Saran (**75**), Institut für Biologie, Gesellschaft für Strahlen-und Umweltforschung, München, Federal Republic of Germany

Donald T. Sawyer (**15**), Department of Chemistry, University of California, Riverside, California 92521

K. Schäfer (**69**), Bereich Strahlenchemie, Hahn-Meitner Institut Berlin, Berlin, Federal Republic of Germany

Karen M. Schaich (**177, 734**), Medical Research Center, Brookhaven National Laboratory, Upton, New York 11973

Andreas Schalhorn (**667**), Ludwig-Maximilians-University of Munich, München, West Germany

Sharon Schmidt (**652**), Experimental Hematology Department, Armed Forces Radiobiology Research Institute, Bethesda, Maryland

F. Schöpfer (**293**), Strahlenzentrum der Justus-Liebig-Universität, Giessen, Germany

Dietrich Schulte-Frohlinde (**61**), Institut für Strahlenchemie im Max-Planck Institut für Kohlenforschung, Stiftstr. 34-36, 4330, Mülheim, Federal Republic of Germany

James J. Shieh (**769**), Department of Chemistry, Boston University, Boston, Massachusetts

F. Y. Shum (**355**), Department of Radiation Oncology, Cross Cancer Institute, Edmonton, Alberta, Canada

Michael G. Simic (**109, 619**), Center for Radiation Research, National Bureau of Standards, Washington, D. C. 20234

Pierre M. Sinet (**45**), Department Biochemie Génétique, Hôpital Necker, Paris, France

Ajit Singh (**461, 736, 739**), Whiteshell Nuclear Research Establishment, Atomic Energy of Canada Limited Research Company, Pinawa, Manitoba, R0E 1L0 Canada

D. Singh (**742**), Department of Medicine, The Ohio State University, Columbus, Ohio 43210

Harwant Singh (**461**), Whiteshell Nuclear Research Establishment, Atomic Energy of Canada Limited Research Company, Pinawa, Manitoba, Canada

Danielle G. Sladdin (**153**), The Memorial Hospital and the Department of Medicine, the University of Massachusetts School of Medicine, Worcester, Massachusetts

F. Lee Smith (**743**), Oklahoma Medical Research Foundation, Oklahoma City, Oklahoma 73104

L. C. Smith (**397**), Department of Organic Chemistry, University of Geneva, Geneva, Switzerland

C. E. Smithen (**329**), Roche Products Limited, Welwyn Garden City, Herts, United Kingdom

Esmond E. Snell (**521**), Department of Microbiology, University of Texas at Austin, Austin, Texas

John R. J. Sorenson (**679**), College of Pharmacy, University of Arkansas, Little Rock, Arkansas

J. D. Spikes (**421, 441**), Department of Biology, The University of Utah, Salt Lake City, Utah 84112

Rajagopalan Sridhar (**363**), Oklahoma Medical Research Foundation, 825 N. E. 13 Street, Oklahoma City, Oklahoma 73104

S. Steenken (**83**), Institut für Strahlenchemie im Max-Planck Institut für Kohlenforschung, Mülheim, Federal Republic of Germany

L. M. Stephenson (**371**), Hydrocarbon Research Institute and Department of Chemistry, University of Southern California, Los Angeles, California 90007

Arnold Stern (**750**), New York University School of Medicine, New York, New York

B. Stevens (**746**), Department of Chemistry, University of South Florida, Tampa, Florida 33620

Reggie H. Stevens (**690**), Radiation Research Laboratory, University of Iowa, Iowa City, Iowa

R. C. Straight (**421, 747**), Medical Research, Veterans Administration Medical Center, Salt Lake City, Utah 84148

Stephen Gene Sullivan (**750**), New York University School of Medicine, New York, New York 10016

Bruce A. Svingen (**147**), Department of Biochemistry, Michigan State University, East Lansing, Michigan

Harold M. Swartz (**161**), University of Illinois College of Medicine, Urbana, Illinois 61801

A. Tallentire (**247**), Department of Pharmacy, University of Manchester, Manchester, M13 9PL United Kingdom

C. Tanielian (**753**), Institut de Chemie, 1 Rue Blaise Pascal, 67008, Université Louis Pasteur, Strasbourg, France

Irwin A. Taub (**769**), Food Engineering Laboratory, U. S. Army Natick Research and Development Command, Natick, Massachusetts

H. H. Thaw (**615, 757**), Department of Experimental Medicine, University of Uppsala, Pharmacia AB, Box 181, Uppsala, S-751 04 Sweden

Eckhardt Thiel (**667**), Ludwig-Maximilians University of Munich, München, West Germany

Michael J. Thomas (**761**), Department of Biochemistry, Bowman Gray School of Medicine, Wake Forest University, Winston-Salem, North Carolina 27109

Ming Tien (**147**), Department of Biochemistry, Michigan State University, East Lansing, Michigan 48824

C. A. Tobias (**716**), Division of Biology and Medicine, Lawrence Berkeley Laboratory, Berkeley, California

John R. Totter (**1**), Institute for Energy Analysis, Oak Ridge Associated Universities, Oak Ridge, Tennessee 37803

T. G. Traylor, Department of Chemistry, University of California at San Diego, La Jolla, California

Richard J. Treinen (**169**), Department of Pathology, The University of Texas Medical Branch, Galveston, Texas

H. Tsuji (**675**), Institute of Physiological Chemistry, The University of The Saarland, Homburg, Federal Republic of Germany

Volker Ullrich (**497, 675**), Department of Physiological Chemistry, 655 Homburg-Saar, Federal Republic of Germany

P. Unrau (**703**), Radiation Biology Branch, Atomic Energy of Canada Limited, Chalk River, Ontario, Canada

J. E. van Lier (**764**), University of Sherbrooke Medical Center, Sherbrooke, Quebec, Canada

Marie E. Varnes (**766**), Division of Radiation Biology, Case Western Reserve University, Cleveland, Ohio 44106

P. Vernole (**641**), Institute of Human Genetics, Catholic University of Rome, Rome, Italy

L. Voituriez (**610**), Laboratoire de Radiobiochimie, Centre d'Etudes Nucléaires de Grenoble, Grenoble, France

Walter H. Waddell (**711**), Department of Chemistry, Carnegie-Mellon University, Pittsburgh, Pennsylvania

S. Y. Wang (**475, 610**), Division of Environmental Chemistry, The Johns Hopkins University, C. O. E .H. - U. S. P. H. S. Hospital, Building 6, 3100 Wyman Park Drive, Baltimore, Maryland

Peter Wardman (**321**), Cancer Research Campaign Gray Laboratory, Mount Vernon Hospital, Northwood, Middlesex, HA6 2RN England

H. Weiss (**263**), Memorial Sloan-Kettering Cancer Center, 1275 York Avenue, New York, New York 10021

Joseph F. Weiss (**622**), Biochemistry Department, Armed Forces Radiobiology, Research Institute, Bethesda, Maryland

Gunnar Westman (**693**), Department of Oncology, University Hospital of Umeå, Umeå, Sweden

D. W. Whillans (**277**), Physics Division, Ontario Cancer Institute, Toronto, M4X 1K9 Canada

Kevin D. Whitburn (**769**), Department of Chemistry, Boston University, Boston, Massachusetts 02215

Richard H. Wiebe (**119**), Atomic Energy of Canada Limited, Whiteshell Nuclear Research Establishment, Pinawa, Manitoba, Canada

Martin D. Williams (**771**), Department of Biochemistry and Biophysics, University of Pennsylvania School of Medicine, Philadelphia, Pennsylvania 19104

Robin Willson (**309**), Biochemistry Department, Brunel University, Uxbridge, England

B. J. Wilson (**507**), Department of Biochemistry, School of Medicine, Vanderbilt University, Nashville, Tennessee

John S. Winterle (**779**), SRI International, Menlo Park, California

Kizashi Yamaguchi (**409**), Department of Chemistry, Faculty of Engineering Science, Osaka University, Toyonaka, Osaka, 560 Japan

Osamu Yamamoto (**781**), Research Institute for Nuclear Medicine and Biology, Hiroshima University, Kasumi, 1-2-3 Hiroshima, 734 Japan

Lawrence J. Young (**787**), Department of Biochemistry, Colorado State University, Fort Collins, Colorado 80523

C. Zannoni (**641**), Institute of Organic Chemistry, University of Bologna, Bologna, Italy

J. D. Zimbrick (**708**), Office of Health and Environmental Research, Department of Energy, Washington, D. C.

J. A. Zuclich (**791**), Technology Incorporated, Life Sciences Division, San Antonio, Texas 78216

Preface

The Conference on Oxygen and Oxy-Radicals in Chemistry and Biology can be considered the second in a series. Following the effective International Conference on Singlet Oxygen and Related Species held at Pinawa, Manitoba in 1977, it was decided to confer again at Austin, Texas in 1980. The scope of this conference was widened to include research not considered at Pinawa, but the unifying theme of the meeting was still the oxygen molecule. Over 160 scientists gathered to discuss research in oxygen and associated radicals and their roles in basic processes in chemistry and biology.

The conference was held 25–29 May 1980 at the Thompson Conference Center on the campus of the University of Texas at Austin, sponsored by the Center for Fast Kinetics Research of the College of Natural Sciences of the University. It was supported financially mainly by the Office of Health and Environmental Research of the U.S. Department of Energy, by the registration fees of the participants, and in part by the Frito-Lay Company of Dallas and the College of Natural Sciences Foundation of the University of Texas at Austin. Direct contributions were made by personnel from the Center for Fast Kinetics Research (supported by the Biotechnology Branch of the Division of Research Resources, NIH) and from the Laboratory of Radiation Biology (supported by General Medical Sciences of NIH and by the Department of Energy).

Professor W. Alfred Noyes, Jr., Ashbel Smith Professor-Emeritus of Chemistry of U.T. Texas at Austin was the honorary chairman of the conference. In what turned out to be his last formal public appearance, he spoke briefly at the convocation session. He died on November 25, 1980.

We, the organizers, are very pleased to acknowledge with most sincere thanks the cheerful and expert assistance of a number of our colleagues and assistants in accomplishing the many tasks associated with this undertaking. Lisa Crumpton and Gloria Tidwell transcribed the discussions, Janet Young did the art work, and Janet Becker and Royce Gehrels gave general assistance. Deborah McWhorter typed the book of abstracts and most of the manuscript corrections, Pamela Bockoven typed the manuscripts, and Linda

Fetters prepared the index. Billy K. Naumann organized the mechanics of the poster sessions. Patrick L. Ogle of the Division of Continuing Education and David W. Odom of the University Accounting Office steered us through the business mazes.

Marian Cross and Peggy Centilli were involved in almost all aspects of organization, staging, and publication. Our sucess is due especially to their efforts.

M. A. J. Rodgers
E. L. Powers
Austin, Texas

SOME REFLECTIONS ON OXYGEN AND OXY-RADICALS IN CHEMISTRY AND BIOLOGY

John R. Totter, Institute for Energy Analysis, Oak Ridge Associated Universities, Oak Ridge, Tennessee 37830

It is a privilege to join in welcoming you to this International Conference on Oxygen and Oxy-radicals in Chemistry and Biology. We are here because our hosts at the University of Texas have recognized the wealth of new knowledge growing out of the many independent and different approaches to the study of oxygen, oxidation, radiation effects in aqueous solutions and their modification, photooxidation, enzymatic oxidation, and metabolism. These approaches have slowly but surely expanded from their isolated origins and are now converging into a recognizably distinct body of information useful to all researchers in these areas, even though very different disciplines are involved. It seemed that a stage of this evolution had been reached which made it appropriate for this diverse array of scientists to get together, get better acquainted, exchange views, and perhaps lay the groundwork for common interpretation of the problems and methods for solving them. Of course, I do not mean to suggest that the number of disciplinary approaches should be narrowed. A large part of the strength of the field lies in its interdisciplinary nature. The purposes of the meeting will surely be met if we all learn of the many approaches to solving the common problems and if some of us learn that there may be better ways than those of which we were aware. I hope that several new collaborations between practitioners of different approaches may result from the meeting. If so, the organizers should feel more than fully repaid for their efforts in bringing the group together.

It is now exactly 80 years since free radicals (as opposed to linked radicals) in organic chemistry were made legitimate by Moses Gomberg's (1) studies of the triphenyl methyl radical. The history of free radicals goes back much further than 1900 but was largely based on conjecture—even though some of the conjectures were surprisingly accurate. Triphenyl methyl radical was recognized because in the absence of oxygen and in the presence of its dimer, hexaphenyl ethane, it appears stable. Furthermore, it is colored, while its dimer is colorless. The reactions of triphenyl methyl, although quite rapid, were sufficiently slow to be followed by methods available 80 years ago. The apparent stability is the result of an equilibrium between the dissociation rate of the dimer and the reassociation rate of the two monomers. Really stable organic radicals have only become well known in the last 20 years or so, and the conditions under which they appear to be stable are limited.

ISBN 0-12-592050-4

Physical methods for measuring organic radicals in low concentration were not very good before about the end of World War II. The most useful was the Gouy balance, with which the presence of paramagnetic radicals could be detected by comparing the apparent weight of a substance in a strong magnetic field with its weight in the absence of the field.

The development of radar during World War II provided new and much more sensitive equipment. The electron spin resonance method was rapidly put into use and the first demonstration of radicals in biological material by this method occurred in 1954 (2). There was, of course, ample reason long before this for believing that free radicals existed in biological material, even if very short lived; nevertheless, the demonstration excited considerable interest and ushered in a new era in radical research.

World War II was also responsible for rapid development in another field. The production of atomic weapons stimulated intense research in the radiation chemistry of materials. Basic work on the radiation chemistry of water was greatly expanded and led in a few decades to rather sophisticated understanding of the interaction of ionizing radiation with water. Early work was, of course, hampered by lack of circuitry and instrumentation for dealing with very short lived phenomenon occurring in 10^{-12} seconds to 10^{-6} seconds. This barrier is now largely overcome, as you all know, and we have accurately determined second-order kinetic constants of the order of 10^{11} $M^{-1}s^{-1}$. Two types of approaches were used here: scavenger techniques and, later, pulse radiolysis.

A somewhat later development has occurred in which photochemistry has blossomed. I am not sure just what the prime stimulus was for this development, other than rapid improvements in instrumentation. The development of photochemistry certainly has accompanied a maturing of physical organic chemistry and has had some impetus from advances in photosynthesis. It has also benefited from enormous improvements in chemical and physical separation methods, some of which stemmed from the development of the nuclear industry during the war. Of course, the current interest in energy from sunlight is a great help also.

Another venerable branch of chemistry has been important in the development of insights into free radical behavior. This is the study of autooxidation, a phenomenon responsible for the hardening of linseed oil used in paints and varnishes. The search for antioxidants to protect rubber from deterioration and to prevent rancidification of fats in foods was a prominent pursuit of many chemists, including myself in the thirties when I was still in college. The decades of the twenties and thirties saw the discovery, isolation, and synthesis of Vitamin E, which is a natural antioxidant found in vegetable oils.

This brings me to the more biological side of oxygen and oxyradicals. With your permission I would like to speculate on what I believe may be a very important consequence of the presence of oxy-radicals in metabolism, namely, their effects on our health and life span.

It may not be surprising even to younger members of the audience that oxygen has long been believed to have a function in aging. I can remember Professor Matill at Iowa University in 1937 feeding rats hydroquinone and trying to see if it could replace Vitamin E or if the life span of the animals would be lengthened.

Hydroquinone was the most common and best known antioxidant easily available at that time. It was well known that it could protect against autooxidative chain reactions that normally took place in animal fats. It was routinely added to commercial cod-liver oil preparations to lengthen their shelf life. (Ordinarily, fresh cod-liver oil begins to rancidify in a few hours.) There were many chemists before and after 1937 who speculated that autooxidation produced cross-linking in tissues much as sulfur treatment of rubber produced a strengthened and more rigid product. There are many chemists today who believe that taking antioxidants may help slow down an aging reaction that they feel may result from autooxidative processes in the lipids of our tissues.

I would like to turn now to the biological handling of oxygen which is also the subject of some of the latter sessions in the symposium. Early life on the planet, it is now widely believed, lived under almost completely anaerobic conditions. Some oxygen was undoubtedly formed by ultraviolet light dissociation of H_2O high in the atmosphere, which was otherwise composed of N_2 and inert or reducing gases. Furthermore, the steady exposure of ferrous iron as its sulfide through erosion of uplifted material must have kept the steady-state level of free oxygen extremely low. Nevertheless, it seems likely that enough oxygen was around so that some adaptation to it could take place. If this had not been the case, the emergence in the pre-Cambrian era of photosynthetic organisms with an oxygen-evolving system may well have resulted in the complete or almost complete extinction of other forms of life. As organisms evolved means to protect themselves and to make use of the oxygen to increase their production of energy from carbohydrate or simpler organic sources, they were able to become more complex. Their ability to move themselves improved and eventually they could emerge from an aqueous medium and adapt to the land. This adaptation did not take place overnight and, I believe, has not yet reached a stage in which all of the deleterious effects of oxygen are successfully eliminated.

The adaptation to oxygen had to take at least two major routes. First, the organism needed to protect itself against unwanted oxidations; second, oxidation of normal substrates needed to be channeled so that it provided chemical energy useful to the organism. The adaptation has resulted in the present situation wherein use is made of oxygen in two ways. The major pathway of oxygen utilization in many organisms is through the cytochrome-cytochrome oxidase system, with an almost complete avoidance of the production of free radicals. The other pathway makes use of the production of oxidizing radicals, possibly to reduce the burden of having a larger number of specific enzymes which would otherwise be needed to help break down compounds that are minor constituents of the animal diet. Also, it is possible that some oxy-radicals can oxidize some xenobiotic compounds as well as some normal metabolites that may have an energy barrier to oxidation too great to be overcome by ordinary enzymatic action.

Some of the normal hydrogen-carrying coenzymes as well as some normal intermediates in metabolism are autooxidizable, and the oxidation usually takes place in two one-electron steps. The organism therefore must be protected from the possible effects of these transient radicals, and an elaborate and effective many-tiered set of defenses has evolved. I believe that the degree to which these defenses have evolved determines the longevity of the species. You will hear quite a bit about the production of the radicals during the symposium, and you will hear some about the defenses; but now I want to stress something about the burden to the animal organism that maintenance of these defenses represents.

The superoxide produced in metabolism is the same as the superoxide produced by exposure to ionizing radiation. It may be converted to more strongly oxidizing substances, perhaps including the ·OH radical, as will be discussed during the next few days. OH radicals are known to produce breaks in the DNA of the cell nucleus. We know the stoichiometry of the conversion, and we know the relation between dose of ionizing radiation and DNA breaks. We also know with somewhat less precision the relationship between acutely delivered X-ray doses and extra cancers. We are therefore in a position to estimate the burden of repair that the body has to carry, assuming that the same number of breaks results in the same number of cancers in the two cases. A reasonable figure for extra cancers from X- or gamma-ray exposures is 200 per 10^6 person rads. The lifetime risk of cancer in the United States is about 20 percent or 200,000 per 10^6. Therefore, the doubling dose is about 1000 rads, or about 13 rads per year for a 75-year lifetime. This is probably a minimum figure because radiation doses protracted this way are likely to affect the cells less than do equivalent exposures delivered acutely. With about 15 single-strand breaks in each cell nuclear DNA per rad, this means about 200 breaks per cell DNA per year. With about 10^{13} cells in the adult body, we are thus called upon to repair each day a minimum of about 5×10^{12} breaks.

I emphasize this to point out that the synthesis and maintenance of the DNA repair enzyme systems and the supply of energy to accomplish the repairs is not an insignificant burden. The work required, of course, is in addition to the not inconsiderable energy used to keep a large supply of superoxide dismutase and catalase in working condition. There are perhaps other defense mechanisms against oxy-radicals or against the results of their actions. This many-tiered defense system must compete with all other demands on our energy supply. If the fraction of the energy supply available for any requirement essential for a survival period sufficient to insure a number of offspring is inadequate, the species will not survive.

We may, for purposes of discussion, divide the major requirements for animal energy into the following categories:

1. energy for maintaining body temperature
2. energy for gathering food and defending territory
3. energy for reproducing and caring for young
4. energy for assuring longevity sufficient to bear and rear an adequate number of young to assure survival of the species

It should be safe to assume that natural selection has operated on each of these requirements in such a way that a near optimum strategy has been developed for surviving species. If two species are in competition with each other, the one which can fulfill its requirements with the least energy expenditure should eventually replace or eliminate the other.

Requirements for energy in the first two categories, maintaining body temperature and fuel, can be reduced by natural selection for a larger-sized organism; this would reduce the ratio of body surface area to volume and therefore also the heat dissipation and relative requirement for food.

As regards the last two categories, numerous reproductive strategies are available and represented in surviving species. It is possible to estimate the energy requirements for each strategy but that is not of interest here. One possible mode of the operation of natural selection for longevity is represented in Fig. 1. Here the assumption is made that a trade-off occurs between energy required for reproduction and that required for maintenance of longevity. (For the sake of simplicity, it is assumed that the size of the animal remains constant.)

In the natural state a great many animals suffer a nearly random loss of life from age-unrelated diseases and predation. If these reduce each cohort to zero before the average age at the end of reproduction, the offspring of the short-lived component of the population will be slightly favored in number because their parents have more energy available to put into reproduction. If this condition remains unchanged over many generations, there will be a slow but steady reduction in the average age at the end of reproduction and average potential longevity.

On the other hand, if, through greater intelligence or some other means, an appreciable proportion of the population remains alive past the average end of the reproductive period, there may be enough reproductive advantage of the older ones to overcome the greater reproductive capacity of the short-lived cohorts. This would initiate selection of a population whose average potential life span is longer than that of the initial population.

It should be obvious that an organism's defense against cancer as an age-related cause of death is multifactorial and includes genetic mechanisms for producing and sustaining the

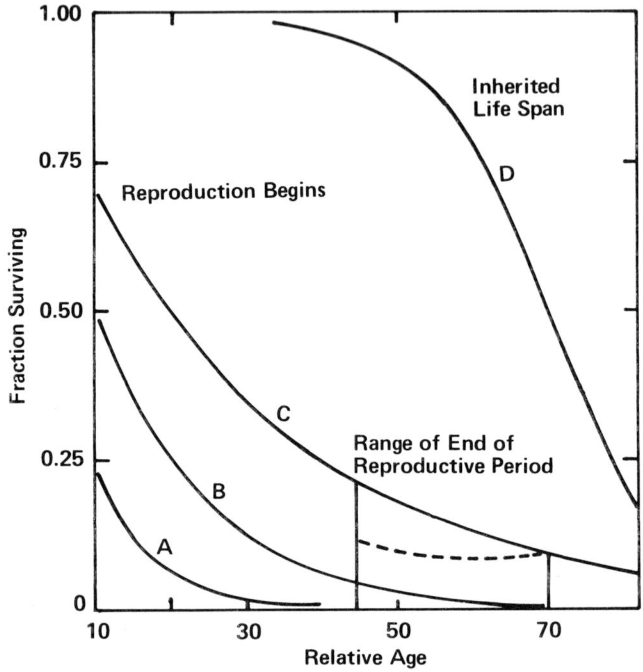

FIGURE 1. Conditions under which selection for increased or diminished internal protection against the deleterious effects of oxy-radicals may occur. Curve A: Selection for reduced longevity owing to selective advantage conferred by increased energy available for reproduction among the short-lived. Curve B: Condition of approximate neutrality for selection. Curve C: Conditions favoring selection for longer life. See text for complete explanation.

enzymatic defense against oxygen and oxy-radicals, numerous DNA repair systems, and probably immune systems. It is also reasonably obvious that if the population has reached equilibrium under some longstanding environmental conditions, we should see a susceptibility or resistance to cancer in the population that is normally or nearly normally distributed.

Vital statistics are available for a large number of countries, and these may be examined for evidence which has a bearing on the hypothesis I have been discussing. It is first necessary to find a mathematical statement which adequately represents the data in a useful way. There are two relationships that are often used for this purpose. The most well known is the Gompertz one (3). This is an empirical equation relating the logarithm of the mortality rate to age. This relationship is more nearly linear above age 40 for crude mortality statistics than it is for the net mortality risk. The latter makes the comparison we wish to use because it is the mortality risk for the given disease in the absence of other risks which compete for the life of the individual (independence assumed).

Another relationship has been used by Armitage and Doll (4) and Burch (5). This makes use of the fact that between about age 40 and 75 the net risks of cancer mortality are very well described as resulting from a target phenomenon in which 4 to 5 "hits" are required. Both of these methods may be described as "double exponential" relationships.

I have used a third double-exponential type of equation in which the normal distribution is used, but the age axis is altered by raising the age to a power, usually differing from one. This method has advantages of simplicity of calculation. It also appears to fit the data from a number of countries somewhat better than the target model and is quite superior to the Gompertz relationship. In Table I the data for net risk of death from cancer [taken from Preston, Keyfitz, and Schoen, 1972 (6)] for males in England and Wales during 1951 is given for ages 45 to 85, in five-year intervals. These data were then used to calculate parameters for the three relationships, and the values obtained from the mathematical expressions were compared with the observed values. It can be seen that the "age-power-normal" curve gives values more consistent with the observed ones than the other two methods give.

TABLE I

NET CANCER RISKS FOR MALES IN ENGLAND AND WALES (1951) CALCULATED ACCORDING TO SEVERAL RELATIONSHIPS

		Calculated Net Risks (R)		
Age	Observed Net Risk	Age-Power Normal	Exponential	Gompertz
45	0.01058	0.01007	0.01149	0.01302
50	0.01807	0.01880	0.02031	0.02049
55	0.03211	0.03292	0.02916	0.03211
60	0.05396	0.05434	0.05387	0.05006
65	0.08549	0.08492	0.08208	0.07757
70	0.12784	0.12615	0.12034	0.11916
75	0.18053	0.17876	0.17032	0.18070
80	0.24422	0.24246	0.23307	0.26874
85	0.31188	0.31581	0.30869	0.38825
Correlation Coefficient		0.9998	0.9994	0.9912

With a good mathematical expression for net risks from each country, it is possible to extrapolate and interpolate with greater confidence and to use the curve parameters for comparisons. The parameters for twenty-five countries (Western and Central European; English-speaking North American countries; and Australia, New Zealand, and Japan) were calculated. A summary of the results for both cancer and cardiovascular disease is given in Table II. The differences in the age power used to convert the data to a normal distribution should be noted. The cancer values for both male and female average at about 1, while there is considerable dispersion in the individual values for each country. For cardiovascular disease there is a difference of almost a unit between the females and the males. This itself may suggest some involvement of the X chromosome in the resistance to cardiovascular mortality.

A value of importance in determining something about the heritability of resistance to cancer is the very low variance in the age at which a risk of 1 percent from cancer mortality in both males and females is reached. The male cardiovascular age is also close to the cancer value, while the female reaches 1 percent risk of cardiovascular mortality at a later average age.

It may be noted that the variance in ages at which higher risks from cancer are reached increases with age, while the variances for similar risks from cardiovascular disease do not. This increase may relate to a greater influence of environment on the cancer risk following the end of the reproductive age. It is the

TABLE II

NET RISK FOR 25 COUNTRIES

		Age at Risk of		
	P	1%	15.87%	50%(μ)
FROM CANCER, MALES 1964				
Mean	1.040	46.166	74.705	96.036
Standard Deviation	0.2986	0.733	2.043	3.268
FROM CANCER, FEMALES 1964				
Mean	1.044	44.374	79.308	106.320
Standard Deviation	0.284	1.402	2.585	4.934
FROM CARDIOVASCULAR DISEASE, MALES 1964				
Mean	2.016	44.006	68.092	81.320
Standard Deviation	0.462	2.643	2.306	2.136
FROM CARDIOVASCULAR DISEASE, FEMALES 1964				
Mean	2.876	49.666	72.978	84.229
Standard Deviation	1.276	3.794	1.641	1.613

end of reproduction which marks a sharp reduction or obliteration of the influence of natural selection.

There is some information on heritability of longevity in humans which suggests that about 20 percent of the variance in longevity is heritable. That is, the standard deviation of the heritable longevity is about 7 to 8 years, while the so-called environmental part is about 15 to 17 years.

In inbred mice kept under "unvarying" environmental conditions, the relationship of the variances to the life span are not much different in proportion than those for humans. This suggests that a very large proportion of the "environmental" portion of the variance in life span is due to inherent biological uncertainty and cannot be further reduced by manipulation of the external environment.

I have in Figs. 2 and 3 a comparison of cancer net risks in males and females (U.S.) for the years 1930 and 1975. You will note that there has been a reduction in the risk for females at all ages, while there has been an increase for males at all ages above 45 or so.

I do not believe there is enough difference in the exposures of men and women to general pollution to account for very much of the difference shown between Figs. 2 and 3. It is possible that smoking habits in the past account for it or at least for some part of it. At any rate it is difficult to attribute much of

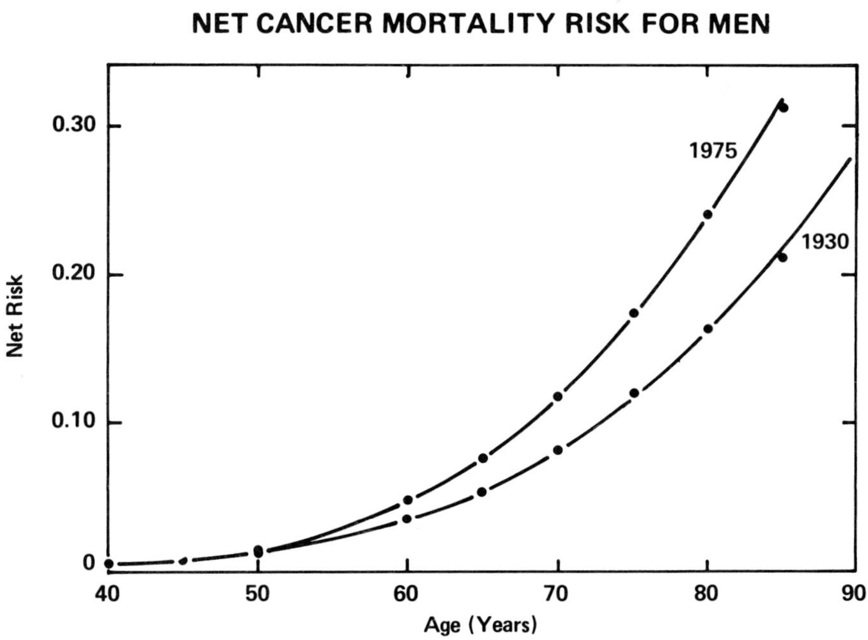

FIGURE 2. Net cancer mortality risk for U.S. men, 1930 and 1975.

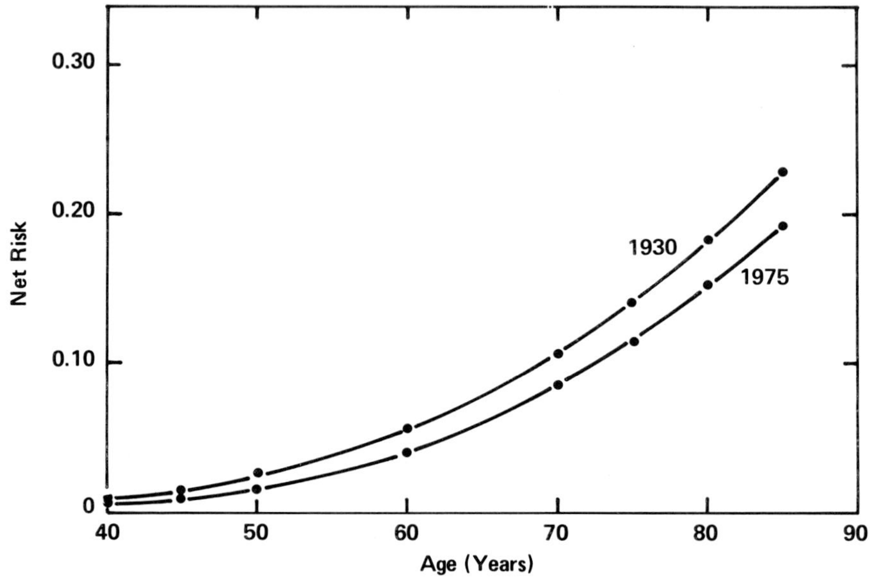

FIGURE 3. Net cancer mortality risk for U.S. women, 1930 and 1975.

modern day cancer mortality to general pollution if the last 40 years or so have shown a reduction in the net risk for females. If this is correct, the causes for cancer must be sought in intrinsic metabolic processes. I believe the normal production of oxy-radicals during metabolism is responsible for a very large part of the cancer produced after the age of 40 or so.

It is clear that there are strong body defenses against the production of cancer from these metabolic causes up to the average age at the end of reproduction. Because selection would operate against all natural intrinsic causes of death up to the end of the reproductive period more or less equally, there would be little direct selection against any particular variety of cancer. Therefore, we might expect the variances in the incidence of individual types of cancer among different populations to be much greater than the variances between the total incidence of cancers. Indirect selective processes may have been at work, however, to produce national or racial characteristics in the mix of cancer types. A lack of appreciation of this possibility has, I believe, led to the suggestion that most cancers are environmental in origin. The very large variance in incidence of individual types of cancer seen in various regions invites the argument that individual types of cancer susceptibility are not heritable. It does not appear to be proper to jump from this conclusion to the more general conclusion that susceptibility to all cancer is not heritable.

This principle may be stated in the more general terms of epidemiology in the following way:

> There is good reason to believe that resistance or susceptibility to cardiovascular disease is entirely independent of resistance or susceptibility to cancer. On the other hand, the resistance or susceptibility to one type of cancer is not necessarily independent of resistance or susceptibility to another type.

The environmental theory of cancer implicitly asserts that susceptibility to any type of cancer is independent of susceptibility to any other type. The expected consequences if either of these hypotheses is correct follow simply. If mortality from cancer types is independent, the removal of the "cause" of one reduces the total mortality from cancer almost as much as that represented in the total. If, however, the susceptibilities show dependence, the removal of the "cause" of one type reduces the total cancer mortality only a fraction of the amount represented by that type. The degree to which this distinction might be blurred by the inherent biological uncertainty is not clear.

I would like to conclude now by reviewing some of the work of the pioneers in radiation biology whose investigations on oxygen toxicity laid the foundation upon which one can construct a speculative scheme such as I have talked about today. A large number of people have contributed to determinations of the oxygen enhancement ratios under various conditions. Some work of this sort is still going on, and I will not attempt to name the people who have been prominent in this research, other than to remark on the late Dr. L. H. Gray's great interest in the subject.

The more direct evidence which I believe to be pertinent to the scheme that has been presented started, I believe, with the work of Alan Conger at Oak Ridge. The work of Conger and Fairchild (7), who showed that exposure to oxygen under pressure resulted in chromosome breaks in the spiderwort (*Trandenscantia*), was quickly picked up by others and extended considerably. Among those who were prominent in directing attention to oxygen toxicity and mutagenicity was Dr. Rebeca Gerschman, who worked for a time with Wallace Fenn at Rochester (8). Dr. Ehrenberg and collaborators (9 and 10) also investigated oxygen effects in plant material, as did Kronstadt, Nilan, and Konzak (11). The first mutational study was conducted by H. L. Plaine (12) at Johns Hopkins University, who showed that *Drosophila* larvae were sensitive to oxygen at certain stages. Recent work has centered around the mutational effects of oxygen under increased partial pressure in the organisms used in the Ames test. These studies were conducted by Burns at Duke University (13) and by Bruyninckx, Mason, and Morse (14).

I should emphasize the work conducted at Duke University first under Dr. Philip Handler's direction and later under Dr. Irwin Fridovich. These researches helped establish that radical-producing reactions take place in the body normally with the involvement of several enzymes and, of course, culminated in the discovery of the very important enzyme superoxide dismutase.

Lengthy, involved research on the mechanism of action of peroxidases and catalase involved many people who made significant advances in understanding the function of these enzymes. I will not take time to name them, and I am sure I would overlook some that should not be omitted.

One other person's work deserves special attention. One must recall Leonor Michaelis' contributions to the redox chemistry of single electron reactions. Tomorrow our sessions begin with a present-day consideration of this type of reaction and then moves to oxy-radicals. On Tuesday the superoxide radical becomes the subject of the first session, and the closely related second section will consider oxygen in radiation biology. On Wednesday we move to excited states of compounds and photooxidation. Thursday's opening topic will be oxygenases, and the closing subject will be gas-phase reactions in the laboratory and in the environment. Please do not forget the poster sessions, which round out the program with several additional topics as well as give material to supplement the formal sessions.

ACKNOWLEDGEMENTS

This work is supported by Contract No. DE-AC05-760R00033 between the U.S. Department of Energy, Office of Health and Environmental Research, and Oak Ridge Associates Universities. The U.S. Government retains a nonexclusive royalty free license to publish or reproduce the published form of this contribution, or allow others to do so, for U.S. Government purposes.

REFERENCES

(1) Gomberg, M., Chem. Ber. 33, 3150 (1900).
(2) Commoner, B., Townsend, J. and Pake, G. E., Nature 174, 689 (1954).
(3) Gompertz, B., Philos. Trans., R. Soc. London, Ser. A. 115, 513 (1825).
(4) Armitage, P. and Doll, R., Brit. Jour. Cancer 8,1 (1954).
(5) Burch, P. R. J., Nature 197,1145 (1963).
(6) Preston, S. H., Keyfitz, N. and Schoen, R., "Causes of Death, Life Tables for National Populations," Seminar Press, New York, N.Y., and London, England, 1972, 787 pages.
(7) Conger, A. D. and Fairchild, L. M., Proc. Natl. Acad. Sci. USA 38, 289 (1952).
(8) Fenn, W. O., Gerschman, R., Gilbert, D. L., Terwilliger, D. E. and Cothran, F. V., Proc. Natl. Acad. Sci. USA 43, 1027 (1957).
(9) Gelin, O., Ehrenberg, L. and Blixt, S., Agric. Horti. Gen. 17, 265 (1959).

(10) Moutschen-Dahman, M., Moutschen, J. and Ehrenberg, L., Hereditas 45, 230 (1959).
(11) Kronstad, W. E., Nilan, R. A. and Konzak, C. F., Science 129, 1618 (1959).
(12) Plaine, H. L., Genetics 40, 268 (1955).
(13) Burns, R. O., U.S. Army Research Office Report No. DRXRO-IP-L-12801-L (1976).
(14) Bruyninckx, W. J., Mason, H. S. and Morse, S. A., Nature 274, 606 (1978).

REDOX CHEMISTRY OF O_2 AND PEROXIDES

Session Chairman, A. J. Bard, Austin, Texas

REDOX CHEMISTRY OF DIOXYGEN SPECIES AND THEIR CHEMICAL REACTIVITY

Donald T. Sawyer and Edward J. Nanni, Jr., Department of Chemistry, University of California, Riverside, California 92521

Because dioxygen (O_2) is vital to many oxidation reactions and biological metabolic/respiration processes, its oxidation-reduction chemistry has been the subject of numerous investigations. However, the mechanisms and energetics for the reduction of O_2 and for the interconversion of its reduction products ($HO_2\cdot$, $O_2^{\cdot-}$, H_2O_2, HO_2^-, O_2^{2-}, $\cdot OH$, $\cdot O^-$, and H_2O) have just begun to be understood, especially in nonaqueous media and when in association with transition metal ions.

The strongly positive potential for the four-electron exothermic reduction of O_2 ($E^{o\prime}$ = +0.82 V vs NHE; 316 kJ/mol at pH 7 relative to the normal hydrogen electrode)

$$O_2 + 4H^+ + 4e^- \rightleftarrows 2H_2O \qquad (1)$$

makes the process attractive for energy transfer and storage, and for use as the cathode in fuel cells (1). However, mechanistic rationale and experience (2,3) preclude realization of this potential and energy via a primary electron-transfer step. Within this electrochemical limit, the reduction of O_2 is a one-electron process which is followed by chemical disproportionation reactions in aqueous media to yield H_2O_2 under acidic conditions (an overall two-electron process) and OH^- under basic conditions (an overall four-electron process). On the basis of pulse radiolysis studies (4) the effective redox potential for the one-electron reduction of O_2 is -0.33 V vs NHE and is independent of pH above pH 5. Hence, mechanistic restrictions on electron transfer to O_2 cause its effective oxidizing energy to be reduced by as much as 444 kJ/mol (relative to the thermodynamic value for eq. 1).

A strongly held view (5) which we share is that apparent concerted multi-electron transfer reactions actually represent atom-transfer processes. As such, their kinetics and mechanisms are strongly dependent upon substrate and catalysts. The corrosion and "rusting" of transition metals by dioxygen is one

ISBN 0-12-592050-4

example of substrate-dependent atom-transfer oxidations (6). Other examples are the biological oxidations by O_2, which are specifically catalyzed by the oxidase, monooxygenase, and dioxygenase enzymes (7). In contrast, the reduction of O_2 by cytochrome oxidase is believed to occur via four sequential one-electron transfer steps with concerted stabilization of the intermediates by the iron and/or copper ions of the enzyme (8). The oxidation of water to O_2 by photo-system II in green-plant photosynthesis, which is catalyzed by a polynuclear manganese cofactor, probably occurs by a four-step electron-transfer sequence that is the reverse of the cytochrome oxidase process (9). At pH 7 the optimal limit for each one-electron step is +0.82 V vs NHE, which requires selective stabilization of the dioxygen intermediates by the transition metal catalyst.

In addition to its role in the formation and decomposition of water, dioxygen is the primary oxidant in most combustion processes. Its reaction with hydrocarbons proceeds via a radical mechanism, with removal of a hydrogen atom in the primary step (10-15).

$$\cdot O_2\cdot + RH \rightarrow R\cdot + HO_2\cdot \qquad (2)$$

This is followed by a number of radical chain processes such as

$$R\cdot + \cdot O_2\cdot \rightarrow RO_2\cdot \qquad (3)$$

$$RO_2\cdot + RH \rightarrow ROOH + R\cdot \qquad (4)$$

These and related radical reactions all involve the reduction of O_2 to either the -1 or -2 oxidation states as well as its incorporation into the oxidized products (16-18).

Hence, beyond the reversible one-electron process at -0.33 V vs NHE (in aqueous media) the effective energetics for oxidations by O_2 are dependent on substrate and catalyst, and frequently are limited by a one- or two-electron rate-controlling step. The over-all thermodynamics due to the reduction of O_2 is determined by its final state, but this is an upper limit which rarely is realized. Fig. 1 summarizes the oxidation-reduction potentials ($\underline{E}^o$) in aqueous media (pH 0, pH 7, and pH 14) for dioxygen and its reduced species (1, 4, 19-23), which can be converted to free-energies ($-\Delta G°$, kJ/mol) for the respective half-reactions by the relation $-\Delta \underline{G}° = \underline{nFE}°$, where $\underline{n}$ is the number of electrons and $\underline{F}$ is the faraday (96,487 C mol^{-1}). Although the redox data for O_2 and its reduction products in aprotic media are limited, the aqueous potentials of Fig. 1 are reasonable approximations for half-reactions that involve cations and neutral species. Because aprotic solvents are poor solvating agents for anions ($O_2^{\cdot -}$, HO_2^-, and OH^-), the redox potentials for half-reaction reductions with anion product species are shifted by at least -0.2 V relative to their aqueous values.

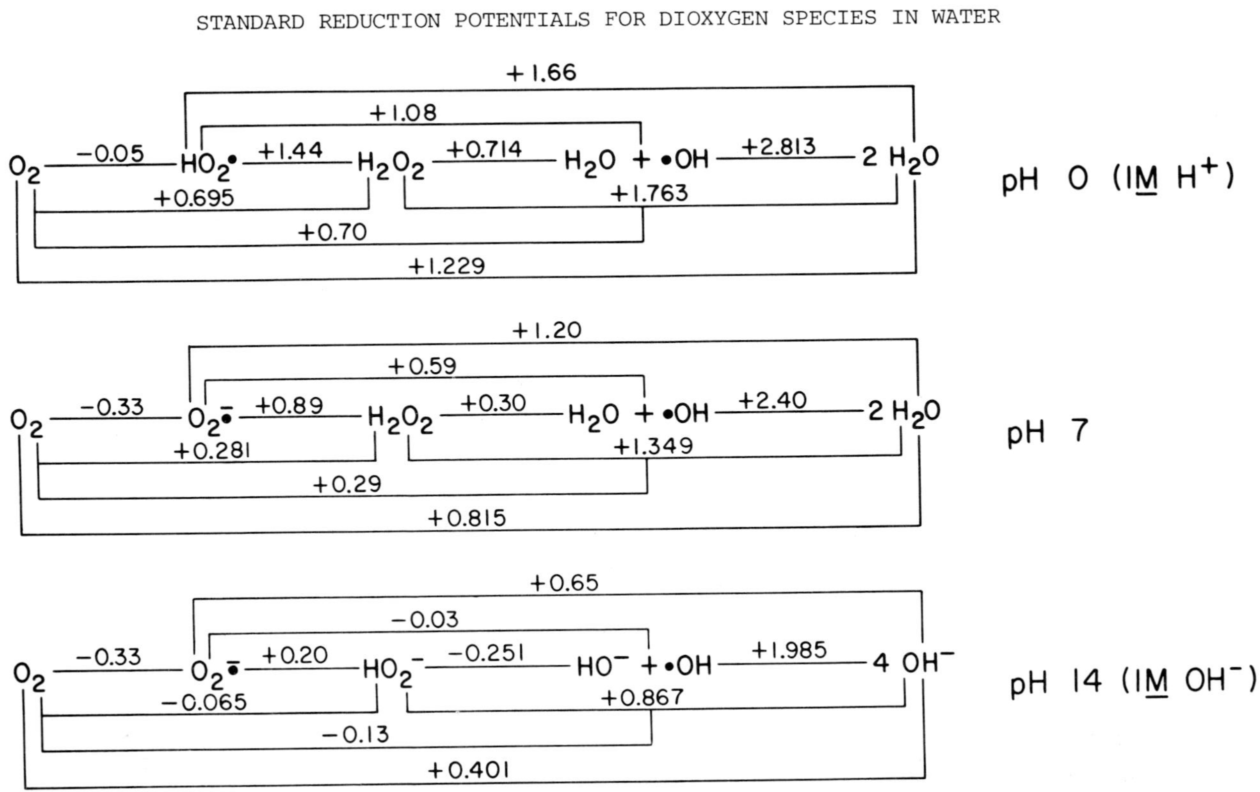

FIGURE 1. Potentials for oxygen redox couples at 25°C; values represent standard potentials versus the Normal Hydrogen Electrode (NHE).

ELECTROCHEMISTRY OF DIOXYGEN

Electron transfer. Fig. 2 illustrates the electrochemical reduction of O_2 under aprotic conditions. The first reduction step proceeds via a reversible one-electron transfer process, which is essentially independent of electrode material and solvent (2,24)

$$O_2 + e^- \rightleftarrows O_2^{\overline{\cdot}} \qquad E^{o\prime}_{pyridine} = -0.60 \text{ V vs NHE} \tag{5}$$

For aqueous conditions the reversible reduction potential of reaction 5 is -0.33 V vs NHE (4); the difference is due to the weak solvation of anions by aprotic solvents.

The second one-electron reduction of O_2 (Fig. 2) is irreversible because the product species reacts with residual water or the solvent (25)

$$O_2^{\overline{\cdot}} + e^- \rightarrow O_2^{2-} \tag{6}$$

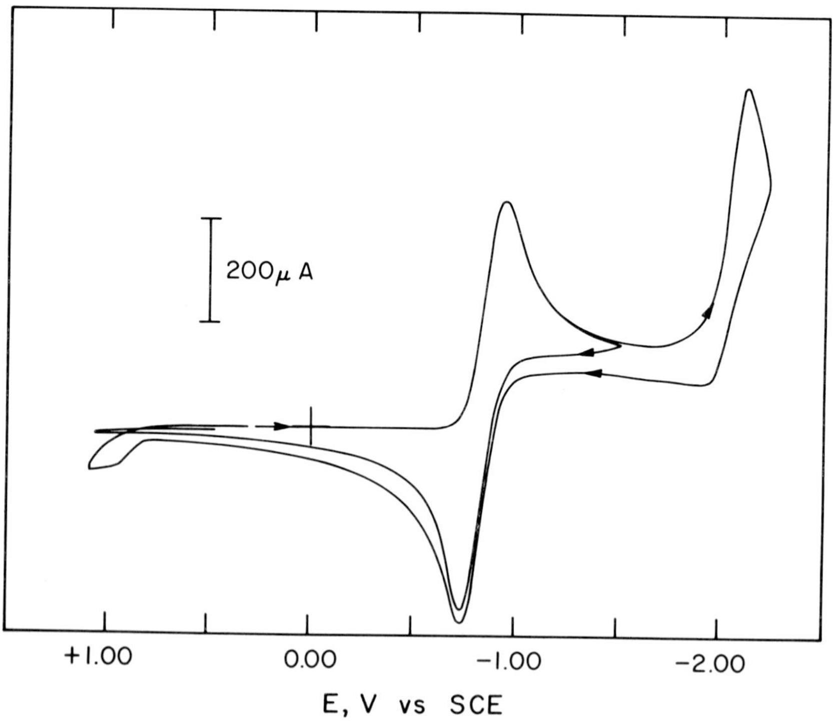

FIGURE 2. Cyclic voltammogram for 5 mM dioxygen (O_2) in pyridine (0.1 M tetrapropylammonium perchlorate at a platinum electrode) (area, 0.23 cm^2). Scan rate, 0.1 V s^{-1}. Saturated calomel electrode (SCE) vs NHE, +0.244 V.

$$O_2^{2-} + H_2O \rightarrow \tfrac{1}{2}H_2 + OH^- \qquad (7)$$

In the case of dimethyl sulfoxide (DMSO) with 0.1 M tetraethylammonium perchlorate, reaction (6) is followed by a series of hydrolytic steps

$$O_2^{2-} + Et_4N^+ \rightarrow Et_3N + CH_2{=}CH_2 + HO_2^- \qquad (8)$$

$$HO_2^- + DMSO \rightarrow DMSO_2 + OH^- \qquad (9)$$

which give dimethylsulfone and hydroxide ion as the stable products (26).

The stoichiometry of the electrochemical reduction of O_2 in aprotic media has been confirmed by controlled potential coulometry as well as by spectroscopic studies of the product species. In an acetonitrile solution with 0.1 M tetrapropylammonium perchlorate, superoxide ion has a single absorption band with a λ_{max} at 255 nm (ε, 1460 $M^{-1}cm^{-1}$) (27,28). The position of this maximum shifts with the supporting electrolyte that is used. Frozen glasses of this same solution at 77°K yield ESR spectra for $O_2^{\overline{\cdot}}$ with $g_\perp = 2.008$ and $g_{\|} = 2.083$ (27,29).

Within experimental error, the reduction of molecular oxygen and the oxidation of $O_2^{\overline{\cdot}}$ are independent of electrode material. The diffusion coefficients for these dioxygen species in DMSO (0.1 M $(C_2H_5)_4NClO_4$) at 25°C also have been determined (24); $D_{O_2} = 3.23 \times 10^{-5}cm^2s^{-1}$, $D_{O_2^{\overline{\cdot}}} = 1.08 \times 10^{-5}cm^2s^{-1}$. The fact that the diffusion coefficient for $O_2^{\overline{\cdot}}$ is one-third that for O_2 indicates there is extensive ion pairing or solvation of the $O_2^{\overline{\cdot}}$ species.

An aqueous solution can be made to approximate an aprotic medium by making it strongly alkaline. With such a solution and a platinum electrode covered with a hydrophobic surfactant, electrochemical reduction of O_2 yields transiently stable superoxide ion (30-32). The standard potential that is calculated from the data for the reduction process is -0.29 V vs NHE (30,31,33), which is in good agreement with the currently accepted value of -0.33 V (4). The formation of transiently stable superoxide ion also is observed in the reduction of oxygen at a carbon paste electrode in alkaline media (34). The reduction and reoxidation on carbon paste occurs at about -0.35 V.

In the presence of protons (or a proton source) superoxide ion participates in a series of reactions (4, 20-22), which for aqueous media at 25°C have the indicated thermodynamic and kinetic constants

$$H^+ + O_2^{\overline{\cdot}} \rightleftarrows HO_2\cdot \qquad pK_a = 4.69 \qquad (10)$$

$$O_2^{\overline{\cdot}} + HO_2\cdot \rightarrow O_2 + HO_2^- \qquad k = 1.0 \times 10^8 M^{-1}s^{-1} \qquad (11)$$

$$HO_2\cdot + HO_2\cdot \rightarrow H_2O_2 + O_2 \qquad k = 8.6 \times 10^5 M^{-1}s^{-1} \qquad (12)$$

$$O_2^{\overline{\cdot}} + O_2^{\overline{\cdot}} \xrightarrow{H_2O} O_2 + HO_2^- \qquad k<0.3\ M^{-1}s^{-1} \qquad (13)$$

The effect of protons on the electrochemical reduction of

oxygen in aprotic media (DMF) is dramatic; the reduction potential becomes more positive and the one-electron process gives way to an apparent irreversible two-electron process (34). This behavior and the spontaneous reaction between superoxide ion and hydrogen ion to form oxygen and hydrogen peroxide are consistent with a reaction pathway that is represented by eqs. 14 and 11 and/or 12 (35).

$$O_2 + H^+ + e^- \rightarrow HO_2\cdot \qquad E^{o}_{H_2O} = -0.03 \text{ V vs NHE} \tag{14}$$

The oxygen formed in the disproportionation of superoxide undergoes reduction by the same pathway. The net reaction is

$$O_2 + 2H^+ + 2e^- = H_2O_2 \tag{15}$$

This pathway accounts for the positive shift of the reduction potential, the production of hydrogen peroxide, and the apparent two-electron stoichiometry; it also avoids reduction processes that require more than a single electron (such as eq. 15) or unstable reactant species, e.g.

$$HO_2\cdot + H^+ + e^- = H_2O_2 \tag{16}$$

even though they are energetically favored.

There is little doubt that superoxide ions react with protons and cause the shift of the reduction potential to more positive values. Further substantiation is provided by similar effects that are observed in the presence of metal cations which form stable superoxide adducts (36-38). The combined information obtained in several solvents indicates that the shift in potential is due solely to complexation or precipitation of superoxide ion. There is no increase in the electron-per-oxygen molecule mole ratio except when catalytic disproportionation of superoxide occurs.

The proposal of a common one-electron reduction mechanism for molecular oxygen is supported convincingly by the data for the polarographic reduction of oxygen in dimethyl sulfoxide-water mixtures (39). Up to a water content of about 8 vol %, the reduction of oxygen is a one-electron process which exhibits ac polarographic reversibility. Above 10% water content, the electron-per-oxygen molecule mole ratio rapidly increases with increasing water content to a maximum value of 2 at about 35% water. A concomitant shift of the reduction potential to more positive values occurs and levels off at the pH-independent value observed in aqueous media.

Reduction of oxygen at inert electrodes in aqueous solutions is first order in oxygen and independent of pH (approximately pH 4-12) (40-42). The reduction potential also is independent of electrode material; i.e., it is the same at mercury and platinum. For these conditions there is extensive evidence (40-43) that the rate-determining step is the one-electron-transfer reduction of oxygen (eq. 5). We believe that the disappearance of superoxide ion follows the identical path it does in aprotic media; i.e., protonation followed by disproportionation, and that it does not involve the electrochemical reduction of superoxide ion or its protonated species, $HO_2\cdot$.

By invoking the same mechanism, we again can explain the formation of hydrogen peroxide at inert electrodes in aqueous media (40-42,44) and preclude electrochemical reduction steps that involve multi-electron transfer or unstable reactants. We, in fact, take issue with any mechanism that requires the electrochemical reduction of superoxide species (40,45); the lifetimes of the superoxides in the presence of protons or bulk water are too short.

For alkaline aqueous conditions, the reduction of O_2 at inert electrodes occurs via reactions 5, 10, 11 and 12, with the overall process represented by

$$O_2 + H_2O + 2e^- \rightarrow HO_2^- + OH^- \qquad E_B^o = -0.06 \text{ V vs NHE} \qquad (17)$$

The superoxide disproportionation reactions are fast over a broad pH range, including moderately alkaline solutions. Between pH 2 and pH 13, for a wide variety of metal electrodes (including platinum, palladium, gold, silver, and nickel (41), mercury (46), and carbon (47)) oxygen is reduced at a potential near +0.05 V vs NHE. The observed potential is positive of the standard aqueous potential for reaction 5 (-0.33 V vs NHE) because the measurement procedure permits most of the $O_2^{\overline{\cdot}}$ at the electrode surface to dismutate (41).

Oxygen atom transfer. In contrast to the preceding electron-transfer mechanism for inert electrodes, the reduction of oxygen at freshly preoxidized metal electrodes is pH dependent, yields negligible amounts of hydrogen peroxide, and occurs at the same potential as that for the metal oxide of the electrode material (41,42) (see Fig. 3). A plausible explanation for these observations is represented by the reaction sequence

$$M(OH)_2 + 2H^+ + 2e^- \rightarrow M + 2H_2O \qquad (18a)$$

$$M(OH)_2 + 2e^- \rightarrow M + 2\ OH^- \qquad (18b)$$

$$2M + O_2 + 2H_2O \rightarrow 2M(OH)_2 \qquad (19)$$

In this mechanism, fresh metal oxide, $M(OH)_2$, is electrochemically reduced and reformed by a chemical oxidation of the surface by the oxygen in the solution. The scheme satisfies the requirement that the reduction potential and rate of reaction be dependent on the electrode material, M, and pH. The net reaction is

$$O_2 + 2H_2O + 4e^- = 4\ OH^- \qquad E_B^o = -0.40 \text{ V vs NHE} \qquad (20)$$

or

$$O_2 + 4H^+ + 4e^- = 2H_2O \qquad E^o = +1.23 \text{ V} \qquad (21)$$

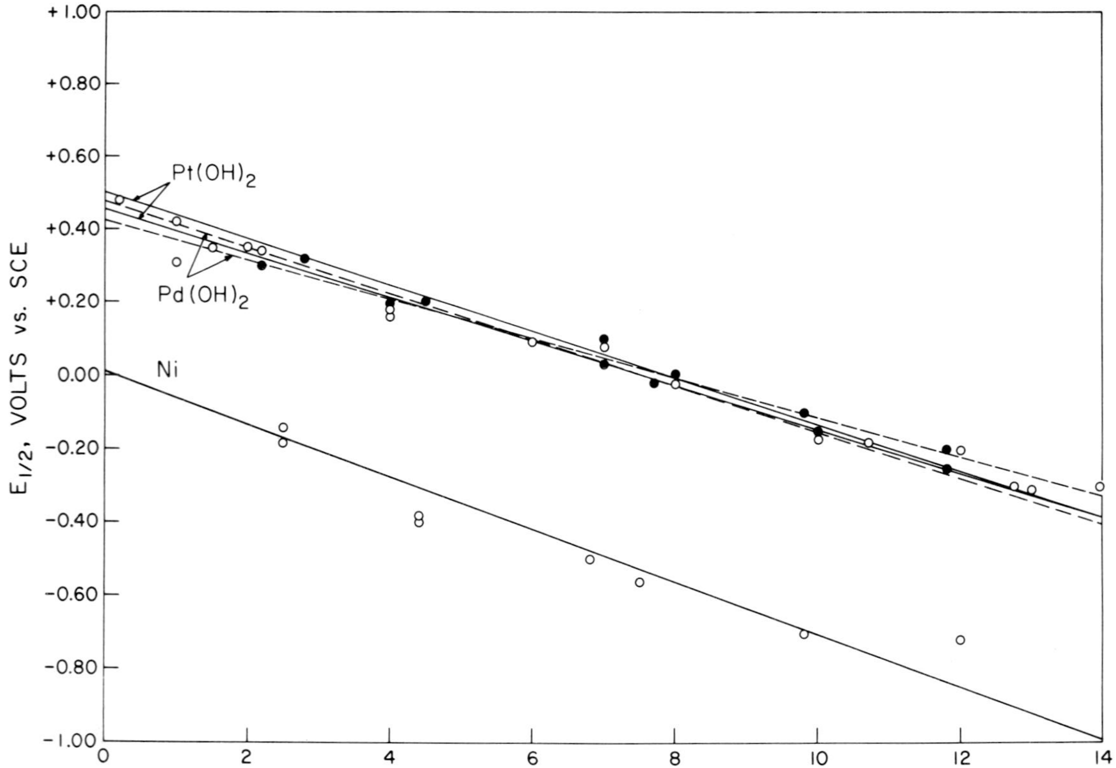

FIGURE 3. Voltammetric (slow-scan) data as a function of pH for the reduction (a) of freshly-formed metal oxides ($Pt(OH)_2$ and $Pd(OH)_2$) (closed circles); and (b) of O_2 (1 atm) with preoxidized electrodes ($Pt(OH)_2$, $Pd(OH)_2$ $Ni(OH)_2$) (open circles). All solutions contained 0.1 M K_2SO_4; H_2SO_4 and NaOH were used to adjust the pH.

and does not involve formation of peroxide. Representative experimental reduction potentials at platinum are +0.65 V vs NHE and -0.05 V at pH 2 and pH 13, respectively (41) (Fig. 3). Although eq. 18 clearly represents the rate-determining step in the reduction of oxygen at peroxidized electrodes, some oxygen may react directly at freshly reduced electrode sites to form trace amounts of peroxide via the electron-transfer mechanism of the preceding section.

Because the electron-transfer reduction of O_2 is a reversible one-electron process, the hope for an electron-transfer catalyst is futile. Atom-transfer catalysts (eq. 18 and 19) can promote more extensive reduction and higher potentials (those which correspond to the reduction of the metal oxide).

The proposition of a one-electron mechanism for the electron-transfer reduction of oxygen and the associated conclusions present significant ramifications relative to the development of improved fuel cells and metal-air batteries. To date the practical forms of such systems have used strongly acidic or basic electrolytes. Such solution conditions normally cause atom-transfer to be the dominant reduction process for molecular oxygen at metal electrodes. Hence, the search for effective catalytic materials should be in this context rather than in terms of a one-electron-transfer process.

Of considerable interest is the third step in the scheme in which the metal is spontaneously oxidized to yield the metal oxide. This is the general equation for the corrosion of metals, which appears to involve the splitting of the oxygen-oxygen bond as oxygen atoms are transferred to the metal. While the mechanism of this reaction is unclear, extensive thermodynamic data are available for the oxidation of metals by molecular oxygen (6).

CHEMICAL REACTIVITY OF SUPEROXIDE ION

Proton-induced dismutation and effective basicity. The kinetics and thermodynamics for the protonation of superoxide ion ($O_2^{\cdot-}$) in water have been evaluated by the pulse radiolysis technique (4,20-22) and are summarized by eq. 10-13 in the preceding section. The data confirm that protons promote a rapid bimolecular process and that the equilibrium in aqueous media at neutral pH is far to the right (Fig. 1)

$$2\ O_2^{\cdot-} + 2H^+ \rightleftarrows H_2O_2 + O_2\ , \qquad K_{pH\ 7} = 4 \times 10^{20} \qquad (22)$$

and even at pH 14 dismutation by H_2O is complete (48).

The effect of acidic substrates upon stable solutions of $O_2^{\cdot-}$ in aprotic solvents has been determined by electrochemical measurements in dimethylformamide (DMF) (49). After addition of the proton source to an $O_2^{\cdot-}$ solution, O_2 and HO_2^- are produced as well as the anion of the substrate. Kinetic measurements indicate that the primary rate controlling step is proton transfer from substrate to $O_2^{\cdot-}$, that the rate law is first-order in HA and first-order in $O_2^{\cdot-}$, and that the dismutation rate is proportional to substrate acidity. The apparent rate constants for the proton-induced dismutation of $O_2^{\cdot-}$ by acidic substrates in dimethylformamide at 25°C are: ascorbic acid ($k >> 2 \times 10^3 M^{-1}s^{-1}$), phenol ($k > 2 \times 10^3 M^{-1}s^{-1}$), α-tocopherol ($k = 8 \times 10^1 M^{-1}s^{-1}$), H_2O ($k = 2 \times 10^{-3} M^{-1}s^{-1}$), and *n*-butanol ($k = 1 \times 10^{-3} M^{-1}s^{-1}$). That the dismutation of $O_2^{\cdot-}$ follows pseudo first-order kinetics implies that a general acid-catalysis mechanism may be involved (50).

On the basis of the kinetic data, a self-consistent mechanism for the proton-induced dismutation of $O_2^{\cdot-}$ in aprotic media has been proposed for acidic substrates (HA) (49)

$$HA + O_2^{\cdot-} \rightleftarrows HO_2\cdot + A^- \qquad (23)$$

$$O_2^{\cdot-} + HO_2\cdot \xrightarrow{k_{24}} [\cdot O_2\text{-}H\text{-}O_2\cdot]^-_{tr} \longrightarrow \cdot O_2\cdot + HO_2^- \qquad (24)$$

$$HO_2^- + HA \rightarrow H_2O_2 + A^- \qquad (25)$$

We envisage that reaction 24 involves a triplet transition-state complex with $O_2^{\cdot-}$ acting as an electron-pair ligand, and that this accounts for the exclusive yield of triplet oxygen (51).

In water, the rate of proton transfer to $O_2^{\cdot-}$ is extremely rapid and the rate-determining step is reaction 24 with $k_{24} = 1 \times 10^8 M^{-1}s^{-1}$ (21). However, in dimethylformamide (DMF) the rate is limited by the transfer of a proton from HA to $O_2^{\cdot-}$, which is controlled by the K_a of the protic substrate.

The tendency of $O_2^{\cdot-}$ to dismutate via abstraction of protons from substrates and solvents is its most dominant reaction characteristic. Although the pK_a for $HO_2\cdot$ in water has a value of 4.69 (which infers that $O_2^{\cdot-}$ is a weak base) (21), combination of the redox data of Fig. 1 yields the net expression

$$2\ O_2^{\cdot-} + H_2O \rightleftarrows O_2 + HO_2^- + OH^- \qquad K = 0.91 \times 10^9 \qquad (26)$$

The value of the equilibrium constant indicates that $O_2^{\cdot-}$ can promote proton transfer from substrates (and solvents) to an extent equivalent to that of the conjugate base of an acid with a pK_a value of approximately 24. Thus, addition of $O_2^{\cdot-}$ to an aqueous solution results in the stoichiometric formation of HO_2^- and OH^-, both strong Brönsted bases. Hence, superoxide ion exhibits the properties of a strong Brönsted base and can effect removal of protons from substrates as weakly acidic as *n*-butanol ($pK_a = 33$ in DMF) (52).

The propensity of $O_2^{\cdot-}$ to remove protons from substrates accounts for its reactivity with acidic reductants and their overall oxidation. Thus, combination of $O_2^{\cdot-}$ with protic substrates (α-tocopherol, hydroquinone, 3,5-di-*tert*-butylcatechol, and L-(+)ascorbic acid) yields products that are consistent with an apparent one-electron oxidation of the substrate and the production of H_2O_2. However, the results of recent electrochemical studies provide clear evidence that these substrates are not oxidized by direct one-electron transfer to $O_2^{\overline{\cdot}}$ (52). The primary step involves abstraction of a proton from the substrate by $O_2^{\overline{\cdot}}$ to give substrate anion and the dismutation products of superoxide, H_2O_2 and O_2. In turn, the substrate anion is oxidized by O_2 in a multi-step process to yield oxidation products and H_2O_2. Thus, by continuously purging the O_2 that results from dismutation of $O_2^{\overline{\cdot}}$ when it is combined with α-tocopherol (by vigorous Ar bubbling through the solution), quantitative yields of substrate anion are obtained without significant oxidation. A mechanism that is consistent with these observations can be outlined by a reaction sequence for an acidic substrate reductant, HA; eq. 23-25 represent the first three steps. In turn, the substrate anion and the O_2 from the proton-induced dismutation of $O_2^{\overline{\cdot}}$ react to give the oxidation products

$$A^- + O_2 \rightleftarrows A\cdot + O_2^{\overline{\cdot}} \qquad (27)$$

$$n\ A\cdot \xrightarrow[(O_2,\ O_2^{\overline{\cdot}},\ H_2O_2)]{} \text{dimeric, dismutation, and redox products} \qquad (28)$$

For aprotic solvents the rate-determining step is proton-transfer from substrate to $O_2^{\overline{\cdot}}$ (eq. 23); the apparent rate is first-order in substrate and is proportional to its acidity.

The reported oxidation of thiols (53,54) and alcohols (53) by $O_2^{\overline{\cdot}}$ presumably occurs by the same initial proton-induced dismutation step.

<u>Nucleophilicity</u>. The reactivity of $O_2^{\overline{\cdot}}$ with alkyl halides via nucleophilic substitution was first reported in 1970 (53,55). These and subsequent kinetic studies (56,59) confirm that the reaction is first-order in substrate, that the rates follow the order $1^o > 2^o >> 3^o$ for alkyl halides and tosylates, and that the attack by $O_2^{\overline{\cdot}}$ results in an inversion of configuration. Thus, the $O_2^{\overline{\cdot}}$-RX reactions occur by an S_N2 mechanism.

In aprotic solvents primary and secondary alkyl halides react to yield primarily dialkyl peroxides with accompanying elimination products (53,58,59), which is consistent with a multi-step S_N2 mechanism

$$RX + O_2^{\overline{\cdot}} \rightarrow RO_2\cdot + X^- \qquad (29)$$

$$RO_2\cdot + O_2^{\overline{\cdot}} \rightarrow RO_2^- + O_2 \qquad (30)$$

$$RO_2^- + RX \rightarrow ROOR + X^- \qquad (31)$$

In dimethyl sulfoxide reaction 30 is followed by attack of the solvent by RO_2^- to give dimethylsulfone and alcohols (55,57,60). The rate constants for these reactions in dimethyl sulfoxide and in pyridine (48) indicate that the nucleophilicity of $O_2^{\overline{\cdot}}$ for such substrates is comparable to the most potent nucleophiles that are known, thiophenoxide and thiocyanate.

The reaction of $O_2^{\overline{\cdot}}$ with esters yields carboxylic acids and alcohols (61-63), and with acyl halides yields diacylperoxides (64). Acyl oxygen scission occurs in the ester reaction on the basis of complete retention of configuration in the alcohol moiety from an optically active ester (61). Further, the order of reactivity for different n-octanoate esters follows typical acyl-oxygen scission patterns.

In pyridine the second-order rate constants for the reaction of $O_2^{\overline{\cdot}}$ with ethyl acetate and with phenyl acetate are 0.011 $M^{-1}s^{-1}$ and 160 $M^{-1}s^{-1}$, respectively (63). Such a ratio of rates (10^4) is consistent with other data for reactions of phenyl and ethyl esters with effective nucleophiles (65,66), and with an S_N2 mechanism. The initial reaction is proposed to occur via nucleophilic addition and then elimination at the carbonyl carbon (63)

$$\text{R-}\overset{\text{O}}{\overset{\|}{\text{C}}}\text{-OR}' + O_2^{\overline{\cdot}} \rightleftarrows \text{R-}\underset{\text{O-O}\cdot}{\overset{\text{O}^-}{\text{C}}}\text{-OR}' \rightleftarrows \text{R-}\overset{\text{O}}{\overset{\|}{\text{C}}}\text{-O-O}\cdot + {}^-\text{OR}' \qquad (32)$$

with subsequent steps to give the net reaction of ester hydrolysis

$$\text{R}\overset{\text{O}}{\overset{\|}{\text{C}}}\text{OO}\cdot + O_2^{\overline{\cdot}} \rightarrow \text{R}\overset{\text{O}}{\overset{\|}{\text{C}}}\text{OO}^- + O_2 \qquad (33)$$

$$\text{R}\overset{\text{O}}{\overset{\|}{\text{C}}}\text{OO}^- + \text{R}\overset{\text{O}}{\overset{\|}{\text{C}}}\text{OR}' \rightarrow \text{R}\overset{\text{O}}{\overset{\|}{\text{C}}}\text{O}\overset{\text{O}}{\overset{\|}{\text{C}}}\text{R} + {}^-\text{OR}' \qquad (34)$$

$$\text{R}\overset{\text{O}}{\overset{\|}{\text{C}}}\text{O}\overset{\text{O}}{\overset{\|}{\text{C}}}\text{R} + 2\ O_2^{\overline{\cdot}} \rightarrow 2\text{R}\overset{\text{O}}{\overset{\|}{\text{C}}}\text{O}^- + 2\ O_2 \qquad (35)$$

Combination of $O_2^{\overline{\cdot}}$ with α-keto, α-hydroxy, and α-halo carbonyl compounds results in nucleophilic addition at the carbonyl carbon followed by oxidative cleavage to give the carboxylic acid that is derived from the α position (67). Likewise, $O_2^{\overline{\cdot}}$ reacts with CO_2 by an initial addition step which is followed by a reduction (a second $O_2^{\overline{\cdot}}$) to give CO_4^{2-} (68).

One-electron reductant. Another reaction characteristic of $O_2^{\overline{\cdot}}$ is its ability to act as a moderate one-electron reducing agent. Voltammetric measurements clearly indicate that O_2 is evolved when $O_2^{\overline{\cdot}}$ is combined with a quinone (DTBQ) and that the major product is 3,5-di-tert-butyl-o-semiquinone anion radical (λ_{max}, 383 nm (ε, 2,250)). The relevant redox potentials in DMF are

$$O_2 + e^- \rightleftarrows O_2^{\overline{\cdot}} \qquad E^{o'} = -0.60 \text{ V vs NHE} \qquad (36)$$

$$\text{DTBQ} + e^- \rightleftarrows \text{DTBQ}^{\overline{\cdot}} \qquad E^{o'} = -0.25 \text{ V vs NHE} \qquad (37)$$

which indicate that the equilibrium constant, *K*, for the reaction of $O_2^{\overline{\cdot}}$ with DTBQ has a value of 0.8×10^6

$$O_2^{\overline{\cdot}} + DTBQ \rightleftarrows DTBQ^{\overline{\cdot}} + O_2 \quad (38)$$

Analogous results are obtained for the reduction of *p*-benzoquinone by $O_2^{\overline{\cdot}}$ in DMF, with a stoichiometric yield of the semiquinone anion radical on the basis of the ESR spectrum.

Electrochemical studies of sulfur dioxide (70) and of molecular oxygen (24) in dimethylformamide indicate that $O_2^{\overline{\cdot}}$ has a sufficiently negative potential (eq. 36) to reduce SO_2 to a limited extent.

$$SO_2 + e^- \rightleftarrows SO_2^{\overline{\cdot}} \qquad E^{o\prime} = -0.58 \text{ V vs NHE} \quad (39)$$

On the basis of these redox potentials the equilibrium constant, *K*, for the reaction of SO_2 with $O_2^{\overline{\cdot}}$ has a value of 2.2

$$SO_2 + O_2^{\overline{\cdot}} \rightarrow O_2 + SO_2^{\overline{\cdot}} \quad (40)$$

However, the reaction goes to completion because of complexation of $SO_2^{\overline{\cdot}}$ by SO_2 and its dimerization to dithionite ion (71).

The electrochemical, spectroscopic, and chemical evidence confirms that $O_2^{\overline{\cdot}}$ is a stronger and more effective reducing agent than dithionite ion in aprotic solvents. This capacity of $O_2^{\overline{\cdot}}$ to act as a strong one-electron reducing agent may represent its most serious biological hazard. For example, the reduction of ferricytochrome c by $O_2^{\overline{\cdot}}$

$$\text{cyt c } (Fe^{3+}) + O_2^{\overline{\cdot}} \rightleftarrows \text{cyt c}(Fe^{2+}) + O_2 \quad (41)$$

has an equilibrium constant with a value of 3.7×10^4 in aqueous media (72).

Superoxide ion also is an effective reducing agent of transition metal complexes; examples include copper (II) (73,74), manganese (III) (75,76), and iron (III) (77). Studies in progress (78) indicate that $O_2^{\overline{\cdot}}$ also reduces ferricinium ion, $Mn_2^{IV}O_2(o\text{-phen})_4^{4+}$, $Co^{III}(o\text{-phen})_3^{3+}$, and $Ir^{IV}Cl_6^{2-}$ by a one-electron process.

Oxidation of $O_2^{\overline{\cdot}}$ to $^1O_2$. Formation of singlet oxygen ($O_2(^1\Delta g)$) has been reported to result from the oxidation of $O_2^{\overline{\cdot}}$ with ferricinium ion (79) and diacyl peroxides (80), and from the dismutation of $O_2^{\overline{\cdot}}$ by proton sources (81). Recent studies in our laboratory support the formation of $^1O_2$ by the oxidation of $O_2^{\overline{\cdot}}$ with ferricenium ion in DMF (51). However, there is little evidence for its formation from the proton-induced dismutation of $O_2^{\overline{\cdot}}$ in DMF.

Conversion of $^3O_2$ to $^1O_2$ is a highly endothermic process (82)

$$^3O_2 \rightarrow {}^1O_2 \qquad \Delta G = +91 \text{ kJ} \quad (42)$$

On the basis of the relationship

$$\Delta G_{rxn} = -nF\Delta E^{o}_{rxn} \tag{43}$$

where ΔG_{rxn} = 91 kJ, n = 1, and F = 96,487C, the value for ΔE^{o}_{rxn} is -0.94 V. In DMF the reversible reduction of $^{3}O_2$ occurs at E^{o} = -0.60 V vs NHE. Subtraction of eq. 42 from eq. 44

$$^{3}O_2 + e^- \rightarrow O_2^{\cdot -} \tag{44}$$

and combination of the appropriate redox potentials yields

$$^{1}O_2 + e^- \rightarrow O_2^{\cdot -} \qquad E^{o} = +0.34 \text{ V vs NHE} \tag{45}$$

Hence, from thermodynamic considerations the oxidation of $O_2^{\cdot -}$ to $^{1}O_2$ requires an oxidant with a redox potential of at least +0.34 V vs NHE.

The results of photon counting experiments indicate that both proton-induced dismutation ($HClO_4 + O_2^{\cdot -}$) and transition-metal ion oxidation ($Fe^{III}(ClO_4)_3 + O_2^{\cdot -}$) of $O_2^{\cdot -}$ yield predominantly triplet oxygen ($^{3}O_2$) (51). This is believed to be the result of a triplet transition state complex for the dismutation process. $[(\cdot O_2:)H(:O_2\cdot)]^{-}_{tr}$, and a triplet state for the redox reaction $[(\cdot O_2:)Fe^{III}(:O_2\cdot]_{tr}$. Wavefunctions for the ground state of $O_2^{\cdot -}$ have been calculated by use of unrestricted open shell LCAO-SCF functions that are based on the INDO formalism (51). The calculations reveal that removal of the highest energy electron from $O_2^{\cdot -}$ will produce a triplet state. The calculations also show that a triplet transition state leads to triplet O_2, while a singlet transition state leads to $^{1}\Delta_g$. Because these reactions yield $^{3}O_2$, triplet transition state complexes are plausible.

The photon counting experiments for the oxidation of $O_2^{\cdot -}$ by ferricenium perchlorate indicate a significant yield of $^{1}O_2$. While both $Fe^{III}(ClO_4)_3$ and ferricenium ion are thermodynamically strong enough one-electron oxidants (E^{o}>+0.34 V vs NHE) to oxidize $O_2^{\cdot -}$ to $^{1}O_2$, only ferricenium ion does so. This can be rationalized on the basis that ferricenium ion is coordinately saturated and may form a singlet transition state complex with $O_2^{\cdot -}$, $[(Cp)_2Fe:O_2:]_{tr}$. Other oxidants which convert $O_2^{\cdot -}$ to $^{1}O_2$ include $Mn^{III}(O_3terpy)_2^{3+}$ and $[Mn^{IV}(o\text{-phen})_2O]_2^{4+}$ (51).

Reactions with metal ions. Metal ions may be classified as Lewis acids and, like the Brönsted acids HCl, NH_4ClO_4, and phenol, affect the reduction of O_2 and the dismutation of $O_2^{\cdot -}$. The reduction potentials of O_2 in DMSO in the presence of a 20-fold excess of $LiClO_4$, $Zn(ClO_4)_2$, $Cd(ClO_4)_2$, $Fe(ClO_4)_2$, $Mn(ClO_4)_2$, and $Co(ClO_4)_2$ are -0.54 V vs NHE, -0.42 V, -0.42 V, -0.27 V, -0.36 V, and -0.41 V, respectively. These metal ions cause the reduction of O_2 to $O_2^{\cdot -}$ to occur at more positive potentials in the order: $Fe^{2+}>Mn^{2+}>Zn^{2+}\sim Co^{2+}>Cd^{2+}>Ni^{2+}>Li^{+}\sim TEA^{+}$ (where TEA^{+} is tetraethylammonium ion, the cation commonly employed as a supporting electrolyte). The overall electron stoichiometry in the presence of Mn^{2+} and Co^{2+} is four electrons per O_2 whereas in the presence of Fe^{2+}, Zn^{2+}, Cd^{2+}, and Ni^{2+} it is two electrons per O_2. When Li^{+} is present, only one electron is consumed in the process.

The observations indicate that O_2 reduction is dependent upon the Lewis acid and that either two- or four-electron reductions of O_2 occur via dismutation reactions after the initial one-electron reduction to $O_2^{\overline{\cdot}}$. As with protons, $O_2^{\overline{\cdot}}$ acts as an electron-pair donor (Lewis base) to the metal ion. In fact, some of the earliest electrochemical studies of $O_2^{\overline{\cdot}}$ in aprotic media have demonstrated this with protons, alkali metal ions, and alkaline earth ions (24,83,84). These superoxy complexes are only transiently stable and decompose to give O_2 and the corresponding peroxide. However, Li^+ ion appears to form a stable $Li(:O_2\cdot)$ complex. ESR and kinetic measurements demonstrate that $O_2^{\overline{\cdot}}$ forms coordination complexes with Ca(II) and Ba(II) in aqueous media as well as DMSO solutions (85).

The electrochemical results indicate that $O_2^{\overline{\cdot}}$ also acts as an electron-pair donor for Zn(II), Cd(II), Mn(II), Fe(II), Co(II), and Ni(II). When Zn^{2+} ion and $O_2^{\overline{\cdot}}$ are combined in aprotic solvents, a two-step reaction occurs with an initial transiently stable bis-coordination complex that decomposes to give O_2 and zinc peroxide (83,84). Again, we envisage this to involve a triplet transition-state complex

$$O_2^{\overline{\cdot}} + Zn^{2+} \rightarrow Zn(:O_2\cdot)^+ \quad (46)$$

$$O_2^{\overline{\cdot}} + Zn(:O_2\cdot)^+ \rightarrow [(\cdot O_2:)Zn(:O_2\cdot)]_{tr} \rightarrow ZnO_2 + \cdot O_2\cdot \quad (47)$$

Analogous mechanisms are proposed for Cd(II), Fe(II), Mn(II), Co(II), and Ni(II). A recent report indicates that $O_2^{\overline{\cdot}}$ produced by pulse radiolysis in water forms a transiently stable electron-pair donor complex with Mn(II) (86). The combination of electrogenerated $O_2^{\overline{\cdot}}$ in acetonitrile with $Mn^{II}(ClO_4)_2$ produces similar results. The complex decomposes to yield O_2 and Mn(III). Based on the overall electron stoichiometry of four electrons for the electrochemical studies, the peroxides of Mn(II) and Co(II) are unstable and dismutate. A possible mechanism for the decomposition of $O_2^{\overline{\cdot}}$ by Mn(II) or Co(II) can be represented by a sequence of dismutation reactions.

$$Mn^{2+} + O_2^{\overline{\cdot}} \rightarrow Mn^{II}(:O_2\cdot)^+ \quad (48)$$

$$Mn^{II}(:O_2\cdot)^+ + O_2^{\overline{\cdot}} \rightarrow [\cdot O_2:Mn^{II}:O_2\cdot] \rightarrow O_2 + Mn^{II}O_2 \quad (49)$$

$$Mn^{II}O_2 + Mn^{II}O_2 \rightarrow 2Mn^{II}O + O_2 \quad (50)$$

$$Mn^{II} + O_2 \rightarrow Mn^{III}(:O_2\cdot)^{2+} \quad (51)$$

$$Mn^{III}(:O_2\cdot)^{2+} + Mn^{II} \rightarrow [Mn^{III}\text{-O-O-}Mn^{III}]^{4+} \quad (52)$$

$$[Mn^{III}\text{-O-O-}Mn^{III}]^{4+} + Mn^{II} \rightarrow [Mn^{III}\text{-O-}Mn^{III}]^{4+} + [Mn^{III}O\cdot]^{2+} \quad (53)$$

$$[Mn^{III}O\cdot]^{2+} + Mn^{II} \rightarrow [Mn^{III}\text{-O-}Mn^{III}]^{4+} \quad (54)$$

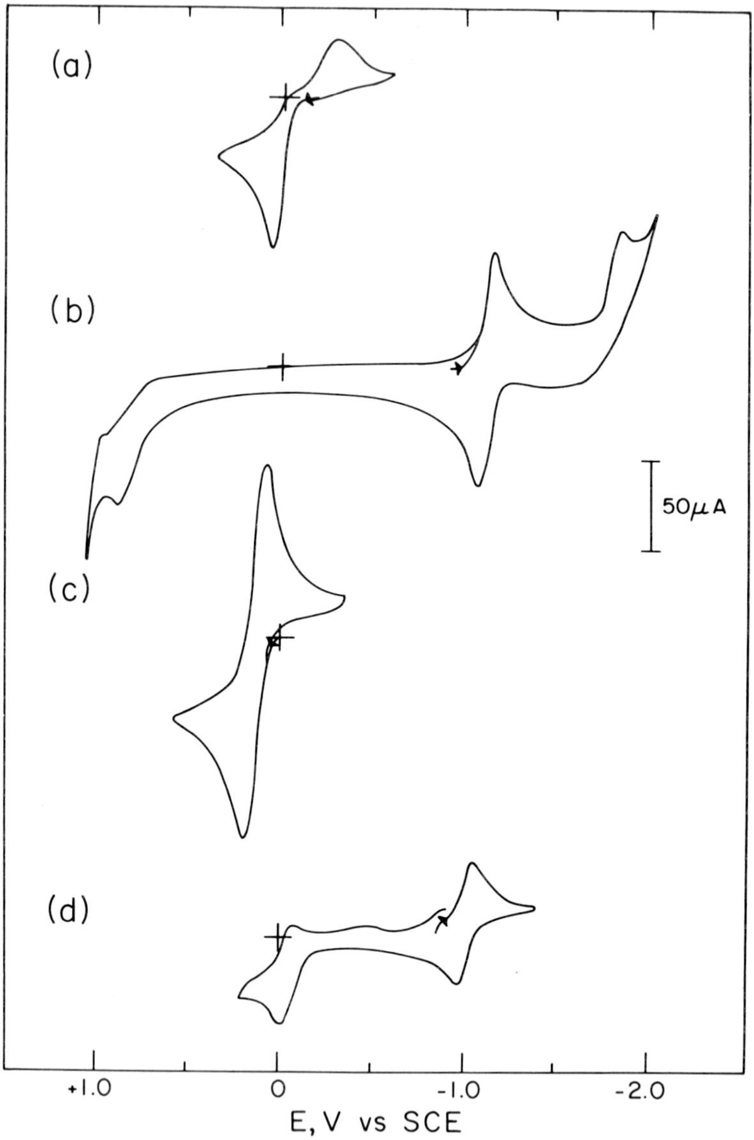

FIGURE 4. Cyclic voltammograms in dimethylformamide (0.1 M tetraethylammonium perchlorate) of (a) 1 mM N,N'-dihydrophenazine; (b) 1 mM N,N'-dihydrophenazine plus 1 mM $O_2^{\overline{\cdot}}$; (c) 1.3 mM N-methyl, N'-hydrophenazine; nad (d) 1 mM N-methyl, N'-hydrophenazine plus 1 mM $O_2^{\overline{\cdot}}$. Measurements were made with a platinum electrode (area, 0.23 cm^2) at a scan rate of 0.1 V s^{-1}; temperature, 25°C.

$$[Mn^{III}O\cdot]^{2+} + [Mn^{III}\text{-}O\text{-}O\text{-}Mn^{III}]^{4+} \rightarrow [Mn^{III}\text{-}O\text{-}Mn^{III}]^{4+} + [Mn^{III}(:O_2\cdot)]^{2+} \quad (55)$$

Oxidant via H-atom transfer from reduced flavins, hydrophenazines, and hydroxylamine. The direct transfer of an electron to $O_2^{\cdot-}$ is an unlikely process in aprotic media because of the extreme instability of the O_2^{2-} species (20,48). As a result, most of the reported oxidations by $O_2^{\cdot-}$ actually represent an initial proton abstraction to give substrate anion and dismutation species, HO_2^- and O_2. The latter oxidize the substrate anion (see eq. 23-27).

Recent studies (87) demonstrate that $O_2^{\cdot-}$ is an effective oxidant of basic reducing substrates with readily transferable hydrogen atoms. Fig. 4 illustrates that dihydrophenazine and N-methyl hydrophenazine (both models for reduced flavin) are oxidized to phenazine and N-methyl phenazine radical, respectively. The reaction stoichiometry in each case is one-to-one, but the net oxidation change is two for dihydrophenazine and one for N-methyl hydrophenazine. The apparent net reactions for hydrophenazines, reduced flavins, hydrazines, and hydroxylamine can be represented by eqs. 56-61.

$$PhenH_2 + O_2^{\cdot-} \xrightarrow{HSol} Phen + \cdot Sol + H_2O + OH^- \quad (56)$$

$$CH_3NPhenH + O_2^{\cdot-} \rightarrow CH_3NPhen\cdot + HO_2^- \quad (57)$$

$$FlH_2 + O_2^{\cdot-} \xrightarrow{HSol} Fl + \cdot Sol + H_2O + OH^- \quad (58)$$

$$C_2H_5NFlH + O_2^{\cdot-} \rightarrow C_2H_5NFl\cdot + HO_2^- \quad (59)$$

$$3\ H_2N\text{-}NH_2 + 4\ O_2^{\cdot-} \rightarrow 3N_2 + 4\ H_2O + 4\ OH^- \quad (60)$$

$$H_2NOH + O_2^{\cdot-} \rightarrow NO + H_2O + OH^- \quad (61)$$

The stoichiometry of the reactions is controlled by the number of reducing hydrogen atoms per substrate molecule. Conversion of such hydrogen atoms to protons provides the means to stabilize the reduction products of $O_2^{\cdot-}$ (O_2^{2-}, $\cdot O^-$, and O^{2-}). Hence, $O_2^{\cdot-}$ is a selective oxidant for those basic substrates that are susceptible to oxidation via a hydrogen-atom transfer mechanism.

Radical-radical coupling with reduced methyl viologen. When reduced methyl viologen is combined with O_2, a stoichiometric yield of $O_2^{\cdot-}$ results

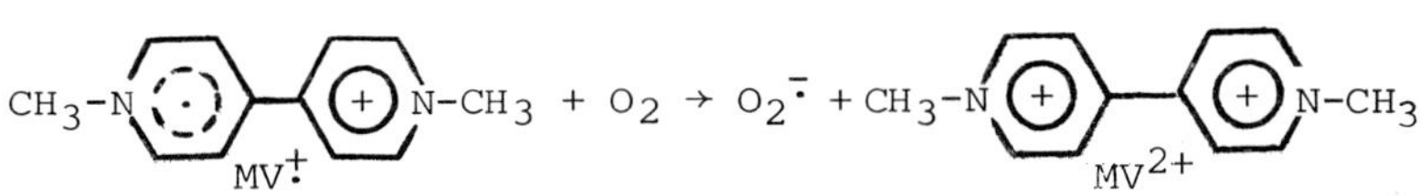

(62)

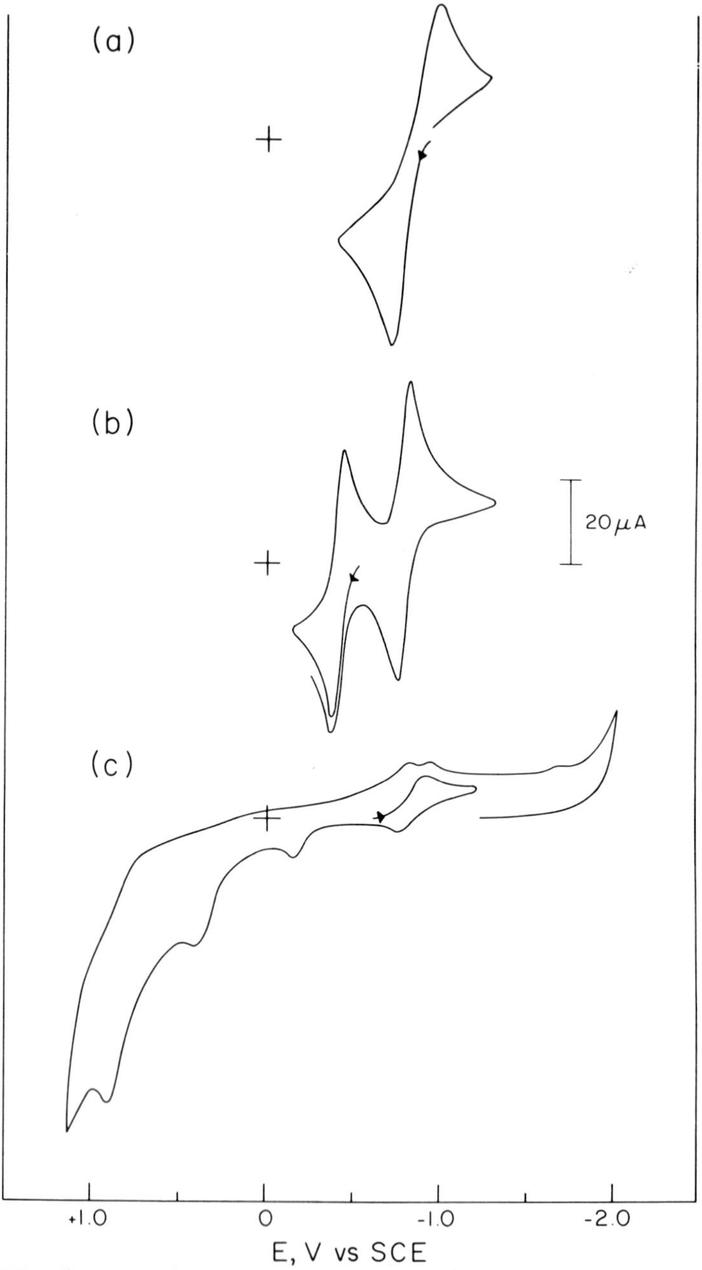

FIGURE 5. Cyclic voltammograms in DMF (0.1 M TEAP) of (a) 1 mM $O_2^{\cdot -}$; (b) 1 mM methyl viologen cation radical ($MV^{\dot{+}}$); and (c) 0.5 mM $O_2^{\cdot -}$ plus 0.5 mM $MV^{\dot{+}}$. Measurements were made with a platinum electrode (area, 0.23 cm^2) at a scan rate of 0.1 V s^{-1}.

which rapidly dismutates in aqueous media as a result of charge neutralization through ion-pairing with MV^{2+}.

$$MV^{2+} + 2\ O_2^{\overline{\cdot}} \rightarrow MV^{2+}(O_2^{\ 2-}) + O_2 \qquad (63)$$

However, combination of DMF solutions of $MV^{\stackrel{+}{\cdot}}$ and of $O_2^{\overline{\cdot}}$ in one-to-one stoichiometry results in the irreversible formation of a dioxygen adduct, $[MVO_2]$ (88).

$$MV^{\stackrel{+}{\cdot}} + O_2^{\overline{\cdot}} \rightarrow [MVO_2] \qquad (64)$$

To the best of our knowledge this is the first example of a radical-radical coupling reaction for $O_2^{\overline{\cdot}}$. Reference to Fig. 5 confirms that the adduct does not have the electrochemical characteristics of the reactants or their redox products. Although $MV^{\stackrel{+}{\cdot}}$ and $O_2^{\overline{\cdot}}$ are paramagnetic and exhibit characteristic ESR spectra, as well as UV absorption spectra, the adduct is ESR-silent and has a new absorption band at 407 nm in DMF. The structural characterization of $[MVO_2]$ is still in progress, but four diamagnetic species that are consistent with the electrochemical and spectroscopic results are likely possibilities

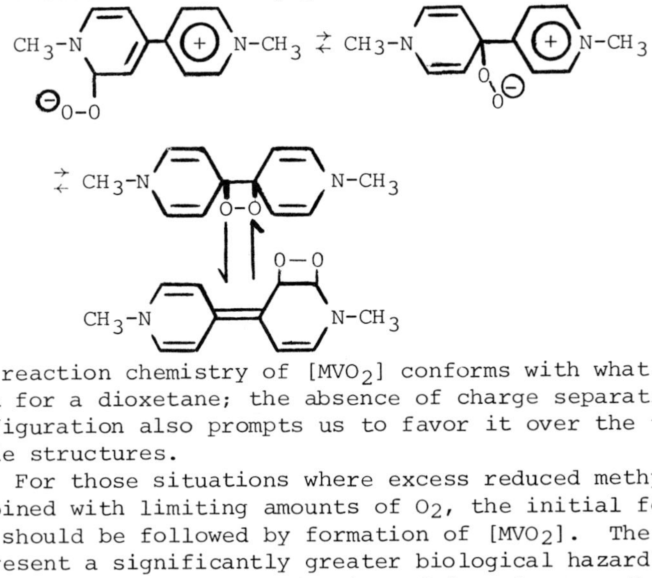

(65)

The reaction chemistry of $[MVO_2]$ conforms with what one would expect for a dioxetane; the absence of charge separation in this configuration also prompts us to favor it over the two alkyl peroxide structures.

For those situations where excess reduced methyl viologen is combined with limiting amounts of O_2, the initial formation of $O_2^{\overline{\cdot}}$ should be followed by formation of $[MVO_2]$. The latter may represent a significantly greater biological hazard than free $O_2^{\overline{\cdot}}$. Furthermore, the formation of $[MVO_2]$ means that methyl viologen (paraquat) probably is specifically involved in toxic reactions rather than acting solely as a reducing agent for the production of $O_2^{\overline{\cdot}}$. This proposition is consistent with the observations that superoxide dismutase (SOD) suppresses the toxic effects of methyl viologen (89). Addition of SOD would remove the $O_2^{\overline{\cdot}}$ that is produced by $MV^{\stackrel{+}{\cdot}}$ in reaction 62 before it could combine with a second $MV^{\stackrel{+}{\cdot}}$ in reaction 64 to produce $[MVO_2]$.

REDOX CHEMISTRY OF H_2O_2 AND HO_2^-

Base-induced dismutation to superoxide ion. When tetraalkylammonium hydroxide is added to H_2O_2 in pyridine, large quantities of superoxide ion are generated (90). Spin-trapping experiments and the subsequent ESR spectra of the products indicate that ·OH also is formed in the reaction. The addition of base to H_2O_2 in acetonitrile yields O_2, but there is no evidence of $O_2^{\cdot-}$. Addition of base to H_2O_2 in water causes a decomposition to water and oxygen (91).

These results and other evidence indicate that superoxide ion is produced by the reaction of peroxide ion with hydrogen peroxide.

$$H_2O_2 + OH^- \rightleftarrows HO_2^- + H_2O \tag{66}$$

$$H_2O_2 + HO_2^- \rightarrow O_2^{\cdot-} + \cdot OH + H_2O \tag{67}$$

Pulse radiolysis data confirm that the reaction of ·OH with pyridine is rapid (92).

$$\cdot OH + py \xrightarrow{k\,=\,3\,\times\,10^9 M^{-1}s^{-1}} py\cdot OH \rightarrow polymer \tag{68}$$

Thus, pyridine traps hydroxyl radicals before they can react any further. Fig. 6 illustrates a mechanistic intermediate for the base-induced dismutation of H_2O_2 in pyridine. The reaction of ·OH with acetonitrile is much slower than with pyridine (92).

$$CH_3CN + \cdot OH \xrightarrow{k\,=\,4\,\times\,10^6 M^{-1}s^{-1}} H_2O + \cdot CH_2CN \tag{69}$$

Thus, reaction 70 becomes competitive with reaction 69.

$$\cdot OH + O_2^{\cdot-} \xrightarrow{k\,=\,1\,\times\,10^{10} M^{-1}s^{-1}} O_2 + OH^- \tag{70}$$

Reactions 66, 67, and 70 appear to represent a viable mechanism for the base-catalyzed decomposition of H_2O_2 to oxygen and water.

The same processes that dominate the base-induced decomposition of H_2O_2 in acetonitrile probably occur in aqueous solution. A second pathway for superoxide decomposition also is likely in aqueous solutions at pH 7 (22).

$$2\,O_2^{\cdot-} + H_2O \xrightarrow{k\,=\,10^7 M^{-1}s^{-1}} HO_2^- + O_2 + OH^- \tag{71}$$

Reaction 67 is presumed to be much faster in pyridine and acetonitrile than in water, due to the general enhancement of the nucleophilic reactivity of HO_2^- in aprotic solvents.

The reaction of H_2O_2 with HO_2^- may prove to be an important pathway to the formation of $O_2^{\cdot-}$ and ·OH in biological systems. Although the rate constant for the decomposition of H_2O_2 in H_2O at pH 7 is only $4 \times 10^{-8} M^{-1}s^{-1}$ (91), the rate is rapid in aprotic media in the presence of base. Thus, for the hydrophobic regions of biomembranes, reaction 67 may become significant and an alternative to the so-called Haber-Weiss reaction (93) as a source of ·OH radical.

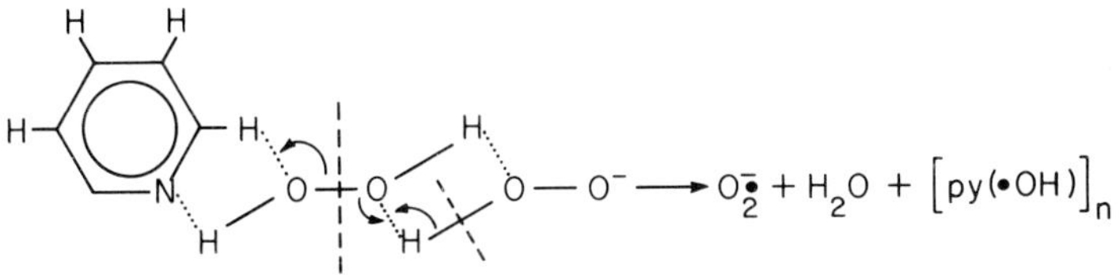

FIGURE 6. Mechanistic pathway for the base-induced dismutation of H_2O_2 in pyridine.

$$H_2O_2 + O_2^{\bar{\cdot}} \rightarrow \cdot OH + O_2 + OH^- \quad (72)$$

Recent reports (94) indicate that the rate constant for reaction 72 is less than 16 $\underline{M}^{-1}s^{-1}$, and probably as low as $10^{-4}\underline{M}^{-1}s^{-1}$; too low a value for this to be a significant source of ·OH radical.

Electron-transfer reduction. The electrochemical reduction of H_2O_2 in pyridine yields superoxide ion (95). That a reduction process generates a species with a higher oxidation state than H_2O_2 is surprising and has prompted a series of experiments. The cyclic voltammogram for the reduction of H_2O_2 in pyridine at a platinum working electrode exhibits a broad cathodic peak at -0.95 V vs NHE and anodic peaks for the reverse scan at -0.50 V and -0.15 V vs NHE. The latter is characteristic of electrolytically generated H_2. Controlled-potential reduction of H_2O_2 in pyridine at -1.0 V vs NHE (with argon degassing) results in a solution that exhibits an anodic cyclic voltammogram that is characteristic of $O_2^{\bar{\cdot}}$. ESR studies of the reduced solution confirm the presence of superoxide ion.

The products and the observed electron stoichiometries for the electrochemical reduction of H_2O_2 are consistent with a mechanism in which the primary step is a one-electron transfer

$$H_2O_2 + e^- \rightarrow HO_2^- + \tfrac{1}{2}H_2 \qquad E^{o\prime} = -1.0 \text{ V vs NHE} \quad (73)$$

followed by a chemical reaction with another hydrogen peroxide molecule (reaction 67). The resultant ·OH is trapped by the pyridine solvent via reaction 68 to yield a stable solution of $O_2^{\bar{\cdot}}$.

There is no evidence of superoxide ion, either by ESR or by cyclic voltammetry, from the electrochemical reduction of H_2O_2 in acetonitrile. This can be explained by the slow rate of reaction of ·OH with CH_3CN (reaction 69) which permits ·OH and $O_2^{\bar{\cdot}}$ to react via reaction 70 to give O_2. The reduction of hydrogen peroxide in aqueous solution appears to be analogous to that in CH_3CN, with the mechanism represented by reaction 73, followed by reactions 67 and 70. Thus, the reduction of H_2O_2 yields H_2 and HO_2^- initially, in a one-electron step. The final products are the result of the reaction of HO_2^- with H_2O_2 and are analogous to those for the base-induced decomposition of H_2O_2.

SUMMARY

The redox chemistry of dioxygen and its reduction products is heavily dependent upon mechanistic pathway, substrate, and solution acidity. For those circumstances that are limited by direct electron transfer, the redox mechanisms for dioxygen species involve one-electron steps. To achieve the full oxidizing potential and energetics for O_2 requires catalysts that facilitate atom-transfer mechanisms and/or stabilize dioxygen intermediates. The dominant characteristic of $O_2^{\cdot-}$ (the primary product from the electron-transfer reduction of O_2) is its ability to act as a Brönsted base. Under aprotic conditions $O_2^{\cdot-}$ is a strong nucleophile, a moderate one-electron reducing agent, a source of $^1O_2$ when oxidized by a strong one-electron oxidant with a closed coordination sphere, and a hydrogen-atom-transfer oxidant of basic substrates such as hydroxylamines and reduced flavins. The base-induced dismutation of H_2O_2 leads to the formation of $O_2^{\cdot-}$ and $\cdot OH$ in aprotic media. The variety of oxidation-reduction, hydrolytic, atom-transfer, and dismutation reaction pathways for dioxygen species is ample evidence of the unique chemical versatility of O_2.

ACKNOWLEDGEMENT

This work was supported by the National Science Foundation under Grant No. CHE-7922040.

REFERENCES

1. Latimer, W. M., "Oxidation Potentials," 2nd ed., Prentice-Hall, New York, N.Y., (1952), p. 38.
2. Sawyer, D. T. and Seo, E. T., Inorg. Chem., 16,499 (1977).
3. Roberts, J. L., Jr., Morrison, M. M. and Sawyer, D. T., J. Am. Chem. Soc., 100,329 (1978).
4. Ilan, Y. A., Meisel, D. and Czapski, G., Israel J. Chem., 12,891 (1974).
5. Williams, R. J. P., private communication, 1978.
6. Pourbaix, M., "Atlas De-Equilibres Electrochimiques a 25°C," Gautheir-Villars et Cie, Paris,(1963).
7. Ochiai, E. I., "Bioinorganic Chemistry, An Introduction," Allyn & Bacon, Boston, Mass.,(1977).
8. Malmstrom, B. G., private communication, 1978.
9. Lawrence, G. D. and Sawyer, D. T., Coordin. Chem. Rev., 27, 173 (1978).
10. Walsh, A. D., Trans. Faraday Soc., 42,269 (1946).
11. Walsh, A. D., Trans. Faraday Soc., 43,297 (1947).
12. Walsh, A. D., Trans. Faraday Soc., 43,305 (1947).
13. Strehlow, R. A., "Fundamentals of Combustion," Internatl. Textbook Co., Scranton, PA,(1968).

14. Kondratiev, V. N. and Azatyan, V. V., "HO_2 Radicals in Combustion Reactions," 14th Symp. on Combustion, The Combustion Institute, Pittsburgh, PA, (1973), p. 37.
15. Lewis, B. and von Elbe, G., "Combustion, Flames, and Explosions of Gases," Academic Press, N.Y., (1951).
16. Harris, E. J. and Egerton, A. C., Proc. R. Soc. London, Ser. A, 168,1 (1938).
17. Harris, E. J., Proc. R. Soc. London, Ser. A, 173,126 (1939).
18. Rice, F. O. and Rodowskas, E. L., J. Am. Chem. Soc., 57,350 (1935).
19. George, P., "Oxidases and Related Redox Systems," (T. E. King, H. S. Mason, and M. Morrison, eds.), John Wiley & Sons, New York, N.Y., (1965), p. 3.
20. Fee, J. A. and Valentine, J. S., "Superoxide and Superoxide Dismutase," (A. M. Michelsen, J. M. McCord, I. Fridovich, eds.), Academic Press, New York, N.Y., (1977), pp. 19-60.
21. Bielski, B. H. J., Photochem. Photobiol., 28,645 (1978).
22. Behar, D., Czapski, G., Rabani, J., Dorfman, L. M. and Schwartz, H. A., J. Phys. Chem., 74,3209 (1970).
23. Hoare, J. P., "Oxidation-Reduction Potentials of Aqueous Solutions: A Selective and Critical Source Book," (A. J. Bard, R. Parsons and J. Jordan, eds.), Internatl. Union of Pure and Applied Chemistry, Chap. 4, In press. (1981).
24. Sawyer, D. T. and Roberts, J. L., Jr., J. Electroanal. Chem., 12,90 (1966).
25. Sawyer, D. T. and Seo, E. T., Inorg. Chem., 16,499 (1977).
26. Goolsby, A. D. and Sawyer, D. T., Anal. Chem., 40,83 (1968).
27. Ozawa, F., Hanaki, A. and Yammamoto, H., FEBS Lett., 74,99 (1977).
28. Fee, J. A. and Hildenbrand, P. G., FEBS Lett., 39,79 (1974).
29. Peover, M. E. and White, B. S., Electrochim. Acta, 11,1061 (1966).
30. Chevalet, J., Rouelle, F., Gierst, L. and Lambert, J. P., J. Electroanal. Chem. Interfacial Electrochem., 39,201 (1972).
31. (a) Kastening, B. and Kazemifard, G., Ber. Bunsenges, Phys. Chem., 74,551 (1970); (b) Divisek, J. and Kastening, B., J. Electroanal. Chem. Interfacial Electrochem., 65,603 (1975).
32. Dubrovina, N. I. and Nekrasov, L. N., Elektrokhimiya, 8,1503 (1972); Sov. Electrochem. (Engl. Transl.), 8,1466 (1972).
33. An arithmetical or typographical error in the original calculation of Chevalet, et al. (30) was corrected by Divisek and Kastening (31).
34. Brezina, M. and Hofmanova-Metejkova, A., J. Electroanal. Chem. Interfacial Electrochem., 44,460 (1973).
35. In a 0.1 $\underline{M}$ ammonium perchlorate DMSO solution the reduction potential is -0.03 V vs NHE and >99% of the superoxide exists as HO_2. (24).
36. Johnson, E. L., Pool, K. H. and Hamm, R. E., Anal. Chem., 39,888 (1967).

37. Iwase, A. and Tada, S., Nippon Kagaku Zasshi, 1828 (1972).
38. Fujinaga, T. and Sakura, S., Bull. Chem. Soc. Jpn., 47,2781 (1974).
39. Fujinaga, T., Izutsu, K. and Akachi, T., Bull. Chem. Soc. Jpn., 42,140 (1969).
40. Appleby, A. J., Mod. Aspects Electrochem., No. 9,369 (1974).
41. Sawyer, D. T. and Interrante, L. V., J. Electroanal. Chem., 2,310 (1961).
42. Sawyer, D. T. and Day, R. J., Electrochim. Acta, 8,589 (1963).
43. Bagotskii, V. S., Nekrasov, L. N. and Shumilova, N. A., Usp. Khim., 34,1697 (1965); Russ. Chem. Rev. (Engl. Transl.) 34, 717 (1965).
44. Sawyer, D. T. and Seo, E. T., J. Electroanal. Chem., 3,410 (1962).
45. Damjanovic, A., Mod. Aspects Electrochem., No. 5, 369 (1969).
46. Kolthoff, I. M. and Miller, C. S., J. Am. Chem. Soc., 63, 1013 (1941).
47. Taylor, R. J. and Humffray, A. A., J. Electroanal. Chem. Interfacial Electrochem., 64,85 (1975).
48. Sawyer, D. T. and Gibian, M. J., Tetrahedron, 35,1471 (1979); at pH 14,2 $O_2^{\cdot-} + H_2O \rightleftarrows HO_2^- + OH^-$, $K = 0.91 \times 10^9$.
49. Nanni, E. J., Jr., Chin, D. -H., Chiericato, G., Jr. and Sawyer, D. T., J. Am. Chem. Soc., (1981) Submitted.
50. Frost, A. A. and Pearson, R. B., "Kinetics and Mechanisms," 2nd ed., John Wiley & Sons, Inc., New York, N.Y.,(1961), p. 229.
51. Nanni, E. J., Jr., Hubbard, L. M., Morrison, M. M., Birge, R. R. and Sawyer, D. T., Inorg. Chem., 20, (1981) In press.
52. Nanni, E. J., Jr., Stallings, M. D. and Sawyer, D. T., J. Am. Chem. Soc., 102,4481 (1980).
53. Dietz, K., Forno, A. E. J., Carcombe, B. E. and Peover, M. E., J. Chem. Soc. B, (1970) 816.
54. Asada, K. and Kanematsu, S., Agric. Biol. Chem., 40,1891 (1976).
55. Merritt, M. V. and Sawyer, D. T., J. Org. Chem., 35,2157 (1970).
56. Magno, F., Seeber, R. and Valcher, S., J. Electroanal. Chem., 83,131 (1977).
57. San Fillipo, J., Jr., Chern, C.-I. and Valentine, J. S., J. Org. Chem., 40,1678 (1975).
58. Johnson, R. A. and Nidy, E. G., J. Org. Chem., 40,1680 (1975).
59. Gibian, M. J., Tangpoonpholvivat, R., Ungermann, T., Morrison, M. M. and Sawyer, D. T., 1978, unpublished data.
60. Gibian, M. J. and Ungermann, T., J. Org. Chem., 41,2500 (1976).
61. San Fillipo, J., Jr., Romano, L. J., Chern, C.-I. and Valentine, J. S., J. Org. Chem., 41,586 (1976).

62. Magno, F. and Bontempelli, G., J. Electroanal. Chem., 68, 337 (1976).
63. Gibian, M. J., Sawyer, D. T., Ungermann, T., Tangpoonpholvivat, R. and Morrison, M. M., J. Am. Chem. Soc., 101,640 (1979).
64. Johnson, R. A., Tetrahedron Lett., 331 (1977).
65. Jencks, W. P., "Catalysis in Chemistry and Enzymology," McGraw-Hill, New York, N.Y.,(1969), pp. 480-487.
66. Bender, M. L., "Mechanisms of Homogeneous Catalysis from Protons to Proteins," Wiley-Interscience, New York, N.Y., (1971), p. 176-179.
67. San Fillipo, J., Jr., Chern, C.-I. and Valentine, J. S., J. Org. Chem., 41,1077 (1976).
68. Hill, H. A. O., 1978, private communication.
69. Sawyer, D. T., Richens, D. T., Nanni, E. J., Jr. and Stallings, M. D., "International Symposium on Superoxide and Superoxide Dismutase", (J. V. Bannister and H. A. O. Hill, eds.), Developments in Biochemistry, Vol. 11A, Elsevier-North Holland, New York, N. Y., (1980) p. 1-26.
70. Martin, R. P. and Sawyer, D. T., Inorg. Chem., 11,2644 (1972).
71. Stallings, M. D. and Sawyer, D. T., J. Chem. Soc. Chem. Commun., 340 (1979).
72. Rao, R. S. and Hayon, E., J. Phys. Chem., 79,397 (1975).
73. Valentine, J. S. and Curtis, A. B., J. Am. Chem. Soc., 97, 224 (1975).
74. Klug-Roth, D. and Rabani, J., J. Phys. Chem., 80,588 (1976).
75. Valentine, J. S. and Quinn, A. E., Inorg. Chem., 15,1997 (1976).
76. Howie, J. K., Morrison, M. M. and Sawyer, D. T., "Electrochemical Studies of Biological Systems," (D. T. Sawyer, ed.) ACS Symp. Ser., 38,97 (1977).
77. Halliwell, B., FEBS Lett., 56,34 (1975).
78. Nanni, E. J., Jr., Univ. of California, Riverside, CA, 1980, Ph.D. Dissertation.
79. Mayeda, E. A. and Bard, A. J., J. Am. Chem. Soc., 95,6223 (1973).
80. Danen, W. C. and Arudi, R. L., J. Am. Chem. Soc., 100,3944 (1978).
81. Mayeda, F. A. and Bard, A. J., J. Am. Chem. Soc., 96,4023 (1974).
82. Guiraud, H. J. and Foote, C. A., J. Am. Chem. Soc., 98,1984 (1976).
83. Johnson, E. J., Pool, K. H., Hamm, R. E., Anal. Chem., 38, 183 (1966).
84. Johnson, E. J., Pool, K. H., Hamm, R. E., Anal. Chem., 39, 888 (1967).
85. Bray, R. C., Mautner, G. N., Fielden, F. M. and Carle, C. J., "Superoxide and Superoxide Dismutases" (A. M. Michelson, J. M. McCord, and I. Fridovich, eds.), Academic Press, New York, N.Y., 1977, p. 61-75.

86. Bielski, B. H. J. and Chan, P. C., J. Am. Chem. Soc., 100, 1920 (1978).
87. Nanni, E. J., Jr. and Sawyer, D. T., J. Am. Chem. Soc., 102, 7591 (1980).
88. Sawyer, D. T., Angelis, C. T., Jr., Dickson, J. and Nanni, E. J., Jr., J. Am. Chem. Soc., (1981) Submitted.
89. Hassan, H. M. and Fridovich, I., J. Biol. Chem., 253,8143 (1978).
90. Roberts, J. L., Jr., Morrison, M. M. and Sawyer, D. T., J. Am. Chem. Soc., 100,329 (1978).
91. Duke, F. R. and Haas, F. W., J. Phys. Chem., 65,304 (1961).
92. Dorfman, L. M. and Adams, G. E., "Reactivity of the Hydroxyl Radical in Aqueous Solutions," NSRAD-NBS46, SD Catalog No. C 13.48:46, U.S. Govt. Printing Office, Washington, D.C., June, 1978.
93. Haber, T. and Weiss, J. J., Proc. R. Soc. London, Ser. A, 147,332 (1932).
94. McClure, G. J. and Fee, J. A., FEBS Lett., 67,294 (1976); Halliwell, B., Ibid, 72,8 (1976); Rigo, A., Stevanato, R., Finazzi-Agro, A. and Rotilio, G., Ibid, 80,130 (1977).
95. Morrison, M. M., Roberts, J. L., Jr. and Sawyer, D. T., Inorg. Chem., 18,1971 (1979).

DISCUSSION

KOPPENOL: You showed in one of your slides a value of 2.4 V for the couple $\cdot OH/H_2O$ at pH 7. I know a theoretical paper by George in 1965 in which this value was calculated (George in "Oxidases and Related Redox Systems, Wiley, New York, 1965, pp. 3-33). However, it does not seem to be very well known that there exists two experimental papers in which the authors come to a much lower value of 1.4 V for the standard reduction potential of the couple $\cdot OH/OH^-$; at pH 7 that gives 1.8 V (Heckner and Landsberg, Z. Phys. Chem. 230: 63-72, 1965 and Stein, J. Chem. Phys. 42: 2986, 1965). These papers were also published in 1965. I was very happy to hear that H_2O_2 and its anion can produce $O_2^{\overline{\cdot}}$ and the $\cdot OH$ radical. Some time ago, I calculated that this reaction is thermodynamically possible, stimulated by a report from a Russian group (Symonyan and Nalbandyan, Biochim. Biophys. Acta 583: 279-286, 1979), who also found evidence for $O_2^{\overline{\cdot}}$ at alkaline pH. I'm glad to hear from you that this reaction really happens.

FRIMER: I would like to make two comments. First of all, most of your work is done in DMSO, pyridine or acetonitrile and you noted the solvent addition products isolated from these systems. Most of our work is done in benzene which is generally considered to be inert and we've been able to isolate biphenyl as a side product in these reactions. If you simply stir crown ether and benzene with potassium superoxide for several days, upon work up of the reaction you can isolate varying amounts of biphenyl. Based on radiation studies, the mechanism of formation probably involves hydroxyl radical formed perhaps by the interaction of $O_2^{\cdot -}$ with crown ether. We've also been doing $O_2^{\cdot -}$ reactions with α,β-unsaturated ketones, and we were interested in the abstraction of an α′-proton. By following the disappearance of starting material in the case of 4,4-disubstituted cyclohex 2-en-1-ones, we've noted that when we use potassium hydroxide, the rate of α′-proton removal is the slowest. Potassium superoxide is faster, and the fastest is potassium t-butoxide. Do you feel that potassium superoxide would be a stronger base than potassium hydroxide?

SAWYER: Yes. Even the thermodynamics would indicate that that reaction goes to the right by 10^8 to produce hydroxide ion via proton abstraction by $O_2^{\cdot -}$ from H_2O. So the tendency for superoxide to abstract protons is greater than that of hydroxide ion.

FRIMER: Do you have any measure of the basicity of potassium hydroxide in an aprotic solvent like benzene?

SAWYER: Yes. But not with benzene. We've made these measurements in dimethylformamide and the pK_a of water is about 32, which is a measure of that. Also, for n-butanol the pK_a is about 34. Thus, the butoxide ion is a slightly stronger base than the hydroxide ion and superoxide ion is stronger than either one of them.

FRIMER: But you wouldn't be surprised that t-butoxide is an even stronger base than potassium superoxide?

SAWYER: No, because we haven't made the measurement. It seems reasonable that it might be comparable.

A. SINGH: Could you comment on the mechanism of disproportionation of $HO_2\cdot$ to H_2O_2 and O_2. Do you think it goes through H_2O_4, or do you think it first gives you an $OH\cdot$ + $HO_3\cdot$? I'm thinking of the reactions of organic peroxy radicals which are supposed to form tetramers or which can give you two alkoxy radicals and O_2. Is there a parallel between the two?

SAWYER: We have not made any measurements which would allow us to give you a reasonable or intelligent answer. I would say that the rate constants in water that the pulse radiolysis people have made, in which the largest rate constant is between the anion and the neutral species, causes us to propose that we have a hydrogen-bridged intermediate or transition state. Certainly a coupling process forming H_2O_4 is a possibility but we do not have any evidence against it. The rate constants can be accounted for just on the basis of the diminished amount of the superoxide anion at lower pH values. I don't know that you can account for it quantitatively. Maybe either Dr. Bielski or Czapski could comment on that.

A. SINGH: If I may ask another question? You had one reaction on one of the slides where RX + $O_2^{\overline{\cdot}}$ goes to $RO_2\cdot$ + X^-. Do you think that's a one step process?

SAWYER: Yes.

A. SINGH: That's very difficult to conceive of.

SAWYER: There's undoubtedly a transition state. A number of studies have looked at this pretty carefully. The Sam Phillipo study at Rutgers has monitored the inversion of alcohols involved in the ester system. The results indicate that it is a classical S_N2 type of nucleophilic displacement process with a transition state that involves the inversion of the center.

A. SINGH: Thank you.

CZAPSKI: I want to comment on the question of dimerization. There is evidence that $O_4^{2=}$ is a very short-lived species; that this is not formed. If it's very short-lived we wouldn't detect it? Evidence from water is that the reaction is not via this intermediate. My question, what I wanted to ask; in one of your slides you showed concerning the mechanism of $O_2^{\overline{\cdot}}$ plus $HO_2^{\cdot}$ shows a hydrogen-bridged intermediate and you mentioned that there is good evidence for triplet oxygen.

SAWYER: The absence of singlet oxygen in our experiments, which are not as sensitive as the ones that have been made by Dr. Bielski, the yield of singlet oxygen when we dismutate superoxide with protons relative to when we use ferricenium is more than 100-fold smaller and is comparable to the yields we get when we use ions like ferric ion to dismutate superoxide ions.

CZAPSKI: So the statement really is that there is no evidence of singlet oxygen.

SAWYER: In our experiments that is right.

BORS: I would like to comment on your reaction of hydroxylamine with $O_2^{\overline{\cdot}}$. We did some experiments with pulse radiolysis and steady state radiolysis in aqueous solution. We found nitrite as the main product. Have you any evidence for the formation of nitrite in your aprotic solution?

SAWYER: Again, the answer is no. We certainly were aware of your work and we looked with great vigor for it because we liked that mechanism and that product, and all I can report is that we cannot find significant quantities of nitrite.

BORS: Thank you.

COHEN: I am interested in the borderline between aprotic and protic media and the biology of what you are presenting, and I wonder if you could give us some feeling for how to think. As I was listening to you speak and listening to the description of the oxidation or the proton abstraction from the catechols I was thinking of those catechols that are present in the nervous system in specific neurons, the catechol amines, and reflecting on the fact that, although they are present in small amounts in the neuron in which they live and are secreted, they are bound in storage vesicles and in those vesicles the vesicle is about 30% by weight catechol. Molarity therefore is in the molar range and that is normal biology; it doesn't mean that there is a tendency for those neurons to become melanized with time, it is sort of a mystery as to how and why that is occurring. Now if superoxide can abstract a proton from a bound catechol in the storage vesicle and perhaps initiate a polymerization or melanization reaction, just as a thought, what degree of water concentration would be permissible before the hydration of superoxide would prevent this reaction?

SAWYER: Well, that is an important question and I wish that I had an important answer. I think an interesting thing is that water at low levels in solvents like dimethylformamide becomes a much weaker acid. The number I cited of 33 for the pK_a as opposed to 15 for bulk water is a dramatic difference in the effective acidity of water. I would think that the value of 33 might well be characteristic of water in membrane systems and that the effect of water there might be much smaller than the bulk-water properties of superoxide. Now in terms of how forgiving are systems involving water <u>vis a vis</u> dismutation of superoxide ion (or other aqueous properties) as opposed to nonaqueous properties, I think that is dependent on the membrane. A DMSO experiment that we did a number of years ago established that CO_2 is reduced to $CO_2^{\cdot -}$ radical stoichiometrically; even up to water contents as high as 30%. Beyond 35% there is zero reduction of CO_2. In acetonitrile if you try to reduce CO_2 with even the minutest amount of water present it fails. The question of how tightly is the water held by membrane materials prevents me from giving an informed response.

NIKI: I just want to comment that the homolytic self reaction of $HO_2\cdot$ is extremely important in the atmosphere. I noticed that the rate constant you are quoting in the liquid phase, however, is at least three orders of magnitude slower than what I know happens in the gas phase; also in the gas phase there is overwhelming evidence that 2 $HO_2\cdot$'s do make an adduct. We don't know the lifetime, however. When we label the O_2 ($^{18}O_2$ and $^{16}O_2$) there is no mixing in the H_2O_2 that ensues from the system. It appears as though the two HO_2's are hydrogen bonded tail-and-head, and then forced apart to H_2O_2. Is there any comment on why the reaction is so slow in the liquid phase?

SAWYER: What is the rate constant in the gas phase?

NIKI: 2.4×10^{-12} $cm^3 molecules^{-1} s^{-1}$; it is roughly three orders faster than your value.

SAWYER: My value is Dr. Czapski's number (and his co-workers), but my opinion as a solution chemist is that things are not soluble unless they are solvated. These species are surrounded by solvent molecules, which slows the reaction. It is good to see that my colleague from San Diego has come all the way to Austin to ask questions.

TRAYLOR: With regard to this bimolecular reaction to hydroperoxy radicals—even the value of 10^5 is a lot faster than for alkylperoxy radicals. I think those bimolecular combination rates are of the order of 10^3 or 10^4. Furthermore, Ray Long in 1958 looked rather carefully at the decomposition of H_2O_2 with catechols and in several other ways. I think uncatalyzed as well, and he found (as mentioned in the gas phase) that in solution he was not able to get any mixing of the labels using doubly labeled and doubly unlabeled H_2O_2. So this evidence would argue against the forming of tetraoxide in solution.

I have a question, however, and that has to do with the reaction of superoxide anion with aryl peroxy radical which you had on one of your slides. I would like to know how that happens. The ester first formed peracid radical prior to reduction to the peracetate ion, if I remember your slide. Are the reduction potentials right for that sort of thing? Is it just an electron transfer?

SAWYER: Apparently.

OXYGEN CONSUMPTION DURING A FENTON-TYPE REACTION BETWEEN HYDROGEN PEROXIDE AND AN IRON-CHELATE (Fe(II)-DTPA)

Gerald Cohen[1] *and Pierre M. Sinet*[2]

[1]*Department of Neurology, Mount Sinai School of Medicine (CUNY), New York, New York 10029*
[2]*Dept. Biochemie Génétique, Hôpital Necker, Paris, France*

1. INTRODUCTION

The Fenton-type reaction between H_2O_2 and iron salts (eq. 1a) or iron-chelates (eq. 1b) generates the highly reactive, hydroxyl free radical (1,2). We were interested in the Fenton-type reaction as it might occur within living cells and, therefore, elected to reexplore the reaction at physiologic pH and with relatively dilute reactants. Most prior studies were performed under acidic conditions and with much more concentrated reactants. In addition, because iron within cells is not "free," but is generally chelated with endogenous cellular constituents, we worked with a "model" iron-chelate: Diethylenetriaminepentaacetic acid (DTPA) was used.

$$Fe(II) + H_2O_2 \longrightarrow Fe(III) + \cdot OH + OH^- \quad (1a)$$

$$Fe(II)\text{-chelate} + H_2O_2 \longrightarrow Fe(III)\text{-chelate} + \cdot OH + OH^- \quad (1b)$$

2. SELECTION OF THE IRON CHELATING AGENT

In oxygenated 50 mM phosphate buffer at pH 7.0, Fe(II) by itself and Fe(II) chelated with either EDTA or ADP were rapidly oxidized to Fe(III) with half-lives of less than 15 seconds (3). In contrast, Fe(II) chelated with DTPA was stable. Fig. 1 illustrates the lack of stability of Fe(II) alone versus the stability of Fe(II)-DTPA, as viewed with an oxygen electrode. The stability of Fe(II)-DTPA permitted further studies with H_2O_2 to be carried out.

3. THE FENTON REACTION: OXIDATION OF IRON AND CONSUMPTION OF OXYGEN

The reaction between Fe(II)-DTPA and H_2O_2 is given below:

$$Fe(II)\text{-DTPA} + H_2O_2 \longrightarrow Fe(III)\text{-DTPA} + OH^- + \cdot OH \quad (1b)$$

$$Fe(II)\text{-DTPA} + \cdot OH \longrightarrow Fe(III)\text{-DTPA} + OH^- \quad (2)$$

ISBN 0-12-592050-4

$$2Fe(II)\text{-}DTPA + H_2O_2 + 2H^+ \longrightarrow 2Fe(III)\text{-}DTPA + 2H_2O \qquad (3)$$

where eq. 3 is the sum of eqs. 1b and 2 written to show formation of water in the buffered medium. Eq. 3 indicates that two moles of Fe(II) can be oxidized for each mole of H_2O_2 added. Less than two moles of Fe(II) are oxidized when a chain decomposition of H_2O_2 with liberation of oxygen occurs (1,4).

When 12.5 µM H_2O_2 was incubated with excess Fe(II)-DTPA under helium, 1.7 moles of Fe(II) disappeared per mole of H_2O_2 added (Table 1), a value close to the expected stoichiometry of 2.0. When oxygen was present, the loss of Fe(II) was doubled (Table 1) and exceeded the theoretic yield. In other experiments performed in an aerobic environment with 50 µM H_2O_2 and either 50 or 200 µM Fe(II)-DTPA, 3.7 moles of Fe(III) were formed for each mole of H_2O_2 that disappeared (3).

Even though Fe(II)-DTPA was stable in the presence of oxygen (Fig. 1), it seemed likely that the extra oxidizing equivalents required to explain the yield of Fe(III) were derived from oxygen. Results in Fig. 2 illustrate that oxygen was indeed consumed in a dose-dependent manner when H_2O_2 was added to Fe(II)-DTPA. Therefore, it is evident that oxygen is activated during the reaction of H_2O_2 with Fe(II)-DTPA.

4. MECHANISMS FOR OXYGEN CONSUMPTION DURING THE FENTON-TYPE REACTION

Three mechanisms can be proposed for oxygen consumption: 1, a reaction of excess DTPA with ·OH; 2, a reaction of the Fe-DTPA chelate with ·OH; and 3, an unspecified reaction in which a reactive intermediate binds oxygen and "activates" it so that both

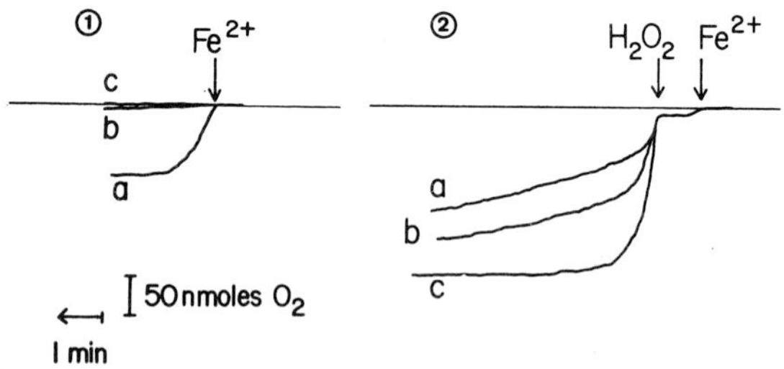

FIGURE 1. (Left) Autoxidation of Fe(II) and stabilization by DTPA. Conditions: 50 mM phosphate buffer, pH 7.0, and 100 µM ferrous ammonium sulfate. The DTPA concentrations were: a, none; b, 100 µM; c, 120 µM (20% molar excess compared to iron).

FIGURE 2. (Right) Oxygen consumption stimulated by the addition of hydrogen peroxide to a stable solution of Fe(II)-DTPA. Conditions: 50 mM phosphate, pH 7.0, 200 µM Fe(II) and 240 µM DTPA (20% molar excess). Hydrogen peroxide was added as follows: a, 20 µM; b, 40 µM; c, 100 µM.

oxygen and Fe(II) are consumed.

Equations 4 and 5 illustrate the first possibility:

$$DTPA + \cdot OH \longrightarrow DTPA\cdot + H_2O \quad (4)$$

$$DTPA\cdot + O_2 \longrightarrow DTPA\text{-}O_2\cdot \quad (5)$$

The peroxy free radical product (eq. 5) or a subsequent organic peroxide could react with Fe(II) to increase the yield of Fe(III). Walling et al. (2) reported an oxidation of EDTA during the reactions of Fe-EDTA with H_2O_2. The second mechanism is demonstrated in experiments in which Fe(II) was added in slight molar excess (10%), so that all of the DTPA was present as an iron-chelate. An initial instability was evident on the oxygen electrode due to the presence of unchelated (i.e., easily oxidized) Fe(II), but the usual rapid increase in oxygen consumption ensued upon addition of H_2O_2. The third mechanism is illustrated by an observation that oxygen consumption (although lesser in rate and in magnitude) occurred upon addition of H_2O_2 to a stable solution of Fe(II) in water (in the absence of both DTPA and phosphate buffer). The latter reaction may involve a reactive intermediate not dependent upon DTPA.

5. THE FENTON-TYPE REACTION WITH BENZENE AS SUBSTRATE

Are increased yields of product obtained from traditional substrates under the same experimental conditions? Experiments were performed with 2-keto-4-thiomethylbutyrate, which yields ethylene as a product, and with benzene, which is oxidized to phenol. Markedly increased yields of products were observed with

TABLE 1. Loss of Ferrous Ions During the Reaction of H_2O_2 with Excess Fe^{2+}-DTPA (100 μM) in the Absence and Presence of Oxygen

Condition	H_2O_2 Added (μM)	Fe(II) Lost μM ± SEM	Fe(II) Lost (μM Corr)	Ratio Fe(II) Lost/ H_2O_2 Added
Helium	0	0		
Helium	12.5	21.0 ± 1.0	(21.0)	1.7
Helium + Oxygen	0	16.5 ± 1.1		
Helium + Oxygen	12.5	58.7 ± 1.8	(42.2)	3.4

Experimental conditions: Sealed 25 ml Erlenmeyer flasks containing 4.0 ml of 50 mM potassium phosphate buffer, pH 7.0, with 200 μM DTPA and 100 μM Fe(II). Where indicated, two ml of oxygen were substituted for two ml of helium. Fe(II) was measured colorimetrically with O-phenanthroline. Results are from 8-13 determinations in three experiments.

both substrates when oxygen was present. Results with benzene are presented in greater detail:

The reaction of benzene with ·OH proceeds by direct addition to yield a free-radical intermediate (5), the hydroxycyclohexadienyl (HCD) radical (Fig. 3). Phenol can be formed from the HCD radical by dismutation or by oxidation, for example, with Fe(III). A second major product, biphenyl (Fig. 3), is the result of dimerization and dehydration (6).

Others (6) had shown that when the yield of phenol was increased by addition of excess Fe(III), the yield of biphenyl was correspondingly decreased. Oxygen can also oxidize the HCD radical (eq. 6):

$$\text{HCD}\cdot + O_2 \longrightarrow \text{phenol} + HO_2\cdot \qquad (6)$$

Therefore, the yields of phenol and biphenyl in the presence and absence of oxygen were examined (Table 2). Although biphenyl production was suppressed by oxygen, the molar yield of biphenyl under helium was low compared to phenol; in the presence of oxygen, the gain in phenol yield could not be explained by a corresponding loss in biphenyl (Table 2). Therefore, oxygen did not act by diverting the HCD radical from biphenyl to phenol.

Several alternate explanations can be excluded:

1, When Fe(III) oxidizes the HCD radical (eq. 7)

$$\text{HCD}\cdot + \text{Fe(III)} \longrightarrow \text{phenol} + \text{Fe(II)} + H^+ \qquad (7)$$

the Fe(II) formed can recycle in the Fenton-type reaction to increase the production of ·OH. But, there is no reason to suspect that this mechanism would become more effective in the presence of oxygen rather than its absence, and there is no explanation for oxygen consumption via this route.

2, Dismutation of $HO_2\cdot$ formed in eq. 6 could increase the availability of H_2O_2. However, the experiments were conducted with a

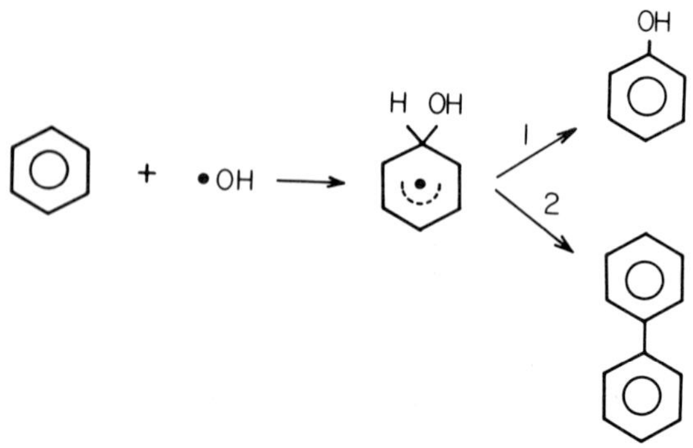

FIGURE 3. Reaction pathways for reaction of hydroxyl radicals with benzene. Oxidizing agents can divert the hydroxycyclohexadienyl radical from pathway 2 to pathway 1.

TABLE 2. Yield of Phenol and Biphenyl During the Fenton-Type Reaction Between H_2O_2 and Fe(II)-DTPA in the Presence and Absence of Oxygen

Atmosphere	*Phenol (μM)*	*Biphenyl (μM)*
	(Mean ± S.D.)	
100% Helium	*20.6 ± 1.7*	*0.3 ± 0.0*
Helium + 10% (v/v) Oxygen	*63.1 ± 3.2*	*Undetectable*

Results from a typical experiment performed in quadruplicate and analyzed by high performance liquid chromatography. Experimental conditions: 50 mM phosphate buffer, pH 7.4, 240 μM DTPA, 200 μM Fe(II) and 400 μM (excess) hydrogen peroxide. Samples were prepared under helium and oxygen was substituted where indicated.

large excess of H_2O_2, which provided optimum yields.

3, Suppression of the formation of other products is a possibility. But, formation of polyhydroxylated products, triphenyl, and opening or fragmentation of the benzene ring appear unlikely under our rather mild conditions, particularly because the concentration of product is low compared to benzene. Moreover, neither catechol nor resorcinol were evident by high performance liquid chromatography.

Conclusion: Other mechanisms must be sought to explain the consumption of oxygen and increased yield of phenol.

6. POTENTIAL SIGNIFICANCE OF OXYGEN CONSUMPTION

The mechanism of oxygen consumption is of considerable interest, but is unresolved. Earlier workers who studied the Fenton reaction with high concentrations of reactants would not have noted small effects of dissolved oxygen because the concentration of oxygen is low in air-equilibrated aqueous solution (only 250 μM). Additionally, the effect of oxygen may be more pronounced at neutral pH and, moreover, may involve some special participation of an iron-chelate. If we assume that Fenton-type reactions between endogenous iron-chelates and metabolically-generated H_2O_2 can occur in living cells, the question should be raised: Can H_2O_2 acting via a concerted mechanism serve to amplify the yield of product in enzymatic reactions that depend upon chelated iron plus oxygen? Biologic mechanisms could involve redox cycling of the chelating agent, or a bound ligand, or formation of an activated complex of iron and oxygen.

7. CONCLUSIONS

During the Fenton-reaction between Fe(II)-DTPA and H_2O_2, oxygen is "activated" and an increased yield of oxidized product is obtained. This reaction pathway may serve as a mechanism for the activation of oxygen by endogenous iron chelates in some biological systems.

ACKNOWLEDGEMENT

Supported in part by USPHS Grant NS-11631 and ERA No. 47, Centre Nationale de la Recherche Scientifique, France. We thank David Lewis and Susan Miller for technical assistance.

REFERENCES

1. Haber, F. and Weiss, J., Proc. Roy. Soc. A, 147,332 (1934).
2. Walling, C., Partch, R. E. and Weil, T., Proc. Natl. Acad. Sci., 72,140 (1975).
3. Cohen, G. and Sinet, P. M., in "Chemical and Biochemical Aspects of Superoxide and Superoxide Dismutase," (J. V. Bannister and H. A. O. Hill, eds.), Elsevier, (1980) p. 27-37.
4. Barb, W. G., Baxendale, J. H., George, P. and Hargrave, K. R., Trans. Faraday Soc., 47,462 (1951).
5. Dorfman, L. M. and Adams, G. E., NSRDS-National Bureau of Standards, 46, Washington, D.C. (1973).
6. Smith, J. R. Lindsay, and Norman, R. O. C., J. Chem. Soc. 3, 2897 (1963).

DISCUSSION

PEISACH: What biological molecule in cells do you envision being able to be bound to Fe and support the reactions that you suggest?

COHEN: Any Fe chelate, let's take one at random.

PEISACH: Let's take hemoglobin.

COHEN: Hemoglobin. I don't know. You are more familiar with hemoglobin than I am, I suspect. What do you think?

PEISACH: If I knew I wouldn't ask.

COHEN: O.K. Well, I don't have a definitive answer to your question. But, hemoglobin can carry out some hydroxylating reactions, such as the oxidation of serotonin to a dihydroxyindole, but under acid conditions where the ferrous-oxygen bond is "destabilized." I don't think the mechanism for that is known.

RICHTER: I have a comment about your finding that you can increase phenol yields in the presence of O_2. Firstly, your suggestion that O_2 oxidizes the hydroxyl radical adduct: I want to comment that we have observed in pulse radiolysis studies that O_2 will react with the hydroxyl radical adduct of dopamine, which is a derivative of benzene, so that reaction seems reasonable to me.

COHEN: What is the hydroxyl radical adduct of dopamine? Do you mean the free radical intermediate of a reaction with dopamine?

RICHTER: The hydroxyl radical adds to the benzene ring of dopamine, so it is very similar to what you have, except that there are other substituents on the ring. This oxidation of the adduct to give phenol should compete very effectively with a side reaction that would normally be depleting the phenol production, which is bimolecular reactions of the adducts to give products besides phenol and biphenyl, products which haven't been identified. Your findings seem very reasonable to me in terms of the reactions which are familiar to radiation chemists.

COHEN: Thank you.

SAWYER: I have two questions. These are both propositions on what might cause or enhance the consumption of O_2. In Fenton chemistry the products include hydroxide ion at the reaction center. When you reduce H_2O_2 you produce one or two hydroxide ions; that changes the localized pH to where the auto-oxidation of ferrous would be enhanced. Now whether it could account for that is one question. As I remember the radiolysis data on benzene plus $\cdot OH$ radicals under neutral or acidic conditions, the dominant product is the benzene radical rather than the adduct. The benzene radical plus O_2 will react cleanly to form the peroxy radical and that would then probably dimerize, dismutate, or something and end up giving you more phenol in a cascade of some sort.

COHEN: I have the distinct impression on reading the literature that the weight of evidence was for addition of hydroxyl radical to the ring, rather than hydrogen abstraction.

SAWYER: In base it is certainly true, but I believe abstraction is found in acidic or neutral conditions.

SIMIC: I just want to add one comment, we have done some work (Dizdaroglu and myself) on the reaction of OH radical with phenylalanine which adds to the benzene ring as in your case. This radical reacts with O_2 to give a peroxy radical, not $O_2^{\cdot -}$.

Furthermore we would like to suggest that phenylalanine is a much better system for picking up hydroxyradicals in biological systems because it is fairly soluble, a protein component, and gives cleaner products.

COHEN: Thank you, yes, you have an abstract here I believe.

FEE: I want to direct the discussion a little bit away from the main thrust of your talk to the rather incredible ability of DETAPAC to prevent the actions of Fe as it is really not well understood. First, if you try to make space-filling models of this compound with Fe you find that it is a very floppy type of a ligand and in doing those constructions we decided that we would try to react the DETAPAC-Fe(II) with $O_2^{\cdot-}$ in an attempt to make a peroxide compound similar to what is observed when one reacts $O_2^{\cdot-}$ with EDTA-Fe(II). We found that at pH 10 a very stable and long-lived purple peroxide compound was formed. Let me make a couple of comments about this purple compound. It is an extremely unreactive form of Fe peroxide. We attempted to react it with a variety of reducing agents including NADH, borohydride, dithionite, etc., and no reaction was found. Perhaps this purple compound is formed in these systems and stabilizing the Fe so as to make it unreactive. It may also be possible that polymerization occurs such that all the coordination positions of the Fe are occupied.

COHEN: Thank you. If there is some way that I can utilize this information to get an answer, if you will describe the experiment privately afterwards, I will try it.

FEE: It was more of a comment, Gerry, than a question.

COHEN: Your reaction is ferrous with superoxide rather than ferric.

FEE: That is true but you can make the same complex starting with ferric and peroxide and it exhibits the same stability.

WILLSON: I am reminded by your studies with the hydroxycyclohexadienyl radical of the pulse radiolysis studies with added ferricyanide which, of course, oxidizes the radical to the phenol. It seems to me that in your system you may have a competition between O_2 and the ferric complex—one to give phenol and the other, the hydroperoxyl radical. Now, with DTPA, I am wondering what is known about the variety of complexes that can be formed between DTPA and Fe and how these complexes oxidize the hydroxycyclohexadienyl radical. Have there been studies, for instance, by pulse radiolysis and could it be that these complexes are competing with O_2? If the ferric form reacts with the intermediate radical, it will give the ferrous complex. Oxygen may actually just oxidize the ferrous complex back to the ferric—and so this system can go on. Is this a possibility?

COHEN: I am sorry I didn't follow some of what you were describing.

WILLSON: One of the key points I think is that several Fe-DTPA complexes undoubtedly exist in your system, do they not? Is it known how rapidly these oxidize?

COHEN: Perhaps Gidi Czapski who has studied the various Fe-EDTA complexes can comment. No, he is shaking his head, no.

WILLSON: If they react with the intermediate, they will leave the ferrous complex. And this, of course, would be oxidized back to the ferric. So you could increase your yields of phenol by O_2 in this particular way.

COHEN: By ferrous being oxidized to the ferric and then the ferric...

WILLSON: ...then the ferric continues to reoxidize the hydroxycyclohexadienyl radical to phenol.

COHEN: So you are setting up a chain in which the Fe is cycling. O.K. That can be ruled out because it is the stability of ferrous-DTPA to autoxidation that provides the basis for the O_2 electrode experiments.

LESHEM: I would like to comment on your opening remarks about the biological implications and the physiological applicability of the Fenton reaction in mammalian or general physiology. Here, as pointed out in your talk, the chelated Fe form is of importance and is perhaps the primary mode in which Fe is incorporated or metabolized. However speaking as a plant physiologist it is known that inorganic Fe plays a major role in vascular bundle translocation and we encounter the problem of chlorosis and its therapy by inorganic Fe, so I think that in plants the Fenton reaction and its development (Cohen, Photochem. Photobiol., 1978) would perhaps be as important as the chelated Fe complex.

COHEN: Thank you.

BARD: Time, I think, for one last question.

BORG: I want to help you find a way of consuming that extra O_2, and I have been searching my recollection trying to remember some ESR flow experiments I did about 18 years ago. My best recollection is that rapid oxidation of phenol by a one-electron oxidant, ceric ion, produced the phenoxy radical at low pH, and we could obtain its ESR signature with our flow system. This was not a deoxygenated system and at higher pH we could also see the p-benzosemiquinone radical, so O_2 had to add on somewhere. Presumably, then under the right conditions some radicals reacted rapidly with O_2. (Perhaps radiation chemists can give us the rates.) Possibly peroxy radicals from O_2 addition which rearranged to semiquinones were what I saw. In any case, hydroxyl radicals can also produce phenoxy radicals and/or semiquinones from phenol, and benzene can be attacked as well. Thus, although the example I have recalled involves O_2 addition to phenol by way of an unspecified free radical mechanism (no hydroxyl was present to form a cyclohexadienyl adduct), a similar kind of O_2 consumption may have occurred in your Fenton reactions with benzene.

OXIDATION OF PURPUROGALLIN

G. Gabor and M. Halmann, Israel Institute for Biological Research, P. O. Box 19, Ness-Ziona, Israel

Pyrogallol (PG) and purpurogallin (PPG) are both oxidized by horse-radish peroxidase (1,2). This enzymatic oxidation is accompanied by chemiluminescence (CL) (1,2). CL is also observed in the non-enzymatic peroxidation, and to a smaller extent in the autoxidation in basic solutions of this compound. Purpurogallin is also obtained as one of the oxidation products of pyrogallol (3), and is a possible intermediate in its enzymatic peroxidation.

Recently, we have observed different reaction mechanisms of these substrates with a series of enzymes, resulting in CL (4). Purpurogallin yields CL with one molecule of HRP, while pyrogallol needs two molecules of enzyme to reach the excited state. With myeloperoxidase (an enzyme having two heme moieties/molecule), both substrates emit light with one molecule of enzyme; two molecules of microperoxidase (small molecular weight) are needed for both substrates for chemiluminescence.

The autooxidation of purpurogallin is a stepwise reaction. H_2O_2, a by-product in the first step, further reacts with other intermediates, producing the excited light emitting species (5,3). Therefore the autooxidation might be also characterized as a peroxidation—at least partially. Even though the end products of the autooxidation and the peroxidation by H_2O_2 reactions are different (Fig. 1), the light emitting species might be the same. Thus the autooxidation process would be defined as a slow peroxidation, due to the small (up to equimolar) quantity of H_2O_2 present in the solution, resulting in a weak chemiluminescence. At higher concentrations of H_2O_2 the production of the light emitting state is accelerated; the light intensity of the CL is increased by several orders of magnitude.

The low rates of the autooxidation permit a detailed study of the various intermediates. We have investigated the pH range 7.5-10.0, which is close to the pH of the enzymatic reaction. In aerated solutions one observes: (1) the decay of the 320nm absorption of purpurogallin; (2) formation of a blue intermediate (610 nm absorbing species); followed by (3) slow decay of the blue intermediate; (4) and simultaneous formation of a yellow intermediate (445 nm absorbing species). This last conversion (610 $\rightarrow$ 445) is a "clean" reaction; the isosbestic points at 386 and 482 nm indicate that there are only two participants in this reaction (that absorb in the 380-700 nm range). The decrease of the 320 nm absorption measures the oxidative decomposition of the purpurogallin. The blue intermediate—identified as a semiquinone radical ion (1,6)—is not the only product of this last reaction; increasing the decay rate of the 320 nm peak is not necessarily

ISBN 0-12-592050-4

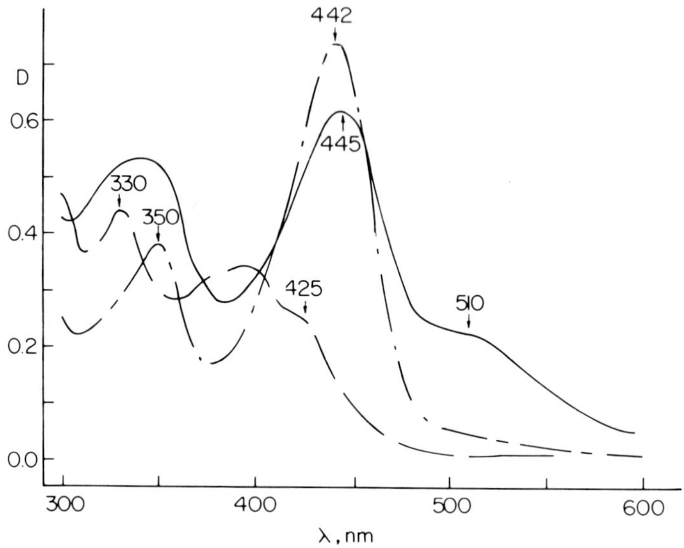

FIGURE 1. Reaction products with (—) and without (--) 0.1M/H_2O_2 [PPG] = $1.36 \cdot 10^{-4}$M, pH 8.5 buffer phosphate 0.1M, 50% ethanol; (-.-.-) same conc. and pH, 4% ethanol.

followed by an accelerated formation of the 610 nm species; e.g. increasing the pH from 7.6 to 9.8, results in a fourfold increase of the 320 nm decay rate, while there is no change in the rate of formation of the 610 nm species (Table 1). The fourfold increase of the 320 nm decay, results from a different pathway at higher pH.

In less polar solutions—more protein like—all reaction rates were considerably reduced (Table 2, Fig. 2). Reaction rates were further reduced on increasing the concentration of purpurogallin (Table 3).

The 445 nm product was stable at room temperature in darkness. It decomposed on exposure to light. The product of this

TABLE 1. Reaction Rates—Expressed as Half Life Times (min)—at Different pH. Phosphate buffer of 0.1M were used. [PPG] = $6.9 \cdot 10^{-5}$M, 4% ethanol

	τ, min			
	320 nm	610 nm		445 nm
pH	Decay	Formation	Decay	Formation
7.7	18	2.3		5
8.7	16	3.5	27	20
9.8	4	3.2	82	52

TABLE 2. Dependence of Reaction Rates (expressed as in Table 1) in Solvent Polarity at pH 8.5 [PPG] = $1.36 \cdot 10^{-4}$

Ethanol Content %	τ, min 610 nm Formation	Decay	445 nm Formation
10	3.0	24	35
50	5.5	46	50

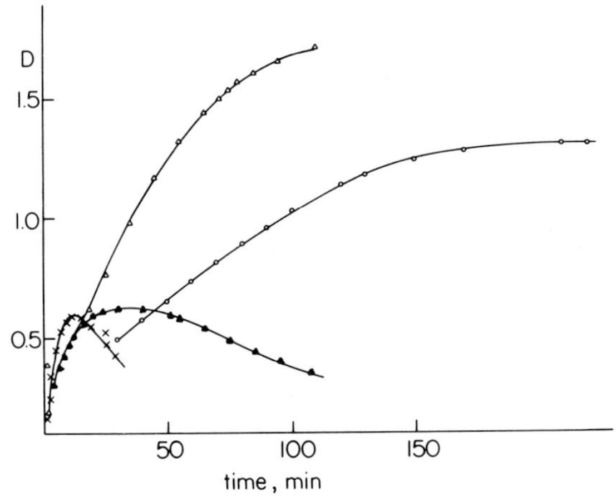

FIGURE 2. Kinetic curves of the production and decay of the blue and the yellow intermediates (X△) - 10% ethanol; (▲○) - 50% ethanol; (X▲) - 610 nm and (△○) - 445 nm species.

TABLE 3. Dependence of Reaction Rates on PPG Concentration at pH 8.5 50% ethanol (expressed as in Table 1)

[PPG] M	τ, min 320 nm Decay	610 nm Formation	610 nm Decay	445 nm Formation
$6.5 \cdot 10^{-5}$		8		41
$1.36 \cdot 10^{-4}$	41	5.5	46	60
$3 \cdot 10^{-4}$	91	9		52.5
$8.6 \cdot 10^{-4}$	144		255	

reaction was not a tropolone derivative as in the H_2O_2 - peroxidation (Fig. 1) (6). The excitation spectrum of the 445 nm intermediate is identical with its absorption spectrum, implying its being the only species that absorbs in the 400-700 range at this stage of the reaction.

In the early phases of the photochemical reaction, a new species absorbing at 420 nm was formed at the expense of the 445 nm species. The relative absorbancies at 420 and 445 nm, summarized in Tables 4 and 5, indicate that the contribution of the 420 nm species increases with light exposure (aging) or by lowering the polarity of the solvent (by adding ethanol) (Fig. 1).

Both the 445 nm and 420 nm's species are fluorescent with maximum at 492 and 462 nm respectively (Fig. 3,4), and, therefore, suitable candidates as the chemiluminescent species. None of these intermediates are radical type, as they yielded no esr signals. Collier (6) has also reported the disappearance of the paramagnetic species (the esr signal), concomitantly with the conversion of the blue radical to the 445 nm absorbing species. Adding 0.1M H_2O_2 at various stages of the 610 nm species' formation results in intense CL, proportional to the quantity of the blue species present.

The blue species as the precursor of the chemiluminescent species in the enzymatic process was suggested by Nilsson (1), and observed by us as well. Its long lifetime under our varying conditions allows its characterization, as well as of other intermediates derived from it.

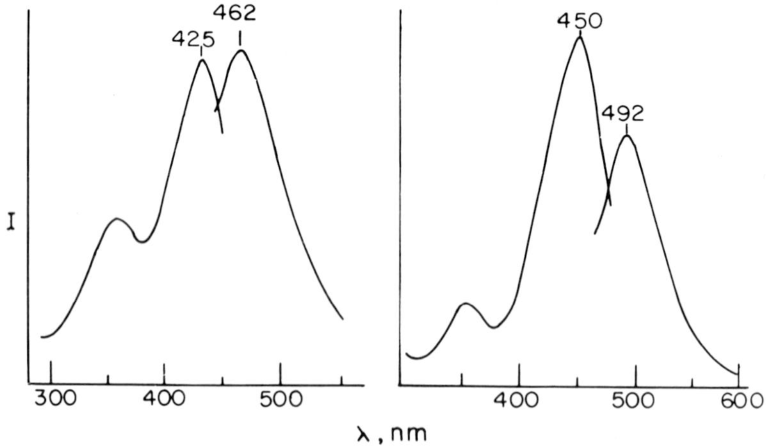

FIGURE 3. Excitation and fluorescence spectra of the 420 nm absorbing species—pH 8.5, 50% ethanol.

FIGURE 4. Excitation and fluorescence spectra of the 445 nm absorbing species—pH 8.5, 10% ethanol.

TABLE 4. Photochemical Decomposition of the 445 nm Absorbing Species as Measured by Relative Absorbances at 410 and 445 nm; [PPG] = $1.36 \cdot 10^{-4}$M, pH 8.5, 50% Ethanol

	A_{410}/A_{445}
At end of normal reaction	*.59*
24 hrs dark reaction	*.61*
20 hrs light	*.79*
60 hrs light	*.95*
120 hrs light	*1.06*

TABLE 5. Relative Absorbances at 410 and 445 nm at the End of Autooxidation as a Function of Solvent Polarity at pH 8.5

[PPG] *M*	*Ethanol: buffer* *v/v*	A_{410}/A_{445}
$6.9 \cdot 10^{-5}$	*1:25*	*.47*
$1.36 \cdot 10^{-4}$	*1:10*	*.55*
$1.36 \cdot 10^{-4}$	*1:1*	*.59*

The enzyme probably has a multiple role (1), it initiates charge transfer at the various stages of the reaction at lower pH (2), determines the end products (3), stabilizes the "correct" light emitting species. We suggest the 420 nm species as the light emitter, as its contribution in non-polar media is greater. Reduced polarity may be provided by the hydrophobic interactions of the protein.

REFERENCES

1. Nilsson, R., Acta. Chem. Scand., 18,389 (1964).
2. Halmann, M., Velan, B., Sery, T. and Schupper, H., Photochem. & Photobiol., 30,165 (1979).
3. Slawinski, J., Photochem. & Photobiol., 13,489 (1971).
4. Halmann, M., Gabor, G. and Velan, B., Photochem. Photobiol. In press.
5. James, T. H. and Weissberger, A., J. Am. Chem. Soc. 68,98 (1938).
6. Collier, P. D., J. Chem. Soc. (C), 2255 (1966); and other references quoted therein.

CHEMISTRY OF O_2 RADICALS

Session Chairman, M. G. Simic, Washington, D.C.

HYDROXYL RADICAL INDUCED OXIDATION OF ACETYLENE IN OXYGENATED AQUEOUS SOLUTION. THE FORMATION OF A HIGHLY ACIDIC INTERMEDIATE.

Dietrich Schulte-Frohlinde, Rudolf Anker and Eberhard Bothe, Institut für Strahlenchemie im Max-Planck-Institut für Kohlenforschung, Stiftstr. 34-36, 4330 Mülheim a. d. Ruhr, Federal Republic of Germany

INTRODUCTION

Gas phase irradiation of acetylene-oxygen mixtures resulted in the formation of carbon dioxide, carbon monoxide and a liquid polymer (1). ^{60}Co-γ-radiolysis of aqueous oxygenated solutions of acetylene (2) has been found to produce mainly glyoxal in a chain reaction (G $\sim$ 14) and hydrogen peroxide (G = 2.6). The influence of pH, concentration of acetylene, and oxygen revealed a rather complex picture and it was not possible to give a mechanism capable of explaining all the experimental results.

In the present work we studied the action of 2.8 MeV-electron pulses on aqueous solutions containing an acetylene-oxygen mixture or an acetylene-oxygen-nitrous oxide mixture in order to investigate the reactions of the β-hydroxy-vinyl-peroxyl radical ($HO\text{-}CH{=}CHO_2^{\bullet}$). This radical is formed from acetylene by reaction with $^{\bullet}OH$ followed by addition of molecular oxygen. The results show that a highly acidic intermediate is formed with a pK value $\leq$ 2. The decay of this intermediate leads to the observed products.

EXPERIMENTAL

Equipment. The pulse irradiation was performed with 1 or 2 μs electron-pulses (dose per pulse 0.1 - 4 krad) from a Van de Graaff accelerator (High Voltage Engineering Corp.; type K, 2.8 MeV). Changes of conductivity during and after the pulse were measured with an apparatus described earlier (3,4).

ISBN 0-12-592050-4

Materials. Triply distilled water from successive alkaline permanganate, acidic dichromate and finally from an all-silica still was used for pulse-radiolysis and for preparative irradiations. The pH of solutions was adjusted using perchloric acid or freshly prepared sodium hydroxide solution. Acetylene gas (acetone free) (Linde AG, Krefeld) was successively washed by sulphuric acid (2 M) and triply distilled water before use. Reagents used were analytical grade.

Performance. Samples were prepared by saturating water with a mixture of acetylene and either N_2O/O_2 (80/20 v/v) or pure O_2. In order to obtain the appropriate mixtures, the flow rates of the gases were controlled by flow meters (Sho-Rate Brooks Instrument, SA, Fribourg, Switzerland). The flow meters were calibrated with the different gases used. The concentration of each gas in the solution was calculated using solubility data from the literature (5), and respective flow rates. For low acetylene concentrations ($< 10^{-3}$ mole/l) also the syringe dilution techniques were applied and gave the same results.

Product Analysis. The product analysis will be described elsewhere.

RESULTS

Product Formation. Solutions containing acetylene and either N_2O/O_2 (80/20 v/v) or pure O_2 were irradiated at dose-rates of 1 krad/μsec (van de Graaff). Below pH = 3 only O_2 was used, because solvated electrons are captured by H_3O^+ and saturation with N_2O is not effective in highly acidic solutions. The main products were: glyoxal, glyoxylic acid, formic acid and hydrogen peroxide. Formaldehyde and an organic peroxide or hydroperoxide are formed as minor products. The list of the corresponding G-values are presented in Table 1. All products show linear yield-dose relationships and no dose rate effects. With the exception of glyoxal no influence of pH ($1 < pH < 6$) and temperature (2 - 50°C) on G(products) was observed.

Conductivity changes after 1 μs electron pulse. In aqueous solutions of acetylene in presence of oxygen from pH 2.5 to 6.5 an increase in conductivity with a risetime of a few microseconds and G-values of 5.4 is observed (initial jump of conductivity) (Figs. 1 and 2). The kinetics of the build-up was of first order (independent of dose rate) with a rate constant which decreased as the concentration of oxygen in the solution was lowered. A slower second build-up of conductivity followed at pH-values ≥ 4. The kinetic of the second increase was of first order with a rate constant of $k = 7 \cdot 10^3\ s^{-1}$ (half-life ∿ 100 μsec) and did also not depend on dose rate (Fig. 2). After approximately 500 microseconds the conductivity reached a maximum value. At times longer than 500 μs the conductivity decayed to a permanent level that was higher than the level prior to the pulse (Fig. 2). The decay of the transient conductivity was predominantly of second

TABLE 1. G-values of products measured after application of 1 μs electron pulses to aqueous solutions containing a gas mixture of acetylene and either N_2O/O_2 (80/20 v/v) or pure O_2 at neutral pH, room temperature and a dose rate of 1 krad per pulse

	N_2O/O_2	O_2
concentration of gases	$[C_2H_2] \sim 8 \cdot 10^{-3}$ M	$[C_2H_2] \sim 9 \cdot 10^{-3}$ M
	$[N_2O] \sim 1.7 \cdot 10^{-2}$ M	$[O_2] \sim 1.1 \cdot 10^{-3}$ M
Products	$[O_2] \sim 2.4 \cdot 10^{-4}$ M	
total acid	5.5 ± 0.2	2.9 ± 0.2
formic acid	2.7 ± 0.3	1.4 ± 0.2
glyoxylic acid	2.8 ± 0.3	1.5 ± 0.2
glyoxal	11 ± 1.5	6.6 ± 0.6
hydrogen peroxide	2.6 ± 0.3	2.9 ± 0.3
organic peroxides	0.3 ± 0.1	0.15
formaldehyde	∿ 0.2	∿ 0.1
succinaldehyde	≤ 0.1	-

order with a rate constant of $2k \sim 1.6 \cdot 10^7\ M^{-1}s^{-1}$ at pH 5.5. The same rate constant has been observed previously for the decay of transient conductivity caused by H^+ and $O_2^{\cdot -}$. Thus we attribute the transient conductivity to the dissociated hydroperoxyl radical. This is confirmed by the fact that the second build-up was not observed at pH-values ≤ 4 since pK ($HO_2^{\cdot}$) = 4.75 (Fig. 2b). The permanent conductivity is due to stable acids. Since permanent conductivity is also formed at pH-values lower than 4 (see Fig. 1) one or more acids with pK-values ≤ 4 are formed. The apparent pK of the permanent conductivity was 3.4 ± 0.2. Since formic and glyoxylic acid are formed as the only permanent acids (Table 1) we attribute the permanent conductivity to their dissociated forms. The apparent pK-value of 3.4 found in the conductivity measurements agrees with the pK value of a mixture of glyoxylic acid and formic acid (pK(CHO-COOH) = 3.3, pH (HCOOH) = 3.7). Fig. 1 shows that the initial jump of conductivity is still present at pH ≤ 3.0, and no second increase but a decrease of conductivity is observed (Fig. 2c).

The G-value of the fast initial increase of conductivity (measured at 20 μs after pulse) as a function of pH can be seen from Fig. 1. The yield does not depend on pH of the solution in the range 2.5 < pH < 6.0. (Value of pH = 2.5 is not shown in Fig. 1). Thus we conclude that the pK-value of the charged intermediate responsible for the fast initial increase of conductivity is ≤ 2.0, i.e. smaller than the pK-value of formic acid and

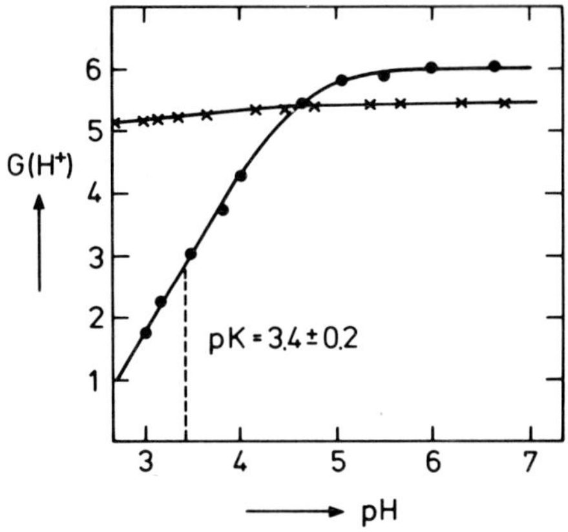

FIGURE 1. G-values for the increase of conductivity 20 μs after pulse (xxx) and for the formation of dissociated stable acids (●●●) observed with aqueous solutions of C_2H_2 (∿ $1 \cdot 10^{-3}$ M) and N_2O/O_2 (N_2O: $2 \cdot 10^{-2}$ M, O_2: $2.8 \cdot 10^{-4}$ M) after 1 μs pulses, 2 krad dose and room temperature. At pH-values ≤ 4.0 a correction was carried out for the reaction: $e^- + H^+ \rightarrow H\cdot$ which reduces G(·OH). For the 20 μs value in addition the neutralization reaction and the $HO_2^{\cdot}$ formation has been taken into account.

glyoxylic acid. The intermediate is unstable and decays in a first order process with a half-life of 90 μsec. The decay rate does not depend on pH within ± 15% in the pH range from 2.5 to 3.5 and no dose rate effects (range 0.5 - 4.5 krad/μs) is observed.

DISCUSSION

Irradiation of aqueous solutions with ionizing radiation gives rise to ·OH radicals, solvated electrons (e^-_{aq}) and H atoms as reactive species with G-values of G(·OH) = 2.7, G(e^-_{aq}) = 2.7 and G(·H) = 0.55, respectively. The hydrated electrons are converted into ·OH radicals by saturating the solution with N_2O. In this case the radicals in the system are 90% ·OH and 10% ·H. The produced ·OH radicals react (6) with C_2H_2 to give β-hydroxy-vinyl radicals **1** (eq. 1).

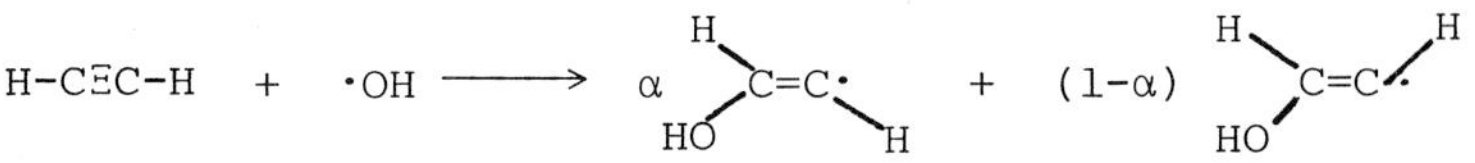

(1)

1 (trans) 1 (cis)

1

It is not known whether the addition of ·OH to acetylene leads to the cis or/and to the trans form of 1 (eq 1).

The rate constant for the reaction of ·OH with acetylene (6) was measured to be $k = 4.5 \pm 0.3 \cdot 10^9\ M^{-1}s^{-1}$. Therefore at concentrations of acetylene larger than $5 \cdot 10^{-4}$ M reaction (eq 1) is practically complete within 1 μs after pulse. Hydrogen atoms can react under our experimental conditions with acetylene or oxygen (reactions (eq 2) and (eq 3)):

$$\cdot H + C_2H_2 \rightarrow H_2C{=}\dot{C}H \qquad k = 2.3 \cdot 10^9\ M^{-1}s^{-1} \qquad (2)$$

$$\cdot H + O_2 \rightarrow HO_2\cdot \qquad k = 2.0 \cdot 10^{10}\ M^{-1}s^{-1} \qquad (3)$$

Reaction (eq 3) leads to $HO_2\cdot$. Due to the pK-value of the hydroperoxyl radical (pK = 4.75) $HO_2\cdot$ dissociated within ∿ 1 μs into H^+ and $O_2^{\bar{\cdot}}$ thus producing a conductivity increase in neutral solution with $G \leq 0.6$.

In the absence of oxygen no change in conductance was observed (6) in the pH-range $3.0 < pH < 9.0$. This shows that radical 1 does not dissociate in this pH-range during its lifetime.

Identification of the acidic intermediate. Since the half-life for the fast initial increase of conductivity is ≤ 10 μs at doses smaller than 0.5 krad, no bimolecular reactions involving two radicals can contribute to the fast formation of conductivity. Particularly, the production of conductivity via tetroxids formed by two peroxyl radicals 2 is excluded, the conducting species are only the product of the reaction of 1 with O_2.

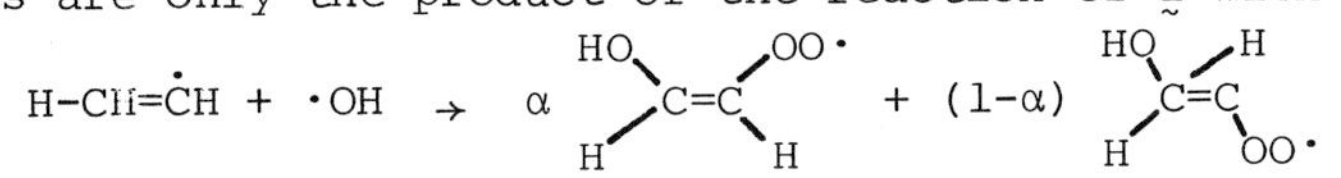

(4)

2 (cis) 2 (trans)

2

Reaction of 2 with a second acetylene molecule is unlikely, because already the rate constant (6) for the addition of 1 to acetylene is only $10^7\ M^{-1}s^{-1}$ and is expected to be smaller for the corresponding reaction of the peroxyl radical 2. Furthermore the concentration of acetylene has no influence on the measured rate constant (Variation of acetylene concentration in the range $8 \cdot 10^{-4}\ M \leq [C_2H_2] < 4 \cdot 10^{-2}\ M$).

We interpret therefore the conductivity results in the following way: In a first rate determining step oxygen adds to 1 forming radical 2 (eq 4). In a fast second step radical 2 deprotonates leading to the anion 3 and a proton (eq 5). Radical 2 is

$$\underset{2}{H(OH)C{=}CHOO\cdot} \longrightarrow H^+ + \underset{3}{H(O^-)C{=}CHOO\cdot} \qquad (5)$$

postulated to be the highly acidic intermediate which produces the initial conductivity change (at 20 μs) down to pK values of 2.5. The acidity of the cis and the trans form of radical 2 is probably very similar and cannot be distinguished by our technique.

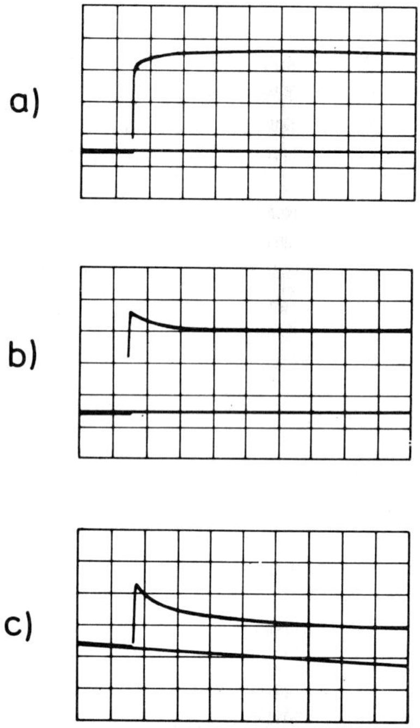

FIGURE 2. Conductivity changes as function of time following 1 μs electron pulses observed with an aqueous solution containing acetylene ($8.8 \cdot 10^{-4}$ M) and N_2O/O_2 (N_2O: $2.1 \cdot 10^{-2}$ M; O_2: $2.8 \cdot 10^{-4}$ M) at room temperature.

FIGURE 2a. Initial jump and transient conductivity obtained at pH ∿ 6.5, after a pulse of 1.2 krad with a sweep of 200 μs/div.

FIGURE 2b. Decrease of transient conductivity and formation of the final conductivity at pH ∿ 6.5 after a 1 μs pulse of 1.2 krad dose with a sweep of 50 ms/div.; ordinate: 100 mV/div.

FIGURE 2c. Decrease of transient conductivity and formation of the final conductivity at pH = 3.0 after a 2 μs pulse of 3.9 krad with a sweep of 200 μs/div.; ordinate: 5 mV/div.

The alternative explanations for the highly acidic intermediate, e.g. electron transfer from 1 to O_2, or elimination of $O_2^{\cdot -}$ and a proton from 2 are unlikely and easily ruled out. Furthermore we were not able to think of a rearrangement which would convert radical 2 to a radical which conceivably has a pK $\leq$ 2. If indeed the formation of the anion 3 produces the initial jump of conductivity the limit for the pK-value allows to estimate the half-life for deprotonation of 2 to be smaller than 7 ns. Therefore for the formation of the initial jump of conductivity, reaction (eq 4) should be rate determining and not reaction (eq 5) in agreement with the $[O_2]$ dependence for the rate of the initial jump of the conductivity build-up. Therefore from the rate of the conductivity build-up as a function of oxygen concentration the rate constant for reaction (eq 3) can be obtained ($k(1 + O_2) = 1.0 \cdot 10^9\ M^{-1}s^{-1}$).

At pH = 3.0 the conductivity decreases monomolecularly as a function of time with a half-life of 90 μs (see Fig. 2c). We assign this half-life to the decay of radical 3. The observed products are formed from this radical since the G-value for this radical is 5.4 ± 0.5. The product forming reactions including the formation of $HO_2\cdot$ will be published elsewhere.

REFERENCES

1. Lind, S. C., Bardwell, D. C. and Perry, J. H., J. Amer. Chem. Soc., 48,1556 (1926).
2. Clay, P. G., Johnson, G. R. A. and Weiss, J., J. Phys. Chem., 63,862 (1959).
3. Klever, H. G., Doctoral Thesis, Ruhr-Universität Bochum, 1974.
4. Bothe, E., Doctoral Thesis, Ruhr-Universität Bochum, 1976.
5. Handbook of Chemistry and Physics, 47th Edition (Robert E. Weast, ed.), Chemical Rubber Co., Cleveland (1966-1967).
6. Anderson, R. F. and Schulte-Frohlinde, D., J. Phys. Chem., 82,22 (1978).

DISCUSSION

A. SINGH: In the last slide you showed, $O^{\cdot -}$ is splitting off from peroxy anion; one might naively ask the question, if that anion can give $O^{\cdot -}$, can HO_2^- split off $O^{\cdot -}$ leaving behind ·OH? Would you care to comment on that?

SCHULTE-FROHLINDE: In our case we have a different energy profile for the reaction because here in contrast to HO_2^- we have the formation of two C=O bonds. This may well shift the reaction to the right side. And this gain in energy is not obtained if you split the HO_2^- into ·OH and $O^{\cdot -}$.

ON THE ADDITION OF MOLECULAR OXYGEN TO SULFUR CENTERED RADICALS

X

K.-D. Asmus, D. Bahnemann, M. Bonifačić, and K. Schäfer, Bereich Strahlenchemie, Hahn-Meitner Institut Berlin, Glienicker Str. 100, 1000 Berlin 39, Federal Republic of Germany

INTRODUCTION

Radicals in general are considered to be quite reactive towards molecular oxygen, and therefore a key role is often attributed to this reaction in a chemical mechanism which involves radicals and takes place under aerobic conditions. Despite the numerous studies in this field, however, many aspects of radical-oxygen interaction and their intermediate products are still not fully understood. The best investigated systems are probably carbon centered radicals, and one is often tempted to extrapolate from these results. Sulfur centered radicals, for example, have also been suggested to be quite reactive towards oxygen since evidently many highly oxygenated stable reaction products are formed in the presence of O_2 (1-8). The only two absolute rate constants measured for the O_2 addition to the thiyl radicals from cysteins (3) and gluthathione (4), $8 \cdot 10^9$ and $1.6 \cdot 10^9$ $M^{-1}s^{-1}$, respectively, seemed in fact to prove a high reactivity. We shall see, however, that these two high rate constants are rather the exception than the general rule. Our discussion will be mainly based on results published in one of our recent papers (9). It will also be restricted to addition reactions and exclude electron transfer reactions which are essentially only a question of the relative redox potentials of the reacting species.

EXPERIMENTAL

The investigations have been carried out with aqueous solutions of various sulfur containing compounds prepared to the usual radiation chemical standards, and using pulse radiolysis techniques. Details on solutions, experimental set-ups, and evaluation of data are to be found in ref. 9.

RATE CONSTANTS FOR THE ADDITION OF O_2 TO SULFUR CENTERED RADICALS

The rate constants for the reaction of sulfur centered radicals measured either directly, or indirectly via competition methods (9) are summarized in the Table.

Only an upper limit can be given for a possible reactivity of the radical cations from disulfides and monosulfides. It

ISBN 0-12-592050-4

TABLE: Rate Constants for the Addition of O_2 to Sulfur Centered Radicals. a) R = any kind of saturated group (9).

Radical	k (O_2 addition), M^{-1} s^{-1}	
$RSSR^{\cdot +}$ a)		
$R_2S^{\cdot +}$ a)	< 10^7	
$(R_2S)_2^{\cdot +}$ a)		
$C_2H_5S^{\cdot}$	$3.4 \cdot 10^8$	
$HOC_2H_4S^{\cdot}$	$2.7 \cdot 10^8$	
$(CH_3)_3CS^{\cdot}$	$7.8 \cdot 10^8$	
Pen $S^{\cdot}$	$4.0 \cdot 10^7$	
Cys $S^{\cdot}$	$8.1 \cdot 10^9$,	$8 \cdot 10^9$ 3)
Glu $S^{\cdot}$		$1.6 \cdot 10^9$ 4)
$CH_2\dot{-}S-CH_3$	$4.4 \cdot 10^8$	
$CH_3CH\dot{-}S-C_2H_5$	$6.2 \cdot 10^8$	
$(CH_3)_2C\dot{-}S-CH(CH_3)_2$	$1.0 \cdot 10^9$	

results from the observation that both the yield and the decay kinetics of these species, which exhibit strong optical absorptions, are not affected at all by the presence of molecular oxygen. For any practical purposes they can therefore be regarded as virtually unreactive towards O_2.

The neutral sulfur centered RS· type radicals, with the exception of the cysteine and to some extent the gluthathione radical, react with rate constants markedly below the diffusion controlled limit.

Somewhat higher rate constants are found for the addition of O_2 to the third category of radicals, namely $\dot{C}-S-$, which are formed by deprotonation of $R_2S^{\cdot +}$ radical cations and hydrogen abstraction from the corresponding sulfide (10). ESR experiments have however shown a spin distribution between sulfur and carbon in these radicals (11) which therefore cannot be regarded as truly sulfur-centered radicals.

DISCUSSION

The results indicate that sulfur-centered radicals are significantly less reactive towards oxygen than, for example carbon centered radicals many of which add oxygen with a diffusion controlled rate constant (12). The experiments with carbon centered radicals have revealed a strong electrophilic character of the O_2 (13). The low, i.e. almost nonexisting reactivity of the sulfur centered radical cations could certainly be explained on this basis. Solely under this aspect it would, however, be quite surprising that the negatively charged $X_2^{\cdot -}$ radical anion complexes

(X = Cl, Br, I, SCN) also do not undergo any addition reaction with O_2. This implies that an additional aspect may have to be considered, and such an aspect emerges from the electronic structure of these species.

All these radical ions contain bonds with an odd number of electrons. In the $(R_2S)_2{}^{\cdot+}$ complex a total of three p-electrons,

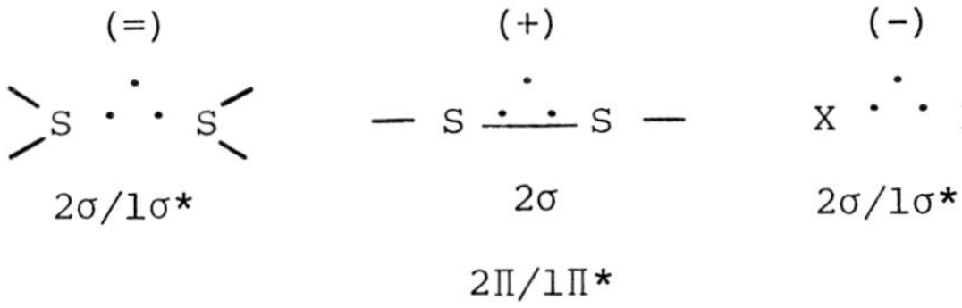

one from the oxidized and two from an unattacked sulfur atom, form a two-electron σ-bond while the third electron is located in an antibonding orbital (10, 14-16). The same electronic structure also exists in the $X_2{}^{\cdot-}$ radical anions where formally a halogen atom is associated with a halide ion. In the disulfide radical cation also no localization of the positive charge is apparent and the unpaired electron therefore in this case is likely to be located in an antibonding Π*-orbital.

Such σ* and Π* electrons are known to be involved in electron transfer reactions (10,17), but they have not been reported to undergo any addition reactions, not even with other radicals. Another example for a three-electron bonded radical species which does not add O_2 (but reduces O_2 in an electron transfer reaction) is the monomer Hg∴CN radical (18). It behaves very differently compared to other ·HgX radicals with X = Cl, Br, SCN, which all readily add O_2 to form $\cdot O_2HgX$ peroxy radicals (18). All these ·HgX species exhibit a different electronic structure with the unpaired electron being located only at the mercury atom rather than between mercury and carbon as in Hg∴CN.

Now, it is interesting to note that O_2 addition generally seems to occur with high rate constants whenever the radical center atom can easily promote its unpaired electron into a hybrid orbital for the oxygen binding. This is the case for the above mentioned ·HgX species where a 6s 6p hybrid is formed but also for phosphorous centered radicals, e.g. $H\dot{P}O_3{}^-$ and $\dot{P}O_3{}^{2-}$ (19), and certainly for carbon centered radicals. It seems obvious that hybrid functions facilitate a much better orbital overlap than other orbital functions, and such a high degree of orbital overlap apparently is necessary to bind molecular oxygen. In fact, this hybrid model may just be the more precise picture of the electrophilicity of O_2.

The lower reactivity of sulfur-centered radicals, and the practically non-existing reactivity of oxygen centered radicals (only exception $O^{\cdot-} + O_2 \longrightarrow O_3{}^{\cdot-}$) thus can find their explanation in a considerably less pronounced tendency of these atoms to establish hybrid bonds (S- and O-atom bonds usually exhibit a smaller s-orbital contribution than, for example, C) (20). In this respect also the singly occupied antibonding orbitals in $(R_2S)_2{\cdot}^+$, $RSSR{\cdot}^+$, $X_2{\cdot}^-$ and Hg∴CN would not be particularly suited for oxygen binding. The above consideration can also be invoked for the trend in reactivity of the >C∸S- type radicals where the spin density at the carbon atom and thus the hybrid orbital probability increases from methyl over ethyl to iso-propyl substitution.

In conclusion, sulfur-centered radicals are certainly able to add molecular oxygen but the respective reaction rates are generally much lower than for carbon centered radicals and below the diffusion controlled limit. In many cases this will allow the S-radicals to successfully undergo competitive reactions such as dimerization or replacement reactions, and the formation of oxygenated stable reaction products may not result from radical-oxygen reaction in the early stages of an oxidation mechanism.

REFERENCES

1. Lal, M., Radiat. Effects 22,237 (1974); Can. J. Chem. 54, 1092 (1976).
2. Al-Thannon, A. A., Barton, J. P., Packer, J. E., Sims, R. J., Trumbore, C. N. and Winchester, R. V., Int. J. Radiat. Biol. 6,233 (1974).
3. Barton, J. P. and Packer, J. E., Int. J. Radiat. Phys. Chem. 2,159 (1970).
4. Quintiliani, M., Badiello, R., Tamba, M., Esfandi, A. and Gorin, G., Int. J. Radiat. Biol. 32,195 (1977).
5. a: Lal, M., Armstrong, D. A. and Wiesner, M., Radiat. Res. 37,246 (1969).
   b: Lin, W. S., Lal, M., Gaucher, G. M. and Armstrong, D. A., Faraday Discussions 63,226 (1977).
6. a: Owen, T. C. and Wilbraham, A. C., Radiat. Res. 50,253 (1972).
   b: Owen, T. C., Wilbraham, A. C., Roach, J. A. and Ellis, D. R., ibid. 50,234 (1972).
   c: Owen, T. C. and Ellis, D. R., ibid. 53,24 (1973).
7. Purdie, J. W., J. Am. Chem. Soc. 89,226 (1967); Can. J. Chem. 47,1029,1037 (1969), ibid. 49,725 (1971), Radiat. Res. 48,474 (1971).
8. Swallow, A. J., J. Chem. Soc. 1334 (1952).
9. Schäfer, K., Bonifačić, M., Bahnemann, D. and Asmus, K.-D., J. Phys. Chem. 82,2777 (1978).
10. Bonifačić, M., Möckel, H., Bahnemann, D. and Asmus, K.-D., J. Chem. Soc. Perkin Trans. II 675 (1975).
11. Gilbert, B. C., Larkin, J. P. and Norman, R. O. C., J. Chem. Soc. Perkin Trans II 272 (1973).
12. a: Adams, G. E. and Willson, R. L., Trans Faraday Soc. 65, 2981 (1969).
    b: Simic, M. and Hayon, E., J. Phys. Chem. 75,1677 (1971).

13. Cercek, B., J. Phys. Chem. 72,3832 (1968).
14. Gilbert, B. C., Hodgeman, D. K. C. and Norman, R. O. C., J. Chem. Soc. Perkin Trans. II 1748 (1973).
15. Symons, M. C. R., J. Chem. Soc. Perkin Trans. II 1618 (1974).
16. Asmus, K.-D., Accounts Chem. Res. 12,436 (1979).
17. Bonifačić, M., Schäfer, K., Möckel, H. and Asmus, K.-D., J. Phys. Chem. 79,1496 (1975).
18. Jungbluth, H., Beyrich, J. and Asmus, K.-D., J. Phys. Chem. 80,1049 (1976).
19. Schäfer, K. and Asmus, K.-D., J. Phys. Chem.,84,2156 (1980).
20. a: Staab, H. A., Einführung in die theoretische organische Chemie," Verlag Chemie, Weinheim 1970.
    b: Cotton, F. A. and Wilkinson, G., "Anorganische Chemie," Verlag Chemie, Weinheim 1970.

DISCUSSION

SCHULTE-FROHLINDE: I like the explanation in your presentation very much because we just finished some work which is not published yet, on the reaction of O_2 with aliphatic carbon radical cations in water. We found a very high reactivity of these radical cations with O_2 and I was puzzled by your result that O_2 did not react with the sulfuric radical cation. But now we have a very nice explanation for this result.

ASMUS: Well, I'm glad to hear that.

A. SINGH: You mentioned that $Br_2^{\bar{\cdot}}$ does not react with O_2, at least there's no evidence for it. The reason may be that no one has looked for that reaction. We have looked at reaction of $\cdot OH$ radicals with sodium azide in which N_6^- or $(N_3)_2^{\bar{\cdot}}$ is formed which would be an analog if $Br_2^{\bar{\cdot}}$ and when we allow O_2 in the system, the decay rate of this species does increase suggesting that there's a reaction between $(N_3)_2^{\bar{\cdot}}$ and O_2. Also when we do gamma radiolysis the decomposition of sodium azide solutions in water in concentrations of $\sim$0.1 molar or greater increases if O_2 is present, which would again indicate that O_2 does take part in such reactions. So in these cases maybe the reaction with O_2 just hasn't been looked for.

ASMUS: Well, you have to be aware of one thing. Your radical ion complex may well look as if it reacts because this complex is in equilibrium with the $N_3^{\cdot}$ radical or with a hydroxyl complex of it and these species may have a different reactivity. Similarly in $Br_2^{\bar{\cdot}}$ the $Br^{\cdot}$ atom could be reactive towards O_2. But these complexes themselves don't react. If you take a $Br_2^{\cdot-}$ or $Cl_2^{\cdot-}$ absorption spectrum and look at the absorption time curve, the addition of O_2 will have no effect. You probably need very high concentrations of O_2 to get $N_3^{\cdot}$ out of the complex. As for the S-centered radicals, by any means, I would not like to say they don't react at all. It's just a much slower reaction than you might extrapolate from the carbon center radicals, as many people did, and the few early results seemed to indicate that you could do this. But the recently measured rate constants really show you that you should think in terms of a comparatively lower reactivity. And this means that O_2 may come into play much later, possibly even on a molecular basis rather than in the initial radical steps of a reaction scheme involving sulfur compounds.

ORGANIC OXYGEN RADICALS IN BIOLOGY
GENERATION AND REACTIONS

Wolf Bors, Christa Michel, and Manfred Saran, Gesellschaft für Strahlen- und Umweltforschung (GSF) Institut für Biologie, Abteilung für Strahlenbiologie, 8042 Neuherberg bei München, Federal Republic of Germany

INTRODUCTION

The doubts which were raised concerning the actual importance of the free hydroxyl and superoxide radicals in biological systems (1-3) led us to suggest the involvement of organic oxygen radicals (4). A verification of this hypothesis requires methods for their specific and selective generation. Recently we investigated (5,6) the radiolytic generation of simple organic alkoxy and peroxy radicals in order to compare their chemical behavior with the inorganic species $^{\cdot}OH$ and $O_2^{\overline{\cdot}}$. The pertinent reactions are:

$$ROOH + e^-_{aq} \longrightarrow RO\cdot + OH^- \quad (1)$$

and

$$RH_2 + \cdot OH \longrightarrow \cdot RH + H_2O \quad (2)$$

$$\cdot RH + O_2 \longrightarrow RHOO\cdot \quad (3)$$

Alternatively, alkoxy radicals were generated by photolytic cleavage of hydro- or dialkyl peroxides. Under these conditions a potential side reaction,

$$ROOH + RO\cdot \longrightarrow ROO\cdot + ROH \quad (4)$$

may occur (7).

To assay the individual processes in detail and to determine their reactivities in aqueous solutions requires specific detection methods. We presently use the bleaching of the water-soluble carotenoid crocin which allows the identification of the various inorganic and organic oxygen radicals (RO· and ROO·) by kinetic spectroscopy (5,6).

Aside from extending these experiments to biologically relevant hydroperoxides of fatty acids, an interesting question concerns the function of antioxidants. Due to the chemical diversity of these compounds, several mechanisms have been proposed (see ref. 8):

(1) preventing the formation of lipid hydroperoxides,
(2) reducing or decomposing hydroperoxides,
(3) scavenging free radical intermediates,

ISBN 0-12-592050-4

(4) scavenging reactive oxygen species, e.g. singlet oxygen,
(5) stabilizing membrane structures.

The present paper describes attempts to determine whether the bleaching of crocin is a genuine radical reaction and to evaluate the efficiency of antioxidants to inhibit crocin bleaching by scavenging alkoxy radicals.

MATERIALS AND METHODS

Crocin was isolated from saffron (9), the free acid crocetin was a gift of Hoffmann-LaRoche, Basel. The peroxides (hydrogen peroxide, tert-butyl hydroperoxide, di-tert-butyl peroxide, cumene hydroperoxide) were obtained from either Merck, Darmstadt or Peroxid-Chemie, München and were used as supplied. The antioxidants and radical scavengers used were diphenylamine, nordihydroguaiaretic acid and propylgallate from Aldrich, butylated hydroxyanisole, butylated hydroxytoluene, mannitol and thiourea from Sigma. Dihydroxy-acetophenone was synthesized from ω-chlorodihydroxyacetophenone (Merck) according to (10).

Photochemical studies, except for product analysis by HPLC, were performed with 254 nm-illumination, intensity 1.75 J/m^2sec. HPLC experiments were performed on a Waters APG 440 instrument with fixed wavelength filters (313, 405 nm).

RESULTS

Pulse-radiolytic experiments (5,6) demonstrated that crocin can be bleached distinctively by the various types of oxygen radicals, yet not by $O_2^{-\cdot}$ (Fig. 1).

To verify whether the bleaching reaction is a genuine radical reaction, we determined the reactivity of H_2O_2 alone both in neutral and alkaline solutions, of H_2O_2 + Fe^{2+}/Fe^{3+} (the Fenton or Ruff reactions, see ref. 11) and of the H_2O_2/Fe^{3+}-systems in the presence of various concentrations of acetonitrile (presumably forming the perferryl complex, see ref. 12). Except for H_2O_2 or HO_2^-, all types of reactions led to a depletion of crocin (data not shown).

The bleaching of carotenoids by radicals probably is due to adduct formation at the polyene moiety. The transient spectra after pulse radiolytic bleaching pointed to new products absorbing at 280-320 nm (Bors et al., submitted). HPLC investigations of the crocin preparation known from the literature (9) revealed that the substrate isolated from saffron actually can be separated into at least six individual peaks. Preliminary photolytic experiments demonstrated a differential sensitivity of the individual peaks towards the various radicals—with ·OH being least selective—and a transient increase in the amount of products absorbing at 313 nm (Fig. 2). The alkoxy radical derived from cumene hydroperoxide yields four different products. To determine

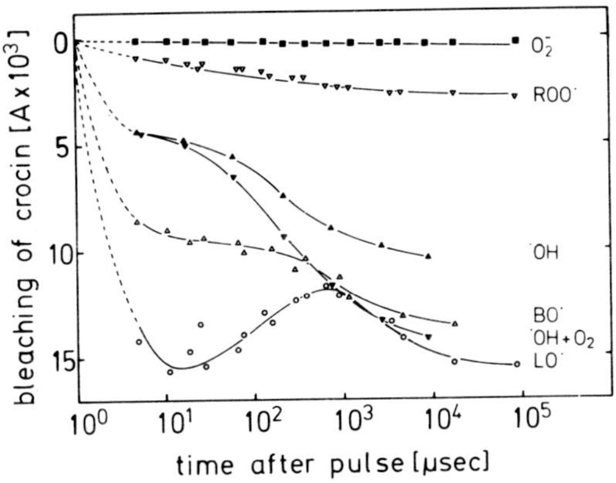

FIGURE 1. Bleaching of crocin by oxygen radicals. Radicals generated by pulse radiolysis (ROO· = $(CH_3)_2$-C(OH)-CH_2OO·, BO· = $(CH_3)_3C$-O·, LO· = alkoxy radical derived from 13-hydroperoxy linoleic acid). For method see text and ref. 5. Ordinate normalized for dose and radical yields.

whether this is due to impurities or differential attack, chromatographic purification or crocin is being pursued.

Contributing to the explanation of the function of antioxidants, the efficiency of these compounds in inhibiting the bleaching of crocin by t-BO· alkoxy radicals generated by the photolysis of t-BOOH, were compared (Fig. 3). Depicted are linear competition plots (13) with the relative reaction rates given in the legend. Absolute values can be obtained only after determining the reaction rate of t-BO· with crocin.

DISCUSSION

The early biochemical evidence on the co-oxidation of carotenoids, among them crocin (14-16), by lipoxygenase already pointed to radical intermediates as reactive species. This is supported by the strong bleaching—as shown in Fig. 1—by simple alkoxy radicals as models for fatty acid intermediates in the lipoxygenase action (17).

One of the arguments in support of the relevance of organic alkoxy radicals in preference to free hydroxyl radicals was the higher selectivity of the former species. The experiments depicted in Fig. 1 and 3 illustrate this point. Mannitol and evidently the gentobiose moiety of crocin do not react with alkoxy radicals. Consequently these radicals react only with the chromophore polyene group and bleach crocin far more effectively than ·OH radicals which also attack the sugar. This evidently represents a test system, which allows to discriminate between ·OH and RO· radicals.

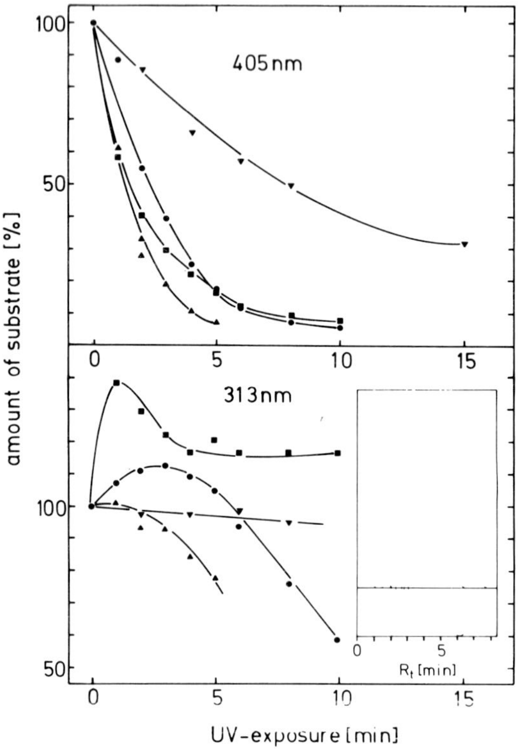

FIGURE 2. HPLC of crocin after UV-exposure in the presence of peroxides. (●) - ·OH radicals from H_2O_2; (▲) - t-BO· + ·OH from tert-butyl hydroperoxide, (▼) - t-BO· from di-tert-butyl peroxide, (■) - CO· from cumene hydroperoxide. Insert: elution profile of unexposed crocin (——— 405 nm, ----- 313 nm).

Whether the bleaching of crocin in the acetonitrile-containing Fenton-system, reportedly generating the perferryl species FeO^{2+} (12), represents an exclusive radical reaction, is open to debate. Zhuravleva and Berdnikov (18), re-interpret the data of Hamilton (19) and favor radical intermediates in this system.

In another attempt to verify the radical nature of the bleaching reactions we tried to determine the adduct respectively fragmentation products. Due to the complexity of crocin and the unresolved difficulties concerning the separation of the free acid crocetin on HPLC, only preliminary data are available. They do suggest, however, that the radicals form covalently bound adducts with the crocin molecule.

Observation of bleaching kinetics as opposed to product identification can be useful to determine rate constants, the

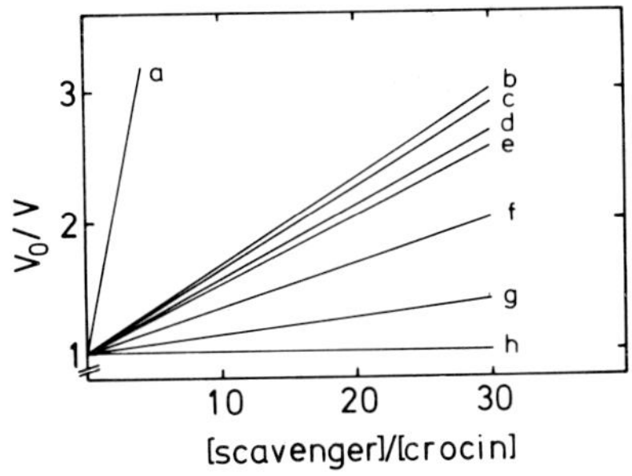

FIGURE 3. Effect of antioxidants and radical scavengers on crocin bleaching. Ordinate: ratio of bleaching rate of crocin by t-BO· in the absence (V_O) vs. presence (V) of scavenger; slope: relative reaction rates - $k_{(scav.\ +\ t\text{-}BO\cdot)}/k_{(crocin\ +\ t\text{-}BO\cdot)}$

(a)	*diphenylamine*	*0.43*
(b)	*propylgallate*	*0.067*
(c)	*dihydroxyacetophenone*	*0.065*
(d)	*butylated hydroxyanisole*	*0.056*
(e)	*butylated hydroxytoluene*	*0.053*
(f)	*nordihydroguaiaretic acid*	*0.035*
(g)	*thiourea*	*0.013*
(h)	*mannitol*	*0.001*

most important physico-chemical parameter to evaluate the biological relevance of the compounds under investigation. Since most known absolute rate constants for organic oxygen radicals have been determined in organic solvents (20), a reference system for aqueous solutions has to be developed.

The importance of these parameters is already apparent when relative reaction rates are compared. For instance, using various chemically diverse antioxidants and/or radical scavengers as inhibitors for the bleaching of crocin by photolytically generated t-BO· radicals reveals quite considerable differences. Two radical scavengers, used preferentially to remove ·OH radicals, i.e. thiourea and mannitol (21-23), were very poor scavengers of the t-BO· radical. Contrarily, diphenylamine which has also been suggested to function as a radical scavenger (24), far surpasses all other compounds in this assay system. Catechol derivatives, in particular with a substituted carbonyl function (propylgallate, dihydroxy-acetophenone), are efficient traps of alkoxy radicals—as they were for the free oxygen radicals ·OH and $O_2^{\overline{\cdot}}$ (25).

This example of assaying a number of compounds, known to interfere in lipid peroxidation reactions, demonstrates the versatility of observing the bleaching of crocin as assay method. The next goal shall be the determination of the absolute rate constants of crocin with various organic oxygen radicals. When these data are known, application of this reference system should help to settle the question, whether organic oxygen radicals are indeed the primarily reactive species in biological systems, not necessarily limited to lipophilic matrices (4).

ACKNOWLEDGEMENT

We appreciate the technical assistance of Ms. M. V. Sanden during the HPLC experiments.

REFERENCES

1. Willson, R. L., "Oxygen Free Radicals and Tissue Damage," CIBA Foundation Symposium 65, new series, 1979, p. 19.
2. Bors, W., Michel, C. and Saran, M., Eur. J. Biochem., 95, 621 (1979).
3. Fee, J. A., "Oxidases and Related Redox Systems," Proceed. Third Int. Sympos., Albany 1979, in press.
4. Bors, W., Saran, M. and Czapski, G., "Biological and Clinical Aspects of Superoxide and Superoxide Dismutase", (W. H. Bannister, J. V. Bannister, eds.), Elsevier/North-Holland, New York, (1980) p. 1.
5. Bors, W., Michel, C. and Saran, M., FEBS Lett., 107,403 (1979).
6. Saran, M., Michel, C. and Bors, W., "Chemical and Biochemical Aspects of Superoxide and Superoxide Dismutase" (J. V. Bannister, H. A. O. Hill, eds.), Elsevier/North-Holland, New York, (1980) p. 38.
7. Funk, M. O., Isaac, R. and Porter, N. A., J. Am. Chem. Soc., 97,1281 (1975).
8. Sies, H., Wendel, A. and Bors, W., "Metabolic Basis of Detoxication" (W. B. Jacoby, J. R. Bend, J. Caldwell, eds.), Academic Press, New York, 1980, in press.
9. Friend, J. and Mayer, A. M., Biochim. Biophys. Acta, 41,422 (1960).
10. Stephen, H. and Weizmann, C., J. Chem. Soc., 105,1046 (1914).
11. Walling, C., Acct. Chem. Res., 8, 125 (1975).
12. Yamamoto, H., Takei, H., Yamamoto, T. and Kimura, M., Chem. Pharm. Bull., 27,789 (1979).
13. Henglein, A., Schnabel, W. and Wendenburg, J., "Einführung in die Strahlenchemie," Verlag Chemie, Weinheim, 1969, p. 172.
14. Dicks, J. W. and Friend, J., Phytochem., 6,1193 (1967).
15. Weber, F., Laskawy, G. and Grosch, W., Z. Lebensm. Unters. Forsch., 155,142 (1974).
16. Weber, F. and Grosch, W., Z. Lebensm. Unters. Forsch., 161, 223 (1976).
17. Vliegenthart, J. F. G., Chemistry Industry, 241 (1979).
18 Zhuravleva, O. S. and Berdnikov, V. M., Russ. J. Phys. Chem., 50,1369 (1976).
19 Hamilton, G. A., J. Am. Chem. Soc., 86,3391 (1964).
20. Howard, J. A., "Advances Free Radical Chemistry" (G. H. Williams, ed.), Logos Press, London, 1972, p. 49.

21. Cederbaum, A. I., Dicker, E. and Cohen, G., Biochemistry, 17,3058 (1978).
22. Cohen, G., Photochem. Photobiol., 28,669 (1978).
23. Ohnishi, K. and Lieber, C. S., Arch. Biochem. Biophys., 191, 798 (1978).
24. Wang, H. P. and Kimura, T., Biochim. Biophys. Acta, 423,374 (1976).
25. Bors, W., Saran, M. and Michel, C., J. Phys. Chem., 83,3084 (1979).

PHENOXYL RADICALS: FORMATION, DETECTION, AND REDOX PROPERTIES IN AQUEOUS SOLUTIONS

P. Neta[1] *and S. Steenken*[2]

[1]*Radiation Laboratory, University of Notre Dame, Notre Dame, Indiana 46556*
[2]*Institut für Strahlenchemie, Max-Planck-Institut für Kohlenforschung, D-4330 Mülheim, West Germany.*

INTRODUCTION

Phenols and quinones are important constituents of many biological systems and have a variety of industrial applications as well. In many cases they participate in redox reactions in which phenoxyl radicals are formed as intermediates. In this article the properties of phenoxyl radicals will be summarized from the chemical point of view, with special emphasis on their involvement in one-electron transfer reactions. Recent measurements of one-electron redox potentials are reported.

FORMATION OF PHENOXYL RADICALS

Phenoxyl radicals are produced from phenols by a variety of methods: chemical oxidation, photoionization, and radiolytic reactions. In the radiolysis of aqueous solutions $\cdot OH$ radicals react rapidly with phenols to form ring adducts which undergo water elimination and yield phenoxyl radicals (1).

$$\cdot OH + C_6H_5OH \longrightarrow HO\dot{C}_6H_5OH \longrightarrow C_6H_5O\cdot \qquad (1)$$

The elimination of water is acid and base catalyzed and in the case of phenol itself the yield of the phenoxyl radical is quantitative. However, with various substituted phenols $\cdot OH$ radicals may add at the ipso position or attack the substituent and the yield of phenoxyl radicals is diminished (2). In order to avoid these side reactions a direct one-electron oxidation of the substituted phenol is carried out. Radicals such as $O^{\cdot-}$, $\cdot CH_2CHO$, $Br_2^{\cdot-}$, $Cl_2^{\cdot-}$, $\cdot SO_4^{-}$, Tl^{2+} are used for this purpose (3). They all oxidize phenols to phenoxyl radicals without any side reactions, e.g.

$$XC_6H_4O^- + \cdot CH_2CHO + H_2O \longrightarrow XC_6H_4O\cdot + CH_3CHO + OH^- \qquad (2)$$

$$XC_6H_4O^- + Br_2^{\cdot-} \longrightarrow XC_6H_4O\cdot + 2Br^- \qquad (3)$$

ISBN 0-12-592050-4

The rate constants for these one-electron transfer reactions are strongly affected by substitution and by the acid-base dissociation of the phenol. Hammett plots of substituent effects have been reported.

ABSORPTION SPECTRA

Phenoxyl radical has a characteristic optical absorption spectrum with a very sharp maximum at 400 nm and two broader peaks at 385 and 370 nm (4). Most substituted phenoxyl radicals have broader maxima in the region of 400-450 nm. These absorptions are used in pulse radiolysis to monitor the rate of formation and decay of these radicals. They are also used to examine the occurrence and the rates of electron transfer reactions between various substituted phenols and phenoxyl radicals.

ESR SPECTRA

The detailed structure of and spin density distribution in phenoxyl radicals has been studied by electron spin resonance (5). In general, the spin density on the ring carbon atoms is found mostly at the ortho and para positions, although substituents such as OH or NH_2 exert strong effects on the spin distribution. When the OH group is dissociated the two oxygens become equivalent, even when they are at positions meta to each other.

ACID-BASE EQUILIBRIA

Protonation of phenoxyl radicals on the oxygen has been studied by ESR and pK_a values in the region of 1 to -5 were measured (6). Acid-base equilibria of hydroxyl or amine substituents on phenoxyl radicals have been studied mainly by optical spectroscopy. A large difference is found between the pK_a of o-, m-, and p-hydroxyphenoxyl radicals (2b).

DECAY OF PHENOXYL RADICALS

Phenoxyl radicals decay to produce dimers through the formation of new C-O or C-C bonds but not O-O (7). In photolytic and radiolytic experiments hydroquinone and catechol were produced from phenol. Under certain conditions, these products were shown to result from the oxidation of dihydroxycyclohexadienyl radicals by phenoxyl radicals (5f).

ELECTRON TRANSFER REACTIONS

Phenoxyl radicals oxidize many substrates by a one-electron transfer process. The oxidation of ascorbate by various substituted phenoxyl radicals has been extensively studied (8). The oxidation of substituted phenols by differently substituted phenoxyl radicals has also been observed (9,10). Certain phenoxyl radicals were shown to reduce quinones by one-electron transfer. The rate constants of the above reactions vary over a wide range and strongly depend on the state of protonation of the substrate. They also depend, of course, on the redox potential difference that drives the reaction.

$$XPhO^{\cdot} + YPhO^{-} \longrightarrow XPhO^{-} + YPhO^{\cdot} \qquad (4)$$

ONE-ELECTRON REDOX POTENTIALS

When the difference in redox potential between the two phenoxyl radicals XPhO· and YPhO· is not very large the electron transfer reaction may reach equilibrium

$$XPhO^{\cdot} + YPhO^{-} \longrightarrow XPhO^{-} + YPhO^{\cdot} \qquad (5)$$

In many cases this equilibrium can be rapidly achieved under pulse radiolysis conditions before the radicals decay to final products. The equilibrium constant K_5 is then determined either from the concentrations of the four species at equilibrium or from the kinetics (k_5/k_{-5}), and the difference in redox potential can thus be calculated. This method was applied to various substituted phenoxyl radicals (10). The redox potential for the hydroquinone/semiquinone pair was used as a reference to establish a set of potentials for the various phenoxyl radicals studied.

This technique has now been used to study the redox properties of several compounds of biochemical interest. First we have determined the potential for the one-electron oxidation of p-dimethylaminophenol (DMAP). This compound was found to be very suitable as reference for determination of redox potentials of several other compounds. The results are summarized in Table I. Norepinephrine, DL-DOPA, 3,4-dihydroxyphenylacetic acid, and 3-hydroxytyramine are all derivates of catechol and their redox potentials are found to be very similar to that of catechol. The oxidation of 5-hydroxyindole and 5-hydroxytryptophan is less favorable by nearly 0.2V. 3,4-Dihydro-2,5,7,8-tetramethylbenzopyran-2-carboxylic acid is a water soluble compound which contains the main ring structure of tocopherol. Therefore, its redox potential (0.19 V) should represent very closely the value expected for tocopherol.

All the above measurements were carried out in alkaline solutions because of experimental limitations on measurements in neutral solutions. Extrapolation of the values in Table I to neutral or acid solutions can be done if all the pK_a values of the compound and its radical are known as was detailed previously (10).

TABLE I. ELECTRON TRANSFER RATES AND EQUILIBRIA AT pH 13.5

$XPhO^-$	$YPhO^-$	λ, nm	$k_5, M^{-1}s^{-1}$	$k_{-5}, M^{-1}s^{-1}$	K^a	$E(XPhO^-), mV^b$
DMAP	hydroquinone	490,430	1×10^8	3×10^5	405	174
DMAP	catechol	490	3×10^7	2×10^5	130	173
DL-DOPA	hydroquinone	430			0.96	22
DL-DOPA	DMAP	490		7×10^7	0.002	14
Norepinephrine	DMAP	490	2×10^5	3×10^7	0.006	44
3,4-dihydroxy-phenylacetic acid	DMAP	490		9×10^7	0.0025	21
3-hydroxytyramine	DMAP	490	3×10^5	9×10^7	0.0022	17
Adrenalone	DMAP	490			∿1.2	∿180
5-hydroxyindole	DMAP	490	3×10^7	3×10^6	5.1	216
5-hydroxytrypto-phane	DMAP	490	3×10^7	2×10^6	3.7	208
Toc[c]	DMAP	490			2.1	192

[a]Determined from the absorbance at equilibrium.

[b]The potentials are for reduction of XPhO· to $XPhO^-$, given vs. NHE using +23mV for hydroquinone as reference.

[c]3,4-Dihydro-2,5,7,8-tetramethylbenzopyran-2-carboxylic acid, a water soluble analog of tocopherol.

ACKNOWLEDGEMENTS

The research described herein was supported by the Office of Basis Energy Sciences of the Department of Energy. This is Document No. NDRL-2103 from the Notre Dame Radiation Laboratory.

REFERENCES

1. Land, E. J. and Ebert, M., Trans. Faraday Soc., 63,1181 (1967) Adams, G. E. and Michael, B. D., Trans. Faraday Soc., 63,1171 (1967).
2. (a) Peter, F. A. and Neta, P., J. Phys. Chem., 76,630 (1972); (b) Steenken, S. and O'Neill, P., J. Phys. Chem., 81,505 (1977); (c) O'Neill, P., Steenken, S., van der Linde, H. and Schulte-Frohlinde, D., Radiat. Phys. Chem., 12,13 (1978); (d) Latif, N., O'Neill, P., Schulte-Frohlinde, D. and Steenken, S., Ber. Bunsenges. Phys. Chem., 82,468 (1978).
3. (a) Adams, G. E., Aldrich, J. E., Bisby, R. H., Cundall, R. B., Redpath, J. L. and Willson, R. L., Radiat. Res., 49,278 (1972); (b) Neta, P. and Schuler, R. H., J. Am. Chem. Soc., 97,912 (1975); (c), Bansal, K. M. and Fessenden, R. W., Radiat. Res., 67,1 (1976); (d) Posener, M. L., Adams, G. E. and Wardman, P., J. Chem. Soc. Faraday Trans. 1,72,2231 (1976); (e) Hasegawa, K. and Neta, P., J. Phys. Chem., 82, 854 (1978); (f) Steenken, S., J. Phys. Chem., 83,595 (1979).
4. Schuler, R. H. and Buzzard, G. K., Int. J. Radiat. Phys. Chem., 8,563 (1976).
5. (a) Atherton, N. M., Land, E. J. and Porter, G., Trans. Faraday Soc., 59,818 (1963); (b) Stone, T. J. and Waters, W. A., J. Chem. Soc., 4302 (1964); (c) Jefcoate, C. R. E. and Norman, R. O. C., J. Chem. Soc. (B), 48 (1968); (d) Neta, P. and Fessenden, R. W., J. Phys. Chem., 78,523 (1974); (e) Dixon, W. T., Moghimi, M. and Murphy, D., J.C.S. Faraday Trans. 2, 70,1713 (1974); (f) O'Neill, P. and Steenken, S., Ber. Bunsenges. Phys. Chem., 81,550 (1977); (g) Dixon, W. T., Kok, P. M. and Murphy, D., J.C.S. Faraday Trans. 2,73,709 (1977).
6. Dixon, W. T. and Murphy, D., J. C. S. Faraday Trans. 2,72, 1221 (1976).
7. Nonhebel, D. C., Tedder, J. M. and Walton, J. C., "Radicals," Cambridge University Press, (1979).
8. Schuler, R. H., Radiat. Res., 69,417 (1977).
9. Schuler, R. H., Neta, P., Zemel, H. and Fessenden, R. W., J. Am. Chem. Soc., 98,3825 (1976).
10. Steenken, S. and Neta, P., J. Phys. Chem., 83,1134 (1979).

DISCUSSION

GROSSWEINER: I would like to mention that the absorption spectrum of the cationic phenoxyl radical from tyrosine was observed by laser flash photolysis in strongly acid solutions and that was published by Baugher and Grossweiner in Photochemistry and Photobiology [Vol. 28, p. 175, 1978].

NETA: Thank you for mentioning this work. But I was referring only to the pK_a determination.

STUDIES OF RADIATION INDUCED PEROXIDATION IN FATTY ACID MICELLES

L. K. Patterson, Radiation Laboratory and the Department of Chemistry, University of Notre Dame, Notre Dame, Indiana 46556

INTRODUCTION

There exists an extensive literature documenting the susceptibility of unsaturated lipids, especially those containing doubly allylic sites, to peroxidative damage (1). Under appropriate conditions of initiation and lipid-lipid interaction such damage may proceed via a radical chain mechanism by which large numbers of molecules undergo oxidation through a single initiate radical. The resulting products of these reactions, e.g. hydroperoxides, have been implicated in a range of pathologies (2). Because such processes may occur in vivo they are of considerable biological importance.

The scheme 1(a)-1(d) outlines the general form of the chain peroxidation process: where RH is the lipid and, in general, the H atom is at a doubly allylic site. Here RO_2H is lipid hydroperoxide, the principal product of the reaction.

$$RH \longrightarrow R\cdot \qquad \text{1a}$$

$$R\cdot + O_2 \longrightarrow RO_2\cdot \qquad \text{1b}$$

$$RO_2\cdot + RH \longrightarrow RO_2H + R\cdot \qquad \text{1c}$$

$$2RO_2\cdot \longrightarrow p \qquad \text{1d}$$

$$2R\cdot \longrightarrow p \qquad \text{1d'}$$

While much of the available information is based on autoxidation studies or initiation by chemical agents, it has been shown that the mechanisms of radical interaction with lipids can be usefully investigated with ionizing radiation. Especially in the study of kinetics for the component steps in above scheme, this latter approach exhibits several advantages. The number of initiating events may be precisely controlled and correlated with the overall product yield. The interrelation between propagation and termination steps depends on the radical concentration which may be manipulated in dose rate dependent studies. Further, with pulse radiolysis techniques it is possible to time resolve several, though not all, of the interactions outlined in the peroxidative scheme.

It is further demonstrated in the literature that the extent of peroxidative damage depends upon the microenvironment in which such reactions occur. In regions of high concentration and lipid

ISBN 0-12-592050-4

close-packing the yield of hydroperoxide per initial event is considerably greater than for regions in which concentrations are low and the reactive sites are not closely arrayed. While peroxidation has been examined in a variety of lipid aggregates, the fatty acid soap micelle in aqueous solution has proven to be a simple but broadly applicable model system in which to investigate the dependence of lipid radical interactions on environmental considerations. From the information gathered on such systems in other laboratories as well as our own, it has become possible to arrive at an elementary picture of the interrelation of primary water radicals, lipid structure and oxygen.

STEADY STATE STUDIES

Gebicki and Allen first demonstrated the dependence of radiation induced hydroperoxide yield upon micelle formation in linoleate solutions by measurements of conjugated diene yield as a function of linoleate concentration (3). They found a sharp rise in the G-value of conjugated diene associated with RO_2H at and above the critical micelle concentration in solutions of varied ionic strength. Raleigh and coworkers applied the more sophisticated technique of liquid chromatography which permitted them to eliminate contributions to diene yield from products other than hydroperoxide (4). Their data obeys the square root relationship of product yield to dose rate expected if the oxidation process is indeed governed by a chain mechanism of the type written above. Deviations from this relationship were found when all diene products are included.

Scavenger experiments in linoleate micellar systems have shown $\cdot OH$ to be the primary water radical responsible for initiation of the chain reaction (4,5). Investigations of radiative degradation in fatty acid soap micelles were also extended from linoleate to both linolenate and arachidonate (5). Even though these latter two compounds contain a greater number of doubly allylic sites and one might intuitively expect enhanced reactivity, the yield of hydroperoxide was instead found to drop. This behavior was interpreted to suggest that the more extensive <u>cis</u>-double bond structure found in linolenate and arachidonate prevents close-packing of adjacent molecules and decreases the probability of the propagation step (1c). Mixed micelles studies using oleate gave results consistent with this interpretation (5). Raleigh and coworkers confirmed these findings with their chromatographic techniques and further refined the measurements by altering the system through the introduction of Cs^+ ions into solution (6). This increased hydroperoxide yield. The Cs^+ acts to reduce headgroup repulsions in the micelle and leads to a more tightly packed structure and enhanced propagation.

Finally, steady state irradiation measurements were carried out with α-tocopherol containing linoleate systems to characterize the protective effect of this antioxidant via:

$$AH + RO_2\cdot \longrightarrow RO_2H + A\cdot \qquad 1c'$$

It was found that 1 molecule in 10^4 of lipid was sufficient to suppress oxidation until approximately two primary radicals had been produced per α-tocopherol (5). Beyond this point substantial prooxidant behavior was exhibited by the products of the α-tocopherol reaction. This is in agreement with findings from autoxidation studies (7).

TIME RESOLVED STUDIES

Information from steady state experiments has been considerably extended using pulse radiolysis techniques to isolate for study initiation and termination kinetics. Complimenting steady state measurements, the overall hydroxyl radical reaction with linoleate in micelles was found to be an order of magnitude slower ($k_{\cdot OH+RH} \sim 10^9\ M^{-1}sec^{-1}$) than with the monomer, indicating that organization of the lipids into extended assemblies not only effects propagation but creates a barrier to ·OH attack from the water phase. In ·OH reactions with methylene-interrupted PUFA, the conjugated radical generated by hydrogen abstraction at the doubly allylic site was found to exhibit an intense absorption spectrum, ($\varepsilon_{max} > 3.0 \times 10^4\ M^{-1}cm^{-1}$) in the region of 280 nm (8). This spectrum provides a highly useful parameter for the measurement of radical formation and disappearance by subsequent reaction. For example, it was found in deaerated solution that the production of conjugated radical following ·OH attack in micellar solution is greater than in monomer systems by a factor of 10. The high reactivity of ·OH dictates that initial radicals will be formed at a number of sites along the hydrocarbon chain not just at the doubly allylic carbon. In solutions containing monomer only some of these with favorable geometry and energetics may produce the conjugated radical by secondary intramolecular ·H abstraction. In micelles, however, the close packing of the lipid molecules enhances the possibility that secondary abstraction may take place by intermolecular means, relaxing the geometric restriction present in intramolecular processes. The second order rate constant for secondary abstraction in micelles was estimated at $10^3\ M^{-1}sec^{-1}$. Decay of the conjugated radical was also found to depend heavily on the distribution of material in micelle and monomeric form. By application of the technique of Henglein and Proske (9) it was further possible to determine rates of radical entry into and exit from the micelles (8).

Rates of radical peroxidation were also determined by absorption at 280 nm using the more selective O^- primary radical to generate the conjugated species, ($k_{O_2+R\cdot} \sim 3 \times 10^8\ M^{-1}sec^{-1}$). Rates of radical-radical interaction were $k_{2RO_2^\cdot} \sim 10^7\ M^{-1}sec^{-1}$ (10). It was further determined in these studies by use of monomeric arachidonate that O_2^- contribution to radical formation is negligible. Extension of these measurements to α-tocopherol containing micelles allowed us to determine the rate of hydrogen

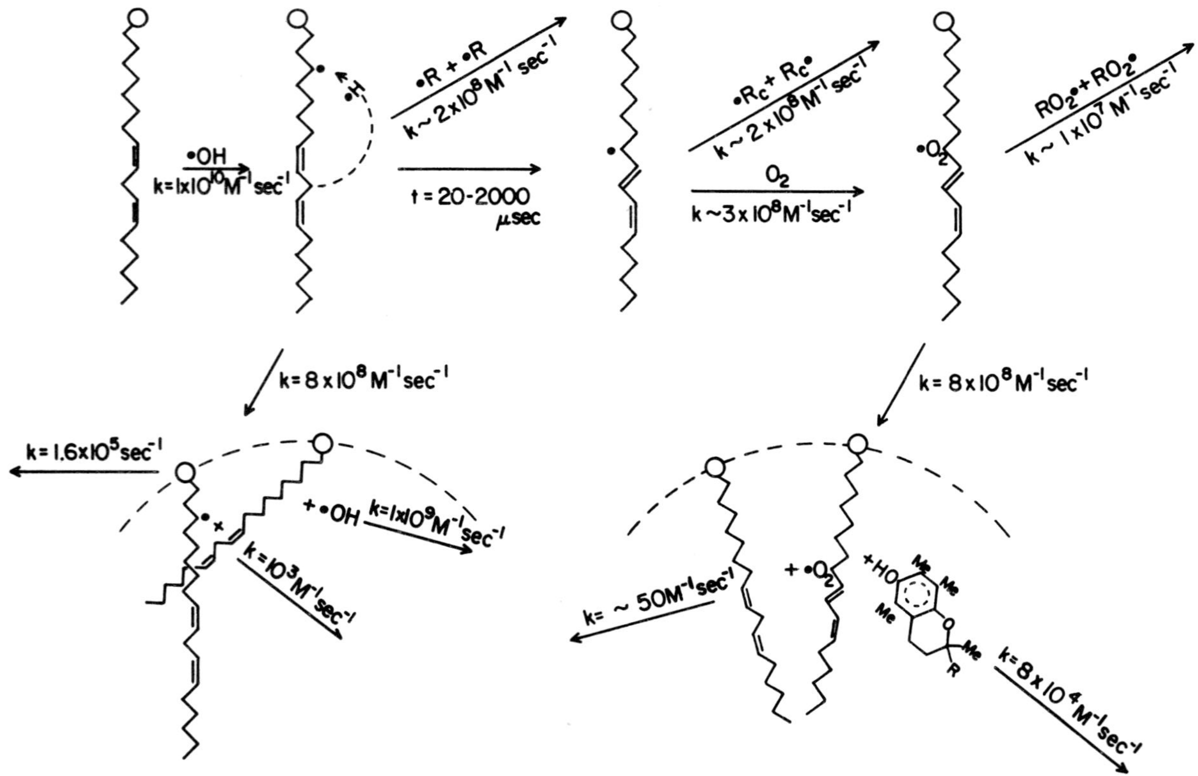

FIGURE 1. Behavior of ·OH induced linoleate radicals in monomeric and aggregated systems.

donation (1d') from the antioxidant, α-tocopherol, to the peroxy radical to block the chain reaction ($k_{RO_2\cdot+HA} = 8 \times 10^4\ M^{-1}sec^{-1}$) and to identify the subsequent α-tocopherol radical spectrum (11). This was found to closely resemble that reported by Willson et al. In mixed solvent systems (12). The scheme for ·OH induced peroxidation of linoleate is outlined in the figure. Because of the low rate constant it is difficult to determine propagation kinetics from pulse radiolysis. Hence the value added here is taken from the work of Bateman (13) on ethyl linoleate.

Because of interest in the potential of alkoxy radicals for initiating peroxidation, we conducted complementary laser measurements on the interaction of the model alkoxy radical, t-BuO·, with a large number of fatty acids (14). The abstraction rate constant can be expressed in terms of contributions proportional to the number of secondary, allylic and doubly allylic hydrogens present in the molecule,

$$k_{RO\cdot + FA} = \{0.07x(\text{no of secH}) + 0.52 \times (\text{no allylic H}) + 2.72 \times (\text{no of doubly allylic H})\} \times 10^6 M^{-1}sec^{-1}$$

and demonstrate the marked specificity of such radicals for the allylic and doubly allylic sites.

In summary, these studies have provided basic information on the kinetics of radical interaction in simple lipids and the role played in such interaction by the degree of assembly in the lipid system. Rate constants for initiation, radical peroxidation, termination, antioxidant behavior and mobility have been obtained by such systems. It is hoped that investigations to date provide information upon which may be built—by further experiments and insight—an enhanced understanding of the mechanisms of lipid peroxidation in complex biological systems.

ACKNOWLEDGEMENTS

The research described herein was supported by the Office of Basic Energy Sciences of the Department of Energy. This is Document No. NDRL-2107 from the Notre Dame Radiation Laboratory.

REFERENCES

1. Mead, S. F., Free Radicals in Biology, Vol. I" (W. A. Pryor, ed.), Academic Press, New York, 1976, p. 51 and references therein.
2. (a) Goldstein, B. D., Lodi, C., Collison, C. and Balchum, O. J., Arch. Environ. Health, 18,631 (1969).
   (b) Tappel, A. L., Fed. Proc., 32,1870 (1973).
   (c) Demopoulos, H. B., Fed. Proc., 32,1859 (1973).
3. (a) Gebicki, J. M. and Allen, A. O., J. Phys. Chem., 73,2443 (1969).

(b) Gebicki, J. M., "Biophysical Aspects of Radiation Quality, I.A.E.A. Vienna," 1971, pp. 229.
4. Raleigh, J. A., Kremers, W. and Gaboury, B., Int. J. Radiat. Biol. 31,203 (1977).
5. Patterson, L. K. and Redpath, J. L., "Micellization, Solubilization and Microemulsions, Vol. 2" (K. L. Mittal, ed.), Plenum Press, New York, 1977, p. 589.
6. Raleigh, J. A. and Kremers, W., Int. J. Radiat. Biol. Relat. Stud. Phys. Chem. ed. 34,439 (1978).
7. Cillard, J. Cormier, M. and Girre, R., C.R. Acad. Sci. Paris, (D), 455 (1975).
8. Patterson, L. K. and Hasegawa, K., Ber. Bunsenges. Phys. Chem., 82,951 (1978).
9. Henglein, A. and Proske, Th., J. Am. Chem. Soc., 100,3706 (1978).
10. Hasegawa, K. and Patterson, L. K., Photochem. Photobiol., 28,817 (1978).
11. Patterson, L. K. and Hasegawa, K., Radiat. Res., 000 (1980).
12. Packer, J. E., Slater, T. F., Willson, R. L., Nature (London), 278,737 (1979).
13. Bateman, L., Quart. Rev. (London), 8,147 (1954).
14. Small, R. D., Jr., Scaiano, J. C. and Patterson, L. K., Photochem. Photobiol. 29,49 (1979).

DISCUSSION

WILLSON: As you know, in simple aliphatic systems where you have a double bond there is a possibility of ·OH addition. Have you any evidence that you get ·OH addition in any of your systems?

PATTERSON: It's a little difficult to measure but if you look at the difference between the rate constants for ·OH attack at the double bonds and ·OH attack at those secondary hydrogens, there are so many secondary hydrogens compared to double bonds that the majority, I think, goes into hydrogen abstraction. Let's say approximately ten to one. I don't know exactly what the numbers are, but some must go to the double bonds.

FLOYD: Since you're doing this in liposomes I assume.

PATTERSON: No. In micelles, not liposomes.

FLOYD: Okay, but it is a two-phase system, is that not correct?

PATTERSON: That's right.

FLOYD: Okay. Could you give us some information as to what happens when you change the amount of water vs the amount of liposome or micelles, and also, something about the temperature effects.

PATTERSON: We have not looked at the temperature effect yet, but because the coiling of the lipid radical for secondary H abstraction should certainly depend on the temperature there should be interesting activation energy effects. As for the ratio between water and the lipid material, there are only about 10^{-5} molar micelles per mole of water so it's really an aqueous system with a very small amount of lipid material. It is much different from the multilamellar systems.

WINTERLE: You mentioned some microviscosity measurements on your micelles.

PATTERSON: That's right.

WINTERLE: Were these determined by some kind of radical coupling?

PATTERSON: No. These are being determined by measurements of fluorescence depolarization, and fluorescence lifetimes of probe-labeled fatty acids. In general, the method of Weber for determining microviscosities is being applied. We're trying to put the fluorescent materials at different points in the micelle by tagging them at various lipid chain positions so we put some of them near the center of the micelle and some of them near the surface.

WINTERLE: Do you have any measurements with numbers at this point?

PATTERSON: I can give you general relative results at present. The oleate is somewhat more fluid than, say, laurate. There's a jump there. There is another jump to linoleate and to linolenate. Those two have about the same microviscosity and a jump downward when you go to arachidonate because you've extended the chain length by 2 carbons. Now these data correspond to a degree with the area per molecule that you would see in monolayer experiments. There's a jump between a saturated chain lipid and a single double bond lipid, another small jump to multiple double bonds with linoleic acid and linolenic acid occupying much the same areas.

THE SPIN TRAPPING OF BIOLOGICALLY GENERATED FREE RADICALS

Gerald M. Rosen and Elmer J. Rauckman, Departments of Pharmacology and Surgery, Duke University Medical Center, Durham, North Carolina 27710

INTRODUCTION

A free radical is a species containing unpaired electrons and as such is paramagnetic. When this free radical is placed in a magnetic field, its magnetic moment is directly detected by an electron paramagnetic resonance (epr) spectrometer. Since epr can only detect free radicals, its use in biological systems is most apparent.

A theoretical lower limit of free radical detection by this technique has been determined (1). For identifying free radicals in aqueous solutions, a practical lower limit is about 10^{-6} M (2). Thus, it is only possible to detect free radicals which can reach this steady state concentration. Nevertheless, there are a number of examples of free radicals being detected in biological systems including: flavin radicals (3), quinone radicals (4), phenothiazine cation radicals (5) and nitroxides (6).

Many free radicals of biological interest, however, are so reactive that they never reach the necessary steady state concentration to be detected directly by epr methodology. An example of this is the hydroxyl radical, which reacts with itself or with most organic molecules at diffusion control rates (7). Thus, direct detection of hydroxyl radical by direct epr measurements is impossible in biological systems.

In theory, the method of spin trapping can overcome many of these difficulties. This technique consists of using a spin trap, i.e., a compound which forms a stable nitroxide by reacting covalently with an unstable radical (Fig. 1). In this way, the unstable radical is "trapped" as a long lived species which can be observed at room temperature using conventional epr equipment. The hyperfine splitting constants of the adduct provide information which can aid in the identification and quantification of the original radical. Since the relatively stable nitroxide free radical accumulates, spin trapping is an integrative method for measuring free radicals, and is inherently more sensitive than procedures which detect only instantaneous, or steady state, levels of free radicals. With this background, we will discuss the spin trapping of free radicals generated during the oxidation of NADPH by mouse hepatic microsomes.

ISBN 0-12-592050-4

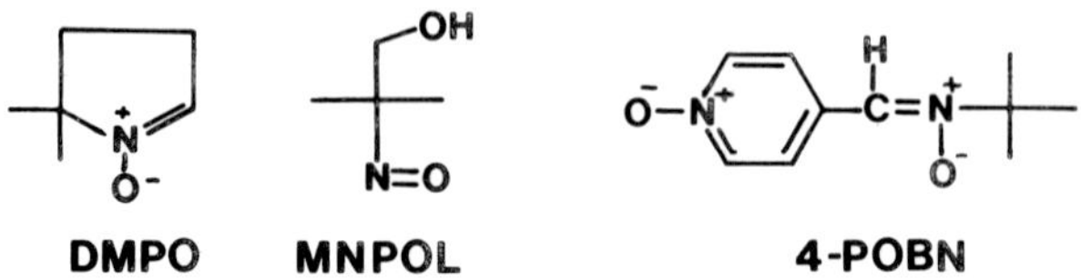

FIGURE 1. The structure of the spin traps.

DISCUSSION

The identification of biologically generated free radicals is quite complicated, especially if one considers the number and complexity of the primary and secondary radicals formed. Reactive primary radicals have the opportunity to interact with a multitude of propitious species to give rise to secondary radicals. This immediately results in a depressed level of primary radicals and leads to the occurrence of what is often an overwhelming level of interfering radical species. An example of this phenomenon is the enzymatic generation of superoxide. Dismutation can lead to hydrogen peroxide formation. If ferrous ion is present, hydroxyl radical can be produced as shown:

$$2\ O_2^{\overline{\cdot}} + 2\ H^+ \longrightarrow H_2O_2 + O_2$$

$$H_2O_2 + Fe^{+2} \longrightarrow Fe^{+3} + HO^- + HO\cdot$$

$$Fe^{+3} + O_2 \longrightarrow Fe^{+2} + O_2$$

$$3\ O_2^{\overline{\cdot}} + 2\ H^+ \longrightarrow 2\ O_2 + HO^- + HO\cdot$$

A further complication may occur in the presence of unsaturated phospholipids. Allylic hydrogen abstraction by hydroxyl radical would produce a lipid radical (L·) which can react with molecular oxygen to give the corresponding lipid hydroperoxy radical (LOO·).

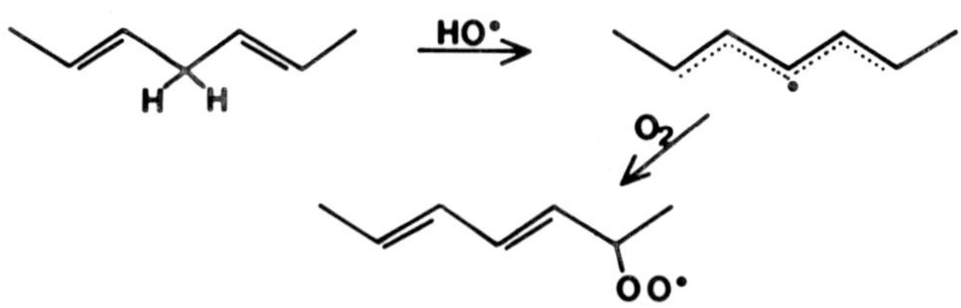

Thus, the production of superoxide can easily result in the generation of at least four distinct radical species: $O_2^{\overline{\cdot}}$, HO·, L· and LOO·. The ability of a technique such as spin trapping to identify any of these radicals depends upon several factors including: the rate of primary free radical production (k_a), the rate of secondary radical formation (k_b), the rate of spin trapping of the different radical species (k_c,k_d) and the rate of spin trapped adduct decomposition (k_e,k_f). If we let X· be the free radical derived from X, this free radical can either form a spin trapped adduct ST$\dot{-}$X or react with another species Y giving the secondary free radical Y$^{\bullet}$. Again, spin trapping this species leads to the spin trapped adduct ST$\dot{-}$Y. Finally, decomposition of the spin trapped adducts ST$\dot{-}$X and ST$\dot{-}$Y can give epr invisible non-radical products.

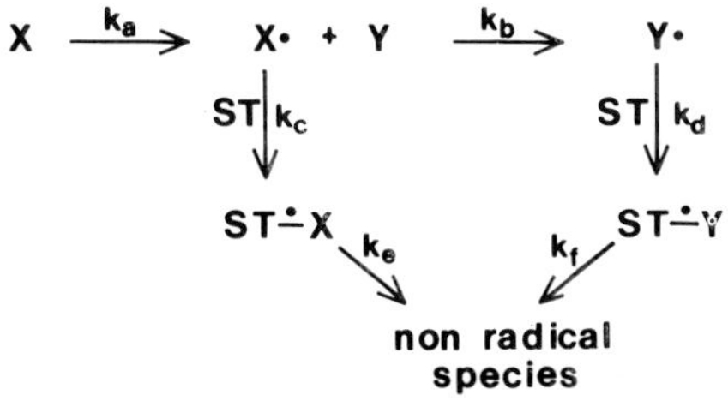

Thus, in order to observe only the primary spin trapped adduct ST$\dot{-}$X, $k_c >> k_b$ and $k_c > k_e$. However, if $k_b \geq k_c$ and k_d and $k_c >> k_f$ and k_e, two different spin trapped adducts can be noted. In biological systems, a variety of secondary radicals can be generated and spin trapped. Frequently, as described herein, the rates of decomposition, k_e and k_f, vary markedly for the various spin traps used, as well as for the different radicals trapped. In fact, it is the disparate decomposition rates (k_e,k_f) that makes the identification of these biologically generated free radicals possible. However, few studies to date have recognized the inherent complexity of the spin trapping technique in biological systems (8). Nevertheless, the spin trapping of certain radicals, in the biological milieu, has been achieved, either by good fortune or by cognizance of the relevant rate constants.

Further complications arise when the spin trapped adduct itself rearranges to give a new spin trapped product. For example, we have found that the cumene hydroperoxyl radical (CuOO·), generated by the action of hematin on cumene hydroperoxide, can be trapped by DMPO forming DMPO-OOCu. Rapid rearrangement at neutral or basic pH gives a second spin trapped adduct, DMPOX (9). If however, the reaction is run in acidic medium (pH=3.0), the initial spin trapped nitroxide, DMPO-OOCu, is quite stable and only slowly undergoes rearrangement to DMPOX.

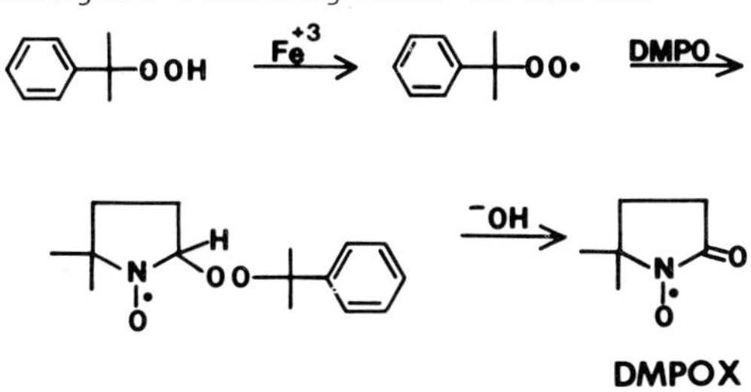

Another example of this phenomenon is best illustrated by the spin trapping of superoxide by DMPO. The ensuing nitroxide, DMPO-OOH, although observable by epr spectrometry, is quite unstable and rapidly decomposes into the more stable DMPO-OH and a non-radical species (10). Because of this phenomenon, misinterpretation of epr spectra are common. This suggested to us (8) that extensive experimentation is essential before one can unequivocally determine the identity of the trapped free radical.

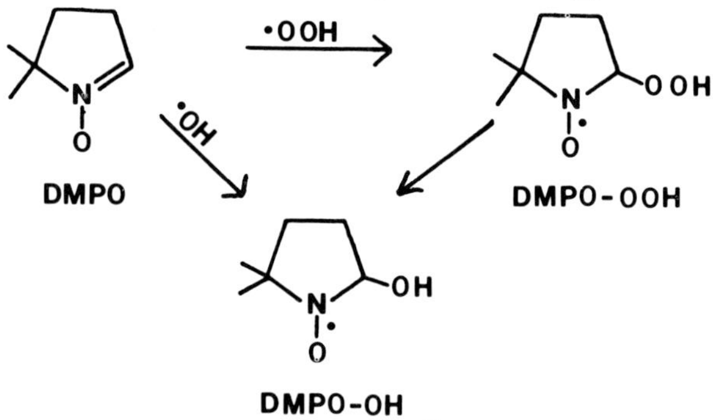

In microsomal preparations, where there are several different enzyme systems capable of generating both superoxide and hydrogen peroxide, as well as a numerous polyunsaturated fatty acid lipids to act as sources of secondary radical formation, the identification of these free radicals by means of spin trapping techniques should be practically impossible. However, we have found that the rate of spin trapped adduct decomposition varies as a function of the spin trap employed and the nature of the radical trapped.

Since the rate of spin trapping of free radicals other than hydroxyl radical is very slow, high concentrations of the probes are essential. For this reason, we have chosen to employ the very water soluble spin traps: DMPO, 4-POBN and MNPOL. When DMPO was incubated with mouse liver microsomes and NADPH, an epr spectrum was obtained (Fig. 2). The spectrum illustrated in this figure is a combination of DMPO-OOH and, to a lesser extent, DMPO-OH. To determine whether the DMPO-OH spectrum is due to the decomposition of DMPO-OOH or to the spin trapping of hydroxyl radical, ethanol (0.95%) was added to the above mixture. It has been well established that the hydroxyl radical reacts with ethanol to produce α-hydroxyethyl radical while superoxide does not undergo such an electron transfer (11). This secondary radical can react with DMPO to produce an adduct with an epr spectrum distinguishable from that of the hydroxyl radical adduct as shown in Fig. 3. Thus, if the production of DMPO-OH is due to the spin trapping of hydroxyl radical, the addition of ethanol should inhibit the production of DMPO-OH and result in the appearance of a new signal due to the spin trapping of the α-hydroxy-ethyl radical. As shown in Fig. 4, a small but finite quantity of

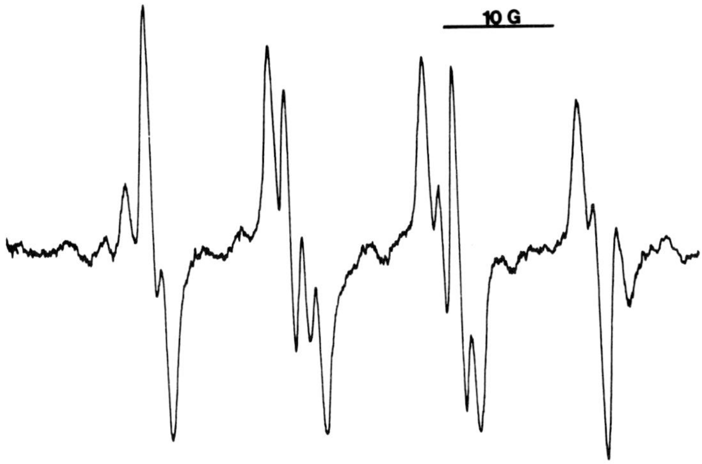

FIGURE 2. Electron paramagnetic resonance spectrum of DMPO-OOH produced by the reaction of DMPO with mouse hepatic microsomes and NADPH at pH 7.4 containing 1 mM DETAPAC, A_N= 14.3 G, A_H = 11.7 G and A_H^{γ} = 1.25 G.

α-hydroxyethyl radical was spin trapped. This suggests that some hydroxyl radical was produced by this microsomal system, probably via an iron catalyzed Haber-Weiss reaction (12) and spin trapped accordingly. Unfortunately, these results using DMPO in microsomal preparations do not allow us to conclude that the carbon centered dienyl lipid radical (L·) and its hydroperoxyl lipid radical species are not spin trapped by DMPO. This may be due in part to the inability of these radical species to approach the site of reaction on the pyrrolidine ring. We have noted that five membered rings, especially containing a double bond, have unusual steric restrictions to one electron reactions (13). To test this hypothesis, we reacted DMPO or 4-POBN with linoleic acid and lipoxygenase at pH 9.0. We found that although 4-POBN was able to spin trap linoleic acid hydroperoxyl radical (Fig. 5), there was no evidence of such free radical trapping using DMPO.

When dealing with a complex mixture of different radical species, instability of certain spin trapped adducts can actually

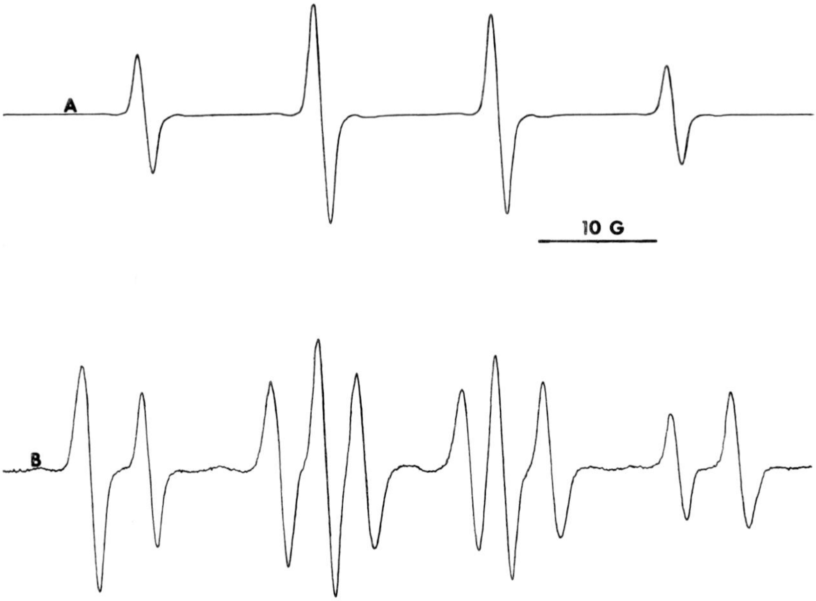

FIGURE 3. The effect of ethanol on hydroxyl radical trapping by DMPO. (A) Hydroxyl radical adduct generated by uv photolysis of H_2O_2, $A_H = A_N = 14.9$ G. (B) Combination of hydroxyl and α-hydroxethyl radical adducts generated by uv photolysis of H_2O_2 in the presence of ethanol, $A_N = 15.8$ G and $A_H = 22.8$ G. (By permission of Academic Press, ref. 8).

be a benefit. For example, the nitroso spin trap MNPOL reacts readily with both carbon and oxygen centered radicals; however the resulting oxygen centered spin adducts decompose too rapidly to be observed by epr spectrometry at ambient temperature. Thus, only carbon centered radical spin adducts are observed when using MNPOL at room temperature. Although this type of "selectivity" simplifies the analysis of the resultant epr spectrum, important radical species may be missed. For example, we observed, as did

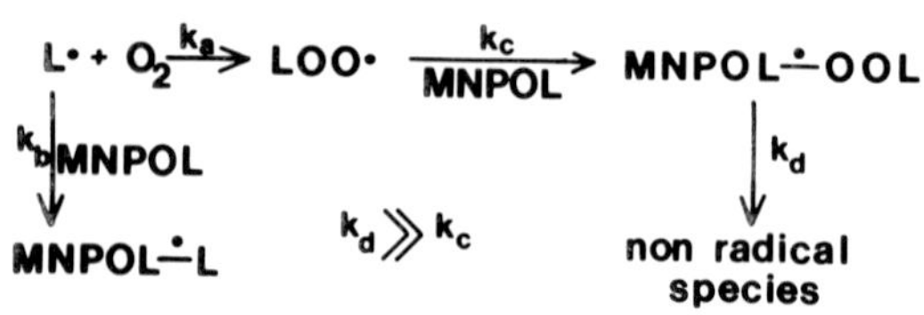

DeGroot, et al. (14) that upon the addition of the spin trap MNPOL to a mixture of lipoxygenase and linoleic acid, linoleic acid free radical (L•) was spin trapped. This species, however, could not be detected until twenty minutes after admixture of the components (Fig. 6). When we substituted 4-POBN in place of MNPOL in the above mixture, an epr spectrum was immediately

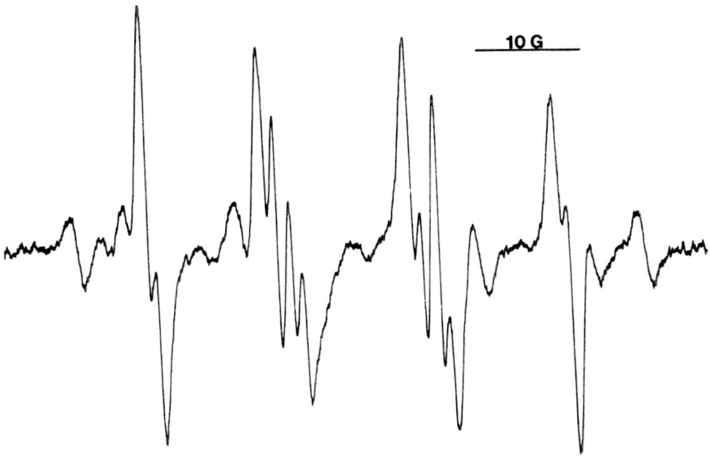

FIGURE 4. Electron paramagnetic resonance spectrum of DMPO-OOH and the α-hydroxyethyl radical adduct produced by the reaction of DMPO and ethanol with mouse liver microsomes and NADPH at pH 7.4 with 1 mM DETAPAC, $A_N = 14.3$ G, $A_H = 11.7$ G and $A_H^{\gamma} = 1.25$ G for DMPO-OOH and $A_N = 15.7$ G and $A_H = 22.8$ G for the α-hydroxyethyl DMPO spin trapped adduct.

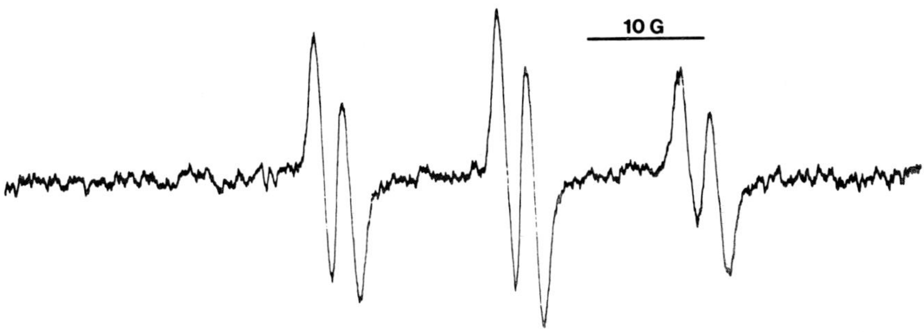

FIGURE 5. Electron paramagnetic resonance spectrum of 4-POBN-OOL obtained by the action of lipoxygenase on linoleic acid in the presence of 4-POBN at pH 9.0, $A_N = 15.8$ G and $A_N = 2.6$ G.

observed (Fig. 5). These results suggest that in the linoleic acid-lipoxygenase system, the linoleic acid free radical, L·, is initially formed, but its reaction with molecular oxygen is faster than its reaction with MNPOL ($k_a >> k_b$). Within fifteen minutes, the reaction mixture becomes anaerobic, as determined by monitoring oxygen consumption as described in the experimental section and in doing so k_a approaches zero. Thus, MNPOL-L is the only product observed by epr. However, in the case when 4-POBN is used in place of MNPOL, no L· is trapped. We arrive at this conclusion based on several observations. Initially, the epr

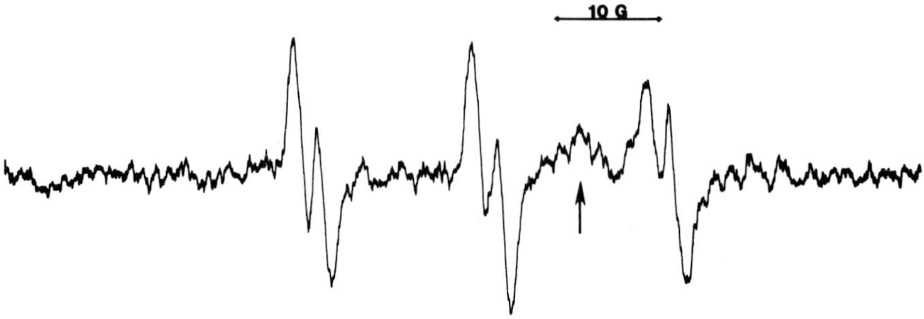

FIGURE 6. Electron paramagnetic resonance spectrum of MNPOL-L obtained by reacting MNPOL with linoleic acid-lipoxygenase system. The spectrum was observed twenty minutes after the reaction was commenced, A_N = 16.6 G and A_H = 2.1 G. The broad peak, as noted by the arrow, is due to an impurity in the epr cell.

spectrum of 4-POBN-OOL is ibtained soon after the reaction is commenced. Secondly, the rate of formation of the spin trapped adduct 4-POBN-OOL approaches zero during the time when the reaction approximates anaerobiosis. Finally, under anaerobic conditions, we do not observe the formation of 4-POBN-L. Although we were surprised at this observation, it is possible that steric hindrance may play a major role in preventing the spin trapping of L· by 4-POBN. Finally, when 4-POBN was incubated with mouse

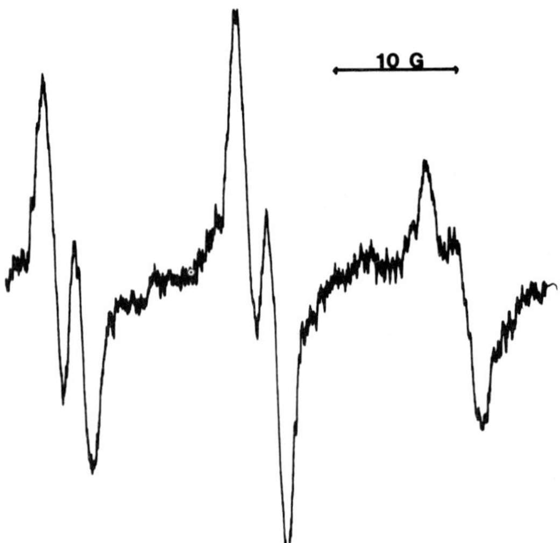

FIGURE 7. Electron paramagnetic resonance spectrum of 4-POBN-OOL obtained by reacting 4-POBN with mouse liver microsomes and NADPH at pH 7.4 with 1 mM DETAPAC, A_N = 15.8 G and A_N = 2.6 G.

liver microsomes and NADPH, we observed only the epr spectrum for 4-POBN-OOL (Fig. 7) and not 4-POBN-OOH or 4-POBN-OH which we have shown are quite unstable and decompose rapidly during the prolonged scan time required to observe the epr spectrum in Fig. 7 (10).

In conclusion, we have found that in mouse hepatic microsomes, a number of different free radical species can be spin trapped depending upon the nature of the free radical and the structural restriction imposed by the spin trap. When using DMPO in this biological system, we can identify the generation of superoxide and hydroxyl radical; hwoever, when 4-POBN is substituted for DMPO only lipid hydroperoxyl radicals are spin trapped. Since the reactions are undertaken aerobically, and since there is no evidence of any lipid radical (L·) being spin trapped, we suggest that once this radical is formed, it reacts with oxygen faster than its reaction with the spin trap.

REFERENCES

1. Bolton, J. R., Borg, D. C. and Swartz, H. M., "Biological Applications of Electron Spin Resonance" (H. M. Swartz, J. R. Bolton and D. C. Borg, eds.), Wiley-Interscience, New York, N.Y.,1972, p. 63.
2. Borg, D. C., "Free Radicals in Biology," Vol. 1 (W. A. Pryor, ed.), Academic Press, New York, N.Y., 1976, p. 69.
3. Beinert, H., "Biological Applications of Electron Spin Resonance" (H. M. Swartz, J. R. Bolton and D. C. Borg, eds.), Wiley-Interscience, New York, N.Y., 1972, p. 351.
4. Yamazaki, I., Mason, H. S. and Piette, L., J. Biol. Chem., 235, 2444 (1960).
5. Borg, D. C., "Biological Applications of Electron Spin Resonance" (H. M. Swartz, J. R. Bolton and D. C. Borg, eds.), Wiley-Interscience, New York, N.Y., 1972, p. 265.
6. Ohnishi, T., Yamazaki, H., Iyanagi, T., Nakamura, T. and Yamazaki, I., Biochim. Biophys. Acta, 1972, 357 (1969).
7. Dorfman, L. M. and Adams, G. E., "Reactivity of the Hydroxyl Radical in Aqueous Solutions," NSRDS, National Bureau of Standards, 46, Washington, D.C., 1973.
8. Finkelstein, E., Rosen, G. M. and Rauckman, E. J., Arch. Biochem. Biophys., 200,1 (1980).
9. Rosen, G. M. and Rauckman, E. J., Mol. Pharmacol. 17,233 (1980).
10. Finkelstein, E., Rosen, G. M., Rauckman, E. J. and Paxton, J., Mol. Pharmacol., 16,676 (1979).
11. Adams, E. G. and Wardman, P., "Free Radicals in Biology," Vol. 3 (W. A. Pryor, ed.), Academic Press, New York, N.Y., 1977, p. 53.
12. Haber, F. and Weiss, J., Proc. Roy. Soc. Ser. A., 147,332 (1934).
13. Rosen, G. M., Rauckman, E. J. and Hanck, K. W., Tox. Lett., 1,71 (1977).

14. DeGroot, J. J., Grott, M. C., Garssen, G. J., Vliegenthart, J. F. G. and Boldingh, J., Biochim. Biophys. Acta, 326,279 (1973).

DISCUSSION

PEISACH: Dr. Rosen, I have some questions of a quantitative nature about your experiments. Specifically, what is the ratio of radical produced in the liver microsomal system compared to NADPH oxidized. What I'm really asking is how leaky is the microsomal electron transport system?

ROSEN: It really depends on how well you make up the microsomes and also upon the concentration of your enzymes. There are at least two major metabolic enzymes in the liver that can produce superoxide. One is the cytochrome P-450 system, both cytochrome P-450 and its reductase will produce superoxide. Also FAD-monooxygenase, a flavoprotein, undergoes a variety of oxidations both on amines as well as sulfur-containing compounds. This enzyme leaks superoxide. So it depends also on the species. For example, in humans as well as hogs and certain mice strains, the FAD monooxygenase is very high in content. Whereas in rats it is exceptionally low so that the amount of free radicals generated varies depending on the species. It is very difficult for me to quantitate exactly how much NADPH is oxidized to superoxide produced, but it is really very little. However it is significant enough so that a variety of peroxidative processes occur, if there aren't any protective mechanisms. In essence, we've washed microsomes so thoroughly that we've removed most of the SOD which is in the cytoplasm as well as the catalase and peroxidases. Thus, there is hardly any catalase left. The glutathione reductase system which is in the cytoplasm of the liver cell is removed; so all these protective mechanisms which occur normally in a cell, we've essentially removed in this *in vitro* system. So I can't completely give you an answer to your question.

PEISACH: By very little do you mean less than 10%, less than 1%, less than 1/1000th of a percent? Based on the NADPH oxidized, what is the order of magnitude?

ROSEN: It's probably in the lower end. It is less than 1%.

ULLRICH: I'm referring to the work of Mr. Kuthan who is doing exactly that work. (See this volume.) You are asking about the efficiency of the microsomal system and we calculated that about 70% of the cytochrome P-450-oxy complex is going to $O_2^{\overline{\cdot}}$. That's a rate of 50 nM $O_2^{\cdot-}$ per mg protein, twice the usual hydroxylation reactions.

FEE: I have a question about the quantitation of the hydroxide radical. It has always bothered me that any strong oxidant might be able to take an electron from the nitrone followed by the addition of water. Could you tell me if that is possible and could it account for any of the free radical signal assigned to hydroxyl radical?

ROSEN: It does not in the DMPO system. Janzen has shown that in his PBN system water will add across the double bond and air oxidation will give you the nitroxide. That does not occur when DMPO is used.

FEE: So it depends on the particular spin trap.

RALEIGH: Are you taking into account the fact that the nitroso can react directly with unsaturated lipids to form nitroxyl free radicals in your trapping experiments with nitroso spin traps?

ROSEN: We have not seen that actually occurring with our model systems. For example, you can take microsomes without NADPH and you will get radicals. We subtract that out. Only the radicals that were actually formed in the presence of NADPH are of significance. One problem you have with this MNPOL is you have to run the reaction in the dark otherwise you end up with free radicals due to dimerization of MNPOL.

SRIDHAR: We've also been able to trap the hydroxyethyl and the hydroxyl free radicals with DMPO when HO· radical is generated in the presence of EtOH. When we trap hydroxyl radical, the trapped radical yield in alcohol is higher than when hydroxyl free radical is trapped under alcohol free conditions. I've got the identical spectrum that you're showing, the only difference is I've got nitroaromatic compounds in an enzyme system which is giving me the hydroxyl radical eventually. But in this particular system, when I add EtOH, the total trapped radical yield is much higher than what I get with just the hydroxyl. I've got a spectrum with me so maybe we can talk later.

ROSEN: What do you want me to answer?

SRIDHAR: Suppose I measure the hydroxyl radical trapped by DMPO, I get a certain amount of radicals trapped. If I do the same reaction in the presence of EtOH, I should be trapping hydroxyl free radicals plus hydroxyl ethyl radicals. But now in this combined system, I'm trapping a lot more free radicals.

ROSEN: Is this in microsomes or is this in the model systems?

SRIDHAR: It is a cytochrome reductase P-450 complex incorporating NADH and an aromatic nitric compound such as m-nitrotoluene.

ROSEN: I do not know the answer.

BORS: There is a conceivable answer. That is the reaction rate of this spin trap with ·OH radicals is less than for EtOH with ·OH. So you're trapping more ·OH radicals with EtOH which then can react with the spin trap and you get an overall higher yield of radicals.

COHEN: Is it clearly established that the spin trap, the DMPO will trap the hydroxyl radical. Is the trapped ·OH reactive? Can you react it with EtOH for example and get the corresponding hydroxyalkyl radical?

VITAMIN E RADICALS

M. G. Simic, Center for Radiation Research, National Bureau of Standards, Washington, D.C. 20234

INTRODUCTION

The extent of publications on vitamin E closely matches its notoriety as a biological antioxidant, anti-aging and anti-cancer agent. In spite of its popularity very little was done on the kinetics of the associated free radical reactions. Direct observation of vitamin E free radicals is possible in general, because of their relative unreactivity. Even the radical-radical reaction is quite slow: $k = 3.5 \times 10^2\ M^{-1}s^{-1}$ as determined by pulse radiolysis in c-hexane (1). The structure of vitamin E free radicals was unequivocally determined by ESR (2) and it was concluded that chromanoxy radicals are formed as expected from mechanistic considerations of vitamin E reactions. Due to the presence of a hetero ring attached to the benzene ring in the chromanols the spin densities at C-5 are about 20% higher than at C-7 (2). The presence of different resonance structures was also demonstrated in the reaction of $\cdot CH_3$ with chromanoxy radical (3). Under these conditions the $\cdot CH_3$ radical adds to C-5 position to give a geminal dimethyl cyclohexadienone and also to the -O· form to give methyl ether of α-tocopherol (3). From these direct observations and previous mechanistic considerations (4-6) it is clear that the most reactive hydrogen is that of the hydroxyl group.

Vitamin E is a two electron donor similar to ascorbic acid. Unlike ascorbic acid the second hydrogen is not well defined, and under different conditions a variety of products can result. Hence in the abbreviated formula, HEOH, both hydrogens are indicated; one in the hydroxy group, and the other unspecified.

EXPERIMENTAL

All experiments were performed at the Center for Fast Kinetics Research using the 5 MeV Van de Graaff pulse radiolysis setup (7). All chemicals used were of the purest commercial grade. Eastman-Kodak α-tocopherol was employed as vitamin E.

FORMATION OF FREE RADICALS

Two media were used in the experiments: (a) aqueous, in which SDS (sodium dodecyl sulfate) and CTAB (cetyltrimethyl-ammonium bromide) micelles were formed ($\sim 1.4 \times 10^{-2}$ M micellar concentration); and (b) pure c-hexane.

ISBN 0-12-592050-4

Peroxy radicals were generated via two distinct mechanisms:

(a) Aqueous media

Radiation generated OH radicals in water were used to produce free radicals on SDS and CTAB molecules which are shown in a general form as P-RH, where P represents the polar head and -RH the hydrocarbon tail.

$$OH + P\text{-}RH \longrightarrow P\text{-}R^{\cdot} + H_2O \qquad (1)$$

and in the presence of oxygen

$$P\text{-}R^{\cdot} + O_2 \longrightarrow P\text{-}ROO^{\cdot} \qquad (2)$$

Under our experimental conditions the P-ROO· radicals are expected to be formed predominantly outside the micelles. The peroxy radicals, irrespective of the origin, will react with vitamin E.

$$P\text{-}ROO^{\cdot} + HEOH \longrightarrow P\text{-}ROOH + HEO^{\cdot} \qquad (3)$$

The contribution of $Br_2^{\overline{\cdot}}$ in the CTAB system has not been determined. From the formation rates of the HEO· absorption bands and taking the average concentration of vitamin E in the bulk the following rate constants were derived for reaction (3): $k = 1.0 \times 10^7$ $M^{-1}s^{-1}$ in SDS and $k = 1.0 \times 10^8$ $M^{-1}s^{-1}$ in CTAB micelles.

The absorption spectra of HEO· in SDS and CTAB micelles are shown in Fig. 1 and Fig. 2. The 423 nm and the 300 nm bands appear simultaneously as demonstrated by time resolved absorption spectra of HEO· in SDS micelles (Fig. 3). Identical response was observed for CTAB micelles.

(b) c-Hexane

In irradiated c-hexane only one type of c-hexane radical is produced—the c-hexyl radical $\cdot C_6H_{11}\cdot$. Unlike in aqueous media this radical is produced directly and in presence of oxygen gives c-hexylperoxy radicals (8).

$$H_2R \longrightarrow HR^{\cdot} + H\ , \qquad (4)$$

$$HR^{\cdot} + O_2 \longrightarrow HROO^{\cdot} \qquad (5)$$

At lower oxygen concentrations (< 1 mM) the H atoms predominantly abstract hydrogen from c-hexane. At higher oxygen concentrations some HOO· radicals will be formed.

The reactions (4) and (5) are followed by

$$HROO^{\cdot} + HEOH \longrightarrow HROOH + HEO^{\cdot} \qquad (6)$$

At 20°C $k_6 = 1.4 \times 10^7$ $M^{-1}s^{-1}$ was measured.

The absorption spectrum of HEO· in c-hexane is in agreement with the previously reported spectrum in the same medium (9,1) and for the sake of comparison with the spectra of HEO· in the micelles it is shown in Fig. 4. Both the 415 nm and the 340 nm bands are generated simultaneously as in Fig. 1 and Fig. 2. The difference between the spectrum of HEO· in a nonpolar medium, Fig.

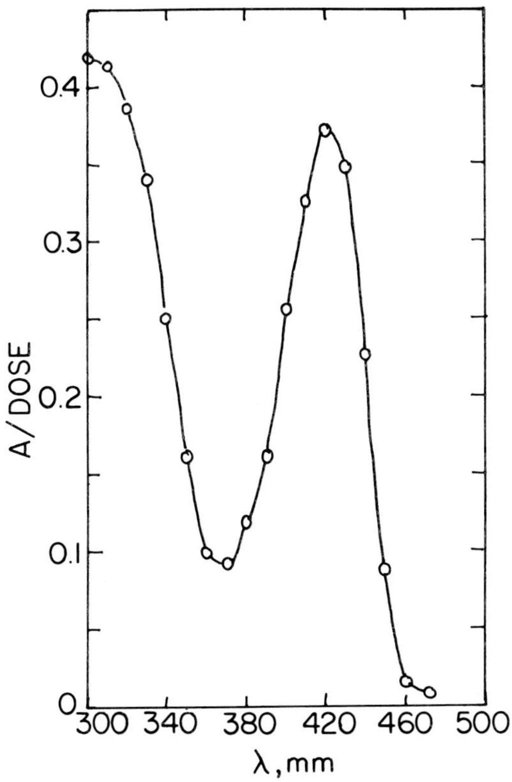

FIGURE 1. Absorption spectrum of α-tocopherol free radical in aqueous SDS micelles. 10^{-1} M SDS, 1.5×10^{-3} M α-tocopherol, air saturated.

4, and the spectra of HEO· in aqueous micellar systems, Fig. 1 and Fig. 2, is obvious and it does not require any comments.

DISCUSSION

The absorption spectra of HEO· in different media are characterized by the 415-425 nm absorption band and resemble the spectra of other phenoxy radicals, ΦO· (10), to which HEO· is related. The strong 340 nm or 300 nm band is most likely a result of the presence of the O-hetero ring in vitamin E.

The spectra of HEO· in aqueous micellar systems are similar to the spectrum of HEO· recorded in 50% isopropanol, 10% acetone, 40% H_2O + CCl_4 system (11). In the latter system the vitamin E radical is produced according to the following mechanism (11),

$$(CH_3)_2\dot{C}OH + CCl_4 \longrightarrow (CH_3)_2CO + Cl_3C\cdot + Cl^- + H^+ \quad (7)$$

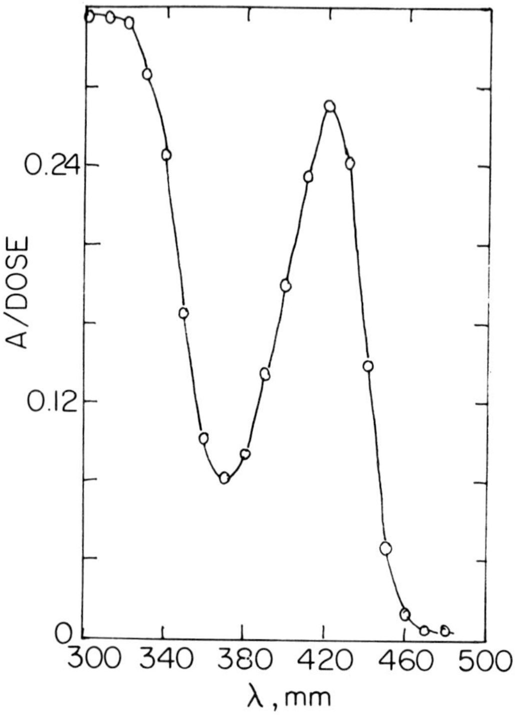

FIGURE 2. Absorption spectrum of α-tocopherol free radical in aqueous CTAB micelles. 10^{-1}M CTAB, 1.5 10^{-3} M α-tocopherol, air saturated.

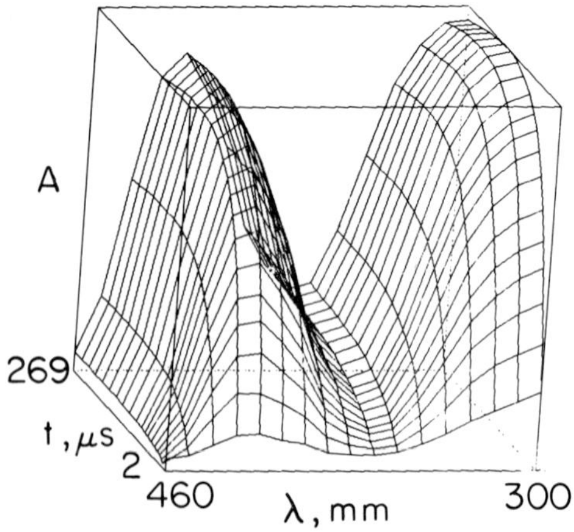

FIGURE 3. Time resolved absorption spectra of α-tocopherol free radical in SDS micelles showing simultaneous formation of the 300 nm and 425 nm bands.

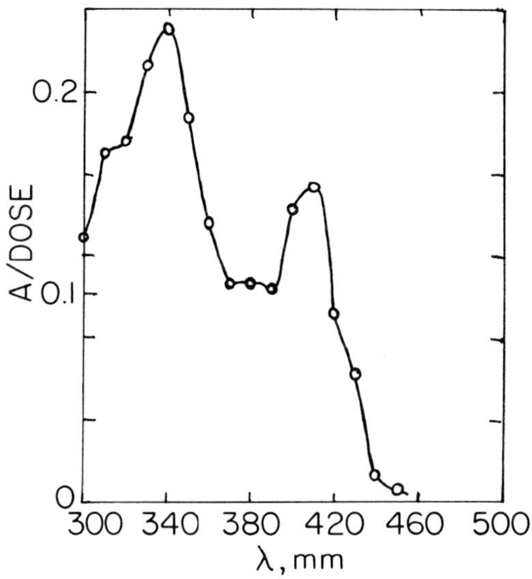

FIGURE 4. Absorption spectrum of α-tocopherol free radical in c-hexane. 2 x 10^{-3} M α-tocopherol, air saturated c-hexane.

$$Cl_3C\cdot + O_2 \longrightarrow Cl_3COO\cdot$$

$$Cl_3COO\cdot + HEOH \longrightarrow Cl_3COOH + HEO\cdot$$

$$k_9 = 5 \times 10^8\ M^{-1}s^{-1}$$

The 340 nm and 415 nm bands observed in a nonpolar medium such as c-hexane (Fig. 4), are shifted in polar media to ≤ 300 nm and 425 nm. This shift in polar media is not inconceivable in view of a highly resonant structure of HEO· and its possible interaction with dipoles.

Extension of these conclusions to aqueous micellar systems infers a polar environment of the HEO· radical, i.e., the chromanoxy part of the free radical is close to or at the surface of the micelle with the hydrocarbon tail of vitamin E ($-C_{16}H_{33}$) buried in the micelle. This proposition is illustrated in Fig. 5. It is far less conceivable that HEO· would have a random position within a micelle.

This premise outlined for micelles can be extended to a question concerning the position of vitamin E and the mechanism of its action in a membrane. If HEOH is to seek ROO· radicals within a membrane the process would be associated with large enthropy changes. The net result would be greatly reduced rate constants for reaction (3) and diminished efficiency of its role as a biological antioxidant. On the other hand if HEOH is on the surface of a membrane with a function to intercept free radical initiators (e.g., peroxy radicals, free radicals in general)

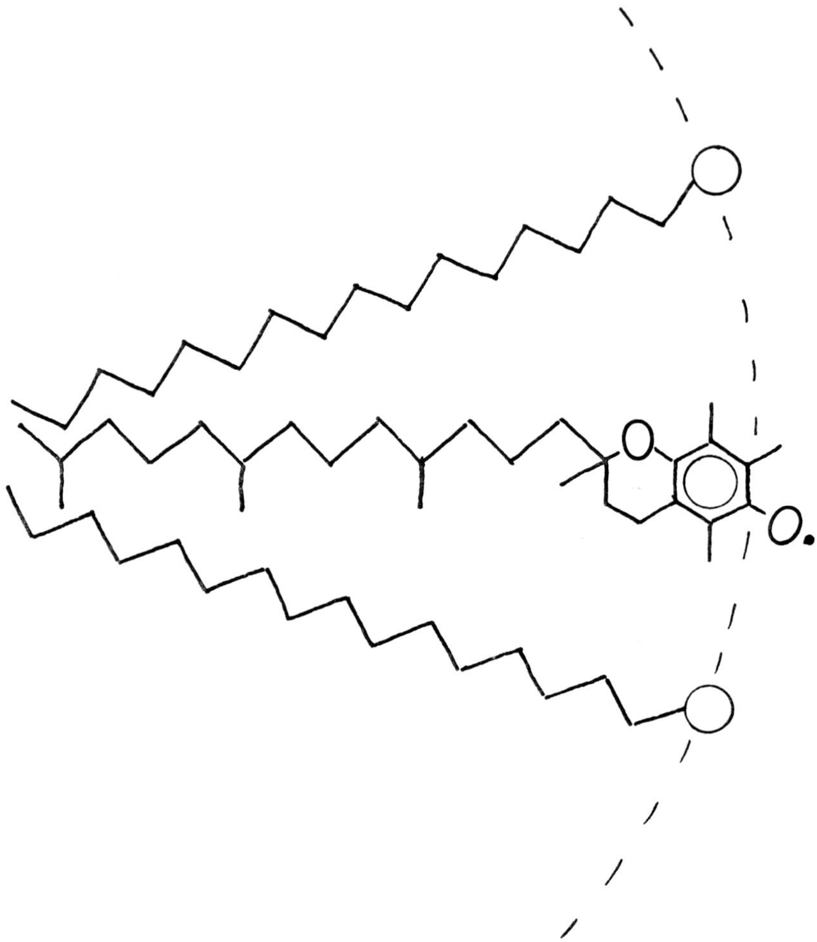

FIGURE 5. Suggested position of α-tocopherol radical (HEO·) in a micelle.

which are generated in a more polar medium, its efficiency as an antioxidant would be greatly increased, as the rate constants for micellar systems suggest ($k = 10^7$ to 10^8 $M^{-1}s^{-1}$).

The presence of the HEO· radical at the surface adds another advantage to the system since it enables recovery of vitamin E through electron (or H) transfer from more polar donors, e.g., ascorbic acid, H_2A, as suggested by Tappel (12).

$$HEO\cdot + H_2A \longrightarrow HEOH + HA\cdot \tag{10}$$

and recently demonstrated to take place in aqueous solutions (11).

ACKNOWLEDGEMENTS

Part of this work was performed at the Center for Fast Kinetics Research at the University of Texas at Austin. The CFKR is supported by NIH grant RR-00886 from the Biotechnology Branch of the Division of Research Resources and by the University of Texas at Austin.

REFERENCES

1. Simic, M. G., "Autoxidation in Food and Biological Systems" (M. G. Simic and M. Karel, eds.), Plenum, 1980, p. 17.
2. Boguth, W. and Niemann, H., Biochim. Biophys. Acta, 248,121 (1971).
3. Urano, S. and Matsuo, M., Lipids, 11,380 (1976).
4. Cook, C. D., Nash, N. G. and Flanagan, H. R., J. Am. Chem. Soc., 77,1783 (1955).
5. Inglett, G. E. and Mattill, H. A., ibid., 77,6552 (1955).
6. Knapp, F. W. and Tappel, A. L., J. Am. Oil Chem. Soc., 38, 151 (1961).
7. Rodgers, M. A. J., IEEE Trans. Nuclear Sci., NS-26, 1733 (1979).
8. Simic, M. and Hayon, E., J. Phys. Chem., 75,1677 (1971).
9. Simic, M. G., NERM 8, Abstracts, Am. Chem. Soc., Boston, p. 47 (1978).
10. See e.g., P. Neta and S. Steenken, This volume, p. 83.
11. Packer, J. E., Slater, T. F. and Willson, R. L., Nature, 278,737 (1979).
12. Tappel, A. L., Geriatrics 23,97 (1968).

DISCUSSION

SINGH: You mentioned that you showed the shift of the spectrum of Vitamin E radical and you speculated that it may be because in one case it is in the hydrocarbon medium and in another case it is polar medium. Is there any other independent information about the shift of this spectrum? I imagine one could produce this free radical in a polar solvent as well as like DMSO or dimethylformamide. Does it shift in that case?

SIMIC: The spectrum of Vitamin E radical reported by Willson was obtained in a predominantly aqueous medium with alcohol added to increase the solubility of Vitamin E. Willson is here, he can comment.

WILLSON: Yes. We've used several aqueous systems up to 50% tertiary butanol.

SAWYER: When we react α-tocopherol with superoxide we can monitor the production of the radical and in DMF the absorption max. is about 430 nm; it is a single absorption peak.

PRYOR: Larry Patterson's paper suggests the answer to something I have been trying to understand for some time: namely, how can one tocopherol molecule per 1000 fatty acid and PUFA molecules in the membrane protect the PUFA from autoxidation. The autoxidative reactions may occur faster than large molecules can undergo lateral transport along the axis of a bilayer biological membrane. Furthermore, the kinetic chain length in autoxidation in a membrane is probably short. (Autoxidation chain lengths are short in vitro and biological systems are protected.) So how can one tocopherol protect 1000 fatty acid molecules? Note that the kinetic chain length is the number of product molecules per primary initiating radical. However, Larry has shown that in a micelle an intermolecular hydrogen atom transfer occurs to pass along the site of damage:

$$(\text{PUFA radical})_1 + (\text{PUFA})_2 \longrightarrow (\text{PUFA})_1 + (\text{PUFA radical})_2$$

If this type of reaction occurs in biological membranes, then many of these reactions occur before a stable product (such as a lipid hydroperoxide) is produced; thus, tocopherol gets many chances to intercept the radical, although this is a single kinetic chain in the sense that one primary radical may give only one product molecule.

PATTERSON: For $RO_2\cdot$ reaction with a PUFA molecule, the rate constant is about 50 $M^{-1}s^{-1}$, while the rate constant for reaction between α-tocopherol and an $RO_2\cdot$ radical in the lipid medium is about $10^4 - 10^5$ $M^{-1}s^{-1}$. The α-tocopherol in kinetic terms can react with $RO_2\cdot$ orders of magnitude more efficiently. So you only really need one α-tocopherol molecule in 10^4 for protection. Some time ago Mead presented levels of protection provided by addition of various tocopherols to pure lipids. When one compares these numbers to the protection suggested by our rate constants plus a peroxidative chain length of 50 and a $RO_2\cdot$-lipid rate constant of 50, the two approaches—steady state and kinetic—fit pretty well. Maybe the reason α-tocopherol is so efficient is because of this massive difference in the rate constants.

There is a limitation to comparing micelles with true membrane bilayers. The micelle is a very dynamic system; and to a first approximation the micelle pseudophase containing α-tocopherol may be considered separate from the water phase with $RO_2\cdot$ simply moving freely between various micelle units of that pseudophase.

BORG: Can I say something to support this? Movement in the plane of cell membranes is very fast, as shown by spin-label experiments of McConnell and others. Hence in a cell membrane α-tocopherol moves around a great deal in the time interval we are considering, and that tends to support the high efficiency of α-tocopherol as an antioxidant in membranes. Based on the rate constants you presented, the rate-limiting step in lipid chain propagation is $RO_2^{\cdot} + R'H \longrightarrow ROOH + R'$; and the lifetimes of $RO_2^{\cdot}$ will be long enough for one α-tocopherol in the membrane to "police" a significant region of membrane for $RO_2^{\cdot}$ to scavenge. That is just a slightly different version of the steady state and kinetic fit you described.

There is one other feature of α-tocopherol to bring to your attention which also relates to what you (Patterson) said this morning. There is more than just redox behavior to take into account. α-Tocopherol, like cholesterol, changes membrane stiffness—that's been shown in red cells and elsewhere—and you pointed out how changes in the packing of the PUFAs make quite a difference in the efficiency of electron transfer in the condensed phase. Hence other roles than redox roles have to be thought of to explain α-tocopherol effects on membrane oxidation.

It still seems to be unrealistic to postulate that tocopherol can move a thousand molecular steps along the membrane to compete with the fast rates of autoxidation of PUFA. The extrapolation of data from homogeneous systems to the biological bilayer system involves great hazards.

PATTERSON: In the micelle you've only got about a hundred molecules and $RO_2^{\cdot}$ can move from one micelle of 100 units to another unit of a hundred. It doesn't have to make a thousand steps. The dynamics are admittedly rather different in the micelles than they are in the membrane. We ought to discuss that in more detail. A separate point concerns the type of radicals produced by α-tocopherol H donation to $RO_2^{\cdot}$. P. Neta and I (mainly P. Neta) looked at the ESR spectrum of the radicals produced and indeed found the phenoxyl radical. Now some people have suggested that there is a mechanism for protection by α-tocopherol other than the donation of phenolic hydrogen, and I guess in none of these studies discussed today do we see that, all we are seeing is the single phenoxyl radical.

WILLSON: I would like to make two points.

First some recent experiments of Andrew Searle at Brunel also indicate that as Dr. Borg says, factors in addition to simple rates of radical reactions are important in determining antioxidant activity. In collaboration with Professor Paoletti and Dr. Malvy at the Institute Gustav Roussy, Paris, we have shown that the compound 6 hydroxy 2,5 dimethyl carbazole (HDC) is a potent inhibitor of microsomal lipid peroxidation. Although the compound reacts rapidly with several free radicals the absolute rate constants are not significantly greater than that for the drug promethazine. Yet in many systems particularly those in which iron is important HDC is some ten times more active. It seems that membrane/iron interactions may play an important role.

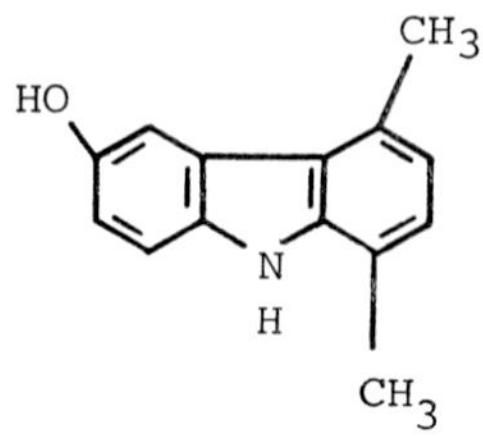

Second, regarding the reactions of peroxy radicals with Vitamin E which Dr. Simic mentioned, we have not undertaken more experiments with biological molecules. In particular, Prof. Trevor Slater, my head of department, who actually suggested the experiment with β carotene and Dr. Packer and Victor Mora-Arellano in our laboratory have shown by pulse radiolysis that the reaction

$$Cl_3COO\cdot + \beta\text{-carotene} \longrightarrow Cl_3COO^- + \beta\text{-carotene}^{\dot{+}}$$

is extremely rapid $k = 1.5 \pm 0.2 \times 10^9\ M^{-1}s^{-1}$.
A rapid bleaching of the ground state β-carotene absorption at 450 nm was observed. In collaboration with Dr. Land at the Christie Hospital Manchester we have shown that a transient absorption characteristic of the β-carotene radical-cation forms simultaneously. Clearly β-carotene is not a specific scavenger for singlet oxygen.

Studies with ascorbate (AH^-) suggest that other peroxy radicals may react with β-carotene similarly albeit somewhat more slowly.

$$Cl_3COO\cdot + AH^- \longrightarrow Cl_3COO^- + AH\cdot \quad k = 2 \times 10^8\ M^{-1}s^{-1}$$

$$ClCH_2OO\cdot + AH^- \longrightarrow ClCH_2OO^- + AH\cdot \quad k = 1 \times 10^8\ M^{-1}s^{-1}$$

$$CH_3OO\cdot + AH^- \longrightarrow CH_3OO^- + AH\cdot \quad k = 2 \times 10^6\ M^{-1}s^{-1}$$

(J. E. Packer, R. L. Willson, D. Bahnemann and K.-D. Asmus, J. Chem. Soc. Perkin II 296, 1980.) Interestingly these reactions do not occur rapidly in the absence of O_2: radicals such as $Cl_3C\cdot$ react at least 100 times more slowly.

Finally β-carotene is closely related to Vitamin A. The protective effect of Vitamin E on Vitamin A which Dr. Totter mentioned earlier could be due to competitive scavenging of peroxy radicals or possibly due to cycling reactions such as those previously observed between Vitamin E and Vitamin C.

SIMIC: Just one more comment, I have also looked at the reactions of cyclohexane and cyclohexane peroxy radicals with β-carotene, just as we did with Vitamin E. They are equally reactive with β-carotene.

KINETIC STUDIES OF PEROXIDES AND PEROXY RADICALS, AND THEIR REACTIONS WITH BIOLOGICAL MOLECULES

Clive L. Greenstock and Richard H. Wiebe, Medical Biophysics Branch, Atomic Energy of Canada Limited, Whiteshell Nuclear Research Establishment, Pinawa, Manitoba, Canada ROE 1LO

INTRODUCTION

Peroxides and peroxy radicals are formed in a variety of xenobiotic and normal biological (1,2), radiation chemical (3), and radiobiological (4,5) processes. Hydrogen peroxide is formed in irradiated aqueous solutions, with a G value of 0.7 (3), and peroxides are important products of irradiated lipids (6), proteins (7) and nucleic acids (8). Because of the ubiquitous nature of these forms of activated oxygen, a growing interest has been expressed in their formation and reactivity, and in the underlying mechanisms of their free-radical-mediated toxicity. Also, there has been some discussion of the possible role of peroxides and peroxy radicals, particularly superoxide ($O_2^{\overline{\cdot}}$) and its protonated form, the hydroperoxy radical (9), in the radiobiological oxygen effect. The realization that the metal ion catalyzed conversion of $O_2^{\overline{\cdot}}$ to the more damaging hydroxyl radical, can occur in irradiated cells in the presence of hydrogen peroxide (Haber-Weiss reaction (10)), may support this hypothesis. Reports of in vitro and in vivo radioprotection (11,12) by the enzyme superoxide dismutase also provides support for the involvement of superoxide in radiation damage to some biological systems.

Since peroxides and peroxy radicals are relatively unreactive species, which are capable of initiating potentially more damaging chain reactions, their involvement in "post-irradiation" effects, or in "damage at a distance" is not an unreasonable assumption (13). The interconversion of peroxides and peroxy radicals, including superoxide, to more reactive forms of activated oxygen is further implicated by the protection afforded mice by superoxide dismutase when administered post-irradiation (14). The persistence of peroxides formed in irradiated biological systems may also have consequences for food sterilization, in terms of their radiotoxicity (15), and even carcinogenicity (16), and their involvement as by-products of multiplicatory oxidative processes.

Peroxides may be harmful, either directly by their reactions with important cellular constituents, or indirectly by their stimulating deleterious free radical processes involving other forms of activated oxygen, as in the Haber-Weiss reaction. Radiation chemical studies of peroxides may improve our understanding of the early processes of radiation inactivation with cellular and

ISBN 0-12-592050-4

multicellular systems. Application of the technique of pulse radiolysis (17) to the study of the reactions of water radiolysis species with organic peroxides and hydroperoxides, and the reactions of peroxy radicals, provides a unique opportunity to examine the free radical mechanisms involved, to determine the important initiating species, and to investigate the kinetics of the redox reactions involving oxygen and activated oxygen.

RESULTS AND DISCUSSION

In the radiolysis of water, three primary species, the hydrated electron (e^-_{aq}), hydrogen atom (H·) and hydroxyl free radical (·OH), are formed with G values of 2.7, 0.6 and 2.8, respectively (3). In aerated solution, $O_2^{\overline{\cdot}}$ is formed with a G value of 3.3 (3) as a result of the fast reaction of e^-_{aq} and H· with oxygen.

$$H_2O \rightsquigarrow e^-_{aq}, H^{\cdot}, \cdot OH, H_2, H_2O_2, H_3O^+, OH^- \qquad (1)$$

$$e^-_{aq}\ (H\cdot) + O_2 \longrightarrow O_2^{\overline{\cdot}} + (H^+) \qquad (2)$$

By the careful use of selective free radical scavengers (18), the reactivity of each of these water radiolysis species towards peroxides can be followed kinetically using pulse radiolysis.

$$e^-_{aq} + N_2O \xrightarrow{H_2O} N_2 + \cdot OH + OH^- \qquad (3)$$

$$\cdot OH(H\cdot) + (CH_3)_3\ COH \longrightarrow (CH_3)_2\ \dot{C}H_2COH + H_2O(H_2) \qquad (4)$$

$$\cdot OH(H\cdot) + HCO_2^- \longrightarrow CO_2^{\overline{\cdot}} + H_2O(H_2) \qquad (5)$$

$$CO_2^{\overline{\cdot}} + O_2 \longrightarrow CO_2 + O_2^{\overline{\cdot}} \qquad (6)$$

(i) Reactions of e^-_{aq} with Peroxides

Rate constants for the reactions of e^-_{aq} with peroxides were measured directly by observing the rate of decay of e^-_{aq} absorption at 650 nm, as a function of peroxide concentration, in deaerated aqueous solution containing 0.5 M t-butanol as a scavenger of H· and ·OH (reaction 4). In all cases, the e^-_{aq} decay rates were exponential and first order in peroxide concentration, yielding the absolute bimolecular rate constants for the reaction of e^-_{aq} with peroxides, given in Table I.

$$e^-_{aq} + ROOH \longrightarrow RO\cdot + OH^- \qquad (7)$$

The e^-_{aq} rate constants were all $\simeq 10^{10}\ M^{-1}\ s^{-1}$, and close to the diffusion controlled limit. The reaction of e^-_{aq} with peroxides (reaction 7), by analogy with that of H_2O_2 (19) is presumed to be a dissociative electron transfer reaction, generating alkoxy or phenoxy-type radicals. In anoxia, peroxides may sensitize by converting e^-_{aq} to RO·. These resulting radicals are reactive and may readily abstract hydrogen from labile hydrogen.

TABLE I. Absolute Rate Constants for the Reaction of Hydrated Electrons with Peroxides and Hydroperoxides

Peroxide	*Rate Constant* $M^{-1}s^{-1} \times 10^9$
Hydrogen peroxide	*9.5*
Ethyl hydroperoxide	*5*
t-Butyl hydroperoxide	*6*
Cumene hydroperoxide	*4*
Dichlorobenzoyl hydroperoxide	*9*
Ascaridole	*4*
Acetyl peroxide	*9.5*
Dichlorobenzoyl peroxide	*5*
Disuccinyl peroxide	*10*

donors such as saturated fats (LH), thus initiating oxidative chain reactions, such as lipid peroxidation (6).

$$\text{initiation } RO\cdot + LH \longrightarrow ROH + L\cdot \quad (8)$$

$$\text{propagation } L\cdot + O_2 \longrightarrow LO_2^{\cdot} \quad (9)$$

$$LO_2^{\cdot} + LH \longrightarrow LOOH + L\cdot \quad (10)$$

(ii) Reactions of ·OH with Peroxides

The rates of reaction of ·OH with peroxides were determined at neutral pH in N_2Osaturated aqueous solution (reaction 3), using the pulse radiolysis competition method (20). In this method, the absorbance at 490 nm of the thiocyanate dimer radical anion $(CNS)_2^{\cdot-}$ product of the reaction of ·OH with CNS^-, was monitored in the absence (A_o) and presence (A) of peroxide:

$$\cdot OH + \text{peroxide} \longrightarrow \text{non-absorbing product} \quad (11)$$

$$\cdot OH + CNS^- \longrightarrow CNS\cdot \xrightarrow{CNS^-} (CNS)_2^{\cdot-} \quad (12)$$

$$A_o/A = 1 + k_{11}\,[\text{peroxide}]/k_{12}[CNS^-] \quad (13)$$

The concentration of CNS^- was kept constant at 2×10^{-3}M, which was sufficient to scavenge ·OH. Typical competition plots show the expected linear variation of A_o/A with increasing peroxide concentration. The rate constants for the reaction of ·OH with peroxides, obtained from the slopes (k_{11}/k_{12}) of competition plots, using a value of $k_{12} = 1.1 \times 10^{10}\ M^{-1}\ s^{-1}$ (21), are given in Table II. Hydroxyl radical rate constants are in the range of $\sim 10^7\ M^{-1}\ s^{-1}$, which is two or three orders of magnitude slower than diffusion controlled. Simple unsubstituted aliphatic hydroperoxides are less reactive towards ·OH than the two peroxides

TABLE II. Absolute Rate Constants for the Reaction of Hydroxyl Radicals with Peroxides and Hydroperoxides

Peroxide	*Rate Constant* $M^{-1}s^{-1} \times 10^7$
Hydrogen peroxide	*2.0*
Ethyl hydroperoxide	*1.5*
t-Butyl hydroperoxide	*<1*
Acetyl peroxide	*6.0*
Succinyl peroxide	*8.0*

tested, acetyl peroxide and succinyl peroxide. None of the products of the reaction of ·OH with peroxides showed any detectable U.V. absorption, making it impossible to ascertain which reaction predominates.

$$\cdot OH + R(H)OOH \rightarrow \cdot ROOH/R(H)O_2^{\cdot} + H_2O \quad (14a)$$

$$\cdot OH + R(H)OOH \rightarrow R(H)O\cdot + H_2O_2 \quad (14b)$$

Hydrogen abstraction by ·OH (reaction 14a), leading to peroxy, alkyl or phenyl radical formation, is expected on the basis of results with H_2O_2 (19), but nucleophilic attack at the 0-0 bond (reaction 14b) is favored energetically, in terms of the lower dissociation energy of the 0-0 bond compared with that of O-H (22).

(iii) Reactions of $O_2^{\bar{\cdot}}$ with peroxides

Attempts to measure the rates of reaction of $O_2^{\bar{\cdot}}$ with peroxides, using a kinetic competition method based on p-benzoquinone as the competitive solute (23) were unsuccessful because of the low reactivity of $O_2^{\bar{\cdot}}$ with peroxides. The one exception to this was acetyl peroxide, which had a measurable rate constant of $2 \times 10^7\ M^{-1}\ s^{-1}$ for its reaction with $O_2^{\bar{\cdot}}$. A steady-state radiolysis competition method using 9,10 dihydroanthracene (9,10 DHA) as the competitive solute, has been developed to estimate an upper limit to the $O_2^{\bar{\cdot}}$ reactivity of peroxides. The assay is based on monitoring the yield of anthracene (An), which is a product of the reaction of $O_2^{\bar{\cdot}}$ with 9,10 DHA (24), in the presence (A) and absence (Ao) of peroxide.

$$O_2^{\bar{\cdot}} + \text{peroxide} \longrightarrow \text{unreactive products} \quad (15)$$

$$O_2^{\bar{\cdot}} + 9{,}10\ \text{DHA} \xrightarrow{k_{16}} X \longrightarrow An \quad (16)$$

$$Ao/A = 1 + k_{15}\ [\text{peroxide}]/k_{16}[9{,}10\ \text{DHA}] \quad (17)$$

This method will only produce linear competition plots with unit intercepts on the ordinate, if the products of $O_2^{\bar{\cdot}}$ with peroxides are unreactive, if A is a direct product of reaction 16, and if k_{16} represents the rate limiting step. Presently, only ethylhydroperoxide has yielded satisfactory results. From the competition plot shown in Fig. 1, the slope of 0.03 indicates that the rate constant for the reaction of $O_2^{\cdot-}$ with ethylhydroperoxide is 30 times lower than that for the reaction of 9,10 DHA. To date, pulse radiolysis studies have shown only that k_{16} is $\leq 10^6\ M^{-1}s^{-1}$ indicating an upper limit for the reaction of $O_2^{\cdot-}$ with ethylhydroperoxide of less than $3 \times 10^4\ M^{-1}\ s^{-1}$.

Although the reactions of $O_2^{\cdot-}$ with peroxides may be very slow, nevertheless, the reactions may be important as a source of reactive free radical products, produced by the interconversion of one form of activated oxygen to another.

$$O_2^{\cdot-} + ROOH \longrightarrow RO^- + O_2 + \cdot OH \quad (18a)$$

$$O_2^{\cdot-} + ROOH \longrightarrow RO^{\cdot} + O_2 + OH^- \quad (18b)$$

These nucleophilic displacement reactions which require metal ion catalysis, are analogous to the Haber-Weiss reaction involving H_2O_2, and, as such, may be an *in vivo* source of hydroxyl radicals (reaction 18a) or equally reactive phenoxy radicals (reaction 18b). In both cases, reaction 18 represents the transformation of a relatively unreactive superoxide to a more reactive free radical, capable of initiating enhanced damage via oxidative chain reactions (reactions 8-10). Peroxides are more powerful oxidants than $O_2^{\cdot-}$, a fact which appears to rule out the possibility of peroxy radical formation following hydrogen abstraction from peroxides by $O_2^{\cdot-}$. In a recent publication (25), reaction 18a was favored over reaction 18b on the basis of electronegativity.

Since the relatively unreactive superoxide can diffuse large distances in cellular systems, reaction 18 represents a possible initiator of "damage at a distance" and of "post-irradiation effects" (13-15). If superoxide is involved in such reactions, then superoxide dismutase, either intracellular or exogenous, should provide a measure of protection (11,12,14) as a free radical defense mechanism against $O_2^{\cdot-}$ mediated damage.

(iv) Reactions of Biological Peroxy Radicals

The reactions of oxygen with radicals and radical anions of pyrimidine bases as model DNA target species, have been investigated, in an attempt to clarity the molecular mechanism of the radiobiological oxygen effect. The rate of the uracil radical anion ($U^{\cdot-}$) formed by simple electron attachment to uracil, was studied in the presence of different concentrations of oxygen:

$$e^-_{aq} + U \longrightarrow U^{\cdot-} \quad (19)$$

$$U^{\cdot-} + O_2 \longrightarrow U + O_2^{\cdot-} \quad (20a)$$

$$U^{\cdot-} + O_2 \longrightarrow (UO_2^{\cdot-}) \quad (20b)$$

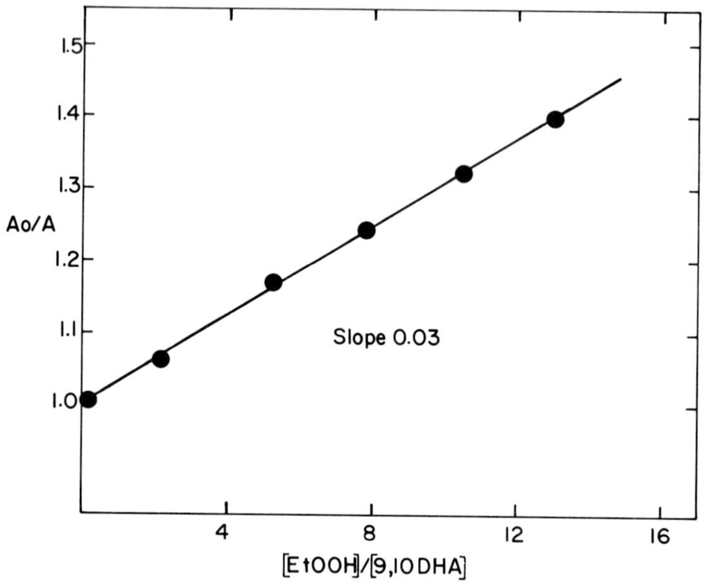

FIGURE 1. Steady state competition between ethyl hydroperoxide and 9,10 dihydroanthracene for superoxide radical anions. Oxygen-free dry DMSO solutions containing 3 mM 9,10 DHA and peroxide were reacted with DMSO solutions containing 20 mM KO_2 and 20 mM 18-dicyclohexyl-6-crown. The yields of anthracene in the presence (A) and absence (Ao) of peroxide were determined using HPLC.

From the decay of the uracil radical anion $U^{\overline{\cdot}}$ absorption at 330 nm (Fig. 2a), which was exponential and first-order in oxygen concentration, a rate constant for reaction 20 was measured to be $1.2 \times 10^9\ M^{-1}\ s^{-1}$ (26,27). To clarify the nature of the product of this reaction, a second experiment was performed in which trace amounts (10-50 μM) of p-benzoquinone were added to aerated solutions of uracil, which were pulse irradiated in the presence of the ·OH scavenger, t-butanol. The result, which is shown in Fig. 2b, indicates that transient absorbance at 410 nm, attributable to the semiquinone radical anion p-BQ$^{\overline{\cdot}}$, is formed with essentially 100% efficiency, and its rate of formation is 1.0×10^9 $M^{-1}\ s^{-1}$ which corresponds closely to the rate of electron transfer from $O_2^{\overline{\cdot}}$ to p-BQ (reaction 21).

$$(U + O_2^{\overline{\cdot}}) + p\text{-}BQ \longrightarrow O_2 + p\text{-}BQ^{\overline{\cdot}} \qquad (21)$$

This result strongly supports reaction 20a in which $U^{\overline{\cdot}}$ transfers an electron to O_2 to form $O_2^{\overline{\cdot}}$. Whether or not $UO_2^{\overline{\cdot}}$ exists as a short-lived intermediate in reaction 20a has not been established, but if it does exist, these experiments indicate that its lifetime must be $\lesssim$ nanoseconds.

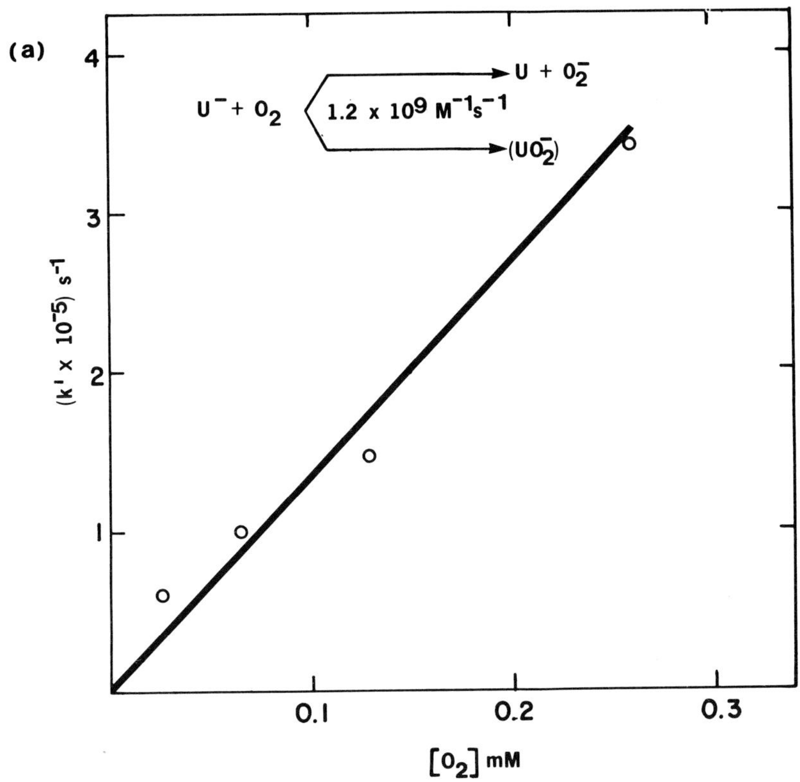

FIGURE 2a. First order plot showing the rate of reaction of the uracil radical anion ($U^{\overline{\cdot}}$) with oxygen. Dose/pulse ~ 1 krad, pH 6, 0.5 M. t-butanol, 330 nm.

In pulse irradiated nitrous oxide saturated solutions of uracil, the ·OH-adduct ·UOH is formed, which absorbs at 380 nm. The reaction of this uracil ·OH-adduct with oxygen has been followed, by observing the rate of decay of ·UOH absorbance as a function of oxygen concentration (28,29). The results in Fig. 3 show that the reaction is first-order in oxygen, with a rate constant of $2.2 \times 10^9\ M^{-1}\ s^{-1}$. No absorbing product was detected, which might have provided information to differentiate between oxygen addition to ·UOH leading to peroxy radical formation (reaction 23a), or electron transfer oxidation forming superoxide (reaction 23b).

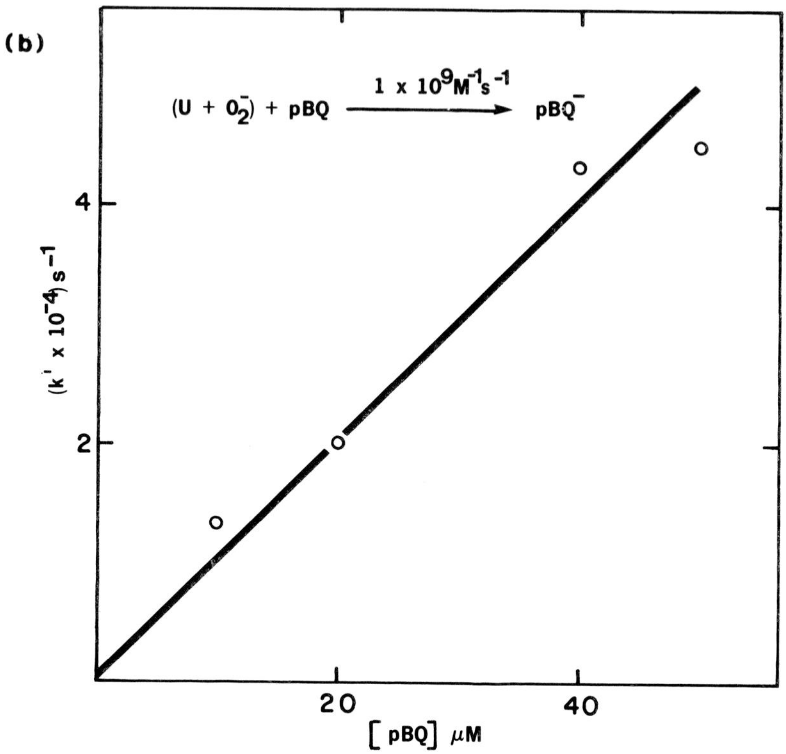

FIGURE 2b. First order plot showing the rate of electron transfer from the product of the reaction of $U^{\bar{\cdot}}$ with oxygen, ($O_2^{\bar{\cdot}}$) to p-benzoquinone. Dose/pulse ~ 0.2 krad, pH 7, 410 nm, oxygen saturated aqueous solution.

$$U + \cdot OH \longrightarrow \cdot UOH \quad (22)$$

$$\cdot UOH + O_2 \longrightarrow \cdot U(O_2)OH \quad (23a)$$

$$\cdot UOH + O_2 \longrightarrow UOH^+ + O_2^{\bar{\cdot}} \quad (23b)$$

However, experiments using trace amounts of p-BQ (10-50 μM) as an electron acceptor probe for $O_2^{\bar{\cdot}}$, produced no evidence for p-BQ$^{\bar{\cdot}}$ formation, strongly supporting the hypothesis that the reaction of ·UOH with oxygen is, in fact, an addition reaction, leading to the formation of a peroxy-hydroxy uracil radical ·U(O_2)OH, which, unlike $O_2^{\bar{\cdot}}$ cannot transfer an electron to p-BQ.

$$\cdot U(O_2)OH + p\text{-}BQ \not\longrightarrow p\text{-}BQ^{\bar{\cdot}} \quad (24)$$

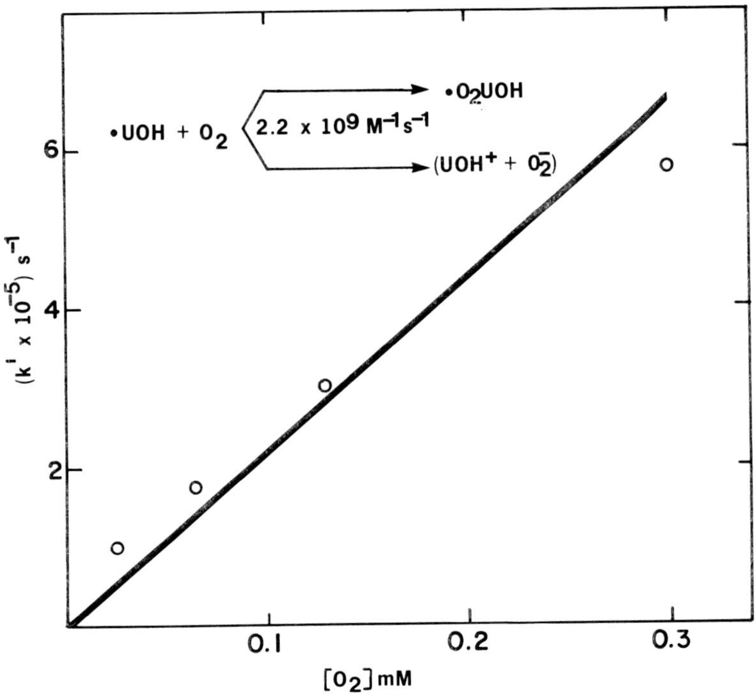

FIGURE 3. First order plot showing the rate of reaction of the uracil OH-adduct (•UOH) with oxygen. Dose pulse ~ 1 krad, pH 7, 380 nm, 2-20% N_2O/O_2 mixtures.

This result is consistent with the observation that hydroperoxides are formed as products in irradiated nucleic acid bases and polynucleotides, including DNA (8).

It is clear from this pulse radiolysis study of the radiation chemistry of peroxide that: a variety of water radiolysis species react with peroxides and hydroperoxides; many of the products of these reactions are, themselves, reactive or unstable species, and will take part in further reactions capable of producing chemical damage of possible relevance to radiobiology; oxygen takes part in primary or secondary reactions in radiation biochemistry, leading to the production of different forms of activated oxygen, including O_2^- and peroxy radicals; interconversion of peroxides and other forms of activated oxygen can be catalysed by ionizing radiation, and these reactions may have serious implications, not only for the radiobiological oxygen effect, but also for autocatalytic chain reactions, as in lipid peroxidation (6), and for the production of damage at a distance (13) and post-irradiation effects (14). In these processes, primary damaging events, occurring during irradiation, may be transformed into a wide variety of secondary effects, some involving only weakly reactive species with long lifetimes and the ability to diffuse large distances of the order of cellular dimensions, to produce damage to remote vulnerable secondary target sites, located some distance away from the site of initiation of the free radical damage.

REFERENCES

1. Paine, A. J., Biochem. Pharmacol., 27,1805, 1978.
2. O'Brien, P. J., Pharmacol. Ther. A., 2,517, 1978.
3. Draganić, I. G. and Draganić, "The Radiation Chemistry of Water," Academic Press, New York, 1971.
4. Alper, T., J. Gen. Microbiol., 11,313, 1954.
5. Adler, H. I., Radiat. Res. Suppl. III, 110,1963.
6. Chipault, J. R., "Lipids and their Oxidation" (H. W. Schultz, E. A. Day and R. O. Sinnhuber, eds.), Avi Publishing Company, 1962, p. 151.
7. Urbain, W. M., "Radiation Chemistry of Major Food Components" (P. S. Elias and A. J. Cohen, eds.), Elsevier Scientific Publishing Company, Amsterdam, 1977, p. 63.
8. Scholes, G., Ward, J. F. and Weiss, J. J., J. Molec. Biol., 2,379, 1960.
9. Bacq, Z. M. and Alexander, P., "Fundamentals of Radiobiology," Butterworths Scientific Publications, London, 1955.
10. Haber, F. and Weiss, J., Proc. Roy. Soc. Lond. 147A,332, 1934.
11. Petkau, A., Chelack, W. S., Pleskach, S. D., Meeker, B. E. and Brady, C. M., Biochem. Biophys. Res. Commun., 65,886 (1975).
12. Nordenson, I., Beckman, G. and Beckman, L., Hereditas, 82, 125, 1976.
13. Tappel, A. L., Fed. Proc. 24,73, 1965.
14. Petkau, A., Chelack, W. S. and Pleskach, S. D., Int. J. Radiat. Biol., 29,297, 1976.
15. Schubert, J., Bull. World Hlth. Org., 41,873, 1969.
16. Van Duuren, B. L., Nelson, N., Orris, L., Palmes, E. D. and Schmitt, F. L., J. Natl. Cancer Inst., 31,41, 1963.
17. Hunt, J. W., Greenstock, C. L. and Bronskill, M. J., Int. J. Radiat. Phys. Chem., 4,87, 1972.
18. "An Introduction to Radiation Chemistry" (J. W. T. Spinks and R. J. Woods, eds.), J. Wiley and Sons, New York, 1976.
19. Currie, D. J. and Dainton, F. S., Trans. Farad. Soc., 61, 1156, 1965.
20. Adams, G. E., Boag, J. W., Currant, J. and Michael, D. B., "Pulse Radiolysis" (M. Ebert, J. P. Keene, A. J. Swallow and J. H. Baxendale, eds.), Academic Press, New York, 1965, p. 131.

21. Willson, R. L., Greenstock, C. L., Adams, G. E., Wageman, R., and Dorfman, L. M., Int. J. Radiat. Phys. Chem., 3,211, 1971.
22. Ingold, C. K., "Structure and Mechanism in Organic Chemistry," Cornell University Press, N.Y., 1953.
23. Greenstock, C. L. and Ruddock, G. W., Int. J. Radiat. Phys. Chem., 8,367, 1976.
24. Tezuka, M., Ohkatsu, Y. and Osa, T., Bull. Chem. Soc. Japan, 48,1471, 1975.
25. Peters, J. W. and Foote, C. S., J. Amer. Chem. Soc., 98, 873, 1976.
26. Greenstock, C. L., Trans. Farad. Soc., 55,2541, 1970.
27. Simic, M., "Fast Processes in Radiation Chemistry and Biology" (G. E. Adams, E. M. Fielden and B. D. Michael, eds.), J. Wiley and Sons, London, 1975, p. 162.
28. Greenstock, C. L., Hunt, J. W. and Ng, M., Trans. Farad. Soc., 65,3279, 1969.
29. Willson, R. L., Int. J. Radiat. Biol., 17,349, 1970.

DISCUSSION

SCHULTE-FROHLINDE: You've found that the ·OH radical reacts four times faster with acetyl peroxide than it did with ethyl hydroperoxide. This is a bit surprising.

GREENSTOCK: Yes. We also do not understand why ·OH as a nucleophilic agent reacts faster with a peracyl acid than with aliphatic hydroperoxide.

SCHULTE-FROHLINDE: Yes, nucleophilic agents should react just the other way around.

GREENSTOCK: We tried to make sense out of this, by comparing reactivity with inductive effects or Hamett sigma constants, but these structure-activity relationships are not well correlated.

SCHULTE-FROHLINDE: Could a possible explanation be that you have dissociation of acetyl peroxide as a function of pH, which alters its reactivity?

GREENSTOCK: That may be possible, but we have no experimental data.

BORS: I think it quite interesting that you find also rather a low reaction rate of O_2^- with ethylhydroperoxide. Did you try any other one? For example, we did an experiment with tertiary butyl hydroperoxide and with linoleic acid hydroperoxide and in each case we came up with an upper limit of less than ten to the third.

GREENSTOCK: Well, our rate constants are all upper limits also, and the true values may be very much lower. I tried to emphasize that. Our steady state competition system using 9,10 dihydroanthracene only gives relative rate constants.

BORS: But was this just an example or were you also investigating other peroxides? That was the question.

GREENSTOCK: We have not looked at any biological hydroperoxides, but I believe their rates with $O_2^{\bar{\cdot}}$ are also quite slow.

BORS: I see.

ASMUS: I'd like to ask you a question concerning the ·OH reaction with the peroxides. You showed a scheme where you have two products either $RO_2^{\cdot}$ radicals or RO· radicals. Now I'm asking for some similarities as to the analogue sulfur system RSSR. In some way this is of course different because there you form radical cations as well as similar radical species and cations are not formed in the case of the peroxides, but otherwise the reactions of ·OH with peroxides match very well with what happens in the case of the disulfides. There we find significant pH dependence determining whether it goes into the RS radical route or RSO radical route, so I just wonder are you aware of any pH dependence in your system? Have these compounds, the $RO_2^{\cdot}$ radical and the RO· radical been measured as a function of pH?

GREENSTOCK: We haven't done any pH studies. We've done some product analysis studies which tend to favor the peroxy pathway although we're not sticking our necks out at this stage. As you know, by analogy with H_2O_2 you might expect the peroxy radical formation; but as on the basis of energetics, OH attack is more likely at a C-H bond of the alkyl or phenyl substituent (R). There seems to be evidence for both pathways but we, at least from kinetics, can't resolve them.

BORS: It would be interesting to see the pH dependence.

GREENSTOCK: That might be one way of getting at the problem.

KALACHANDRA: In peroxide systems I would like to know the concentration ranges of the reacting species. And my second question is that you report that the rate constants are involved $\sim 10^7$ $M^{-1}s^{-1}$. Are you calling them diffusion controlled?

GREENSTOCK: No, no. The hydroxyl radical reactivity with

peroxides is much less than diffusion controlled. The concentrations we generally work in are around 10^{-3} - 10^{-4} molar.

KALACHANDRA: Oh. That is the lower limit.

GREENSTOCK: Within that range.

RADICAL-INDUCED OXIDATIONS IN BIOLOGY

Session Chairman, W. A. Pryor, Baton Rouge, Louisiana

AUTOXIDATION IN BIOLOGICAL SYSTEMS. A NOVEL LOW-TEMPERATURE METHOD FOR SPIN TRAPPING AND THE MECHANISM OF REACTION OF OZONE WITH POLYUNSATURATED FATTY ACIDS.

William A. Pryor, Department of Chemistry, Louisiana State University, Baton Rouge, Louisiana 70803

Oxygen reacts with biological molecules in living organisms both in enzyme-controlled processes and in a reaction called autoxidation in which a slow, flameless oxidation by O_2 occurs (1-9). Autoxidation is a chain process that involves the formation of free radicals and propagation reactions involving both carbon-centered and oxygen-centered radicals. Some of the enzyme-controlled processes also produce radicals, often in reactions that involve hydrogen peroxide and the superoxide ion-radical. For these reasons, all organisms that live in an atmosphere of air have developed a variety of metabolic activities to protect themselves against the damage that free radicals, and particularly oxygen-centered radicals, can cause to biological materials. It also is now clear that cells often utilize hydrogen peroxide and superoxide for vital cellular functions, despite the toxic nature of these species.

Radicals produced either in enzyme-controlled reactions or in non-enzymatic reactions can react with susceptible biomolecules; often when this occurs, the autoxidation of the material is initiated. A number of types of biomolecules are easily oxidized and these materials are particularly sensitive targets of radicals; these compounds include the polyunsaturated fatty acids (PUFA), ascorbate and tocopherol and other oxidizable cofactors, steroids and other unsaturated organic compounds, and thiol-containing compounds such as glutathione and thiol-containing enzymes.

One of the most intriguing questions in free radical biology asks: What is the source of the primordial radicals that initiate radical-mediated damage to living systems? It is becoming increasingly clear that the most important sources of these primary radicals probably are processes that are enzyme-controlled

ISBN 0-12-592050-4

and that often involve hydrogen peroxide and superoxide. However, some of the radical-producing processes that are important in vivo do not involve enzymes. (i) The first of these that might be considered is the spontaneous reaction of oxygen with organic materials, eq 1; although this may contribute to the very slow deterioration of rubber and plastics in air, it is too slow to be important in the cell (5, 10, 11).

$$RH + O_2 \rightarrow R^{\cdot} + HOO^{\cdot} \quad (1)$$

(ii) Similarly, the unimolecular, unassisted homolysis of either H_2O_2 or of the lipid hydroperoxides, LOOH, that are produced in the autoxidation of PUFA is far too slow to be important in vivo. (iii) However, the molecule-assisted homolysis (MAH) of either H_2O_2 or LOOH may be important; it is known that both H_2O_2 and lipid peroxides decompose many orders of magnitude faster in olefinic solvents than in saturated solvents (12, 13), and most of the LOOH molecules also contain double bonds and may undergo an intramolecular MAH reaction (5). (iv) In certain situations, light or ionizing radiation can produce damaging radicals. (v) Recently, it has been suggested that complexes of oxygen or of superoxide ion with chelated iron ions can react with organic materials such as PUFA to produce LOOH directly (14). (vi) Many xenobiotics cause the formation of radicals, some in non-enzymatic reactions (13). Most notable in this list is NO_2, which is itself a free radical, and ozone, which although not a radical causes the formation of radicals (15-18). We will return to a discussion of the mechanism of radical production by ozone below.

Despite this plethora of radical-producing reactions that do not involve enzymes, it is probable that most of the radicals in the cell result from enzyme-controlled reactions. Table I lists a few of the many systems that have been studied.

The preceding remarks are meant to provide an all too brief introduction to the topics that will be addressed by the speakers that follow me in this session of this symposium. In the short time that is left to me, I wish to turn my attention to a summary of our recent studies of the reaction of ozone with PUFA, a reaction that is known to cause the formation of oxy-radicals in vivo.

THE MECHANISM OF THE REACTION OF OZONE WITH PUFA

Ozone is one of the most troublesome pollutants of modern life; it is present in urban smog and has been cited by the W.H.O. as one of the three most damaging pollutants worldwide.

Ozone is extremely reactive; it reacts even with hydrocarbons at temperatures as low as -78°C, and its reaction with olefins and some other types of organic molecules can only be studied by rapid kinetic techniques.

The effects of ozone in living systems are known to be at least partly mediated through free radical reactions. Table II

TABLE I. Enzyme-controlled reactions that produce radicals

Source of the radical	Notes	References
Respiratory electron transport chain	H_2O_2 is produced	19
Some processes involving monoxygenases (e.g., cytochrome P-450)	H_2O_2 and/or $O_2^{\cdot-}$ may be produced	20-23
Metabolism of QH_2-type compounds such as hydroxydopamine and dialuric acid	H_2O_2 and $O_2^{\cdot-}$ are postulated to be responsible for the cytotoxicity of some of these compounds	24,25
Phagocytosis and the chemistry of neutrophils	H_2O_2 and $O_2^{\cdot-}$ are involved	26,27
Metabolism of certain precarcinogens such as N-hydroxyacetylaminofluorene	A stable nitroxide radical is postulated	28,29
Synthesis of prostaglandins	Feedback control is postulated to involve radical destruction of the synthetase enzyme system	30,31
CCl_4 in liver	Generates $\cdot CCl_3$ that initiates PUFA autoxidation	32,33
Alcohol oxidation	Apparently involves the $HO\cdot$ radical	34
Steroid oxidation	Products can be toxic	35

gives some of the evidence that supports this statement. Despite the general acceptance of this postulate, the mechanism by which ozone reacts with typical organic target molecules is not known.

The most direct technique to probe the mechanism of radical production is by electron spin resonance (ESR), a technique that detects radicals directly. In 1968, Goldstein et al. (36) reported that ESR signals are observed when ozone is allowed to bubble through neat linoleic acid. This publication is often cited as proof that ozone does, in fact, cause the formation of radicals in PUFA; however, we have had reservations about this pioneering and extremely important publication for some time for reasons we have summarized elsewhere (37).

Therefore, we repeated these experiments, and we also performed a number of experiments modified from the original design.

TABLE II. *Evidence that Ozone Causes the Formation of Radicals in Living Systems*

Observation	*Reference*
The induction period for the autoxidation of PUFA is greatly shortened by O_3	*17*
Ethane and/or pentane is evolved	*8*
The TBA test is positive	*17*
Vitamin E protects animals and model systems	*15,16*

Our method involved bubbling 40 ppm ozone in an oxygen stream at 20-40 ml/min through the substrate for 3 to 5 hours in an ESR tube held in the ESR cavity. We also bubbled ozone through the sample outside the cavity, quickly froze it in a dry ice bath, and then cooled it to liquid nitrogen temperatures; the sample was then allowed to slowly thaw in the cavity of the ESR instrument. For neither linoleic acid nor methyl linoleate was any ESR absorption observed. We conclude, therefore, that the previous results appear to be due to an artifact.

STUDIES OF RADICAL PRODUCTION AT LOW TEMPERATURES USING A SPIN TRAP TECHNIQUE

The method of spin trapping can very often be used to demonstrate the production of radicals when direct ESR methods fail (38), and we therefore explored the reaction of ozone with PUFA using spin traps. It was necessary to develop a methodology that would allow the use of spin traps, which of course are extremely reactive, in the presence of ozone, a very strong oxidant. Our technique involves the following steps. (i) Ozone at 40-300 ppm in oxygen is allowed to bubble through the organic substrate in Freon-11 (or similar suitable solvent) held in an ESR tube at -78°C. (ii) The sample is then thoroughly flushed with argon to remove all ozone. At this point no ESR signal can be observed. (Control experiments show that ozone reacts with spin traps to produce ESR signals; however, the argon flush removes all ozone.) (iii) A solution of the spin trap in Freon-11 at -78°C is then added so as to produce a final concentration of the spin trap of 0.1M. At this point, no ESR signal (or only a very weak signal) is observed. (iv) The sample is allowed to slowly warm in the cavity of the ESR instrument.

When methyl linoleate is studied in this way, the ESR signal of a spin trap is observed beginning at approximately -50 to -40°C. We have studied this reaction using three common spin traps (37,38). Our results are shown in Table III.

We also have performed some pilot experiments with model compounds using DMPO and the same methodology. Hexane gives no

TABLE III. Spin Adducts from Ozone and Methyl Linoleate

Spin Trap	*hfsc of the spin adduct* a_N	a_H	*Identification of radical(s) trapped*
PBN	*13.7*	*1.8*	*R·, RO· and/or ROO.*
DMPO	*13.9*	*20.4*	*R· and a weaker background signal that could be due to the less stable RO· adduct.*
NtB	*15.2*	*1.8*	*An alkyl radical with one hydrogen, suggestive of the allylic L· radical.*

spin adduct, indicating that saturated alkanes do not react with ozone at -78°C in 3-5 hours. The analogue of a PUFA, 1,4-pentadiene gives strong signals both from an alkyl radical spin adduct and the HOO· spin adduct. Since we reasoned that the HOO· spin adduct might result from a process such as eq 2, we studied the reaction of some compounds that are known to react with ozone to form hydrotrioxides.

$$RH + O_3 \rightarrow ROOOH \rightarrow RO\cdot + HOO\cdot \qquad (2)$$

Dioxolane gives a very strong HOO· spin adduct, eq 3(R=H) ·

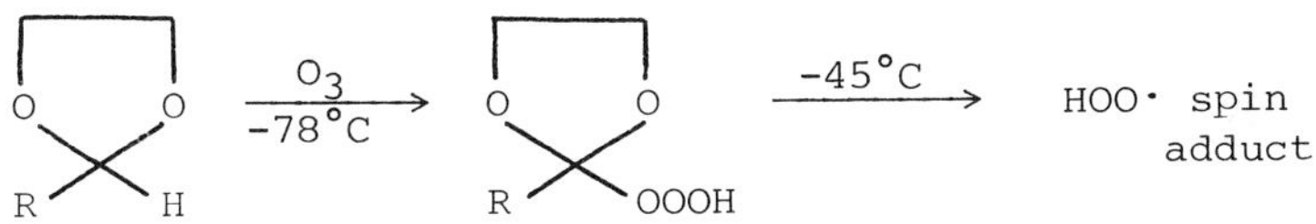

(3)

(R = H or Ph)

2-Phenyldioxolane reacts similarly, but 2,2-dimethyldioxolane gives no ESR signal, indicating that the hydrogens between the two oxygens are the reaction center. We also have shown that 2-phenyldioxolane forms the expected hydrotrioxide, shown in eq 3, by low temperature NMR analysis, and that the amount of hydrotrioxide is roughly consistent with the amount of spin adduct formed.

Thus, our experiments with the dioxolanes indicate that a hydrotrioxide is the precursor of the HOO· spin adduct. The precursors of the alkyl and oxy-radical spin adducts formed with linoleate and with pentadiene are less clear. However, it is known that dioxide species are far too stable to decompose at a temperature of -50°C where we see spin adducts formed, and tetroxides such as RO_4R or RO_4H decompose at -80°C and are far too unstable to give radicals at -45°C. In addition, the 1,2,3-trioxolane (the "molozonide") of most olefins also is far too unstable to give radicals at -50°C. However, RO_4H or RO_4R type species are known to decompose at temperatures in the range of -50 to

-40°C (37). Thus, our current hypothesis is that ozone reacts with PUFA ultimately to form either ROOOR or ROOOH type compounds, and that these materials decompose at about -50°C to produce the spin adducts that we observe (37).

These observations appear to prove conclusively that ozone reacts with PUFA to form radicals, despite the failure to observe ESR signals directly from ozone-PUFA reaction mixtures at room temperature. The detailed mechanism for the reactions of ozone with PUFA, the nature of the intermediates that are formed, the pathways of their decomposition, and the nature of the radicals that are formed are the subject of our current investigations.

REFERENCES

1. Pryor, W. A., "Free Radicals in Biology," Vol. I, Academic Press, New York, N.Y. (1976).
2. Pryor, W. A., "Free Radicals in Biology," Vol. II, Academic Press, New York, N.Y. (1976).
3. Pryor, W. A., "Free Radicals in Biology," Vol. III, Academic Press, New York, N.Y. (1977).
4. Pryor, W. A., "Free Radicals in Biology," Vol. IV, Academic Press, New York, N.Y. (1980).
5. Pryor, W. A., "The Role of Free Radical Reactions in Biological Systems," Chapter 1 in reference 1.
6. Pryor, W. A., "Free Radical Mechanisms of Lipid Damage and Consequences for Cellular Membranes," Chapter 2 in reference 1.
7. Fridovich, I., "Oxygen Radicals, Hydrogen Peroxide, and Oxygen Toxicity," Chapter 6 in reference 1.
8. Tappel, A. L., "Measurement of and Protection from <u>in Vivo</u> Lipid Peroxidation," Chapter 1 in reference 4.
9. Witting, L. A., "Vitamin E and Lipid Antioxidants in Free-Radical-Initiated Reactions," Chapter 9 in reference 4.
10. Pryor, W. A., Fed. Proc., 32,1862 (1973).
11. Pryor, W. A., Medicinal Chemistry, V, 331 (1973).
12. Pryor, W. A., "Free Radicals," McGraw-Hill Book Co., New York, N.Y. (1966).
13. Pryor, W. A., "Organic Free Radicals," American Chemical Society, Washington, D.C. (1978).
14. Svingen, B. A., O'Neal, F. O., and Aust, S. D., Photochem. Photobiol., 28,803 (1978).
15. Menzel, D. B., "The Role of Free Radicals in the Toxicity of Air Pollutants (Nitrogen Oxides and Ozone)", Chapter 5 in reference 2.
16. Tappel, A. L., "Pathological Aspects of Cell Membranes," Volume I, Academic Press, New York, N.Y. (1972).
17. Pryor, W. A., Stanley, J. P., Blair, E., and Cullen, G. B., Arch. Environ. Health, 21,201 (1976).
18. Pryor, W. A., "Molecular Basis of Environmental Toxicity," Ann Arbor Science, Ann Arbor, Michigan (1980).

19. Chance, B., Sies, H., and Boveris, A., Phys. Revs. 59,527 (1979).
20. Griffin, B. W., Peterson, J. A., and Estabrook, R. W., "The Porphyrins," (D. Dolphin, ed.) Academic Press, Chapter 7, Vol. 7B (1979).
21. Griffin, B. W., FEBS Letters, 74, 139 (1977).
22. Ullrich, V., Fortsch. Chem. Forschung, 83, 68 (1979).
23. Hamilton, G. H., "Molecules Mechanisms of Oxygen Activation," (O. Hayaishi, ed.) Academic Press (1974).
24. Borg, D. C., Schaich, K. M., Elmore, J. J., and Bell, J. A., Photochem. Photobiol., 28, 887 (1978).
25. Cohen, G., Photochem. Photobiol., 28,669 (1978).
26. Tauber, A. I., and Babior, B. M., Photochem. Photobiol., 28, 701 (1978).
27. Levine, P. H., Weinger, R. S., Simon, J., Schoon, K. L., and Krinsky, N. I., J. Clin. Invest., 57, 955 (1976).
28. Floyd, F. A., "Free Radicals in Arylamine Carcinogenesis," Chapter 6 in reference 4.
29. Ts'O, O. P., Caspary, J., and Lorentzen, R. J., "The Involvement of Free Radicals in Chemical Carcinogenesis," Chapter 7 in reference 3.
30. Porter, N. A., "Prostaglandin Endoperoxides," Chapter 8 in reference 4.
31. Egan, R. W., Marnett, L. J., et al., Prostaglandins, 16,861 (1978).
32. Reynolds, E. S. and Moslen, M. T., "Free Radical Damage in Liver," Chapter 2 in reference 4.
33. Recknagel, R. O., Glende, E. A., Jr., and Hruszkewycz, A. M., "Chemical Mechanisms in Carbon Tetrachloride Toxicity," Chapter 3 in reference 3.
34. Cohen, G., and Cedarbaum, A. E., Arch. Biochem. Biophys., 199, 438 (1980).
35. Imai, H., et al., Science, 207,651 (1980).
36. Goldstein, B. D., Balchum, O. J., Demopoulos, H. B., and Duke, P. S., Arch. Environ. Health, 17,46 (1968).
37. Pryor, W. A., Prier, D. G., and Church, D. F., Environ. Res. In press.
38. Janzen, E. G., "A Critical Review of Spin Trapping in Biological Systems," Chapter 4 in reference 4.

THE ROLE OF HEMEPROTEINS IN AUTOXIDATION PROCESSES

Brenda Walker Griffin, Biochemistry Department, The University of Texas Health Science Center, Dallas, Texas 75235

INTRODUCTION

The autoxidation of unsaturated fatty acids stimulated by hemeproteins requires low concentrations of the fatty acid hydroperoxides, which form spontaneously in the presence of O_2. These "peroxidase" activities are very low, non-specific, and result in destruction of the catalyst, consistent with a heme-initiated radical decomposition of the hydroperoxide (1). By contrast, the true hemeprotein peroxidases exhibit unusually large catalytic activities only with H_2O_2 and certain other peroxidic agents, which are not reduced to radical species in these reactions (2). However, certain oxidase, or oxygen-consuming, reactions catalyzed by peroxidases generate oxy radicals which can sustain a non-enzymatic radical chain oxidation (3,4). In this presentation, recent results from several laboratories will be summarized which define more clearly the fundamental chemical properties of these distinct peroxide-dependent activities of hemeproteins.

PROPERTIES OF HORSERADISH PEROXIDASE

The peroxidase isolated from horseradish root (HRP) has been extensively characterized with respect to isozyme composition and catalytic function (2). Three distinct states of the hemeprotein participate in the dehydrogenation of many electron donors by H_2O_2:

$$\text{HRP}\ (Fe^{3+}) + H_2O_2 \xrightarrow{k_1} \text{Compound I} \qquad (1a)$$

$$\text{Compound I} + AH \xrightarrow{k_2} \text{Compound II} + A\cdot \qquad (1b)$$

$$\text{Compound II} + AH \xrightarrow{k_3} \text{HRP}\ (Fe^{3+}) + A\cdot \qquad (1c)$$

$$A\cdot + A\cdot \longrightarrow A - A \qquad (1d)$$

The essential features of this cycle are: 1) two-electron oxidation of the ferric hemeprotein by H_2O_2 to Compound I; 2) successive one-electron transfers to oxidized HRP which regenerate the ferric enzyme and produce radical species of the electron donor substrate (AH); and 3) non-enzymatic reaction of these radicals, yielding dimers or other products. Electron donor substrates of HRP include phenols, aromatic amines, ascorbic acid, hydroquinone, and iodide. The latter compound effects a two-electron reduction of Compound I, without detectable formation of Compound

ISBN 0-12-592050-4

II (2). Several aromatic substrates of HRP bind weakly near the heme group, but not directly coordinated to the cation (5). The rate constant (k_1) for oxidation of ferric HRP by H_2O_2 has a value of 1.4×10^6 M^{-1} sec^{-1} for the A (acidic) isozymes and is about 10-fold larger for the C (basic) isozymes (2). The rate constant for reduction of Compound I depends on the electron donor, but is normally intermediate between k_1 and k_3 for the same donor. Thus, the rate of reduction of Compound II is generally rate-limiting in the oxidation cycle. The redox potentials of Compounds I and II are approximately 1 V in acidic solution (6); however, both species have been characterized by several spectroscopic techniques under conditions where each is relatively stable. The results have indicated that the oxidation state of the cation in both oxidized species of HRP is Fe(IV) and that the second oxidizing equivalent of Compound I is associated with a porphyrin *pi*-cation radical (2).

Other hemeprotein peroxidases, such as catalase and cytochrome *c* peroxidase, are also rapidly oxidized by stoichiometric H_2O_2, Eq. (1a). The physical properties of Compound I of catalase (2) are very similar to those of HRP Compound I, but the corresponding oxidized form of cytochrome *c* peroxidase is clearly different, indicating the presence of an amino acid radical species instead of a porphyrin *pi*-cation radical (7). Unlike HRP, catalase and cytochrome *c* peroxidase show high specificity for their respective electron donor substrates, H_2O_2 and ferro-cytochrome *c*, and the specific oxidation of both substrates proceeds without detectable formation of enzyme intermediates analogous to Compound II of HRP. Although this implies a concerted two-electron transfer from the reducing substrate to the oxidized hemeprotein, ferro-cytochrome *c* is a functional one-electron donor. The magnitude of the rate constant for association of cytochrome *c* peroxidase with its protein substrate, 5×10^8 M^{-1} sec^{-1}, may preclude detection of the enzyme species presumed to be an intermediate in catalysis (7).

In addition to dehydrogenation reactions, HRP catalyzes the H_2O_2-supported halogenation of certain substrates by iodide or bromide, but the mechanism of these reactions is unclear (8). Halogenation, however, appears to be the physiologically important catalytic function of mammalian peroxidases. The oxidase reactions of HRP, in which O_2 is consumed during the oxidation of certain electron donors, are especially relevant to the topic of this session. A detailed kinetic analysis of two such reactions has recently confirmed that they are variations on the general mechanism of HRP catalysis [cf. Eq. (1)], in which, however, the intermediate substrate radicals react readily with oxygen. For example, as shown in Fig. 1, the HRP-catalyzed oxidation of NADH produces the NAD· radical which rapidly reduces O_2 to $O_2^{\cdot -}$; the latter species can oxidize NADH at a slower rate, forming another NAD· (3). On the other hand, the substrate, indole-3-acetic acid, yields a radical species which adds to O_2, with simultaneous

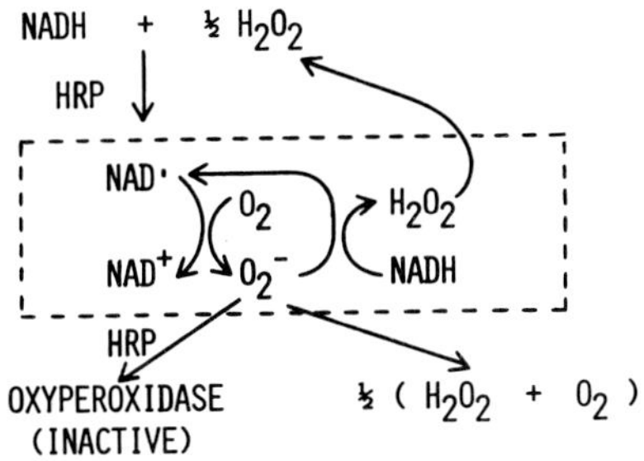

FIGURE 1. Principal reactions in the aerobic oxidation of NADH catalyzed by HRP, with the radical chain reaction enclosed by the broken line (from Ref. 3, by permission American Chemical Society).

decarboxylation, to form a peroxy radical; this species is converted by electron abstraction to a hydroperoxide which oxidizes peroxidase C to Compound I at a substantial rate (4). The much greater reactivity of this organic hydroperoxide with peroxidase C than with the A isozyme probably relates to specific hydrophobic interactions of the oxidant with the enzyme, consistent with other observed differences in reactivity of these isozymes.

The primary catalytic function of HRP in these oxidase reactions is the generation of substrate radicals having high reactivity with molecular oxygen. Either $O_2^{\cdot-}$ or a peroxy radical will be produced, depending on the chemical properties of the substrate radical; these oxy radicals will be subsequently reduced in a non-enzymatic radical chain reaction. Thus, HRP does not directly activate molecular oxygen in these reactions. As expected, superoxide dismutase affects the NADH, but not the indole-3-acetic acid, oxidase activity of HRP. Because several distinct free radical species may be formed during these reactions, many side reactions are possible, such as the formation of inactive oxyperoxidase shown in Fig. 1. Under specified experimental conditions the significance of these reactions will depend, in a complex manner, on many rate constants and on the relative concentrations of the various enzyme and substrate species.

HEMEPROTEIN-CATALYZED N-DEMETHYLATION REACTIONS

Our laboratory is investigating the mechanism of oxidation reactions catalyzed by liver microsomal cytochrome P-450 in the presence of either NADPH and O_2 (monooxygenase activity) or in the presence of hydroperoxides (peroxidase activity). It has been previously proposed that both activities of cytochrome P-450 involve the same Compound I-like enzyme species, which transfers an oxene (two-electron deficient oxygen atom) directly from the oxidant to the electron donor (9). Since such an activity has not been previously demonstrated for HRP Compound I, we have investigated the ability of peroxidase and other hemeproteins to catalyze oxidations considered to be typical cytochrome P-450 activities. We have shown that the N-demethylation of aminopyrine occurs quite generally by the following radical pathway in several hemeprotein-hydroperoxide systems, including HRP-H_2O_2 (10):

$$R_2\ N\text{-}CH_3 \xrightarrow{-e^-} R_2\ N\text{-}CH_3^{+}\cdot \xrightarrow{-H\cdot} R_2\ N^+ = CH_2 \xrightarrow{H_2O} R_2\ NH_2^+ + H_2C = O. \quad (2)$$

This reaction sequence is consistent with proposed mechanisms of chemical N-demethylation reactions and with other radical oxidations catalyzed by HRP, but does not support the oxygen-transfer mechanism proposed for cytochrome P-450 peroxidase activities. The unusually stable aminopyrine radical does not couple with another such species or add to oxygen, but loses an electron, either by enzymatic oxidation or by dismutation.

The oxidation of aminopyrine catalyzed by HRP has been clearly shown to involve catalytic cycling of the hemeprotein, as indicated by Eq. (1), consistent with a measured catalytic activity of 2.5×10^4/min (10). However, the rate of oxidation of aminopyrine in the metmyoglobin-cumene hydroperoxide system is three orders of magnitude smaller (11), quite comparable to rates reported for the cytochrome P-450-cumene hydroperoxide system. Since both of these systems oxidize aminopyrine to the free radical, the question of the identity of the one-electron oxidant arose. Although H_2O_2 can oxidize metmyoglobin to a species similar to HRP Compound II, which oxidizes electron donors very slowly, the reaction of metmyoglobin or cytochrome P-450 with cumene hydroperoxide could not be shown to produce similar hemeprotein species. Indeed, concentrations of this hydroperoxide required to cause any measurable reaction destroyed the heme group. The possibility that free radicals derived from cumene hydroperoxide are the oxidizing species in these systems was established by the use of nitroso compounds as spin trap probes (11). As shown in Fig. 2, at least two radical species, cumyloxy and methyl, have been implicated as the functional oxidants of substrates, as well as the electron-rich heme group. Thus, heme destruction by cumene hydroperoxide can be inhibited by electron donors and by spin traps.

In summary, the cumene hydroperoxide-supported activities of hemeproteins and heme-dependent autoxidation of unsaturated fatty acids appear to be examples of a rather general radical decomposition of organic hydroperoxides initiated by the heme group. By contrast, the true hemeprotein peroxidases effect a stoichiometric two-electron reduction of specific hydroperoxides and are quite unreactive with most organic hydroperoxides. In all of these systems, however, specific chemical properties of the electron donors and the presence of O_2 dictate the course of

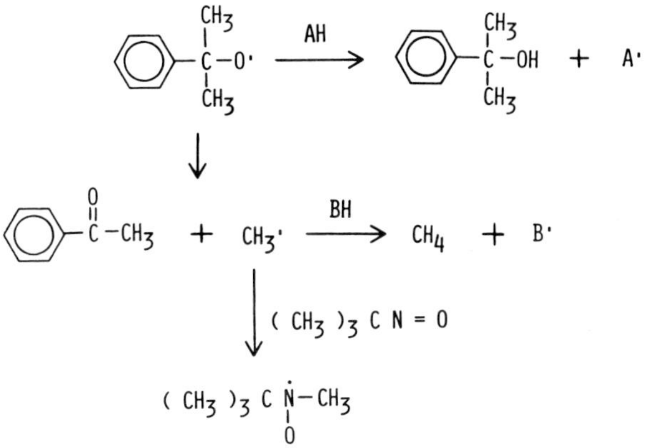

FIGURE 2. Reactions of radical species arising from cumene hydroperoxide with electron donors (AH and BH) or with a spin trap, 2-methyl-2-nitrosopropane.

oxidation, which may be very complex. The radical species of certain electron donors can react with, i.e., "activate," molecular oxygen without direct participation of the enzyme. This is a consequence of the greater probability of reaction of triplet O_2 with a radical species than with a singlet molecule. We speculate that the activation of enzyme-bound O_2 may occur similarly, involving enzyme-promoted development of radical character on the bound electron donor molecule which can, thus, participate in activation of the oxygen molecule. The role of radical species in enzymatic oxygenation reactions has been generally discounted because the high reactivity of free radicals is incompatible with the specificity of such reactions. However, the unique molecular architecture of an enzyme active site where both oxygen and the reducing substrate are specifically bound may provide a means of controlling the reactivity and ultimate fate of radical intermediates of the substrate molecules. We are planning experiments to test this hypothesis and hope to report those results at the next meeting of this conference.

ACKNOWLEDGEMENT

Supported by NIH Grant No. AM 19027 and Grant No. I-601 of The Robert A. Welch Foundation.

REFERENCES

1. Tappel, A. L., Arch. Biochem. Biophys., 44, 378 (1953).
2. Hewson, W. D., and Hager, L. P., "The Porphyrins, Vol. VII, Biochemistry, Part B" (D. Dolphin, ed.) Academic Press, New York, N.Y. (1979), p. 295.

3. Yokota, K. and Yamazaki, I., Biochemistry, 16,1913 (1977).
4. Nakajima, R., and Yamazaki, I., J. Biol. Chem., 254,872 (1979).
5. Morishima, I., Ogawa, S., Inubushi, T., and Yonezawa, T., FEBS Letters, 80,177 (1977).
6. Hayashi, Y., and Yamazaki, I., J. Biol. Chem., 254,9101 (1979).
7. Yonetani, T., Adv. in Enzymol., 33,309 (1970).
8. Morrison, M., and Schonbaum, G., Ann. Rev. Biochem., 45, 861 (1976).
9. Rahimtula, A. D., and O'Brien, P. J., Biochem. Biophys. Res. Commun., 60,440 (1974).
10. Griffin, B. W., and Ting, P. L., Biochemistry, 17,2206 (1978).
11. Griffin, B. W., "Fundamental Research in Homogeneous Catalysis, Vol. 3" (M. Tsutsui, ed.), Plenum Pub. Corp., New York, N.Y. (1979), p. 773.

INITIATION OF LIPID PEROXIDATION BY PERFERRYL COMPLEXES

Ming Tien, Bruce A. Svingen and Steven D. Aust, Department of Biochemistry, Michigan State University, East Lansing, Michigan 48824

Except for probably a limited amount of photochemically activated oxygen, the forms of oxygen involved in lipid peroxidation must arise from superoxide and/or hydrogen peroxide or redox reactions involving transition metals and oxygen. Four model lipid peroxidation systems have been studied in order to determine which mechanism of lipid peroxidation may prevail in biological systems. These model systems include liver microsomal NADPH dependent lipid peroxidation, xanthine oxidase dependent lipid peroxidation, hydroxyl radical (Fenton's reagent) dependent lipid peroxidation, and lipid hydroperoxide dependent lipid peroxidation. The first two systems have been studied by many investigators. The hydroxyl radical system was developed because the hydroxyl radical has been proposed to be involved in the first two systems. Lipid hydroperoxide dependent lipid peroxidation was studied because it also occurs in most lipid peroxidation systems being investigated. It also involves chelated forms of iron but not those forms involved in the initial formation of the lipid hydroperoxides.

Kellogg and Fridovich (1,2) studied xanthine oxidase promoted lipid peroxidation and found that both superoxide dismutase and catalase inhibited activity. Hydroxyl radical traps had very little effect, but singlet oxygen traps were very effective inhibitors. They therefore concluded that both hydroxyl radical and singlet oxygen, produced by the Haber-Weiss reaction, were responsible for initiation of lipid peroxidation. While several reports have also proposed that singlet oxygen, generated by the non-enzymatic dismutation of superoxide, is responsible for initiation of lipid peroxidation (3-5), several others have been unable to demonstrate the participation of singlet oxygen (6,7). More recently, Foote and coworkers (8) demonstrated that less than 8 x 10^{-4} percent of the molecular oxygen formed by non-enzymatic dismutation of superoxide is in the singlet state. In addition, the possible role of singlet oxygen produced via the Haber-Weiss reaction has been questioned since the rate of the Haber-Weiss reaction should be insufficient to account for the rates of lipid peroxidation observed.

Fong and coworkers (9) investigated microsomal NADPH dependent lipid peroxidation and xanthine oxidase promoted lipid peroxidation and concluded that both systems generated hydroxyl radical by an iron catalyzed Haber-Weiss reaction. These investigators proposed that iron chelates, reduced by superoxide, react

ISBN 0-12-592050-4

with hydrogen peroxide to yield the hydroxyl radical. This is not in accord with the results of Svingen and coworkers (7,10) who also investigated NADPH dependent lipid peroxidation and xanthine oxidase promoted lipid peroxidation. These investigators concluded that both systems initiated lipid peroxidation by an ADP-ferrous iron oxygen complex (ADP-Fe^{+3}-O_2 $\rightleftarrows$ ADP-Fe^{+2}-O_2). This complex, referred to as the perferryl ion complex by Ingraham (11), is proposed to exist and be involved in lipid peroxidation because anionic chelators such as ADP or pyrophosphate are required in these lipid peroxidation systems (12) and they are required for stabilization of the complex (13). In microsomal NADPH dependent lipid peroxidation, this complex is generated by reduction of ADP-Fe^{+3} to ADP-Fe^{+2} and subsequent complexation with oxygen. In xanthine oxidase promoted lipid peroxidation the complex is generated by superoxide complexing with ADP-Fe^{+3}.

Both NADPH and iron are required for maximal rates of NADPH dependent lipid peroxidation (Table I,A). Chelation of Fe^{+3} by ADP caused a 4 fold increase in activity. The direct addition of ADP-Fe^{+2} alone to microsomes caused a rate of lipid peroxidation comparable to that of NADPH and ADP-Fe^{+3}. These results are in agreement with the results obtained by Svingen and coworkers (10), indicating that NADPH dependent lipid peroxidation is initiated by NADPH cytochrome P450 reductase catalyzed reduction of ADP-Fe^{+3} to ADP-Fe^{+2}. The rate of lipid peroxidation is first order in ADP-Fe^{+2}. Log-log plots of rates of MDA formation versus ADP-Fe^{+2} concentration are quite linear over an 8 fold change in ADP-Fe^{+2} concentration with slope of 0.88. The lack of inhibition by 50 mM Tris suggest that hydroxyl radical was not one of the radicals involved. The large amounts of contaminating catalase in microsomes (14) would also minimize the production of hydroxyl radical from hydrogen peroxide, and washing microsomes to remove contaminating catalase has no effect on the rate of lipid peroxidation.

The role of the hydroxyl radical was also investigated in xanthine oxidase promoted peroxidation of isolated microsomal phospholipid liposomes (Table I,B). The generation of superoxide in the absence of iron did not result in lipid peroxidation. These results confirm the conclusion that neither superoxide nor hydrogen peroxide or a reaction product of the two initiate lipid peroxidation. Chelation of Fe^{+3} by ADP greatly stimulated activity. The system is not inhibited by catalase but is completely inhibited by SOD. These results tend to rule out a role for the hydroxyl radical and singlet oxygen produced by the Haber-Weiss reaction.

It is also important to point out that the iron catalyzed decomposition of lipid hydroperoxides produces radicals which can initiate lipid peroxidation (15,16). In the case of microsomes, ferric cytochrome P450 reacts with lipid or other organic hydroperoxides to promote the further peroxidation of microsomal lipid. In any system lacking substances like cytochrome P450 the

TABLE I. Model Lipid Peroxidation Systems

Systems	Additions	nmole MDA/min/ml
A. Microsomal	none	0.01
	NADPH	0.42
	ADP-Fe^{+3}	0.11
	NADPH, Fe^{+3}	0.81
	NADPH, ADP-Fe^{+3}	3.79
	ADP-Fe^{+2}	2.63
	NADPH, ADP-Fe^{+3}, Tris	3.62
B. Xanthine Oxidase	none	0.01
	Fe^{+3}, EDTA-Fe^{+3}	0.01
	ADP-Fe^{+3}, EDTA-Fe^{+3}	0.03
	X.O.	0.05
	X.O., Fe^{+3}, EDTA-Fe^{+3}	1.00
	X.O., ADP-FE^{+3}, EDTA-Fe^{+3}	3.70
	X.O., ADP-Fe^{+3}, DPTA-Fe^{+3}	3.24
	X.O., ADP-Fe^{+3}, EDTA-Fe^{+3}, catalase	3.80
	X.O., ADP-Fe^{+3}, EDTA-Fe^{+3}, SOD	0.06
C. Lipid Hydroperoxide	none	0.01
	ADP-Fe^{+3}	0.02
	EDTA-Fe^{+3}	0.01
	ADP-Fe^{+2}	0.60
	Fe^{+2}	4.98
	EDTA-Fe^{+2}	2.44
D. Fenton's Reagent	none	0.01
	Fe^{+2}	0.15
	H_2O_2	0.10
	Fe^{+3}, H_2O_2	0.10
	Fe^{+2}, H_2O_2	5.20
	Fe^{+2}, H_2O_2, catalase	0.25
	Fe^{+2}, H_2O_2, mannitol	0.12
	ADP-Fe^{+2}	0.20
	ADP-Fe^{+2}, H_2O_2	5.10
	ADP-Fe^{+2}, H_2O_2, mannitol	4.85

A. Microsomal reaction mixtures contained 0.5 mg microsomal protein/ml in 0.25 M NaCl, pH 7.5 at 37°C. The additions were 0.1 mM NADPH, ADP-Fe^{+3} (1.0 mM ADP + 0.1 mM $FeCl_3$), ADP-Fe^{+2} (1.0 mM ADP + 0.1 mM $FeCl_2$), and 50 mM Tris.

B. Xanthine oxidase dependent lipid peroxidation reaction mixtures contained liposomes (1.0 mM lipid phosphate) in 0.03 M NaCl and 0.33 mM xanthine, pH 7.5 at 37°C. The additions were 0.1 unit xanthine oxidase/ml, 0.1 mM $FeCl_3$, ADP-Fe^{+3} (0.2 mM ADP + 0.1 mM $FeCl_3$), EDTA-Fe^{+3} (0.11 mM EDTA + 0.1

mM $FeCl_3$), and DPTA-Fe^{+3} (0.11 mM DPTA + 0.1 mM $FeCl_3$).

C. *Lipid hydroperoxide dependent lipid peroxidation reaction mixtures contained liposomes (1.0 mM lipid phosphate), 0.2 mM phospholipid hydroperoxides in 0.03 M NaCl pH 7.5 at 37°C. The additions were ADP-Fe^{+3} (1.7 mM ADP + 0.1 mM $FeCl_3$), EDTA-Fe^{+3} (0.11 mM EDTA + 0.1 mM $FeCl_3$, ADP-Fe^{+2} (1.7 mM ADP + 0.1 mM $FeCl_2$), 0.1 mM $FeCl_2$ and EDTA-Fe^{+2} (0.11 mM EDTA + 0.1 mM $FeCl_2$).*

D. *Fenton's Reagent dependent lipid peroxidation reaction mixtures contained liposomes (1.0 mM lipid phosphate) in 0.04 M NaCl, pH 7.5 at 37°C. The additions were 0.2 mM $FeCl_2$, 0.2 mM $FeCl_3$, 0.1 mM H_2O_2, ADP-Fe^{+2} (1.0 mM ADP + 0.2 mM $FeCl_2$), 10 mM mannitol and 1.0 unit catalase/ml.*

rate of lipid peroxidation would be must less. Rates of lipid peroxidation in these systems would then be dramatically affected by certain forms of chelated iron. The decomposition of lipid hydroperoxides by unchelated ferric iron has a pH optimum at 2.2 (17) so at physiological pH, lipid hydroperoxides are very slowly decomposed by ferric iron and therefore will be associated with very little lipid peroxidation. At physiological pH, ferrous iron will readily decompose lipid hydroperoxides and consequently promote further lipid peroxidation. However, in most biological lipid peroxidation systems reducing equivalents are required to reduce the iron and chelators affect the ability of the iron to be reduced. Thus, the inclusion of EDTA or DPTA chelated iron greatly stimulates iron reduction and lipid peroxidation in systems that have a mechanism for forming the initial hydroperoxides (Table I,C).

In order to further investigate the role of the hydroxyl radical in lipid peroxidation, a lipid peroxidation system dependent upon the hydroxyl radical was developed. The system was greatly affected by buffers, presumably because many buffers are iron chelators. Thus lipid peroxidation dependent on the hydroxyl radical was generated in unbuffered sodium chloride utilizing a Fenton's reagent of ferrous chloride and hydrogen peroxide (Table I,D). This system was inhibited by catalase and mannitol. Addition of ADP decreased activity somewhat but more importantly the activity was no longer sensitive to hydroxyl radical traps such as mannitol. We would conclude that the hydroxyl radical readily promotes lipid peroxidation but in the presence of iron chelators, which are extremely prevalent in biological systems, the mechanism of lipid peroxidation probably does not involve the hydroxyl radical.

In conclusion, our results with both microsomal and xanthine oxidase promoted lipid peroxidation suggest that both systems initiate lipid peroxidation by the same mechanism. In their model, ADP chelated iron is either reduced by NADPH cytochrome P450 reductase, in the case of microsomes, or by superoxide in the case of xanthine oxidase dependent lipid peroxidation (Table I, B). The degree of reduction of the complex is actually unknown. One electron reduction would yield what we term the perferryl ion. It would be stabilized by the complexing anion (13). Two additional electrons would give the ferryl ion:

$$ADP\text{-}Fe^{+2}O_2 \xrightarrow[2e-]{2H^+} ADP\text{-}[FeO]^{+2} + H_2O$$

This can be considered equivalent to the reaction of ferrous iron with hydrogen peroxide (18):

$$Fe^{+2} + H_2O_2 \longrightarrow [FeO]^{+2} + H_2O$$

Perhaps this is the predominant species formed by the reaction of ADP chelated ferrous with hydrogen peroxide. This system readily promotes lipid peroxidation but it is not sensitive to inhibition by mannitol (Table I,D). The possibility exists that this is lipid peroxidation by a form of the hydroxyl radical which is not susceptible to inhibition by the classical hydroxyl radical traps. Neither microsomal nor xanthine oxidase dependent lipid peroxidation are inhibited by catalase suggesting that the ferryl ion cannot be formed by the reaction of ferrous with free hydrogen peroxide. This does not, however, rule out direct two electron reduction of the perferryl ion by either superoxide or NADPH-cytochrome P450 reductase. However, lipid peroxidation appeared first order with respect to ADP-Fe^{+2}, which argues against this possibility.

ACKNOWLEDGEMENTS

Supported in part by NSF Grant PCM 79-15328. Michigan Agricultural Experiment Station Article number 9449.

REFERENCES

1. Kellogg, E. W., III and Fridovich, I., J. Biol. Chem., 250, 8812 (1975).
2. Kellogg, E. W., III and Fridovich, I., J. Biol. Chem., 252, 6721 (1977).
3. Bus, J. S., Aust, S. D., and Gibson, J. E., Biochem. Biophys. Res. Commun., 58, 749 (1974).
4. Khan, A. U., Science, 168,476 (1970).
5. Mayeda, E. A. and Bard, A. J., J. Am. Chem. Soc., 96,4023 (1974).
6. King, M. M., Lai, E. K. and McCay, P. B., J. Biol. Chem., 250,6496 (1975).
7. Svingen, B. A., O'Neal, F. O. and Aust, S. D., Photochem. Photobiol. 28,803 (1978).
8. Foote, C. S., Shook, F. C. and Abakerli, R. A., J. Am. Chem. Soc., 102,2503 (1980).

9. Fong, K., McCay, P. B., Poyer, J. L., Keele, B. B. and Misra, H., J. Biol. Chem., 248,7792 (1973).
10. Svingen, B. A., Buege, J. A., O'Neal, F. O. and Aust, S. D., J. Biol. Chem. 254,5892 (1979).
11. Ingraham, L. L., "Comprehensive Biochemistry," (M. Florkin and E. H. Stotz, eds.)Vol. 14, Elsevier, Amsterdam (1964), p. 424.
12. Hochstein, P., Nordenbrand, K. and Ernster, L., Biochem. Biophys. Res. Commun. 14,323(1964).
13. Weiss, J., Experientia, 9,61 (1953).
14. Hildebrandt, A. G. and Roots, I., Arch. Biochem. Biophys., 171,385 (1975).
15. O'Brien, P. J., Can. J. Biochem., 47,485 (1969).
16. Maier, V. P. and Tappel, A. L., J. Am. Oil Chemists' Soc., 36,8 (1959).
17. Aust, S. D., Tien, M. and Svingen, B. A., J. Am. Oil Chemists' Soc., 57,186 (1980).
18. Walling, C., Acc. Chem. Res., 8,125 (1975).

EFFECTS OF OXYGEN RADICALS ON PLATELET FUNCTIONS

Norman I. Krinsky,[1] *Danielle G. Sladdin*[2] *and Peter H. Levine*[2]

[1]*Department of Biochemistry and Pharmacology, Tufts University School of Medicine, Boston, Massachusetts 02111*
[2]*Blood Research Laboratory, The Memorial Hospital and the Department of Medicine, the University of Massachusetts School of Medicine, Worcester, Massachusetts 01605*

Polymorphonuclear leukocytes (PMNL) can be activated by a variety of stimuli including bacteria, latex spherules, zymosan-activated plasma, aggregated IgG or a single component of the complement system, C5a, to initiate metabolic and structural changes. The metabolic response includes the generation and release of oxygen radicals or products derived therefrom such as $O_2^-\cdot$, H_2O_2, $OH\cdot$ (1) and other as yet unidentified oxidants (2). The structural changes include the process of phagocytosis, when particulate stimuli are used (3), or degranulation by exocytosis when the particles are too large to be ingested by the cells (4). In both cases, the structural changes are associated with the release of active oxygen species capable of destroying ingested bacteria (1) or of damaging associated cells (5,6).

We have been concerned with the effects of active oxygen species on platelet function (7,8) and have demonstrated that activated PMNL release a sufficient quantity of H_2O_2 in their environment to affect nearby platelets (5). The present study was initiated to determine the impact of the two other reactive oxygen species elaborated by activated PMNL ($O_2^-\cdot$ and $OH\cdot$) on platelet function.

MATERIALS AND METHODS

Purification of Platelets: Human platelet-rich plasma (PRP) and gel-filtered platelets (GFP) were prepared fresh as described earlier (5). In the case of GFP that were to be exposed to $O_2^-\cdot$ *in situ*, the buffer contained 1 mM sodium formate. When GFP were to be exposed to exogenous $O_2^-\cdot$ solutions, the pH was adjusted to pH 8.0 one-half min before the addition of the $O_2^-\cdot$, which changed the pH to 9.8. After 2.5 min, the preparation was readjusted to pH 7.2. For $OH\cdot$ exposure, GFP were suspended in buffer which had been equilibrated with N_2O prior to irradiation.

Measurement of $O_2^-\cdot$: The superoxide dismutase (SOD)-inhibitable reduction of cytochrome *c* was followed at 550 nm to determine $O_2^-\cdot$ production (9).

Irradiation of GFP: Samples of GFP, under air, O_2, or N_2O, were irradiated at 24°C in a Co-60 research irradiator, model NRI-300

ISBN 0-12-592050-4

(Nuclear Materials and Equipment Corp., Apollo, Pa) as described earlier (10,16)

Production of $O_2^-\cdot$ Solutions: Steady-state radiolysis of aqueous solutions was carried out as described by Simic and Taub (12) and a modification of the technique of Bielski (13) was employed for the production of solutions of $O_2^-\cdot$ (10).

Measurement of Platelet Aggregation: Aggregation studies were performed on GFP using the turbidometric method of Born (14) as modified by Mustard *et al.* (15). The aggregating agent (10-20 μM ADP) was added after a 1 min incubation at 37°C and the percent aggregation measured 5 min after adding the ADP.

Double Labeling of GFP: PRP was incubated with ^{3}H-serotonin and ^{14}C-adenine and subsequently gel-filtered to produce doubly-labeled GFP, as described earlier (16).

Materials: Milk xanthine oxidase (XO), catalase, superoxide dismutase (SOD), cytochrome *c*, ADP, and human serum albumin were obtained from Sigma Chemical Co., St. Louis, MO. Perchloric acid (70% ultrapure) and pentaerythritol were obtained from Alfa Division, Ventron Corp., Danvers, MA. All other chemicals were reagent grade obtained from Fisher Scientific Co., Pittsburgh, PA. All buffers were prepared with centrally distilled Milli-Q treated water (Millipore Corp., Bedford, MA). Reagents to be used in the irradiation procedures were purified further to remove metal ions, according to the methods of Bielski and Allen (17).

RESULTS

Although XO and xanthine were used originally as a source of $O_2^-\cdot$, it soon became apparent that the XO could interact with the GFP in the absence of any other added substrate. As can be seen in Table I, XO causes extensive platelet aggregation as well as the release of cytochrome *c*-reducing material, which was identified as $O_2^-\cdot$ by the inhibition of cytochrome *c* reduction upon the addition of SOD. It should be noted that the added SOD was without effect on the XO-induced aggregation. From the results presented in Fig. 1, it would appear that XO causes a lytic phenomenon to occur, for ^{3}H-serotonin (a granule marker) and ^{14}C-adenine (a cytoplasmic marker) are released in parallel when doubly-labeled GFP are exposed to increasing concentrations of XO.

In order to by-pass the enzymatic production of $O_2^-\cdot$, which in the case of XO may be affecting platelets *independently* of $O_2^-\cdot$ production, we have utilized a radiolytic process to generate $O_2^-\cdot$. When GFP were exposed to 10 krad of γ-irradiation under O_2 in the presence of 1 mM sodium formate, a significant inhibition of subsequent ADP-induced aggregation was observed (Fig. 2). The addition of SOD (200 μg/ml) significantly increased the action of γ-radiation on GFP, whereas catalase (1000u/ml) protected the GFP.

In an attempt to avoid the possibility of short-lived radicals interacting with the GFP during irradiation, we resorted to adding stable $O_2^-\cdot$ solutions, prepared by γ-irradiation of oxygen-

TABLE I. The Effect of SOD on Xanthine Oxidase-Induced Aggregation, on Cytochrome c Reduction, and on $O_2^-\cdot$ Production of Gel-Filtered Platelets

Additive to GFP	% Aggregation ± SEM	Cytochrome c reduced	$O_2^-\cdot$ produced
		nmoles/2.5 x 10^8 platelets/20 min	
None	2 ± 0.29	4.55 ± 0.92	0.47
Xanthine oxidase	67 ± 3.63	9.20 ± 0.99*	5.12*
Xanthine oxidase + SOD	69 ± 3.22	5.58 ± 0.99	1.50
SOD	2 + 0.42	4.08 + 1.02	0

$p<.001$/ all other additives not significantly different from buffer control. The $O_2^-\cdot$ production represents the difference, in nmoles cytochrome c reduced, between the gel-filtered platelets + SOD sample and the other samples. Xanthine oxidase was used at 50 μg/ml. SOD was used at 20 μg/ml.

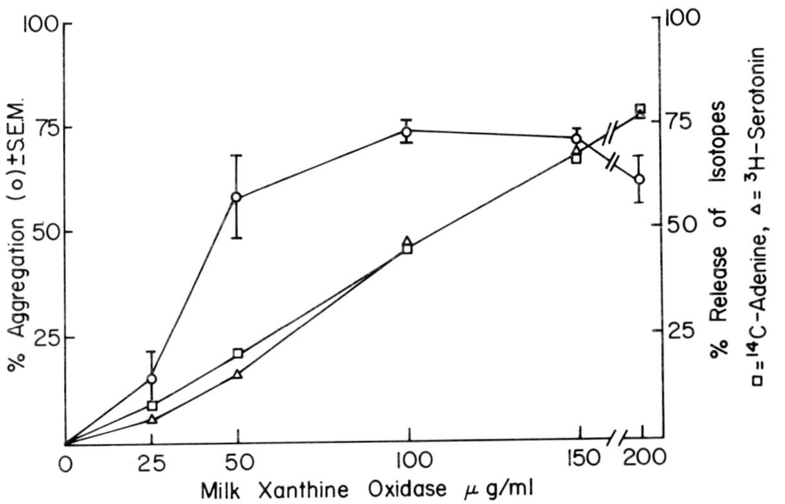

FIGURE 1. The effects of varying concentrations of XO on either spontaneous aggregation (o) in GFP or on the appearance in the suspending medium of ^3H-serotonin (Δ) and ^{14}C-adenine (□). The % aggregation and analyses of serotonin and adenine were determined at 5 min after the XO had been added (10), by permission F. K. Schattauer Verlag.

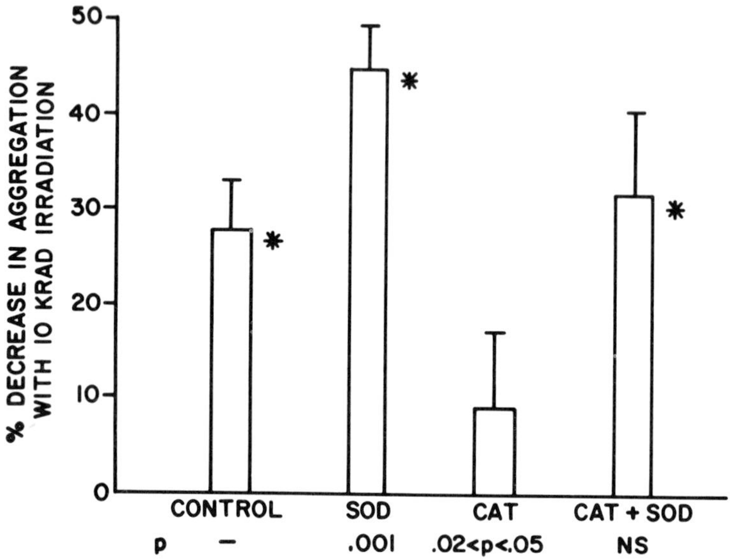

FIGURE 2. The effect of $O_2^{\bar{\cdot}}$ on ADP-induced aggregation of GFP. Platelets in calcium-free modified Tyrode's buffer pH 7.2, 1 mM sodium formate, were exposed to 10 krad γ-radiation. Additives were tested at the following levels: SOD, 200 μg/ml; catalase (CAT), 1000 u/ml. The samples which show a significant decrease (p <.02) in platelet aggregation attributable to the effects of radiation are designated by an asterisk, (11), by permission F. K. Schattauer Verlag.

saturated solutions of 10 mM sodium formate in 0.1 mM EDTA, pH 11.9, to our GFP preparations. As seen in Table II, the addition of exogenous $O_2^{-}\cdot$ (27.1 μM) to GFP results in a significant inhibition of ADP-induced aggregation. SOD is without effect in this system, whereas catalase, in the presence or absence of SOD, completely abolishes the effects of added $O_2^{-}\cdot$.

Irradiation of GFP under N_2O also resulted in a significant decrease in ADP-induced aggregation (Fig. 3A). This inhibition of platelet function could be partially overcome by the addition of either pentaerythritol (80mM) or catalase, and completely reversed by a combination of both agents. As seen in Fig. 3B, a moderate dose-related lysis was also observed when GFP were irradiated either under air or under N_2O, and again, pentaerythritol was able to significantly block this radiation-induced lysis.

DISCUSSION

Human platelet function can be altered by exposure to either exogenous H_2O_2 (7,8,18,19) or by exposure to phagocytosing PMNL (5). Although catalase eliminated most of the inhibitory effects of PMNL, some activity remained which we suggested might be due to OH·, $O_2^{-}\cdot$, or another oxygen-related species (5). Since then,

TABLE II. The Effect of Exogenous $O_2^{-}\cdot$ Solutions on the Aggregation of GFP

Additive	*% decrease in aggregation*	*p values*
None	*12 ± 5.2*	*.02<p<.05*
SOD	*10 ± 3.1*	*<.01*
CAT	*-4 ± 5.4*	*N.S.*
SOD + CAT	*-4 + 7.3*	*N.S.*

Gel-filtered platelets, adjusted to pH 8.0 in Ca^{2+}-free modified Tyrode's buffer, were exposed to a 27.1 µM solution of $O_2^{-}\cdot$ at pH 9.8 for 2.5 min. The results are expressed as the difference in % aggregation between control samples similarly treated with un-irradiated sodium formate solution and the experimental samples (n=10). The p values reflect the effects of the experimental treatment.

Handin et al. (20) reported that treatment of platelets with XO and xanthine resulted in platelet aggregation, attributable to a direct effect of $O_2^{-}\cdot$ on platelets. However, based on our observations that XO alone causes platelet aggregation which is not prevented by the addition of SOD (Table I) and that XO appears to lyse platelets (Fig. 1), it would seem that the effect of XO is not related to its ability to generate $O_2^{-}\cdot$ (11). This same conclusion has been reached by Patscheke et al (21) who concluded that the XO-induced aggregation and release reaction of platelets occurred independently of XO enzymic potency.

Since the irradiation of oxygenated water in the presence of dilute sodium formate generates $O_2^{-}\cdot$ almost exclusively (22) with a G-value of 6.2, we irradiated GFP in 1 mM sodium formate and observed an inhibition of ADP-induced aggregation (Fig. 2). This inhibitory effect is not associated with platelet lysis or release reaction and can be prevented by the addition of catalase, but not SOD (Fig. 2). We have developed a technique (10) for infusing relatively stable solutions of $O_2^{-}\cdot$ produced according to the procedures of Bielski and Allen (17) into suspensions of GFP under conditions in which the $t_{1/2}$ for $O_2^{-}\cdot$ was 46.1 sec (10). As seen in Table II, this treatment also results in a significant decrease in ADP-induced aggregation. Again, the addition of catalase completely abolished this effect whereas SOD was without effect. We therefore conclude that $O_2^{-}\cdot$, either generated radiolytically in situ or added as a relatively stable solution is without direct effect on GFP, but can dismutate to H_2O_2 which inhibits platelet function.

Finally, we have investigated the role of radiolytically-generated OH· on platelet function (16). When GFP are irradiated under air or N_2O, their subsequent response to ADP is inhibited (Fig. 3A). This inhibition can be partially reversed by the

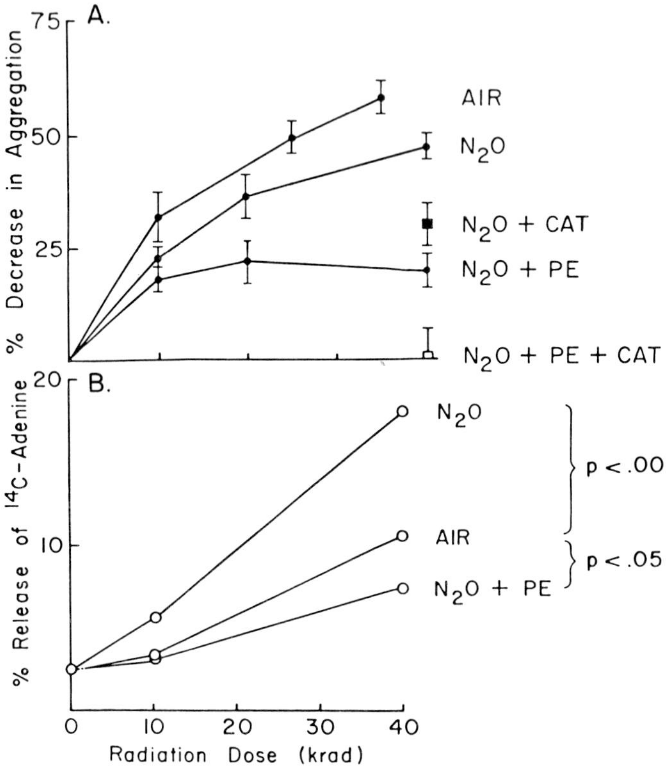

FIGURE 3. A. The effect of irradiation on ADP-induced aggregation of GFP, plotted as the % decrease in aggregation ± SEM. GFP were irradiated either in air, or with N_2O saturation. The GFP irradiated in N_2O were also supplemented with either 80 mM pentaerythritol (PE), 1000 u/ml catalase (CAT), or with both PE and CAT at these same concentrations. B. Percent release of ^{14}C-adenine from either control or irradiated GFP. The results of statistical significance are shown as p values for each set of comparisons indicated by a bracket. Each point represents the mean of 10 experiments. Pentaerythritol (PE) was used at a final concentration of 80 mM.

addition of pentaerythritol, an OH· quencher (23), as well as by catalase. The combination of both pentaerythritol and catalase completely eliminates the inhibiting effects of γ-radiation in a N_2O atmosphere, suggesting that both OH· and H_2O_2 are active under these circumstances. In addition, γ-radiation under N_2O-saturation also resulted in platelet lysis, which effect could be significantly inhibited by pentaerythritol (Fig. 3B).

Wörner et al. (24) have observed a release reaction when platelets are exposed to K_3CrO_8, which can be blocked by the addition of mannitol, DMSO, EDTA or histidine, and is attributed to the formation of OH· during the decomposition of K_3CrO_8. These observations, together with ours (16) would suggest that another PMNL radical species, OH·, may also play a role in modulating platelet function <u>in vivo</u>.

SUMMARY

O_2^{-}· does not appear to affect GFP directly, but exerts an action through its dismutation product, H_2O_2, which can inhibit platelet function. In addition, OH· also inhibits ADP-induced platelet aggregation.

ACKNOWLEDGEMENTS

We are indebted to Drs. B. H. J. Bielski, M. G. Simic and I. A. Taub for their help and use of their laboratories. Supported in part by a grant from the Hemostasis Research Fund, The Memorial Hospital, Worcester, MA.

REFERENCES

1. Babior, B. M., New Engl. J. Med., 298,659 (1978).
2. Kuehl, F.A., Jr., Humes, J. L., Ham, E. A., Egan, R. W. and Dougherty, H. W., Adv. Prostag. Thrombox. Res., 6,77 (1980).
3. Klebanoff, S. J. and Clark, R. A., "The Neutrophil: Function and Clinical Disorders," North-Holland, Amsterdam, 1978.
4. Henson, P. M., Am. J. Pathol., 68,593 (1972).
5. Levine, P. H., Weinger, R. S., Simon, J., Scoon, K. L. and Krinsky, N. I., J. Clin. Invest., 57,955 (1976).
6. Sacks, T., Moldow, C. F., Craddock, P. R., Bowers, T. K. and Jacob, H. S., J. Clin. Invest., 61,1161 (1978).
7. Canoso, R. T., Rodvien, R., Scoon, K. and Levine, P. H., Blood, 43,645 (1974).
8. Rodvien, R., Lindon, J. N. and Levine, P. H., Brit. J. Haemat., 33,19 (1976).
9. McCord, J. M. and Fridovich, I., J. Biol. Chem. 244,6049 (1969).
10. Krinsky, N. I., Sladdin, D. G., Levine, P. H., Taub, I. A. and Simic, M. G., Thromb. Haemos. In press. (1981).
11. Levine, P. H., Sladdin, D. G. and Krinsky, N. I., Thromb. Haemos. In press. (1981).
12. Simic, M. G. and Taub, I. A., Faraday Disc. Chem. Soc., 63, 270 (1977).
13. Bielski, B. H. J., Photochem. Photobiol., 28,645 (1978).
14. Born, G. V. R., Nature (London), 194,927 (1962).
15. Mustard, J. F., Hegardt, B., Roswell, H. C. and MacMillan, R. L., J. Lab. Clin. Med., 64,548 (1964).

16. Levine, P. H., Krinsky, N. I. and Sladdin, D. G., manuscript submitted (1981).
17. Bielski, B. H. J. and Allen, A. O., J. Phys. Chem., 81,1048 (1977).
18. Higashi, O., Kikuchi, Y. and Konno, K., Tohuku J. Exp. Med., 106,399 (1972).
19. Clark, R. A. and Klebanoff, S. J., J. Clin. Invest., 63,177 (1979).
20. Handin, R. I., Carabin, R. and Boxer, G. J., J. Clin. Invest. 59,959 (1977).
21. Patscheke, H., Paschen, W. and Wörner, P., Hoppe-Seyler's Z. Physiol. Chem., 359,933 (1978).
22. Hart, E. J., J. Am. Chem. Soc., 76,4198 (1954).
23. Simic, M. G., Taub, I. A., Tocci, J. and Hurwitz, P. A., Biochem. Biophys. Res. Commun., 62,161 (1975).
24. Wörner, P., Patscheke, H. and Paschen, W., Hoppe-Seyler's Z. Physiol. Chem., 360,559 (1979).

THE ROLE OF ASCORBIC ACID ON RADICAL REACTIONS IN VIVO

Harold M. Swartz[1] *and N. J. F. Dodd*[2]

[1]*University of Illinois College of Medicine, Urbana, Illinois*
[2]*Paterson Laboratories, Manchester, England*

INTRODUCTION

Ascorbic Acid and its oxidation products play an important role in redox reactions in biological systems, but the roles of the principal components remain unsettled. This paper summarizes some of the data that lead to the conclusion that ascorbic acid is a metabolically active form while the first product of its oxidation, the ascorbyl radical, is relatively inert, even though it is a free radical. As a consequence, the ascorbyl radical is a principal radical species observed in complex biological systems.

Ascorbic acid (Vitamin C) exists in biological systems as negatively charged lactone (pK $\sim$ 4.2) (Fig. 1). It oxidizes in two steps to dehydro-ascorbic acid (pK $\sim$ 9) which is mostly unionized under physiological conditions. The intermediate in the redox reaction is the ascorbyl radical which is negatively charged (pK $\sim$ 0) and has the electronic structure shown in Fig. 2 (1). The resonance structure of the radical may contribute to its low reactivity.

EXPERIMENTAL DETERMINATION OF THE ASCORBYL RADICAL

Data on the ascorbyl radical in solutions have been obtained by pulse radiolysis (2,3) (Fig. 3) and by electron spin resonance (ESR) (1,4) (Fig. 4). The pulse radiolysis technique is useful especially for quantitative determination of reaction rates while ESR provides direct and specific information on the radical and its electronic structure. The ESR spectrum of the ascorbyl radical in solution is quite characteristic and therefore it can often be identified more readily than most of the radicals that occur in biological systems.

Recently we have also studied the ascorbyl radical in the solid state, using ESR techniques (5,6). With conventional methods, the ESR spectrum of the ascorbyl radical in solid systems is nonspecific. At high microwave frequencies (e.g., 35 GHz), the ESR spectrum of the ascorbyl radical is sufficiently specific that the spectrum, in combination with suitable experimental conditions, permits detection of the ascorbyl radical in dried complex biological materials (Fig. 5).

ISBN 0-12-592050-4

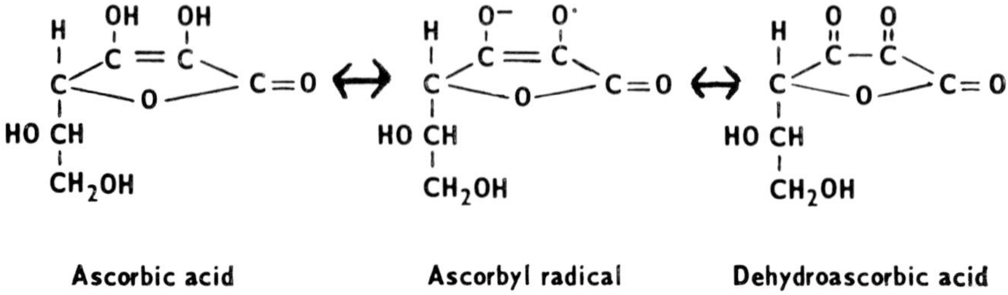

FIGURE 1. Structural formulae of ascorbic acid redox system.

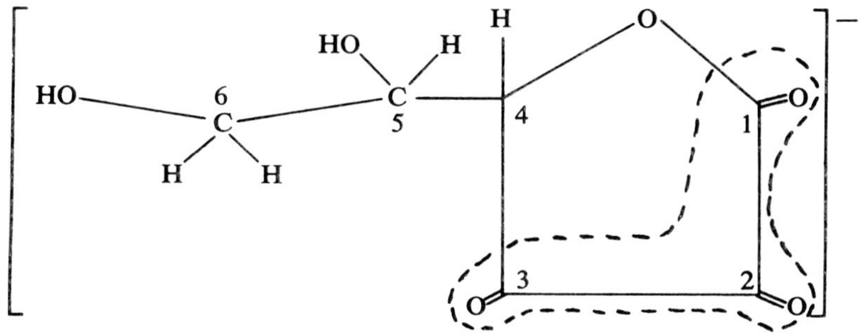

FIGURE 2. The ascorbyl radical.

OCCURRENCE OF THE ASCORBYL RADICAL IN COMPLEX BIOLOGICAL MATERIALS

The characteristic ESR spectrum of the ascorbyl radical in aqueous solutions led to its unambiguous identification in very complex biological materials such as tissues. It is the principal radical observed in tissues under some conditions. It occurs in relatively high concentrations in some tumors (7,8). High doses of ionizing radiation greatly increase its concentration in some tissues (9,10). Attempts to correlate the intensity of the ascorbyl radical in wet tissues with some biological properties (e.g., malignancy or radiation sensitivity) were not successful (9,11,12). The intensity also does not correlate with total concentration of ascorbic acid (12).

We recently reinvestigated the nature of the ESR signals seen in solid dried (lyophilized) tumors (13,14). Such tumors have a relatively intense ESR signal. Although it appears that the ESR signal in such preparations is an artifact (in the sense that it is induced and/or changed in intensity by the lyophilization process), its intensity seems to have some relation to the state of the tumor (15,16) and therefore we attempted to

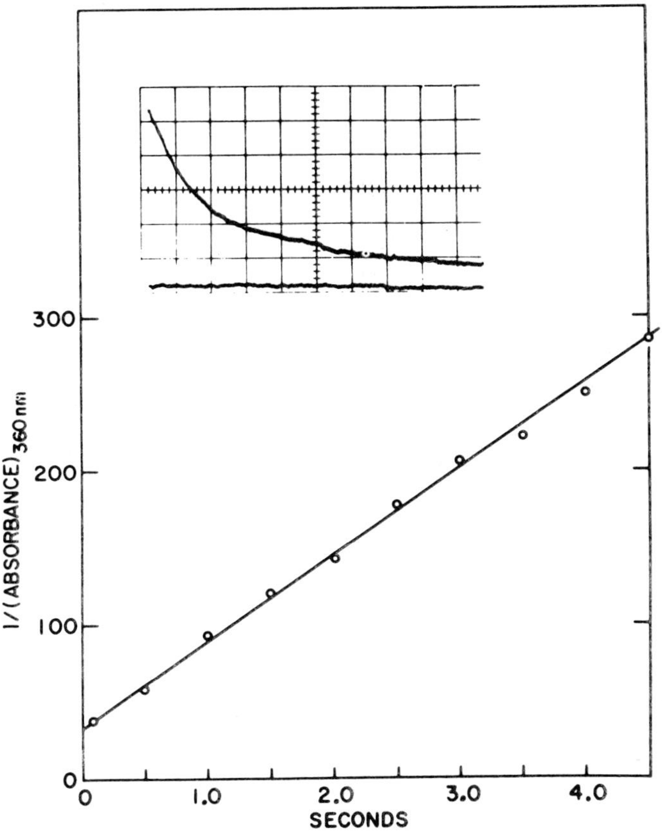

FIGURE 3. Second-order decay of ascorbate radicals. A nitrous oxide-saturated solution of ascorbate was irradiated and mixed with a nitrogen-saturated phosphate buffer at pH 8.7. The solution after mixing contained 2.65 μM ascorbate radicals, 47.35 μM ascorbic acid, 50 mM sodium phosphate, 11 mM nitrous oxide, and 0.32 mM nitrogen. The change in absorbance (inset) at 360 nm (0.005 absorbance units/cm) with time (0.5 sec/cm) was determined with an optical light path of 2 cm. The measurements were taken at 23°C (from ref. 3).

determine its molecular basis. We concluded that much of the ESR signal in lyophilized tumors is due to the ascorbyl radical and that its intensity is related to the breakdown of normal intracellular barriers (5). This interpretation is consistent with previous observations of the ascorbyl radical in wet tissue and was based, in part, on our parallel studies of nonlyophilized tissues.

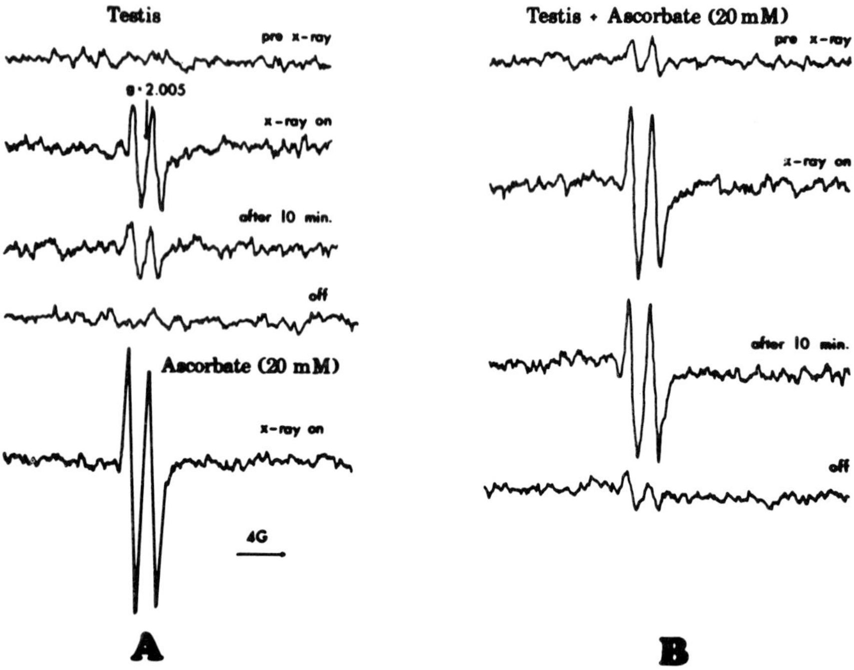

FIGURE 4. ESR spectra of testis tissue homogenate taken during x-irradiation. A: Testis homogenate prepared as described in text. X-irradiation intensity 3,000 Rad/sec. The four upper traces from the top down show the ESR spectra before x-irradiation, immediately after the start of x-irradiation, after ten minutes of x-irradiation, and then after x-rays are switched off. The lower trace shows the ESR spectra during x-irradiation of 10 mM ascorbate in 0.5 M Tris pH 7.5. Modulation amplitude 1 gauss, sweep speed 1 G/3 sec, temperature 15°, frequency 8.754 GHz. B: Same experiment as that seen in (A) except that 20 mM ascorbate was added to the tissue homogenate (from ref. 10, by permission N. Y. Academy of Sciences).

OCCURRENCE OF THE ASCORBYL RADICAL AND BIOCHEMICAL EQUILIBRIA

Recently Borg and coworkers (17) summarized a large body of data in which a number of different redox systems were studied alone, and in combination with other redox systems including ascorbic acid. In virtually all cases, the primary species observed was the ascorbyl radical. This indicates that ascorbic acid was readily reduced and that its product, the ascorbyl radical was the most stable radical component (Table 1). Other reducing agents such as sulfhydryl compounds did not produce stable radical products.

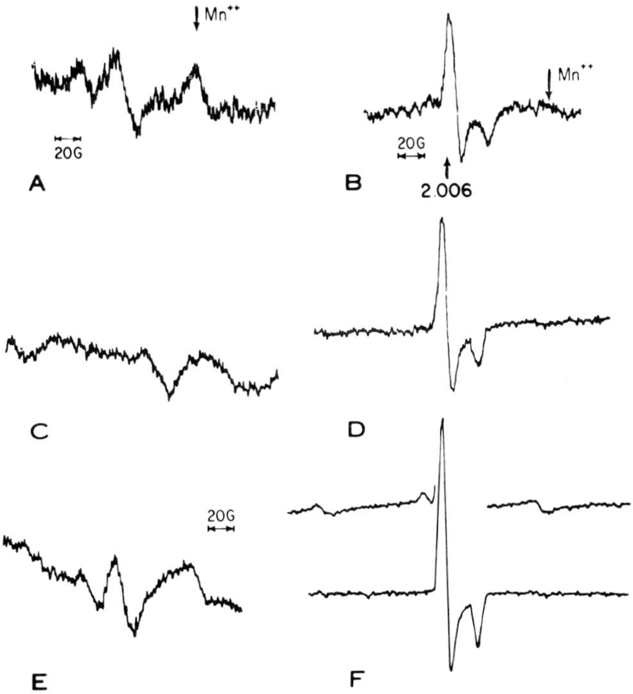

FIGURE 5. ESR spectra of lyophilized smaples at 35 GHz affect exposure to air for 45 min. A. 7, 12-Dimethylbenzanthracene induced rat breast tumor (before exposure to air). B. The same tumor after exposure to air. C. Bovine serum albumin. D. Bovine serum albumin plus 10^4 m ascorbic acid. E. Rat liver microsomes (1 ml contained microsomes from 0.5 g liver). F. Rat liver microsomes plus 10^4 m ascorbic acid. ESR conditions similar for all spectra except for amplification; temperature 150°C, incident microwave power 0.06 mW, modulation amplitude 8 G. In F a spectrum obtained at 60 mW is also shown to indicate the position of the second, third, and fourth peaks of Mn^{2+}. The portion of the fourth Mn^{2+} peak is also indicated in A and B (from Swartz, Submolecular Biology and Cancer (Ciba Foundation Symposium No. 67) 107, 1979, Elsevier/Excerpta Medica, Amsterdam).

DIRECT MEASUREMENT OF THE REACTIVITY OF THE ASCORBYL RADICAL

Bielski et al. (3) studied the reactivity of the ascorbyl radical using pulse radiolysis. This experimental procedure enabled them to generate the ascorbyl radical without simultaneously producing other active species and they therefore obtained unambiguous data, some of which are summarized in Table 2. These data indicate that the ascorbyl radical reacts very sluggishly

TABLE 1. Principal Radical Observed for $QH^{\cdot} + DH_2 \rightleftharpoons QH_2 + DH^{\cdot}$

QH_2	$DH^{\cdot}$ Ascorbate	$DH^{\cdot}$ Glutathione
6-OHDAcid	+	−
6-OHDA	+	±
5-OHDAcid	+	
6-ADAcid	+	−
6-ADA	+	±
6-ADSuccinamide	+	−
Dialurate	+ and −	±
Di-OH-maleate	+	−
Triose Reductone	+	−
Reductate	+	

Abbreviations used: 6-OHDAcid = 2,4,5-trihydroxyphenlpropionic acid; 5-OHDAcid = 3,4,5-trihydroxyphenlpropionic acid; 6-ADAcid = 2-amino-4,5-dihydroxyphenylpropionic acid; 6-ADSuccinamide = N-(2-amino-4,5-dihydroxyphenylethyl)-succinamide. From Borg et al. (ref. 17), by permission of Pergamon Press.

Symbols used: + = radical of ascorbate or glutathione
− = radical of QH_2 observed
± = small amount of $DH^{\cdot}$ observed

with most substrates, while the dismutation reaction is relatively rapid. Thus, ascorbyl radicals in many systems would dismutate, regenerating ascorbate and dehydroascorbate, rather than reacting with other types of molecules.

IMPLICATIONS OF THE DATA

On the basis of the data summarized in this paper, it seems reasonable to conclude that under many conditions, the ascorbyl radical is the principal long-lived radical that is produced in biological systems. Its stability and the data on its reactivity indicate that it is a relatively nontoxic product. This is consistent with ascorbic acid having a significant antioxidant role in biological systems; such a role requires that its oxidative product (the ascorbyl radical) be nontoxic.

SPECULATION ON POSSIBLE TOXIC EFFECTS FROM OXIDATION OF ASCORBIC ACID TO THE ASCORBYL RADICAL

Although it seems likely that generation of the ascorbyl radical by oxidation of ascorbic acid generally does not lead to biological damage, there may be circumstances when damage does occur.

*TABLE 2. Interaction of Ascorbate Radicals ($A^{\cdot}$) with Various Biochemical Compounds**

Reaction	($A^{\cdot}$)obs (μM·)	(Q) or (QH_2) (mM)	pH	Rate Constant (M^{-1} sec^{-1})
$A^{\cdot}$ + $A^{\cdot}$	2.7		8.7	$(2.8\pm0.2)x10^5$
$A^{\cdot}$ + dopamine	5.8	10-50	8.4	$(3.6\pm0.4)x10^2$
$A^{\cdot}$ + cytochrome c	6.4	0.20	7.4	$6.6x10^3$
$A^{\cdot}$ + O_2	4.7	0.63	8.6	$<5x10^2$
$A^{\cdot}$ + methanol	3.7	7.40	8.8	<0.1
$A^{\cdot}$ + lactate	6.6	50.00	8.6	<10
$A^{\cdot}$ + pyruvate	6.2	50.00	8.6	<10
$A^{\cdot}$ + fumarate	6.4	50.00	8.7	<10
$A^{\cdot}$ + L-α-ketoglutarate	8.6	20.00	9.7	<10

(Q) and (QH_2) represent the final concentrations of the given scavenger after mixing. The ascorbate radical was generated by radiation in a nitrous oxide saturated solution at pH 9.2 and 23°C. The pH of the reaction mixture is given in column 4. From Bielski et al., ref. 3.

One possibility is that the substance reduced by ascorbic acid becomes potentially damaging. An obvious example is the reduction of ferric ions in the presence of hydrogen peroxide, leading to production of hydroxyl radicals via a Fenton-type reaction. Although damage is likely to result in such a situation, the ascorbyl radicals would have no direct role in the damage.

Under some circumstances the ascorbyl radical itself could react in a damaging way. Although its reaction rate is low, it is nonzero for at least some substrates. This low reactivity could enable ascorbyl radicals to migrate relatively long distances and eventually react in fairly specific ways (e.g., with cytochrome C). These effects would be limited by the rate of the dismutation reaction. The effects presumably would occur only in targets in the hydrophilic regions of the cells, because of the negative charge of the radical.

Finally, it is possible that the ascorbyl radical could serve as a relatively inert form for transporting ascorbic acid to sites that are normally inaccessible to it because of the presence of intervening substances that would be reduced by ascorbic acid. When two ascorbyl radicals dismutate, ascorbic acid and dehydroascorbic acid are produced at that site and these could then react with nearby molecules.

REFERENCES

1. Laroff, G. P., Fessenden, R. W. and Schuler, R. H., J. Amer. Chem. Soc., 90,9062 (1972).

2. Bielski, B. H. J. and Allen, A. O., J. Amer. Chem. Soc. 92, 3793 (1970).
3. Bielski, B. H. J., Richter, H. W. and Chan, P. C., Ann. N.Y. Acad. Sci. 258,231 (1975).
4. Yamazaki, T., Mason, H. S. and Piette, L. H., J. Biol. Chem. 235, 2444 (1960).
5. Dodd, N. J. F. and Swartz, H. M., submitted.
6. Swartz, H. M., "Submolecular Biology and Cancer," Ciba Foundation Series 67,107 (1979).
7. Dodd, N. J. F., Br. J. Cancer 28,257 (1973).
8. Duke, P. S., Exp. and Mol. Pathol. 8,112 (1968).
9. Vaughan, W. M., Henry, J. T. and Commoner, B., Biochem. Biophys. Acta Report 329,159 (1970).
10. Floyd, R. A., Brondson, A. and Commoner, B., Ann. N.Y. Acad. Sci., 222,1077 (1973).
11. Dodd, N. J. F. and Giron-Conland, J. M., Br. J. Cancer 32, 451 (1975).
12. Silcock, J. M. and Dodd, N. J. F., Br. J. Cancer 34,550 (1976).
13. Gutierez, P. L. and Swartz, H. M., Br. J. Cancer 39,24 (1979).
14. Gutierez, P. L., Swartz, H. M. and Wilkinson, E. G., Br. J. Cancer 39,330 (1979).
15. Swartz, H. M. and Gutierez, P. L., Science 198,936 (1977).
16. Emanuel, N. M., Q. Rev. Biophys. 9,283 (1976).
17. Borg, D. C., Schaich, K. M., Elmore, J. J., Jr. and Bell, J. A., Photochem. and Photobiol. 28,887 (1978).

CARBON TETRACHLORIDE: FREE-RADICAL HEPATOTOXIN

Edward S. Reynolds, Mary Treinen Moslen and Richard J. Treinen, Department of Pathology, The University of Texas Medical Branch, Galveston, Texas 77550

Free-radical mechanisms of liver injury have been proposed for a number of synthetic compounds foreign to biological systems. The proposed mechanisms generally involve biotransformation of the compound to a reactive free-radical intermediate capable of injuriously interacting with cell macromolecules.

Proof that such mechanisms operate requires substantial evidence both that the toxin is converted *in vivo* to a free-radical intermediate and that interactions between this free-radical intermediate and cell components initiate a process destructive to the liver.

Experimental evidence indicative of the involvement of free-radical intermediates in the pathogenesis of a chemical toxin can be either direct or indirect (1). *Direct* evidence would be detection of radical intermediates of the toxin in tissues. However, direct evidence is not easy to obtain because radicals, as reactive species with short half-lives, are present in low concentrations. Fortunately, abundant *indirect* evidence can be obtained in three major forms:

(a) Identification of reaction products indicative of specific and predictable interactions between the proposed radical intermediate and cell components.

(b) Modulation of the toxin's injurious response by altering its "site" of free-radical activation or components which "contain" an injurious reaction of the radical intermediate.

(c) Demonstration of a structure-activity correspondence within closely related chemicals between propensity for activation to a free-radical and relative hepatotoxic effect.

In this presentation we will briefly examine the evidence indicative of a free-radical mechanism for the hepatotoxin CCl_4, and discuss our recent metabolic studies which suggest a role for the trichloromethylperoxy radical in the hepatotoxicity of CCl_4.

HISTORY AND PROGRESS OF FREE-RADICAL INTERMEDIATE MECHANISM FOR CCl_4

In 1961, Butler (2) proposed that the hepatotoxicity of CCl_4 was related to its biotransformation to a free-radical; this proposal was based on his demonstration that CCl_4 was reduced *in vivo* and *in vitro* to $CHCl_3$. He considered the hydrogen to be

ISBN 0-12-592050-4

acquired after a C-Cl bond in CCl_4 had undergone an "unusual" homolytic fission reaction, possibly with needed activation energy furnished by an unknown enzymatic system. Butler considered that the other metabolites of CCl_4, namely CO_2 and urea which had been detected by McCollister et al. (3) using ^{14}C-labeled CCl_4, could have arisen from the biologically "conventional" heterolytic fission of C-Cl bonds.

Butler proposed that the free-radical intermediate of CCl_4 could cause serious biochemical consequences by reacting with tissue components, e.g., alkylation of a sulfhydryl group. He mentioned protective experiments involving treatments with sulfhydryl amino acids and drugs, but found the then present knowledge inadequate for a "confident interpretation." However, he considered a survey of the literature supportive for his proposed mechanism, since the survey indicated that CCl_4 was far more hepatotoxic than other chloromethanes. He reasoned that these other compounds, which had stronger C-Cl bonds that CCl_4, would be less readily activated to a toxic free-radical species.

In his classic presentation, Butler mentions each of the three major forms of indirect evidence for a free-radical intermediate: specifically, reaction products indicative of the proposed radical, modulation of hepatotoxic severity by conditions which affect potentially injurious reactions of the radical, and structure-activity correspondence. Subsequently, other investigators have applauded, amplified, slightly modified, or provided additional evidence to substantiate Butler's proposed mechanism.

Direct evidence, of at least presumptive nature, for the in vivo formation of $\cdot CCl_3$ from CCl_4 was recently reported by Poyer et al. (4). They used phenyltertbutylnitrone (PBN) as a spin trapping agent in vivo to isolate a stable, lipid-soluble radical-adduct from livers of rats fed CCl_4 and PBN two hours previously. This adduct has an esr spectrum nearly identical to that formed when CCl_4, PBN and NADPH are incubated with liver microsomes in vitro, or when PBN and CCl_4 are irradiated with u.v. light anaerobically, as well as to that simulated by computer for a PBN-CCl_3 radical adduct.

Other reaction products identified in vivo are hexachloroethane (5) which is the annihilation product of two trichloromethyl radicals, and trichloromethylated fatty acids which are putatively formed by the addition of the trichloromethyl radical to polyunsaturated fatty acids (6). Also several known consequences of radical-initiated lipid peroxidation have been detected in animals poisoned with CCl_4, including an increase in the lipid diene conjugate contents of liver (7), and increased exhalation of the stable volatile breakdown products, ethane (8) and pentane (9).

Modulation studies provide support for the radical mechanism (see 1,5). For example, the hepatotoxicity of CCl_4 is usually predictably altered by pretreatments which increase or decrease cytochrome P-450. This hemoprotein, which can accept or donate single electrons, is the most likely site of CCl_4 activation to $\cdot CCl_3$. Also dietary deficiencies of Vitamin E render animals vulnerable to CCl_4 injury. This lipid soluble antioxidant plays a major role in protecting liver components from injury by free-radical species generated by endogenous lipid peroxidation systems.

Structure-toxicity studies of the trichloromethanes indicate a rank order of $CBrCl_3 > CCl_4 > CHCl_3 > CFCl_3$ for both propensity to form a $\cdot CCl_3$ intermediate (1) and relative hepatotoxicity (10).

We believe there is substantial evidence for conversion of CCl_4 to a free-radical intermediate. However, the evidence that its free-radical intermediate initiates a process destructive to the liver is not as clear cut.

Recknagel and colleagues (11) have proposed that the lipid peroxidation process initiated by interaction of $\cdot CCl_3$ with membrane lipids is the critical process in CCl_4 injury. Reaction product studies provide support for this concept. For example, dose-response studies indicate a linear relationship between levels of lipid conjugated dienes *in vivo* and CCl_4 dose (12). However, time-course studies indicate lipid diene conjugates have a transient existence (12); therefore the magnitude of their formation is difficult to determine under *in vivo* conditions. Sagai and Tappel (9) demonstrated a log-linear relationship between CCl_4 dose and exhalation of the stable lipid peroxidation product pentane. Hafeman and Hoekstra (8) found a concurrence between exhalation of the stable peroxidation product ethane and the magnitude of liver injury in animals made more vulnerable to CCl_4 injury by Vitamin E deficiency. Yet because the amounts of lipid diene conjugates detected and of pentane and ethane recovered are very small, we wonder how relevant lipid peroxidation is to the process of cell injury!

Others including Castro and colleagues (13) have suggested that the alkylation or covalent binding of $\cdot CCl_3$ to cell macromolecules is the critical process in cell injury. Relationships between magnitude of covalent binding and injury have been reported under some conditions and at some times, but not under other conditions or at other times (see 1). Since the liver is a dynamic system, many of the covalently bound macromolecules formed in putatively injurious processes are subject to further alterations or are rapidly removed by ongoing repair processes. Therefore the magnitude of their formation is difficult to assess *in vivo*.

Frankly, during our years of studying this model hepatotoxin we have ascribed relevance to both the peroxidation and binding processes. Probably both contribute to the ensuing cell injury. Butler's original proposal that the reduction of CCl_4 to $\cdot CCl_3$ underlies its hepatotoxicity cannot be disputed. But is the oxidation of CCl_4 necessarily as conventional as he proposed, and thus innocuous?

Little is known about the interrelationship between the pathways responsible for the reduction of CCl_4 to $CHCl_3$ and for its oxidation to CO_2. For example, it is not known if the two pathways are competitive or are sequential, nor is it known if both pathways proceed from the $\cdot CCl_3$ radical. Packer et al. (14) have reported high avidity of $\cdot CCl_3$ for O_2.

In order to clarify the relevance of these pathways, our laboratory is now looking carefully at both the time-course and dose-response relationships of CCl_4 metabolism to CO_2 and to $CHCl_3$. We administer $^{14}CCl_4$ to rats by gavage, place the rats in individual metabolism chambers, and monitor $CHCl_3$ and $^{14}CO_2$ exhalation for up to 24 hr. Animals are sacrificed at 24 hr and the extent of liver injury assessed by measuring liver derived transaminases (GOT and GPT) in serum.

Fig. 1 illustrates $^{14}CO_2$ and CHCl3 exhalation following 0.3 mmoles $^{14}CCl_4$/kg. We have confirmed Butler's intuition that CO2 is the major metabolite and $CHCl_3$ is a relatively minor product. Note the relatively complete temporal relationship between $^{14}CO_2$

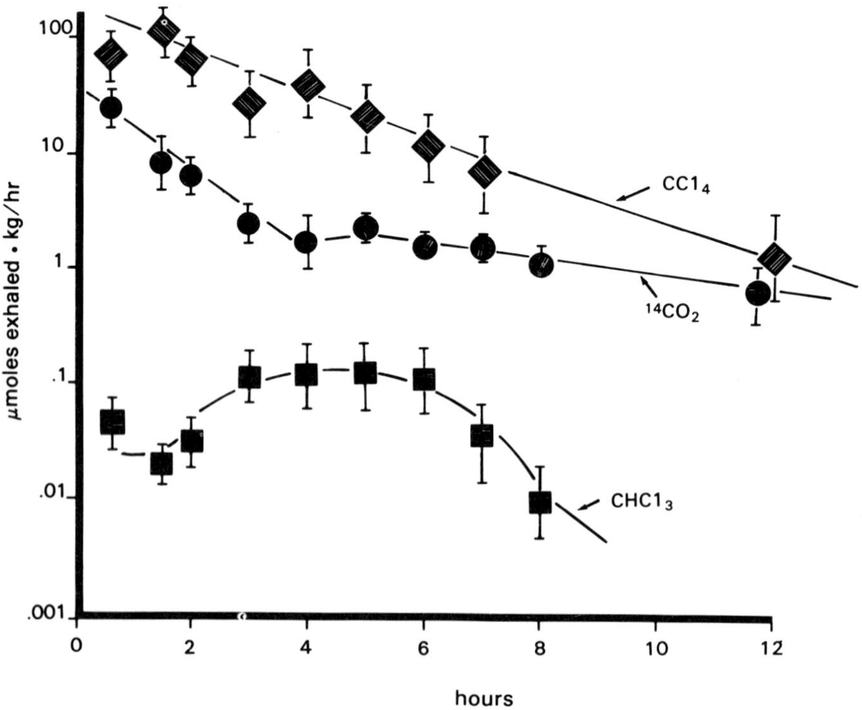

FIGURE 1. Time-course of CCl_4, $^{14}CO_2$ and $CHCl_4$ exhalation following 0.3 mmoles $^{14}CCl_4$/kg.

and $CHCl_3$ exhalation. Most $^{14}CO_2$ metabolite is recovered in the first few hours when only minor amounts of $CHCl_3$ metabolite are exhaled. Peak exhalation of $CHCl_3$ occurs between 4-6 hr.

We have similarly monitored the metabolism of one smaller, less hepatotoxic dose and 5 larger, more hepatotoxic doses of $^{14}CCl_4$. Increases in dose progressively decrease the total fraction of dose recovered as metabolite (from 30% with the smallest 0.1 mmole dose to 1% with the largest 26 mmole dose). However, the amounts of $^{14}CO_2$ and $CHCl_3$ metabolite exhaled progressively increase. The time-course of CO_2 metabolite exhalation was not appreciably altered as the dose increased. In contrast, there were dose-dependent alterations in $CHCl_3$ metabolite exhalation including a shortening of the interval before and lengthening the duration of peak exhalation. It is possible that the proportionally greater amounts of CCl_4 are reduced to $CHCl_3$ at later times because tissue oxygen content is diminished by injury.

Since $CHCl_3$ is itself oxidized to CO_2, we also examined the fate of $^{14}CO_2$ and $^{14}CHCl_3$ given by gavage. Following administration of 2 μmoles $Na_2\,^{14}CO_3$ per kg, almost all $^{14}CO_2$ was rapidly exhaled as $^{14}CO_2$ with an initial $t_{1/2}$ of 8 minutes. Three quarters of the dose administered was exhaled within 30 minutes and more than 90% by the end of the first hour.

Two dose levels of $^{14}CHCl_3$ were given, 0.1 and 0.3 mmole/kg. Neither dose caused liver injury. However the larger $CHCl_3$ dose corresponds to the maximum amount of CCl_4 metabolized following the largest (26 mmole) dose of CCl_4 we studied. More than two-thirds of both doses of $CHCl_3$ were rapidly metabolized to $^{14}CO_2$ within 6 hr. Less than 5-10% was recovered as $CHCl_3$, and $CHCl_3$ was not detected in the breath beyond 6 hr.

Analysis of the sequential relationship between the exhalation of $CHCl_3$ and its oxidized metabolite $^{14}CO_2$ at various time points up to 6 hr after $CHCl_3$ administration indicate a linear relationship over a thousand-fold range of $CHCl_3$ in the breath (0.003 to 3 mmoles $CHCl_3$/kg.hr). At any one time this relationship approximates $[CO_2] = 20\ [CHCl_3]^3$. On this basis we propose that the amount of $^{14}CO_2$ derived from the $^{14}CHCl_3$ metabolite of CCl_4 can be estimated.

At times beyond 2 hr following CCl_4 administration, the amounts of $^{14}CO_2$ metabolite recovered can be accounted for on the basis of their secondary derivation from $CHCl_3$ metabolite. However, the large amounts of $^{14}CO_2$ metabolite produced during the first 2 hr following CCl_4 poisoning cannot be accounted for on the basis of secondary oxidation of $CHCl_3$. We have found, as shown in Fig. 2, a highly significant regression correlation between early $^{14}CO_2$ metabolite exhalation and injury at 24 hr following CCl_4 over a dose range from 0.1 to 26 mmole CCl_4/kg.

Based on these admittedly complex findings which suggest oxidation of CCl_4 could be relevant to the injurious mechanism of CCl_4, we propose the metabolic scheme illustrated in Fig. 3. The radical formed by electron transfer from cytochrome P-450 could

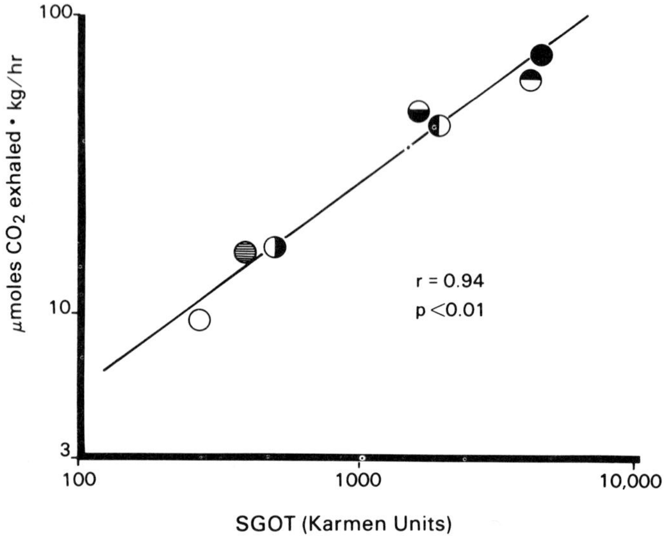

FIGURE 2. Regression correlation between injury (SGOT) and $^{14}CO_2$ exhalation during first 30 minutes following CCl_4. From left to right symbols represent mean values from 3-5 animals given 0.1, 0.3, 1, 2, 4, 10, or 26 mmoles CCl_4/kg.

covalently bind to cell components, be reduced by ·H abstraction to form $CHCl_3$, or combine with O_2 to form a peroxy radical. Such a peroxy radical could abstract a ·H and then decompose to liberate highly reactive ·OH radicals. We consider this peroxy pathway likely to yield more injurious oxidative species of CCl_4 than the alternative oxidation pathway which is secondary to $CHCl_3$ formation. Even though $CHCl_3$ is extensively oxidized to CO_2, at least in part through the proven reactive intermediate phosgene (15), on a mole for mole basis, $CHCl_3$ is far less hepatotoxic than CCl_4. Therefore, if oxidation of CCl_4 is relevant to the

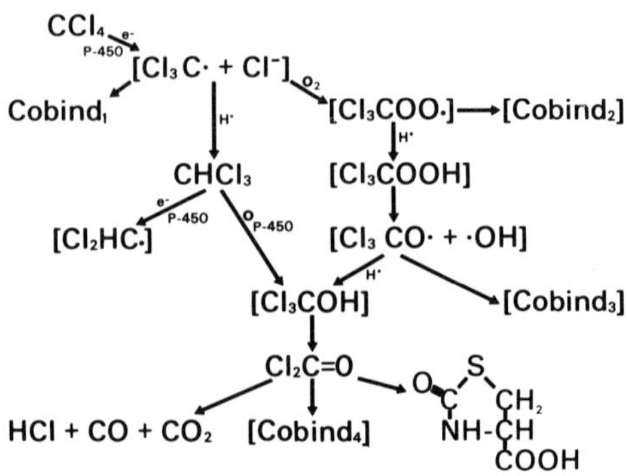

FIGURE 3. Proposed metabolism scheme for CCl_4.

injurious mechanism of CCl_4, it is not likely to be via a species in common with $CHCl_3$, but via some other species such as the trichloromethylperoxy radical.

ACKNOWLEDGEMENTS

This work was supported by NIH grant AM 19814.

REFERENCES

1. Reynolds, E. S. and Moslen, M. T., "Free Radicals in Biology," Vol. IV, (W. A. Pryor, ed.) Academic Press, New York, N.Y. (1980), p. 99.
2. Butler, T. C., J. Pharmacol. Exp. Ther., 134,311 (1961).
3. McCollister, D. D., Beamer, W. H., Atchison, G. J. and Spencer, H. C., J. Pharmacol. Exp. Ther., 102,112 (1951).
4. Poyer, J. L., Floyd, R. A., McCay, P. B., Janzen, E. G. and Davis, E. R., Biochim. Biophys. Acta., 539,402 (1978).
5. Fowler, J. S. L., Br. J. Pharmacol., 37,733 (1969).
6. Gordis, E., J. Clin. Invest., 48,203 (1969).
7. Recknagel, R. O. and Ghoshal, A. K., Lab. Invest., 15,132 (1966).
8. Hafeman, D. G. and Hoekstra, W. G., J. Nutr., 107,656 (1977).
9. Sagai, M. and Tappel, A. L., Toxicol. Lett., 2,149, (1978).
10. Burdino, E., Gravella, E., Ugazio, G., Vannini, V. and Calligaro, A., Agents Actions, 4,244 (1973).
11. Recknagel, R. O. and Glende, E. A., Jr., CRC Crit. Rev. Tox., 2,263 (1973).
12. Klaassen, C. D. and Plaa, G. L., Biochem. Pharmacol., 19, 2019 (1969).
13. de Toranzo, E. G. D., Diaz Gomez, M. I. and Castro, J. A., Res. Commun. Chem. Pathoo. Pharmacol., 19,347 (1978).
14. Packer, J. E., Slater, T. F. and Willson, R. L., Life Sci., 23,2617 (1978).
15. Mansuy, D., Beaune, P. Cresteil, T., Lange, M. and Leroux, J. P., Biochem. Biophys. Res. Commun., 79,513 (1977).

AUTOXIDATION AND CYTOTOXICITY

Donald C. Borg, Karen M. Schaich and John J. Elmore, Jr., Medical Research Center, Brookhaven National Laboratory, Upton, New York 11973

We essay the presentation of a comprehensive synthesis, or reaction schema, to relate autoxidations of non-lipid compounds to lipid chain peroxidation in vivo. As at Pinawa three years ago, we do this in the context of cytotoxic autoxidation reactions, and now—as then (1)—we conclude that hydroxyl radicals, $OH^{\cdot}$, produced by iron-dependent Fenton reactions serve as both primary toxicants and as sources of secondary toxicants. The latter stem from lipid chain peroxidation initiated by the Fenton-derived $OH^{\cdot}$, which we visualize as the obligate coupling step linking enzyme-dependent and non-enzymic autoxidations to potentially toxic outcomes.

HYDROXYL RADICALS AND MEMBRANE LIPID PEROXIDATION

At Pinawa we cited evidence implicating the generation of $OH\cdot$ in "dark" biochemistry (i.e., not photochemistry or radiation chemistry) (1), and a surge of new data in support of this contention has appeared in the intervening few years, as recently reviewed by Borš, Saran and Czapski (see Table III of ref. 2). We proposed (1) that in addition to the recognized radiomimetic cytotoxicity of the extremely reactive (and, hence, short-lived and short-ranged) hydroxyl radicals themselves, their initiation of lipid peroxidation in membrane lipids could slowly (hours) "broadcast" damage to important ("target") biomolecules at considerable distances, mediated by reactive intermediates and products of lipid chain peroxidation, both free radical (e.g., lipid alkoxy radicals) and non-radical (e.g., carbonyl compounds (3)) in nature.

AN INTEGRATED REACTION SCHEMA BASED ON AUTOXIDATION

A reaction schema justified by the arguments reviewed in detail in our previous paper (1) is depicted by Fig. 1. In the figure, electron flow (i.e., the transfer of redox equivalents) from the critical autoxidation step (beaded arrow) through the Fenton-like iron cycle to yield $OH^{\cdot}$ reacting with [membrane] lipid, LH, is emphasized by the set of heavy arrows.

Although the capability of $HO_2\cdot/O_2^{\cdot-}$ directly to reinitiate lipid peroxidation chains from metastable lipid hydroperoxides, LOOH (the reaction denoted by a question mark in Fig. 1), has been challenged by contrary findings (4,5), and Winterbourn has

ISBN 0-12-592050-4

claimed that the reduction of ferric iron by ascorbate (looping arrow given by long dashes and connecting the ultimate electron donor, DH_2, with complexed iron [Fe--], in Fig. 1) is faster than its reaction with $O_2^{\overline{\cdot}}$ and probably more important in vivo (6), the schema implied by our earlier paper remains otherwise intact (Fig. 1).

The critical feature of driving the cycling iron-dependent generation of OH· or of lipid alkoxy radicals (LO·) is the provision of an autoxidizing substrate that reacts spontaneously with dioxygen to yield $O_2^{\overline{\cdot}}/HO_2$ (beaded arrow in Fig. 1). The basic schema of Fig. 1 relates this autoxidation to the linked reactions of quinol-type cytotoxins which catalyze electron donation via an iron redox cycle from ascorbate or other ultimate reductants to H_2O_2 or to LOOH, because autoxidation reactions of those quinols were the focus of our concern at Pinawa. The H_2O_2, in turn, can be provided endogenously by spontaneous or dismutase-catalyzed disproportionation of the superoxide afforded by the autoxidation step, Fig. 1. However, we emphasize here other versions of the schema that are conceptually equivalent and differ only in the way that the autoxidizing substrate is provided.

In the cases of cytotoxins such as 6-hydroxydopamine, 6-aminodopamine and their congeners or of alloxan (which is reversibly reduced to dialuric acid in vivo), we contended that it was their quinolic forms (or quinolamine equivalents) which were the active substrates for autoxidation (1). The reactions summarized and integrated by the cycles to the right of the autoxidation reaction (beaded arrow) in the basic schema of Fig. 1 were invoked to explain the potentiation of the cytotoxicity of these agents in vivo by ascorbic acid, their catalysis of ascorbate oxidation in vitro (7), and certain other observations relating to these interactions (see Charts 1 and 2 of ref. 1). However, in cases where electron affinic drugs (such as nitrofurans, nitrobenzenes, etc.) or carcinogens (such as 4-nitroquinoline-1-oxide, 4-nitropyridine-1-oxide, etc.) catalyze the oxidation of ascorbate (8) or interact with non-protein thiols in cells (9), it appears that spontaneous reduction of these compounds by ascorbate can occur to afford nitro radical anions as reactive species which can also autoxidize in the presence of dioxygen (8,10). A version of the reaction schema that incorporates these observations and affords semiquinones rather than quinols (hydroquinones) as the autoxidizable substrates in non-enzymic reactions is given by Fig. 1A.

SCHEMA FOR AUTOXIDATION DEPENDING UPON ENZYMIC REDUCTION

When the reduction of ultimate electron donor to autoxidation substrate is enzymic in vitro or in vivo, the schema of Fig. 1B applies. Moreover, this generalized pathway appears consistent with many biochemical and metabolic findings regarding the detection of $O_2^{\overline{\cdot}}$, H_2O_2, or OH· (the latter directly by spin-trapping or inferentially by product analysis) or of products associated with lipid peroxidation from metabolizing or NAD(P)H-supported cells, microsomes, and solubilized microsomal enzyme systems. Recent findings (11) show that even nuclei and nuclear membrane preparations undergo iron-stimulated, NADPH-supported lipid peroxidation.

In general, any compound that can enter cells and accept electrons from an appropriate carrier to give rise to an autoxidizable form could be expected to enter into a reaction schema like that in Fig. 1B, producing $O_2^{\overline{\cdot}}$ and H_2O_2 (12) and—as an ultimate toxin—perhaps OH·. Indeed, substances such as paraquat, methyl viologen, etc. (12) and various nitro (10,13) and azo compounds (14) do increase intracellular production of $O_2^{\overline{\cdot}}$ and H_2O_2, and the toxicity of some of them increases as pO_2 rises (12).

Experimental evidence suggests that within cells most of the autoxidizable intermediates formed by enzymic action are associated with microsomal oxidases. There is some indication that reduced forms of microsomal cytochrome reductases (15,16) and certain valence states of P-450-type cytochromes themselves (J. A. Peterson, this conference) are autoxidizable and that P-450 cytochromes may directly effect one-electron reduction of some substrates (most substrates are oxidized or hydroxylated). However, the weight of evidence appears to favor the direct enzymic reduction of substrates to autoxidizable products by the flavoprotein cytochrome reductases associated with the microsomal and mitochondrial cytochrome P-450 oxidase/monoxygenase systems.

ROLE OF FLAVOPROTEIN REDUCTASES

In support of electron flow from reduced pyridine nucleotides through flavoprotein cytochrome reductases to compounds that are thence converted into autoxidation substrates are observations that solubilized, purified preparations of microsomal reductase will support OH· generation (detected by spin-trapping) and lipid peroxidation (malonaldehyde production) when stimulated by chelated iron (Fe^{2+}-EDTA) and provided reducing equivalents by NADPH (17). Comparable behavior of intact rat liver microsomes also appears to be independent of cytochrome P-450 when the latter is inhibited by metyrapone (16,18). In a similar vein, rat liver cytochrome P-450 reductase without the oxidase present catalyzes reduction by NADPH of tumoricidal quinone anthracycline antibiotics to rapidly autoxidizable anion free radicals (19). Also indicative of the competence of flavocompounds alone to mediate NADPH-supported reduction of substrates to autoxidizable products is the report that a model system of nitro compounds, NAD(P)H, and flavin mononucleotide (FMN) can form autoxidizable nitro dianion free radicals with no heme present (15).

As another line of evidence supporting the participation of flavoproteins, hepatic microsomal lipid peroxidation is itself independent of cytochrome P-450, and even when the lipid oxidation is inhibited by drug substrates of the cytochrome P-450

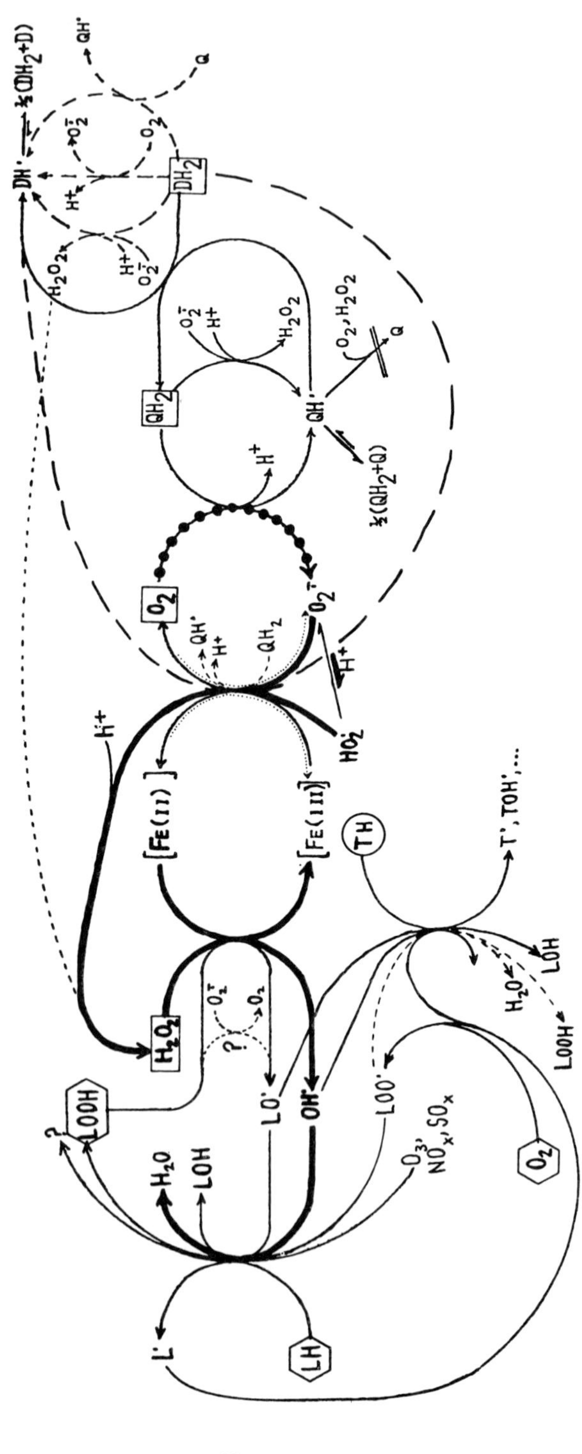

DH·
½(DH2+D)
QH·
Q
O2
DH2
H+
H2O2
QH2
QH·
½(QH2+Q)
O2, H2O2
O2
HO2·
Fe(II)
Fe(III)
TH
T·, TOH·,...
H2O2
LOOH
LO·
OH·
LOO·
H2O
LOH
O3, NOx, SOx
LOOH
O2
L·
LH

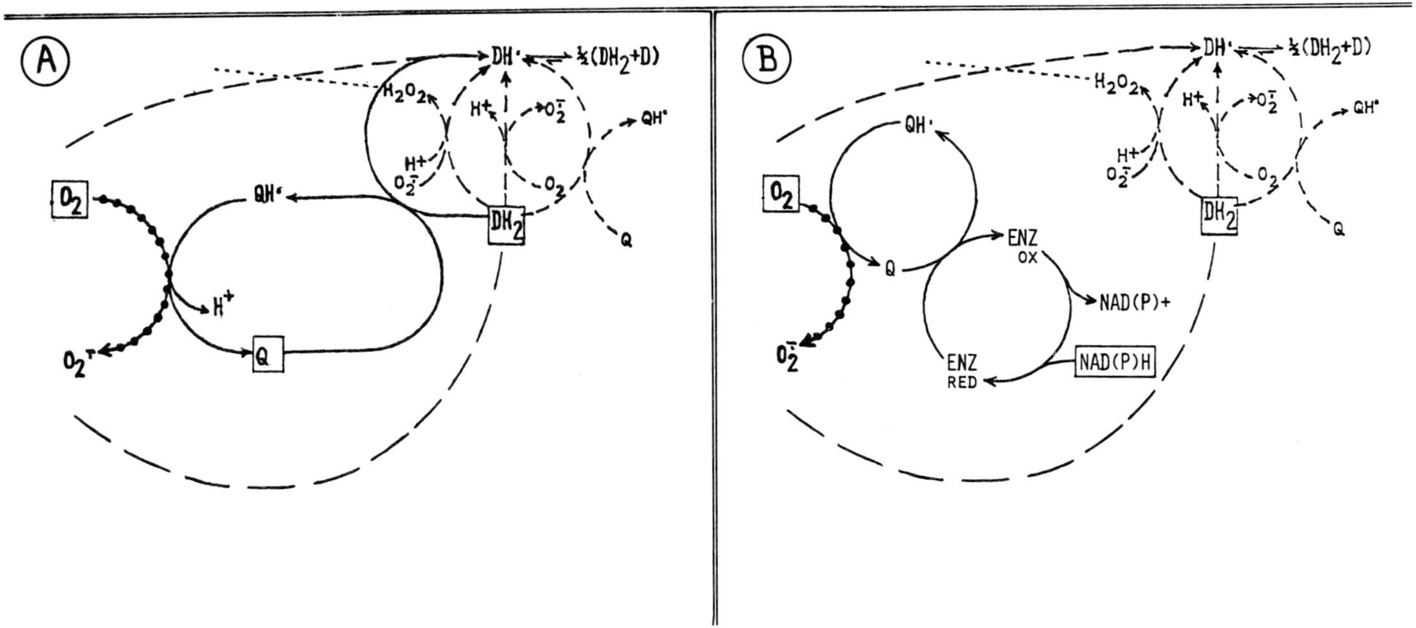

FIGURE 1. Iron-dependent, biological autoxidation schema of quinolic or quinolamic compounds (Qtt_2). Inserts A and B depict autoxidation of reactive free radicals (Qtt·) formed nonenzymically and enzymically, respectively (see text). Non-lipid substrates: When autoxidation catalyst is present, DH_2 = ultimate electron donor (ascorbate anion, etc.) in its hydroquinone quinol) form, with DH· its free radical (semiquinone) form and D its quinone form; QH_2(QH·,Q) = autoxidation catalyst or ultimate electron donor in the absence of DH_2; TH = target ("essential") biomolecule; [Fe--] = iron complexed to ADP, ascorbate, EDTA, etc. in vitro and to undetermined ligands in vivo. Lipid substrates: LH = polyunsaturated lipid; L = lipid alkyl radical; LO· = lipid alkoxy radical; LOO· = lipid peroxy radical; LOOH + lipid hydroperoxide.

mixed function oxidase in intact microsomes, the inhibition fails to correlate with drug metabolism. Probably this is so because direct suppression of lipid chain peroxidation results from the antioxidant properties of the drug substrates and does not reflect competition between drug metabolism by the oxidase and lipid peroxidation for NADPH reducing equivalents (20).

The oxidation of ethanol by rat liver microsomes is also dissociated from drug metabolism, and it is yet another metabolic process that depends on OH· mediation. Microsomal oxidation of alcohol is not correlated with mixed function oxidation of drug substrates in intact microsomes and even proceeds in the absence of cytochrome P-450 in reconstituted systems (21). Rather than depending on cytochrome oxidation, the microsomal production of OH· from endogenous H_2O_2 and the subsequent oxidation of primary aliphatic alcohols reveals patterns of product yields and of inhibition consistent with a Fenton system (21,22).

Additional support for the nonessentiality of cytochrome P-450 for enzyme-generated autoxidation in cells comes from detergent-solubilized superoxide-forming enzyme from human polymorphonuclear leukocytes (PMN's), which is associated with the well-known elaboration of $O_2^{\overline{\cdot}}$ and associated reactive forms of oxygen that accompany the bactericidal and cytocidal components of the "oxidative burst" of activated PMN's and macrophages. This purified enzyme requires only NADPH and a flavin (flavin adenine dinucleotide is best) for its activity, with no need for cytochrome P-450 (23).

ADEQUACY OF FENTON-SUPPORTED INITIATION OF LIPID PEROXIDATION

Although we argue that Fenton-supported lipid peroxidation can be mediated by OH·, as our schema shows (Fig. 1), Aust and his colleagues hold that Fenton-generated OH· is incapable of initiating lipid peroxidation in biological membranes and that "perferryl" iron-oxygen complexes, such as those alleged to form with iron chelated by ADP, are the obligate "primordial" initiators of lipid chain peroxidation therein (24,25). Above and beyond our disagreement concerning the nomenclature (see discussion, these proceedings), we remain unconvinced. We cannot readily fit a cycling reaction involving the implied iron-oxy complex, $Fe(II)\text{-}O_2 \rightleftharpoons Fe(III)\text{-}O_2^{\overline{\cdot}}$, into our integrated reaction schema, but what is far more compelling is that we note that Fenton-type reactions in homogeneous reaction systems seem largely oblivious of whether ADP is present or not, even when lipid peroxidation is involved, whereas most—or all—of the systems that fail to follow expected Fenton behavior are heterogeneous, multiphasic ones with some compartmentalized and membrane-bound reactants.

With lipids in membranes or vesicles, such as the microvesicular extracted microsomal lipids used by Aust et al. (24,25), the rate constants for the reaction of OH· with lipid may be lowered by nearly an order of magnitude (26,27). Hence, even Fenton initiation would be expected to be less efficient than with solubilized lipids. Considering the short diffusion range of reactive OH· radicals, we have the notion that ADP may serve to distribute the redox-active iron so as to increase OH· concentrations locally when peroxidation of vesicular lipids is initiated _in vitro_ by Fenton reactions in the suspending medium. OH· production _in vivo_ at cell surfaces or intracellularly, therefore, may represent an altogether different situation.

Failure of inhibitor patterns in biological, compartmentalized systems to follow expectations based upon Fenton reactions in solution (especially ineffectiveness of catalase in some cases (28,29)) may also be explained by a speculation of Green _et al_. (29). They propose that within cells H_2O_2 may be inaccessible to catalase if the H_2O_2 remains coordinated to a metal ion (represented by brackets in the equations below). Then OH· _would appear to be_ formed directly from $O_2^{\bar{\cdot}}$ by a disproportionation reaction due to the set of reactions occurring within the locus of the reducing metal (such as Fe^{2+}), and no "free" H_2O_2 would exist:

$$H^+ + O_2^-\cdot \longrightarrow HO_2\cdot$$

$$HO_2\cdot + H^+ + O_2^-\cdot \longrightarrow [H_2O_2] + O_2$$

$$[Fe^{2+}] + [H_2O_2] \longrightarrow OH\cdot + [Fe^{3+}] + OH^-$$

$$\text{Net:}\quad 2O_2^-\cdot + [Fe^{2+}] + 2H^+ \longrightarrow OH\cdot + OH^- + O_2 + [Fe^{3+}]$$

Contrary to Aust's claims, with peroxidation of lipids in intact microsomes OH· does appear to be present, because spin trapping reveals authentic OH· adducts (16,18). Furthermore, as we asserted above and document below, Fenton-generated OH· seems to be an effective initiator of lipid peroxidation in homogeneous reactions, and the presence of ADP makes little difference (provided that iron is otherwise complexed to avoid precipitation). Even in the presence of ADP the relatively "hydroxyl-specific" spin trap, 4-pyridyl-1-oxide-N-_tert_butyl nitrone, forms its characteristic OH· adduct in a Fenton reaction (30). As an another example: reducing agent inhibition of Fenton-catalyzed linoleic acid peroxidation is no different whether Fe^{2+} that contaminates the buffer is used, additional $FeSO_4$ is provided, or ADP-Fe^{2+} is added (20).

Even enzymically-induced lipid oxidation can be initiated by OH· in homogeneous reaction systems. Thus with detergent-solubilized methyl linoleate, cytochrome P-450 reductase, NADPH, and $FeCl_3$ (even when excess EDTA is present), lipid peroxidation occurs and reveals an inhibitor pattern expected of a Fenton system with OH· intermediation (17,31).

Our own studies also support the concept of OH· initiation of lipid oxidation in Fenton-like reactions. Previously we reported a pattern of oxygen consumption with 6-hydroxydopamine or 6-aminodopamine catalysis of iron-enhanced oxidation of ascorbate in the presence of arachidonic acid that was interpreted as most consistent with $OH^{\cdot}$ initiation of lipid oxidation (1). Fig. 2 reveals that with Fe^{2+} and an excess of ascorbate to keep the iron reduced and to provide net reducing power (as in Fig. 1), there is a background rate of O_2 consumption by metal-catalyzed ascorbate autoxidation. Addition of exogenous H_2O_2 to complete a a Fenton system has little further effect <u>except</u> in the presence of linoleic acid, which increases O_2 consumption. The additional O_2 consumption, in turn, is taken to be a measure of peroxy

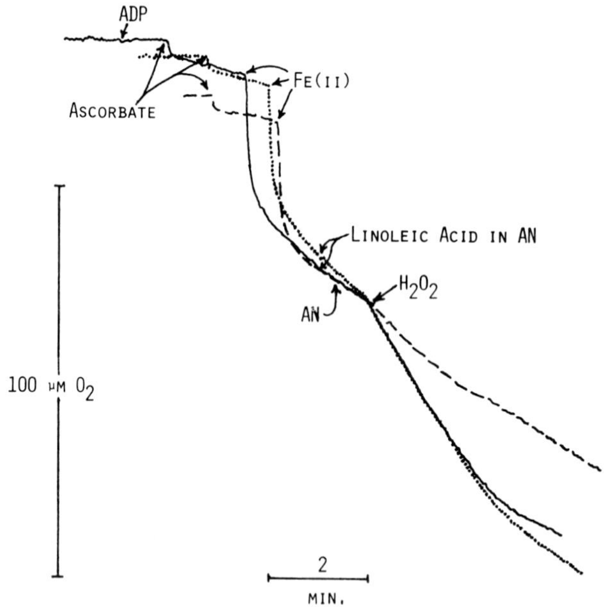

FIGURE 2. Oxygen consumption at 37°C and pH 7.4 (0.1 M phosphate buffer). Arrows indicate times of addition of various reactants in 10-30 μl aliquots into 3 ml total volume. Final concentrations: Adenosine diphosphate (ADP) = 0.2 mM; Ascorbate = 3.3 mM; Fe (II), as $Fe(NH_4)_2(SO_4)_2$ = 0.1 mM; Acetonitrile (AN) = 67 mM; Linoleic acid in AN = 0.3 mM; H_2O_2 = 0.3 mM. Tracings have been normalized (superimposed) at the point where H_2O_2 is added to complete the Fenton reaction system. The dashed line (---) represents O_2 consumption in the absence of lipid. For the other two tracings linoleic acid was added before the Fenton reaction commenced, either with (solid line,____) or without (dotted line, ...) ADP present.

radical formation (see LOO· in Fig. 1) as the lipid alkyl radicals, L·, produced by reaction with OH· rapidly react with O_2 (32). The addition of ADP does not alter this pattern of O_2 consumption (Fig. 2), regardless of whether ADP is present before Fe^{2+} is added, as in Fig. 2, or whether ADP and Fe^{2+} are premixed anaerobically.

Furthermore, Fig. 2 represents a more generalized behavior, because linoleic acid can be replaced by linolenic acid, and enhanced O_2 consumption from lipid oxidation cum peroxidation is also seen when Ti^{3+} is substituted for iron/ascorbate as the Fenton reductant. Iron sequestration quenches O_2 consumption, as expected, and OH· scavengers (including the spin trap 4-phenyl-N- tertbutyl nitrone) eliminate the enhanced O_2 uptake that is due to the reaction of Fenton OH· with lipid (work in progress).

In summary, we conclude that our proposed schema provides a useful integration into a concerted toxicological pathway of various cellular and laboratory redox processes that have been studied as isolated reaction systems. Guided by the schema, which will require revision as new data appear, we may now begin to study cellular cytotoxic processes for evidence that the connected reactions actually occur in vivo.

ACKNOWLEDGEMENTS

This work was supported by the U.S. Department of Energy, under contract DE-AC02-76CH00016.

REFERENCES

1. Borg, D. C., Schaich, K. M., Elmore, J. J., Jr. and Bell, J. A., Photochem. Photobiol., 28,887 (1978).
2. Bors, W., Saran, M. and Czapski, G., "Proceedings of Second International Symposium on Superoxide and Superoxide Dismutases," in press.
3. Benedetti, A., Casini, A. F., Ferrali, M. and Comporti, M., Biochem. Pharmac., 28,2909 (1979).
4. Hasegawa, K. and Patterson, L. K., Photochem. Photobiol., 28,817 (1978).
5. Bors, W., Michel, C. and Saran, M., FEBS Lett. 107,403 (1979).
6. Winterbourn, C. C., Biochem. J., 182,625 (1979).
7. Heikkila, R. E. and Cohen, G., Mol. Pharmac., 8,241 (1972); Ann. N. Y. Acad. Sci., 258,221 (1975).
8. Biaglow, J. E., Jacobson, B., Varnes, M. and Koch, C., Photochem. Photobiol., 28,869 (1978).
9. Varnes, M. E. and Biaglow, J. E., Cancer Res., 39,2960 (1979).
10. Mason, R. P. and Holtzman, J. L., Biochem. Biophys. Res. Commun., 67,2367 (1975).

11. Baird, M. B., Birnbaum, L. S. and Sfeir, G. T., Arch. Biochem. Biophys., 200,108 (1980).
12. Hassan, H. M. and Fridovich, I., Arch. Biochem. Biophys., 196,383 (1979).
13. Sealy, R. C., Swartz, H. M. and Olive, P. L., Biochem. Biophys. Res. Commun., 82,680 (1978).
14. Mason, R. P., Petersen, F. J. and Holtzman, J. L., Biochem. Biophys. Res. Commun., 75,532 (1977); Mol. Pharmac., 14,665 (1978).
15. Mason, R. P. and Holtzman, J. L., Biochem., 14,1626 (1975).
16. Lai, C. S. and Piette, L. H., Biochem. Biophys. Res. Commun., 78,51 (1977).
17. Lai, C. S., Grover, T. A. and Piette, L. H., Arch. Biochem. Biophys., 193,373 (1979).
18. Lai, C. S. and Piette, L. H., Arch. Biochem. Biophys., 190, 27 (1978).
19. Bachur, N. R., Gordon, S. L., Gee, M. V. and Kon, H., Proc. Natl. Acad. Sci. U.S.A., 76,954 (1979).
20. Miles, P. R., Wright, J. R., Bowman, L. and Colby, H. D., Biochem. Pharmac., 29,565 (1980).
21. Cederbaum, A. I., Miwa, G., Coyen, G. and Lu, A. Y. H., Biochem. Biophys. Res. Commun., 91,747 (1979).
22. Cohen, G. and Cederbaum, A. I., Arch. Biochem. Biophys., 199,438 (1980).
23. Babior, B. M. and Kipnis, R. S., Blood, 50,717 (1977).
24. Pederson, T. C., Buege, J. A. and Aust, S. D., J. Biol. Chem., 248,7134 (1973).
25. Svingen, V. A., Buege, J. A., O'Neal, F. O. and Aust, S. D., J. Biol. Chem., 254,5892 (1979).
26. Barber, D. J. W. and Thomas, J. K., Radiat. Res., 74,51 (1978).
27. Patterson, L. K. and Hasegawa, K., Ber. Bun. Phys. Chem., 82,951 (1978).
28. Rosen, H. and Klebanoff, S. J., J. Clin. Invest., 64,1725 (1979).
29. Green, M. R., Hill, H. A. O., Okolow-Zuhkowska, M. J. and Segal, A. W., FEBS Lett., 100,23 (1979).
30. Jansen, E. G., Wang, Y. Y. and Shetty, R. V., J. Am. Chem. Soc., 100,2923 (1978).
31. Kameda, K., Ono, T. and Imai, Y., Biochim. Biophys. Acta, 572,77 (1979).
32. Kolthoff, I. M. and Medalia, A. I., J. Am. Chem. Soc., 71, 3784 (1949).

DISCUSSION

PRYOR: This has been a brief but excellent presentation of data from a number of biological systems where oxy-radicals are involved. I see a number of persons with questions.

PEISACH: I have a general question that has to do with spin-trapping reagents. These have been used in two of the lectures this evening. In all of these discussions, we never told about the stoichometry of these reactions nor are we told what the overall yields are or how meaningful these spin-trapped radicals are in the overall reaction scheme.

PRYOR: It is extremely difficult and at the same time extremely important to quantify spin adducts in our systems. I think Ed Janzen should make some remarks about this.

JANZEN: That is a very important question, and I was going to make some remarks about the chain length, which is related. We wanted to know how much is happening per initiating radical. We measured three reactions several years ago using the phenyl radical, the t-butoxyl, and the benzoyloxyl radical. With enough spin-trap, one can quantitatively trap all radicals produced by a given initiator. However, to date, those are the only three radicals that have been studied quantitatively.

PRYOR: How much is "enough" spin trap?

JANZEN: In the cases we studied, anywhere from 0.05 to 0.1 M in benzene solvent. The hydrogen peroxide system, which produces hydroxyl radicals, is a much more complex system. One has to find an initiator that produces a known number of radicals per unit time. That has not been done, although there has now been published an initiator which will give hydroxyl radicals thermally. We have not studied that one as yet, but it looks like a source one could use to obtain quantitative results. But surely at lower concentrations of spin trap, radicals as reactive as hydroxyl will do many other things. This is simply a competitive reaction, and the spin trap competes with all other reactions in the system.

PRYOR: Quantifying radicals in complex systems such as those that are dealt with in our ozone work or in complex biological systems is much more difficult than in a simple, clean system in which one initiator gives a high yield of one (or at most two) radicals. In many biological systems and our ozone systems, the overlapping absorption spectra of several spin adducts occurs. When one has overlapping derivative peaks, you can not use peak heights or peak areas either.

JANZEN: The only way to sort that out is to do some extensive computer simulation of model mixtures and from those get an estimate. Spin adducts are not stable molecules either. If one is going to trap hydroperoxyl radicals, the spin adduct is a hydroperoxide of some kind and it will be reduced. In the case of DMPO, which is an excellent spin trap (it is very fast, and it gives very good information), the reactivity of the spin adducts is very high. In fact, some of its spin adducts cannot be detected. The phenyl adduct of DMPO for example is barely detectable under any condition.

PRYOR: Spin trapping is a very useful technique. However, it has been applied to biological systems for so short a time that we have not yet done all the studies and controls that need to be done. Spin adducts have not been quantified in any biological system as yet; to a certain extent, results must be taken with a grain of salt until radical yields are quantified. This definitely applies to our ozone spin-trapping work.

BORG: One comment just to second what you said about the reactivity of DMPO spin adducts. When we look at spin traps in a fast (millisecond region) ESR flow system, we often see changes in the profiles of radical adducts within seconds or fractions of a second. This depends on the reactants, their concentrations, and pH of course, but there are apt to be changes in very short times indeed. It's certainly a complicated picture we see.

In answer to Jack Peisach's question, there's a practical, useful thing one can do that is fairly obvious. It's hard to quantify the ESR in situ, as everyone just said, but in many cases something is known about the reaction rates, so what you can do is treat the spin adduct (if it is supposed to be a competitive reactant for a free radical intermediate) like a competitive inhibitor. Look for the expected changes in the relative intensities of the spin adduct signal and look for other kinetic evidence that the free radical being trapped is in the main reaction path. Gerry Rosen in a recent review that was cited this morning (Arch. Biochem. Biophys. 200, 1 (1980)) listed some other useful techniques to help check that the adduct that is attributed to a certain source is, in fact, obtained in that way. He presented some examples in the case of the hydroxyl radical.

PEISACH: This raises another question related to what Don Borg said. If we consider spin trapping reagents as competitive inhibitors, does that mean that if they are ineffectual in an enzymatic reaction, then the radical which is supposedly formed is not part of the reaction sequence? If a spin trap has no effect on an enzymatic reaction can you argue that the radical produced is not formed in the enzymatic pathway?

PRYOR: A spin trap is a radical inhibitor, and its presence may change the system.

FEE: It seems that one of the things overlooked here is the reactivity of O_2 with biological molecules, and I would like to point out one of the many possible cases where that leads to a reaction deleterious to the metabolism of the cell. The second enzyme in purine metabolism (an iron-sulfur containing protein), for reasons which are not understood, reacts directly with molecular oxygen and loses its activity. There are a large number of iron-sulfur centers in cells and these generally exist at low redox potentials and many of them are sensitive to oxygen. I believe they probably are sensitive to direct reaction with oxygen. When the redox status of the iron-sulfur center controls the enzymatic activity, the metabolism of the cell is going to be changed upon its autoxidation. Once you have changed the metabolism of the cell to throw it out of the fine-tuned coordination in which it normally exists, a general malaise sets in, iron will be released from these iron-sulfur centers, etc., and then you are in the area of rancidification where free radical reactions occur. I hear a great deal of discussion here about rancidification, and I believe that iron and other metals are essential for the process. I am not yet convinced, however, that iron is essential for the "primordial" reactions of oxygen toxicity.

PRYOR: Any events that decrease the organization of the cell produce increased sensitivity to autoxidation. For example chaotropic chemicals do this (Y. Hatefi and W. G. Hanstein, Arch. Biochem. Biophys., 138,73 (1970)).

BORS: I would like to make a general comment on the nomenclature of this perferryl species. For example, I am at a loss as to whether the Fe^{3+} $O_2^{\bar{\cdot}}$ is not merely $O_2^{\bar{\cdot}}$ chelated and stabilized by ferric iron.

In the case of cytochrome P-450, when there is a second electron transfer, you may have a formal splitting or the oxygen molecule. After water is formed, then you obtain a FeO^{2+} species, with only one oxygen atom left in the iron complex. There is some supporting evidence for that, when iodosobenzene is used as an artificial substrate. Iodosobenzene eliminates one oxygen atom that then stimulates or participates in cytochrome P-450 reactions. Thus, the FeO^{2+} species is a completely different intermediate that has more similarity to free $OH^{\cdot}$ than to $O_2^{\bar{\cdot}}$ chelated to ferric iron. That is an alternative that should be considered.

PRYOR: Table A Below recommends nomenclature for iron-oxygen species.

AUST: I agree that I do not know exactly what terminology to use for what we call a perferryl complex [Fe^{2+} O_2 $\longleftrightarrow$ Fe^{3+} $O_2^{\bar{\cdot}}$]. The formal oxidation state of iron is plus 6, but it cannot remove 4 electrons and come up with 2 waters from oxygen, so I'd rather not talk about the formal oxidation state. What we were calling the ferryl complex is ferrous plus H_2O_2, which may be kinetically the same as the hydroxyl radical.

WILLSON: We've been doing some experiments with nitro-compounds that may be relevant to iron-sulfur complexes. If you take a de-aerated solution of cysteine and ferrous ion you can obtain the iron-sulfur complex $Fe^{2+}(RS^-)_2$. In the presence of a trace of a nitro-compound a larger complex forms rapidly:

$$Fe^{2+}(RS^-)_2 + RNO_2 \longrightarrow [Fe^{3+}(RS^-)_2 \text{ --- } RNO_2 \cdot^-]$$

This complex is very similar to Dr. Aust's.

FEE: Oxygen toxicity may start with the autoxidation of key metabolic enzymes, and this leads to the disruption of metabolism. Once this disruption occurs, rancidification sets in. I think what Dr. Willson was referring to are possible mechanisms whereby iron-cysteine sulfur complexes could produce the strong oxidizing agents that we all know to be involved in this process. Maybe Jack Peisach could tell us about bleomycin.

PEISACH: Before I talk about bleomycin, let's get something straight in terms of nomenclature and the problem of redox and oxidation states. We all agree that the ferryl oxidation state kinetically resembles hydroxy radical; in other words it is a ferric and hydroxy radical, although they are not kinetically equivalent. For what you call the perferryl state you are willing to consider oxidation state six if we can assume a formal oxidation state which is at the same oxidation state as oxyhemoglobin—that's also oxidation state six. What has not been considered in Dr. Aust's mechanism is something which is undoubtedly operative in cytochrome P-450 mechanisms and also in the peroxidase mechanisms, and that is oxidation state five....a one electron reduced form from the state six form. Is this a possible intermediate in some of these reactions? After all, Dr. Aust has shown that you need two different iron compounds to demonstrate activity.

AUST: The difference is in forming the initial lipid hydroperoxide. This does not have anything to do with EDTA-ferrous except that EDTA-ferrous reacts with preformed lipid hydroperoxides to initiate lipid hydroperoxide-dependent lipid peroxidation. What you are talking about though is something that Bill Pryor and I have discussed: can you trap the lipid alkyl radical? Perhaps you can in the absence of oxygen, but in the presence of oxygen, no. (Just as with carbon tetrachloride, for example.) We have tried to add just ferrous, anaerobically, and looked for the alkyl radical. We cannot find it. The question then is, does the oxygen combine with reduced iron in some complex and that is the initiating species?

PEISACH: What I am suggesting is that even though you make a state six oxygenated ferrous complex, that may not be the active species in the peroxidation reaction. What I am suggesting is an intermediate at oxidation state five must be produced, analogous to protein peroxidation.

AUST: Let me say one thing. The rate of reaction is first order with ferrous. That's the only evidence against another electron's being supplied by ferrous, and, as I said earlier, we have no evidence that it comes from the lipid.

FEE: It seems to me that what may happen with the ferric-$O_2^{\overline{\cdot}}$ complex, or ferric peroxide complex is simply that electrons are fed into it until the formal level of Fe^{3+}-OH· is achieved. This is what I always refer to as the Fenton reagent. There is a great deal of controversy about whether it is a free hydroxyl radical or whether it is a metal-bound oxene, or something like that. Nevertheless, the free energy change in going from hydroxyl to water is enormous, and this is where the energy comes from to abstract hydrogens or to add to difficult carbon-carbon bonds, etc.

TRAYLOR: I just want to discuss the ferrous dioxygen complex. This is a very unreactive species. It will react with phenol, perhaps to abstract a hydrogen. If it did that it would produce the precursor to Compound I. That is a very strong oxidizing agent. Thus, by supplying to horseradish peroxidase an electron as such (or a hydrogen atom which is the same thing), you can produce the iron (V) FeO^+ species by splitting the O—O bond. This is an extremely strong oxidizing agent. But we can make a protoheme oxygen complex in solvents such as DMSO or ethers that is very stable at room temperature to attack by solvents. It is not a very good oxidizing agent. In fact, it reacts with dithionite with a rate constant of only 10^4 M^{-1} sec^{-1}. I would suggest that a dioxygen-iron complex would react with polyolefins too slowly to be significant.

AUST: It would have to get that electron from lipid.

TRAYLOR: It will not react with things like lipid double bonds. I'm pretty sure of that.

AUST: Then how do we get our activity?

TRAYLOR: You need a source of electrons or easily extractable hydrogen atoms.

AUST: You need that other source of electrons if you start out with, for example, just linoleic acid and ADP ferrous and oxygen. And you get activity. Okay, where do you get the iron (V)? You are going to have to get that electron from lipid. Right? Where is the alkyl radical?

TRAYLOR: There are several ways that these oxygen complexes can be decomposed. For example deoxy Fe(II) and the oxygen complex can come together to give an iron peroxide that splits into the equivalent of horseradish peroxide compound II, also an oxidizing agent. There are ways at getting at these other oxidation states. But I doubt very much that you are going to get a direct reaction of an oxygen complex of iron II with a simple olefin or diene. It is just not that reactive.

AUST: Would the reaction be first order with respect to ferrous?

TRAYLOR: I do not know. It is second order with respect to ferrous if it goes through the iron four system but I do not know what to expect in your system.

KOPPENOL: This scheme (see Koppenol, this volume, Fig. 3) is a little bit more simplistic than that shown by Don Borg but I think it is very similar, except that it shows another reaction between R· and O_2. We start on the left with the hydrogen peroxide from $O_2^{\overline{\cdot}}$ or from enzymes, and then we have a metal ion (e.g. iron (II)) react with it. We form water and, I hope I say it correctly, the ferryl radical. Reduction of H_2O_2 does not require a metal ion: it could be ascorbic acid or a semiquinone, in which case you get the real OH· radical. Now MO^{2+} or OH· oxidizes RH to an R· radical. What is different from Borg's scheme is that R· can combine with oxygen and give $O_2^{\overline{\cdot}}$ back and that starts a Fenton-driven Haber-Weiss reaction. The reaction of R· + O_2 and R^+ apparently does not occur in the case of lipids. Instead you get what is shown in the lower right part, lipid peroxidation, and the trivial combination of two R· radicals. This scheme shows that superoxide dismutase comes in quite handy here because it can stop this whole mechanism either at top left or in the center itself, in the Fenton-driven Haber-Weiss cycle. But you can still reduce H_2O_2 with ascorbic acid and for that reason you need catalase or a peroxidase. It was said by Steve Aust that no effect was observed by ordinary OH· scavengers in some of his experiments. If indeed the ferryl radical is formed and not OH·, it might be that FeO^{2+} is more selective in its reaction and that you need something else than mannitol to scavenge it and to observe an effect.

BORG: I would like to comment on Hal Swartz's paper. The fact that you can see the monodihydroascorbyl radical in tissue and resolve its 1.4 gauss ESR splitting—even in tissues—means that you're seeing a motionally narrowed ESR spectrum with only anisotropic hyperfine splitting detected, because the ESR is not broadened at all. You showed yourself, how the ascorbyl radical when it is immobilized or in powder spectra is seen — as in lyophilized tissue samples—as a big broad signal. Well that is further evidence, if we need more evidence, that the ascorbic acid really is in the extracellular fluid and not in membranes for the most part: not fixed in tissue. At least the radical that we see is freely mobile in a solution or sol. That issue was brought up in the discussion earlier today, when the question was raised whether reactions that go in homogeneous systems with ascorbate interacting with vitamin E and other antioxidants and hydrophobic reducing substances, will actually occur in cells, because ascorbate is mostly outside. Some of you biologists know that one of the tricks used to determine whether a spin label is on the outside of a membrane or on the inside is to see whether ascorbate will quence (reduce) it; if so, it is on the outside of the membrane, because ascorbate is outside. So, the question of whether ascorbate contacts the other reactants with which it could react in homogenous phase—whether it meets them in the cell—I think is an open one, and the evidence from the narrow ESR signal suggests that ascorbate is really outside the cell in the extracellular fluid. Do you want to comment on that?

SWARTZ: I do not understand why you could not get the same signal if it was in the cytoplasm?

BORG: If it is free and not in a gel you might.

SINGH: Don Borg, I do not see in your slide where you need OH· radicals and in what concentration. In initiating autoxidation, obviously there is an energy requirement, depending on where the attack is going to be and where you want to abstract hydrogen. Do you really require OH· or can some less active radical also start it? Once the process gets started, you form lipid hydroperoxides or lipid peroxides and they can then give alkoxy radicals that are themselves very reactive. So I think as starting point, do we really need ·OH radicals, and if so, in what concentrations? Even in case of in vivo systems, the natural radiation background or earth gives something like 10^{-20} molar (that is mole dm^{-3}) hydroxyl radicals all the time, in steady state. These might be enough to initiate lipid peroxidation.

PRYOR: Where did that 10^{-20} come from?

SINGH: The 10^{-20} molar OH· is due to earth's natural background due to normal radiation. Another thought that goes through my mind is that in photolysis there is a process called energy pooling which we are all familiar with, where two exited molecules can come together and then with the total energy bring about reactions for which either one of them did not have enough energy. Do we possibly have a biological analog of this? Where one single enzyme in dealing with the substrate may not have enough energy to bring about a reaction, but two molecules of the enzyme with two molecules of the substrate can pool their energy and give you a reaction analogous to OH· radical reaction?

PRYOR: I doubt if background radiation is important in the initiation of radical reactions in vivo. Some years ago ("Advances in Free Radical Chemistry," Vol. V (G. H. Williams, ed.), Academic Press, New York, 1975) I reported on the very slow, self-initiated polymerization of styrene. This is a free radical process called "thermal" polymerization that goes on at an extremely slow rate at 25°C but at 0.1% per hour at 60°C and 2% per hour at 100°C. Willard Libby thought this might be initiated by the earth's natural radiation background. I was sure it was not, since the rate of this process is very sensitive to the structure of the styrene (or other monomer) that is used. Nevertheless, while I was at UCLA in collaboration with Libby, we did an experiment in which we compared the rate of polymerization of styrene on the surface and in a salt cave deep underground. (The cave is near China Lake in California.) Salt has a very low concentration of radiation-emitting isotopes (unlike granite). As I had anticipated, background radiation caused no measurable increase in the rate, even of this very slow free radical process. We also showed by calculations based on G values that this should be true. On the other hand, I believe that the background levels of ozone (about 0.02 ppm) are sufficient to initiate autoxidation of polyunsaturated lipids in a significant way. We have published experimental evidence for this [Arch. Environ. Health, 31, 201 (1976)] and Zelac et al. [Environ. Res., 4, 262,325 (1971)] also have suggested that background concentrations of ozone are far more biologically significant than are background levels of radiation.

NIKI: In connection with the formation of free radicals in ozonolysis reactions, mention has been made of the O'Neal-Blumstein mechanism. This mechanism was originally designed to account for certain anomolous products of the gas-phase reactions of ozone with simple olefins. Subsequent experimental work has shown, however, that this mechanism does not account for the major features of the reactions. Work in our laboratories indicates that the Criegee mechanism for solution-phase ozonolysis is also applicable in the gas phase. The main difference between the solution-phase and gas-phase reactions lies in the subsequent chemistry of the Criegee intermediate. In the gas phase the Criegee intermediate can react in a unimolecular or bimolecular mode leading to a very complex sequence of free radical reactions. In the condensed phase these kinds of free radical reactions are unlikely to take place to any significant extent. The current status of the gas-phase ozonation problem is discussed by me in "Chemical Kinetic Data Needs for Modeling the Lower Atmosphere," (J. T. Henson, R. S. Huie, and J. A. Hodgeson, eds.), NES Special Publication 557, U.S. Government Printing Office, Washington, D.C. 20402 (1979).

PRYOR: We have had a lengthy discussion of the nomenclature of iron-oxygen species during this session. In order to aid in future discussions of the chemistry of such species by workers in the field, and also to help neophytes with this difficult nomenclature, a number of us will get together after this session and produce a list of recommended names and structures. Those of you that wish to contribute to this effort should see me in the coming several days. (See Table A below.)

I might say in closing this most interesting session that free radical biology and its implications on complex phenomena such as aging, carcinogenesis, electron transport, etc., can no longer be regarded as a set of untenable, weird, or unscientific postulates. We now are in the happier state of doing solid science and asking sophisticated questions that have strong implications on biology. Although we are far from having answers to all the questions we ask, we clearly are on the road to getting answers to some extremely intriguing, perplexing, and important problems in biology and medicine.

I thank all the participants and this patient and knowledgeable audience.

In response to the lengthy discussion of nomenclature of iron-oxygen complexes, we decided to review the nomenclature and make several recommendations. We suggest that the names listed below for the oxidation states listed. We also recommend that the name "perferryl" not be used.

TABLE A. A LIST OF RECOMMENDED NAMES AND POSSIBLE STRUCTURES FOR VARIOUS IRON-OXYGEN COMPLEXES.

by H.B. Dunford and W.A. Pryor

in collaboration with S. Aust, W. Bors, I. Fridovich, W.H. Koppenol, T. Traylor and V. Ullrich

FORMAL OXIDATION STATE [1]	POSSIBLE STRUCTURES [2]	SOURCES	RECOMMENDED NAME(S)
6	$[Fe^{II}\ O_2]^{2+}$, $[Fe^{III}\ O_2^{\overline{\cdot}}]^{2+}$	[3] Oxyhemoglobin. Oxymyoglobin. Peroxidase compound III.	Oxy-ferrous complex. Dioxygen-ferrous complex.
5	$[Fe^{III}\ O_2^{=}]^{+}$, $[Fe^{II}{-}O{-}O^{-}]^{+}$	Possibly non-heme iron-dependent monooxygenases.	Peroxo complex.
	$[Fe^{III}{-}\underline{\overline{O}}\ ]^{3+}$, $[Fe^{IV}{-}\underline{\overline{O}}\cdot]^{3+}$, $[Fe^{V}{-}\underline{\overline{O}}|]^{3+}$, $[Fe^{IV}{-}\underline{\overline{O}}{-}H,\ R\cdot]^{3+}$, $[Fe^{IV}{-}\underline{\overline{O}}|^{-},\ \overset{+}{\cdot}R{-}H]^{3+}$ [4]	Peroxidase compound I.	Compound I.
	$[RS^{-}\ Fe^{III}{-}\underline{\overline{O}}\ ]^{2+}$, $[RS^{-}\ Fe^{IV}{-}\underline{\overline{O}}\cdot]^{2+}$, $[RS^{-}\ Fe^{V}{=}\underline{\overline{O}}\ ]^{2+}$, $[RS{-}Fe^{IV}{-}\underline{\overline{O}}\cdot]^{2+}$, $[RS{-}Fe^{IV}{=}\underline{\overline{O}}\ ]^{2+}$	Cytochrome P_{450}	Oxenoid complex. (Oxene is less desirable.)
4	$[Fe^{III}\cdot OH]^{3+}$, $[Fe^{III}{-}\dot{O}H]^{3+}$, $[Fe^{IV}\ OH^{-}]^{3+}$	Myoglobin + H_2O_2. Peroxidase compound II.	Ferryl complex or compound II.

[1] Formal oxidation state is here defined as the number of electrons necessary to reduce the iron and oxygen species to atomic iron and water respectively.

[2] When calculating the total charge on the complex, the two negative charges on the porphyrin ring are neglected as is the nature of any ligand in the 5th coordination site (except for the P_{450} oxenoid complex).

[3] See D.A. Summerville, B.D. Jones, B.M. Hoffman and F. Basolo, J. Chem. Ed., 56, 157 (1979).

[4] R is a radical species in a metal environment.

ROLE AND TOXICITY OF SUPEROXIDE IN CELLULAR SYSTEMS

Session Organizer, G. Czapski, Jerusalem, Israel

Session Chairman, D. C. Borg, Brookhaven, U.S.A.

THE BIOLOGY OF SUPEROXIDE AND OF SUPEROXIDE DISMUTASES - IN BRIEF

Irwin Fridovich, Department of Biochemistry, Duke University Medical Center, Durham, North Carolina 27710

We shall here briefly review current knowledge of the biology of superoxide radical and of the superoxide dismutases, so that those new to this field may easily gain some appreciation of its recent history and current status. Several more complete reviews are available and no attempt will be made to provide full documentation here. Readers needing references not provided in this discussion are directed to these reviews (1-5).

RECENT HISTORY

Generation of the superoxide radical ($O_2^{\overline{\cdot}}$) during the aerobic action of milk xanthine oxidase was deduced from the ability of this enzyme to: 1. initiate the free radical chain oxidation of sulfite, 2. cause the dioxygen-dependent reduction of cytochrome c, and 3. induce the chemiluminescence of aminophthalhydrazide. Proof of the release of $O_2^{\overline{\cdot}}$ from xanthine oxidase was finally provided in 1968 (6) and was confirmed by esr the following year (7).

Superoxide dismutase (SOD) activity was discovered during the course of studies of the production of $O_2^{\overline{\cdot}}$ by xanthine oxidase. A preliminary report establishing the catalytic activity of SOD and describing its isolation from bovine blood was made in 1968 (8). The full report appeared the following year (9). The SOD isolated from erythrocytes proved to be identical to the hemocuprein previously isolated on the basis of its copper content (10) and extensively studied during the following decades, without knowledge of its enzymatic activity.

ISBN 0-12-592050-4

MULTIPLICITY OF SUPEROXIDE DISMUTASES

Oxygenation of the biosphere, as a consequence of photosynthesis, must have applied a common selection pressure to a varied biota. It is not surprising that independent evolution of more than one SOD occurred. There are, in fact, three kinds of SODs and these fall into two independent evolutionary families. The copper and zinc-containing SOD (CuZnSOD) is characteristically found in the cytosol of eukaryotes. This enzyme is a dimer, whose subunit molecular weight is close to 16,000. The second variety of SOD to be isolated was the manganese-containing type (MnSOD) and it was obtained from *Escherichia coli* (11). *E. coli* also contain an iron-SOD (FeSOD) (12). SOD has been isolated from a wide range of bacteria, blue-green algae, fungi, plants, fish, birds and mammals. The CuZnSOD, MnSOD and FeSOD are the only types of SOD thus far noted. Amino acid sequence analyses have exposed an evolutionary relationship between the MnSOD and the FeSOD (13,14), while the CuZnSOD represents an independent line.

All of these varieties of SOD catalyze the same reaction, i.e.

$$O_2^{\bar{\cdot}} + O_2^{\bar{\cdot}} + 2H^+ \leftrightarrow H_2O_2 + O_2$$

and do so at a rate which is close to the diffusion limit. It has been possible to reversibly resolve all classes of SOD. The apoenzymes were, in all cases, devoid of catalytic activity and restoration of the metal prosthetic group was paralleled by regain of activity. Substitution of the native metal by other metals has been possible; however in each case the enzyme is active only when the catalytic center is occupied by the metal found in the native enzyme. Thus CuZnSOD is only active with Cu in the active site, MnSOD only with Mn and FeSOD only with Fe. It is instructive of the specificities of these active sites that MnSOD, substituted with Fe, is inactive (15) even though these enzymes show extensive amino acid sequence homologies (13,14). The Zn in the CuZnSOD plays some secondary role and it can be replaced by a variety of metals without compromising the activity of the copper site (16,17).

MECHANISM

The catalytic cycle of all of the SODs appears to involve alternate reduction and reoxidation of the active site metal, during successive encounters with $O_2^{\bar{\cdot}}$ (18-21). In the case of CuZnSOD the two half reactions are equally rapid. The structure of the active site of the bovine erythrocyte CuZnSOD is available from studies of X-ray crystallography (22,23). The two metals are bridged by an imidazolate ring and the other ligands to both Cu and Zn are known. There is an arginine residue in the solvent access channel to the Cu and its modification, by αβ-diketones, markedly decreases catalytic activity (24).

ACTIVITY STAINS

It is possible to localize SOD activity on polyacrylamide gel electropherograms. Two methods are currently available. One of these gives achromatic bands against a purple background (25), while the second gives brown bands against an unstained background (26). These activity stains can be applied to electropherograms prepared from crude extracts of cells. When this is done with a variety of extracts, a small number (sometimes only one) of activity bands are seen. In all cases, which have been thoroughly studied, each of the bands, thus visualized, corresponds with a specific SOD isozyme. In extracts of aerobically grown *E. coli* one sees a band due to FeSOD, another due to MnSOD and a third due to the hybrid composed of one subunit of each of these. Many bacteria show only one band of SOD, which is due to MnSOD in some cases and to FeSOD in others. Extracts of mammalian tissues ordinarily show a band due to the mitochondrial MnSOD and one or more bands due to the cytosolic CuZnSOD; which sometimes occurs as a family of closely related electrophoretic variants. In the case of chicken liver, there are four electrophoretically-distinct forms of the CuZnSOD and they are so closely related that they copurify (27).

The important point to be stressed here is that one sees SOD activity only at positions corresponding to isolatable and known types of SOD. Dialysis does not eliminate any of these bands. If there were low molecular weight metal complexes within cells capable of exerting significant SOD activity, they would be visualized on these electropherograms. Similarly, if other metalloproteins exhibited SOD activity, they would also be seen. We must conclude that the dismutation of $O_2^{\cdot-}$, within cells, is a specific function of the superoxide dismutases, rather than an activity common to many metal complexes or metalloproteins.

INDUCTIONS AND BIOLOGICAL SIGNIFICANCE

Circumstances which increase the intracellular production of $O_2^{\cdot-}$ have often been seen to cause increased synthesis of SOD. Such induction of an enzyme by its substrate is a common feature of metabolic control and enzymes which are under such control are called inducible. Other enzymes, whose synthesis appears to be independent of the presence of their substrate, are called constitutive. In *E. coli* the FeSOD is constitutive, while the MnSOD is inducible (28). Thus, anaerobically grown *E. coli* contain only the FeSOD and exposure to increasing levels of dioxygen calls forth progressively greater synthesis of MnSOD. In *E. coli*, increased total intracellular SOD correlates with increased resistance towards the lethality of high pressure oxygen. This was the case whether the intracellular SOD level was manipulated by aeration (28), by variation of growth rate in chemostat culture (29) or by changing the metabolism from primarily fermentative to primarily respiratory (30).

Particularly instructive correlations, between levels of SOD and resistance to the lethal effects of $O_2^{\cdot-}$, were achieved by artificially increasing the rate of intracellular production of $O_2^{\cdot-}$. This can be done with compounds which can enter cells and there be converted, by a diaphorase, to a reduced form capable of direct and rapid univalent reduction of dioxygen. Such compounds thus undergo a redox cycle which effectively shunts electron flow from the water-producing cytochrome pathway to an $O_2^{\cdot-}$-producing truncated pathway. In *E. coli*, methyl viologen, pyocyanine, juglone, plumbagin and streptonigrin, are among the compounds which can function in this way (31-34). One indication of the extent of diversion of electron flow by these compounds is the increase in cyanide-resistant respiration which they cause.

E. coli responds to the presence of such compounds by a profound increase in the rate of synthesis of MnSOD. If this were truly a response to intracellular $O_2^{\cdot-}$, then such inductions should be absolutely dependent upon both dioxygen and an electron source. This is the case. Thus, methyl viologen increases the production of SOD in *E. coli* only when both dioxygen and a rapidly metabolizable substrate are present. Increased synthesis of SOD, under such circumstances, is an adaptive response and provides protection against what would otherwise be a lethal stress. Thus, if *E. coli* are in a medium which will support rapid induction of SOD, they will easily tolerate mM methyl viologen. In contrast, if rapid induction is prevented, by puromycin or by removing amino acids from the medium, the cells will succumb to 0.01 mM methyl viologen. The toxic effect of methyl viologen is, in any case, entirely dependent upon both dioxygen and an electron source. It is clear that an increased intracellular production of $O_2^{\cdot-}$ can kill *E. coli* and that increased intracellular SOD can decrease this threat to cellular integrity.

Although most thoroughly studied in *E. coli*, inductions of SOD have been observed in a variety of cells including *Streptococcus faecalis* (35), *Photobacterium leiognathi* (36), *Saccharomyces cerevisiae* (37), rat lung (38,39) and potato (40). An interesting induction was seen in the fungus *Dactylium dendroides* which, when deprived of Cu and unable to synthesize CuZnSOD maintained its net level of SOD by a compensatory increase in MnSOD (41).

THE DANGERS OF $O_2^{\cdot-}$

That an intracellular flux of $O_2^{\cdot-}$ can be deleterious and even lethal and that SOD protects, has already been clearly shown. We do not, however, know the mechanisms of the damaging actions of $O_2^{\cdot-}$. In aqueous buffers, $O_2^{\cdot-}$ has been seen to oxidize catechols, catecholamines, thiols, ascorbate and pamoate. Its rate of reaction with a given molecule can be greatly influenced by the microenvironment of the target. Thus, $O_2^{\cdot-}$ does not react, at a significant rate, with NADH in free solution, but NADH, which is bound to lactic dehydrogenase, is very reactive (42). When tested against a variety of small molecules, including metabolic intermediates and common buffers, $O_2^{\overline{\cdot}}$ was found unreactive (43), yet it rapidly oxidized the 1,2-dihydroxyethylthiamine pyrophosphate intermediate of the transketolase reaction (44). It is not unlikely that living systems, in their enormous complexity, contain multiple critical targets for $O_2^{\overline{\cdot}}$. In the main, however, they are yet to be identified. One target of importance has been noted. Thus, normal human plasma contains a component, which reacts with $O_2^{\overline{\cdot}}$ to generate a powerful neutrophil chemotoxin, which appears to play a role in the inflammatory process (45).

THE HABER-WEISS REACTION

An interaction of $O_2^{\overline{\cdot}}$ with H_2O_2, which generated a powerful oxidant, was noted a decade ago (46). In this particular case the xanthine oxidase reaction, which generates both $O_2^{\overline{\cdot}}$ and H_2O_2, was seen to cause the co-oxidation of methional to ethylene. SOD prevented this co-oxidation, implicating $O_2^{\overline{\cdot}}$ and catalase did likewise, implicating H_2O_2. Since low levels of either SOD or catalase could inhibit ethylene production nearly completely and since heat-inactivated controls had no effect, it was clear that both $O_2^{\overline{\cdot}}$ and H_2O_2 were required for the co-oxidation of methional. The need for H_2O_2 was independently shown by an exponential increase in ethylene production, during the first minutes of the reaction, due to the accumulation of H_2O_2. This lag in ethylene production was eliminated by exogenous H_2O_2, which had no effect on methional in the absence of the enzymic flux of $O_2^{\overline{\cdot}}$.

It appeared reasonable that the oxidant capable of attacking methional had been generated by an interaction of $O_2^{\overline{\cdot}}$ with H_2O_2. Since Haber and Weiss had invoked such an interaction in their explanation of the catalytic decomposition of H_2O_2 by iron salts (47), we used their proposal in explaining our results (46). It was thus supposed that $O_2^{\overline{\cdot}}$ had reduced H_2O_2 to OH^- plus $\underline{OH}\cdot$ and that $OH\cdot$ might have been the species directly attacking methional. In that case compounds capable of scavenging $OH\cdot$, but unreactive towards either $O_2^{\overline{\cdot}}$ or H_2O_2 should also have inhibited the co-oxidation of methional. Ethanol and benzoate have the required properties and did indeed inhibit ethylene production. The potent oxidant, generated from $O_2^{\overline{\cdot}}$ plus H_2O_2, was also detected in terms of its ability to oxidize ferrocytochrome _c_ and once again it was intercepted by ethanol or benzoate and its production was prevented by SOD or catalase (46).

The production of $OH\cdot$, or something of comparable reactivity, from $O_2^{\overline{\cdot}} + H_2O_2$, has been repeatedly observed. Enzymic or cellular sources of $O_2^{\overline{\cdot}}$ plus H_2O_2 have been seen to depolymerize hyaluronate (48), oxidize linolenate (49), lyse erythrocytes (50) and vesicles formed from washed erythrocyte stroma (51) and kill bacteria (52) and neutrophils (53). $O_2^{\overline{\cdot}}$ and H_2O_2 produced by activated granulocytes have been seen to release ethylene from methional (54), methane from DMSO (55) and CO_2 from benzoate (56). Yet, the direct reduction of H_2O_2 by $O_2^{\overline{\cdot}}$ appears to be a very slow reaction (57,58). Catalysis by iron appears to provide the answer to this apparent paradox. The observations which support the production of OH· from $O_2^{\overline{\cdot}}$ plus H_2O_2 were made with reaction mixtures certainly contaminated with μmolar levels of iron, whereas direct attempts to measure the rate of the Haber-Weiss reaction were not. This catalytic effect of iron has been demonstrated (59-64). EDTA-iron was an effective catalyst but diethylenetriaminepentacetic acid-iron was not.

Several mechanisms can be proposed for the iron catalysis of the Haber-Weiss reaction, but the data now available do not allow a clear choice of the most likely. From the biological point of view, the important question is whether an iron-catalyzed Haber-Weiss reaction has reality within living systems and that question cannot yet be answered.

In summary, it is clear that $O_2^{\overline{\cdot}}$ presents a serious threat to living cells and that superoxide dismutases evolved to provide a defense. In contrast, the mechanisms by which $O_2^{\overline{\cdot}}$ exerts its deleterious effects are not yet known.

REFERENCES

1. Fridovich, I., Accounts Chem. Res., 5, 321 (1972).
2. Fridovich, I., Advan. Enzymol., 41, 35 (1974).
3. Fridovich, I., Ann. Rev. Biochem., 44, 147 (1975).
4. Halliwell, B., Cell Biol. Int. Res., 2, 113 (1978).
5. Fridovich, I., Science, 201, 875 (1978).
6. McCord, J. M. and Fridovich, I., J. Biol. Chem., 243, 5753 (1968).
7. Knowles, P. F., Gibson, J. F., Pick, F. M. and Bray, R. C., Biochem. J., 111, 53 (1969).
8. McCord, J. M. and Fridovich, I., Fed. Proc., 28, 346 (1968).
9. McCord, J. M. and Fridovich, I., J. Biol. Chem., 244, 6049 (1969).
10. Mann, T. and Keilin, D., Proc. Roy. Soc. Ser B, 126, 303 (1939).
11. Keele, B. B., Jr., McCord, J. M. and Fridovich, I., J. Biol. Chem., 245, 6176 (1970).
12. Yost, F. J., Jr. and Fridovich, I., J. Biol. Chem., 248, 4905 (1973).
13. Harris, J. I. and Steinman, H. M., in "Superoxide and Superoxide Dismutases (Proc. EMBO Workshop) 1976," (A. M. Michelson, J. M. McCord, and I. Fridovich), eds.) Academic Press, London, 1977, p. 225.
14. Steinman, H. M., J. Biol. Chem., 253, 8708 (1978).
15. Brock, C. J. and Harris, J. I., Biochem. Soc. Trans., 5, 1537 (1977).

16. Forman, H. J. and Fridovich, I., J. Biol. Chem., 248, 2645 (1973).
17. Beem, K. M., Rich, W. E. and Rajagopalan, K. V., J. Biol. Chem., 249, 7298 (1974).
18. Rotilio, G., Bray, R. C. and Fielden, E. M., Biochim. Biophys. Acta, 268, 605 (1972).
19. Klug, D., Fridovich, I. and Rabani, J., J. Amer. Chem. Soc., 95, 2786 (1973).
20. Pick, M., Rabani, J., Yost, F. and Fridovich, I., J. Amer. Chem. Soc., 96, 7329 (1974).
21. Lavelle, F., McAdam, M. E., Fielden, E. M., Roberts, P. B., Puget, K. and Michelson, A. M., Biochem. J., 161, 3 (1977)
22. Richardson, J. S., Thomas, K. A. and Richardson, D. C., Biochem. Biophys. Res. Commun., 63, 986 (1975).
23. Richardson, J. S., Thomas, K. A., Rubin, B. H. and Richardson, D. C., Proc. Natl. Acad. Sci. U.S.A., 72, 1349 (1975).
24. Malinowski, D. P. and Fridovich, I., Biochemistry, 18, 5909 (1979).
25. Beauchamp, C. and Fridovich, I., Anal. Biochem., 44, 276 (1971).
26. Misra, H. P. and Fridovich, I., Arch. Biochem. Biophys., 183, 511 (1977).
27. Weisiger, R. A. and Fridovich, I., J. Biol. Chem., 248, 3582 (1973).
28. Hassan, H. M. and Fridovich, I., J. Bacteriol., 129, 1574 (1977).
29. Hassan, H. M. and Fridovich, I., J. Bacteriol., 130, 805 (1977).
30. Hassan, H. M. and Fridovich, I., J. Bacteriol. 132, 505 (1977).
31. Hassan, H. M. and Fridovich, I., J. Biol. Chem. 252, 7667 (1977).
32. Hassan, H. M. and Fridovich, I., J. Biol. Chem. 253, 8143 (1978).
33. Hassan, H. M. and Fridovich, I., Arch. Biochem. Biophys., 196, 385 (1979).
34. Hassan, H. M. and Fridovich, I., J. Bacteriol., 141, 156 (1980).
35. Gregory, E. M. and Fridovich, I., J. Bacteriol. 114, 543 (1973).
36. Puget, K. and Michelson, A. M., Biochem. Biophys. Res. Commun., 58, 830 (1974).
37. Gregory, E. M., Goscin, S. A. and Fridovich, I., J. Bacteriol., 117, 456 (1974).
38. Crapo, J. D. and Tierney, D. L., Amer. J. Physiol., 226, 1401 (1974).
39. Stevens, J. B. and Autor, A. P., J. Biol. Chem., 252, 3509 (1977).
40. Boveris, A., Sanchez, R. A. and Beconi, M. T., FEBS Lett., 92, 333 (1978).

41. Schatzman, A. R. and Kosman, D., Biochim. Biophys. Acta, 544, 163 (1978).
42. Bielski, B. H. J. and Chan, P. C., J. Biol. Chem., 251, 3841 (1976).
43. Bielski, B. H. J. and Richter, H. W., J. Amer. Chem. Soc., 99, 3019 (1977).
44. Asami, S. and Akazawa, T., Biochemistry, 16, 2201 (1977).
45. McCord, J. M., Fed. Proc., 39, 566 (1980).
46. Beauchamp, C. and Fridovich, I., J. Biol. Chem., 245, 4641 (1970).
47. Haber, F. and Weiss, J., Proc. Roy. Soc., London A 147, 332 (1934).
48. McCord, J. M., Science, 185, 529 (1974).
49. Kellogg, E. W., III and Fridovich, I., J. Biol. Chem., 250, 8812 (1975).
50. Kellogg, E. W., III and Fridovich, I., J. Biol. Chem. 252, 6721 (1977).
51. Lynch, R. E. and Fridovich, I., J. Biol. Chem. 253, 1838 (1978).
52. Babior, B. M., Curnutte, J. T. and Kipnes, R. S., J. Lab. Clin. Med., 85, 235 (1975).
53. Salin, M. L. and McCord, J. M., J. Clin. Invest., 56, 1319 (1975).
54. Weiss, S. J., Rustagi, P. K. and LoBuglio, A. F., J. Exp. Med. 147, 316 (1978).
55. Repine, J. E., Eaton, J. A., Anders, M. W., Hoidal, J. R. and Fox, R. B., J. Clin. Invest., 64, 1642 (1979).
56. Sagone, A. L., Decker, M. A., Wells, R. M. and DeMocko, C., Biochim. Biophys. Acta, 628, 90 (1980).
57. Ferradini, C. and Seide, C., Internat. J. Radiat. Phys. Chem., 1, 219 (1969).
58. Ferradini, C., Foos, J., Hovee, C. and Pucheault, J., Photochem. Photobiol., 28, 697 (1978).
59. Fong, K.-L., McCay, P. B., Poyer, J. L., Misra, H. P. and Keele, B. B., Chem.-Biol. Interactions, 15, 77 (1976).
60. McCord, J. M. and Day, E. D., Jr., FEBS Lett., 86, 139 (1978).
61. Halliwell, B., FEBS Lett., 92, 321 (1978).
62. Buettner, G. R., Oberley, L. W. and Chan-Leuthauser, S. W. H., Photochem. Photobiol., 28, 693 (1978).
63. Halliwell, B., FEBS Lett., 96, 238 (1978).
64. Gutteridge, J. M. C., Richmond, B. and Halliwell, B., Biochem. J., 184, 469 (1979).

IS SUPEROXIDE TOXIC AND ARE SUPEROXIDE DISMUTASES ESSENTIAL FOR AEROBIC LIFE?

James A. Fee, Biophysics Research Division and Department of Biological Chemistry, The University of Michigan, Ann Arbor, Michigan 48109

It is the purpose of this article to show reasonable cause for negative replies to the two questions expressed in the title.

IS SUPEROXIDE TOXIC?

An examination of the early literature of radiobiology reveals that this question has been discussed since the early 1950's. In 1952, Alexander and Fox (1) proposed that superoxide was the responsible factor in the x-ray induced depolymerization of polymethacrylic acid. In subsequent papers Alexander and his co-workers (2-6) emphasized the importance of $HO_2(O_2^{\overline{\cdot}})$ in effecting radiation damage of both model and biological systems. However, Alexander's view that HO_2 was more reactive than OH· was inconsistent with chemical facts, and the idea that superoxide played a significant role in oxygen enhanced radiation damage fell into disfavor (7).[1] Indeed, reviews of the mechanisms of radiation damage to biological systems either do not mention superoxide or allude to its unreactivity (8-11).

---

[1]*It is ironic that Joseph Weiss' name has been so closely tied with recent attempts (66) to confer toxicity on superoxide: Weiss made the following statement at the Colloquium on Organic Peroxides Formed by Radiations and Their Role in Radiobiology, Paris, January 9 and 10, 1957 (7).*

"...it is a great pity that two or three years ago there was a tendency, by certain quarters, to make the HO_2 responsible for almost everything which happens in radiobiology. If we look at the reactions of HO_2 in vitro, it is obvious that this radical is relatively unreactive towards saturated organic systems, where it does not do anything or only very rarely, but, on the other hand, it can act as an oxidising agent towards strongly reducing substances such as ferrous ions and as a reducing agent, in the form of its anion $O_2^{\overline{\cdot}}$, towards e.g. ceric salts. However, there seems to be no justification to explain almost everything in radiobiology by the action of HO_2, as was done particularly by Alexander. It is unfortunate that so much time still has to be wasted now to point out that these views were really of very little consequence."

ISBN 0-12-592050-4

Superoxide was discussed very little by radiobiologists during the 60's and 70's, but a great deal of work was done on its physical and chemical properties, and it can now be considered to be a chemically well defined species. The properties of superoxide have been extensively reviewed (12-14) and mention will be made here only of its benign chemical properties in aqueous media.

Superoxide, like fluoride, is very strongly hydrated by water molecules and has an ionic size comparable to chloride. It is a facile one-electron reductant possessing a standard redox potential of -0.33 V for the half-reaction, $O_2 + e^- \rightleftarrows O_2^{\overline{\cdot}}$. It is a strong oxidant from the thermodynamic view but cannot accept an electron directly because of the instability of O_2^{2-} (13). Therefore its capacity to oxidize organic molecules is greatly restricted. It can, however, be an effective one-electron oxidant of metal ions where reaction presumably proceeds by way of a metal-superoxo complex. The types of substances with which aqueous superoxide has been shown by direct demonstration to react are listed in Table I, while Table II indicates substances with which superoxide has been shown, by the same criterion, not to react. No convincing evidence has been obtained for nucleophilic displacement or addition to non-activated double bonds,[2] or for H-atom abstraction (13). Of singular importance here is that no reaction of aqueous superoxide has yet been discovered which could reasonably by considered deleterious to a cell.

However, since the discovery of the superoxide dismutase activity of erythrocuprein (16), superoxide has been indirectly implicated in processes which lead to the destruction of biological molecules and indeed even to the killing of various types of cells. These manifestations of superoxide toxicity are listed in Table III. Examination of Tables II and III should convince the reader that it is not superoxide itself which is responsible for the degradative processes, but there is no doubt that scavenging superoxide by means of added superoxide dismutase interferes with the degradation of the substance under consideration.

Beauchamp and Fridovich (66) were the first to recognize the necessity of both H_2O_2 and $O_2^{\overline{\cdot}}$ for these degradative phenomena since both (or either) superoxide dismutase or catalase slowed or stopped the degradation. They proposed that superoxide reduced H_2O_2 to form a hydroxyl radical

$$O_2^{\overline{\cdot}} + H_2O_2 + H^+ \rightarrow O_2 + H_2O + OH\cdot \qquad [1]$$

and that OH· was responsible for the actual degradation. There are two problems with this explanation: first, Reaction [1] does

---

[2]*When stabilized in aprotic, weakly solvating solvents such as dimethyl sulfoxide, superoxide, like fluoride, is a powerful nucleophile. Indeed, its chemical behavior in such solvents is quite different from that observed in aqueous medium.*

TABLE I. Some Representative Reactions of Aqueous Superoxide[a]

As a reducing agent	*References*
$C(NO_2)_4 \rightarrow C(NO_2)_3^- + NO_2 + O_2$	12
Quinones → Semiquinones	15
Cytochrome $c^{III} \rightarrow$ *Cytochrome* c^{II}	16,17,18
Peroxidase Compound I → Compound II	19
Fe(III)-EDTA → Fe(II)-EDTA	20,21
Fe(III) → Fe(II)	12
$HClO + O_2^{\overline{\cdot}} \rightarrow OH\cdot + O_2 + Cl^-$	22
As an oxidizing agent	
Ascorbate	23
NADH - Lactate dehydrogenase	24
Fe(II)-EDTA	20
Metal-Complex formation	
General	13
$Mn^{2+} + O_2^{\overline{\cdot}} \rightarrow Mn^{2+}{-}O_2^{\overline{\cdot}}$	25
Addition to an activated double bond	
Nitrones + $O_2^{\overline{\cdot}} \rightarrow$ *hydroperoxynitroxides (spin trap)*	26

[a]*There are other examples in each class but these were selected because the experiments involved direct observation of superoxide rather than non-stoichiometric flux-type experiments.*

not occur; and second, in no case mentioned in Table III (Ref. 28 excepted, cf. Table II) was a serious attempt made to account for and control trace element contamination.

The present author has pointed out that the in vitro manifestations of superoxide toxicity (Table III) can all be rationalized with known superoxide chemistry (69,70). In the following scheme M is a metal ion and MO is a metal-oxygen complex having properties analogous to the hydroxyl radical.

$$M^{n+} + O_2^{\overline{\cdot}} \rightarrow M^{(n-1)+} + O_2 \quad [2]$$

$$M^{(n-1)+} + H_2O_2 \rightarrow MO(\equiv OH\cdot + M^{n+}) + H_2O \quad [3]$$

$$MO + RH \rightarrow R^{\cdot} + H_2O \quad [4]$$

$$R^{\cdot} \rightarrow \text{degradation products} \quad [5]$$

TABLE II. Substances with which Aqueous Superoxide has Been Shown Not to React[a]

Substance	*Condition or Type of Experiment*	*References*
ΦX - 174 DNA	*Radiolysis/Centrifugation*	27
Col E1 DNA	*Xanthine/Xanthine Oxidase/Electrophoresis*	28
Nuclear bases	*Gamma irradiation/Product analysis*	29
Poliovirus	*KO_2 dissolution/virulence*	30
Cholesterol	*Product analysis*	31
Formate, fumarate, α-ketogluterate, pyruvate, oxalate, imidazole, Tris, EDTA	*Combined pulsed radiolysis and stopped-flow/H_2O_2 and O_2 analyses*	32[b]
ClO_4^-, ClO_3^-, ClO_2^-	*Pulsed radiolysis*	22
H_2O_2	*KO_2 dissolution/O_2 analysis*	33
	Stopped flow kinetics	34
	H_2O_2 inhibition of NBT reduction by $O_2^{\cdot-}$	35
	Pulsed radiolysis	36
	Search for aromatic hydroxylation by $O_2^{\cdot-}$ and H_2O_2	37,38
ROOR[c]		39
ROOH[c]		40
Linolenic acid (dispersion)	*Stopped-flow kinetics*	*(Fee and McClune, unpublished)*

[a]*Representative only*
[b]*Several other related compounds were also found not to react with O_2^-.*
[c]*In acetonitrile*

Note that superoxide acts only as a one-electron reductant in this scheme, a role to which it is well suited. One is thus forced to the conclusion that any form of superoxide toxicity must be mediated by metal ions.[3]

The ramifications of this conclusion are best discussed in the light of Professor Fridovich's hypothesis that oxygen toxicity is mediated by superoxide. This theory which was presented

---

[3]*For a complete discussion of these ideas the reader is directed to References 69, 70, and 73.*

TABLE III. In vitro Degradative Processes in which Superoxide has been Shown to Participate[a]

Process	*References*
DNA Degradation/depolymerization	*28, 41-44*
Lipid Peroxidation	*45-52*
Depolymerization of polysaccharides	*53,54*[b]
Hydroxylation of aromatic substances	*55-60*
In vitro cell killing (bacteria and myoblasts)	*61-65*
Ethylene formation from methional[c]	*66-68*

[a]*Representative. The participation of superoxide has generally been ascertained by an inhibition of the overall process by superoxide dismutase.*
[b]*Contains no experimental data supporting superoxide involvement.*
[c]*Including compounds related to methional.*

under the title "An Enzyme Based Theory of Obligate Anaerobiosis: The Physiological Function of Superoxide Dismutase" (71) has three essential axioms:

(1) Superoxide is formed by biological systems.
(2) Superoxide is in some way toxic. - Metal
(3) Superoxide dismutases remove superoxide by catalyzing the reaction

$$2\ O_2^{\overline{\cdot}} + 2H^+ \rightarrow H_2O_2 + O_2 \qquad [6]$$

and are thus essential for any life form producing $O_2^{\overline{\cdot}}$.

A vast amount of evidence suggests that small, but not trivial, amounts of superoxide are formed by biological systems: living cells, flavoproteins, iron/sulfur proteins, and a variety of other common cellular components properly disposed react with oxygen to form superoxide. Thus, the first point of Fridovich's theory is firmly established.

Returning to the second point of the hypothesis, three reactions have been proposed as processes whereby a toxic species could be formed from superoxide, now generally recognized as chemically benign.

$$O_2^{\overline{\cdot}} + H_2O_2 + H^+ \rightarrow {}^3O_2 + H_2O + OH\cdot \qquad [1]$$

$$O_2^{\overline{\cdot}} + H_2O_2 + H^+ \rightarrow {}^1O_2({}^1\Delta g) + H_2O + OH\cdot \qquad [7]$$

$$2O_2^{\overline{\cdot}} + 2H^+ \rightarrow {}^1O_2({}^1\Delta g) + H_2O_2 \qquad [8]$$

As pointed out above, Reaction [1] does not occur probably because of the inability of H_2O_2 to accept an additional electron.

For the same reason, the analogous Reaction [7] (47), although not directly tested, probably does not occur. And, after considerable controversy reaction [8] has definitely been shown not to occur (72). We are thus left with metal ion catalysis of Reaction [1] (Reactions [2-3]) to generate a toxic species from $O_2^{\overline{\cdot}}$, and this would seem to adequately account for the in vitro observations of Table III as well to be an acceptable explanation for many workers who favor the superoxide hypothesis (73). However, a requirement of metal ions for oxygen toxicity represents a significant expansion of the original theory and the implications of this modification need to be examined.

According to this modified hypothesis, oxygen toxicity is due to Reactions [2-5] and is to be understood as a simple variant of Fenton chemistry, previously dubbed superoxide mediated Fenton chemistry (69,70). Two important questions arise:

(1) Can $O_2^{\overline{\cdot}}$ effectively compete with all the other one-electron reductants, ΣA^-, in a cell to regenerate the reduced metal ion?

$$\Sigma A^- + M_n \rightarrow \Sigma A + M^{(n-1)+} \qquad [9]$$

(2) Are trace metals within a cell so disposed that Reactions [2-4] can occur to such an extent so as to account for the observables of oxygen toxicity?

Concerning the first of these questions, several in vitro tests of the capacity of physiological reductants such as ascorbic acid and glutathione to compete with superoxide in Reactions [2-5] have been made and are reviewed in Refs. 69 and 70. It is clear that ascorbic acid at levels well below those found in some tissues can effectively out-compete $O_2^{\overline{\cdot}}$ being formed by the action of xanthine oxidase. For example, in the absence of ascorbic acid, Winterbourn (74) showed ethylene production from methional was inhibited by superoxide dismutase while the presence of as little as 50 μM ascorbic acid obviated the ability of dismutase to retard production of ethylene. It would thus seem unlikely that $O_2^{\overline{\cdot}}$ could effectively compete against reductants normally present in viable cells and tissues.

The second question is impossible to answer directly, but some reasonable conjecture is warranted. Reaction [4] has been harnessed by cells (catalase, peroxidase, cytochrome P-450)[4] and therefore it would be likely that the cell has also "learned" to control this process so as to avoid its unnecessary occurrence. In addition, one should recall the exacting control of metal ion distribution exhibited by biological systems. For example, the difficulty with which bacteria colonize avian eggs has been attributed in part to their inability to extract iron from conalbumin (75).

In summary, there appears to be no reasonable argument for

---

[4]*Porphyrin associated* Fe^{3+}, *for example, need not be reduced prior to reaction with peroxide to form a reactive MO complex.*

superoxide toxicity and failure of researchers to find any supporting evidence over a period of approximately 50 years would seem to make the answer to the first question a resounding, No!

WHAT ARE SUPEROXIDE DISMUTASES AND ARE THEY NECESSARY FOR AEROBIC LIFE?

Superoxide dismutases catalyze Reaction [6], and there are three classes of such proteins based on the metal ion cofactor necessary for catalytic activity (Table IV). The Zn/Cu protein is found principally in eucaryotic systems, the iron protein in procaryotes, and the manganese protein in the mitochondria of eucaryotes; there are exceptions to each of these statements. Eucaryotic cells thus contain the Zn/Cu- and Mn- superoxide dismutases which appear to exist in different cell compartments; E. coli has both iron and manganese proteins present in approximately equal amounts in the cytosol (76); Paracoccus denitrificans (77) and Photobacterium leigonathi (78,79) possess both iron and Zn/Cu dismutases.

The formulation of the theory of superoxide mediated oxygen toxicity was largely predicated on:

(1) The existence of proteins having superoxide dismutase activity.

(2) A survey of 26 microorganisms in which relatively high dismutase activity was found in strict and facultative aerobes, lesser amounts in aerotolerent organisms, and none in strictly anaerobic species (71). And, further, indirect support for the theory has been taken from the additional observations.

TABLE IV. General Properties of Superoxide Dismutases

Essential Co-factor	*Molecular Weight*	*Number[a] Subunits*	*Metals/ Subunit[b]*	*Source*
Cu[c]	*32,000*	*2*	*1 Zn, 1 Cu*	*eucaryotic cells*
Fe	*38,000*	*2*	*1 Fe*	*procaryotic cells*
Mn	*40,000/80,000*	*2/4*	*1 Mn*	*pro- and eucaryotic cells*

[a]*Subunits appear to be identical.*

[b]*Idealized, generally fractions of a metal ion per subunit are found in purified preparations. It is not known whether this is due to incomplete occupancy of the metal binding site or to loss during purification.*

[c]*The function of Zn in this protein is not known, but it is not necessary for superoxide dismutase activity.*

(3) Increased oxygen tension resulted in an induction of the manganese containing dismutase in *E. coli*, and it is now established that both the presence of the activity (80) and the synthesis of the protein (81) are absolutely dependent on the presence of oxygen.

(4) Certain redox dyes such as methylviologen effect a dramatic induction of the manganese dismutase in *E. coli* and it has been presumed that this induction is due to the reactions

$$\text{Cellular reducing agents} + MV^{2+} \rightarrow MV^{+} \quad [10]$$

$$MV^{+} + O_2 \rightarrow O_2^{\overline{\cdot}} + MV^{2+} \quad [11]$$

which increase the concentration of $O_2\cdot$ within the cell (82-86).

(5) Cells so adapted, either by raising the pO_2 or treatment with methylviologen, are much more resistant to the bactericidal effect of a combined treatment with hyperbaric oxygen and interference in protein synthesis as with puromycin or chloramphenicol.

(6) Certain mutants lacking manganese dismutase have been shown to be unusually sensitive to oxygen poisoning (87).

Taken alone and at face value the above observations, all emanating from Prof. Fridovich's laboratory, constitute a fairly strong suggestion that superoxide dismutase is a pre-requisite of aerobic life. However, these observations constitute only indirect evidence, and conflicting results have now begun to appear from other laboratories (see below).

Of critical importance to the arguments presented here are the recent findings of two quite different, strictly aerobic, oxygen requiring and consuming microorganisms which have been shown to be totally lacking in superoxide dismutase (88,89). *Neisseria gonorrhoeae* and related strains (Table V) are strictly aerobic, oxidase positive organisms. It can be seen that several of the pathogenic strains completely lack superoxide dismutase activity as assessed by normal assays while the nonpathogenic strains have normal though variable levels. An interesting fact here is that the pathogenic strains require many more substances for growth in a chemically defined medium than the non-pathogenic strain (90,91). A similar situation exists in the case of the Mycoplasma (Table VI). All are strict aerobes and consume oxygen but those lacking dismutase require a more complex growth medium than those possessing this activity (89). Some mention should also be made of the lactic acid bacterium, *Lactobacillus plantarum*. McCord *et al*., (71) found this aerotolerant organism to have no superoxide dismutase but claimed that it consumed no oxygen so therefore had no need for dismutase activity. Elstner and coworkers (92,93) re-examined this organism, confirmed that it had no dismutase activity but in contrast to the report of McCord *et al* (71) observed that, under their conditions, it actively consumed O_2 and formed H_2O_2. These findings refute the general concept that superoxide dismutase activity is necessary for aerobic

TABLE V. Absence of Superoxide Dismutase in Some Strictly Aerobic Strains of Niesseria Gonorrhoeae[a]

Pathogenic strains[b]	*Dismutase activity*
CS-7	1.5
JW-31	0.5
WP (T4)	0.5
WP (T2)	0.4
F62 (T4)	0.4
F62 (T2)	0.4
2686 (T4)	0
2686 (T2)	0
1384	0
2024	0
Non-pathogenic Neisseria species	
N. subflava	2
N. mucosa	22
N. lactamica	18
N. flavesens	22
N. flava	18
N. perflava	5

[a]*Taken from Ref. 88.*
[b]*All strains had catalase: ∿4 nmoles O_2/mg protein-min.*

life and strongly support a continued survey of microbial species for possible clues to the molecular differences among aerobic organisms which require these proteins and those which do not.

The absence of superoxide dismutase in aerobic organisms is not really surprising in view of the lack of superoxide toxicity. However, it allows the second question of the title to also be answered in the negative.

AN ALTERNATIVE HYPOTHESIS

We are now faced with finding alternative explanations for the observations enumerated in the above section which have previously been taken as supportive of the superoxide mediated theory of oxygen toxicity. The following alternative hypothesis is suggested:

The superoxide dismutase activity possessed by the various metallo-proteins presently called superoxide dismutases is a trivial, artifactual one, and it does not represent the true biological function of these gene products.

TABLE VI. *Activities of Superoxide Dismutase and of Catalase in Extracts of Mycoplasma Species and of Acholeplasma Laidlawii*[a]

Species	Superoxide Dismutase Assayed in:			Catalase Assayed in:		
	Solution		Gels	Solution		Gels
	Specific Activity units/mg	μg Protein Assayed		Specific Activity Units/mg	μg Protein Assayed	
M. hominis	0.04*	1880	-	0	940	-
M. pulmonis	0.1*	740	-	0	370	-
M. arthritidis	0.02*	1780	-	0	890	-
M. gallisepticum	0.1	1500	-	0	750	-
M. fermentans	0.2	1460	-	0	730	-
M. synoviae	0.01*	500	-	0	250	-
A. laidlawii						
Strain M728	32.8	38	+	0	1900	-
Strain 1305	32.9	32		0.004	1580	
M. pneumoniae						
Strain Jose	0, 0.01*	540-730	-	0	270-730	-
Strain TW 48-5	0*	200	-	0	100	
Strain 1428	0*	478		0	236	

[a]Taken from reference 89.

*Values marked with an asterisk were not significantly different from the uninhibited rate and should probably be considered zero.

The first question is whether it is possible that the superoxide dismutase activity is artifactual. As discussed elsewhere (94,95) the structure-function requirements for metal ion catalysis of superoxide dismutation are very limited with no apparent requirement for protein mediated events. Moreover, a few small complexes of Cu, Fe, and Mn have been found to have dismutase activity which ranges from approximately four times more than the proteins, $Cu^{2+}aq$, to extremely weak. Thus, artifactual activity is a possibility (Cf. Ref. 95 for a fuller discussion of mechanistic details).

As regards the probability that the dismutase activity of the proteins described in Table IV is artifactual, some support comes from the presence of dismutase activity in strictly anaerobic organisms. More extensive surveys for dismutase activity than that conducted by McCord et al., (71) have revealed the presence of superoxide dismutase in a wide variety of strictly anaerobic organisms at levels ranging from those similar to anaerobically grown E. coli to undetectable (96-101). Asada and others (102) have demonstrated that the dismutase activity of these anaerobes is due to a protein having physical properties very similar to the iron containing dismutase isolated from E. coli. Some interesting "explanations" have been put forth for the presence of dismutase activity in a variety of strict, aerointolerant organisms: protection against transient exposure to oxygen (97), protection against superoxide formed by cosmic radiation (103), and use as a virulence factor by pathogenic anaerobes (97); none of these appears very reasonable. The presence of superoxide dismutase in organisms grown in the absence of oxygen is not readily explained in terms of the normal efficiency of living systems to limit the production of unnecessary enzymes (104), and cannot but suggest the possibility that this protein may have some other, essential function for the anaerobic cell.

Turning now to the manganese protein, observations, 3, 4, and 5 above: In essence, Fridovich and co-workers have shown that mild hyperoxia or treatment with the representative redox dye, methylviologen, induce the manganese protein many-fold. Cells thus treated have been claimed to be highly resistant to killing by oxygen or by the enhancement of oxygen toxicity by Reactions [10,11], and this resistance has been assigned to the increased levels of manganese dismutase. These conclusions are misleading, partly because Hassan and Fridovich (82,83,85) fail to recognize some important, well-documented facts in the interpretation of their observations:

(1) Neither treatment with hyperbaric oxygen (105) nor with aerobic methylviologan (Cf. 84,110) is bactericidal—they are purely bacteristatic! It is only when the cells are treated (82, 83,85) with an additional, bacteristatic agent which interferes with protein synthesis, e.g., puromycin or chloramphenicol, that cell killing occurs and the adapted cells exhibit resistance to the bactericidal mixture. Since the two bacteristatic agents act

synergistically to produce a bactericidal condition, any simple interpretation would seem to be precluded.

(2) It is not simply the levels of dismutase activity which determine the resistance to bactericidal, puromycin/oxygen mixtures. Thus, Simons et al. (106), showed that when the dismutase levels of E. coli were varied a factor of ∿4 by manipulating the concentration of Fe in the culture medium, there was no difference in the resistance of the organisms to a bactericidal mixture of puromycin, methylviologen, and air.

This latter result strongly implies that an adaptation is occurring upon treatment with mild hyperoxia or methylviologen and air. Indeed, a survey of most of the proteins of E. coli by a quantitative two-dimensional technique (107) has revealed that treatment with aerobic methylviologen results in a dramatic change in the levels of many cellular proteins (81). Very recently, Seither and Brown (108) have found that bacteristatic hyperoxia invokes stringency (109). This phenomenon, which is under the control of the rel genes, is known to be a pleitropic response, i.e., the whole of cellular metabolism is subject to dramatic change. Moreover, it was also shown that rel A^- cells were unable to adapt to hyperoxia in this fashion and for these cells oxygen poisoning led to death.[5] While it is not yet known whether manganese dismutase is under the control of the rel genes and would therefore be a part of the stringency response, experiments to test this are in progress. Other studies by Brown and co-workers have shown that oxygen poisoning includes specific and irreversible inhibition of certain biosynthetic enzymes (111).

In summary, treatment with mild hyperoxia or air and certain redox dyes induces the stringent response and manganese dismutase. It would appear that resistance to bactericidal mixtures of puromycin and oxygen is related to the adaptive stringent response of which dismutase may be a part, but resistance is not conferred only by increased dismutase levels. Indeed, to assign such resistance to a single protein would appear rationally impossible in the face of the pleiotropic response of the cell.

Finally, the various mutants described by Hassan and Fridovish (87) strongly suggest an important, perhaps vital, role for the manganese protein in the aerobic life of E. coli, but they offer no evidence that superoxide dismutation is that essential function. It is important to note that the amount of dismutase activity in these mutant organisms was decreased at most by a factor of 2-3 (in some cases less) due to the presence of unchanged levels of the iron superoxide dismutase. It is improbable that reducing the amount of any enzyme in the cell by

---

[5] *Sokawa and Sokawa (110) have shown that chloramphenicol can interfere in the onset of the stringency response; perhaps puromycin functions in this manner.*

these amounts would be fatal.

CONCLUSIONS

Reasonable arguments have been made that superoxide does not mediate oxygen toxicity and that superoxide dismutase activity is not a prerequisite of aerobic life. It has been suggested that all available facts present a more coherent picture if the proteins presently termed superoxide dismutases actually have biological functions unrelated to superoxide dismutation.

ACKNOWLEDGEMENTS

GO TO 277 -

The author wishes to thank Joan Valentine, Gregory McClune, and John Groves for helpful discussions of the chemistry of iron/oxygen systems and Phillip Bloch, Frederick Niedhardt, and Olen Brown for aid in understanding the complexities of microbial physiology. Particular thanks go to Olen Brown for providing manuscripts prior to publication.

This work was supported by a grant from the United States Public Health Service, GM 21519.

REFERENCES

1. Alexander, P. and Fox, M., Nature, 169, 572 (1952).
2. Alexander, P., British J. Radiol., 26, 413 (1953).
3. Alexander, P., Fox, M. and Hitch, S. F., British J. Radiol., 27, 130 (1954).
4. Alexander, P., Bacq, Z. M., Cousens, S. F., Fox, M., Herve and Lazar, J., Radiat. Res., 2, 392 (1955).
5. Alexander, P., "Radiobiology Symposium 1954," (Z. M. Bacq and P. Alexander , eds.), Academic Press, New York, 1955, p. 49.
6. Alexander, P., Radiat. Res., 6, 653 (1957)
7. Latarjet, R., "Organic Peroxides in Radiobiology," Pergamon Press, London (1958).
8. Nakken, K. F., "Current Topics in Radiation Research", (M. Ebert and A. Howard, eds.), North Holland, Amsterdam (1965) 1, 49.
9. Adams, G. E., Ibid, (1967) 3, 35.
10. Teoule, R. and Cadet, J., Extrait du Bulletin d'Information Scientifiques et Techniques du Commissariat à l'Energie Atomique No. 145, p. 1970.
11. Czapski, G., Ann. Rev. Phys. Chem., 22, 171 (1971).
12. Bielski, B. H. J. and Gebicki, J. M., Adv. Radiat. Chem.,2, 177 (1970).
13. Fee, J. A. and Valentine, J. S., "Superoxide and Superoxide Dismutases", (A. M. Michelson, J. M. McCord and I. Fridovich, eds.), Academic Press, London (1977), p. 19.
14. LeeRuff, E. Chem. Soc. Revs., 6, 195 (1977).
15. Patel, K. B. and Willson, R. L., J. C. S. Faraday 1,814(1973).

16. McCord, J. M. and Fridovich, I., J. Biol. Chem., 244, 6049 (1969).
17. Shafferman, A. and Stein, G., Biochim. Biophys. Acta, 416, 287 (1975).
18. Simic, M. G., Taub, I. A., Tocci, J. and Hurwitz, P. A., Biochem. Biophys. Res. Commun., 62, 161 (1975).
19. Bielski, B. H. J. and Gebicki, J. M., Biochim. Biophys. Acta, 364, 233 (1974).
20. McClune, G. J., Fee, J. A., McClusky, G. A. and Groves, J. T., J. Am. Chem. Soc., 99, 5220 (1977).
21. Ilan, Y. A. and Czapski, G., Biochim. Biophys. Acta, 498, 386 (1977).
22. Long, C. A. and Bielski, B. H. J., J. Phys. Chem., 84, 557 (1980).
23. Bielski, B. H. J. and Richter, H. W., J. Am. Chem. Soc., 99, 3019 (1977).
24. Bielski, B. H. J. and Chan, P. C., J. Biol. Chem., 251, 3841 (1976).
25. Bielski, B. H. J. and Chan, P. C., J. Am. Chem. Soc., 100, 1920 (1978).
26. Harbour, J. R., Chow, V. and Bolton, J. R., Can. J. Chem., 52, 3549 (1974).
27. Blok, J., "Radiation Research", (G. Silini, ed.), North Holland, Amsterdam, p. 423 (1967).
28. Brawn, K. and Fridovich, I., Fed. Proc., 39, Abs. No. 1945 (1980).
29. Cadet, J. and Teoule, R., Photochem. Photobiol., 28, 661 (1978).
30. Pelouz, Y., Nofre, C., Cier, A. and Colobert, L., L'Inst. Pasteur Anal., 102, 6 (1962).
31. Smith, L. L., Kulig, M. J. and Teng, J. I., Chem. Phys. Lipids, 20, 221 (1977).
32. Bielski, B. H. J. and Richter, H. W., J. Am. Chem. Soc., 99, 3019 (1977).
33. George, P., Disc. Faraday Soc., 2, 196 and 219 (1947).
34. McClune, G. J. and Fee, J. A., FEBS Letters, 67, 294 (1976).
35. Halliwell, B., FEBS Letters, 72, 8 (1976).
36. Czapski, G. and Ilan, Y. A., Photochem. Photobiol. 28, 657 (1978).
37. Fee, J. A. and Hildenbrand, P. G., FEBS Letters, 39, 79 (1974).
38. Rigo, A., Stevanato, R., Finazzi-Agro, A. and Rotilio, G., FEBS Letters, 80, 130 (1977).
39. Peters, J. W. and Foote, C. S., J. Am. Chem. Soc., 98, 873 (1976).
40. Gibian, M. J. and Ungermann, T., J. Am. Chem. Soc., 101, 1290 (1979).
41. Van Hemmen, J. J. and Meuling, W. J. A., Biochim. Biophys. Acta, 402, 133 (1975).

42. White, J. R., Vaughn, T. O. and Yeh, W. S., Fed. Proc. Abst. #537 (1971).
43. Lorentzen, R. J. and Tso, P. O. P., Biochemistry, 16, 1467 (1977).
44. Cone, R., Hasan, S. K., Lown, J. W. and Morgan, A. R., Can. J. Biochem., 54, 219 (1976).
45. Fee, J. A. and Tietelbaum, H. D., Biochem. Biophys. Res. Commun., 49, 150 (1972).
46. Lynch, R. E. and Fridovich, I., J. Biol. Chem., 253, 1838 (1978).
47. Kellog, E. W., III and Fridovich, I., J. Biol. Chem., 250, 8812 (1975).
48. Tyler, D. D., Biochim. Biophys. Acta, 396, 335 (1975).
49. Pederson, T. C. and Aust, S. D., Biochem. Biophys. Res. Commun., 48, 789 (1972).
50. Bus, J. S., Aust, S. D. and Gibson, J. E., Biochem. Biophys. Res. Commun., 58, 749 (1974).
51. Tyler, D. D., FEBS Letters, 51, 180 (1975).
52. Noguchi, T. and Nakano, M., Biochem. Biophys. Acta, 368, 446 (1974).
53. McCord, J. M., Science, 185, 529 (1974).
54. Isbell, H. S. and Frusch, H. L., Carbohydrate Res., 59, C25 (1977).
55. Strickland, S. and Massey, V., "Oxidases and Related Redox Systems," (T. E. King, H. S. Mason and M. Morrison, eds.), University Park Press, p. 189, 1973.
56. Goscin, S. A. and Fridovich, I., Arch. Biochem. Biophys., 153, 778 (1972).
57. Halliwell, B., Eur. J. Biochem., 55, 355 (1975).
58. Halliwell, B., and Ahluwalia, S., Biochem. J., 153, 513 (1976).
59. Halliwell, B., Biochem. J., 163, 441 (1977).
60. McCord, J. M. and Day, E. D., Jr., FEBS Letters, 86, 139 (1978).
61. Lavelle, F., Michelson, A. M. and Dimitrijevic, L., Biochem. Biophys. Res. Commun., 55, 350 (1973).
62. Michelson, A. M. and Buckingham, M. E., Biochem. Biophys. Res. Commun., 58, 1079 (1974).
63. Gregory, E. M., Yost, F. J. and Fridovich, I., J. Bact., 115, 987 (1973).
64. DeChatelet, L. R., Shirley, P. S., Gordson, P. R. and McCall, C. E., Antimicrobial. Agents and Chemother., 8, 146 (1975).
65. Van Hemmen, J. J. and Mauling, W. J. A., Arch. Biochem. Biophys., 182, 743 (1977).
66. Beauchamp, C. and Fridovich, I., J. Biol. Chem., 245, 4641 (1970).
67. Weiss, S. J., Rustagi, P. K. and LoBuglio, A. F., J. Expt. Med., 147, 316 (1978).
68. Weiss, S. J., Turk, J. and Needleman, P., Blood, In press.

69. Fee, J. A., "Third International Dymposium on Oxidases and Related Oxidation-Reduction Systems",(T. E. King, M. Morrison and H. S. Mason, eds.), in press.
70. Fee, J. A., "Metal Ion Activation of Dioxygen," (T. G. Spiro, ed.), John Wiley and Sons, New York, 1980, p. 209.
71. McCord, J. M., Keele, B. B., Jr. and Fridovich, I., Proc. Natl. Acad. Sci., 68, 1024 (1971).
72. Foote, C. S., Shook, F. C. and Abakerli, R. A., J. Am. Chem., Soc., 102, 2503 (1980).
73. Proceedings of the Second International Conference on Superoxide and Superoxide Dismutases, Msida, Malta, September, 1979.
74. Winterbourne, C. C., Biochem. J., 182, 625 (1979).
75. Bullen, J. J., Rogers, H. J. and Griffiths, E., "Microbial Iron Metabolism," (J. B. Neilands , ed.), Academic Press (1974), p. 517.
76. Britton, L. and Fridovich, I., J. Bact., 131, 815 (1977).
77. Vignais, P. M., Henry, M. F., Tereech, A. and Chabert, J., "The Significance of Superoxide and Superoxide Dismutases," (J. V. Bannister and H. A. O. Hill , eds.), Vol. 1, Elsevier, In press.
78. Puget, K. and Michelson, A. M., Biochemie, 56, 1255 (1974).
79. Puget, K. and Michelson, A. M., Biochem. Biophys. Res. Commun., 58, 830 (1974).
80. Hassan, H. M. and Fridovich, I., "Strategies of Microbial Life in Extreme Environments," (M. Shilo , ed.), Verlag Chemie Weinheim (1979), p. 179.
81. Fee, J. A., Lees, A. C., Bloch, P. L. and Neidhardt, F. C., Unpublished results.
82. Hassan, H. M. and Fridovich, I., J. Biol. Chem., 252, 7667 (1977).
83. Hassan, H. M. and Fridovich, I., J. Biol. Chem., 253, 8143 (1978).
84. Fee, J. A., Lees, A. C., Bloch, P. L. and Neidhardt, F. C., "Biochemical and Clinical Aspects of Oxygen," (W. S. Caughey, ed.), Academic Press, New York, 1979, p. 635.
85. Hassan, H. M. and Fridovich, I., J. Biol. Chem., 254, 10846 (1979).
86. Hassan, H. M. and Fridovich, I., Arch. Biochem. Biophys., 196, 385 (1979).
87. Hassan, H. M. and Fridovich, I., Rev. Inf. Diseases, 1, 357 (1979).
88. Norrod, P. and Morse, S. A., Biochem. Biophys. Res. Commun., 90, 1287 (1979).
89. Lynch, R. E. and Cole, B. C., "The Significance of Superoxide and Superoxide Dismutase," (Bannister, W. H., and Bannister, J. V., eds.), Vol. 2, Elsevier, In press.
90. Catlin, B. W., J. Inf. Diseases, 128, 178 (1973).
91. Kenny, C. P., Ashton, F. E., Diena, B. B. and Greenberg, L., Bull. World Health Org., 37, 569 (1967).

92. Götz, F., Sedewitz, B. and Elstner, E. F., Arch. Microbiol., In press.
93. Götz, F., Elstner, E. F., Sedewitz, B. and Lengfelder, E., Arch. Microbiol., In press.
94. Fee, J. A., "Superoxide and Superoxide Dismutases," (A. M. Michelson, J. M. McCord and I. Fridovich, eds.), Academic Press, London (1977) p. 173.
95. Fee, J. A. and McClune, G. J., "Mechanisms of Oxidizing Enzymes," (T. P. Singer and R. N. Ondarza, eds.), North Holland, Amsterdam (1978) p. 273.
96. Hewitt, J. and Morris, J. G., FEBS Letters, 50, 315 (1975).
97. Tally, F. P., Goldin, B. R., Jacobus, N. V. and Gorbach, S. Z., Infect. and Immunity, 16, 20 (1977).
98. Lindmark, D. G. and Müller, M., J. Biol. Chem., 249, 4634 (1974).
99. Hatchikian, E. C. and Henry, Y. A., Biochemie, 59, 153 (1977).
100. Carlsson, J., Wrether, J. and Geckman, G., J. Clin. Microbiol., 6, 280 (1977).
101. Gregory, E. M., Moore, W. E. C. and Holdeman, L. V., Appl. and Environ. Microbiol., 35, 988 (1978).
102. Kanematsu, S. and Asada, K., FEBS Letters, 91, 94 (1978).
103. Lumsden, J. and Hall, D. O., Nature, 257, 670 (1975).
104. Lehninger, A. L., "Biochemistry," Worth, New York, 1975, Chapter 1.
105. Brown, O. R., Howitt, H. F., Stees, J. L. and Platner, W. S., J. Lipid Res., 12, 692 (1971).
106. Simons, R. S., Jackett, P. S., Carroll, M. E. W. and Lowrie, D. B., Toxicol. and Applied Pharmacol., 37, 271 (1976).
107. Bloch, P. L., Phillips, T. A. and Neidhardt, F. C., J. Bact., 141, 1409 (1980).
108. Seither, R. L. and Brown, O. R., Submitted for publication.
109. Nierlich, D. P., Ann. Rev. Microbiol., 32, 393 (1978).
110. Sokawa, J. and Sokawa, Y., J. Biochem., 83, 1699 (1978).
111. Brown, O. R. and Yein, F., Biochem. Biophys. Res. Commun., 85, 1219 (1978).

DISCUSSION

BORG: Acrimony we will avoid, and we won't vote on who the winner will be. I did say that in trying to follow Gidi's (Czapski) plans we'd have the two principal speakers address each other's arguments in rebuttal before the rest of us got into the act. So I would like to ask Irwin (Fridovich) if he would like to make some comments about what he considers to be the nubbin of the controversy, perhaps noting some of the similarities as well as the major differences between the two positions. After he has had his five or ten minutes, Jim (Fee) will have time to comment again.

FRIDOVICH: In his attempts to demolish the superoxide theory of oxygen toxicity Dr. Fee has constructed an argument based primarily upon areas of current ignorance. The theory is not yet complete and there is still much we do not know; but ignorance does not negate knowledge, it is rather an opportunity for the further expansion of knowledge. I will here try to comment on the major points made by Dr. Fee.

1. Do reductants, present within cells, efficiently scavenge $O_2^{\cdot}$ and so obviate the need for superoxide dismutases? Oxidation of any reductant by $O_2^{\cdot}$ begets a free radical of the reductant. That free radical, be it derived from ascorbate, glutathione or some phenol, may itself be damaging to the cell. Furthermore, scavenging of $O_2^{\cdot}$ with a reductant consumes the reductant, whereas catalytic scavenging by SOD consumes only $O_2^{\cdot}$ and produces only O_2 plus H_2O_2. H_2O_2 is itself deleterious, but we have catalases and peroxidases to deal with H_2O_2.

We can illustrate the virtues of SOD, as a catalyst of $O_2^{\cdot}$ dismutation, by considering a relatively simple cell, i.e., the mature human erythrocyte. This cell contains almost 5 mM hemoglobin and only 0.01 mM SOD. Hemoglobin can react with $O_2^{\cdot}$, indeed hemoglobin can catalyze the dismutation of $O_2^{\cdot}$. Why then do we need SOD in the red cell? The relevant reactions and their rate constants are:

a) $HbO_2 + O_2^{\cdot} + 2H^+ \rightarrow$ Met Hb $+ O_2 + H_2O_2$ } $k = 4 \times 10^3\ M^{-1}\ sec^{-1}$

b) Met Hb $+ O_2^{\cdot} \rightarrow HbO_2$ } $k = 5.7 \times 10^3\ M^{-1}\ sec^{-1}$

c) $O_2^- + O_2^{\cdot} + 2H^+ \xrightarrow{SOD} H_2O_2 + O_2$ } $k = 2 \times 10^9\ M^{-1}\ sec^{-1}$

The rate of reaction of $O_2^{\cdot}$ with SOD is 10^6 fold more rapid than with hemoglobin. Allowing for the 500-fold excess of hemoglobin over SOD, we see that the SOD will intercept over 1000 molecules of $O_2^{\cdot}$ for every one intercepted by hemoglobin. Moreover, the SOD is unchanged in the process, whereas the hemoglobin is converted to methemoglobin. Were we willing to walk around with a 50% methemoglobinemia and thus to sacrifice half of the oxygen-carrying capacity of our blood, we could dispense with 0.1% of the SOD present in our erythrocytes. Clearly this would be a poor bargain!

2. Why do some anaerobes contain SOD and why can an oxygen-tolerant organism like Lactobacillus plantarum do without a classical SOD? Here is one of those areas begging to be filled with knowledge. The anaerobic lifestyle is a precious one on an

oxygenated planet. Anaerobes do not actively grow in the presence of oxygen, but they must survive temporary exposure to it, enroute from one anaerobic niche to the next. How else could they get from one gut to another? The SOD present in some of these organisms probably ensures their survival during transient exposure to oxygen.

Bacteroides are considered to be strict anaerobes. Dr. E. M. Gregory has noted that they contain a low level of SOD, when examined after anaerobic growth. If a culture growing anaerobically is exposed to just 2% oxygen, they cease growing. They also increase their SOD content five-fold and survive this oxygen exposure. The increase in SOD thus appears to be an adaptive response in the anaerobe Bacterioides as it is in the facultative *Escherichia coli*.

Lactobacillus plantarum is a peculiar organism. It contains no iron, makes no hemes and has no cytochrome chain. Most germane to this discussion, it can respire in certain media, is reasonably oxygen-tolerant, yet contains no classical SOD. Does this mean that $O_2^{\overline{\cdot}}$ is benign and SOD a luxury? Hardly! Rather, it means that there are many ways to skin a cat and that evolution, given almost infinite time, has tried them all. Thus, Dr. F. Archibald has found that *L. plantarum* accumulates Mn(II) until its intracellular level approximates 20 mM. This huge amount of Mn(II) serves to scavenge $O_2^{\overline{\cdot}}$. Since earlier studies of this organism looked for SOD activity in dialyzed cell extracts and used SOD assays which included EDTA in the reaction mixture, the activity of this Mn(II) was overlooked. These cells may contain mechanisms for reducing the Mn(III) or the MnO_2^+ generated by the $O_2^{\overline{\cdot}}$ + Mn(II) reaction. If so, then they can use their manganese catalytically. In any event they do possess a mechanism for scavenging $O_2^{\overline{\cdot}}$ and they can be killed if exposed to artificially increased rates of intracellular $O_2^{\overline{\cdot}}$ production. Most organisms use micromolar levels of SOD to scavenge $O_2^{\overline{\cdot}}$; *L. plantarum*, for reasons of its own uses millimolar levels of Mn(II) to achieve the same end.

3. Dr. Fee notes that Cu(II) can itself catalyze the dismutation of $O_2^{\overline{\cdot}}$ and uses this to argue that SOD is therefore not necessary. Why build a metalloprotein when the metal will do, by itself? Free Cu(II) is intolerable. It catalyzes too many oxidations and it inhibits too many enzymes. That is probably why there is no free Cu(II) inside cells and that is why we need copper enzymes, which so channel the reactivity of the copper that it will catalyze only a single reaction or only a single class of reactions. The fact is that an extract of *E. coli* or of liver exhibits only a few bands of SOD activity on polyacrylamide gel electropherograms and each of these bands corresponds to an SOD. We see no bands due to Cu(II) or to other low molecular weight components in *E. coli* or in liver or in most other places. *L. plantarum*, as already explained, is a horse of a different color.

MnO_2 will catalyze the decomposition of H_2O_2 and so will triethylene tetramine-Fe(III). Do we use this to argue that catalases are superfluous?

There are superoxide dismutases containing Cu + Zn or Mn or Fe and all are equally efficient as catalysts of the dismutation of $O_2^{\bar{\cdot}}$. Moreover the Mn and Fe enzymes show extensive homologies even when the enzymes compared are from sources as diverse as E. coli and chicken liver mitochondria. The Cu + Zn enzymes appear totally unrelated to the Mn or the Fe enzymes. If these enzymes really served functions in the cell unrelated to the dismutation of $O_2^{\bar{\cdot}}$, as Dr. Fee believes, would these homologies not be exceedingly unlikely? Why should a Mn-protein isolated from one source be homologous to a Fe-protein isolated from another source, unless they catalyze the same reaction and have evolved from a common ancestral protein? These Mn or Fe enzymes were isolated based upon their abilities to dismute $O_2^{\bar{\cdot}}$. Their extensive amino acid sequence homologies indicate that this is, and has long been, their biological function.

There are a number of papers which erroneously attribute SOD activity to various metal complexes. Thus Fe-EDTA does not catalyze the dismutation of $O_2^{\bar{\cdot}}$ to any significant extent and neither does Cu-pennicillamine. Cu-complexes, such as the salicylate or the amino acid complexes, are active to the extent that they dissociate to free Cu. The SOD-like activities of such complexes are uniformly eliminated by EDTA, which has no effect on SOD. Superoxide dismutases are in fact carefully evolved catalysts. The E. coli MnSOD is not active when the manganese at its active site has been replaced by iron. The E. coli FeSOD is conversely not active when its active site is occupied by manganese.

4. Dr. Fee claims that with O'Farrell gels he can visualize each and every one of the proteins of E. coli. O'Farrell gels will resolve at most between 200 and 400 proteins. E. coli certainly contain at least several thousand proteins and possibly several tens of thousands of proteins. More to the point, Dr. Fee uses these gels to demonstrate that methyl viologen, which can cause a profound induction of MnSOD, also can increase or decrease the levels of several proteins resolvable on O'Farrell gels. He concludes that the correlations we have made, based on changes in MnSOD, are therefore arbitrary. In the case of MnSOD the induction by methyl viologen was absolutely dependent upon molecular oxygen and upon a source of electrons. Was that true for the proteins tracked by Dr. Fee? In the case of MnSOD, elevated levels correlated with protection against the lethality of hyperbaric oxygen and against the oxygen-enhanced lethality of methyl viologen. Was that true for the proteins tracked by Dr. Fee? E. coli contains a certain amount of protein. A massive induction of any one protein will ordinarily be associated with decreases in other proteins. Metabolism is aptly described as being a complex net of interrelated reactions. Any change in conditions is apt to change several enzyme levels.

Consider that methyl viologen exerts its effect because it can be reduced inside *E. coli* by NADPH, under the influence of a soluble diaphorase, and because the reduced methyl viologen can, in turn, reduce O_2 to $O_2^{\cdot-}$. Adding methyl viologen will therefore lower [NADPH] and raise [$NADP^+$]. These changes might induce the biosynthesis of the diaphorase. They would probably also induce the synthesis of the enzymes of the hexose monophosphate shunt; because this pathway is responsible for maintaining the level of [NADPH]. Inducing the biosynthesis of MnSOD might also require induction of some Mn-transport protein. The argument based upon O'Farrell gels thus clearly reflects a lack of appreciation of the complexity of metabolism.

5. Dr. Fee questions the experiments based upon the lethality of hyperbaric oxygen or of methyl viologen. In particular he objects to the use of puromycin in many of these experiments. *E. coli* exposed to an increased intracellular flux of $O_2^{\cdot-}$, whether caused by raising pO_2 or by adding electron-shunting compounds such as methyl viologen, respond adaptively by making more SOD. Puromycin inhibits protein synthesis and so prevents this increase in SOD and exposes the cells to the full onslaught of the $O_2^{\cdot-}$. Of course, controls are needed to demonstrate that any observed lethality was not due to the puromycin. Such controls were done and have been reported. For example, puromycin does not kill *E. coli* grown anaerobically and kept anaerobic; because there is then no exposure to $O_2^{\cdot-}$. And, puromycin does not kill *E. coli* grown aerobically and kept aerobic; because they then already contain a high level of SOD to handle the $O_2^{\cdot-}$ made aerobically. But, puromycin applied to anaerobically-grown cells then exposed to air does cause cell death; because it prevents the cells from increasing their content of SOD to that level appropriate to the aerobic rate of production of $O_2^{\cdot-}$. All of this should be very clear. Moreover, one can prevent rapid protein synthesis by the stringent response in media devoid of amino acids. This also prevents rapid induction of SOD and exposes the cells to the lethality of intracellular $O_2^{\cdot-}$. In these cases puromycin was not present. Clearly, an intracellular production of $O_2^{\cdot-}$, which exceeds the scavenging capacity of the intracellular SOD, kills the cells and puromycin has nothing to do with this deleterious effect of the oxygen radicals.

FEE: My only purpose in this debate is to bring the superoxide theory of oxygen toxicity out of the shadow of conjecture into the light of experimental fact. It is only under such scrutiny that we will be able to decide if indeed it has real value in guiding our efforts to understand the molecular basis of oxygen toxicity. The fundamental deficiencies of the theory, as I see them, are described in my paper and will not be repeated here. However, I would like to address the several points of rebuttal raised by Dr. Fridovich.

1. The chemistry of superoxide is not an area of ignorance.

We know from many careful studies over a period of some 40 years that aqueous superoxide is extremely unreactive toward essential functional groups of cellular constituents, and the efforts of the last decade have confirmed and extended this understanding of superoxide chemistry. Therefore my assertion that superoxide is not cytotoxic is drawn from a large base of knowledge not from an area of ignorance.

Since it is a basic assumption of the theory that superoxide leads directly or indirectly to cell damage, the theory would be greatly strengthened if there were at least one chemical reaction involving superoxide which all workers in the field could agree leads to the formation of cell degradation. The "biological Haber-Weiss reaction" should be carefully examined in this regard.

My statements regarding the presence of reducing agents within cells and their relation to the concept of superoxide driven Fenton Chemistry appear to be misunderstood. Consider the following reactions which compose this type of chemistry:

$$Fe^{3+} + O_2^{\overline{\cdot}} \longrightarrow Fe^{2+} + O_2 \qquad [1]$$

$$Fe^{2+} + H_2O_2 \longrightarrow Fe^{3+} + OH\cdot + H_2O \qquad [2]$$

$$OH\cdot + RH \longrightarrow \text{degradation} \qquad [3]$$

The reducing equivalent of superoxide is carried to peroxide via the metal, here represented by Fe, and this is done by reducing the metal. The point I wish to make is that cells are known to contain rather high concentrations of other reducing agents, such as ascorbate, that are kinetically and thermodynamically capable of reducing metal ions, as shown in Reaction [4], and thus compete with superoxide

$$AH + Fe^{3+} \longrightarrow Fe^{2+} + A^{\dot{-}} + H^{+} \qquad [4]$$

I have argued that the quantitative contribution of superoxide to this type of Fenton chemistry is the ratio of the velocity of Reaction [1] to the sum of the velocities of Reactions [1] and [4]. Thus,

$$\text{Superoxide Contribution} = V_1/(V_1 + V_4) \qquad [5]$$

Because the concentration of the organic reductant is so high (∿mM) compared to superoxide (∿μM, even if dismutase were absent) and the rate constants for reduction of the metal are probably similar for both types of reductants, the contribution of superoxide to the Fenton chemistry must be negligible. Indeed, there are several examples of *in vitro* reactions where 50 micromolar ascorbate completely abolishes the contribution of a significant flux of superoxide to Fenton degradation.

It should not be *a priori* assumed that metal ions will be

available for Fenton chemistry within the cellular milieu. The simple fact is that we know nothing about the intracellular concentrations of "free" transition ions. In summary, a case might be made that Fenton chemistry is the source of superoxide toxicity and therefore oxygen toxicity, but the logical derivation of that position from present knowledge would be very tenuous indeed.

2. L. plantarum utilizes oxygen differently from E. coli, and it does grow quite happily without a superoxide dismutase. The original argument for the oxygen tolerance was that L. plantarum consumed no oxygen and therefore needed no dismutase, but Elstner later showed that L. plantarum can respire quite rapidly, found that extracts had a weak capacity for scavenging superoxide, and confirmed that the organism had no true dismutase activity. Since I have not had access to the data regarding the possible use of Mn(II) as a superoxide trap, it is not possible for me to comment on this assertion.

But let us not dwell on this case as it constitutes only one of four distinct aerobic cell types which do not possess superoxide dismutases. These are adipocytes (Peeters-Joris et al, Biochem. J. 150, 31-39 (1975), not mentioned in my manuscript), L. plantarum, Neisseria gonorrhoeae, and Mycoplasma. The latter two are strict aerobes possessing an oxygen supported electron transport system and utilizing it for obtaining metabolic energy. While further work needs to be done on these systems they stand as living contradictions of the theory that superoxide dismutase is required for aerobic life.

Other examples where the biological distribution of superoxide dismutases is not consistent with the concept that its presence is necessary for superoxide dismutation are available. Consider two of these: a. Of the water soluble protein of certain seeds, such as peas and other legumes, 2.5% is Zn/Cu dismutase! It is very difficult to imagine that much protein being needed for protection against superoxide. b. Many strictly anaerobic organisms are rapidly killed by exposure to oxygen. Yet several species contain significant amounts of superoxide dismutase but there is no general correlation between dismutase activity and the ability of these organisms to withstand oxygen.

It would seem that as our areas of ignorance are filled in with further facts the validity of the idea that superoxide mediates oxygen toxicity becomes less tenable.

3. Concerning dismutase activity of low molecular chelates of Cu, Fe, and Mn. Here again, my comments appear to be misunderstood. My point is simple: If Cu(II) alone can catalyze superoxide dismutation at rates comparable to Cu-protein complexes this type of catalysis is inherent in the chemistry of the metal and needs no help from a protein. This simply opens the possibility that the Cu bound to the protein may catalyze superoxide dismutation because this aspect of its chemistry is not suppressed in the Cu-protein complex. Indeed, many Cu-complexes (those

which have at least one unoccupied coordination position) are effective catalysts. By contrast, if one suppresses this chemistry by occupying all the coordination positions of the Cu, as with EDTA, activity is lost. It is therefore argued that the dismutase activity of these proteins may be a trivial one. I am not arguing that small complexes exist in cells for the purpose of dismuting superoxide, although, that is a position Dr. Fridovich has adopted to explain the accumulation of Mn(II) by L. plantarum.

The sequence homology between the Fe- and Mn-proteins is indeed interesting, and it does suggest a common metabolic link between the two proteins. It does not add anything to the argument that the metabolic function is known.

The statement that Fe-EDTA does not catalyze superoxide dismutation is incorrect, and adds nothing to this discussion.

4 and 5. The conclusions drawn by Hassan and Fridovich in their experiments demonstrating induction of Mn-dismutase by treatment of E. coli with aerobic methylviologen and from the adaptation of these cells to the bactericidal mixture of methylviologen, oxygen, and puromycin are not unique. They are also confusing because they fail to integrate facts known from other people's work into the interpretation of their observations.

Let me first deal with the uniqueness question. When one examines essentially all the proteins of E. coli using quantitative O'Farrell gels (Cf. Ref. 107) many proteins, including the Mn-dismutase, undergo dramatic changes in concentration after treatment with aerobic methylviologen. In this investigator's opinion, the choice of a particular protein to account for the adaptation of the cell to this type of treatment, if not arbitrary, is at least highly selective. The most one can conclude from this type of experiment is that the Mn-dismutase changes its concentration during the adaptation of the cell to this type of treatment. Whether this change is a primary or a secondary one cannot be stated.

The conclusions drawn are confusing because neither aerobic methylviologen nor hyperoxia are bactericidal toward E. coli; they are both bacteriostatic. Immediately upon removing the poison all the cells present in the culture begin to reproduce. It is only when the cells are confronted with the additional poison, puromycin, that they die. Perhaps a trained microbiologist can shed some light on the behavior of organisms subjected to multiple bacteriostatic agents.

The second important experimental fact that has not been incorporated into these conclusions is that when the levels of the two dismutases are manipulated over a four-fold range by simply increasing or attenuating the concentrations of metals in the culture medium the cells show no difference in their resistance toward the bactericidal mixture of puromycin/methylviologen/oxygen (Cf. Ref. 106). These results strongly suggest that the dismutases are not central to the adaptation and that puromycin does not simply act to prevent the synthesis of superoxide dismutase. All facts considered, it is apparent that the cells undergo a complex adaptation to hyperoxia and to aerobic methylviologen which involves many proteins one of which is the Mn-dismutase. However, simply increasing the Mn-dismutase (or total dismutase) does not protect against the bactericidal mixture. These experiments lend little support to the idea that superoxide promulgates oxygen poisoning, but do suggest that we have no satisfactory understanding of the underlying chemistry of the phenomenon.

BORG: As we agreed earlier, I am going to ask the four members of the panel that Gidi (Czapski) selected, other than myself, to make comments or to ask questions to introduce a broader discussion. I have some questions myself, but I will save them until last. I do feel that there is some common ground here; and although the two main speakers have given different emphasis to what one of them called superoxide-driven Fenton chemistry, there still is a considerable area of agreement. Let us hear what the rest of you have to say about that.

COHEN: I would like to make some general comments about the debate we have heard. There are points of agreement, but I am more struck by the disagreement. The argument boils down to the following: Either superoxide is toxic and SOD protects or superoxide is unreactive and, therefore, enzymes with SOD activity are unnecessary. The superoxide-SOD field is over ten years old. Let's look at where we were 10 years ago and where we are now.

One question that had to be answered was: Does the protein in question really dismute superoxide? Rabani and others showed unequivocally that it did by pulse radiolysis study. Once that question was answered, a powerful research tool was placed in the hands of the biochemist, and the second question could be asked: Does superoxide exist in biological systems? Well, now we know that it does. So, superoxide does exist in biology and SOD also exists and can, moreover, be used as an experimental tool. We have seen much progress since then.

Fridovich and his associates have directed our attention towards one point of view which is O_2 toxicity. That may or may not be the full story for superoxide; there may be other things going on. But the insights and experiments have been good ones, and they have provided provocative new data. We have seen new ways to provoke lipid peroxidation, learned about cell toxins like methyl viologen (Hassan), lysis of red blood cells (Lynch) and about intercellular attractants for phagocytes (McCord). Other workers such as Autor, Oberley and Babior have provided new information about a host of topics concerning O_2 toxicity, cancer and phagocytosis. Much of the information is too recent to be fully evaluated. A point I want to make is that when we look at the logical construct from the sum of all of these experiments, we should ask if the concept that superoxide is toxic provides a guide for experimentation. Is the superoxide theory useful? It is! It is a powerful guide to experiments. I think it is too late to go back and say that superoxide does not exist in biology or that it doesn't react. There is now too much evidence that superoxide can effect a lot of changes in biological systems. In my own personal experience I have found it a valuable guide leading to ways to prevent 6-hydroxydopamine hemotoxicity and the destruction of β-cells by alloxan.

The main point I want to make is that when superoxide is present, a lot of changes, many oxidative, but poorly understood, happen. So, the problem is to understand how superoxide does what it does. Here, there is a lot of confusion. But, certainly both superoxide and superoxide dismutases can play important roles in biology and toxicology. It is too late to look on SOD merely as a Cu-binding protein. That would mean giving up too powerful a research tool.

BIELSKI: I would like to add to Gerry's (Cohen) comments that first of all we should define under what conditions $O_2^{\cdot-}$ is harmful. While it does not react with many compounds in aqueous solutions, as our studies at Brookhaven have shown, it does react with some of these compounds in aprotic solvents at considerable rates. Hence the question arises, do we expect $O_2^{\cdot-}$ to react with compounds of biological importance without the aid of some catalyst. There are a few examples which suggest the need for a catalyst or modification of a given compound for reaction to take place with $O_2^{\cdot-}$. For example, $O_2^{\cdot-}$ does not react with free NADH, the rate is less than 27 $M^{-1}s^{-1}$, but attached to lactate dehydrogenase (E-NADH) it reacts with $O_2^{\cdot-}$ at a rate of 10^5-10^6 $M^{-1}s^{-1}$ in a chain mechanism. Also, while horseradish peroxidase does not react with $O_2^{\cdot-}$ its reaction product with H_2O_2, that is Compound 1 reacts at a rate of 10^6 $M^{-1}s^{-1}$. In view of these observations I suggest that many of the so-called unreactive compounds should be re-investigated under biological conditions; that is, in the presence of appropriate enzymes.

Also, the diagnostic methods of adding SOD and catalase to $O_2^{\cdot-}$ generating systems should be used with caution as we have some preliminary results which suggest that in the presence of formate, $O_2^{\cdot-}$ reacts with one of the forms of catalase, most likely Compound 1. In this system, the rate of $O_2^{\cdot-}$ disappearance appears to be dependent upon the peroxide concentration and the overall mechanism of the disappearance is very complex. It is well known that catalase Compound 1 is a relatively strong oxidizing agent which oxidizes alcohols to aldehydes, formate to CO_2 and NO_2^- to NO_3^-. If Compound 1 of catalase reacts similarly to Compound 1 of horseradish peroxidase, I would expect it to react with $O_2^{\cdot-}$. One can assume that when one adds SOD and catalase, or catalase alone to an $O_2^{\cdot-}$ generating system, that a very complex situation ensues because of the intermediate states of catalase and the numberous cross-reactions which can occur. Since

SOD and catalase are often found together in biological systems, one cannot exclude the possibility that they act in unison to protect the living cell or control some yet unknown metabolic pathways.

I should like to add that the removal of $O_2^{\overline{\cdot}}$ in a highly organized cell could hardly occur via Cu complexes or free Cu since the latter would be strongly bound by a variety of terminal groups of proteins. Hence from an organizational point of view, the need of Cu protein like SOD appears at present the logical answer.

HEMPFLING: I am the trained microbiologist that Dr. Fee asked for. Unfortunately, I don't think I can respond to his question in regard to this nutritional effect, so let me go on to something that I hope to have some more familiarity with. First, to congratulate Dr. Fridovich for his contribution to microbial physiology, even though he didn't set out to make that contribution. I believe the unifying notion that there could be a general explanation for obligate anaerobiosis was most stimulating to microbial physiologists. I would like to point out, however, that the only thing that ought to appeal more to a scientist than the construction of a beautiful hypothesis is its destruction, so I would like to congratulate Dr. Fee for his approaches in that direction. I think that the dialectical result of those two opposing views ought to provide us with more strongly supported models in the future.

I would like to make a couple of points though. It seems to me that the notion that SOD is the overriding consideration in the ability to grow under aerobic conditions lacks the experimentum crucis. Dr. Fridovich addressed himself to that all too briefly when he brought up observations about mutations that had led to the inability to grow under aerobic conditions. I thought "Aha, here it is," but I then heard that the mutation appears to affect specifically the Mn SOD.

Does it affect the Fe SOD as well? The next consideration is the association among catalase, peroxidase and Mn SOD. Are you suggesting that there is an operon?

FRIDOVICH: Of course E. coli happened to be an unfortunate choice for this search for a mutant because, in fact, the mutant which could not make Mn SOD still could and did make the Fe SOD. These experiments are now being repeated with organisms which have only one SOD. Dr. Larry Britton is in the audience here, he is doing studies like this with Streptococcus fecalis; others are undertaking similar studies and we'll see what their results are. It should be of course, a much more clean-cut experiment in an organism which has but one SOD.

HEMPFLING: And what I think we are all driving at is to choose an organism that has one SOD, to knock out the SOD without knocking out anything else. Which means that we ought to affect the structural gene for SOD formation as opposed to some pleiotropic effect and see what the outcome is. And I would trust that conditional mutants are being sought and all that sort of business.

I was somewhat disturbed to see that there appears to be a controversy as to the penultimate effect of paraquat on organisms. Is it bacteriostatic or does it actually kill the organisms? I would hope that some accommodations could be reached with regard to the measurements of such effects between the opposing forces.

Finally, I would like to know what the targets of superoxide are in the cell that lead to these effects. I think maybe that is a rhetorical question while I hope that it would go beyond rhetoric.

Perhaps by focussing on the more dramatic question of life or death in the presence of O_2 we are missing an equally important phenomenon: the apparent repression of synthesis of many cellular components in the presence of atmospheric O_2, the mechanisms of which are completely unknown. Examples of importance are found in the bacteriorhodopsin content of the "purple" membrane of *Halobacterium*, branched respiratory chain oxidases in coliform bacteria (that is, cytochromes *d* and a_1) and in other respiratory components. I urge consideration of these questions utilizing the idea that oxygen radicals may be exerting a controlling effect at the level of transcription, and that the activity of SOD might affect, say, bacteriorhodopsin biosynthesis by adjusting the level of superoxide within the cell. Such more subtle effects seem to me to be more consistent with the low concentrations of superoxide present than does cytotoxicity.

MITCHEL: I think that Dr. Bielski started out in the right direction when he said that we should think about what is important, and that means what is important to the cells. Perhaps we should harken back to our keynote address where it was pointed out that, from an evolutionary standpoint, there are just two things that a cell is interested in, and that is survival to the point of reproduction, and then a successful replication of an unaltered cell. To that point, measuring the survival of the cell, or a genetic alteration, is perhaps the end point that we really should be looking at. Since he is not here to quote for himself, perhaps I could refer to Dr. Czapski's recent paper where he used radiolytic techniques to produce and modulate superoxide in *E. coli* and arrived at the conclusion that superoxide from the standpoint of lethality was unimportant to the *E. coli*. I would like to ask if there is any evidence if induced *E. coli*, which have high levels of SOD, are any more resistant to radiolytic cell death. Or in fact if there is any evidence that there is any change in mutation.

FRIDOVICH: The E. coli always contain an Fe SOD, even if grown anaerobically. An experiment was done by Steve Goscin in my lab, and published some years ago, which compared the OER in E. coli which had been grown aerobically with those grown anaerobically so that there was a difference of about three-fold in the level of total SOD. There was no difference in their radiosensitivity. On the other hand, other experiments have been done by Dr. Hara Misra, and others, in which SOD has been added to the suspending medium. And in that case, radicals, presumably produced in the medium by ionizing radiation, were intercepted by the SOD and there was a substantial diminution in the OER.

Clearly more work needs to be done. I am well aware of Dr. Czapski's results and we have ourselves seen that with some strains of bacteria there is no effect, but with other strains there is a large protective effect of the SOD. This may have to do with the form of Fe which is available to catalyze that Haber-Weiss reaction among the $O_2^{\bar{\cdot}}$ and peroxide species produced in the suspending medium. I think we have to concentrate on $O_2^{\bar{\cdot}}$ produced and peroxide produced in the medium and not that produced in the cell because the cell contains (even the anaerobically grown cell) SOD and some catalase. I hope that answers your question.

BORG: It does seem to me from this conference and from preceding conferences that there are some generalities that hold. One, despite Ben Bielski's disclaimer that one shouldn't over-read the rate constant data that come from his laboratory, is that we're really not sure that there aren't certain situations, such as intermediate reactive states of some enzymes requiring pyridine nucleotide cofactors, where $O_2^{\bar{\cdot}}$ reactivity may not be important. In general however, there is widespread agreement that $O_2^{\bar{\cdot}}$ is surprisingly unreactive; and that is one part of the problem. It has led to another broadly accepted conclusion, however: what Jim Fee chooses to call the superoxide-driven Fenton reactions. Even if that is Fee's phrase, I think Irwin (Fridovich) agrees that the biologically important reactants may form downstream from superoxide production and that they are Fenton-like. The question, then, of looking at the toxicity of $O_2^{\bar{\cdot}}$ is to say - as Bohrs and Saran and Gidi Czapski put it in their review from the Malta conference—that it serves primarily as a transport vehicle to transmit radical damage over a wider distance. There still seems to be a problem.

Jim Fee recognizes that Fenton reactions may be important, but looking at the sources of reactants kinetically and semiquantitatively, he feels they are probably not significant, in fact, in living systems. Irwin Fridovich also accepts the possibility that Fenton-like products may be ultimate cytotoxins. However, I have a question, in that regard, about the protective effect of SOD. If Fenton reactions are an important outcome of $O_2^{\bar{\cdot}}$ production in living systems, how does SOD protect? It just postpones the moment of truth. If, in fact, it is the H_2O_2 from the $O_2^{\cdot-}$ that is feeding into a Fenton cycle, the H_2O_2 production is going to occur somewhere with or without SOD. Is it the role of SOD, then, primarily to regulate where that happens? Is the association with catalase or with glutathione peroxidase important for SOD to afford protection from $O_2^{\cdot-}$ effects? The total amount of H_2O_2 to come from the $O_2^{\cdot-}$ is going to be the same with or without SOD, but the intracellular distribution will be different.

FRIDOVICH: The point is that if $O_2^{\cdot-}$ is made in water, in the absence of other reactions, it will dismute and it will make peroxides, so the same amount of H_2O_2 will be made, but the tremendous difference is that when SOD is present, the steady state level of $O_2^{\cdot-}$ will be many orders of magnitude lower than it would be in the absence of the enzyme. So, if there were a process that depended upon cooperative interaction, a Fenton-driven reaction, or what I prefer to call the iron-catalyzed Haber-Weiss reaction ; its rate depends upon the steady state concentrations of both of those reactants, then of course, that process will be depressed many orders of magnitude by the presence of SOD. As you saw in the case of ethylene production or of the nicking of the DNA, SOD protected completely and catalase protected completely, because both $O_2^{\cdot-}$ and H_2O_2 were required.

BORG: But where has the H_2O_2 from the dismutation catalyzed by SOD gone?

FRIDOVICH: It has gone into the solution, but H_2O_2 by itself, as the controls indicated, was not capable of attacking DNA or liberating ethylene from methional. So yes indeed, peroxide is going to be made, and there are other enzymes that have been known for a hundred years (catalases, peroxidases, and glutathione peroxidase more recently) which take care of the H_2O_2.

MISRA: Yes, it's very interesting that hydroxyl radicals are produced in a great amount, but at the same time we should know that the half-life of the hydroxyl radical is very short compared to the superoxide radical or H_2O_2. Hydrogen peroxide and superoxide radicals can crawl through and they can be transported to the site of injury where they can combine with the cell components or the metals or anything that will catalyze the reaction to make hydroxyl radicals. But if the hydroxyl radical is produced inside the cell, there are a lot of proteins—a lot of other components that can take care of hydroxyl radicals too. And, at the same time since its half-life is very short and it is very reactive, they cannot diffuse through the cell wall even to cause the damage. So that explains that SOD and catalase are important to take care of this Fenton reaction.

BORG: Right, I agree. Maybe Wolf Bors would like to comment, since I am really citing his preprint. In talking about SOD as a regulatory enzyme did you mean that there is still more cellular detoxification to be done with the resultant H_2O_2?

BORS: Actually I showed in my presentation yesterday on the first slide some alternative proposals which may help to explain the toxicity of O_2. They take into account the delayed effect and the higher specificity of a number of biochemical oxidation processes which cannot be explained by a reaction of $\cdot OH$ radicals. We think one important aspect is the transporting of the species, $O_2^{\bar{\cdot}}$ being a transport vehicle or a storage pool of radical properties. We derive this from the fact that $O_2^{\bar{\cdot}}$ is easily formed by a number of enzymatic and biological processes but is not in itself a very active agent.

Since I am here at the microphone I might as well continue because I have further arguments for this discussion. Dr. Fridovich was suggesting there might be other unknown targets in the cells which do react with $O_2^{\bar{\cdot}}$ and he was questioning whether pulse radiolysis was actually capable of coming up with such targets. I have an example.

We recently determined the reactivity of $O_2^{\bar{\cdot}}$ with a number of catechol derivatives. Catechols are very important biological substrates and we found some activated catechols which have reaction rates of 10^7 $M^{-1}s^{-1}$ for $O_2^{\bar{\cdot}}$ which is extremely high considering that normally the rate constants for $O_2^{\bar{\cdot}}$ do not exceed 10^4 $M^{-1}s^{-1}$.

Another point I want to make pertains to Dr. Fee's talk. He was suggesting that since metal ions or small metal complexes are capable of imitating or mimicking SOD, they might be considered as an alternative, so there would be no need for a protein. I consider this not a very reasonable argument. I would rather suggest that the protein increases the specificity of the reaction.

Copper complexes have been studied in a number of biochemical systems as substitutes of SOD. However, recently Estabrook, et al. could show that the inhibition of the cytochome P-450 reaction by Cu tyrosine does not involve scavenging of $O_2^{\bar{\cdot}}$ Similarly, fatty acid oxidation by lipoxygenase is not inhibited by SOD (Grosdy, pers. commun.) but is inhibited by the Cu penicillamine complex. Thus, small Cu complexes do react far less specifically with $O_2^{\bar{\cdot}}$ and interfere differently in these reactions. In total, these arguments favor the conventional theory.

FEE: I repeat that there seems to be some misunderstanding. I am not saying that these small complexes are used in cells as dismutases. What I am saying is that because they have this activity, it means that the dismutase activity of the protein may possibly be an artifactual one and therefore the protein was not made for this function.

SIMIC: I just want to comment along the line of inability of SOD to protect from radiation. First of all, SOD is not distributed homogeneously in a cell. Radiation-induced free radicals will be distributed throughout the cell in the areas with and without SOD. Secondly, I don't like the expression "cell killing" because that is a very misleading term to begin with. You have to define what you mean by the effect of radiation, you have to specify it, because if you are talking about loss of reproductive ability then we are probably talking about DNA and a possible reaction of $O_2^{\bar{\cdot}}$ with DNA; and as you know there is no SOD around DNA. Therefore, that experiment probably was doomed from the beginning. I want also to ask one question about blue-green algae which make a great deal of H_2O_2. What is going on with the Fenton system in these cells? They should be dying like flies.

OBERLEY: I am a firm believer in the superoxide theory of toxicity but I believe that there is one very important paradox that remains to be understood. I have been studying tumor cells for the last five or six years and tumor cells are very unusual in that almost all of them have low levels of SOD. At the same time they have normal or higher than normal levels of superoxide. Thus, you would expect tumor cells to be very sensitive to O_2. However, they are not sensitive to O_2, and in fact, they are much more resistant than a normal cell with high levels of SOD. However, I do not believe that this is evidence against the superoxide theory of toxicity. I think it points to the importance of the hydroxyl radical in O_2 toxicity. Tumor cells are also very abnormal in Fe metabolism and I believe this abnormality in Fe metabolism enables the tumor cell to withstand the high levels of superoxide because of low hydroxyl radical production. Now this is a system that has been studied very well and I would urge people that are looking at O_2 toxicity to consider the tumor cell as an ideal system because it has low SOD and high superoxide. The forms of Fe itself are abnormal in a tumor cell; the same as in a normal cell but the chelated forms are different. For instance, ferritin is very abnormal in tumor cells and the levels of picolinic acid may be much different.

KOPPENOL (post-session contribution): Fee's remark that the reaction of $O_2^{\bar{\cdot}}$ with SOD is a fortuitous event is refuted elsewhere in this volume, page 671.

CZAPSKI: (Editor's note: These comments were submitted after the session. Dr. Czapski was unable to attend the session.) I am adding some remarks to this session. I was supposed to chair it, but according to some participants, the SOD level in my heart was not right so I checked in to the intensive care unit rather than chairing a session on the role of SOD. I can see why with a wrong level of SOD I was not allowed to be in this session. As these faults are apparently repaired by now, I would like to add these few remarks and express some of my opinions on the subject of this discussion.

The first point may be a stubborn approach on a semantic problem—a name of a reaction, but as it was repeatedly raised in the session I'll add my comments.

Many of the explanations how $O_2^{\bar{\cdot}}$ is toxic suggest that it is a precursor of OH·, via the reactions.

$$O_2^{\bar{\cdot}} + Fe^3 \longrightarrow O_2 + Fe^{2+} \quad [1]$$

$$Fe^{2+} + H_2O_2 \rightarrow Fe^{3+} + OH^- + OH\cdot \quad [2]$$

This sequence of reactions gives us the stoichiometric equation.

$$O_2^{\bar{\cdot}} + H_2O_2 \longrightarrow OH^- + OH\cdot + O_2 \quad [3]$$

This last reaction 3 was proposed long ago by Haber and Wilstätter (1) and is called by Fridovich (3,4) and others the "Haber-Weiss reactions" (4).

Reaction 3 was originally proposed by Haber and Willstätter three years prior to the "Haber-Weiss paper" (1,2).

This reaction was ruled improbable already in 1951 by Barb et al (5) and more recently was shown to be far too slow (6,7,8,9).

If the mechanism is via reaction 1 and 2, then the role of $O_2^{\bar{\cdot}}$ is just to reduce Fe^{3+} and reaction 2 is the known Fenton reaction or Fenton reagent (10). Therefore I feel the name "Haber-Weiss" is very inappropriate and, if at all, one should call it "the Fenton reaction, initiated by reduction of Fe^{3+} to Fe^{2+} with $O_2^{\bar{\cdot}}$."

The Haber-Weiss paper dealt with the mechanism of the catalysis of the decomposition of H_2O_2 by iron salts. This mechanism of $O_2^{\bar{\cdot}}$ toxicity has no relevance to the catalytic decomposition of H_2O_2. It apparently is due to the formation of OH· through reaction 2 and should therefore be called "Fenton reaction" driven by $O_2^{\bar{\cdot}}$ reduction of Fe^{3+}. If one wants to use reaction 3—the stoichiometric equation, then as this reaction was found and proposed by Haber and Wilstätter (1), it should bear their names. To call it "the Haber-Weiss reaction" is wrong, and Jim Fee quoted how unjust it is to Weiss, who thought that $O_2^{\bar{\cdot}}$ has nothing to do with O_2 toxicity (11).

Now let me make a few remarks on the problem of $O_2^{\bar{\cdot}}$ and SOD and toxicity. I believe that several facts are accepted by everyone:

1) That $O_2^{\bar{\cdot}}$ is formed in cells and is even an intermediate in one of the pathways (a minor one apparently) of O_2 reduction in the O_2 metabolism.
2) That SOD exists in most cells.
3) That SOD catalyzes $O_2^{\bar{\cdot}}$ dismutation.
4) There is also sufficient experimental evidence that suggests that $O_2^{\bar{\cdot}}$ is toxic in some cases and SOD seems to protect from this toxicity.

There is some controversy as to the role of SOD. I'm not willing to go into this controversy, where Fee and Fridovich represent the extreme positions. Bors, Saran and my group (12) looked for other reactions of SOD. We failed to find any. On the other hand, the fact that an enzyme exists that promotes the metabolism of an intermediate ($O_2^{\cdot-}$) is not surprising regardless of whether $O_2^{\cdot-}$ is toxic or not. Nor is it surprising that you can induce an increase in SOD level by various means.

I myself believe that there are sufficient experimental data indicating that $O_2^{\cdot-}$ seems to act as a toxic agent and SOD has protective effects. A lot of the experimental data are not conclusive, some are apparently even wrong, but enough is left for me to make the above statement.

The role of SOD was discovered as being effective, as a "miracle drug," long before SOD was known and clearly no bias has influenced it. On the other hand I doubt very much in the mechanism proposed for the toxicity of $O_2^{\cdot-}$. I would like to raise the following difficulties that should be solved before one can accept the "$O_2^{\cdot-}$- SOD" theory.

If the addition of SOD protects cells or even whole animals, from ionizing radiation damage (13), does SOD penetrate into cells? From experiments done by Q. Mehdi and myself the answer is negative. Even SOD in a 0.01-5.0 M solution of DMSO does not cause any SOD to get into *E. coli*. (This phenomena was found with various strains of *E. coli*.).

If this is the case, then protection is occurring in the medium external to the cell. In this case, $\cdot OH$ has hardly any chance to reach cell membranes unless formed within 10-30A° of them.

In our experiments we showed that *E. coli* irradiated in air-saturated water containing 1mM formate is not sensitized as compared to cells without formate (14). On the contrary, 1mM formate protects. In this system all radicals outside the cell form $O_2^{\cdot-}$ and hence one would expect sensitization if $O_2^{\cdot-}$ is toxic.

I also repeated my studies to see if SOD protects various strains of *E. coli* being irradiated in aerated suspension.

No protection was observed (15). These results, together with older radiation biological studies, seem to show that if $O_2^{\cdot-}$ is formed by irradiation both inside and outside the cell it is not toxic, and SOD does not protect (16).

I have no explanation for these observed differences which seem to be contradicting, but as long as they are not explained, there seems to be some question whether $O_2^{\cdot-}$ is toxic. In the irradiation experiments, I see no reason why the Fenton mechanism should operate less than in the other experiments. As to the Fenton-driven mechanism, if it is responsible for toxicity, I think vitamin C and many other reducing agents should be as efficient and should show toxicity, which they do not.

To conclude, there is a lot of evidence for $O_2^{\cdot-}$ toxicity and for the protective role of SOD but there are also serious data which show that the mechanism and the interpretation and even the toxicity of $O_2^{\overline{\cdot}}$ may be wrong. If $O_2^{\overline{\cdot}}$ is toxic I believe the mechanism cannot be reaction 1-3 in this simplified form. The more probable mechanism, as proposed by Bors, Saran and myself in the Malta meeting (17), is that $O_2^{\overline{\cdot}}$ forms complexes at essential sites, that these complexes are stable and could then react to yield OH· much later at this site. A similar mechanism accounts for the bleomycin toxicity.

REFERENCES

1. Haber, F. and Wilstätter, R., Ber. 64, 2844 (1931).
2. Haber, F. and Weiss, J., Proc. Roy. Soc. A., 147, 332 (1934).
3. Fridovich, I., "Free Radicals in Biology I," (Pryor, W. A., ed.), Academic Press, NY., 1976, p. 239.
4. Misra, H. P., and Fridovich, I., Arch. Biochem., 176, 577 (1976).
5. Barb, W. G., Baxendale, J. H., George, P. and Hargrave, K. R., Trans. Far. Soc., 47, 462 (1951).
6. Czapski, G., and Ilan, Y. A., Photochem. Photobiol., 28, 651 (1978).
7. Halliwell, B., FEBS Letters, 72, 8 (1976).
8. McClune, G. J., and Fee, J. A., Ibid 67, 294 (1976).
9. Maiboroda, V. D., Petrayaer, E. P., Byakov, V. M., and Ivashonevich, L. E., Khim. Vys. Energ., 8, 284 (1974).
10. Fenton, H. J. H., J. Chem. Soc., 65, 899 (1894).
11. Fee, J. A., This volume, p. 205.
12. Bors, W., Saran, M., Ilan, A. Y., and Czapski, G., unpublished results.
13. Petkau, A., Chelack, W. S., and Plestach, S. D., Int. J. Radiat. Biol., 29, 297 (1976).
14. Mehdi, Q., and Czapski, G., unpublished results.
15. Samuni, A., and Czapski, G., Radiat. Res., 76, 624 (1978).
16. Czapski, G., to be published.
17. Bors, W., Saran, M., and Czapski, G., Proc. 2nd Inter. Sym. on SOD, Malta 1979, in print.

O_2 AND CELLULAR RADIATION SENSITIVITY

Session Chairman, E. L. Powers, Austin, Texas

SENSITIZATION OF CELLS TO RADIATION BY OXYGEN: A BRIEF HISTORY

E. L. Powers, Laboratory of Radiation Biology, Center for Fast Kinetics Research, The University of Texas at Austin, Austin, Texas 78712

Molecular oxygen is the most physiological and operationally one of the best, if not the best, sensitizer of cells fo radiation. The earliest evidence of its sensitizing ability was recorded in 1909 by Schwarz (1) who demonstrated that the induction of skin erythema required larger doses in the absence of air over the skin than in its presence. But it was not until 1933 that the phenomenon was addressed in formal manner when Crabtree and Cramer (2) saw diminished radiation sensitivity of tumor cells in the absence of oxygen. Naturally, they wondered whether the enzymatic systems concerned with oxidative processes were involved in whatever mechanisms were operating. A variety of respiratory poisons was tested. This study was followed by several other investigations that were directed to the biochemical and metabolic processes involving oxygen. The results were variable. Cyanide with O_2 was shown to be either without effect, or sensitizing, in Crabtree and Cramer's experiments, but it abolished the oxygen component of radiation sensitivity in bacteria (3). Other enzyme poisons in other systems had no detectable effects.

Thoday and Read (4) with the bean root gave indirect evidence that the oxygen effect is a physico-chemical rather than a metabolic set of actions and reactions. Then entered L. H. Gray and his colleagues. Gray is responsible for the gathering together of several lines of evidence from the work of his laboratory and that of others, and presenting the case formally that it is truly in the physico-chemical domain that explanations for the sensitizing effects of oxygen must be sought.

Some of the facts were:

1. The action of oxygen in sensitizing Ehrlich ascites cells to X-rays is the same at 35°, 18°, and 3°C; but, over this temperature range, there must be a very large change in enzymic reaction rates and in efficiencies of respiratory systems. Its action is non-enzymatic.

ISBN 0-12-592050-4

2. The relationship between oxygen concentration and radiation sensitivity in the ascites system is such that 4 μM/l of O_2 half saturates the O_2 sensitizing effect, whereas a Q_{O_2} for the same cells is at saturation values at concentrations approximately 1/10th that.
3. The evidence of Thoday and Read (4) was that the bean root is sensitive to oxygen at low LET radiations but not for high, and this, even at that time, was according to expectations from the chemical standpoint.
4. The aerobic and anaerobic sensitivities of three microorganisms were independent of a mutation that caused a hemin deficiency in the respiratory enzyme systems (5).
5. Gray's group examined (6) the relevance of oxygen sensitization as a factor in radiation therapy, proposing that cells in the tumor with less oxygen would be less radiation sensitive and could be restored to sensitivity by increasing the partial pressure of oxygen. This notion of the concentration gradient of oxygen arising from its diffusing through metabolizing tissues was used by Gray to explain the variable results reported by others in different systems where these gradients occurred. When this factor was taken into consideration, a metabolic or biochemical action of oxygen could be eliminated, from the initial stages at least, as an important factor in the action of oxygen as a radiation sensitizer.

The focus, then, was almost entirely on molecular oxygen's acting as a chemical intervening species, and certain mechanisms for the action of oxygen were proposed. In 1942 Laterjet and Loiseleur (7) had suggested that the action of oxygen in radiobiological systems could be due to either of two mechanisms: the taking up of hydrogen to form hydrogen peroxide, or the direct action of the oxygen molecule with an elementary particle that had been damaged by radiation. We note that at that time that the concept of free radicals as chemical entities was in its early developmental stages. A second early contribution was made by Rebeca Gerschman and colleagues in 1954 (8). They proposed that oxygen is a poisonous substance, and that there is significance in the similarity in oxygen poisoning and the effects of X-radiation. Their language was quite direct with respect to the species involved saying that "it would appear that radiation and oxygen poisoning produces some of their lethal effects through at least one common mechanism, possibly that of the formation of oxydizing free radicals." Later Gerschman (9) became much more explicit about oxygen and its characteristics pointing out that oxygen becomes "activated" with respect to energy release when an "oxidizing free radical" is formed and she suggested $R\cdot + O_2 = RO_2^{\cdot}$ as an initial step.

In her 1956 paper Alper (10) examined the oxygen effect and introduced the term "metionic reaction" that allows, among other things, reaction with radiation-induced radicals and peroxides.

She suggested that if oxygen intervenes in the metionic reaction "it can form a product with the target which makes it incapable of restoration." Note that this allows for back reaction to a restored target (repair!), but of a physico-chemical kind.

At the same time (about 1955), and in closely proximate institutions, Howard-Flanders and Alper with bacteria (11) and Deschner and Gray with mammalian cells (12) worked on the relationship between O_2 concentration and radiation sensitivity. Both groups found that the data could be fitted (albeit quite loosely) by a Michaelis-Menten type competition expression which, as Gray (13) points out, can be applied to many chemical, biological or medical situations and can have many interpretations. The demonstration from these sometimes loose fits with this algebraic expression was that the concentration dependence of the "O_2 effect," judged by the values of $\underline{m}$ and $\underline{K}$ is remarkably similar in a variety of experimental biological systems.

Our first paper dealing with these questions was published in 1960 (14). The first contribution is the method of analysis of numbers arising from our studies on dried spores. Since first order survival behavior is observed with our system under a variety of conditions, it follows that the total response consists of a number of parts, each probably independent of the others. The radiation response is divisible into sections or components and to understand one we must control the others.

Thus, if N_N is the number of cells that survive irradiation by dose D given to an initial number N, and N_O is the number surviving after irradiation in high concentrations of O_2, both are first order:

$$N_N/N = e^{-k_N D} \qquad \text{----------} \quad (1)$$

$$N_O/N = e^{-k_x D} \qquad \text{----------} \quad (2)$$

in which k_x is larger than k_N. Since the factors determining k_N operate whether or not O_2 is present, k_x contains both k_N and k_O, and equation 2 is more precisely written as

$$N_O/N = e^{-k_N D} \cdot e^{-k_O D} = e^{-(k_N + k_O)D} \text{--------} \quad (3)$$

The notation developed in 1960 is that k_N is written k_I, and since k_O, as defined experimentally above, was shown to consist of two components, it is described by k_{II} and k_{III} and equation 3 is most properly written

$$N_O/N = e^{-(k_I + k_{II} + k_{III})D} \qquad \text{----------} \quad (4)$$

The factors determining k_{II} operate only if O_2 is present during irradiation, while k_{III} represents the formation of free radicals that have extended life-times in the dried spore, and that can combine with O_2 to become lethal to the spore, even at long times after irradiation. Or,

$$R^{\cdot} + O_2 \longrightarrow RO_2^{\cdot} \text{ (lethal)}$$

is an experimentally demonstrated fact in this system.

So, the "O_2 effect" here is not a singular oxygen effect, but rather two oxygen effects, and each of these, as does k_I, the anoxic component, has its own unique set of properties, saying that at least three sets of physico-chemical processes operate in bringing about the radiation-induced death of the spore. Here no single explanation of the action of O_2 is permissible. This alerts us to the possibility that more than one action of O_2 might be seen in other systems also.

In the fully wet spore system these two O_2 effects cannot be demonstrated by experiments used to describe them in the dried system; and this is to be expected since water drastically reduces free radical life-times.

But the existence of these two O_2 effects in wet systems was discussed and allowed for by us (15), and, later, in experiments in which O_2 concentration was varied (16), evidence for two O_2 actions in the spore in aqueous suspension was found. Even later evidence confirmed and extended these results (17) to suggest that there is an additional O_2 effect at low [O_2] that can be reversed by alcohols, indicating an action of ·OH. Subsequently, the complex, two-step relation between [O_2] and radiation sensitivity was described for mammalian cells (18); earlier evidence from fast-flow mixing experiments with Chinese hamster cells had indicated two resolvable time-dependent O_2 effects (19).

Also, experiments in which O_2 was presented to spores and vegetative cells at various times before and after irradiation, firm evidence for two actions of O_2 was discovered. In the spore (20), two O_2-dependent components with first order decay kinetics were recognized, a fast one with a rate of 0.077/s, and a slow one with 0.0058/s. And, subsequently, in cells of *Serratia marcescens* also two first-order O_2-dependent components were revealed (21), one decaying at 2000/s and the other at 170/s. The results in both studies support the proposition that there are two O_2 effects.

I call attention to two other historical notes. First, in his 1957 paper (13) Gray proposed that the total radiation sensitivity S is equal to the anoxic component S_n added to the aerobic component S_o when each is the reciprocal of the dose to bring about the effect. This antecedes our 1960 statement to that effect. And, second, in their study of chromosomal abnormalities, Deschner and Gray (12) super-corrected the ascites data for varying [O_2] caused by metabolism, and these numbers (see Fig. 7 of ref. 12) fit the step function described years later for spores (16,17), and then mammalian cells (18), much better than the Michaelis-Menten type line drawn.

A summary of this brief presentation of early history is as follows: the "oxygen effect" had received little attention until the 1930's when detailed studies began that sought explanations in the biochemical and metabolic actions of oxygen in the cell. It became apparent over the next ten years that there are perhaps few reasons for thinking that the "oxygen effect" as "it" relates to very early stages is anything but physical and chemical in nature, and we must look to assistance from the physical and chemical sciences for help in understanding the "oxygen effect" in radiation biology. The search for mechanisms in these terms were initiated separately and independently by several sets of workers. Latarjet and Loiseleur, Gerschman and colleagues, and Alper each made suggestions that in retrospect are quite appropriate and consistent with some of our current understandings. The importance of the dry spore work, followed by the wet spore work, was the recognition of independently operating components of radiation damage, and the necessity of describing their properties separately and independently. So, we see the anoxic component on the one hand, and the oxic on the other, and then the immediate recognition at that point that there is not one but two oxygen effects in the dry spore, and one of these is demonstrably the $R\cdot + O_2 \longrightarrow RO_2^{\cdot}$ reaction proposed earlier by others. In the water saturated spore, two O_2 effects were delineated, these were confirmed, and then a third, and a possible fourth, were described. Then in vegetative bacterial cells and in mammalian cells evidence for two actions of O_2 was discovered.

The O_2 black box is now open. A variety of techniques (physical, chemical and biological) allows now investigation of O_2 participation in radiation-induced biological changes with precision that permits recognition of fine structure and small differences. The "oxygen effect" is now plural rather than singular, certainly in some cell systems. Even though the implications of this, especially with reference to the manipulation of the radiation sensitivity of tumor cells, are now not clear, the kind of work that must be done is.

ACKNOWLEDGEMENTS

Research in these laboratories is supported by the U.S. National Institutes of Health (GM-13557 and RR-00886) and by the Department of Energy (EY-76-05-3408).

REFERENCES

1. Schwarz, G., Münch. Med. Wochnschr., 56,1217 (1909).
2. Crabtree, H. G. and Cramer, W., Proc. Roy. Soc. London, B 113,238 (1933).
3. Laser, H., Nature, 174 (1954).
4. Thoday, J. M. and Read, J., Nature, 163 (1949).
5. Moustacchi, E., Annal. Inst. Pasteur, 94,89 (1958).
6. Gray, L. H., Conger, A. D., Ebert, M., Hornsey, S. and Scott, O. C. A., Brit. J. Radiology, 26,638 (1953).
7. Latarjet, R. and Loiseleur, J., Compt. Rendu Soc. Biol., 136,60 (1942).
8. Gerschman, R., Gilbert, D. L., Nye, S. W., Dwyer, P. and Fenn, W. O., Science, 119,623 (1954).

9. Gerschman, R., "Proceedings of XXI International Congress of Physiological Sciences," Buenos Aires (1959), p. 222.
10. Alper, T., Radiat. Res., 5,573 (1956).
11. Howard-Flanders, P. and Alper, T., Radiat. Res., 7,518 (1957).
12. Deschner, E. E. and Gray, L. H., Radiat. Res., 11,115 (1959).
13. Gray, L. H., "Lectures on the Scientific Basis of Medicine," Athlone Press, London Volume VIII (1957), p. 314.
14. Powers, E. L., Webb, R. B. and Ehret, C. F., Radiat. Res. Suppl. 2,94 (1960).
15. Tallentire, A. and Powers, E. L., Radiat. Res., 20,270 (1963).
16. Tallentire, A., Jones, A. B. and Jacobs, G. P., Isr. J. Chem., 10,1185 (1972).
17. Ewing, D. E. and Powers, E. L., Science, 194,1049 (1976).
18. Millar, B. C., Fielden, E. M. and Steele, J. F., "Radiation Biology and Chemistry: Research Developments" (H. E. Edwards, S. Navaratnam, B. J. Parsons and G. O. Phillips, eds.), Elsevier, Amsterdam (1979), p. 397.
19. Shenoy, M. A., Asquith, J. C., Adams, G. E., Michael, B. D. and Watts, M. E., Radiat. Res., 62,498 (1975).
20. Stratford, I. J., Maughan, R. L., Michael, B. D. and Tallentire, A., Int. J. Radiat. Biol., 32,447 (1977).
21. Michael, B. D., Harrop, H. A. and Maughan, R. L., "Proc. 6th International Congress of Radiation Research," Tokyo (1979), p. 288.

OXYGEN-DEPENDENT DAMAGING PROCESSES IN IRRADIATED CELLS

A. Tallentire, Department of Pharmacy, University of Manchester, Manchester, M13 9PL, U.K.

INTRODUCTION

The radiation scientist, who seeks with the aid of experiments done on intact cells to describe the events and processes responsible for induction of ionizing radiation damage in cells, sets himself an almost insuperable task. His test system is physically and chemically of the greatest complexity which, for the most part, is uncharted; the measures of damage that he is able to make are mostly non-specific; he has no direct means of detecting or identifying transient species or products so he cannot therefore measure yields of these, nor can he directly follow reaction rates. Despite this, he is undeterred, and it is proper that he is. For, the only unequivocal evidence of the importance and effectiveness of a given induced event, species, product, reaction, or process is that which is generated using a cellular test system.

By analogy with radiation-induced events known to occur in model chemical systems, and from direct observation of irradiated cells, it is generally believed that cellular radiation damage develops sequentially. The stages of this development, pertinent to present considerations, are shown schematically in Fig. 1. The experimentalist working with cells measures then an observable response to radiation that is the end result of a sequence of events; typical of such a response is the inability of an irradiated cell to reproduce itself and survive. Enhancement of response by oxygen presence at the time of irradiation is thought to occur as a consequence of physical or chemical intervention of this agent at one or more stages of the sequence in a way (or ways) which produces damage additional to that induced in the absence of oxygen. The arrows in Fig. 1 indicate those stages in the sequence where, from studies performed on non-cellular systems, oxygen might be expected to intervene, altering yields or fates of chemical species or products induced by radiation in cells and thereby increasing observed response. Clearly, enhancement of response could well be a net consequence of several different induced processes, whose courses are under the influence of oxygen.

Using cells, how might these different processes be distinguished and recognized? In the absence of direct means of following radiation-induced events in cells, it is necessary to rely on an indirect approach. The usual experimental methodology is

ISBN 0-12-592050-4

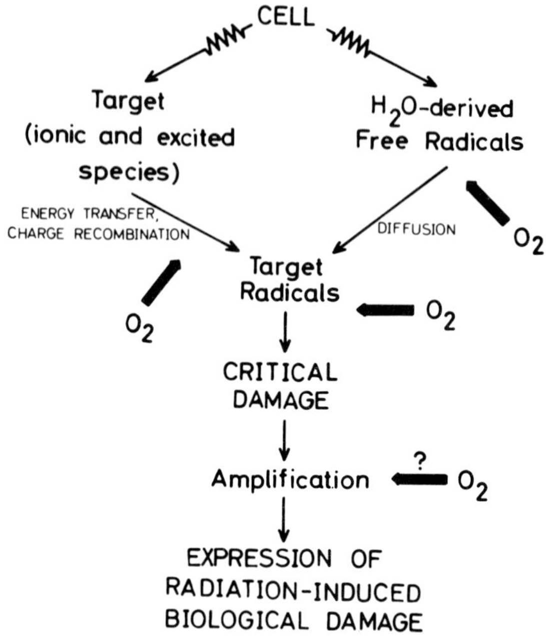

FIGURE 1. Schematic representation of the development of observed damage induced by radiation in cells. Arrows indicate points in this development where oxygen might intervene.

to make intentional changes in the physical or chemical state of the irradiated cell in a manner expected to change the yields of potentially damaging species, the courses of reaction of these species, and hopefully the amounts of critical damage; from corresponding changes in observed responses, the nature of damaging events is inferred. It may well be, of course, that underlying the choice of the physical or chemical condition imposed on the cell, is its known influence on reactions or processes recognized under ideal circumstances in model systems, but in the end, the interpretations put on findings from experimentation with cells are based wholly on inference. It is important to appreciate this point at the outset; all too often in the past in the area of mechanistic cellular radiation biology, the tendency has been for inference and supposition, stated at length and with certitude, to become accepted fact.

Very recently, Ewing and Powers (1) have examined inferential evidence on the nature and properties of chemical mechanisms involved in oxygen-dependent sensitization of cells irradiated in suspension. Their examination is comprehensive, interpretive and balanced, and so there is little to be gained from repeating the exercise here. Rather, the purpose of the present paper is to describe the latest data on sensitization by oxygen in wet cells which augments the Ewing/Powers' survey and allows a rationalization of certain oxygen-dependent events seen in dried spores with those occurring in wet cells.

COMPONENTS OF OXYGEN-DEPENDENT DAMAGE

The proposal that oxygen sensitization is a consequence of one kind of oxygen intervention in the damage sequence, made from a consideration of early findings on changes in cell sensitivity seen with changing oxygen concentration (2-4), is no longer tenable. The fact that certain parts of the oxygen-dependent damage in cells are specifically influenced by variables such as time, chemical environment, temperature and oxygen concentration, is taken as evidence that oxygen sensitization in cells consists of more than one component. Furthermore, the particular behavior of the inferred components in respect of changing experimental conditions is suggestive that oxygen's sensitizing action is brought about by several different physico-chemical or chemical processes. The foundation work for the characterization and ultimate identification of these processes was done some 20 years ago by Powers and his collaborators (5-10) and parts of this work are considered below in the light of recent findings.

TIME RESOLVED COMPONENTS

Dried cells. Powers, Webb and Ehret (5) were the first to demonstrate that sensitization with oxygen could be operationally divided into at least two components. Using dried *Bacillus megaterium* spores, they were able to distinguish three distinct levels of X-ray sensitivity depending upon the gaseous conditions prevailing at the time of and after irradiation (Fig. 2). A low baseline sensitivity occurred with spores irradiated in nitrogen and treated with nitric oxide after irradiation prior to exposure to oxygen (symbolized by $N_2 \rightarrow NO$); a higher level of sensitivity was seen for spores also irradiated in nitrogen but exposed to oxygen within minutes after irradiation without any intervening treatment ($N_2 \rightarrow O_2$); and the highest sensitivity occurred with spores irradiated in oxygen and held in this gas after irradiation ($O_2 \rightarrow O_2$). The existence of these three levels, clearly dependent on the state of oxygenation of the spores during and after irradiation, allowed division of radiation damage in dried spores into three components or classes (8), *viz*: Class I—that described by k_I in the algebraic expression of Powers above that is wholly oxygen-independent; Class III—that described by k_{III} that is developed only if oxygen is presented post-irradiation to spores following irradiation in nitrogen (called the oxygen-dependent 'postirradiation' component); and Class II—that described by k_{II} that is seen only when spores are irradiated in the presence of oxygen (designated the oxygen-dependent 'immediate' component). The two classes of damage II and III, resolvable on the basis of time, account for the overall sensitization by oxygen in dried spores.

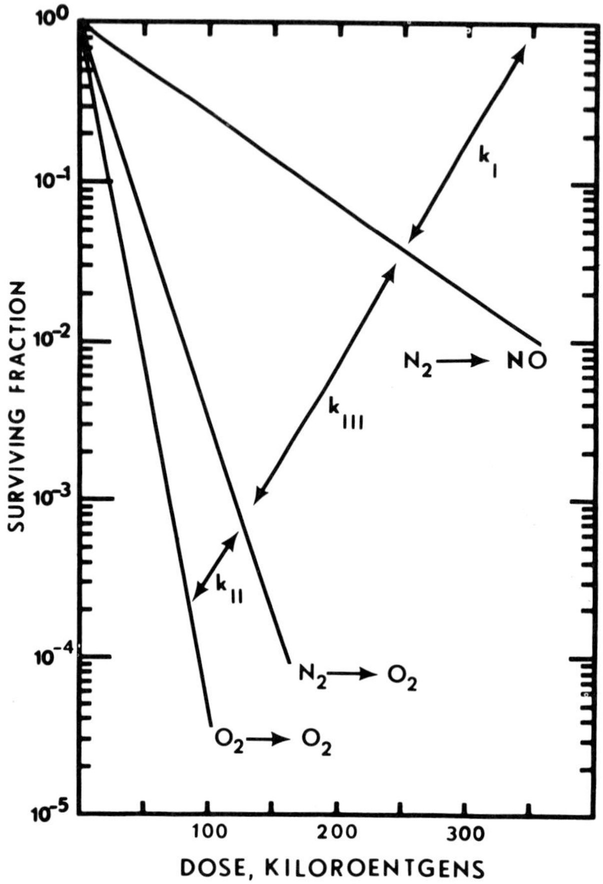

FIGURE 2. Survival of dried Bacillus megaterium spores X-irradiated in three experimental conditions. The symbols are described in the text.

The operational separation of the oxygen effect into two components, 'immediate' and 'postirradiation,' led to the proposal that there are two kinds of oxygen-dependent X-radiation-induced species in the dried spore system, one with a short life and one with a long life, reaction of oxygen with either leading to lethality (5). The fact that the 'post-irradiation' component was prevented from operating by treatment of the spores with the radical scavenging agent, nitric oxide, was taken to mean that the long lived species is a free radical. To this day, this is the only unequivocal demonstration of the involvement of free radicals in the mechanisms of the oxygen effect in cells. While not revealed specifically by these experiments, the damaging interaction between the induced radical and oxygen was suggested as being the formation of a peroxy radical (9).

Since the original observations on dried B. megaterium spores were made, the existence of two time resolvable components of the oxygen effect has been found with every other cell type that has been examined in a manner similar to that used for bacterial spores. These include vegetative cells of Staphylococcus aureus (11) and Streptococcus faecium (12), conidiospores of the eucaryote, Aspergillus nidulans (13), and endospores of other Bacillus species (14,15). There is therefore good reason to believe that the phenomenon is a general one.

Wet cells. The 'postirradiation' or 'free radical' component of the oxygen effect is prevented from operating in dried cells by anoxic treatment with the gas nitric oxide (7) or hydrogen sulphide (8) or with heat prior to exposure to oxygen (6). It is also prevented by anoxic exposure to water, which presumably acts by hastening the decay of the induced free radicals to a form in which they are no longer reactive and therefore harmless when oxygen is admitted to cells after irradiation (10). This modulating action of water supposedly determines why anoxic wet cells given X or γ-radiation show no postirradiation oxygen effect. However, in wet cells oxygenated at the time of irradiation, it is conceivable that damaging events occur which are similar in kind to those operating in the postirradiation oxygen effect, as recognized in dried cells. With oxygen and water present together in cells, the harmful reactions of radicals with oxygen will be expected to compete successfully with harmless decay reactions giving a level of response to radiation greater than that seen with cells in water without oxygen.

Recently a number of studies has been aimed at characterization of the competing harmful and harmless reactions, as they occur in intact wet cells. These studies have used the so-called 'fast-mix' techniques in which intense pulse irradiation is coupled with a means of changing the level of oxygenation within the cell at precisely known short times after irradiation. Two such studies done on different cell types provide evidence that oxygen-dependent damage in wet cells also has time resolvable components.

Fig. 3 gives the basic findings of Michael et al. with wet vegetative bacterial cells of Serratia marcescens (16). Each point represents the survival of cells exposed to a fixed radiation dose, given over a very short time interval, and to oxygen at the time indicated. Changes in survival seen with changing time of exposure to oxygen have been used to deduce the kinetics of decay of the oxygen-dependent damage (presumably free radical in nature). The best description of these kinetics is obtained when it is assumed that decay occurs by two first order processes

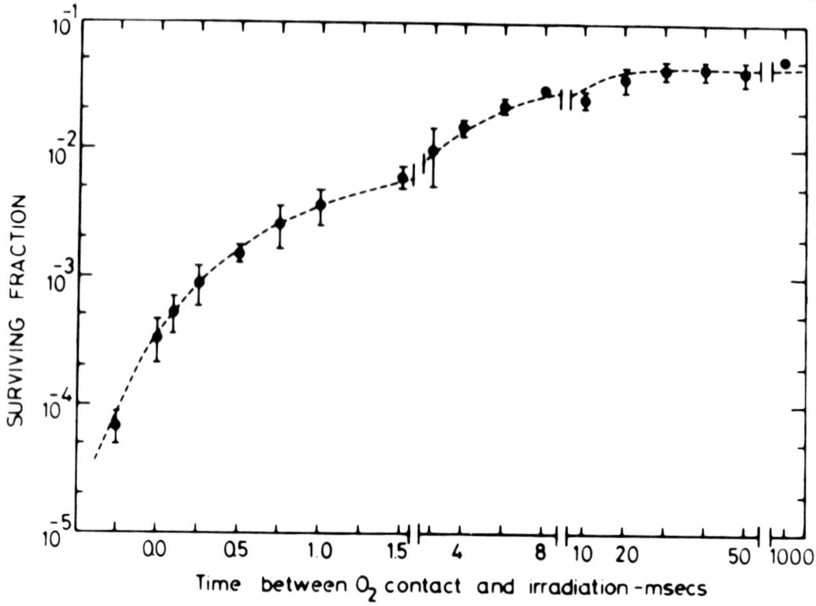

FIGURE 3. Time course for the action of oxygen on the survival of Serratia marcescens cells irradiated with a single 175 krad pulse of electrons of 5 ns duration (reproduced by permission of B. D. Michael).

operating in parallel at appreciably different rates. The rate of the 'fast' decaying component of oxygen-dependent damage in these cells is $2000s^{-1}$ and that of the 'slow' component $170s^{-1}$ (the dashed line in Fig. 3 is derived from this dual exponential decay model). It should be noted that the decay rates of these two inferred components show about a twelve fold difference.

The findings with Serratia cells closely parallel, qualitatively, those made originally with wet bacterial spores handled in a like manner (17). Again the best fit to the data requires the simultaneous decay of two oxygen-dependent damages having different first order rates, the faster process taking a value of $0.079s^{-1}$ and the slower a value of $0.0058s^{-1}$. Clearly, these values are very much less than the estimates for decay corresponding to 'fast' and 'slow' components of oxygen-sensitive damage in the vegetative cell test system. The reason for the slow decay in bacterial spores is presently not known, but the reduced water content of spores over that of vegetative cells could be a contributing factor. However, we note that the relatively slow rates associated with the spore system also differ by a factor of around twelve, giving further support to the idea that different cells have common mechanisms of oxygen sensitization.

The relatively long time-scale of decay of damage in spores means that with this test organism there is an opportunity for experimental manipulation of the damage in a way that may not be possible with cell types in which diffusion rates of solutes into the cell exceed decay rates (16). For instance, the susceptibility of O_2-dependent damage to oxidizing agents other than oxygen is open to study by the use of a simple fast liquid-mix technique. Recently, we have initiated experiments of this kind (18). Instead of admitting oxygen in solution to spores at various times after pulse irradiation, hydrogen peroxide in solution has been given to them under anoxia. Apparently hydrogen peroxide acts on the 'slow' component of oxygen sensitization, giving a response comparable with that of oxygen. This result could be a pointer to another possible mechanism of sensitization, partly mimicking oxygen, which may become significant in circumstances where hydrogen peroxide yields are known to be high, for example, with high LET irradiation and with high dose rates.

The most recent thrust of the experimental work employing time to separate components of oxygen-dependent damage has involved modification of individual components. For example, it has been found that the presence of oxygen at the instant of delivery of the pulsed irradiation removes preferentially the 'slow' component of damage (16,17); the presence of certain agents belonging to the electron-affinic class of radiosensitizers apparently also removes this component only (19,20), but with a sulphydryl compound (dithiothreitol) present, the decay rates of both the 'fast' and 'slow' components are affected equally (16). These findings are highly encouraging since, with the availability of means of selectively influencing time-resolved components, there are real prospects of identifying the chemical species and processes involved in these forms of damage.

OXYGEN CONCENTRATION DEPENDENT COMPONENTS

A division of the overall oxygen sensitization in wet cells has now been recognized on the basis of the dependency of sensitivity on oxygen concentration; again the bacterial spore was the test organism in which the phenomenon was first discovered.

An exact measurement of the relationship of the sensitivity of B. megaterium spores to gamma-rays as a function of oxygen concentration in aqueous suspension (21) yielded datum points which cannot be represented properly by a simple hyperbolic function of the type described by Alper and Howard-Flanders (2). Rather, the curve describing best the relationship exhibits a definite discontinuity at an oxygen concentration in the gas phase of around 2% (equivalent to $\simeq$ 26 $\mu mol/dm^3$ in solution), indicating that, at oxygen concentrations below this level, sensitization is being determined by one form of oxygen concentration dependency and that, at concentrations above it, a second and different form of oxygen dependency operates. The simplest interpretation of this precise finding is that oxygen sensitization in wet spores is a consequence of at least two oxygen-dependent components of damage, a 'low' oxygen concentration component amounting to about 40% of the overall oxygen effect and a 'high' component representing about 60%. Since the first finding was made, similar divisions in oxygen sensitization in aqueous cells have been reported for spores irradiated with soft X-rays, for which the ratio of 'low' and 'high' components was found to be close to 70:30 (22), and gamma-irradiated V79 Chinese hamster cells which gave again a 'low':'high' component ratio of 40:60 (23). A likely origin of the difference in values of this ratio is considered below.

A collation of the results of six oxygen concentration studies, covering three different cell types whose absolute sensitivities differ 500 fold, has recently been prepared by Ewing and Powers (1) in a way that allows direct comparison of the different cells' response to changing oxygen concentration. The close agreement within the data argues, according to the authors, for an oxygen effect that is the same in the three cell systems and perhaps too for a unity for the mechanisms involved. We note, however, that without further information there is no justification for equating the processes separated on a temporal basis in dried or wet cells with the two distinguished on a concentration basis in the wet cells.

TEMPERATURE DEPENDENT COMPONENTS

Recently, another way of dividing the oxygen effect in wet spores has been found. The basis of this separation is the behavior of oxygen sensitive damage with changes in temperature at the time of irradiation.

Fig. 4 (Gaszo and Tallentire, unpublished results) shows how the responses of spores in nitrogen saturated (solid line) and in oxygen saturated (dashed line) water suspensions vary with

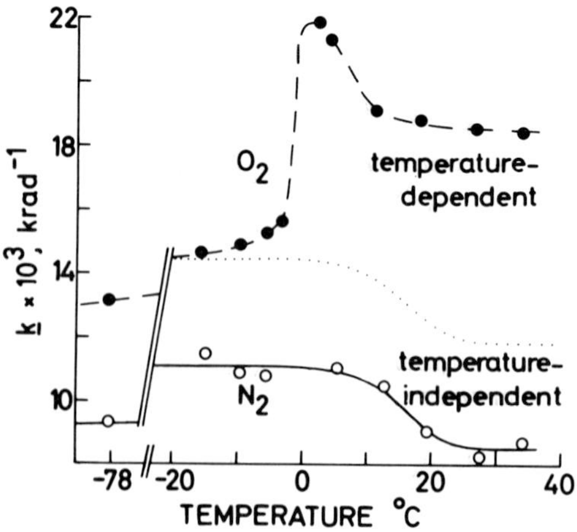

FIGURE 4. Responses of γ-irradiated Bacillus megaterium spores in N_2-saturated (o) and O_2-saturated (•) water suspensions held at various temperatures during irradiation.

temperature. The difference between the responses in nitrogen and in oxygen at any one temperature represents maximal sensitization by oxygen at that temperature. It is noted that over the temperature range 35° to 2°C the extent of oxygen sensitization is invariant; at around 0°C oxygen-dependent damage falls sharply to a level which is about 40% of that seen above 0°C; and for the range -4° to -78°C, the oxygen effect holds constant at this low level. Our interpretation of these findings is that the oxygen effect in wet spores at ambient temperature has two components possessing different properties in respect of temperature change, one component representing some 40% of the overall effect that is temperature independent and another that is wholly sensitive to a critical temperature change corresponding to a transition in the phase state of the spore suspension. While we choose at present to take no stand regarding the significance of this correspondence, the likelihood is that it will figure in future experiments designed to explore mechanisms involved in the temperature dependent, oxygen-dependent component.

RATIONALIZATION OF THE VARIOUS COMPONENTS RECOGNIZED OPERATIONALLY

An obvious next step in this inferential approach is to look for interrelationships among the various components of the oxygen effect resolved by different experimental manipulations. First results from work of this kind are presented in Fig. 5. This gives the responses to variations in temperature of nitrogen equilibrated (solid line) and 100% oxygen equilibrated (dashed line) spores taken from the previous figure, together with datum points for spores equilibrated with 0.5, 2.2 and 4.1% oxygen in nitrogen mixtures. These oxygen concentrations span the critical part of the oxygen dependency curve, as seen under ambient conditions, so that the effects of temperature on the 'low' and 'high' oxygen concentration components may be examined. From Fig. 5 we see that, for oxygen concentrations at which the 'low' component partially or wholly operates (0.5 and 2.2% oxygen in the gas phase), the extent of sensitization by oxygen is independent of the temperature during irradiation over the range tested, and, for an oxygen concentration at which the 'low' component plus part of the 'high' component potentially operate (4% oxygen in the gas phase), sensitization shows a dependence upon temperature. Furthermore, both datum points generated with spores in equilibrium with 4% oxygen at temperatures below 0°C fall on a level of sensitivity representing the operation of only the temperature independent component in maximally oxygenated spores. While it is obvious that more data are needed before categorical statements can be made, the results obtained so far are consistent with the notion that the temperature independent component of the oxygen effect is analogous to that component responding to low oxygen concentrations.

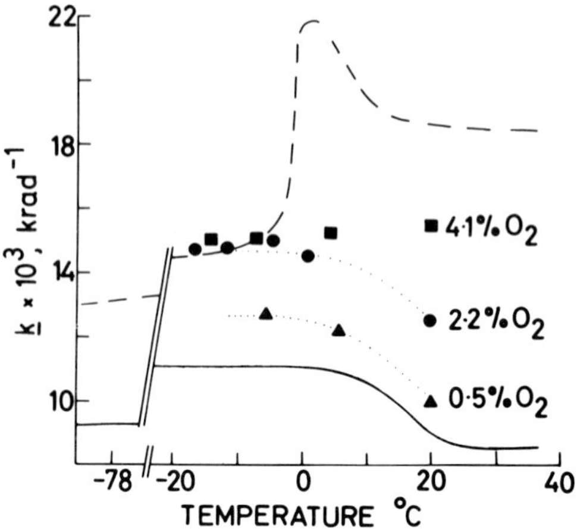

FIGURE 5. Responses of γ-irradiated Bacillus megaterium spores in water suspensions equilibrated with different O_2/N_2 mixtures during irradiation at various temperatures. The solid and dashed lines represent responses for N_2-saturated and 100% O_2-saturated suspensions respectively, taken from Figure 4.

The recognition in wet spores of a component of the oxygen effect that is temperature independent means that there are now grounds for rationalizing, at least in part, the components separated on a time basis in dried cells and those of wet spores. The early work done with dried spores on the relationship of sensitivity and temperature under specifically controlled conditions during and after irradiation (5) revealed that the two components resolved by time behave differently in respect of temperature (Fig. 6). The 'postirradiation' or 'free radical' component, shown bounded by curves 1 and 3/4 in Fig. 6, is unaffected by the temperature during irradiation within the range extending from just above room temperature to -200°C [the loss of this component at high temperatures has been ascribed to thermal annealment of potentially effective free radicals during anoxic irradiation (6)], whereas the 'immediate' oxygen-dependent component, shown bounded by curves 2 and 1, shows strong temperature dependence for temperatures between ambient and -150°C, ultimately taking a value of zero at the low end of the temperature range. The similarity in behavior, in respect of temperature independence, of the 'free radical' component of dried spores and of one of the components of wet spores implies that the two are linked. Taking this notion a stage further, we can propose that the component of the oxygen effect exhibiting a low oxygen concentration dependency in wet cells is equal to that showing temperature independency in wet spores and that this component has involved in its damaging process a cellular free radical/oxygen interaction. Evidently, for wet spores and for V79 Chinese hamster cells exposed to

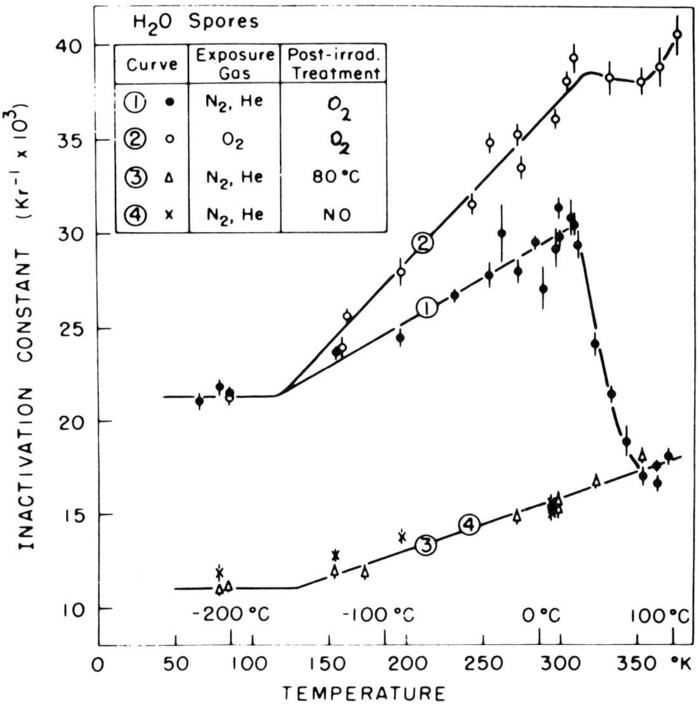

FIGURE 6. Relationship between temperature and radiation sensitivity of Bacillus megaterium spores held under specified gaseous conditions during and after irradiation [taken from Powers, Webb and Ehret (5), by permission Academic Press.]

gamma-radiation, this 'free radical' component is responsible for a little less than a half of the overall oxygen effect.

INVOLVEMENT OF WATER RADIOLYSIS PRODUCTS IN OXYGEN SENSITIZATION

So far, attention has been focused on a mechanism of oxygen sensitization involving free radical/oxygen interactions in the damaging event. This mechanism, it is inferred, accounts at least in the wet spore and in certain mammalian cells for a part only of the oxygen effect, so it is proper to ask which other processes contribute to the oxygen effect. For example, there is the process responsible for the 'immediate' component recognized operationally in dried cells; what is the nature of this process and is it involved in oxygen sensitization in wet cells? To date, findings from experiments with cells have not provided an answer to these questions, nor in fact even supplied clues. Another question is, in progressing from dry to wet cells, do other oxygen-dependent processes that are water dependent come into operation (10)? As about 70% of the weight of the wet cell is water, this seems a reasonable question to pose. Also there are unconnected pieces of evidence that hint indirectly at water involvement in oxygen-dependent processes; typical of these is the large change in the magnitude of the oxygen-dependent damage seen at the phase transition point in temperature dependency experiments [above and references (24) and (25)].

A possible way for a water dependent process of oxygen sensitization to act is via the transient species formed from the radiolysis of the cell's water. A means of testing for the involvement of the species so formed is to examine how and to what degree a given chemical additive, capable of scavenging selectively a particular transient, influences the oxygen-dependent components of the oxygen effect. In the light of such information, inferences may be drawn regarding the species involvement in the damaging process. There exists already a body of information of this kind implicating the hydroxyl radical as a damaging species [for reviews of this see references (1) and (26)] and here it is proposed to re-examine briefly the cogent evidence in the context of multi-component sensitization by oxygen.

The apparently decisive set of experiments linking OH radicals and oxygen sensitization is that of Ewing and Powers (22) in which it was shown that the OH radical scavenging agent, t-butanol, protects against oxygen-dependent soft X-ray damage in water suspended spores over a specific range of oxygen concentrations. The Ewing-Powers findings and these authors' own re-analysis are shown in Fig. 7. With no t-butanol and increasing oxygen concentration in suspension (open points), the behavior of oxygen-dependent damage is qualitatively similar to that found earlier with spores gamma-irradiated in phosphate buffer (21), but with the difference that the level at which the intermediate plateau occurs is about 70% of the total oxygen effect rather than 40%. With 0.1M t-butanol present (solid points), there is no effect on sensitization at high oxygen concentrations, but at low and intermediate concentrations, a part of the sensitization apparently due to oxygen is removed. Ewing and Powers (1) have called that portion removed by t-butanol the OH component of oxygen-dependent damage and proposed that it results from OH radical damage. Similar experiments to those of Ewing and Powers, run recently with spores in water and with gamma-rays, show that in this system the OH radical scavengers, t-butanol and KSCN, have very little effect on oxygen sensitization; under these irradiation conditions, the presence of either scavenger gives a barely significant reduction in oxygen sensitization at high and intermediate oxygen concentrations, and no reduction at low oxygen concentrations. Certainly there is not a pronounced OH component as observed with soft X-rays. These differences in findings for soft X-rays and for gamma-rays could well be indicative of differences in yields of effective water-derived species or products at given levels of oxygen concentrations for the two radiation

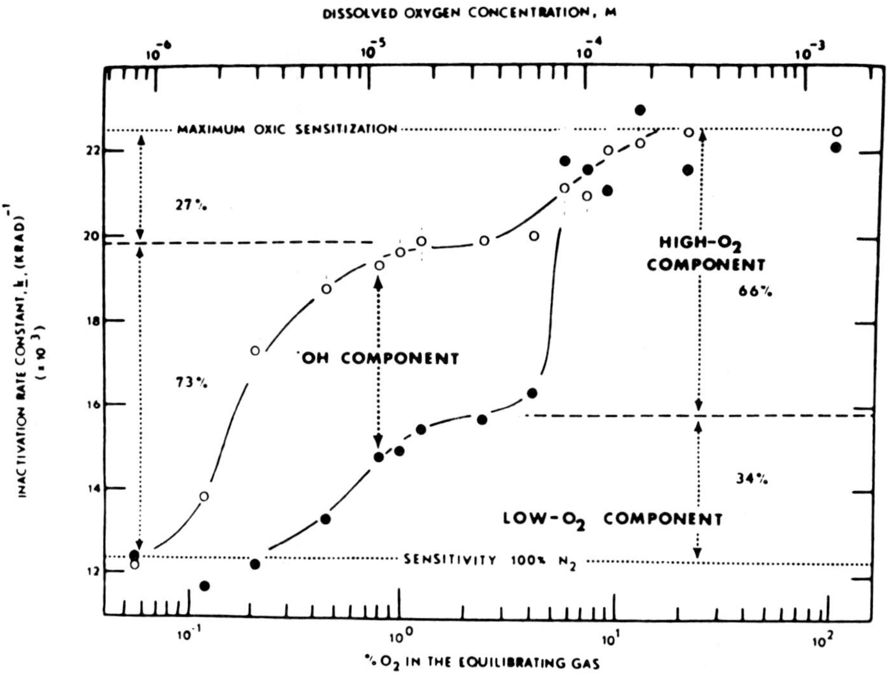

FIGURE 7. Radiation sensitivity as a function of O_2 concentration for Bacillus megaterium spores irradiated with 50 kVp X-rays while suspended in water alone (o) or in water + 0.1M t-butanol (●) [taken from Ewing and Powers (1), by permission American Association for the Advancement of Science].

types.

A particularly interesting comparison, in the context of the different findings noted above, is one involving data for soft X-irradiation with t-butanol and gamma-ray data with no t-butanol. In Fig. 8, the data for these two sets of conditions, which yield different absolute sensitivities, have been normalized by plotting the degree to which different oxygen concentrations sensitize within the maximal effect of oxygen. From the locations of the datum points, it can be concluded that all are describing a single relationship of sensitization and oxygen concentration. One interpretation that can be put on this conclusion is that the form of the relationship for gamma-rays is the basic response and that t-butanol presence with soft X-rays causes a shift in the X-ray relationship, making the two equal. In other words, the t-butanol can be thought of as negating an OH radical dependent process arising out of differences between exposure with soft X-rays and gamma-rays.

In addition to t-butanol, Ewing (27) has tested separately the abilities of methanol, ethanol, allyl alcohol and t-amyl alcohol to remove part of the oxygen-dependent damage caused by soft X-rays and found that these correlate well with OH radical

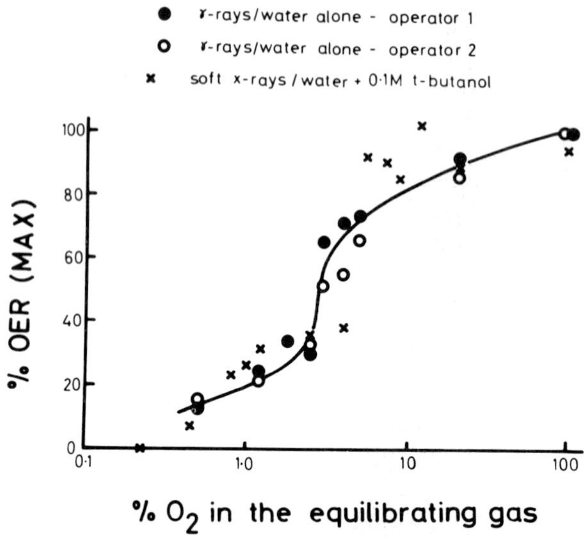

FIGURE 8. A comparison of the relationships between sensitization and O_2 concentrations for Bacillus megaterium spores irradiated under two sets of conditions. The solid line is fitted to the circles, which represent data for spores γ-irradiated in water; the crosses are the findings of Ewing and Powers (22) for spores in 0.1M t-butanol given 50 kVp X-rays.

scavenging efficiency. This finding supports strongly the involvement of OH radicals in the induction of this particular component of damage, but it argues against their involvement as the damaging species operating within the cell. The concentrations of scavengers in these experiments (and, in fact, in all others in which chemical additives are used) were extracellular concentrations and these nominal concentrations were used to derive values of scavenger efficiency. If the correlation recognized by Ewing is to reflect the scavenging ability towards OH radicals effective inside the cell, then it has to be allowed that either the concentrations inside and outside the cell do not differ (i.e. there is no concentration gradient across the spore coat), or that they differ by a proportion that is common to all additives. With five such additives, possessing widely varying physicochemical properties to this author it is unlikely that either of these requirements holds. Clearly then, serious consideration has to be given to an alternative possibility, namely, that the functional radicals in respect of the OH component of oxygen sensitization with soft X-irradiation are produced outside the cell.

SUMMARY

Studies on cells demonstrate, with the degree of certainty that an inferential investigational approach allows, that multiple components exist in oxygen-dependent radiation sensitization. Two components can be distinctly recognized on the basis of the behavior of oxygen sensitization towards each of (i) the time elapsing between irradiation of cells and exposure to oxygen, (ii) the concentration of oxygen surrounding them at the instant of irradiation, and (iii) their temperature during irradiation. The clear indications are then that oxygen sensitizes cells by more than one chemical or physical process.

Application of combinations of experimental variables allows relationships to be drawn between components of the oxygen effect distinguished by imposing a single variable. A 'free radical' component, first recognized 20 years ago with dried bacterial spores and involving cellular free radical/oxygen interaction, accounts for about 40% of oxygen sensitization in gamma-irradiated wet cells. Evidence is provided that suggests that the OH component of oxygen sensitization seen with soft X-irradiation of wet spores is a characteristic of this radiation type. The proposal is made that the water-derived OH radicals implicated in this component are functioning extracellularly.

REFERENCES

1. Ewing, D., and Powers, E. L., "Radiation Biology in Cancer Research" (R. E. Meyn and H. R. Withers, eds.), Raven Press, New York, N.Y. (1980), p. 143.
2. Alper, T. and Howard-Flanders, P., Nature, 178,978 (1956).
3. Alper, T., Radiat. Res., 5,573 (1956).
4. Howard-Flanders, P., "Advances in Biological and Medical Physics," Vol. VI, Academic Press, New York, N.Y., 1958, p. 554.
5. Powers, E. L., Webb, R. B. and Ehret, C. F., Radiat. Res., Suppl. 2,94 (1960).
6. Webb, R. B., Powers, E. L. and Ehret, C. F., Radiat. Res., 12,682 (1960).
7. Powers, E. L., Webb, R. B. and Kaleta, B. F., Proc. Nat. Acad. Sci., 46,984 (1960).
8. Powers, E. L. and Kaleta, B. F., Science, 132,959 (1960).
9. Ehret, C. F., Smaller, B., Powers, E. L. and Webb, R. B., Science, 132,1768 (1960).
10. Tallentire, A. and Powers, E. L., Radiat. Res., 20,270 (1963).
11. Webb, R. B., "Physical Processes in Radiation Biology" (L. Augenstein, R. Mason and B. Rosenburg, eds.), Academic Press, New York, N.Y., 1964, p. 267.
12. Mavinkurve, S. S., Ph.D. Thesis, University of Manchester, 1973.

13. Wilson, J. B. and Powers, E. L., Radiat. Res., 43,698 (1970).
14. Ley, F. J., "Ionising Radiation and the Sterilization of Medical Products," Taylor and Francis Ltd., London, 1964, p. 51.
15. Al-Shaickly, M. A. S., Jones, A. B. and Tallentire, A., "Spore Research 71," (A. N. Barker, G. W. Gould and J. Wolf, eds.), Academic Press, London, 1971, p. 264.
16. Michael, B. D., Harrop, H. A. and Maughan, R. L., Proceedings of the 6th International Congress of Radiation Research, Tokyo, 1979, p. 288.
17. Stratford, I. J., Maughan, R. L., Michael, B. D. and Tallentire, A., Int. J. Radiat. Biol., 32,447 (1977).
18. Tallentire, A., Barber, D. J. W., Maughan, R. L. and Michael, B. D., "Radiation Biology and Chemistry: Research Developments" (H. E. Edwards, S. Navaratnam, B. J. Parsons and G. O. Philips, eds.), Elsevier Scientific Publishing Co., Amsterdam, 1979, p. 283.
19. Tallentire, A., Stratford, I. J., Maughan, R. L., and Michael, B. D., Br. J. Cancer, 37, Suppl. III, 34 (1978).
20. Michael, B. D., Harrop, H. A., Maughan, R. L. and Patel, K. B., Br. J. Cancer, 37, Suppl. III, 29 (1978).
21. Tallentire, A., Jones, A. B. and Jacobs, G. P., Isr. J. Chem., 10,1185 (1972).
22. Ewing, D. and Powers, E. L., Science, 194,1049 (1976).
23. Millar, B. C., Fielden, E. M. and Steele, J. J., Int. J. Radiat. Biol., 36,177 (1979).
24. Stapleton, G. E. and Edington, C. W., Radiat. Res., 5,39 (1956).
25. Wood, T. H. and Taylor, A. L., Radiat. Res., 6,611 (1957).
26. Powers, E. L., "Fast Processes in Radiation Chemistry and Biology" (G. E. Adams, E. M. Fielden and B. D. Michael, eds.), Wiley and Sons, London, 1975, p. 329.
27. Ewing, D., Radiat. Res., 73,121 (1978).

DISCUSSION

POWERS: If we should begin thinking that the radicals involved in the ·OH component are produced and function external to the cell, then there must be a species produced from ·OH (a very reactive molecule) with low chemical activity, and, therefore, a long lifetime, because in molecular terms "It's a long, long way" to DNA—from outside the spore. Also, how then do we think about the oxygen-dependent radicals that may involve ·OH seen to act at variable and very low water concentrations that are required to exist <u>inside</u> the spore in the dried spore experiments.

THE EFFECT OF OXYGEN, CHEMICAL AGENTS, DOSE RATE AND TEMPERATURE ON THE RADIATION RESPONSE OF WET SPORES

H. Weiss and A. Santomasso, Memorial Sloan-Kettering Cancer Center, 1275 York Avenue, New York, New York 10021

While we are interested in the effect of oxygen on radiation survival, the focus of the work in our laboratory is upon the more fundamental question of determining the nature of the lesions induced by ionizing radiation. Knowledge of these lesions will permit proposals to be made on how they interact with various agents including, for example, oxygen. The work I will describe involves a post-irradiation sensitizing effect of oxygen. Wet spores of Bacillus megaterium equilibrated with 100% nitrogen gas are irradiated and held in nitrogen in the sample chamber for various times before exposure to oxygen delivered at high pressure. Our irradiation source is a field emission accelerator capable of delivering doses up to megarads in single pulses of 450 keV electrons of 3 nsec duration. It is appropriate to recall that Powers and co-workers (1) pioneered this area and approach about 20 years ago using dry spores of B. megaterium for which damage development is slow. Michael, also in this session, and co-workers in 1973 (2) described a gas exchange apparatus similar to ours but used vegetative bacterial cells as the test organism.

Fig. 1 shows the surviving fractions after single pulses of electrons providing a dose of 430 krads to anoxic spores in a thin aqueous layer on a membrane filter. Each point on the figure represents an average of about 40 separate samples. It is seen in these measurements at room temperature that storage for a time after irradiation before exposure to oxygen yields increased survival. The ability of oxygen to sensitize the spores after irradiation decreases with increasing time, reaching zero after 10-15 min. in nitrogen. These results indicate the existence of a slowly decaying radiation-induced component or lesion which when combined with oxygen is lethal to the wet spore. Using the terminology in the literature of spore radiobiology (1) the damage resulting after the long decay times is denoted class I, and the damage due to reaction with oxygen after irradiation in nitrogen is denoted as class III (oxygen independent and oxygen dependent, respectively). With the assumption that the concentration of the decaying component decreases at the same rate as the time change of radiation sensitivity, various models were examined to determine the kinetics of the decay. It was found that second order kinetics of the form $dR/dt = \lambda R^2$ integrating to $R_o/R = \lambda R_o t + 1$ predicts the decay and therefore also the

ISBN 0-12-592050-4

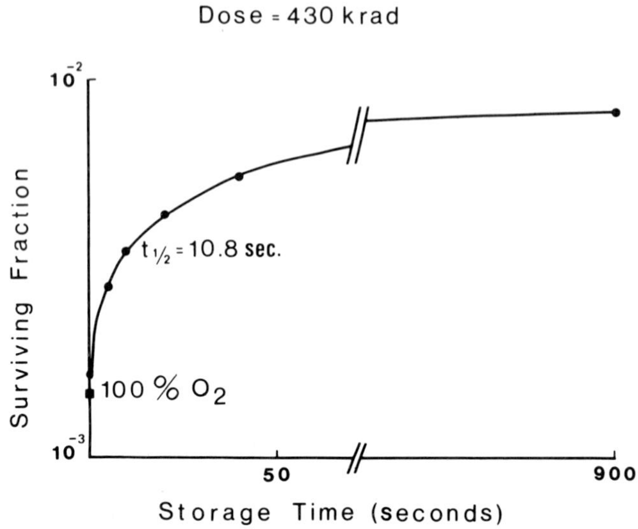

FIGURE 1. Post-irradiation effect of oxygen. Closed circles are surviving fractions for wet spores irradiated and stored in N_2 and then exposed to O_2. Closed square is wet spores irradiated in O_2.

survival. The second order rate constant is λ, and R_o is the initial concentration of the decaying component.

A consequence of second order kinetics is that the rate of decay will be dose dependent due to the R_o term $(=G(R) \cdot D)$. As seen in Fig. 2a, when the inverse fraction of the measured radiosensitivities under various conditions are plotted against storage time in nitrogen and before delivery of oxygen, the data fit a straight line with a unity intercept. The lower line in Fig. 2a is the data and second order fit shown in Fig. 1 and the next line up is a similar experiment with the dose at 650 krad. The slopes of the lines are found to be .0925 sec^{-1} at 430 krad and .135 sec^{-1} at 650 krad. Within the experimental error the ratio of the slopes are in the ratio of the doses (1.50).

Spores were also irradiated at room temperature in anoxic aqueous suspensions of ethyl, allyl or t-butyl alcohol. All three alcohols are hydroxyl radical scavengers. Ethanol and allyl alcohol are also reducing agents. All three alcohols offer a small degree of protection to spores irradiated and stored in nitrogen for minutes after the irradiation pulse (i.e. a reduction of class I damage). The amount of maximum class III damage is unaffected. Only when the reducing agent alcohols are present during irradiation is the half life shortened. The implication is that a hydrogen atom will repair this radiation-induced lesion leading to class III damage, thus indicating that the lesion is due to H atom abstraction from a vital molecule.

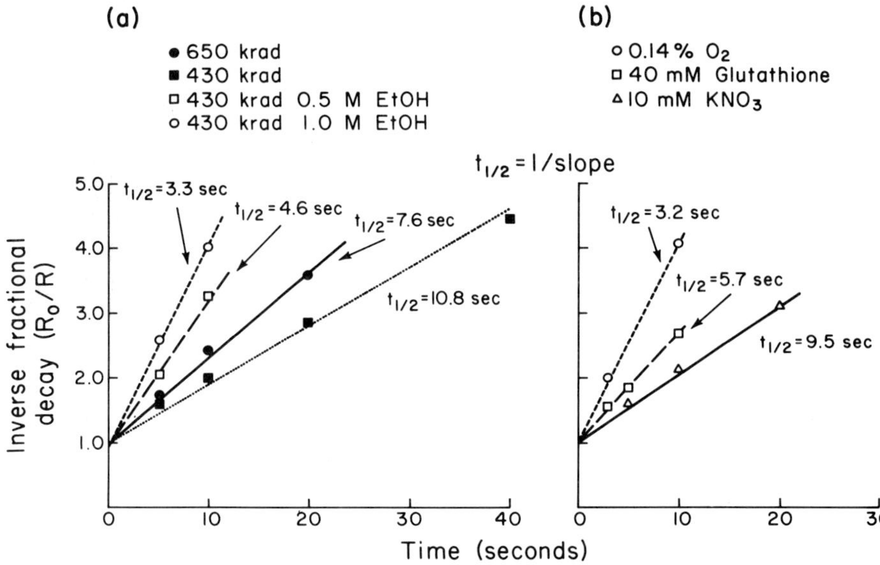

FIGURE 2. Inverse fractional decay (R_O/R) of the oxygen sensitive component.

Irradiations using the gas exchange apparatus were also carried out with spores in aqueous suspensions of other agents; KNO_3 and glutathione. Results with glutathione are that the oxygen independent component is protected to a degree greater than by the alcohols, and glutathione is more effective on a concentration basis than the alcohols in reducing the half-life of the decay of the oxygen sensitive lesion.

Irradiations were also carried out at room temperature with spores in aqueous suspensions but the 100% N_2 gas was replaced with 0.14% O_2 in nitrogen. Pure oxygen was then delivered to spores equilibrated with 0.14% O_2 and stored in 0.14% O_2 for various times after the pulse. These results are shown in Fig. 2b.

This experiment indicates that the decay of the lesion is due to competition between repair and reaction with 0.14% O_2. The equation for this reaction is $R_O/R = [(\lambda R_O + \lambda' O_2) \exp(\lambda' O_2 t) - \lambda R_O]/\lambda' O_2$, where R, R_O and λ have the same meaning as in the previous equation and λ' is the bimolecular rate constant for the reaction of R with oxygen. This equation will also yield an intercept of unity and for short times it will approximate a straight line. The equation also applies to the results with EtOH, allyl alcohol and glutathione with their concentrations replacing O_2. Fitting this equation to the results with 0.14% O_2 yields a value for the rate constant for the reaction of the radiation induced lesion with oxygen of $\lambda' = 1.5 \times 10^5\ M^{-1}s^{-1}$.

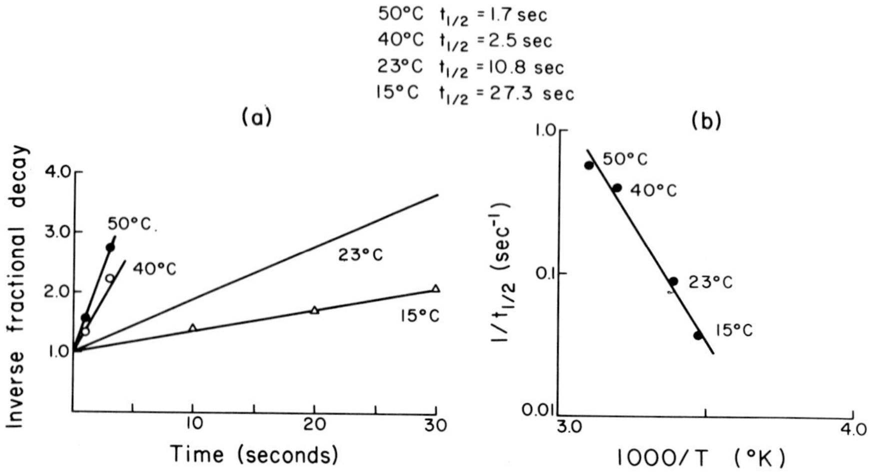

FIGURE 3. Effect of temperature of the sample on the decay of the oxygen sensitive component.

(a) Inverse fractional decay (R_o/R)

(b) Arrhenius plot of log λR_o vs T^{-1}

Fig. 3a shows the results of irradiation at 430 krad when the sample chamber is held at the various temperatures indicated. Using thermoelectric devices the temperature of the sample chamber can be varied over the range of 0°C to 70°C. The half life of the decay of the lesion is strongly dependent on temperature. In Fig. 3b, a plot of the log of slope (the inverse of the half-life) against 1/T (in degrees Kelvin) yields a straight line. This indicates a fit to an Arrhenius of Eyring (3) type of reaction with an activation energy determined from the slope of the graph of 15 kcal/mole for the decaying radiation induced lesion. This is to be compared with the activation energy of 15.9 kcal reported for dried spores by Webb et al. (4).

Fig. 4 shows the effect of temperature on class I and class III damage. Inactivation constants (radiosensitivities) are normalized to the value for class I damage at room temperature. These results preliminarily reported (5) show that increasing temperatures reduces class I damage, an unexpected result. The reverse is true in dry spores (1). Water somehow reverses the effect.

One further note from Fig. 1. Surviving fractions measured with the spores in 100% O_2 are very nearly the same as with spores irradiated in 100% nitrogen with oxygen delivered within a few msec. Similar results were observed under other irradiation conditions with alcohols, glutathione and other temperatures. This implies that the production and nature of the lesion induced by the radiation is the same whether or not oxygen is present, and the diffusion of oxygen into the spore is rapid.

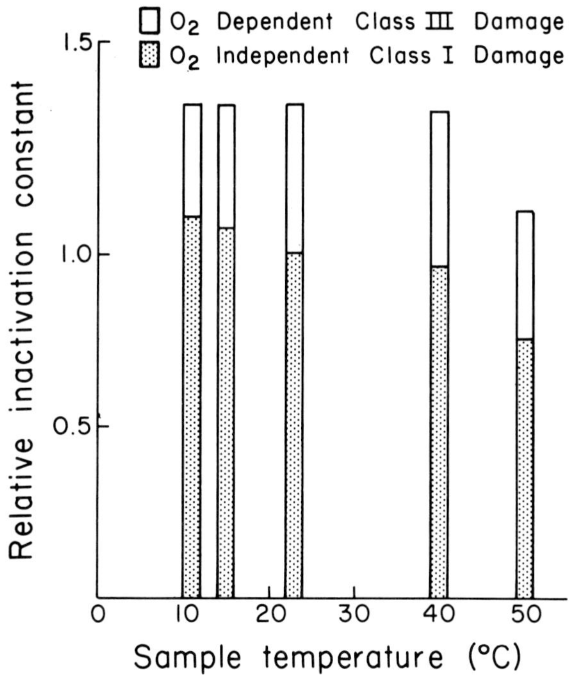

FIGURE 4. Variation of inactivation constants for wet spores irradiated at various temperatures.

The results presented here have the appearance and are described by kinetics that could be taking place in a laboratory reaction vessel. In this case the reaction vessel with its contents is a spore. While this may be impressive a qualification must be made, that is, methods and reaction equations from dilute solutions have been used. The spore is not a dilute solution. Additionally, although we know the concentration of the various added agents outside the spore we do not know them inside. However, it is useful to carry over the techniques to learn whatever we can, so long as this qualification is kept in mind.

ACKNOWLEDGEMENTS

Support in part by NCI grant No. 18179 and American Cancer Society grant No. PDT-59.

REFERENCES

1. Powers, E. L., "Cellular Radiation Biology," Williams and Wilkins Co., Baltimore, 1965, p. 286.

2. Michael, B. D., Adams, G. E., Hewitt, H. B., Jones, W. B. G. and Watts, M. E., Radiat. Res., 54,239 (1973).
3. Johnson, F. H., Eyring, H., and Polisan, M. J., "The Kinetic Basis of Molecular Biology," John Wiley and Sons, Inc., New York, 1954, p. 5.
4. Webb, R. B., Powers, E. L. and Ehret, C. F., Radiat. Res., 12,682 (1960).
5. Weiss, H. and Santomasso, A., Spore Radiobiology Meeting, Tokyo, Int. J. Radiat. Biology 37,557 (1980).

HYDROXYL RADICAL DAMAGE IN IRRADIATED BACTERIA: A RE-EVALUATION

David Ewing, Department of Radiation Therapy and Nuclear Medicine, Hahnemann Medical College and Hospital, 230 N. Broad Street, Philadelphia, Pennsylvania 19102

INTRODUCTION

When suspensions of bacterial spores are exposed to ionizing radiation, damage from hydroxyl radicals (·OH) is oxygen-dependent: only when the dissolved O_2 concentration is within the range 10^{-6} - 10^{-4}M will ·OH removal protect (1). Some additives, which react well with ·OH, can also reduce the response outside that O_2 concentration range, but the basis for this effect is not simple ·OH removal. Instead, after reacting with ·OH, these scavengers form hydroxy radicals, and there is evidence that these reducing radicals are probably the actual protectors (2,3).

This situation with spores is different from that reported from studies with bacteria (4,5) and mammalian cells in culture (6). In both these latter kinds of cells, the conclusion is that simple ·OH removal will protect when these cells are irradiated under anoxic or well-oxygenated conditions.

However, some of the additives used in the studies with bacteria and mammalian cells form reducing radicals, and it is important to know whether or not the protection observed with these cells was correctly attributed to simple ·OH removal. Accordingly, we have used Serratia marcescens to retest the conclusions reached earlier with bacteria.

MATERIALS AND METHODS

Stationary phase Serratia marcescens were washed twice by centrifugation, and then resuspended in doubly-distilled de-ionized water at ∿ 2 x 10^7 cells/ml. Unirradiated controls were taken before and after each experiment to monitor cell viability, and in no case were these two control values significantly different.

Irradiation was with 260 kVp X rays, delivered at 21.7 krad/min (measured by ferrous sulfate dosimetry). Difco's MR-VP was the culture medium, and the plates were incubated overnight at 29°C for colony development.

Survival curves were fitted to the equation $\ln (S) = \ln (n) - kD$, where S is the fractional survival after dose D; n is the extrapolation number and k is the inactivation rate constant. Values of k were used as indices of radiation sensitivity.

ISBN 0-12-592050-4

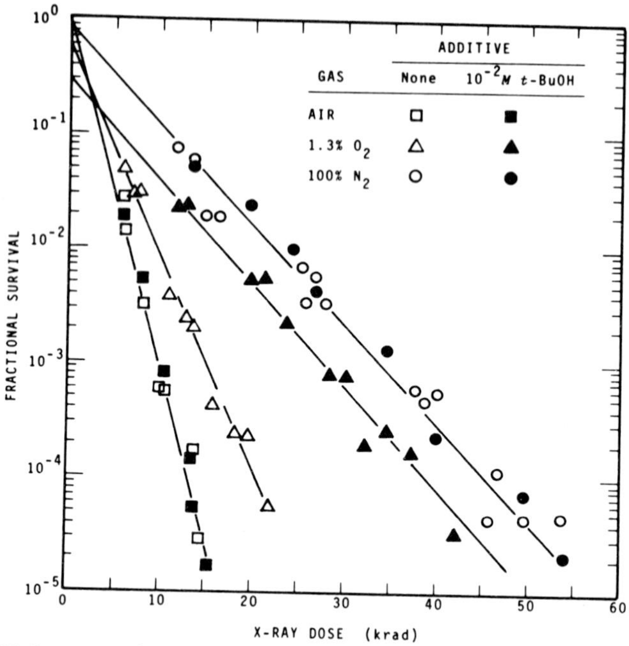

FIGURE 1. Survival curves for Serratia marcescens irradiated with 260 kVp rays. Gases and the additive concentration are given in the figure.

RESULTS

Fig. 1 presents survival curves obtained when Serratia were equilibrated with air, 1.3% O_2, or with 100% N_2 and then irradiated. The figure also shows the results when these three curves were repeated with 10^{-2}M t-butanol present. [The additive was chosen because it cannot form a reducing radical after it reacts with ·OH, and, therfore, it cannot protect through this pathway. The specific concentration of the additive was chosen because, according to previous work (4,5), it should be ∿ 100X too low to protect via ·OH removal alone.] As the figure shows, the responses in air and in 100% N_2 were unaffected, but in 1.3% O_2 this t-butanol concentration eliminated all sensitization by O_2.

We have also run a series of experiments to see if the protection in 1.3% O_2 could be attributed to OH radical scavenging. For these, we held the O_2 concentration at 1.3% and measured the response when different OH radical scavengers were present. We used t-butanol, t-amyl alcohol, and dimethylsulfoxide (DMSO) over a range of concentrations, and, in Fig. 2, we plotted the responses against the ·OH Scavenging Efficiency. These data form a single relationship when plotted in this way; this is good evidence that simple ·OH removal by the additives was the process responsible for this protection.

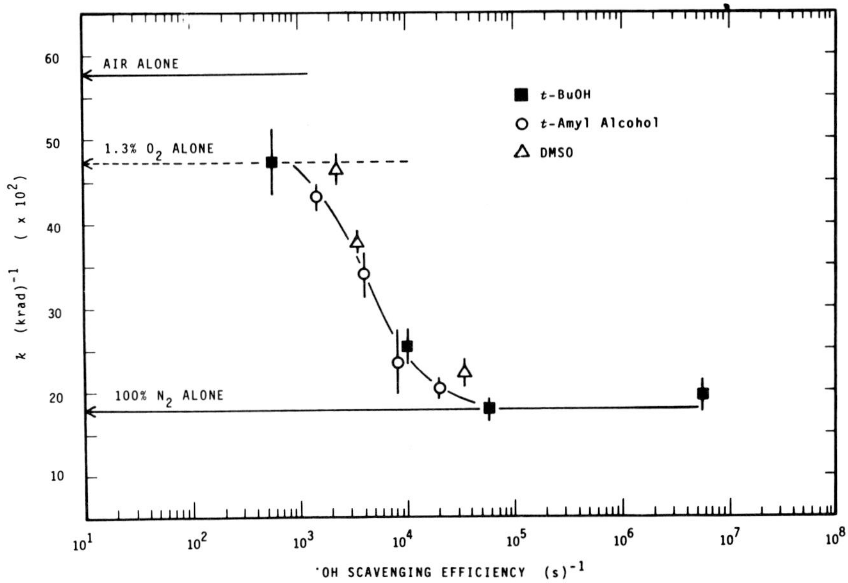

FIGURE 2. The relationship between rediation sensitivity (k, plotted on the ordinate) and the ·OH Scavenging Efficiency for different concentrations of three additives. [The ·OH Scavenging Efficiency is the product of the additive's concentration and the rate constant for the reaction between ·OH and the additive; the rate constants are given in Table 1.] In these experiments with Serratia, the O_2 concentration was constant at 1.3%.

We also tested the additives in air and in 100% N_2 at the highest concentrations used for the set of experiments in Fig. 2. These results (Table 1) show clearly that these additives change the response only in 1.3% O_2.

Throughout these experiments, the values of k were highly reproducible; however, we had little control over the values of n, which fluctuated widely and sometimes dropped as low as 0.2. We could not eliminate this problem, and we are at present running all these experiments with another bacterium.

DISCUSSION

Radiation protection by added chemical agents has been studied by Johansen and Howard-Flanders (4) using air-equilibrated and Sanner and Pihl (5) with N_2-equilibrated bacteria. Both these studies attributed protection to ·OH radical scavenging by the added chemicals, and both studies concluded that ·OH cause about 50%-60% of the lethal damage from exposures of ionizing radiation. Johansen and Howard-Flanders (4), in fact, suggested a specific two-step sequence to relate damage from ·OH radicals and

TABLE 1. Tests of the highest concentrations of the additives used in Fig. 2 on the radiation sensitivity of Serratia. *The data show that the response is changed only in 1.3% O_2, not in air or in 100% N_2.*

	RADIATION SENSITIVITY*		
ADDITIVE	100% N_2	1.3% O_2	AIR
NONE	17.6 ± 0.7	42.9 ± 1.3	57.3 ± 6.1
t-BUTANOL @ $10^{-2}M$	16.9 ± 0.2	19.5 ± 7.4	63.6 ± 2.6
t-AMYL @ $10^{-5}M$	18.5 ± 0.5	20.9 ± 1.0	69.1 ± 5.2
DMSO @ 6 X $10^{-6}M$	18.2 ± 1.6	21.3 ± 2.6	68.4 ± 1.8

*DATA ARE VALUES OF k, MULTIPLIED BY 10^2 KRAD ± 1 STANDARD ERROR

the radiation sensitization by O_2:

$$\text{DNA} + {}^{\cdot}\text{OH} \longrightarrow {}^{\cdot}\text{R} \quad \text{(a)}$$

$$^{\cdot}\text{R} + \text{O}_2 \longrightarrow \text{"inactivate product"} \quad \text{(b)}$$

Chapman and his colleagues have suggested that these reactions also apply to mammalian cells (6).

Our results presented here are incompatible with these conclusions reached earlier with bacteria. A reduction in the O_2 concentration would lower the overall response by reducing the efficiency of reaction (b), but a reduced O_2 concentration should have no effect whatever on the $^{\cdot}$OH Scavenging Efficiency needed to protect by blocking reaction (a). From Fig. 2, we can estimate that an $^{\cdot}$OH Scavenging Efficiency of $\sim 5 \times 10^3$/s will eliminate half this kind of damage. This value is very near that found with bacterial spores in similar experiments, $\sim 10^4$/s (7).

Although the large variation in the values of n during these experiments indicate the need for caution, our data suggest that ·OH radicals do not cause damage when bacteria are irradiated under well-oxygenated or anoxic conditions. With bacteria, as with bacterial spores, damage from $^{\cdot}$OH seems to be observed only when the O_2 concentration is suitably low. If these results are confirmed, we must conclude that the protection observed by Johansen and Howard-Flanders (4) and by Sanner and Pihl (5) was incorrectly attributed to ·OH radical removal.

ACKNOWLEDGEMENTS

These studies were supported by the National Cancer Institute, DHEW, through Grant Number CA 14043. My sincere thanks to Charles Dort and Martin Brubaker for their expert help with these experiments.

REFERENCES

1. Ewing, D. and Powers, E. L., Science, 194,1049-1051 (1976).
2. Ewing, D., Radiat. Res., 68,459-468 (1976).
3. Ewing, David, Int. J. Radiat. Biol., 30,419-432 (1976).
4. Johansen, Ivar and Howard-Flanders, Paul, Radiat. Res., 24, 184-200 (1965).
5. Sanner, T. and Pihl, A., Radiat. Res., 37,216-227 (1969).
6. Chapman, J. D., Dugle, D. L., Reubers, A. P., Gillespie, C. J. and Borsa, J., "Radiation Research: Biomedical, Chemical, and Physical Considerations" (O. Nygaard, H. Adler and W. Sinclair, eds.), Academic Press, New York, 1975, pp. 752.
7. Ewing, D., Radiat. Res., 73,121-136 (1978).
8. Dorfman, L. M. and Adams, G. E., "Reactivity of the Hydroxyl Radical in Aqueous Solutions," Washington, D.C.: U.S. Government Printing Office, NSRDS-NBS #46, 1973.
9. Reubers, A. P., Greenstock, C. L., Borsa, J. and Chapman, J. D., Int. J. Radiat. Biol., 24,533-536 (1973).

DISCUSSION

SINGH: Dr. Ewing, in the work you have described, and all the other previous work described, it is not clear to me—how does the differential solubility in the aqueous phase in the cell and in the organic phase come into play? You see, in the organic solvent, the solubility of O_2 is about 10 times greater as compared to that in water. The same would apply in a cell, inside of the membrane and in the regions which are predominantly organic; your O_2 concentration is going to be about 10 times greater than in the aqueous portions. When you go to lower concentrations of O_2, the same differential would apply. How do you think this affects the expression of the radiation effect?

EWING: I am sorry I have no idea. Do you think it might influence the scavenging ability of the added $\cdot$OH scavengers?

SINGH: In your case, I cannot say. But I was wondering earlier on whether the biphasic O_2 effect that has been seen, has something to do with this difference in solubility. Because, certainly the additive efficiency of O_2 (to radicals) depends on concentration. So, in the organic phase this would be greater, because you have more O_2; and in the aqueous phase it would be lower.

EWING: I am not sure. The work with spores that you have seen is all done by washing the spores as free as we can of extracellular debris and then irradiating the spores in water. For the results that I have shown you with Serratia, the cells were also washed and irradiated in water. For the results with E. coli, the cells were washed but put into phosphate buffer. The qualitative results here are all the same.

POWERS: Singh has raised a question of general importance, one that was, I think, touched on in the previous session when someone wondered what were the concentrations inside cells when they are in suspension of dissolved solute. Usually no one knows. Much less is known concerning the distribution of the additives in the various locales within the cells. The best that can be done now is to fix the concentration at various levels outside the cell and see whether there is a smooth relationship between concentration and effect.

EWING: May I make one other comment? Tallentire and I talked very briefly yesterday afternoon; and on the slide that both Tallentire and I showed, he mentioned that either the extraspore and intraspore concentrations of the additives have to be the same, or else they have to all be excluded in the same amount or —as he suggested—we are dealing with things that are outside the spore. As I told him yesterday, there is now very good evidence that the hydroxyl radicals that are involved in the spore ·OH component of damage are actually outside the cell. And this damage also involves H_2O_2. That perhaps takes care of the spore's ability to concentrate these additives, since it looks like we are dealing with something on the outside of the spore. Now, whether that means the target is on the outside of the spore or not is uncertain; but the ·OH's that are important at least start in the bulk solution.

POWERS: It is clear that the critical target is inside the spore.

EWING: Everyone knows that it is.

SWARTZ: My comment relates to the points under discussion. I think we know what is the target and where it is. Therefore, I wonder if we are dealing with an interesting artifact in the papers under discussion. The target at doses of interest in mammalian cells is clearly DNA; the data supporting this are overwhelming. I have two principal problems with accepting the data as pertinent to mammalian radiobiology. One is that the diffusion distance of the hydroxyl radical is very short, so the pertinent species are not hydroxyl radicals in general; but rather, only those that are within diffusion distance of the DNA. The second problem I have is that although butanol is a good hydroxyl radical scavenger, so are most components of the cell, and therefore in these experiments one is adding what seems to be a small amount to an almost infinite number of hydroxyl scavengers which are already existing in the cell. When one puts these together, I wonder whether or not you are looking at a component of radiation damage which may not be pertinent to mammalian cell killing at doses used clinically or in the usual laboratory experiments. The doses used in these experiments may lead to killing via non-DNA targets, but such killing probably occurs only at very high doses.

EWING: I think a partial resolution to what you are saying here is what we just covered. The hydroxyl radicals are outside the cell, and for years we have routinely made the spore suspensions as clean as possible; so the amount of organic material, while not zero, is certainly low. In this way, our adding μM concentrations of OH scavengers is important.

POWERS: Whatever one's intuition is, we must deal with results. The hard fact is that varying the concentration by small amounts brings about the effect.

SWARTZ: Well, another possibility is that in all of these systems we are dealing with rather high doses, compared to mammalian cells, so we may in these systems be dealing with a component of damage which is related to DNA but which is a relatively small component of damage and one which won't carry over into mammalian radiation biology.

EWING: No. Wait. It's not a small component of damage. It is a large component of damage. And may I just remind all of us, since we keep forgetting, that for many years Alper has said that the target for O_2-dependent damage is not the DNA. She says it is the cell membrane.

ROOTS: I would just like to add that we have also done some experiments with hydroxyl radical scavengers under hypoxic conditions; and we find that, whereas the hydroxyl radical may be responsible for about 50% of the damage in aerobic mammalian cells, in anaerobic cells this part of the damage is reduced to about 20%. And actually most publications agree that in hypoxic systems, the hydroxyl radical mediated damage is much reduced. But it is not absent.

EWING: Dr. Roots, in the system that you talked about, is this a cell assay measurement or was this breaks in the DNA? The one that you just mentioned.

ROOTS: This is cell killing.

MECHANISMS OF OXYGEN RADIOSENSITIZATION IN CHO CELLS

D. W. Whillans, Physics Division, Ontario Cancer Institute, 500 Sherbourne Street, Toronto, Canada M4X 1K9

It is not many years since it was first recognized that irradiation in the presence of oxygen greatly increases the sensitivity of a wide variety of biological systems. Attempts have been made to relate these changes to the well-known interactions between oxygen and many radiation-induced biochemical species. In 1957, Howard-Flanders and Alper (1) interpreted the shape of the oxygen response curves ("K-curves") of irradiated bacteria in terms of a competition between damage fixation and repair. Such a model is shown below where, for our purposes, rate constants have been assigned to each of the "repair" and "fixation" pathways. Such a model predicts (i) that oxygen should be dose-modifying and (ii) that if the oxygen enhancement ratio (OER (O_2)) is

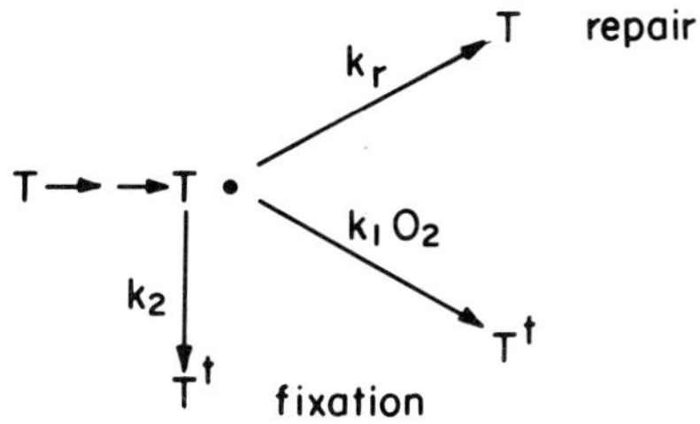

defined as the ratio of the slopes of the survival curves under oxygenated and anoxic conditions, a simple relationship can be derived which has the same form as the K-equation proposed by

$$\mathrm{OER}(O_2) = \frac{m[O_2] + K}{[O_2] + K} = \frac{\mathrm{OER}_{max}[O_2] + \frac{k_2}{k_1}\mathrm{OER}_{max}}{[O_2] + \frac{k_2}{k_1}\mathrm{OER}_{max}} \qquad (1)$$

Alper. The second term in both numerator and denominator is the "K-value," the concentration at which the increase in OER is at one-half its maximum value. In this relationship, it can be shown simply that

$$\mathrm{OER}_{max} = 1 + \frac{k_r}{k_2} \qquad (2)$$

$$K = \frac{k_r + k_2}{k_1} \qquad (3)$$

ISBN 0-12-592050-4

This analysis, which is similar to the simplest case considered by Kieffer (2), predicts that an increase in the repair path, such as has been postulated for addition of a radioprotective sulphydryl, should increase OER_{max} in proportion to k_r', the new repair rate constant, and also increase K with k_r'. Conversely, any radiosensitizer which mimics the action of oxygen, as has been proposed for a number of nitroimidazole derivatives, simply increases k_2, so that the OER_{max} decreases as $1/k_2'$ but K increases as k_2'. We have tested these predictions in mammalian CHO cells making use of a Clark-type oxygen probe (3) with defined conditions to ensure that oxygen is not depleted by radiation or cellular consumption (4,5), and so oxygen levels are known with accuracy. Levels here are quoted relative to a saturation value at 22°C of 1.3 mM.

In a complementary study, we have used the technique of rapid-mixing (6,7,8) to measure the rate of development of oxygen sensitization in these same cells. By a variation of this rapid-mixing approach, we also have measured independently the rate of diffusion of oxygen into these cells.

MATERIALS AND METHODS

In all experiments, Chinese hamster ovary (CHO) cells were irradiated in suspension at room temperature in α medium containing 0.2% foetal calf serum, using techniques already described (4,5). Survival was defined as the ability to form colonies after 8 days' incubation. The rapid-mixing system, similar to that of Shenoy et al. (7), had a minimum time resolution of 2.7 ± 1.7 ms.

RESULTS

K-curves. When CHO cells are irradiated, the effect of oxygen is strictly dose-modifying and the upper solid line of Fig. 1 shows that the oxygen response is well-fitted by equation 1 down to 0.1% where oxygen depletion in our suspension system becomes significant. In contrast to the 2-component response reported for V79-753B cells (9), there is no evidence here for more than one target or one mechanism. The effects of the radiosensitizer, misonidazole, and the protectors, cysteamine and DMSO, on this response were determined. Fig. 2a shows that in the presence of 10 mM cysteamine, a concentration which gives no overt toxicity (10) during the period of the experiment up to 43% oxygen, a protection which appears dose-modifying is observed, weaker in the presence of oxygen than in nitrogen. Fig. 1 shows that the maximum OER is greatly increased as predicted, but that this increase occurs only at very high oxygen concentrations. In fact, there is no sensitization observed until about 5% O_2 and, since at high oxygen concentrations the approach of overt toxicity due to oxidation products (data not shown) must be considered, it is

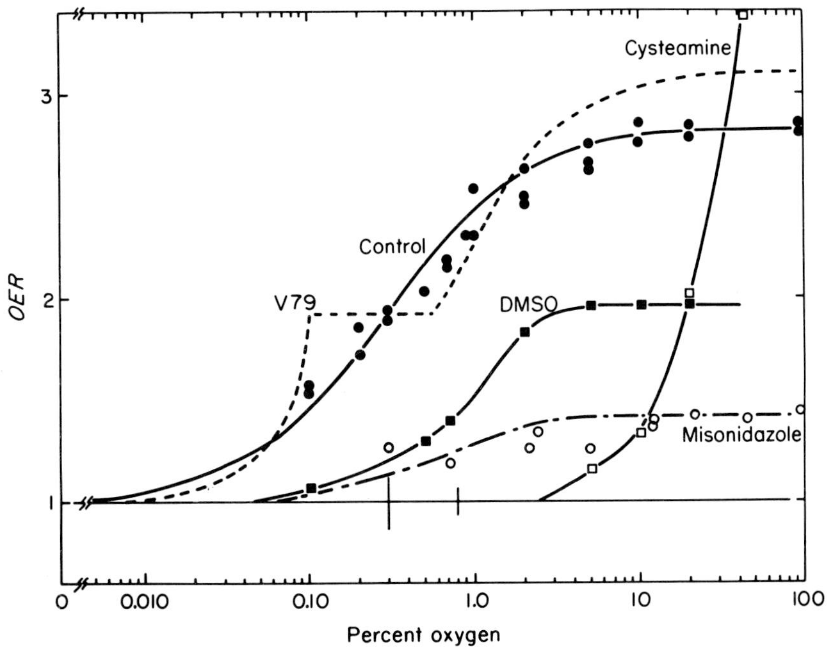

FIGURE 1. K-curves for CHO cells in α medium containing 0.2% foetal calf serum for ^{60}Co irradiation in the presence of a number of radiation sensitizers and protectors. ● no added modifiers; ■ 1M DMSO; ○ 2 mM misonidazole; and □ 10 mM cysteamine. The upper solid line represents the Alper reaction for m=2.85 and K=0.30%. The dashed line represents the two-component curve seen with V79-753B cells by Millar et al (9); the dash-dot line is a theoretical fit to the misonidazole OER_{max} of 1.45. The small bar indicates the K-value of 0.8% for DMSO.

doubtful whether this simple model applies.

Fig. 2b shows the sensitization of hypoxic but not aerobic cells in the presence of 2 mM misonidazole. Fig. 1 demonstrates that, while the data are somewhat scattered, the response to oxygen concentration is not inconsistent with the model, though clearly more precise determinations are needed. Also given in Fig. 1 is a K-curve obtained in the presence of 1M DMSO, a dose-modifying radioprotector which was proposed to act not in a competition for T·, but by reducing its yield through scavenging of a major precursor of cellular damage, the hydroxyl radical (11). Thus it would be predicted to have no effect on the shape of the K-curve. However, as can be seen, the OER_{max} is markedly reduced to ∿ 2.0, and, in addition, the K-value is raised to about 0.8%. It has been proposed, however, that indirect attack by the water radiolysis ·OH species accounts for only part of the radiation damage (11).

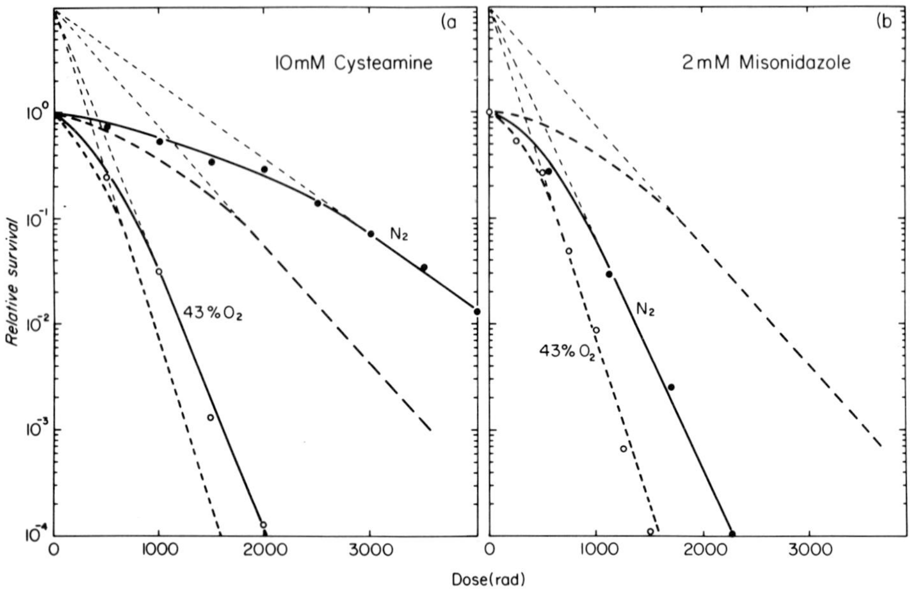

FIGURE 2. Survival curves for CHO cells irradiated in the presence of (a) 10 mM cysteamine, and (b) 2 mM misonidazole. In both cases, the dashed curves represent hypoxic and 43% O_2 controls.

Rapid-mixing. When hypoxic cells are incubated with oxygen for only milliseconds before irradiation, their sensitivity is increased (7). Fig. 3 shows survival data for CHO cells irradiated 6 ms after contact with oxygen, in comparison with hypoxic and aerobic controls. In this short time, the OER is already about 2.0. Fig. 4 gives the complete time response after mixing hypoxic cells to a concentration of 2.6% oxygen. Full sensitization is observed within 10 ms, in contrast to observations with V79-GL1 cells (7) where two distinct components were resolved in time. When hypoxic cells labelled with an oxygen-sensitive fluorescent probe were mixed with oxygen, the rate of quenching of the fluorescence indicated a diffusion of oxygen into the cells with a coefficient of 4 x 10^{-6} cm^2s^{-1} (12). Fig. 4 includes a theoretical response based on a central target, the above diffusion coefficient and the steady-state oxygen response of Fig. 1. Clearly, penetration into these cells is sufficiently rapid to allow sensitization generally throughout the cell in times as short as those observed. In this rapid response, there is no evidence for more than one component in CHO cells.

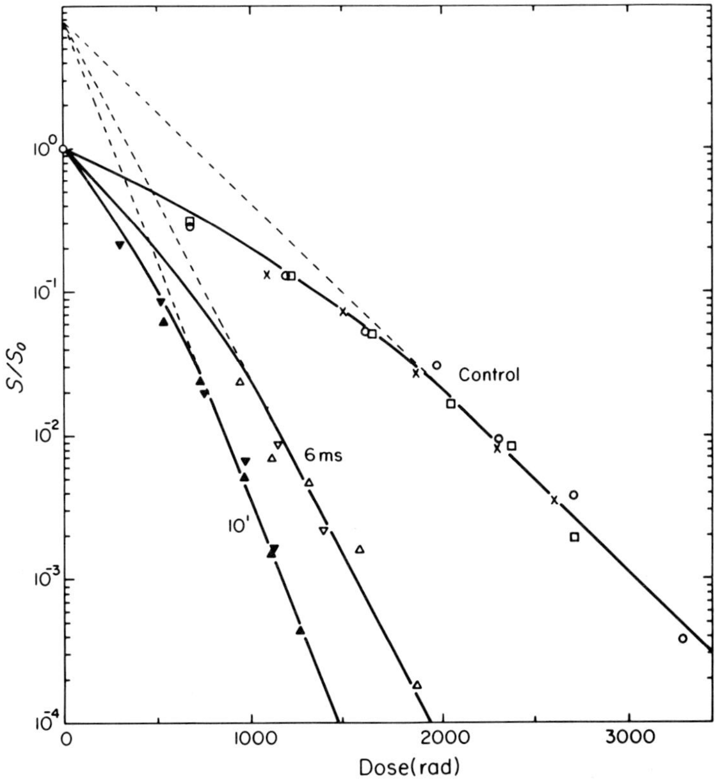

FIGURE 3. Survival curves derived following irradiation by 25 MeV electrons at high dose rate in the rpaid-mixing apparatus. Hypoxic cells were mixed with either hypoxic medium (control) or α medium containing 5.2% oxygen, 6 ms or 10' before irradiation.

DISCUSSION

O Enhancement Ratio.

Neither the dependence of OER on oxygen concentration nor the development of radiosensitivity with time of incubation in oxygen gives evidence in CHO cells for two components of sensitization, indicative of two sites or two mechanisms of action, as seen in some V79 sublines. It is possible that in CHO cells, the mechanisms may be so similar in these respects as not to be resolved by our techniques. However, Ling et al. (13) using a double-pulse technique with even better time resolution also failed to observe two components. What is clear, however, is that sensitization by oxygen occurs too quickly to involve anything but a direct interaction with a radiation-induced target species, such as that proposed by Howard-Flanders and Alper.

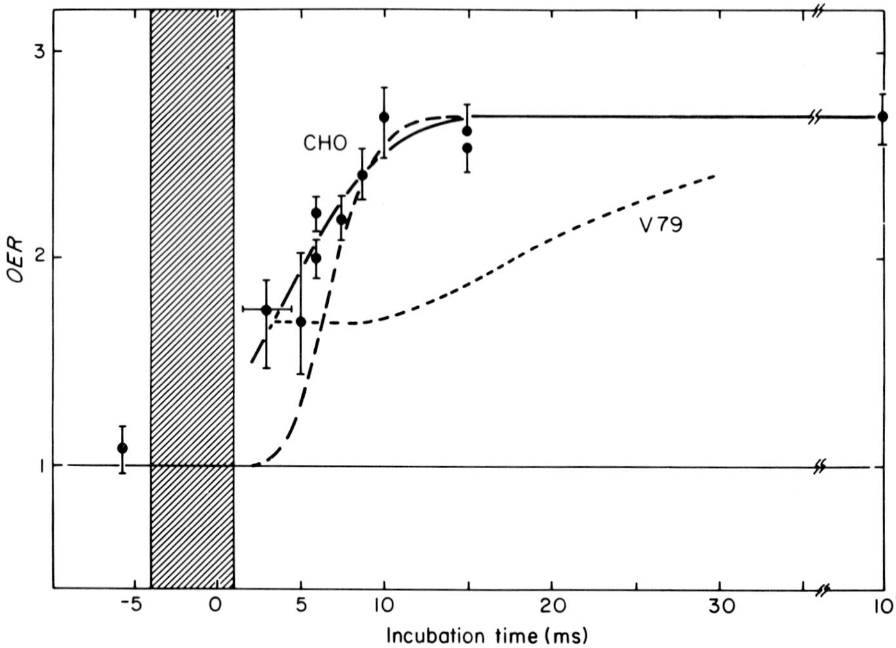

FIGURE 4. Complete time response of the development of oxygen sensitization in CHO cells. Each point is derived from a ratio of slopes of survival curves as in Fig. 3. The hatched area represents a time region not available for observation. The short-dashed curve is the V79 response taken from reference 7. The long-dashed curve is a theoretical response based on diffusion to a central target as described in the text.

Our preliminary studies of the modification of the shapes of K-curves lends some support to this model. In our hands, the effects of oxygen on CHO cells fit such a model very well, and the limited misonidazole data are also consistent. Addition of the radioprotector cysteamine leads to an increase in OER_{max} as expected, but only at much higher oxygen concentrations than predicted, so that we must conclude that other mechanisms are involved. Interestingly, a high concentration of DMSO also fits the model well qualitatively, but not quantitatively and it is tempting to interpret these data as a K-curve for direct radiation damage. In summary, it is clear that the concepts of a radiation-chemical model for the radiation inactivation of mammalian cells provide a useful basis for further studies. In future, however, it would also be fruitful to extend this approach to those cell lines which appear to show two clearly-resolvable components of oxygen-sensitizable damage.

REFERENCES

1. Howard-Flanders, P. and Alper, T., Radiat. Res., 7,518 (1957).
2. Kieffer, J., "Radiation Research: Biomedical, Chemical and Physical Considerations" (O. Nygaard, H. Adler and W. Sinclair, eds.), Academic Press, New York, p. 1025, 1975.
3. Koch, C. J. and Kruuv, J., Anal. Chem. 44,1258 (1972).
4. Whillans, D. W. and Rauth, A. M., Canc. Clin. Trials 3,63 (1980).
5. Whillans, D. W. and Rauth, A. M., Radiat. Res., 84,97 (1980).
6. Adams, G. E., Cooke, M. S. and Michael, B. D., Nat., 219, 1368 (1968).
7. Shenoy, M. A., Asquith, J. C., Adams, G. E., Michael, B. D. and Watts, M. E., Radiat. Res. 62,498 (1975).
8. Whillans, D. W. and Hunt, J. W., Br. J. Cancer 37 Supp. III 38 (1978).
9. Millar, B. C., Fielden, E. M. and Steele, J. J., Int. J. Radiat. Biol. 36,177 (1979).
10. Vergroesen, A. J., Burke, L. and Vos, O., Int. J. Radiat. Biol. 13,77 (1967).
11. Chapman, J. D., Reuvers, A. P., Borsa, J. and Greenstock, C. L., Radiat. Res. 56,291 (1973).
12. O'Loughlin, M. A., Whillans, D. W. and Hunt, J. W., Radiat. Res., 84,477 (1980).
13. Ling, C. C., Michaels, H. B., Epp, E. R. and Peterson, E. C., Radiat. Res. 76,522 (1978).

DISCUSSION

BIAGLOW: I was wondering why you have used pyrene as a fluorescent probe when nicotinamide adenine dinucleotide reduced (NASD(P)H) are probably more sensitive.

WHILLANS: Well, initially what we were looking for was some sort of a DNA binding probe because we wanted to have a probe at the location of the proposed target. This work was done by Mary Ann O'Loughlin in the laboratory. She examined a very large number of probes; and at least when they were introduced into cells, very few of them had lifetimes long enough to give you large changes of intensity in air as opposed to hypoxia. In other words, the lifetimes were so short that you couldn't quench them. Although pyrene really distributes only throughout the cytoplasm and is not an ideal probe, at least it allowed us to measure diffusion through that part of the cell. It was simply that none of the others worked.

BIAGLOW: But the cellular NAD(P)H fluorescence as used by B. Chance definitely shows a very fast response to O_2. This occurs because the mammalian cytochrome chain involving electron transfer from NADH to cytochromes is extremely sensitive to O_2.

WHILLANS: Were these measurements made in whole cells?

BIAGLOW: Inside cells, yes.

EPP: We have made some K-curve measurements for CHO cells in a thin layer and also in the presence of medium using 50 KV x-rays in the former and 250 KV x-rays in the latter. We seem to see the same type of thing that you do, in other words, no evidence of plateauing for CHO cells in the response.

WHILLANS: I have seen your data. In fact with the thin layer technique you can reliably go to less than .1%, and I think with that region it's even better evidence that only one component seems to be involved, or seems to be resolved at least in that system.

TIMESCALE AND MECHANISMS OF THE OXYGEN EFFECT IN IRRADIATED BACTERIA

B. D. Michael, H. A. Harrop and K. D. Held, Cancer Research Campaign, Gray Laboratory, Mount Vernon Hospital, Northwood, Middlesex HA6 2RN, England

INTRODUCTION

In the course of our studies of the fast kinetics of the actions of radiosensitizers and radioprotectors, we have examined the oxygen effect extensively (1-5). This paper summarizes some of our findings with bacteria and relates the fast time-dependent behavior of oxygen to the concentration-dependent effects that are determined in steady-state irradiation experiments.

It is generally held that the concentration dependences of the effects of modifying agents represent their participation in competing reactions in which lethal and/or non-lethal products are formed. Sometimes the concentration dependences are complex and the same agent (e.g. O_2) may show multiple (6-8) or even opposing (9) effects over different concentration ranges.

The aim of the fast kinetics approach is to study these competing processes as functions of time as well as of concentration and so gain further information about the mechanisms of the reactions that are involved. This has stimulated the development of fast response methods that can be used with intact cells (1,10, 11).

FAST KINETICS OF THE OXYGEN EFFECT

In this work we have used the gas explosion technique (1,4) in which cells are irradiated with a 5-nanosecond pulse of electrons and are exposed to a rapid ($\sim$ 100 μs) transition from a hypoxic to a well-oxygenated environment at a preset time prior to or after irradiation.

a) The Kinetics of Chemical Repair of Oxygen-dependent Damage

The kinetics of the repair of oxygen-dependent damage in hypoxic Serratia marcescens are shown by the top curve in Fig. 1. Each point represents the geometric mean of the survival of several different membrane filter-mounted preparations of the bacteria. Each preparation was rendered hypoxic by flushing with N_2 for 2 mins. and was then exposed to a single 175-gray pulse of electrons at time t = 0 and to oxygen contact at the time indicated. Thus t = +2 ms represents O_2 contact 2 ms after the electron pulse and similarly negative times represent O_2 contact

ISBN 0-12-592050-4

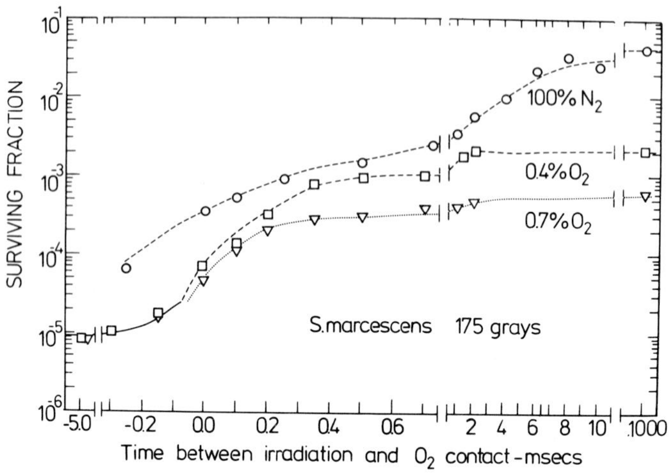

FIGURE 1. The influence of small concentrations of O_2 in the pre-irradiation flushing gas, as indicated, upon the decay kinetics of O_2-dependent damage in S. marcescens. The shot of 100% O_2 is delivered at the indicated times.

before the electron pulse. Almost the full degree of sensitization is produced by O_2 contact just before irradiation, but with increasing delay of O_2 contact after the electron pulse the survival rises progressively towards the hypoxic level which corresponds to the survival shown for t = +1000 ms.

As will be described later, the oxygen shot reacts with the damage on a time scale that is short compared with that of the curves in Fig. 1. Thus when O_2 contact occurs as a certain preset time after the electron pulse, it reacts with and renders lethal the amount of oxygen-dependent damage remaining unrepaired at that instant. In this way we have been able to determine the kinetics of the repair under hypoxia of oxygen-dependent damage in a number of systems and some of these are summarized below.

i) S. marcescens. Here there appear to be two stages of exponential decay of damage, a "fast" stage, amounting to 70% of the overall effect, that decays with a pseudo-first order rate of 2100 s^{-1} and a "slow" stage, the remaining 30%, that decays at 170 s^{-1} (2,4). A two-component time-exponential expression has been used to generate the line drawn through the points on the top curve in Fig. 1.

ii) Escherichia coli. Comparisons of a wild type E. coli and some related mutants (12,13), deficient in the glutathione (GSH) synthesis pathway have been made (5,14). The behavior of four

TABLE 1. Comparison of Fast Kinetics and Concentration Dependences of the Oxygen Effect in Various Types of Bacteria

Bacteria	*Relevant Phenotype*	*First Half Life of Decay of O_2-Dependent Damage,ms.*	*Measured K-value % O_2*	*Calculated K-value (see text) % O_2*
E. coli				
K12 AB 1157 (a)	*Wild type*	*0.2*	*0.71*	--
" 830 (a)	*GSH synthetase deficient*	*0.45*	*0.26*	--
" 821 (a)	*γ-glutamyl cysteine synthetase-deficient*	*2.6*	*0.06*	--
" 7 (a)	*"*	*1.6*	*0.06*	--
S. marcescens				
NCTC 1377	*Wild type*	*0.6*	*0.33-0.41*	*0.34*

(a) See ref. 13 for the derivation of strains.

such related strains is compared in Table 1. Here, the oxygen-dependent damage decay rates have been shown in terms of the first half lives of the decay. It is evident that the damage half lives are much longer in the GSH-deficient mutants than in the wild type and we conclude that in the wild type E. coli the repair of O_2-dependent damage under hypoxia is specifically attributable to GSH. There is direct competition between O_2 and GSH for the same damage site.

b) The Kinetics of the Reaction of O_2 with O_2-dependent Damage in S. marcescens

The reaction rate of oxygen has been determined by a modification (2,4) of the gas explosion experiment in which, instead of rendering the bacteria anoxic by nitrogen flushing, we equilibrate them with a gas mixture containing a small proportion of O_2 and then expose them to the electron pulse and to the 100% O_2 shot. The results are shown in Fig. 1. The survival levels at t = +1000 ms fall below that of the top curve (100% N_2 flush) because of sensitization by the levels of O_2 present in the flushing gas. The lines through the points have been generated with

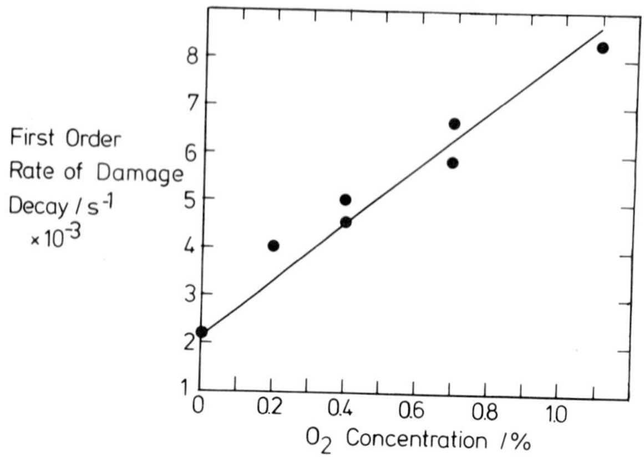

FIGURE 2. First order rate of decay of the fast component of O_2-dependent damage in S. marcescens as a function of O_2 concentration in the pre-irradiation flushing gas.

two-component exponential expressions. The effect of increasing the concentration of O_2 in the flushing gas is to shorten the lifetime of the damage. In Fig. 2 is shown the dependence of the rate of decay of the major, fast component on O_2 concentration. As with our other studies with sensitizers and protectors (4), there appears to be a linear relationship, indicating simple bimolecular reaction kinetics. The slope of the line corresponds to a rate of $5 \times 10^8 dm^3\ mol^{-1}\ s^{-1}$ if it is assumed that the solubility of O_2 in the region of the sensitive target is the same as that in water. This differs from our previously reported value of $2.8 \times 10^8\ dm^3 mol^{-1} s^{-1}$ (4) which was based partly on preliminary data with more scatter. Clearly, the high rate of reaction of O_2 will mean that, at the high concentrations present in the gas shot (up to 100%), O_2 will react with the damage in a time of the order of 10 μs. This is short compared with the kinetic decay times such as those shown in Fig. 1 and Table 1, and contact with the O_2 shot is therefore quite a precise fast probe to detect the amount of damage remaining unrepaired at each instant after the electron pulse. The time needed for O_2 to diffuse through the sensitive regions of the cell does limit the ability of the technique to resolve reaction kinetics in mammalian (3) and yeast (4,15) cells, but with bacteria diffusion is much faster because of their smaller size, and approximately 80% of the damage decay can be fitted kinetically.

COMPARISON OF THE CONCENTRATION DEPENDENCE WITH THE FAST KINETICS OF THE OXYGEN EFFECT

We have also examined the relationships between O_2 concentration and radiosensitization in the bacterial strains shown in Table 1. In the cases of the E. coli strains these were determined from full survival curves obtained by γ-irradiating the bacteria in suspensions, vigorously bubbled with gas mixtures containing the appropriate concentrations of O_2. Full survival curves for S. marcescens were obtained by exposing them to various single-pulse doses of electrons in the gas explosion apparatus flushed with the appropriate concentration of O_2, but without applying the O_2 shot. The concentration dependence of O_2 sensitization of bacteria has been studied by Alper and Howard-Flanders (16) who showed that the behavior could be characterized by parameters m, corresponding to the maximum oxygen enhancement ratio (OER) in the system and K, corresponding to that concentration of oxygen which gives an OER of (m + 1)/2. Their model was described further in a later publication (17). The significance of K is that it represents the concentration of O_2 at which competition between repair and fixation proceed at equal rates. Thus,

$$K = \frac{\text{first order rate of repair of damage}}{\text{second order rate of fixation of damage by } O_2}$$

The K-values for the E. coli strains were measured as described and are listed in Table 1. The values correlate inversely with the damage half lives in approximate conformity with the above equation if, as might be expected, the rate of reaction of O_2 is the same in all of these related E. coli strains. For S. marcescens, from the ratio of the rate of repair of the major (70%) fast component of damage (2100 s^{-1}) to the rate of its reaction with O_2 (5×10^8 dm^3 mol^{-1} s^{-1}) we have calculated a K-value of 0.34% (4.2×10^{-6} mol dm^{-3}) which is within the range 0.33-0.41% that was measured as described above. These values are also shown in Table 1.

Clearly the above analysis is only approximate because it does not take into account the two time-resolvable components of damage and because the errors of measurement, which are partly due to the systematic variations in the responses of these vegetative bacterial systems, are appreciable.

CONCLUSIONS

The above comparisons do show that in these systems the concept of oxygen acting as a radiosensitizer by fixing damage in competition with its repair by intracellular free SH is applicable to the oxygen effect phenomena, whether they are observed as functions of concentration or of time.

The ability of sulphydryl compounds in polymeric systems to repair free-radical damage by H atom donation was pointed out by Alexander and Charlesby (18) and they also showed that oxygen could block this repair. Powers and Kaleta showed that in dry bacterial spores similar competition to that described above occurred between reactions of radiation-induced damage with the simplest sulphydryl, H_2S, and with O_2 (19). Recently, Cullen and co-workers have shown a correlation between GSH levels in mammalian cells and the oxygen K-value (20). Similarly, it has been found that radiosensitization by the electron-affinic compound, misonidazole, is also dependent on the intracellular level of GSH (5,14).

ACKNOWLEDGEMENT

The authors wish to acknowledge the support of the Cancer Research Campaign. KDH is supported by a National Science Foundation Needs Post-doctoral Fellowship.

REFERENCES

1. Michael, B. D., Adams, G. E., Hewitt, H. B., Jones, W. B. G. and Watts, M. E., Radiat. Res., 54,239 (1973).
2. Michael, B. D., Harrop, H. A., Maughan, R. L. and Patel, K. B., Br. J. Cancer, 37, Suppl. III, 29 (1978).
3. Watts, M. E., Maughan, R. L. and Michael, B. D., Int. J. Radiat. Biol., 33,195 (1978).
4. Michael, B. D., Harrop, H. A. and Maughan, R. L., "Radiation Research," Proceedings 6th Int. Cong. Radiat. Res., Tokyo (S. Okada, M. Imamura, T. Terashima and H. Yamaguchi, eds.), Jap. Assoc. for Radiat. Res., 288 (1979).
5. Michael, B. D. and Harrop, H. A., "Proceedings of Key Biscayne Conference on Combined Modalities in Cancer Treatment: Radiation Sensitizers and Protectors", (L. W. Brady, ed.), Masson, New York, 14, (1980).
6. Tallentire, A., Jones, A. B. and Jacobs, G. P., Isr. J. Chem., 10,1185 (1972).
7. Ewing, D. and Powers, E. L., Science, 194,1049 (1976).
8. Millar, B. C., Fielden, E. M. and Steele, J. J., Int. J. Radiat. Biol., 36,177 (1979).
9. Held, K. D. and Powers, E. L., Int. J. Radiat. Biol., 36, 613 (1979).
10. Adams, G. E., Cooke, M. S. and Michael, B. D., Nature, 219, 1368 (1968).
11. Epp, E. R., Weiss, H., Kessaris, N. D., Santomasso, A., Heslin, J. M. and Ling, C. C., Radiat. Res., 54,171 (1973).
12. Fuchs, J. A. and Warner, H. R., J. Bacteriol., 124,140 (1975).
13. Apontoweil, P. and Berends, W., Biochem. et Biophys. Acta, 399,10 (1975).
14. Harrop, H. A., Held, K. D., Crompton, N. E. A. and Michael, B. D., in preparation.
15. Kiefer, J., Harrop, H. A. and Michael, B. D., unpublished data.
16. Alper, T. and Howard-Flanders, P., Nature, 178,978 (1956).
17. Howard-Flanders, P. and Moore, D., Radiat. Res., 9,422 (1958).
18. Alexander, P. and Charlesby, A., "Radiobiology Symposium 1954" (Z. M. Bacq and P. Alexander, eds.), Butterworths, London, 49 (1955).

19. Powers, E. L. and Kaleta, B. F., Science, 132,959 (1960).
20. Cullen, B. M., Michalowski, A., Walker, H. C. and Revesz, L., Int. J. Radiat. Biol., 38,525 (1980).

DISCUSSION

EPP: Barry, could you in fact make a comment about the two components in your point one of the summary?

B. MICHAEL: Well, if I can just show a further slide. This is for *Serratia marcescens* and shows $1/D_o$ against O_2 contact time in experiments where we determined full survival curves. We find that the behavior cannot be fitted by the single time exponential that might be expected from the Alper and Howard-Flanders model, but it can be fitted by two exponentials. We associate these with a fast component of O_2-dependent damage amounting to about 70% and a slow component which is the remaining 30%. In certain DNA repair deficient mutants of *E. coli*, we only see one such component.

BIAGLOW: Barry, have you considered that the lack of glutathione inhibits glutathione peroxidase and that your radiation effect is due to H_2O_2 production? Have you tried adding catalase? This enzyme would destroy H_2O_2.

B. MICHAEL: We have not tried deliberately adding catalase. I do not see how one could get a fast enough rate of reaction to explain these millisecond kinetics that we observe.

BIAGLOW: Well, glutathione peroxidase catalyzes the reaction between H_2O_2 and glutathione ($H_2O_2 + 2GSH = H_2O + GSSG$). The rate of this reaction is $10^8 M^{-1} s^{-1}$.

B. MICHAEL: Yes, but what we feel is happening here is that there is radical damage induced at some site in the cell and that by a diffusion process this has to get together with the repairing substance which in the case of *E. coli* we are attributing to glutathione. There is a limit consequent from diffusion kinetics, the rate of the collision process. The concentration of the peroxidase would have to be quite high, even with a reaction rate of 10^8, to account for the kinetics we observe.

BIAGLOW: Well, that is the second question. Are cell thiols (NPSH) really different in concentration in the cell types? Are you attributing the radiation response to just the glutathione? You still have the same non-protein thiol (NPSH) concentration.

B. MICHAEL: Yes. That is rather complex. I couldn't really go into this. We obtained these strains from Apontoweil and Berends in the Netherlands. The strain that we call GSH^- is the one defective in the final step of glutathione synthesis. It produces about 12% of the wild type glutathione, and it also accumulates the precursor γ-glutamylcysteine up to a concentration comparable with the GSH content of the wild type. In the other two strains, which were deficient in the second from last step of GSH synthesis, they only have about 1% of the wild type GSH level; and they accumulate cysteine which is subject to feedback inhibition which makes the situation more complicated.

SAGONE: I wonder if you think that there is an intermediate between the GSH and DNA such as ascorbate which directly repairs the DNA or whether you think that there is enough GSH in the nucleus to account for the protection. I guess, since bacteria really do not have a distinct nucleus, a direct reaction of GSH in the cytosol with DNA might occur without difficulty.

B. MICHAEL: Well, I think that there are two things there. We are not making any assumptions of what we are looking at really as DNA damage. This came up in discussion this morning, and there are various proponents of the idea that DNA is involved in these fast processes and others that it is not. We are currently working with Held, who is here, doing some work on transforming DNA in an in vitro system; and we are hoping to throw some light on that. Regarding the second matter, I think that what you suggested is a three-body reaction, that is, between the target damage the ascorbate and the glutathione—am I correct? Now kinetically this is going to take a lot of time for these things to get together. I do not think that this can account for the sort of rates we are observing experimentally. Have I interpreted you correctly?

SAGONE: Yes, the real question is whether, for example, GSH reacts rapidly with a macromolecule causing repair and then is regenerated more slowly in irradiated cells through the hexose monophosphate shunt pathway. The question is whether or not the GSH in the region of the DNA is in high enough concentrations to explain the repair in bacteria or whether you have to postulate some intermediate which can react very rapidly and then be regenerated by the GSH through the HMPS pathway.

B. MICHAEL: This was a procaryote—there is no nucleus.

SAGONE: We have done some work with human lymphocytes. We have irradiated them under air and hypoxic conditions and found paradoxically that the hexose monophosphate shunt pathway was stimulated better under hypoxic conditions than aerobic conditions. We postulated that this reaction reflects a capacity of GSH to repair DNA as your data suggests in the bacterial system.

THE KINETICS OF THE OXYGEN EFFECT IN YEAST IRRADIATED IN DRY AND WET CONDITIONS

J. Kiefer, F. Schöpfer, and J. Luggen-Hölscher, Strahlenzentrum der Justus-Liebig-Universität, Giessen, Germany

INTRODUCTION

There is a number of indications that the radiobiological oxygen-effect consists of more than one component: in dry bacterial spores three different classes can be separated depending on exposure- and after-treatment conditions (1), the dependence of the oxygen-enhancement ratio (OER) on oxygen concentration shows breaks in various systems (2,3), and it has been suggested that "type O" and "type N" damage are localized in different parts of the cells (4). We studied these questions in the simple eucaryot Saccharomyces cerevisiae using two approaches: the dependence of OER on oxygen tension was determined both for survival and mutation induction. Since a forward mutation was used a haploid strain had to be employed in this case. In order to assess whether also in yeast cells more than one component may exist, the techniques originally developed for bacterial spores were adapted for dried diploid yeast cells. The results show that the dependence on oxygen concentration is the same for survival and mutation within error limits, implicating DNA as the main target and that also in our system three classes of oxygen dependent damage exist.

MATERIALS AND METHODS

Strains and culture conditions: Saccharomyces cerevisiae haploid wild type 211-1a and diploid wild type 211 were grown on nutrient agar plates to stationary phase and washed off for further manipulations. Survival was determined in the usual way by counting macrocolonies after 3-4 days incubation at 30°C. The rate of forward mutations to canavanine resistance was estimated by plating on nutrient agar plates containing 40 mg/l canavanine.

Dried cells: Diploid cells were collected on membrane filters and carefully dried in a two-step procedure. At first they were kept for about 30 minutes in air, and then desiccated in vacuo in the presence of P_2O_5. Control survival was about 70% under these conditions. After exposure the cells were either plated immediately on nutrient agar or kept in dry air for various periods of time at 30°C before plating. To test also the ability for delayed plating recovery aliquots were plated on non-nutrient agar, incubated for 4 days at 30°C in air and then transferred to nutrient conditions. Irradiation conditions: The

ISBN 0-12-592050-4

oxygen dependence was measured with 150 kV X-rays. Different concentrations were maintained by bubbling appropriate air/N_2-mixtures through aqueous suspensions. The dose-rate of 0.9 GY/min was low enough to exclude radiation-induced oxygen depletion.

Dry cells were exposed to ^{241}Am-α-particles in dry air or nitrogen.

RESULTS

Fig. 1 shows the OER as a function of oxygen partial pressure for mutation and survival. The difference between the two curves is not statistically significant, and they can be fitted to a common "Alper-curve" with parameters: m = 1.9, K = 24.2 μmol dm^{-3}. Although we tried carefully to find out whether an

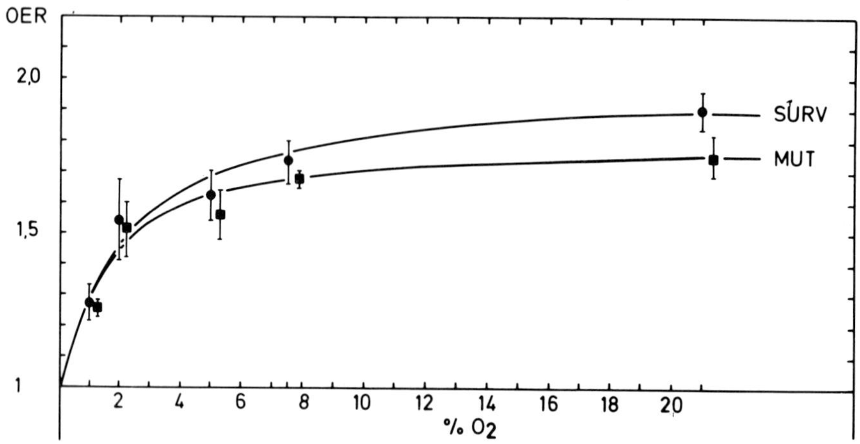

FIGURE 1. The dependence of OER on oxygen concentration for survival and mutation with 150 kV X-rays. Ordinate: Oxygen enhancement ratios as a function of oxygen concentration; abscissa: oxygen concentration.

intermediate plateau exists as reported for other systems (2,3) the resolution of our method does not seem to be sufficient to give a clear-cut answer. There may be an indication at 6% O_2, but the significance is doubtful. We shall explore this point further in future experiments.

An example of an experiment with the dry system is given in Fig. 2. In this case exposure to ^{241}Am α particles took place in N_2, and the cells were either plated immediately or subjected to a delayed plating procedure. One notes a substantial sensitivity difference between wet and dry conditions which is surprising since high-LET-irradiation (LET_∞ = 160 keV/μm) was used where the indirect effect is expected to be of minor influence. It has to be taken into account that also other factors—e.g. shrinking of intracellular targets—may play a role. There is also substantial DP-recovery after exposure in dry conditions.

Space does not permit presentation of all survival curves,

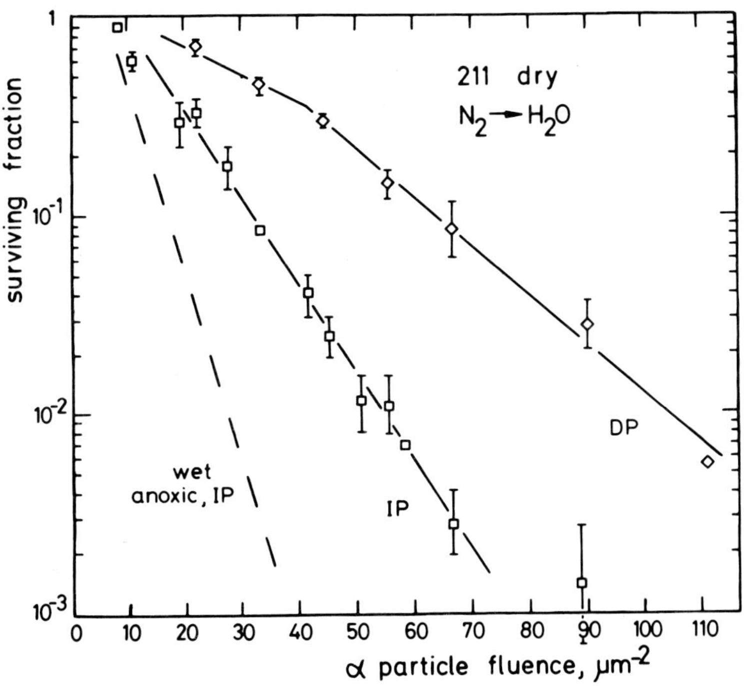

FIGURE 2. Survival of diploid yeast exposed to ^{241}Am-α-particles in wet and dry state under nitrogen. IP: Immediate plating on nutrient agar, DP: Incubation on non-nutrient agar for four days before final plating. Ordinate: surviving fraction; abscissa: α-particle fluence.

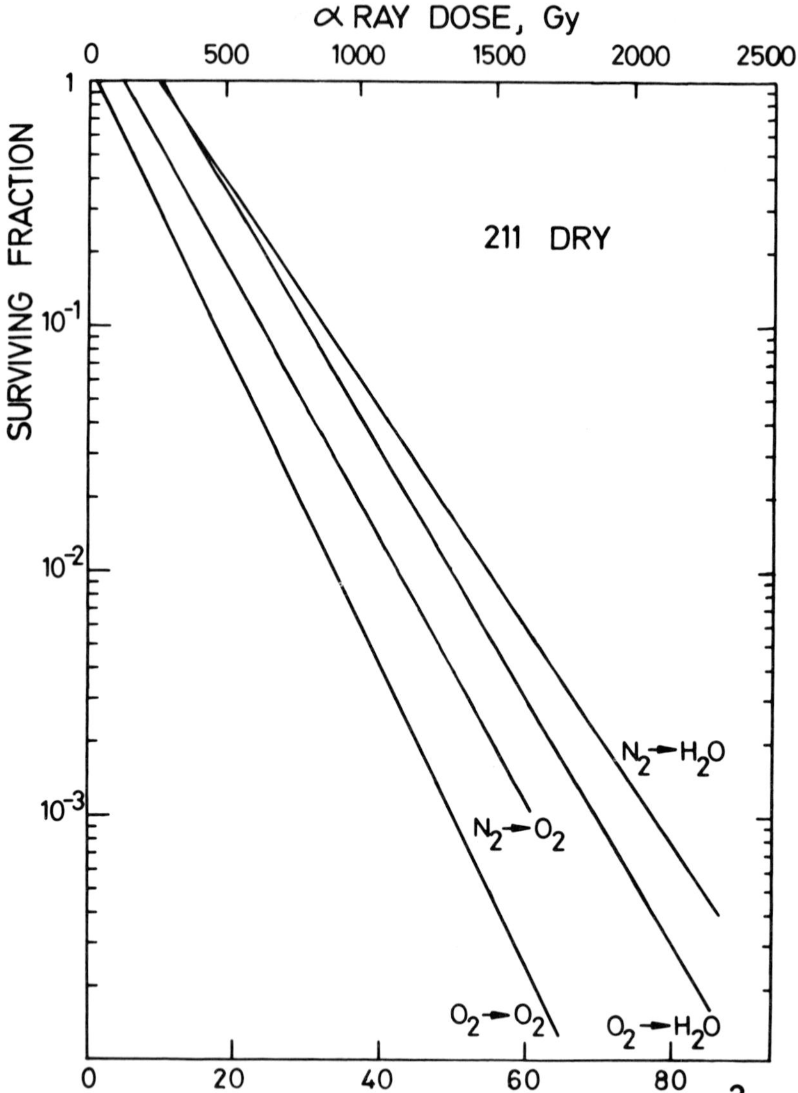

FIGURE 3. Summary of results with dried yeast (data points omitted for clarity). First letter designates exposure conditions, H_2O: immediate plating on nutrient agar, O_2: 40 hours after treatment with dry air at 30°C before plating. Axes as in Fig. 2. The α-dose was calculated assuming a track average LET of 160 keV/μm since the particles were slowed down by traversal through air.

and they will be published in detail elsewhere. The main features are summarized in Fig. 3. In all cases the sensitivity is increased if the cells are kept in air for about 40 hours after exposure compared to immediate plating. The kinetics of this process are depicted in Fig. 4. The "half-lives" for damage fixation after irradiation in air or N_2 are 17-20 hours and 6-7 hours, respectively. We have to conclude from this that the underlying mechanisms are different.

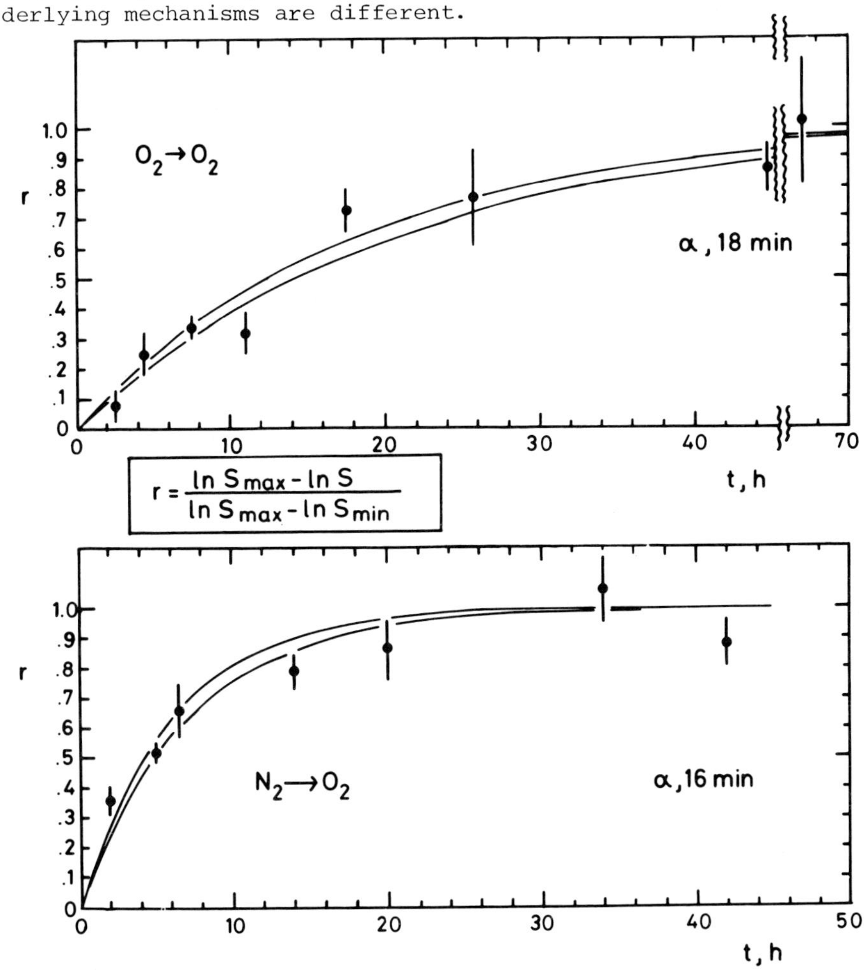

FIGURE 4. Kinetics of the after-effect by treatment with air. The points were fitted by simple exponentials with different "half-lives" (upper curve: 20 and 17,5 hours, lower curve: 6 and 7 hours) to illustrate the degree of variation. S indicates the actual sensitization factor, S_{max} and S_{min} the respective maximal and minimal values. Ordinate: r as given by the formula; abscissa: duration of posttreatment before plating.

DISCUSSION

There is clear oxygen effect if dried cells are plated immediately after exposure in N_2 (Component I). The sensitivity is increased by after-treatment with O_2, irrespective of exposure conditions. Since the time course is different, we have to assume two further components. The conclusion is therefore that also in our eucaryotic system the oxygen effect consists of at least 3 distinct classes.

The question remains why this does not show up in the dependence on oxygen concentration in wet cells. Apart from the poor resolution, it may be possible that in our cells the time-courses of the respective reactions overlap in such a way that the single pathways may not be discernible.

ACKNOWLEDGEMENTS

We thank K. Stark, C. Stühmeier, B., Huckenbeck and H. Pickl for technical assistance and Dr. E. L. Powers for helpful suggestions. This study was partly supported by the "Bundesministerium für Forschung und Technologie" and "Gesellschaft für Schwerionenforschung."

REFERENCES

1. Powers, E. L., Webb, R. B. and Ehret, C. F., Radiat. Res., 2,94 (1960).
2. Ewing, D. and Powers, E. L., Science 194,1049 (1976).
3. Millar, B. C., Fielden, E. M. and Steele, J. J., Int. J. Radiat. Biol. 36,177 (1979).
4. Alper, T., Phys. Med. Biol. 8,365 (1963).

DISCUSSION

MITCHEL: We have also used diploid yeast and looked at the same sort of thing with varying O_2, and we had the same idea to look at both survival and actually a DNA damage (we were using gene conversion as a measure of DNA damage), and we got a plateau just where your one aberrant point was at around 50 μM in both cases. We did see biphasic curves in both cases, and the plateau was the same in both cases. The ratio I noted, if I could guess, was about 60-40 perhaps?

KIEFER: Yes.

MITCHEL: And that is what we got too in both cases: 60-40 ratio.

KIEFER: I have seen your poster. I was curious to see that. I think it is very difficult to really resolve the plateau. I think you still have a fair amount of scatter in the data.

MITCHEL: That is true, and in any one experiment I would be skeptical too, but we have done it many times now, and it always comes out the same, and I believe it now.

KIEFER: O.K. I think we are going to do more experiments, and then we will see what we believe.

MITCHEL: A point we could perhaps make to David Ewing's suggestion is that some of the radicals are formed outside the cell. Since those two curves for cell survival and DNA damage are so similar, it would tend to argue that all the O_2 component is due to DNA damage and if something is formed outside the cell it has to get to the DNA somehow.

KIEFER: Yes, I think so; and I don't think it is really conceivable, particularly in the yeast cell, with not only a cell membrane but also a very thick cell wall. It is not conceivable to assume that something from outside the cell really comes to the DNA in the cell to do the mutation. I don't think so, at least not in our system.

POWERS: The fact that you do not see the plateau there does not mean that you have shown that it is not there.

KIEFER: No. I did not want to state that.

THE EFFECTS OF OXYGEN ON SURVIVAL AND SINGLE STRAND DNA BREAKS IN IRRADIATED MAMMALIAN CELLS

Barbara C. Millar and E. Martin Fielden, Radiobiology Unit, Physics Division, Institute of Cancer Research, Sutton, Surrey, U.K.

INTRODUCTION

The survival of cells irradiated in vitro varies according to the concentration of oxygen present at the time of irradiation. Also several authors have shown that oxygen increases the yield of single strand breaks (ssb) in the DNA of bacteria (1,2) and mammalian cells (3,4,5,6) following irradiation. In bacteria, the increase in ssb production occurs at lower oxygen concentrations than are required to increase X-ray killing in E. coli (1) suggesting that ssb do not contribute appreciably towards cell death. Similarly, using a rapid mix technique it has been shown that the presence of oxygen is required for a much shorter time period prior to irradiation for the maximal effect on ssb production in mammalian cells than for the full OER for cell survival (7). In this work the radiation survival response of exponentially growing Chinese hamster cells at various oxygen concentrations has been closely examined and compared with the initial yield of ssb determined by the rapid lysis technique.

MATERIALS AND METHODS

Cell culture. Chinese hamster cells, line V-79-753B, were used throughout the work. Cell culture procedures were carried out by methods described previously (8,9). Survival curves were determined for cells attached to glass petri dishes using a Co^{60} γ source at a dose rate of 280 rad/min. The basic techniques and method of controlling gas tension have been described elsewhere (8,9). Gas mixtures of certified composition were supplied by British Oxygen Corporation and checked using a "Thermox 1" meter (Thermox Instruments Inc., Pittsburgh, Pennsylvania).

Rapid lysis procedures and ssb determinations. The preparation of cells for rapid lysis has been described previously (6). Cells labelled with 0.6 µCi/ml ^{3}H-Thymidine (20-30 mCi/mM) were suspended in Dulbecco's PBSA under controlled gas tension at 4°C. Following irradiation (duration 20 msec) the cells were lysed within 0.2 sec and the mean molecular weight (Mn) of the DNA determined by velocity sedimentation (6).

ISBN 0-12-592050-4

RESULTS AND DISCUSSION

Fig. 1 shows the slope ratio obtained from the survival curves of cells irradiated in the presence of different concentrations of oxygen (8). Between the concentrations 0.4 to 1.48 μM there is progressively more sensitization as the concentration of oxygen is increased. However, between 1.48 and 7.0 μM the amount of sensitization is constant, corresponding to an enhancement or slope ratio (SR) of 1.9.

Further sensitization is achieved when the concentration of oxygen is increased above 7 μM. However, the oxygen enhancement ratio exhibited by cells irradiated in air (∿ 280 μM) is not exceeded when the oxygen concentration is further increased (1395 μM).

In the rapid lysis experiments the plots of dose vs. 1/Mn were linear, indicating that oxygen depletion was not a problem in the dose range used. Values for the yields of ssb at each concentration of oxygen were calculated from the slope of these plots. At a concentration of 1.5×10^{-6}M oxygen the yield of ssb was $1.7(2) \times 10^{-9}$ daltons^{-1} krad^{-1}, similar to that found for cells irradiated in pure O_2 ($1.7(3) \times 10^{-9}$ daltons^{-1} krad^{-1}). However, the slope ratio (SR) for cell killing was 1.9 for cells irradiated in 1.5×10^{-6}M oxygen compared with the maximum SR of 3.1 for cells irradiated in 1.48×10^{-3}M oxygen. Fig. 2 shows the relationship between the yield of ssb at 0.2 sec after irradiation and the slope ratio for survival. For convenience the

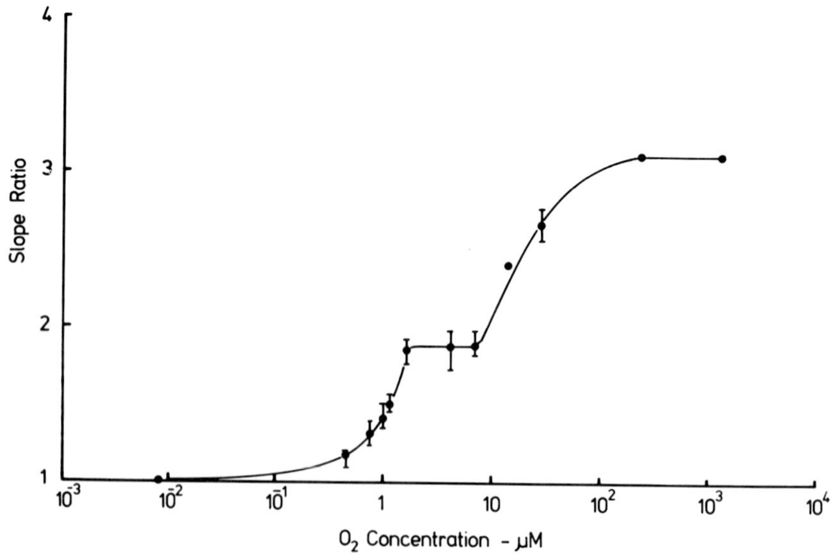

FIGURE 1. Changes in the slope ratio of survival curves for Chinese hamster cells, V-79-753B, as a function of oxygen concentration (8), by permission Taylor and Francis, Ltd.

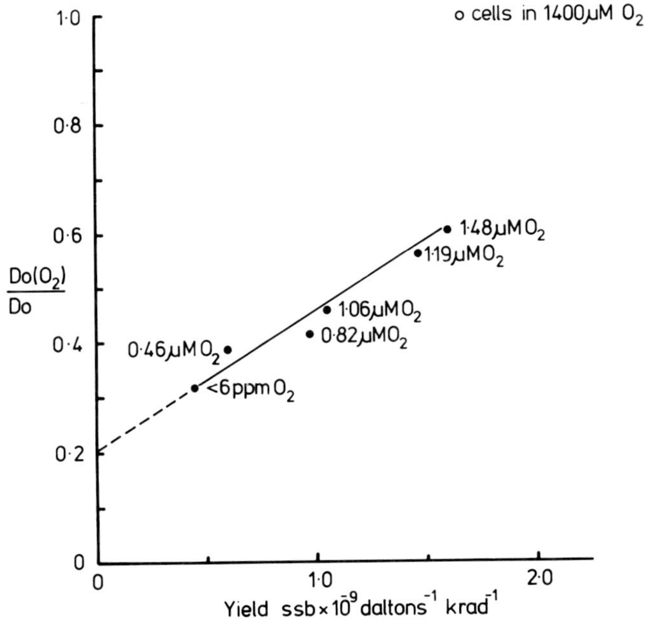

FIGURE 2. The relationship between cell survival and ssb production in Chinese hmaster cells, V-79-753B (6), by permission Academic Press.

SR values have been normalized to a value of 1 in pure oxygen by taking the ratio of D_o values. Figs. 1 and 2 show that there is a discontinuity in both the survival response and initial yield of ssb as a function of oxygen concentration. Thus a simple competition model cannot explain either the response of ssb or survival to oxygen. There appears to be a linear relationship between cell survival and the initial yield of ssb over the low oxygen tension region (0.46 to 1.5 µM O_2). Further work with radiosensitizers indicates that this is probably not a causal relationship however (Millar and Fielden, to be published).

A biphasic response to radiation as a function of oxygen concentrations has been noted previously in bacterial spores (10). In mammalian cells and bacteria there is evidence that the oxygen effect on survival can be divided into at least two components (11,12), although no previous evidence has been presented to suggest that they may be separated on the basis of oxygen concentration. Using a rapid-mix technique, where an anoxic suspension of mammalian cells was mixed with oxygen at various concentrations prior to irradiation, an enhancement ratio of 1.7 was found after the shortest possible contact time (4 msec). Further enhancement of sensitization required longer contact times (11).

CONCLUSIONS

Oxygen acts in at least two ways in enhancing cell killing and exerts a maximal effect on DNA single strand breaks at a concentration that gives only 40% (SR = 1.9) of the full OER for cell killing. A similar discontinuity in SR vs concentration curves has been noted with several hypoxic cell sensitizers (13).

ACKNOWLEDGEMENTS

We would like to thank the CRC and MRC for financial support for this research.

REFERENCES

1. Johansen, I., Gulbrandsen, R. and Petterson, R., Radiat. Res., 58,384 (1974).
2. Sapora, O., Fielden, E. M. and Loverock, P. S., Radiat. Res., 64,431 (1975).
3. Palcic, B. and Skarsgaard, L. D., Int. J. Radiat. Biol., 21,417 (1972).
4. Dugle, D. L., Chapman, J. D., Gillespie, C. J., Borsa, J., Webb, R. G., Meeker, B. E. and Reuvers, A. P., Int. J. Radiat. Biol., 22,545 (1972).
5. Roots, R. and Smith, K. C., Int. J. Radiat. Biol., 26,467 (1974).
6. Millar, B. C., Fielden, E. M. and Steele, J. J., Radiat. Res., 83,57 (1980).
7. Adams, G. E., Agnew, D. A., Stratford, I. J. and Wardman, R., Fifth Symposium on Microdosimetry, Verbania Pallanza (Italy), September 22-26 (1975).
8. Millar, B. C., Fielden, E. M. and Steele, J. J., Int. J. Radiat. Biol. 36,177 (1979).
9. Cooke, B. C., Fielden, E. M., Johnson, M. and Smithen, C. E., Radiat. Res. 65,152 (1976).
10. Tallentire, A., Jones, A. B. and Jacobs, G. B., Israel J. Chem., 10,1185 (1972).
11. Shenoy, M. S., Asquith, J. C., Adams, G. E., Michael, B. D. and Watts, M. E., Radiat. Res., 62,498 (1975).
12. Sapora, O., Fielden, E. M. and Loverock, P. S., Radiat. Res., 69,293 (1977).
13. Millar, B. C., Fielden, E. M. and Jenkins, T. C. This volume, p. 329.

DISCUSSION

BIAGLOW: Dr. Fielden, it looks like you are titrating something. Do you think that there could be a respiratory component; have you tried respiratory inhibitors to see if they will follow that break?

FIELDEN: No. We have not tried any specific respiratory inhibitors but they may well be worth trying. However, I cannot see how respiratory effects could produce the "plateau" under our experimental conditions. To elaborate on radiation induced O_2 consumption—it really is rather surprising that one can put in 25 krads and apparently not see any O_2 consumption in the SSB vs dose curves. There could be a couple of explanations for that. One, the O_2 may be transiently bound—that is during this short time scale of these experiments, O_2 can be converted into superoxide and other reactive oxy radicals which may still be able to be involved in the chemistry of the O_2 effect. A further explanation could be that as far as DNA single strand breaks are concerned, the thing that determines the pattern or level of breaks would be the O_2 status of the cell at the time of irradiation or the biochemical redox status, if you like, which would be set by the initial concentration of O_2.

EWING: Martin, you answered my question almost exactly since I was also concerned with radiolytic O_2 binding. But, may I suggest one other thing? Not only do we need to worry about the rate of radiolytic O_2 binding and the rate at which O_2 can rediffuse into the cells, but the rates at which O_2 reacts to cause damage are also important here. If the rates for oxygen's reactions which cause damage are slow enough, then it does not matter how fast the rate of O_2 binding is or how slow the rate of O_2 rediffusion is, so long as the radical can stay there long enough for the O_2 to get there. Perhaps that is also involved.

FIELDEN: Yes, but I don't quite see how a slow phenomenon like you have proposed is helped by rapid O_2 binding.

EWING: It would overcome it; then it wouldn't make any difference about the rapidity of O_2 binding.

MICHAELS: I would like to report that I did some work in my laboratory on the very same cells (V79-753B) that Martin has been talking about here. These were cells that were actually sent from Sutton to my lab for the express purpose of seeing if we could observe the same OER vs O_2 concentration relationship, i.e. the K-curve, with the thin layer technique which we use. However, in a series of careful experiments, we were not able to detect the behavior described by Millar and Fielden. In particular, we see no plateau in the K-curve, over the O_2 concentration range that Martin has reported, using basically the same experimental conditions. Dr. Fielden and I have discussed these results, and we have yet to discover a clear explanation for the discrepancy.

FIELDEN: At Hammersmith, Dr. Cullen has found the same effect as we did for the K-curve with the same cell line. I suppose my thoughts for the discrepancy with the Boston group is probably that the biochemical status of the cell could be different. If you are using a different medium or growth conditions the cells could have different SH levels and these may well be involved in the separation of the components. Thus if they have moved sufficiently close together you will see a single curve.

MICHAELS: I agree that the two groups are certainly using different medium and different serum for these experiments. Perhaps this is the explanation. On another point, Dr. Fielden, I also am surprised that you observe no O_2 depletion effects under your conditions, since you are giving such a large dose in a very short time. I have also considered the point made in your suggestion that perhaps there is a transient O_2 binding step involved, and I will be referring to this possibility this afternoon in Robin Willson's session. I will be showing that there can be significant effects on the O_2 depletion process on this very fast time scale, because of the presence of highly reactive molecules and radical species in the cell; this deserves further consideration.

WHILLANS: Some resolution of these controversies may be in differences in techniques which did not seem significant at the time. I would just like to support Martin's last suggestion and say that one of the differences between the two rapid-mixing experiments, where one showed two components and the other only one, was that in the former case the cells were hypoxic—you could call it acutely hypoxic—for only a few minutes whereas in our system they had been hypoxic for perhaps an hour. There is a difference in the way we had achieved hypoxia. It is a real possibility, I think.

EPP: Dr. Fielden has shown evidence for the plateau in the O_2 K-curve for V79-753B cells, but work done in our laboratory by Howard Michaels and Clift Ling, and by several other groups, indicate that this effect is not observed for CHO cells. One likes to look for generalized phenomena. Now either the phenomenon is not generalized, i.e. different cells, or different cell lines, under different conditions may have different shaped K-curves; or perhaps the phenomenon is generalized, but the experimenter's Nemesis, namely an artifact, may be lurking about somewhere in one or other of the experiments. I think that it would be a good idea if we could jointly try to encourage other groups to take a very hard look with their mammalian cell lines to try to nail down the responses in detail. One has to do this in detail if one wants good results, of course, before we go too far in deciding whether it is one component or whether it is two components. Until the data we have are clearly reproducible by various groups, I think investing the time experimentally might be the best bet at this time.

FIELDEN: That is right. It is of course, as you know, extremely time consuming and not easy to do this carefully controlled O_2 tension work. Also, I didn't put it into the talk as it will be coming this afternoon. We find this "plateau" effect with radiosensitizers of the O_2 mimetic, electron-affinic type.

POWERS: Dr. Epp, you do not doubt that there are two O_2 effects?

EPP: I am just saying that with respect to the O_2 k-curve: the question of the shape of the K-curve has been around for a long time as you well know.

POWERS: The K-curve has been fitted with very bad data throughout the history of the use of K-curve. The data scatter badly; the K-curve is usually tightly compressed, very close to the 0 point.

EPP: Let me just get specific about mammalian cells. Clearly Millar and Fielden have looked in detail for the response for their type of mammalian cells and Michaels has done the same more recently, so I do not think your comment applies to those.

POWERS: No it does not.

EPP: So I am just saying that we should try to nail the K-curve shape down experimentally—that is all I am really saying.

WEISS: One question, two comments. At which temperature were these experiments done with the measurements of the K-curves: were they all at room temperature?

FIELDEN: They were all at room temperature: 25-27°C.

WEISS: Dr. Epp, was that the same for yours too?

EPP: Yes.

WEISS: Also, I observed a second order decay and Barry Michael observed (this has nothing to do with the present paper) a first order decay. I don't think that there is any disagreement really between the two because we are delivering doses on the order of at least a factor of 10 greater, so that you can see a second order decay in one case and a linear decay in the other. My second comment does not have to do with science. This is a conference on O_2. I am guilty of misuse of language myself—and particularly in this case. There are hypoxic cells, there are anoxic cells, there are not oxic cells—if you look it up in a dictionary you cannot find it. It is the same thing as emic. If a person is healthy he is not emic, but he can be anemic.

MICHAEL: Just because you do or do not see an effect does not mean it is not there. I think from the studies that we have been engaged in, and other people, it is plain that there are competing fast reactions involved in the O_2 effect; and the rates of those reactions are crucial in the concentration dependences that are observed. We think that the rates of those reactions are (from the Stokes-Einstein relationship) strongly dependent upon the viscosity within the cell and particularly in the microenvironment of the target. And it may be that the observation or nonobservation of these biphasic behaviors represent differences of reaction rate reflecting perhaps local viscosity changes under different concentrations.

ROOTS: I would just like to add that, when you look at cells that have been in N_2 for an hour or so vs cells that have been in N_2 for only a short time, certainly the intracellular concentrations will vary a great deal and the concentration of substances which may compete with O_2 may be quite different in these two cases.

OXYGEN-LIKE RADIATION SENSITIZERS: MECHANISMS OF ACTION

Session Chairman, R. L. Willson, Uxbridge, U. K.

FREE RADICALS, OXYGEN AND RADIOSENSITISING DRUGS: A VERY BRIEF INTRODUCTION

Robin Willson, Biochemistry Department, Brunel University, Uxbridge, England

Radiation therapy continues to be a major form of cancer treatment. Results, however, still vary widely. When control of even the primary tumour fails, it is thought that sometimes this is due to some cells being poorly supplied with oxygen at the time of treatment. The efficient capillary network of normal tissue ensures that all cells are adequately supplied with oxygen. In tumours, network development may fail to keep up with the expanding tissue mass and as the intercapillary distance increases, some cells may become so removed that all the oxygen is consumed by intervening cells before it reaches them. These cells die forming an area of necrosis. Cells adjacent to this area however may just remain viable but contain very low amounts of oxygen. Such hypoxic cells are well-known to be relatively resistent to radiation killing and should they survive treatment, they may then serve as foci for regrowth of the tumour.

In 1955 hyperbaric oxygen chambers were introduced in the hope that whole body exposure to 3 atmospheres of oxygen would increase the level of oxygen leaving the capillaries and thereby increase the distance it would diffuse before being metabolised. If hypoxic cells were present, their number might be reduced. By the mid 1960's however it was clear that any improvement in patient survival was unlikely to be more than marginal (1). Several research groups, particularly those of Howard-Flanders at Yale and Scott and Adams at Mount Vernon Hospital England, turned their attention to the possibility of combining conventional treatment with drugs which sensitised only hypoxic cells. Any natural bias towards normal rather than tumor tissue destruction might thus be overcome (2,3).

In 1973 the first clinical trial of a hypoxic radiosensitising drug was begun. Patients with brain tumours (notoriously difficult to cure) were given the drug metronidazole (Flagyl) before radiation treatment. Although the median survival was increased only from 15 to 26 weeks and questions were raised as to

ISBN 0-12-592050-4

the suitablity of the radiation dose regimen used, the results were encouraging to both research workers and clinicians alike (4,5). The rationale for using Flagyl and later the closely related nitroimidazole, misonidazole stemmed from a wealth of fundamental clinical and biological data concerning the action of ionizing radiation, and the role of oxygen free radicals, that had accumulated over half a century.

A wide assortment of compounds chosen for many different reasons had been examined for sensitising activity, particularly using bacterial or in vitro mammalian cell systems. Occasionally with bacterial systems (for example Serratia marcescens irradiated in buffered saline) plots of the logarithm of the fraction of cells surviving against radiation dose (survival curves) are linear. In these instances the sensitising efficiency, (expressed in terms of the enhancement ratio (E.R.), the factor that the dose can be reduced for the same level of survival) is constant and the chemical is said to be dose modifying (Fig. 1a). With other bacterial systems and with mammalian cells the situation is

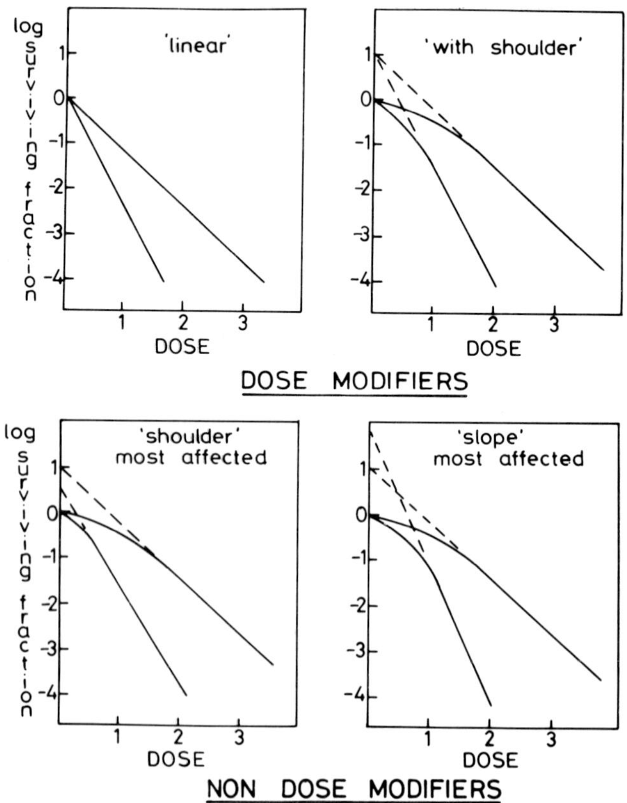

FIGURE 1. Typical survival curves.

often more complicated. Survival curves are indeed often curved consisting of a 'shoulder' followed by a linear 'slope'. When a particular chemical is 'dose modifying' the linear portion of the curve will extrapolate to the same point on the zero dose ordinate as the control curve and the ratio of the slopes will again correspond to the enhancement ratio (Fig. 1b). Where chemicals are not dose modifying and affect the slope and the shoulder by different amounts (Figs. 1c and d) the manner in which sensitising efficiency is expressed must be carefully defined.

The activity of dose modifying chemicals is usually assessed from plots of the enhancement ratios against concentration (Fig. 2). Although in some cases mathematical equations have been fitted to such plots and the activity has been expressed in terms of a mathematically defined parameter it has been recent practice to express activity simply in terms of the concentrations of the chemical required to produce a particular enhancement ratio (7-9).

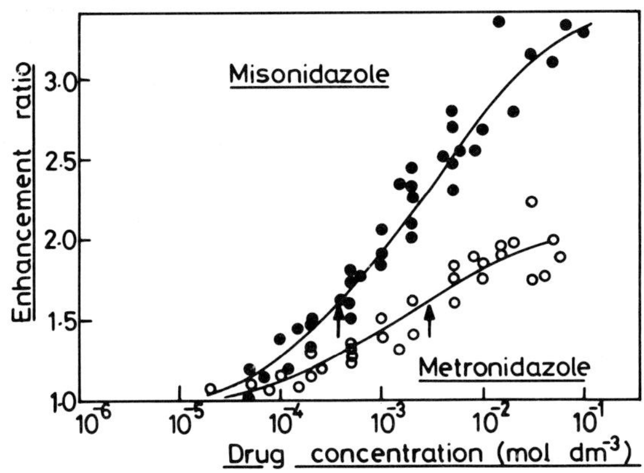

FIGURE 2. Plot of enhancement ratio against sensitiser concentration for mammalian cells in vitro. (Courtesy I. Stratford ref. 6).

Of the many compounds tested to date few have achieved more than a passing interest from clinicians. Some are too toxic, others pharmacologically unsuitable, some are also likely to sensitise well oxygenated cells and hence normal tissue. Fortunately, however, the studies of the effects of such chemicals continue to be undertaken in both chemical and cellular systems and are providing very useful mechanistic information. Indeed, without these studies it is unlikely that Flagyl or misonidazole would have ever reached the clinic. Other compounds that have been studied include dyes such as fluorescein (10) and trypan blue (11), the respiratory poison, iodoacetate (12), mitotic inhibitors such as menadiol phosphate (13) paramagnetic compounds for example nitric oxide (14, 15, 16) and several stable free radicals (17, 18), sulphydryl reagents such as N-ethyl maleimide (19), and diamide (20), conjugated carbonyls, for example diacetyl (21), metal ions and complexes including those of iron (22) copper (23) silver (24) platinum (25) and cobalt (26), nitro-compounds including nitrobenzene (27), nitrofuran (28) and other two and 5 nitroimidazole derivatives (29), membrane active compounds such as pyridinium salts and chlorpromazine (30, 31) and DNA interactive agents such as 8-methoxypsoralen (32).

By the late 1940's it was well known that free radicals played a role in the damaging action of ionising radiation and that sulphydryl compounds could sometimes afford protection to chemicals as well as cells.

In 1953 Alexander and Charlesby proposed (33) that this might be due to radicals being repaired by hydrogen donation from the sulphydryl group, the resulting product radical R'S· being relatively unimportant. Soon afterwards Alper and Howard-Flanders suggested (34) that the sensitising action of oxygen could be due to its competing with the sulphydryl compound for the vital radical, so fixing it and preventing its repair

$$R\cdot + R'SH \longrightarrow RH + R'S\cdot \qquad \text{REPAIR}$$

$$R\cdot + O_2 \longrightarrow \text{products} \qquad \text{FIXATION}$$

The concept of these competing reactions was to provide the basis for many of the later studies on radiosensitisation. It was quickly appreciated that if oxygen acted in this manner then other compounds that reacted with free radicals might behave similarly. The sensitising and protective effects of nitric oxide was soon demonstrated (14, 15) and led to studies with several stable nitroxide free radicals, which continue today (16, 17). The alternative possibility that sensitisation might be achieved by reducing the levels of intracellular sulphydryl compounds was also apparent and in 1960 Bridges reported sensitisation of bacteria by the well-known sulphydryl binding agent, N-ethyl maleimide (18).

Looking back it seems particularly fortunate that this new free radical approach to radiosensitisation was to coincide with the introduction of the fast reaction technique of pulse radiolysis. Many of the hitherto somewhat speculative reactions of free radicals were soon to be demonstrated directly. In particular the primary reducing species produced on the radiolysis of water, the hydrated electron (e^-_{aq}), was observed for the first time causing considerable excitement. The kinetics and the absorption spectra of the products of its reaction with a host of compounds began to be measured. Early data indicated that the electron adduct of compounds containing conjugated double bonds were relatively long-lived compared to e^-_{aq} itself. In 1963 Adams and Dewey pointed out the N-ethyl maleimide possessed such bonds and suggested that these may contribute to its sensitising action (21). They argued that "damage resulting from attachment of electrons at a particular site would depend upon the competition of non-damaging processes occurring elsewhere in the system". "A molecule combining a high electron affinity with a structure suitable for delocalisation of the attached electron" might act as an electron carrier thereby extending its original sphere of activity and increase the probability of it reacting at a critical site. (Fig 3).

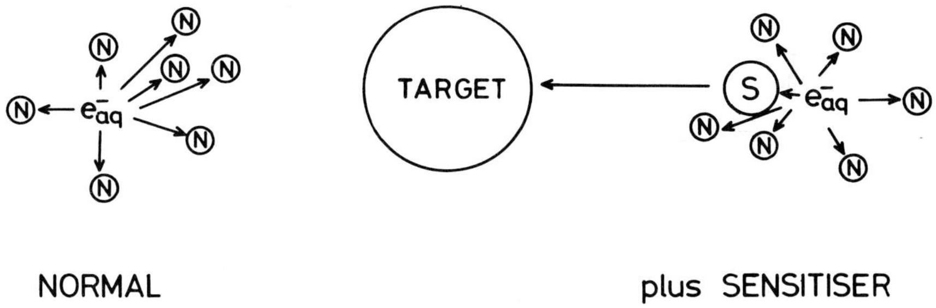

FIGURE 3. Adams and Dewey (1963) electron carrier model of sensitisation.

Although this hypothesis proved a considerable stimulus for sensitiser research, contradictory evidence appeared. In particular, if, as was widely accepted, DNA was a prime target, then the hypothesis could not accommodate either the damaging role of electrons or the sensitising action of oxygen or other so-called electron affinic sensitisers. Steady state studies with DNA and related molecules, particularly by Scholes and co-workers, (35, 36), indicated that the hydroxyl radical, rather than the electron, was likely to be the major damaging species. The electron adduct of oxygen, the superoxide radical $O_2^{\overline{\cdot}}$, reacted only slowly if at all with DNA constituents. In fact on the contrary, the DNA electron adduct was more likely to react with oxygen. Pulse radiolysis studies (37, 38) supported these findings and showed that N-ethyl maleimide as well as other newly described electron affinic sensitisers such as benzophenone and menadione behaved similarly:

$$\text{thymine}^{\overline{\cdot}} + O_2 \longrightarrow \text{thymine} + O_2^{\overline{\cdot}}$$

$$\text{thymine}^{\overline{\cdot}} + S \longrightarrow \text{thymine} + S^{\overline{\cdot}}$$

Yet compounds chosen on the basis of their electron affinity did indeed sensitise. In 1968 Adams wrote (3):

"The best sensitiser on a molar basis is oxygen itself, a molecule of considerable electron affintiy. However, the electron adduct $O_2^{\cdot-}$ is a relatively unreactive species and, therefore, electron transfer from $O_2^{\cdot-}$ to a potential target in a biological system is a suspect mechanism. It is in this respect that the original sensitiser hypothesis of electron transport to a target molecule is seriously deficient.

The following alternative hypothesis involving electron transfer from a damaged target molecule to an electron affinic compound would remove the oxygen anomaly. In this model, a direct ionisation produces some molecular site. If subsequent electron transfer from this site to a more electron-affinic molecule could occur, the possibility of any internal neutralisation process taking place would be eliminated. Thus, electron transfer would result in a higher probability of irreversible chemical change since the target molecule would then be an odd electron system." (Fig. 4).

Electron affinity thus continued as the criterion for the choice of many of the compounds screened for sensitising activity. By the end of the 1960's, however, no appreciable sensitisation of hypoxic mammalian cells by electron affinic compounds other than oxygen had been observed (39).

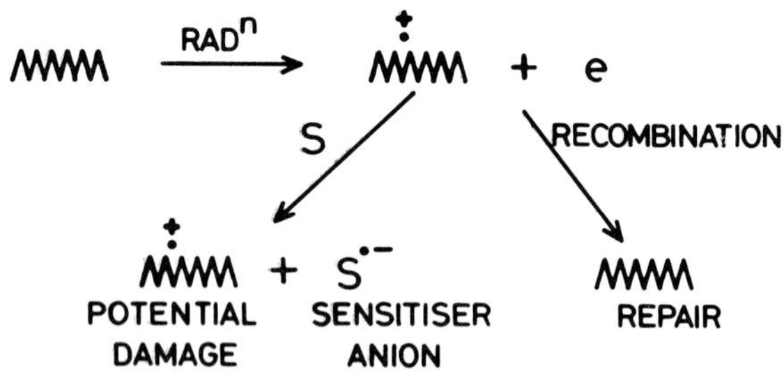

FIGURE 4. Adams (1968) direct action model of sensitisation.

In 1969 during a series of pulse radiolysis studies of electron transfer reactions, the nitro-compound, paranitroacetophenone (PNAP) was found to react rapidly, not only with the thymine electron adduct but, like nitro benzene studied earlier (40), with simple organic radicals in neutral solution (41, 42). This raised the possibility that it might also sensitise by entering into 'fixation' reactions similar to those proposed for oxygen, nitric oxide and the nitroxyl free radicals.

$$(CH_3)_2\dot{C}OH + PNAP \longrightarrow (CH_3)_2CO + PNAP^{\bar{\cdot}} + H^+$$
electron transfer

$$\dot{C}H_2OH + PNAP \longrightarrow (PNAP - CH_2OH)\cdot$$
adduct formation

In 1970 appreciable sensitisation of mammalian cells in vitro by PNAP was reported (43). Studies with other nitro compounds, including several nitrofuran derivatives, led eventually to the discovery of the in vitro and the in vivo sensitising activities of metronidazole, Flagyl (44-47) and misonidazole (9, 48). Their usefulness in radiation treatment remains to be seen. However the fact that they are presently undergoing clinical trials as the result of basic free radical and cellular research, has led to considerable interest in mechanisms of sensitising action generally (49). At the molecular level studies being undertaken include measurements of:

i) one-electron oxidation potentials (E^1_7) of sensitisers using the pulse radiolysis equilibrium technique (50-54)
ii) rates and extent of adduct formation and electron transfer reactions taking place when sensitisers interact with organic radicals (55-65)
iii) effects of sensitisers on damage to DNA constituents (52, 66-72)
iv) radiation induced binding of sensitisers to DNA (73, 74)
v) reactions of sensitiser radicals with biomolecules (62, 75, 76).

Studies at the biochemical level have been made into the reaction of sensitisers with cellular constituents, for example sulphydryl compounds and flavins (77-79) and on enzyme processes particularly respiration (80, 81).

At the cellular level studies have included experiments into the effect of:

i) sensitiser concentration E^1_7 and lipophilicity on survival (10, 29, 82, 83)
ii) various sensitisers on the extent of DNA damage (84-88)
iii) oxygen, sensitiser and protective agent combinations (17, 46, 89-99)
iv) different time intervals between chemical addition and irradiation using rapid mix and double pulse techniques (95-101)
v) cell cycle time using synchronised cells (9, 102-104)

From the information obtained it now appears that several processes may be involved in sensitisation.

a) interference with charge recombination due to a sensitiser having a high electron affinity (3)
b) an increase in the yield of oxidising hydroxyl radicals by 'electron sequestration': a sensitiser which reacts rapidly with e^-_{aq} may at high concentrations scavenge this reducing

species and so lessen the possibility of $OH^{\cdot}$ being rapidly reduced to hydroxide ion (105, 106)

c) interference with radical combination reactions due to a sensitiser having a high one electron oxidation potential (49, 52)

d) oxidation of organic radicals so 'fixing' them

e) formation of products which are toxic

f) changes in the biochemistry of the cell.

The relative importance of each remains a matter for debate. It is hoped that the references cited here and in the following papers will be useful in enabling those newly interested in the radiobiologist's search for 'oxygen-like' drugs to become acquainted with some of the recent literature.

REFERENCES

1. Duncan, W., Brit. Med. Bul. 29, 33 (1973) and ref. cited.
2. Emerson, P. T. and Howard-Flanders, P., Radiat. Res., 26, 54 (1965).
3. Adams, G. E., "Radiation Chemistry of Aqueous Solutions" (G. Stein, ed.) Interscience, London, New York, Sydney; p. 241 (1968).
4. Urtasun, R., Band, P., Champman, J. D., Feldstein, M. L., Meikle, B. and Fryer, C., New Eng. J. Med. 294, 1364 (1976).
5. Belli, J. A. and Hellman, S., ibid 294, 1399 (1976).
6. Stratford, I. J., "Radiosensitizers of hypoxic cells" (A. Breccia, C. Rimondi and G. E. Adams, eds.) Elzevier/North Holland (1979), p. 109.
7. Alper, T. and Howard-Flanders, P., Nature, 178, 978 (1956).
8. Tallentire, A. and Jacobs, G. P., Int. J. Radiat. Biol., 21, 205 (1972).
9. Asquith, J. C., Watts, M. E., Patel, K., Smithen, C. E. and Adams, G. E., Radiat. Res. 60, 108 (1974).
10. Adams, G. E., Clarke, E. D., Flockhart, I. R., Jacobs, R. S., Sehmi, D. S., Stratford, I. J., Wardman, P. and Watts, M. E., Int. J. Radiat. Biol. 35, 133 (1979).
11. Mottram, J. C., Brit. Med. J. (1929) p. 149; Russ, J. and Scott, G. M., Proc. Roy. Soc. B, 118, 316 (1935).
12. Franks, W. R., Shaw, M. M. and Dickson, W. H., Am. J. Cancer, 22, 601 (1934).
13. Mitchell, J. S., Brit J. Cancer 2, 351 (1948).
14. Howard-Flanders, P. Nature 180, 1191 (1957).
15. Powers, E. L., Webb, R. B. and Kaleta, B. F., Proc. Nat. Acad. Sci. 48, 984 (1960).
16. Dale, W. M., Davies, J. V. and Russell, C., Int. J. Radiat. Biol. 4, (1961).
17. Emmerson, P. T. and Howard-Flanders, P., Nature 204, 1005 (1964).
18. Millar, B. C., Fielden, E. M. and Smithen, C. E., Radiat. Res. 69, 489 (1977).
19. Bridges, B. A., Nature 188, 415 (1960).

20. Harris, J. W. and Power, J. A., Radiat. Res. 56, 97 (1973).
21. Adams, G. E. and Dewey, D. L., Biochem. Biophys. Res. Comm. 12, 473 (1963).
22. Moroson, H. and Tenney, D., Experentia 24, 1041 (1968).
23. Cramp, W. A., Nature, 206, 636 (1965).
24. Richmond, R. C. and Powers, E. L., Radiat. Res. 58, 470 (1974).
25. Richmond, R. C. and Powers, E. L., J. Clin. Hematol. Oncol. 7, 580 (1976).
26. Richmond, R. C., Simic, M. and Powers, E. L., Radiat. Res. 63, 140 (1975).
27. Champman, J. D., Raleigh, J. A., Borsa, J., Webb, R. G. and Whitehouse, R., Int. J. Radiat. Biol. 21, 475 (1972).
28. Chapman, J. D., Reuvers, A. P., Borsa, J., Petkau, A. and McCalla, D. R., Cancer Res. 32, 2616 (1972).
29. See Wardman, P., Current Topics in Radiat. Res. Quarterly 11, 347 (1977).
30. Redpath, J. L. and Patterson, L. K., Radiology 118, 725 (1976); Anderson, R. F. and Patel, K. B., Int. J. Radiat. Biol. 32, 471 (1977).
31. Shenoy, M. A., George, K. C., Singh, B. B. and Gopal-Ayengar, A. R., Int. J. Radiat. Biol. 28, 519 (1975).
32. Redpath, J. L., Brit. J. Cancer 37, Supp. III, 64 (1978).
33. Alexander, P. and Charlesby, A., "Radiobiology Symp. Liege" (Z. M. Bacq and P. Alexander, eds.) Butterworths, London (1955) p. 49.
34. Alper, T. Radiat. Res. 5, 573 (1956); Howard-Flanders, P., Nature, 186, 485 (1960).
35. Scholes, G., Ward, J. F. and Weiss, J., J. Mol. Biol. 2, 379 (1960).
36. Blok, J., Luthjens, L. J. and Roos, A. L. M., Radiat. Res. 30, 468 (1967).
37. Greenstock, C. L., Adams, G. E. and Willson, R. L., "Radiation Protection and Sensitisation" (H. L. Moroson and M. Quintiliani, eds.) Taylor and Francis, London (1970) p. 65.
38. Adams, G. E., Greenstock, C. L., Van Hemmen, J. J. and Willson, R. L., Radiat. REs. 49, 85 (1972).
39. Adams, G. E. adn Cooke, M. J., Int. J. Radiat. Biol. 15, 457 (1969).
40. Asmus, K. D., Wigger., A. and Henglein, A., Ber. Buns. Phys. Chem. 70, 862 (1966).
41. Adams, G. E., Asquith, J. C., Dewey, D. L., Foster, J. L., Michael, B. D. and Willson, R. L., Int. J. Radiat. Biol. 19, 575 (1971).
42. Adams, G. E. and Willson, R. L., J. Chem. Soc. Farad. I. 69, 719 (1973).
43. Asquith, J. C., Willson, R. L. and Adams, G. E., Abstracts 4th In. Cong. Rad. Res., Evian (1970) p. 11.
44. Foster, J. L. and Willson, R. L., Brit. J. Radiol. 46, 234 (1973).

45. Asquith, J. C., Foster, J. L. and Willson, R. L., Brit. J. Radiol. 46, 648 (1973).
46. Asquith, J. C., Foster, J. L., Willson, R. L., Ings, R. M. and McFadzean, J. A., Brit. J. Radiol. 47, 474 (1974).
47. Begg, A. C., Sheldon, P. W. and Foster, J. L., Brit. J. Radiol. 47, 399 (1974).
48. Sheldon, P. W., Foster, J. L. and Fowler, J. F., Brit. J. Radiol. 30, 560 (1974).
49. Willson, R. L., "Metronidazole", S. M. Finegold Excerpta Medica (1977) p. 147.
50. Patel, K. B. and Willson, R. L., J. Chem. Soc. Farad. I., 69, 814 (1973).
51. Wood, P. M., FEBS Letters, 44, 22 (1974).
52. Willson, R. L., Cramp, W. A. and Ings, R. M. J., Int. J. Radiat. Biol. 26, 557 (1974).
53. Meisel, D. and Neta, P., J. Amer. Chem. Soc. 97, 5198 (1975).
54. Wardman, P. and Clarke, E. D., J. Chem. Soc. Farad. I., 72, 1377 (1976).
55. Adams, G. E., Michael, B. D. and Richards, J. T., Nature, 215, 1248 (1967).
56. Emmerson, P. T. and Willson, R. L., J. Phys. Chem. 72, 3669 (1968); Willson, R. L. and Emmerson, "Radiation Protection and Sensitisation", (H. L. Moroson and M. Quintiliani, eds.) Taylor and Francis, London (1970) p. 73.
57. Willson, R. L., Trans. Farad. Soc., 67, 3008 (1971).
58. Emmerson, P. T., Fielden, E. M. and Johansen, I., Int. J. Radiat. Biol., 19, 229 (1971).
59. Roberts, P. B. and Fielden, E. M., Int. J. Radiat. Biol., 20, 363 (1971).
60. Hayon, E. and Simic, M., Radiat. Res. 50, 464 (1972).
61. Simic, M. and Hayon, E., Int. J. Radiat. Biol., 22, 507 (1972).
62. Greenstock, C. L. adn Dunlop, I., Radiat. Res. 56, 428 (1973).
63. Whillans, D. W. and Neta, P., Radiat. Res. 56, 428 (1973).
64. Asmus, K. D., Nigam, S. and Willson, R. L., Int. J. Radiat. Biol. 29, 211 (1976).
65. Nigam, S., Asmus, K. D. and Willson, R. L., J. Chem. Soc. Farad. I. 72, 2324 (1976).
66. Greenstock, C. L., Raleigh, J., McDonald, E. and Whitehouse, R., B-ochem. Biophys. Res. Comm. 52, 276 (1973).
67. Raleigh, J. A., Greenstock, C. L., Whitehouse, R. and Kremers, W., Int. J. Radiat. Biol. 24, 595 (1973).
68. Varghese, A. J., Int. J. Radiat. Biol. 28, 477 (1975).
69. Whillans, D. W. and Adams, G. E., Int. J. Radiat. Biol. 28, 501 (1975).
70. Michaels, H. B., Rasburn, E. J. and Hunt, J. W., Radiat. Res. 65, 2509 (1976).

71. Hissung, A. and von Sonntag, C., Int. J. Radiat. Biol. 35, 449 (1979).
72. Asmus, K. D., Deeble, D. J., Garner, A., Idriss-Ali, K. M. and Scholes, C., Brit. J. Cancer 37 Supp. III, 46 (1978).
73. Johansen, I., Ward, J. F., Siegel, K. and Sletten, A., Biochem. Biophys. Res. Comm. 33, 949 (1968).
74. Brustad, T., Jones, W. B. G. and Wold, E., Int. J. Radiat. Biol. 24, 33 (1973).
75. Simic, M. G., Taub, I. A., Rocci, J. and Hurwitz, P. A., Biochem. Biophys. Res. Comm. 2, 161 (1975).
76. Biaglow, J. E., Jacobson, B., Varnes, M. and Koch, C., Photochem. Photobiol. 28, 869 (1978).
77. Willson, R. L. and Searle, A. J. F., Nature 255, 498 (1975).
78. Bahnemann, D., Basaga, H., Dunlop, J. R., Searle, A. J. F. and Willson, R. L., Brit. J. Cancer 37 Supp. III, 16 (1978).
79. Clarke, E. D. and Wardman, P., to be published.
80. Biaglow, J. E., Greenstock, C. L. and Durand, R. E., Brit. J. Cancer 37, Supp. III, 145 (1978).
81. Koch, C. J. and Biaglow, J. E., ibid. 37, Supp. III, 163 (1978).
82. Adams, G. E., Flockhart, I. R., Smithen, C. E., Stratford, I. G., Wardman, P. and Watts, M. E., Radiat. Res. 67, 9 (1976).
83. Anderson, R. F. and Patel, K. B., Brit. J. Cancer 29, 705 (1979).
84. Dugle, D. L., Chapman, J. E., Gillespie, C. J., Borsa, J., Webb, G., Meeker, B. E. and Reuvers, A. P., Int. J. Radiat. Biol. 22, 545 (1972).
85. Johansen, I., Radiat. Res. 58, 398 (1974).
86. Fielden, E. M., Sapora, O. and Loverock, P. S., Radiat. Res. 75, 54 (1978).
87. Held, K. D., Synek, R. W. and Powers, E. L., Int. J. Radiat. Biol. 33, 317 (1978).
88. Held, K. D. and Powers, E. L., Int. J. Radiat. Biol. 38, 293 (1980).
89. Skov, K. A., Palcic, B. and Skarsgard, L. D., Radiat. Res. 79, 591 (1979).
90. Redpath, J. L. and Willson, R. L., Int. J. Radiat. Biol., 23, 51 (1973).
91. Chapman, J. D., Reuvers, A. P. and Borsa, J., Radiat. Res. 56, 391 (1973).
92. Greenstock, C. L., Champman, J. D., Raleigh, J. A., Sheirman, E. and Reuvers, A. P., Radiat. Res. 556 (1974).
93. McNally, N. J. and De Ronde, J., Brit. J. Cancer 37, Supp. III, 90 (1978).
94. Johansen, I., Gulbrandsen, R., Fielden, E. M. and Sapora, O., Radiat. Res., 70, 597 (1977).
95. Harris, J. W., Power, J. A. and Koch, C. J., Inst. J. Radiat. Oncol. Biol. Phys., 1, 487 (1976).

96. Dewey, D. L. and Michael, B. D., Biochem. Biophys. Res. Comm., 21, 392 (1965).
97. Adams, G. E., Cooke, M. S. and Michael, B. D., Nature, 219, 1368 (168).
98. Epp, E. R., Weiss, H., Kessaris, N. D., Santomasso, A., Heslin, J. and Ling, C. C., Radiat. Res., 54, 171 (1973).
99. Shenoy, M. A., Asquith, J. C., Adams, G. E., Michael, B. D. and Watts, M. E., Radiat. Res., 62, 498 (1975).
100. Whillans, D. W. and Hunt, J. W., Brit. J. Cancer, 37, Supp. III, 38 (1978).
101. Michael, B. D., Harrop, H. A. and Maughan, R. L., "Proc. 6th Int. Cong. Rad. Res.", (1979) p. 288.
102. Chapman, J. D., Webb, R. G. and Borsa, J., Int. J. Radiat. Biol., 19, 561 (1971).
103. Han, A., Sinclair, W. K. and Kimler, B. F., Radiat. Res. 65337 (1976).
104. Pettersen, E. O., Radiat. Res. 73, 180 (1978).
105. Powers, E. L., Isr. J. Chem. 10, 1199 (1972).
106. Powers, E. L., "Fast Processes in Radiation Chemistry and Biology (G. E. Adams, E. M. Fielden and B. D. Michael, eds.) John Wiley and Sons, New York (1975) pp. 329-340.

OXYGEN-LIKE RADIOSENSITIZING DRUGS: THE IMPORTANCE OF FREE-ENERGY RELATIONSHIPS

Peter Wardman, Cancer Research Campaign Gray Laboratory, Mount Vernon Hospital, Northwood, Middlesex HA6 2RN, England

INTRODUCTION

Most groups currently involved in the search for clinically valuable hypoxic cell radiosensitizers have adopted a medicinal-chemical approach which is not primarily designed to elucidate the molecular mechanisms involved. The vast majority of the literature discussing mechanisms of radiosensitization at the chemical level is now over five years old. It therefore seems timely to outline the mechanistic implications of some recent work in the development of oxygen-like radiosensitizing drugs.

ACTIVITY OF NITROAROMATIC COMPOUNDS AS RADIOSENSITIZERS

Watts and her colleagues in the Gray Laboratory have now measured the activity of over 150 nitroaromatic compounds in radiosensitizing hypoxic mammalian cells in vitro (1-6). The extracellular concentrations required to achieve significant radiosensitization range over at least 5 orders of magnitude, depending on molecular structure. 8-Nitrocaffeine is more efficient than oxygen (3). A given compound may demonstrate some activity at concentrations 100 to 1000-fold smaller than that required to achieve maximal sensitization. Of the typical nitroaromatic sensitizers shown in Fig. 1, metronidazole and misonidazole are now being evaluated in the radiotherapy clinic.

OXYGEN AND OTHER ALTERNATIVE ELECTRON-ATTRACTING GROUPS

Demonstration of the activity of 4-nitroacetophenone (7) was the key factor leading rapidly to these clinical developments, but many other alternative electron-attracting groups had been examined previously (e.g. (8)) following the proposal of the 'electron-affinic' hypothesis (8,9). Some of the active structures are also shown in Fig. 1; chemists (with the benefit of hindsight) will immediately recognize the common feature they share with the nitroaryl compounds, of their ability to accept an electron in a one-electron reduction to yield a radical which is relatively stable. Oxygen, too, has this ability and this raises the important question as to mechanistic similarities between the radiosensitizing action of oxygen and other electron-affinic compounds. The title of this paper reveals my own prejudices: I suggest that nitro compounds (for example) replace not only a

ISBN 0-12-592050-4

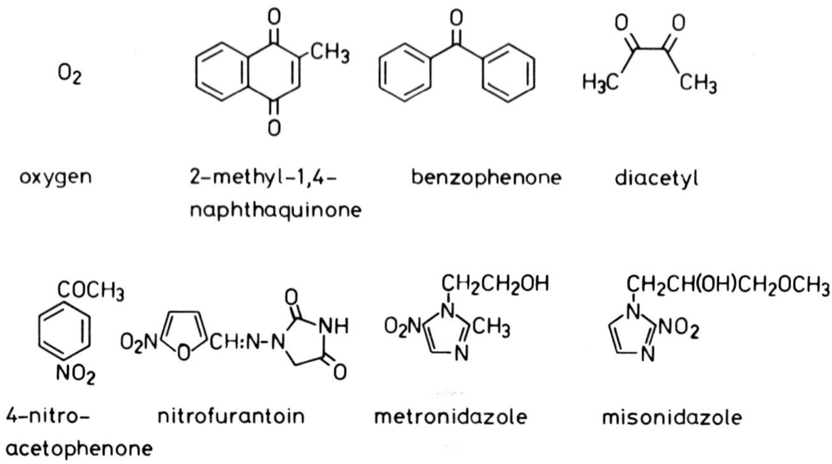

FIGURE 1. Some typical electron-affinic hypoxic cell radiosensitizers

major part of the radiobiological effect of oxygen, but that a common mechanism is involved.

An answer might be sought by studying the effects of mixtures of oxygen and e.g. misonidazole. Alper (10) has reviewed some such experiments and concluded that "...the mechanism of radiosensitization by oxygen is mimicked only in part, perhaps only to a minor extent, by the nitroaromatic compounds..." However, let us consider some of the evidence from the contrary viewpoint of the two scavengers present together, both reactive towards a species in a reaction ultimately defining cell death. With n multiple scavengers each characterized by a rate constant k_i and an extracellular concentration C_i, the combined biological response r will be set by the summation of the products: $\sum_{i=1}^{n} (k_i g_i C_i)$ where g_i is a factor, different for each compound, which quantifies the concentration gradient within the cell so that $g_i C_i$ is the concentration at the site of action. The lipophilicity and charge on the molecule will be the most important parameters controlling g_i. To predict the effect of combinations, we can convert one individual response (e.g. the lower) to an equivalent concentration C_i of the other solute, to add to the concentration of this co-solute and then read off the combined response from the appropriate curve. Such a procedure compensates for differences in both k_i and g_i and is precisely that used by Adams (11) in discussing the clinical implications.

From Fig. 2 we can see that C_{miso} = 100 µmol dm^{-3} ($r_{miso} \simeq$ 1.20) is 'equivalent' to $C_{O_2} \simeq$ 1.2 µmol dm^{-3}. A combination of 100 µmol dm^{-3} misonidazole with 13 µmol dm^{-3} oxygen would be expected to yield a combined response r_{comb} appropriate to that for ∿ 14.2 µmol dm^{-3} oxygen alone, i.e. an increase over oxygen alone

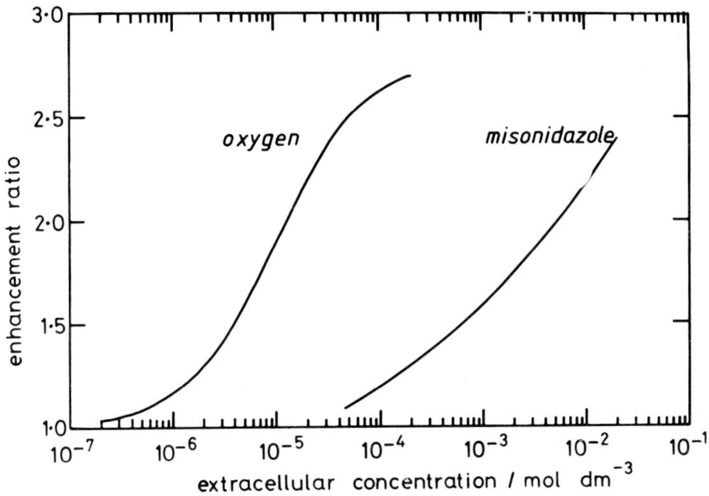

FIGURE 2. Typical response vs. concentration curves for radiosensitizers of hypoxic mammalian cells in vitro.

of only 0.04. Since such small differences are well within the experimental uncertainty, it is not surprising a conclusion of no 'addivity' was drawn. Even this oversimplified treatment (see below) using elementary chemical concepts quantitatively explains most of the 'additivity' experiments discussed by Alper, and the results for the combination of two chemically virtually identical nitro compounds discussed by McNally and de Ronde (12). One must not use the term 'no additivity,' with its mechanistic implications, when the additivity was simply too small to readily detect.

FREE-ENERGY RELATIONSHIPS BETWEEN ACTIVITY AND STRUCTURE

Possibly the first semiquantitative correlation between sensitizing efficiency and chemical properties was made by Tallantire et al. (13), using bacterial spores and four ketones whose relative one-electron affinities had been established from pulse radiolysis observations. Hammett sigma substituent constants are known to reflect differences in reduction potential of substituted nitrobenzenes, and Raleigh et al. (14) showed the radiosensitization response at a fixed concentration of these compounds was closely dependent upon the Hammett free-energy function. Simic and Powers (15) showed how general the relationship was between reduction potential and sensitizing efficiency, using data for bacterial spores with sensitizers of widely varying chemical type covering an extremely wide range of potentials; a result which has major mechanistic implications.

To subject such correlations to quantitative mechanistic evaluation it is preferable to follow the conventional medicinal-chemical approach (16) and measure the extracellular concentration C_r required to achieve a fixed response r, to set up the free-energy relationship ideally of the form:

$$\log(1/g_i C_r) = b_0 + b_1 . E \qquad (1)$$

where E is the reduction potential. Since g_i is not known, its effect can be included by the addition of lipophilicity terms:

$$\log(1/C_r) = b_0 + b_1 . C + b_2 . \log P + b_3 . (\log P)^2 \qquad (2)$$

where P is e.g. the octanol:water partition coefficient. Adams *et al.* (1,2) measured E as the one-electron reduction potential, from the positions of free-radical one-electron transfer equilibria (17). After showing that the effects of P on response were much smaller than those of E for radiosensitization of mammalian cells *in vitro* by 38 nitroaromatic compounds, Adams *et al.* fitted their data to the simplified form of eqn. (1), setting $g_i = 1$. Extension of this approach (4) to include data for 54 nitroaromatic compounds gave a coefficient $b_1 = 8.93 \pm 0.48\ V^{-1}$ relating radiosensitizing efficiency to reduction potential.

EQUILIBRIUM- OR RATE-LIMITING: THERMODYNAMIC OR KINETIC CONTROL

Irreversible electron-transfer from an electron-adduct or other nucleophilic free radical to a sensitizer, thus preventing repair by a recombination mechanism, was invoked in the original Adams mechanism for electron-affinic sensitization (8). Incorporation of nitroaromatic sensitizers in DNA increases the radiolytic yield of guanine radical-cations (18), supporting this proposal. We now know that many likely electron-donor sites react with most sensitizers with rate constants approaching diffusion-controlled, whereas sensitization efficiency covers 5 orders of magnitude in concentration terms. The redox relationship led Willson *et al.* (19) to extend this 'prevention of recombination' hypothesis to involve an electron-transfer *equilibrium* between a recombining nucleophilic radical and the sensitizer. Some features of the Adams/Willson model are illustrated in Fig. 3, in a schematic form as a basis for discussion: thus the biomolecule B_2 involved in recombination is not necessarily the same as that involved in reaction with the sensitizer, nor is the recombining cationic centre $B_1^{+\cdot}$ necessarily the species reacting with RSH. However, in both cases a coupling of the reagents is indicated.

From this scheme, assuming a one-electron transfer equilibrium was controlling an overall response, I originally (perhaps naively) expected the slope b_1 of eqn. (1) would be close to $nF/2.3RT = 16.9\ V^{-1}$. For example, the overall rate of 4-electron reduction of nitro compounds by reduced flavins is probably controlled by a one-electron transfer equilibrium, and log(rate) vs. E *does* have a slope of this order (20). However, comparison with other biological free-energy relationships is informative. We recently drew attention (4) to the fact that not only do different biological manifestations of nitro compound toxicity—cell survival, cellular or microsomal reduction rate, inhibition of DNA synthesis, production of DNA strand breaks, and mutation frequency—all have a similar redox dependence (see e.g. (21)), this dependence was quantitatively similar to the coefficient $\underline{b}_1$ for radiosensitization. Since these non-radiation effects of nitro compounds all presumably involve interaction with the cellular electron-transport system in a one-electron transfer equilibrium, these similarities might not be coincidental.

Competitive 'fixation' vs. repair of bioradical B· is also a feature of the earlier model of the oxygen effect in radiobiology developed by Alexander, Charlesby, Alper and Howard-Flanders, which for our purposes can be represented by eqns. (3)-(5):

$$B\cdot \longrightarrow \text{lethal damage, rate constant } \underline{k}_3 \quad s^{-1} \qquad (3)$$

$$B\cdot \longrightarrow \text{repair,} \quad " \quad " \quad \underline{k}_4 \quad " \qquad (4)$$

$$B\cdot + O_2 \longrightarrow \text{lethal damage,} \quad " \quad " \quad \underline{k}_5\,[O_2] \quad " \qquad (5)$$

(Since all three reactions are first order, we do not need to stipulate the nature of the fixation or repair reactions (3) or (4).) It can be readily shown (10) that the concentration $\underline{C}_r$ of O_2 needed to achieve an enhancement ratio $\underline{r}$ is given by:

$$\underline{C}_r = [(\underline{k}_3 + \underline{k}_4)/\underline{k}_5][(\underline{r} - 1)/(\underline{r}_\infty - \underline{r})] \qquad (6)$$

where $\underline{r}_\infty$ is the maximum that can be achieved. Hence if the mechanism were generally applicable to other sensitizers:

$$\log(1/\underline{C}_r) = \text{constant} + \log\underline{k}_5, \qquad (7)$$

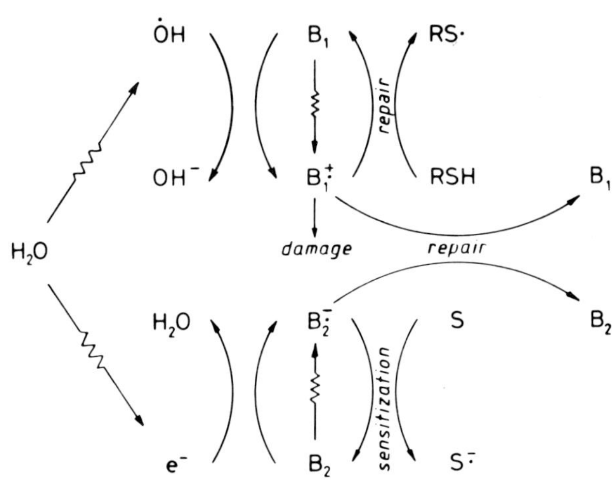

FIGURE 3. Schematic representation of the Adams/Willson model for electron-affinic radiosensitization.

and using eqn. (1) with $g_i = 1$:

$$\log k_5 = \text{constant} + b_1 . E \qquad (8)$$

I have illustrated (22) how the Marcus theory of electron-transfer reactions can be used to predict a dependence of rate constant, log k_5 on E which is approximately linear over the range of interest and which has the appropriate value of slope $b_1 \simeq 10\ V^{-1}$.

MECHANISTIC CONSTRAINTS

Both these approaches—involving either thermodynamic or kinetic control of radiosensitizing efficiency—appear plausible, but must be subjected to two exacting tests. First, any model has to be capable of quantitatively predicting a dependence of efficiency on redox properties covering 5 orders of magnitude in concentration terms. Secondly, any model has to be applicable to different types of electron-affinic molecule, in a quantitative manner defined at least principally by E. The important study of Simic and Powers (15) supporting the latter conclusion has already been noted, and we reported (1,2) that oxygen and a quinone radiosensitized mammalian cells with an efficiency predictable from their one-electron reduction potentials and the relationship (1) established for nitroaromatic compounds.

Both the above approaches are now seen to have drawbacks. It is difficult to accept that an equilibrium condition will operate over 5 orders of magnitude in concentration terms, particularly when there are other, competing fast reactions which may prevent equilibrium being established. On the other hand, linear free-energy relationships for biochemical reactions cover almost as wide a range. Variation in rate constant by over 5 orders of magnitude, in a quantitatively similar manner for several different types of one-electron acceptor is less difficult to accept, but still requires experimental demonstration. Of course, both approaches represent gross oversimplifactions since we already know conclusively (see e.g. (23)) that there are at least two mechanisms of radiosensitization either by oxygen or other electron-affinic molecules (or two sites of action so that concentration gradients define the response; the 10-fold higher lipophilicity of oxygen over misonidazole should be noted). The above constraints are equally applicable to other possible models, such as the Adams irreversible electron-trapping scheme with a kinetically controlled electron-transfer rate. Both oxygen and nitro compounds add to some carbon-centered sigma-radicals, and competition between electron-transfer and addition will be redox-related. Some nitroxyls are highly electron-affinic but undergo irreversible electron-capture (24), of obvious relevance to the above discussion.

ACKNOWLEDGEMENTS

This work is supported by the Cancer Research Campaign.

REFERENCES

1. Adams, G. E., Flockhart, I. R., Smithen, C. E., Stratford, I. J., Wardman, P. and Watts, M. E., Radiat. Res., 67,9 (1976).
2. Adams, G. E., Clarke, E. D., Flockhart, I. R., Jacobs, R. S., Sehmi, D. S., Stratford, I. J., Wardman, P., Watts, M. E., Parrick, J., Wallace, R. G. and Smithen, C. E., Int. J. Radiat. Biol., 35,133 (1979).
3. Wardman, P., Clarke, E. D., Jacobs, R. S., Minchinton, A., Stratford, M. R. L., Watts, M. E., Woodcock, M., Moazzam, M., Parrick, J., Wallace, R. G. and Smithen, C. E., Cancer Clinical Trials, in press (1980).
4. Watts, M. E., Anderson, R. F., Jacobs, R. S., Patel, K. B., Wardman, P., Woodcock, M., Smithen, C. E., Moazzam, M., Parrick, J. and Wallace, R. G., Cancer Clinical Trials, in press (1980).
5. Smithen, C. E., Clarke, E. D., Dale, J. A., Jacobs, R. S., Wardman, P., Watts, M. E. and Woodcock, M., Cancer Clinical Trials, in press (1980).
6. Adams, G. E., Ahmed, I., Clarke, E. D., O'Neill, P., Parrick, J., Stratford, I. J., Wallace, R. G., Wardman, P., and Watts, M. E., Int. J. Radiat. Biol., 38,613 (1980).
7. Adams, G. E., Asquith, J. C. and Willson, R. L., British Empire Cancer Campaign for Research, 47th Annual Report, 1969, pp. 167,168.
8. Adams, G. E. and Cooke, M. S., Int. J. Radiat. Biol., 15,457 (1969).
9. Adams, G. E. and Dewey, D. L., Biochem. Biophys. Res. Commun., 12,473 (1963).
10. Alper, T., "Cellular Radiobiology," Cambridge University Press, Cambridge,(1979).
11. Adams, G. E., "Cancer, A Comprehensive Treatise, Vol. 6" (F. F. Becker, ed.), Plenum Press, New York,(1977), p. 181.
12. McNally, N. J. and de Ronde, J., British J. Radiol., 52,589 (1979).
13. Tallentire, A., Jones, A. B. and Jacobs, G. P., Israel J. Chem., 10,1185 (1972).
14. Raleigh, J. A., Chapman, J. D., Borsa, J., Kremers, W. and Reuvers, A. P., Int. J. Radiat. Biol., 23,377 (1973).
15. Simic, M. and Powers, E. L., Int. J. Radiat. Biol., 26,87 (1974).
16. Hansch, C., "Drug Design, Vol. 1" (E. J. Ariëns, ed.), Academic Press, New York,(1971), p. 271.
17. Meisel, D. and Neta, P., J. Am. Chem. Soc., 97,5198 (1975).

18. Gräslund, A., Ehrenberg, A. and Rupprecht, A., Int. J. Radiat. Biol., 31,145 (1977).
19. Willson, R. L., Cramp, W. A. and Ings, R. M. J., Int. J. Radiat. Biol., 26,557 (1974).
20. Clarke, E. D., Goulding, K. H. and Wardman, P., Biochem. Pharmac., 29,2684 (1980).
21. Olive, P. L., Cancer Clinical Trials, in press (1980).
22. Wardman, P., Curr. Topics in Radiat. Res. Q, 11,347 (1977).
23. Millar, B. C., Fielden, E. M. and Steele, J. J., Int. J. Radiat. Biol., 36,177 (1979).
24. O'Neill, P. and Jenkins, T. C., J.C.S. Faraday I, 75,1912 (1979).

NOTE ADDED IN PROOF:

References 3, 4, 5 and 21 are now published in: Brady, L. W. (ed.), "Radiation Sensitizers: Their Use in the Clinical Management of Cancer", Masson, New York (1980).

EVIDENCE FOR THE DUAL ACTION OF THE NITROXYL AND NITROAROMATIC MOIETIES IN A MIXED FUNCTIONAL RADIOSENSITIZER

Barbara C. Millar,[1] *T. C. Jenkins,*[1] *E. Martin Fielden,*[1] *and C. E. Smithen*[2]

[1]*Radiobiology Unit, Physics Department, I.C.R., Sutton, Surrey*
[2]*Roche Products Limited, Welwyn Garden City, Herts, U. K.*

INTRODUCTION

Both "electron affinic" and nitroxyl free radical radiosensitizers reduce the survival of irradiated hypoxic mammalian cells in an oxygen-like manner (1,2). We have previously shown that electron affinic agents such as misonidazole and the free radical compound NPPN enhance the sensitization of Chinese hamster cells, line V79-753B, irradiated <u>in vitro</u>, in the presence of submicromolecular concentrations of oxygen (3). However, TMPN, a free radical compound which does not sensitize appreciably in this cell line, decreases the amount of sensitization produced by oxygen or misonidazole in certain conditions (3). Thus, TMPN either acts as a radioprotector in these systems or competes with oxygen or misonidazole for radion-induced damage to produce products that are non-lethal or capable of undergoing repair. The present paper is concerned with an investigation of a mixed functional compound Ro. 03-9454 (1-(1-oxyl-2,2,6,6-tetramethylpiperidin-4-yl)amine-3-(2-nitroimidazol-1-yl)propan-2-ol, which has both a nitroaromatic (electron affinic type) and a nitroxyl (free radical type) moiety to determine the contributions of the separate functions in determining the efficacy of this compound as a radiosensitizer. A modified form of this compound has also been examined in which the free radical function has been eliminated by the addition of a cyanopropanol group.

COMPOUNDS

Misonidazole and Ro. 03-9454 were kindly supplied by Roche Products Ltd. TMPN and the monofunctional derivative of Ro. 03-9454 viz. ((1-(1-2-cyanoprop-2-yl)oxy-2,2,6,6-tetramethylpiperidin-4-yl)amine-3-(2-nitroimidazol-1-yl)propan-2-ol), RSU 4013 were synthesized in our laboratories. There was no appreciable toxicity from any of the compounds at the concentrations used, when oxic or hypoxic cells were exposed to the drugs for 2 hr at room temperature.

ISBN 0-12-592050-4

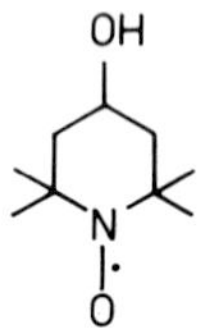

TMPN

NPPN

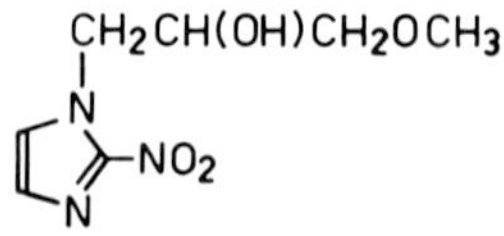

Misonidazole

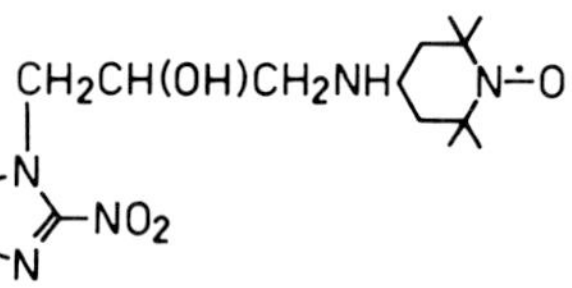

Ro 03 9454

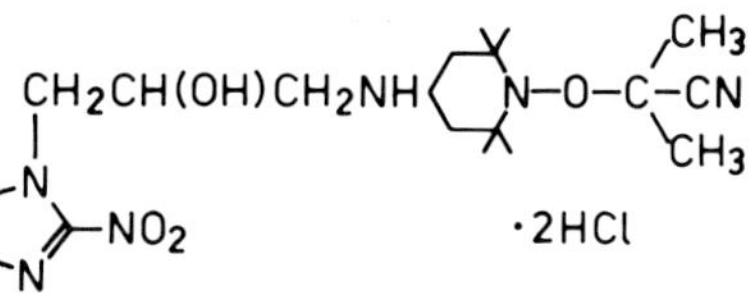

RSU 4013

CELL CULTURE

Asynchronous cultures of Chinese hamster cells, line V79-753B, were used throughout the work. Routine handling of cells was carried out by methods described previously (4). For experimental purposes cultures were trypsinized and plated on to 61 mm glass petri dishes in 2.0 ml of growth medium (Earle's based MEM supplemented with 15% foetal calf serum and 20 mM HEPES buffer). After attachment, the growth medium was aspirated from the cells and replaced with 2.0 ml of Dulbecco's phosphate buffered saline (PBSA) with or without drug at the required concentration. Triplicate plates of each sample were rendered hypoxic by gassing the samples in sealed "Dural" containers with O_2-free N_2 (BOC white spot) for 1 hr prior to irradiation. Irradiation was carried out using a Cobalt-60 source and a dose rate of approximately 400 rad min^{-1}. All irradiations were done at room temperature. After irradiation the PBSA was removed from the cells and replaced with 2 ml of fresh growth medium. Cultures were incubated in an atmosphere of 5% CO_2-95% air for 6 days to allow colony formation. At this time the medium was removed, the colonies fixed with ethanol, stained with methylene blue and counted.

All experiments were repeated at least twice and data was accumulated from complete survival curves. In experiments where mixtures of two compounds were carried out, each experiment consisted of a control survival curve, survival curves for cells irradiated in the presence of each separate compound and a survival curve for cells irradiated in the presence of the mixture. Pooled data for each experiment were analyzed using the multitarget equation as described previously (5) to give errors on the survival curve parameters. The multitarget equation was used to remove subjective bias from the experimental data and not as an indication of the underlying validity of this model (5). The survival parameters obtained from the computed data were used in a second programme (J. L. Miller, unpublished) to obtain theoretical curves from the data. Some of those curves are shown in the text with the observed experimental survivals superimposed.

RESULTS AND DISCUSSION

Fig. 1 shows survival curves determined in the presence of various drugs and drug combinations. In all cases the effect of the drugs was essentially dose modifying and the ratio of the slope of the exponential portion of the survival curves relative to that

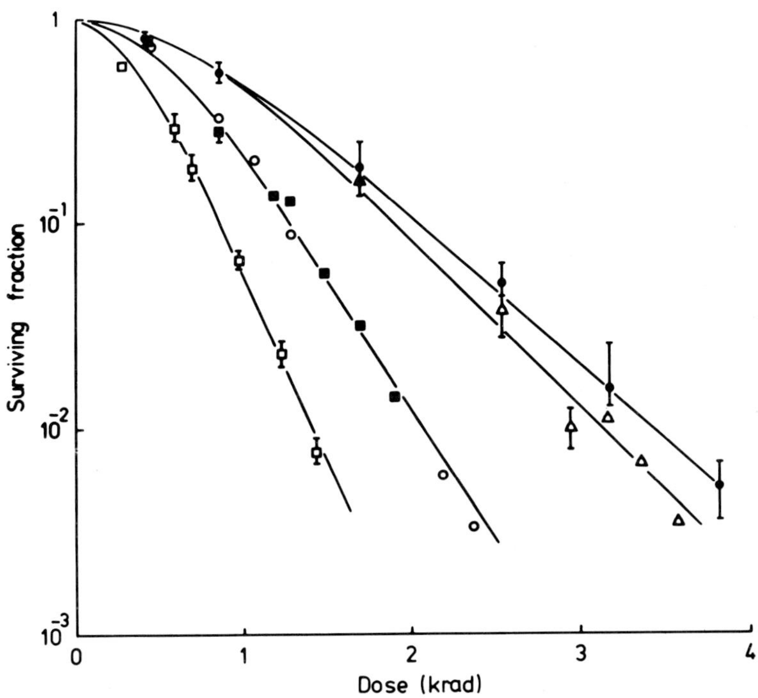

FIGURE 1. Comparison of the effect of 0.5 mM RSU 4013 in combination with 0.5 mM TMPN with 0.5 mM RO.03-9454 on the survival of hypoxic Chinese hamster cells, V79-753B ● *N_2 control, D_o 630 ± 20;* ▲ *1.0 mM TMPN, D_o 540 ± 20;* □ *0.5 mM RSU 4013, D_o 240 ± 8;* ○ *0.5 mM Ro.03-9454, D_o 336 ± 11;* ■ *0.5 mM RSU 4013 + 0.5 mM TMPN, D_o 336 ± 9.*

of the anoxic controls (SR) was calculated.

Both the mixed functional compound (9454) and the modified monofunctional derivative (4013) were more effective sensitizers than misonidazole on a concentration basis. The SR versus concentration curve for 9454 and 4013 shows that the amount of sensitization remains almost constant at SR = 1.9 for an appreciable range of drug concentrations within the extremes of 0.05 and 0.5 mM. Similar behavior (Fig. 2) is shown by misonidazole (3) and oxygen (6). In the case of 4013 the "plateau" is maintained over a 4-fold concentration range but in the case of 9454 the range is at least 10-fold in concentration (c.f. a 3-fold range for misonidazole (3) and 4.5-fold for O_2 (6)). At the highest concentration of 4013 tested (0.5 mM) there was greater sensitization than for cells irradiated in the presence of a similar concentration of the mixed functional compound. It was not possible to test higher concentrations of RSU 4013 because of toxicity at concentrations greater than 0.5 mM. The effect of an equimolar concentration of TMPN (0.5 mM) in combination with RSU 4013 can be compared to the effect of 0.5 mM Ro. 03-9454 alone, in Figs. 1 and 2. TMPN reduces sensitization by the monofunctional compound and the amount of sensitization for the combination was similar to that seen in hypoxic cells irradiated in the presence of 0.5 mM Ro. 03-9454 alone. However, when the mixed functional compound was used in combination with an equimolar concentration of TMPN (Fig. 2) there was no appreciable change in the amount of sensitization. We have previously shown that TMPN can protect against sensitization by oxygen and misonidazole in certain conditions (3), which

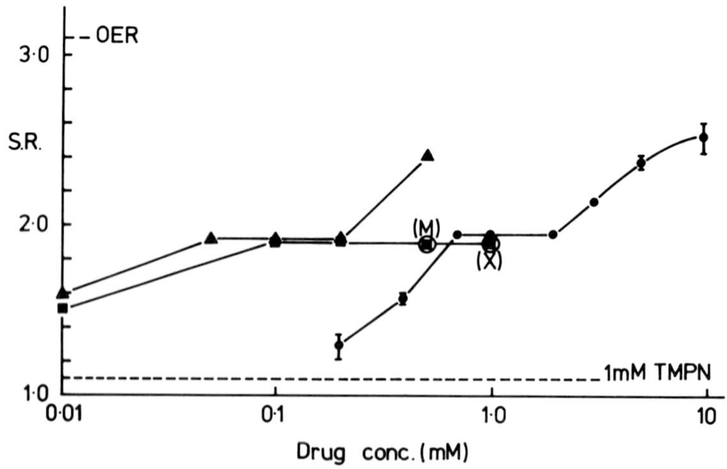

FIGURE 2. Comparison of the effects of Ro.03-9454, RSU 4013 and misonidazole on the survival of hypoxic Chinese hamster cells, V79-753B. ● *Misonidazole;* ■ *Ro.03-9454;* ▲ *RSU 4013;* ○ *(M) 0.5 mM RSU 4013 + 0.5 mM TMPN;* ○ *(X) 1.0 mM Ro.03-9454 + 1.0 mM TMPN, by permission of Academic Press.*

suggests that electron affinic and free radical moieties can compete for the same radiation induced lesions. The present data support these observations and show that competition reactions can occur between the two types of moiety, when contained in the same molecule. This competition limits the efficacy of the mixed functional compound as a radiosensitizer because removal of the nitroxyl group results in enhanced sensitization. Because of the wide plateau on the SR vs concentration curve for 9454 it would presumably need considerably more than 1 mM TMPN to reduce the SR for the combination below 1.9.

One reason for investigating the mixed bifunctional compound 9454 was to test whether this compound could be effective in removing or reducing the shoulder of survival curves as had been shown for the nitroxyl biradicals (4,7). In the present case no significant changes in extrapolation number were seen so that shoulder modification cannot be a common feature of bifunctional agents. Indeed, the two active moieties of 9454 evidently act independently and no co-operative phenomenon is observed.

CONCLUSIONS

1) The data present further evidence that "electron affinic" and nitroxyl free radical compounds may operate by different mechanisms and that these mechanisms involve competition for the initial radiation-induced damage. The efficacy of such competition would thus determine the resultant sensitization.

2) The mixed functional compound Ro. 03-9454 and the modified form of this compound, RSU 4013, in which the nitroxyl function has been eliminated by the addition of a cyanopropyl group are more effective sensitizers than misonidazole at low concentrations.

3) Like misonidazole, the two sensitizers contain a nitroimidazole group, 9454 and 4013 both show a plateau at a SR of 1.9 over a range of concentrations.

ACKNOWLEDGEMENTS

We should like to thank the CRC and MRC for financial support for this programme.

REFERENCES

1. Adams, G. E., Asquith, J. C., Darcy, D. L. Foster, J. L., Michael, B. D. and Willson, R. L., Int. J. Radiat. Biol. 19,575 (1971).
2. Parker, L., Skaarsgaard, L. D. and Roe, P. T., Radiat. Res. 38,493 (1969).
3. Millar, B. C., Fielden, E. M. and Steele, J. J., Radiat. Res. In press. (1980).
4. Cooke, B. C., Fielden, E. M., Johnson, J. and Smithen, C. E., Radiat. Res. 65,162 (1976).

5. Millar, B. C., Fielden, E. M. and Millar, J. L., Int. J. Radiat. Biol. 33,599 (1978).
6. Millar, B. C., Fielden, E. M. and Steele, J. J., Int. J. Radiat. Biol. 36,177 (1979).
7. Millar, B. C., Fielden, E. M., Huntley, S. E. and Smithen, C. E., Radiobiol. Res. and Radiotherapy 1,239 (1977).

EXPLORING THE TIME SCALE OF RADIOSENSITIZATION

Howard B. Michaels, Edward R. Epp and C. Clifton Ling, Division of Radiation Biophysics, Department of Radiation Medicine, Massachusetts General Hospital, Harvard Medical School, Boston, Massachusetts 02114*

INTRODUCTION

In recent years, considerable information on the time scale of the processes involved in radiation sensitization has been obtained through various novel fast time response techniques in radiation chemistry and radiobiology. The radiation chemical reactions which give rise to the initial damage of biological molecules in the living cell occur within $\sim 10^{-6}$ s after radiation energy deposition. In general, it is the reactions of these initial damage sites with sensitizer or protector species, occurring subsequently in the time range from 10^{-6} - 10^{-2} s, which are responsible for the fixation of such damage into harmful or lethal products, or, alternatively, for their repair (1-3). That oxygen sensitization occurs at times shorter than $\sim 10^{-2}$ s was demonstrated over 20 years ago by Howard-Flanders and Moore (4), in an elegant experiment using a rapid-transfer technique. In recent years, the time scale of radiosensitization in cells has been demonstrated experimentally to be much shorter than 10^{-2} s, using such methods as double pulse (5-9), gas explosion (10,11), rapid liquid mixing (12,13) and rapid lysis (14), all of which make use of pulsed radiation sources. These techniques have provided information on the diffusion of oxygen and other sensitizers, and also some protectors; on the lifetime of radiation-induced oxygen-dependent damage in cells; on direct and indirect effect processes; on the action of hydroxyl radical scavengers; and on the production of DNA strand breaks. The time relationship of these, and other experimental approaches to the processes involved in cellular radiation damage is shown in Fig. 1. The accumulated evidence implies that the reactions responsible for sensitization and protection take place on the microsecond to millisecond time scale, and that free radical reactions are probably involved (3). No attempt will be made in this brief presentation to review all the different experimental methods and results; the following is a summary of recent work on certain aspects of the problem as studied using the thin layer and pulse irradiation techniques developed in this laboratory.

**Present address: Division of Radiation Oncology and Biophysics, George Washington University Medical Center, Washington, D.C. 20037*

ISBN 0-12-592050-4

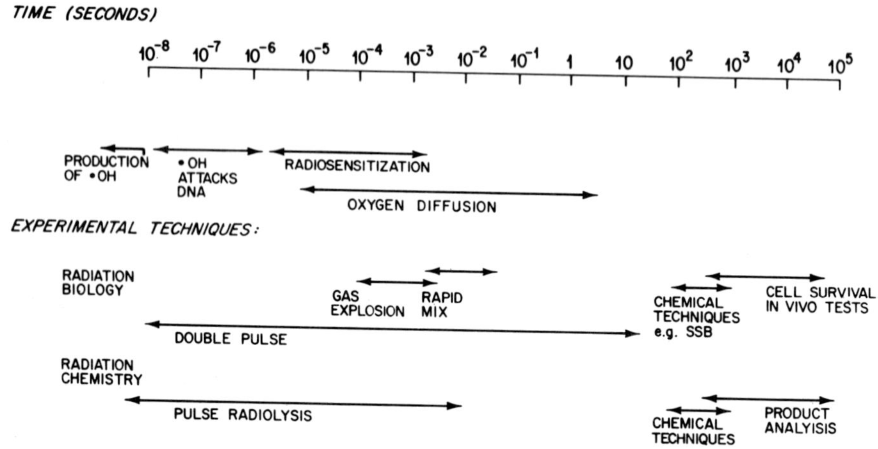

FIGURE 1. A simplified time scale for radiation damage in respiring cells at room temperature, showing the approximate time ranges for various experimental techniques.

BREAKING SURVIVAL CURVES

It has been established that when cultured mammalian cells are equilibrated with a low level of oxygen and irradiated with short pulses of electrons, a breaking survival curve is observed, due to the cells exhibiting an initial sensitized response, and then, at higher doses, an anoxic one (6,7). This phenomenon, first studied in bacteria (15,16), is attributed to radiochemical depletion of the intracellular oxygen. Cells responding to doses greater than the depletion dose display an anoxic sensitivity, since oxygen is unable to diffuse into the cells fast enough to affect the response, resulting in the "breaking" nature of the survival curve. An example of this effect in mammalian cells is shown in Fig. 2. Chinese hamster ovary (CHO) cells, plated onto a thin coverslip and equilibrated with 0.44% O_2 + 99.56% N_2, were irradiated with 3 ns electron pulses from a field emission source (instantaneous dose rate $\sim 10^{12}$ rad/s); cell survival was assayed by colony formation. Complete details of this thin layer cell handling technique and irradiation procedure have been described previously (7). While the survival curves for air and for nitrogen equilibration are similar to those observed at conventional dose rates, the curve for 0.44% O_2 clearly demonstrates breaking behavior. The "breakpoint dose", i.e., the dose required to deplete the oxygen in these cells, was $\sim$1200 rads, implying a g-value of 1% O_2/2.7 krad for the radiolytic oxygen depletion process (7), a value about the same as that obtained for bacteria (15,16).

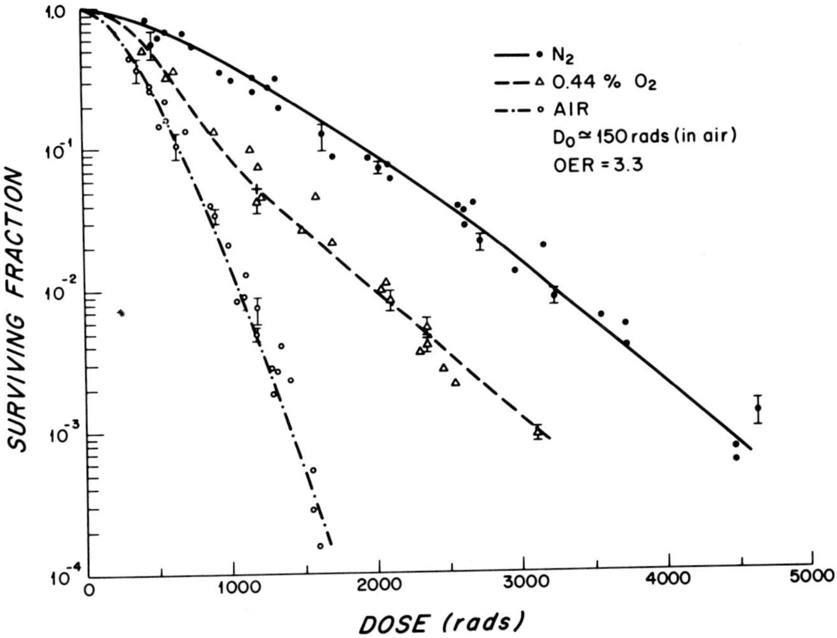

FIGURE 2. The breaking survival curve observed for CHO cells in 0.44% O_2, obtained using the thin layer procedure with pulsed electron irradiation. Survival curves for air and for nitrogen are also shown (7), by permission Academic Press.

EFFECT OF NITROIMIDAZOLE SENSITIZERS

It has been postulated that the electron-affinic radiosensitizers, including the 2-nitroimidazoles misonidazole (Ro-07-0582) and Ro-05-9963, and the 5-nitroimidazole metronidazole (Flagyl), act as oxygen mimics in their mechanism of action as sensitizing agents. These drugs are dose-modifying sensitizers of hypoxic mammalian cells both at conventional and at ultrahigh dose rates (9,17). The response of cells containing these compounds and a low level of oxygen, and irradiated with single 3 ns electron pulses, was studied in detail. As illustrated in Fig. 3, the breaking survival curve normally observed for cells equilibrated with 0.44% O_2 is altered by the presence of a relatively small concentration of either misonidazole or metronidazole (9). The curves measured in the presence of the nitroimidazole sensitizers clearly do not exhibit breaking behavior; similar results are obtained with Ro-05-9963. In addition, the final slope of each non-breaking survival curve is substantially steeper than that expected if the nitroimidazole, at the concentration used, were simply replacing oxygen as the radiosensitizer in the cells after the 0.44% O_2 had been depleted radiochemically. The implication is that the oxygen has not been depleted, and is continuing to sensitize the cells (9,17). These data, in conjunction with the

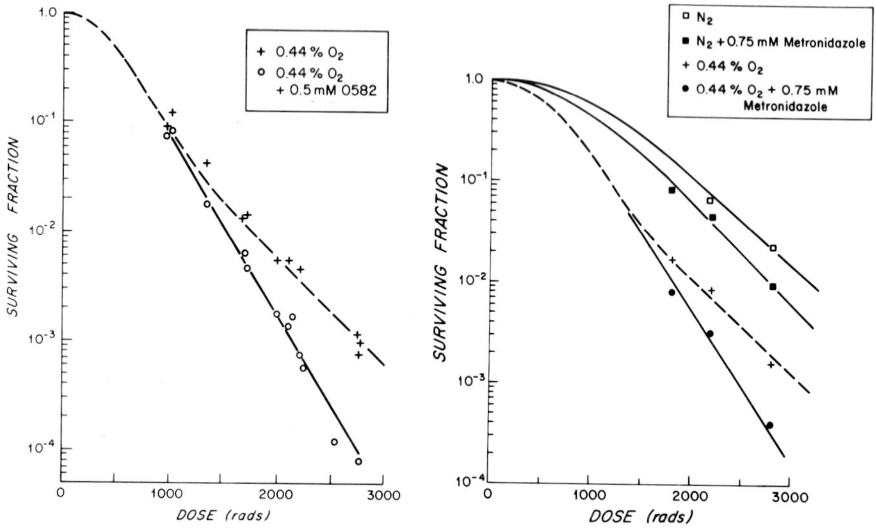

FIGURE 3. The effect of nitroimidazole sensitizers on the breaking survival curve for cells in 0.44% O_2: (a) misonidazole; (b) metronidazole, Also included in (b) are survival curves for cells equilibrated in N_2 (9), by permission Academic Press.

results of other experimental work discussed in detail (9), support the hypothesis that interference by nitroimidazoles in the radiochemical oxygen depletion reactions at ultrahigh dose rates is responsible for the prevention of the breaking survival curve.

The observed results can be explained by the competition of the nitroimidazole compounds with oxygen for the radiation-induced O_2 binding sites, which, though non-lethal to the cell, serve to remove oxygen from participation in radiosensitization reactions. It can be postulated that the very same properties which make the nitroimidazoles effective oxygen mimics in radiosensitization also enable the drugs to be oxygen mimics in terms of radiolytic depletion reactions. Then, the nitroimidazole could act as a substitute for oxygen in these reactions, and if its concentration were sufficiently high, the sensitizer could essentially block oxygen consumption by this pathway, leaving oxygen free to sensitize. The reactions involved in oxygen depletion, and the interference of nitroimidazoles in these reactions, probably involve radiation-induced free radical species and occur at microsecond times; the mechanism for oxygen depletion has yet to be elucidated (6,7).

DOUBLE PULSE EXPERIMENTS

Both the gas explosion technique of Michael and co-workers (10,11) and the double pulse method developed in this laboratory

(6,8) have been employed to investigate the kinetics of oxygen diffusion in bacterial and in mammalian cells and to obtain information about the lifetime of radiation-induced oxygen-dependent damage. While there are some differences in the results obtained by the two groups for bacterial cells, it is clear that, for mammalian cells, oxygen diffusion takes place at millisecond times, and that the half-life for the decay of oxygen-dependent damage in irradiated mammalian cells is not longer than a few milliseconds (8,18). The double pulse method employs the thin layer cell handling procedure, and irradiation is performed with two sequential electron pulses separated by an accurately-controlled, variable, interpulse time Δt; complete details of the rationale and procedures have been published (8). The results of two types of double pulse experiment with CHO cells are shown in Fig. 4. In the left panel, cells were equilibrated with 0.21% O_2 + 99.79% N_2. The first pulse D_1 depleted the oxygen; the second pulse D_2 was of variable dose. If the interpulse time Δt was zero ($\leq 10^{-8}$ s), a breaking survival curve was observed (open triangles), as though a single pulse experiment similar to that shown in Fig. 2, had been performed. The breakpoint dose and the spacing of the parallel anoxic portions of the breaking survival curve for 0.21% O_2 are less than those measured with 0.44% O_2 because these both depend on the initial oxygen concentration (7). When Δt = 60 s, allowing for complete re-equilibration of the cells with the 0.21% O_2 by diffusion, a re-breaking survival curve is generated (closed triangles), as described previously (8).

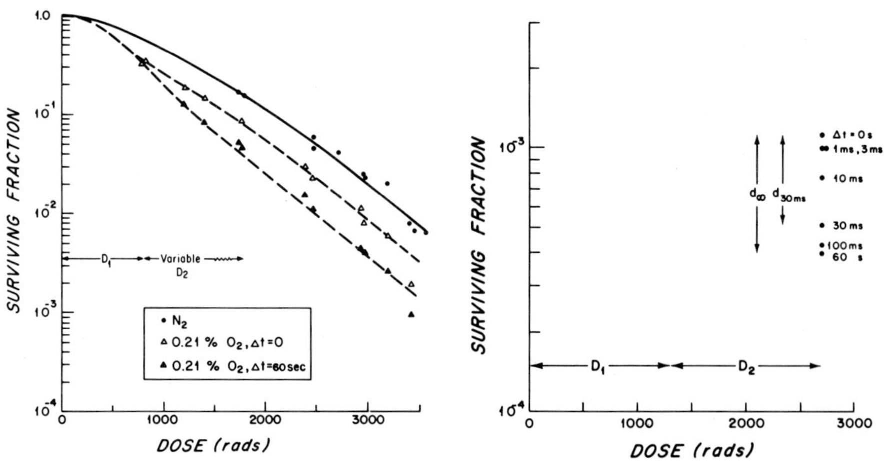

FIGURE 4. Two types of double pulse experiment (8). (a) A rebreaking survival curve is generated when the interpulse time is essentially infinite (Δt=60 sec) and variable second pulse doses are used. (b) With a fixed second dose, survival decreases with increasing interpulse time, because of diffusion of oxygen into the cells. By permission Academic Press.

In the right panel of Fig. 4, a double pulse experiment was performed with D_1 and D_2 fixed, but with Δt varied from 0 to 60 s; here the cells were initially equilibrated with 0.44% O_2. In this type of experiment, the kinetics of oxygen diffusion can be examined. As discussed in detail (8), as the interpulse time is increased in the millisecond time range, survival decreases due to oxygen diffusing into the cells from the surrounding atomsphere of 0.44% O_2, and sensitizing the cells to the second pulse D_2. From the measured distances of change in surviving fraction with Δt, labeled in the figure, a time-diffusing curve for oxygen can be derived (8). This is shown in Fig. 5. In this figure, it is seen that the experimental data fit the calculated theoretical diffusion curve in one dimension, using a distance of 4μ (approximately the cell radius) and a diffusion coefficient for oxygen of 2×10^{-5} cm^2 s^{-1}, the same value as that in pure water. From this time-diffusion curve, it can be concluded that a radiobiologically significant amount of oxygen diffuses into the cells within 3×10^{-3} s after the first pulse. This time, then, represents an upper limit to the lifetime of the radiation-induced oxygen-dependent damage species, since otherwise breaking survival curves would not have been observed for single pulse irradiation.

The double pulse method can also be applied to test the hypothesis proposed to explain the apparent interference of nitroimidazole sensitizers with radiolytic oxygen depletion; namely, that the sensitizer protects against oxygen depletion (9). In Fig. 6,

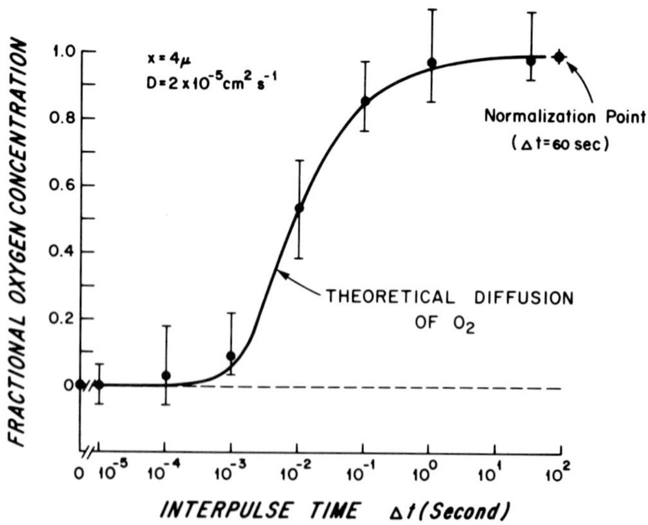

FIGURE 5. The diffusion curve for oxygen in CHO cells, measured by the double pulse method (8). The experimental data can be fit by the one-dimensional diffusion equation. By permission Academic Press.

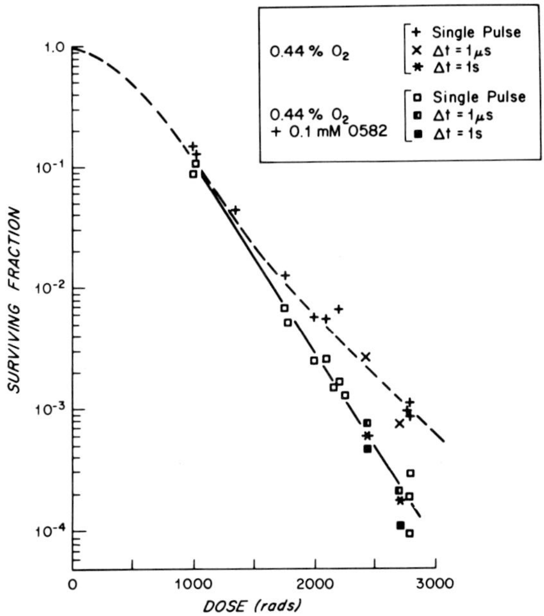

FIGURE 6. The effect of 0.1 mM misonidazole on the breaking survival curve for cells in 0.44% O_2, observed in single and double pulse experiments (8), by permission Academic Press.

single and double pulse experiments with 0.1 mM misonidazole are shown. The single pulse data indicate that the presence of misonidazole prevents breaking behavior, similar to Fig. 3. When cells are irradiated by two pulses, there is only a slight difference in survival for $\Delta t = 1$ s (essentially equivalent to $\Delta t = 60$ s) compared with $\Delta t = 1$ μs (essentially $\Delta t = 0$), when the misonidazole is present. This result suggests that a small amount of oxygen had been depleted, and was replenished by diffusion during the 1 s interpulse time. However, that the survival is not greatly changed as Δt is varied from 0 to 1 s is evidence that most of the oxygen had not been depleted at all by the first pulse. The data from this, and other double pulse experiments (9), therefore support the hypothesis that sensitizers are capable of interfering with the oxygen depletion process on this fast time scale.

MECHANISMS OF SENSITIZATION

It is beyond the scope of this paper to evaluate the current understanding of the mechanisms for radiosensitization for the many classes of sensitizing agents now known. There are many factors involved, and it is likely that there are several pathways

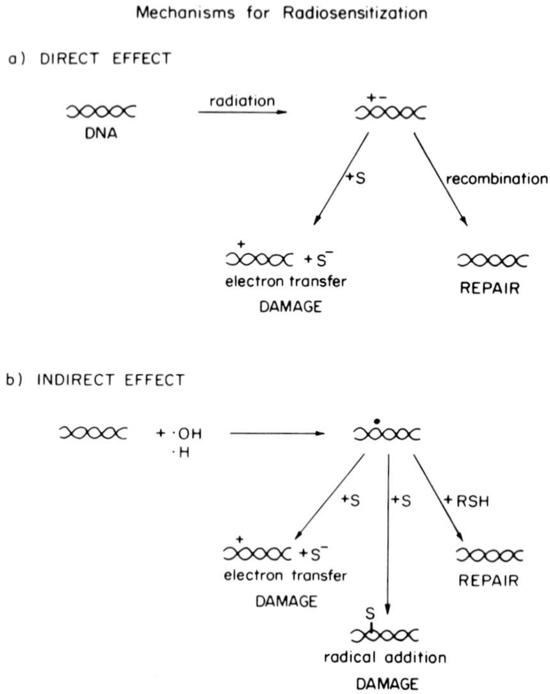

FIGURE 7. A simplified schematic of possible mechanisms of sensitization by oxygen and electron-affinic sensitizers. In this diagram, S represents a sensitizer; RSH a sulfhydryl compound; and ·OH and ·H the hydroxyl radical and hydrogen atom, respectively.

leading from radiation-induced initial damage to lethality. An attempt to depict a simplified picture of possible mechanisms of action for electron affinic compounds is presented in Fig. 7. Oxygen probably acts by the radical addition pathway, giving rise to DNA hydroperoxides (the oxygen fixation hypothesis); some chemical sensitizers, e.g., nitrofurans, nitroxyls, and nitroimidazoles, may also act by radical addition (1,19). An electron transfer process is appealing as a mechanism for highly electron affinic sensitizers such as p-nitroacetophenone and other nitroaromatic compounds. However, demonstration of electron transfer in vitro and in chemical systems have not indicated unambiguously the role of this mechanism (20). Both of these pathways, leading to enhanced radiation damage, and also the competing repair processes of ion recombination and H donation by sulfhydryl compounds, are consistent with a time scale for these events of microseconds to milliseconds. Diffusion of sensitizers and protectors also occurs in millisecond times. Experimental techniques of fast time response represent a favorable approach toward our gaining a more complete understanding of the complex interaction of these important processes in radiobiology.

REFERENCES

1. Chapman, J. D., Dugle, D. L., Reuvers, A. P., Meeker, B. E. and Borsa, J., Int. J. Radiat. Biol. 26,383 (1974).
2. Michaels, H. B. and Hunt, J. W., Radiat. Res. 74,23 (1978).
3. Quintiliani, M., Int. J. Radiat. Oncol. Biol. Phys. 5,1069 (1979).
4. Howard-Flanders, P. and Moore, D., Radiat. Res., 9,422 (1958).
5. Epp, E. R., Weiss, H., Kessaris, N. D., Santomasso, A., Heslin, J. and Ling, C. C., Radiat. Res. 54,171 (1973).
6. Epp, E. R., Weiss, H. and Ling, C. C., Curr. Top. Radiat. Res. Q. 11,201 (1976).
7. Michaels, H. B., Epp, E. R., Ling, C. C. and Peterson, E. C., Radiat. Res. 76,510 (1978).
8. Ling, C. C., Michaels, H. B., Epp, E. R. and Peterson, E. C., Radiat. Res. 76,522 (1978).
9. Michaels, H. B., Ling, C. C., Epp, E. R. and Peterson, E. C., Radiat. Res., 85, In press (1981).
10. Michael, B. D., Adams, G. E., Hewitt, H. B., Jones, W. B. G. and Watts, M. E., Radiat. Res. 54,239 (1973).
11. Michael, B. D., Harrop, H. A. and Maughan, R. L., "Radiation Research: Proceedings of the 6th International Congress of Radiation Research" (S. Okada, M. Imamura, T. Terashima, H. Yamaguchi, eds.), Tokyo (1979), p. 288.
12. Shenoy, M. A., Asquith, J. C., Adams, G. E., Michael, B. D. and Watts, M. E., Radiat. Res. 62,498 (1975).
13. Whillans, D. W. and Hunt, J. W., Br. J. Cancer 37, Suppl. III,38 (1978).
14. Fox, R. A., Fielden, E. M. and Sapora, O., Int. J. Radiat. Biol. 29,391 (1976).
15. Dewey, D. L. and Boag, J. W., Nature 183,1450 (1959).
16. Epp, E. R., Weiss, H. and Santomasso, A., Radiat. Res. 34, 320 (1968).
17. Michaels, H. B., Ling, C. C., Epp, E. R., Gerweck, L. E. and Peterson, E. C., Radiat. Res. 74,494 (1978). [Abstract]
18. Watts, M. E., Maughan, R. L. and Michael, B. D., Int. J. Radiat. Biol. 33,195 (1978).
19. Chapman, J. D., Greenstock, C. L., Reuvers, A. P., McDonald, E. and Dunlop, I., Radiat. Res. 53,190 (1973).
20. Michaels, H. B., Rasburn, E. J. and Hunt, J. W., Radiat. Res. 65,250 (1976).

DISCUSSION

WILLSON: Could I start the discussion by asking Dr. Wardman why he chose 1.6 in his activity/oxidation potential study? Would you get the same results if you used, say, 1.4?

WARDMAN: The value of enhancement ratio 1.6 was chosen for purely practical reasons. Adams et al., Int. J. Radiat. Biol. 35,133 (1979) gave the data for different values of response and showed the slope b. in the Hansch equation did indeed vary from about 6 to 9 V^{-1}, depending on the enhancement ratio taken. In contrast to the theoretical predictions, experimental measurements of the rate constant for the reaction of the nitro radical-anion with O_2 gave a slope for log k vs E which was rather less—about 5 V^{-1} [P. Wardman and E. D. Clarke, Biochem. Biophys. Res. Comm. 69,942 (1976)]. So we are at the stage now where we are just playing with figures. However, this application of the Marcus theory might be relevant to the question of competing radical addition and electron transfers. Dr. Willson and I were discussing this last night and he suggested that the choice between electron transfer and radical addition may occur at a transition state. I think that the theoretical treatment, e.g. by Marcus, of electron-transfer reactions probably has also something to tell us about radical addition, because the theory deals with the equilibrium between reactants and the transition state. So some of my discussions about the control of rate constants by electron-affinity is also relevant to radical addition.

RALEIGH: When you are talking about radical addition I think you have to consider two possibilities. One is where the radical adds to the aromatic ring and leads to a stable adduct; the other where the radical adds to the oxygen of the nitro group with the possibility of an oxygen transfer process occurring. Either of these seem to be equally possible. Of course electron transfer can also occur. In 1973 we published a paper concerning the cleavage of phosphate esters sensitized by nitroaromatic compounds. I think we could explain our results best by invoking a carbonium ion intermediate, formed by the oxidation of the intermediate radical on the carbon-bearing phosphate group [D. W. Whillans and G. E. Adams, Int. J. Radiat. Biol. 28,501 (1975)].

WILLSON: Regarding the point about the possibility of oxidation of DNA radicals that Howard Michaels brought up in his talk, he said that it seemed that sensitizers did not oxidize the radicals very readily. I think this stems from work with the thymine-hydroxyl adducts. Hydroxyl radicals react with thymine to form an adduct. You can see this aduct by pulse radiolysis; you can observe its absorption. If you have O_2 present then the absorption decays very rapidly. We, and also Dr. Whillans and Dr. Michaels, have found that if you take some "nitro" sensitizers and try to observe this reaction that we see with O_2—you cannot. Interestingly, with guanine you do not ever see the reaction with O_2. It seems that the product from guanine, a different species, it might be a radical cation or even the protonated adduct—whatever it is, it does not seem to react with O_2.

CADET: We have reported several steady state studies on thymine and guanine irradiated in the presence of TAN, a nitroxyl free radical sensitizing agent [Cadet et al., Int. J. Radiat. Biol. 30, 1 (1976); J. Radiat. Res. 18,93 (1977); Tetrahedron 35,2743 (1979)]. We identified two products which are in fact very stable. For guanine, we do not see any adduct or any long-lived product so our results support completely your proposal.

ROOTS: I just want to ask Dr. Michaels about the prevention of the break in the survival curve by the sensitizer. Is that because the sensitizer reacts with the hydrated electron so that O_2 is not depleted that way?

MICHAELS: It is not known what the mechanism is for O_2 depletion in the first place. Certainly, the reaction of hydrated electrons with O_2 to form $O_2^{\overline{\cdot}}$ which is relatively unreactive in any of these systems would account for the removal of O_2 from participation in sensitization reactions. Now what I think happens in my experiments is that the sensitizer, being an O_2 mimic, i.e., almost equally reactive as O_2 with electrons and free radicals, is capable of reacting rapidly with the $O_2^{\overline{\cdot}}$ binding species. These species may be solvated electrons or they may be solvated electron adducts that are formed by the addition of electrons with other reactants in the cell. Now if the sensitizer reacts with the $O_2^{\overline{\cdot}}$ binding species, and, of course, the sensitizer is present in higher concentration than that of the O_2 in the cells (about 5μM), then those $O_2^{\overline{\cdot}}$ binding species would be prevented from reacting with O_2. This would leave O_2 free to react with DNA radicals and, therefore, sensitize. We do not know whether this binding of the sensitizer to the supposed radiation-induced $O_2^{\overline{\cdot}}$ binding species is transient or permanent. It could be transient. The only restriction on the time for this transient phenomenon is that the sensitizer must bind to the $O_2^{\overline{\cdot}}$ removing species long enough to allow O_2 to react with the DNA radicals (or whatever it does react with) to cause sensitization.

WILLSON: I think that if the sensitizer does react with the radical and the sensitizer is a stable molecule like the nitro compounds beforehand, then if an adduct is formed, this must itself be a radical so you might expect that this would not be stable. But, of course, in the case of the nitroxyl free radicals of which Dr. Fielden spoke, then you well might get a stable adduct. It will not be a free radical.

MICHAELS: I would just like to make one other comment regarding radicals, namely ·OH-adducts on thymine and some of the other bases, and their being able to react with O_2 and other sensitizers in ways detectable by pulse radiolysis. Everything is correct that you said regarding my own work with DNA base radicals and with PNAP. In addition, I would like to add that Eric Rasburn and I also did some conductivity measurements with a pulse conductivity set up [H. B. Michaels, E. J. Rasburn and J. W. Hunt, Radiat. Res. 65,250 (1976)] where we looked for electron transfer from hydroxyl radical adducts of bases, sugars, and alcohols to sensitizers, specifically to PNAP. In the case of the bases, such as thymine and uracil, we saw no electron transfer; we cannot tell if there is addition or not since we cannot see this with conductivity. However, in the case of the sugars, we did see substantial electron transfer. So this means that hydroxyl radical attack on the sugars could lead to electron transfer reactions with the sensitizer, or could lead to addition. In the case of ·OH addition products on the DNA bases, such as thymine, we think that addition is likely.

WILLSON: Yes, I agree with you entirely. If you get attack on the sugar and you get hydrogen abstraction to give a sugar radical, and this could behave like any other simple alcohol radical; it could react by electron transfer or form an adduct as in the case of O_2. We also showed in fact, though, that if you take the thymine ·OH-adduct and you have a compound that is sufficiently electron-affinic like menadione then reaction can indeed occur [R. L. Willson and P. T. Emmerson, Radioprotection and Sensitization (H. Moroson and M. Quintiliani, eds.), Taylor and Francis, London, 1970, p. 73].

FIELDEN: I'd also like to comment on the reaction between O_2 and the thymine ·OH-adduct and the possibility of O_2's reacting with the ·OH-guanine product. For a long time it was believed that O_2 did not react with the DNA·OH-adduct although we have not demonstrated a fast reaction. The problem was that optical detection pulse radiolysis is just not the best means of looking at these species, in that there is quite a lot of spectral overlap between the hydroxyl radical adducts and the product of O_2 added to the ·OH-adducts. This is particularly so in the case of DNA. My colleague, Dr. O'Neill, has demonstrated the spectral similarity of a range of DNA derived radicals involving ·OH, O_2 and even nitroxyl addition. Thus the fact that a reaction between O_2 and G(·OH) is not "seen" by pulse radiolysis does not mean that it does not occur.

WARDMAN: I would like to ask Dr. Fielden about the 2-nitroimidazole compound with the nitroxyl terminating the side chain, RO 03-9454. Dr. Smithen presented the Gray Laboratory data for sensitization by this compound last year at Key Biscayne. Like you, we found it to be about an order of magnitude more efficient than misonidazole. This increased efficiency is precisely similar to the behavior of some 30 or more other (2-nitro-1-imidazolyl) propanolamines which are related to this compound. The enhanced activity arises from the secondary amine function which separates the nitroxyl residue from the 2-nitroimidazole. If nitroxyls protect against sensitization by nitroimidazoles why do you not, in effect, get the molecule's protecting against itself more? Why is the efficiency as high as that of other (2-nitro-1-imidazolyl) propanolamines?

FIELDEN: As to your comment on the effectiveness of base moieties in the side chain, I absolutely agree, I really did not want to go into the business of sensitizer efficiency here, but compound RSU 4013 is much more effective than misonidazole for the reasons that you stated. Our contention is that 9454 is not as effective as 4013 because it has the nitroxyl function which is acting independently in a subtractive manner, as can be demonstrated by the addition of the simple nitroxyl TMPN.

BIOCHEMICAL INTERACTIONS OF RADIOSENSITIZERS

John E. Biaglow, Division of Radiation Biology, Case Western Reserve University, 2064 Abington Road, Cleveland, Ohio 44106

Aerobic and anaerobic reduction of nitro compounds involves electron transfer from cellular flavoproteins such as NADPH-Cytochrome c reductase, NAD(P)H dehydrogenase, xanthine oxidase and aldehyde oxidase to the nitro group (1). Recent data obtained with cyanide inhibited cells, with purified Ehrlich microsomes and with the S-9 fraction of rat liver indicate that the NADPH-cytochrome c reductase may be the major electron donating flavoprotein (1,2). As seen schematically in Fig. 1, the reduction of nitro compounds via this microsomal flavoprotein is linked with glucose oxidation (1). NADPH, supplied via oxidation of glucose in the pentose cycle, donates electrons to the flavin moiety of the enzyme forming flavin radical. It has been demonstrated that flavin radicals can react with nitro compounds. In the presence of O_2

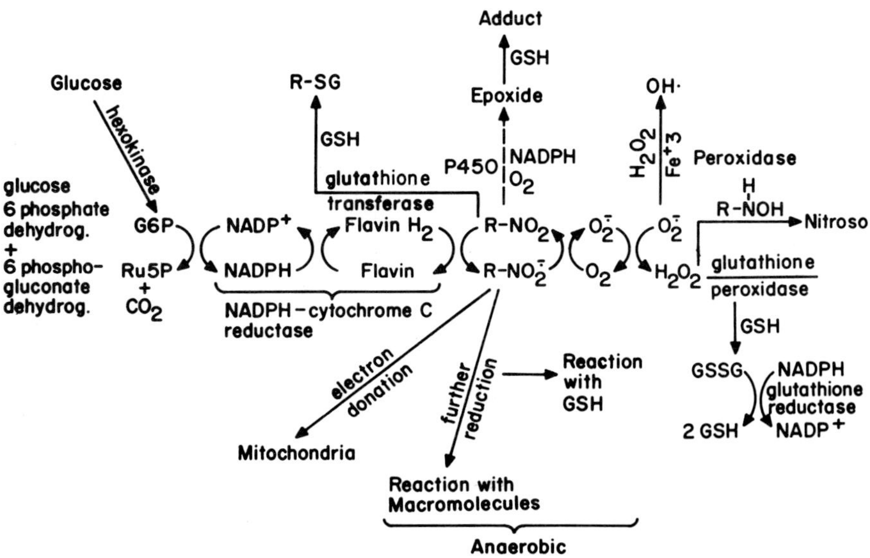

FIGURE 1. Reaction Scheme for cellular activation of nitro compounds. The abbreviations are: ATP, adenosine triphosphate; $NADP^+$, nicotinamide adenine dinucleotide phosphate; NADPH = reduced; GSH, glutathione; P450, cytochrome P450; GSSG, oxidized glutathione; $R-NO_2$, nitro compound; R-NO, nitroso intermediate; $R-NO_2^{\cdot-}$, nitro radical; R-NHOH, hydroxyl-amine; $R-NH_2$, amine; $O_2^{\cdot-}$, superoxide.

ISBN 0-12-592050-4

the resultant nitro radicals react with O_2 producing superoxide and peroxide (1). In the absence of O_2, reduction beyond the one-electron state occurs. Some nitro compounds such as 4-nitroquinoline-1-oxide(4-NQO), dinitrobenzonitrile (DNBN) and dinitropyridinium chloride (DNPC) are reduced, even in the presence of oxygen, beyond the radical state to form more stable intermediates such as hydroxylamine (1). Other sensitizers like metronidazole and misonidazole are reduced in the absence of oxygen to amines (3). In addition to the above mentioned flavoprotein reductions the nitro compounds are reactive with a variety of biochemicals such as thiols, ascorbate and NADPH or NADH plus riboflavin (1). Like NADPH cytochrome c reductase, these electron donors produce oxygen reactive nitro intermediates in air and reduce the nitro compounds to amines in the absence of oxygen (1).

PRODUCTION OF PEROXIDE

With intact cells, in the presence of oxygen (Fig. 1), the formation of oxygen reactive intermediates results in production of H_2O_2 which can be conveniently measured in the presence of N-ethylmaleimide (NEM) and sodium azide. As determined with HEp-2 cells (Fig. 2), azide will inhibit catalase and electron transfer to the cytochromes (Fig. 1). NEM inhibits glutathione peroxidase activity by removing endogenous glutathione (GSH) necessary for reduction of H_2O_2 (Fig. 1). As seen in Fig. 2, left panel, with HEp-2 cells, there is no peroxide produced in the presence of azide until NEM has been added. In the presence of NEM alone, a small amount of H_2O_2 is produced and a continuous generation of H_2O_2 occurs only after azide is added. The upper curve shows the production of H_2O_2 when both NEM and azide is present and the production of it continues for a longer period of time. The panel on the right shows the effect of misonidazole concentration on H_2O_2 production by NEM and azide inhibited cells. No H_2O_2 production was detected in the absence of glucose. H_2O_2 production increases as the electron affinity of the nitro compounds increases. Some nitro compounds react with endogenous thiols *via* conjugation, either spontaneously or through the enzyme glutathione-S-transferase, and therefore do not require NEM for maximal H_2O_2 production (4). Other nitro compounds, such as the nitrofurans, also exhibit a concentration dependency for H_2O_2 production (5).

OTHER OXYGEN CONSUMING REACTIONS

Not all of the oxygen consumption results in the production of H_2O_2. For example, the hydroxylation of the parent ring as well as adjoining ring structures on various nitro compounds can also consume oxygen. In addition, reactions of H_2O_2 with protein and nonprotein thiols result in less measurable H_2O_2 but a net disappearance of O_2 (Fig. 1). We have found that H_2O_2 and 4-hydroxyl-amino-quinoline-1-oxide will react *via* a peroxidase

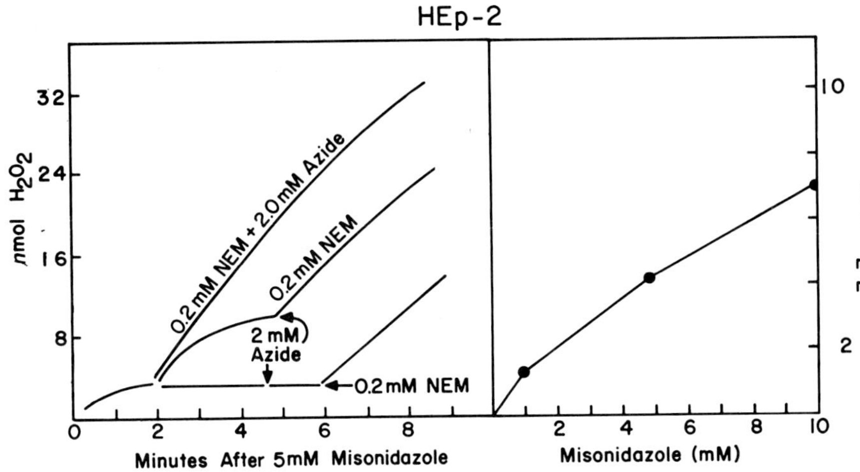

FIGURE 2. On the left side is the effect of 2 mM azide and 0.2 mM N-ethylmaleimide on 5 mM misonidazole-linked H_2O_2 production by 10^7 HEp-2 cells suspended in 10 mM glucose, 0.05 M Hepes buffered physiological saline, pH 7.4, 37°C. Peroxide was continuously monitored with a Yellow Springs oxidase probe and meter. The values represent the nmol of H_2O_2 produced 2 min after the addition of 15 µmoles misonidazole. The right side represents misonidazole concentration effect on the rate of H_2O_2 production.

catalyzed reaction to produce a nitroso intermediate. A similar reaction could account for the increased cytotoxicity of aerobic cells exposed to the hydroxylamine or amine reduction products of misonidazole (1).

ANAEROBIC REDUCTION OF NITROCOMPOUNDS AND REACTION WITH THIOLS

Thiols protect against the cytotoxic effects of misonidazole and other more electron affinic nitro compounds such as 4-NQO (1). In addition, we have found that prior removal of thiols by oxidizing agents such as diamide will enhance the radiosensitizing properties of misonidazole (6). Thiol removal by anaerobic incubation with misonidazole and washing off the drug also enhances the cellular radiation response (6). The mechanism of thiol protection against cytotoxic effects may involve the thiol's ability to react with either the nitro radical or nitroso intermediate before it reacts with macromolecules such as DNA (1). As seen in Fig. 1 there are a number of possible ways that thiols can influence the anaerobic metabolism of nitro compounds leading to nitro, nitroso or nitroxyl radicals. We have determined the ability of nitro compounds to deplete the NPSH of Ehrlich, V79 and CHO cells (7). As seen in Fig. 3B increasing concentration of misonidazole alters both the NPSH and protein thiol levels of Ehrlich cells. The ratio of protein thiol to NPSH in Ehrlich cells is approximately 10:1 (7). Thus 5 mM misonidazole reacted more with protein thiols

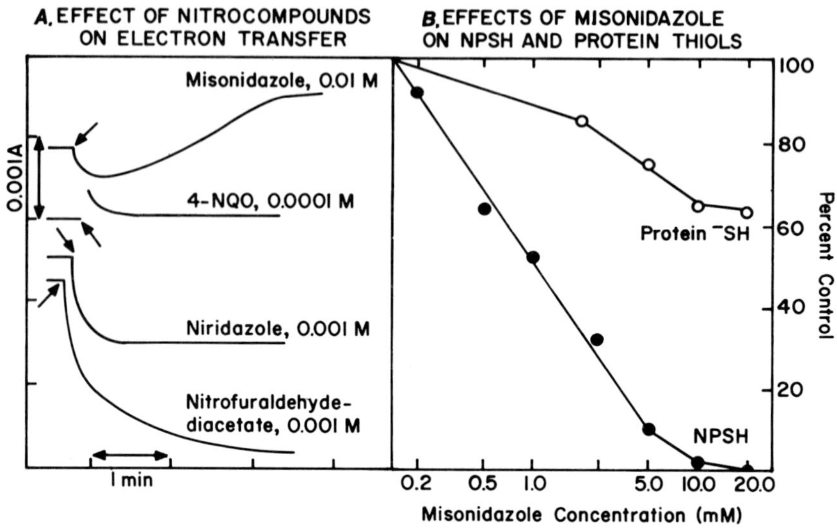

FIGURE 3A and 3B. Figure 3B show the effect of misonidazole on the non-protein and the protein thiol content of Ehrlich ascites tumor cells. Cells (10^7/ml) were incubated anaerobically in 0.5 M PBS and 10 mM glucose at 37°C, pH 7.3. Thiol determinations were performed 15 min after anaerobic conditions had been achieved. Fig. 3A shows the effect of nitro compounds on electron transfer to cytochrome c + c_1 (550-540 nm) in Ehrlich ascites tumor cells. The 10^7 cells/ml were buffered with 0.05 M PBS at pH 7.4, 37°C. The drug concentrations were as indicated. The upward deflection in absorbancy change indicates reduction and a downward deflection indicates oxidation.

than with non-protein thiols on a molar basis. This suggests that one of the reduced drug intermediates is non-specific in its reaction with thiols. We have previously found that misonidazole-linked thiol reaction is enhanced by glucose and ascorbate (8) and is decreased by inhibitors of the pentose cycle. CHO and Ehrlich cells treated with 5 or 10 mM misonidazole for 30 minutes are unable to regenerate reacted NPSH. This data (8) suggests that the reactive nitro compound forms an adduct that is metabolically inert.

EFFECT OF NITRO COMPOUNDS ON ELECTRON TRANSFER IN EHRLICH ASCITES CELLS

Nitro compounds may stimulate or inhibit cellular electron transfer (1,2,6). These effects can be monitored by using a sensitive dual beam spectrophotometer to measure intracellular cytochrome c + c_1 (550-540 nm) changes within the mitochondrial structure. We have chosen nitro compound concentrations that previously were found to produce the most dramatic effects. The 4-NQO concentration was based on previous results with rotenone inhibited cells. The data in Fig. 3A indicate that misonidazole at 10 mM will increase the steady state reduction of cytochromes $c + c_1$, suggesting that it has a slight stimulatory effect on cellular oxygen utilization. 10^{-4}M 4-NQO had little effect, but produces a dramatic effect on cytochrome $c + c_1$ if the electron transfer is first inhibited with rotenone (1). Niridazole, a nitro compound with a similar redox potential as misonidazole and a radiosensitizer (1), was also found to inhibit electron transfer. The most potent inhibitor was nitrofuraldehyde diacetate (NFDA) (lower curve, Fig. 3A).

Stimulation of cellular electron transfer and oxygen consumption usually occurs with the more electron affinic nitro compounds such as 4-NQO, DNBN and DNPC (1). Blockage of the cytochrome chain with cyanide or azide inhibits this reaction (cf. Fig. 1 and Ref. 1). Inhibition by NFDA (Fig. 3A) involves removal of essential thiols that are necessary for the electron transfer (1). Inhibition by less electron affinic nitro compounds such as nitrotoluene, metronidazole and 2-nitropyrrole may involve a direct alteration of electron transfer at a site prior to cytochrome $c + c_1$ in a manner similar to that reported for the classical inhibitor rotenone. Whatever the mechanism for inhibition of electron transfer and cellular oxygen utilization is, it is important to note that the resulting oxygen sparing effect is an important means for sensitizing multicellular systems to radiation <u>in vitro</u> and may be as important <u>in vivo</u> (1,2,9).

ACKNOWLEDGEMENTS

This investigation was supported by Grant CA 13747, awarded by the National Cancer Institute, Department of Health, Education, and Welfare.

REFERENCES

1. Biaglow, J. E., Varnes, M. E., Koch, C. J. and Sridhar, R., "Free Radicals and Cancer", (R. Floyd ed.) Marcel Dekker, N. Y. In press. (1981).
2. Jacobson, B., Biaglow, J. E., Fielden, E. M. and Adams, G., Cancer Clin. Trials, 3,47 (1980).
3. Mohindra, J. and Rauth, A. M., Cancer Res., 36, (1976).
4. Schroy, C. B. and Biaglow, J. E., Biochem. Pharmacol. Submitted. (1981).
5. Varnes, M. E. and Biaglow, J. E., this volume, p. 766.
6. Biaglow, J. E., J. Chem. Ed. In press. (1981).
7. Varnes, M. E. and Biaglow, J. E., Cancer Res., 39,2960 (1979).

8. Varnes, M. E., Biaglow, J. E., Koch, C. J. and Hall, E. J., Proceedings on the Conference on Combined Modality Cancer Treatment: "Radiation Sensitizers" Vols of Cancer Management, (L. Brady, ed.) Masson Press, New York, (1980) p. 121.
9. Biaglow, J. E., Radiation Res. (1981) In press.

OXYGEN MIMETIC REACTIONS OF NITROAROMATIC RADIOSENSITIZERS ACTIVATED BY NITROREDUCTASES IN THE PRESENCE OF MODEL MEMBRANES

J. A. Raleigh and F. Y. Shum, Radiobiology, Department of Radiation Oncology, Cross Cancer Institute, 11560 University Avenue, Edmonton, Alberta, Canada

INTRODUCTION

In principle, nitroaromatic radiosensitizers ($ArNO_2$) can react with cellular free radicals ($R\cdot$) in three ways (reactions 1-3). Nucleophilic aromatic substitution (1)

$$R\cdot + ArNO_2 \longrightarrow RArNO_2{}^{\cdot} \qquad (1)$$

Electron transfer oxidation (2,3)

$$R\cdot + ArNO_2 \longrightarrow R^+ + ArNO_2{}^{\overline{\cdot}} \qquad (2)$$

Oxygen transfer (4)

$$R^{\cdot} + ArNO_2 \longrightarrow (RO{-}\overset{O\cdot}{\overset{|}{N}}{-}Ar) \longrightarrow RO\cdot + ArNO \qquad (3)$$

Reactions 2 and 3 may be considered oxygen mimetic and have been studied in a variety of systems including models of the polar head group in phospholipids (3,5) and of the internucleotide bond in nucleic acids (6-8). Of current concern, however, is radiosensitizer toxicity and, in particular, peripheral nerve toxicity (9, 10). Metabolic activation of nitroaromatics has been proposed as the source of toxicity. Nitroreductase reduction of nitroaromatics produces arylhydroxylamines (11) and arylamines (12) from which the formation of precursor arylnitroso intermediates can be deduced (13,14). The formation of arylnitroso compounds as freely diffusible entities, however, has not previously been demonstrated. Nitroso compounds may be considered to be isoelectronic with oxygen (O_2) and to react with olefins, including unsaturated lipids, in an "ene" reaction in a way analogous to singlet oxygen (Scheme 1) (15,16). This reaction of nitrosoaromatics has been proposed as a possible source of the carcinogenic action of arylnitroso compounds (17) and is readily detectable by virtue of the formation of a stable and easily identifiable nitroxyl radical ($>N{-}O\cdot$) as one of the products. We have found that nitroxyl radicals are formed when nitrobenzene radiosensitizers (5) are reduced by the nitroreductase, xanthine oxidase, in the presence of the phospholipid vesicles and fatty acid micelles.

MATERIALS AND METHODS

Nitrosobenzene, nitrobenzene, dinitrobenzonitrile (DNBN) and 4-oxo-2,2,6,6,-tetramethylpiperidinoxy were purchased from Aldrich

ISBN 0-12-592050-4

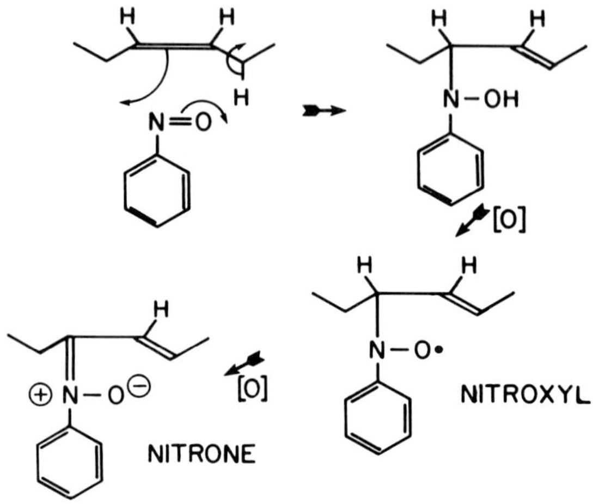

SCHEME 1. Mechanism proposed for the formation of nitroxyl radicals in the reaction of arylnitroso compounds with unsaturated lipids.

Chemical Company. The unsaturated lipids, dioleoyl phosphatidyl choline (DOL) and linoleic acid (18:2), were obtained from NuChek Prep. Sigma Chemical Company was the source of xanthine and xanthine oxidase. Electron spin resonance (ESR) spectra were recorded with a Varian V3600-12 magnet with Mark I Fieldial controller and Bruker ER400 ESR consol. High performance liquid chromatography (HPLC) was carried out on Tracor equipment incorporating a 950 pump and a 970A variable wavelength detector. A post-column Technicon Artoanalyzer was used to analyze phospholipid components specifically in a non-digestive, continous flow analysis (Ruddock and Raleigh, in preparation).

In a typical nitroreductase experiment, 100 mg of linoleic acid or dioleoyl phosphatidyl choline were dispersed in 25 ml of doubly distilled water, buffered with tris (pH 8.5), by sonication for 20 minutes under nitrogen with an Artek Sonic Dismembrator Model 150. Nitrobenzene (50 m molar) and xanthine (96 m molar) were added and the solution degassed for 10 minutes with nitrogen. Xanthine oxidase (12 units) was added and the tube sealed under nitrogen. The reaction was allowed to proceed overnight at which time the mixture was extracted with an equal volume of chloroform/methanol (2/1). In the case of the fatty acid experiment, the mixture was acidified with concentrated sulfuric acid before extraction. The concentrated extract was chromatographed on a silica gel column (2.5 x 35 cm) with the nitroxyl-containing lipid fraction eluting with $CHCl_3$/MeOH (2/3) in the case of DOL and $CHCl_3$/MeOH (99/1) in the case of 18:2. The lipid containing

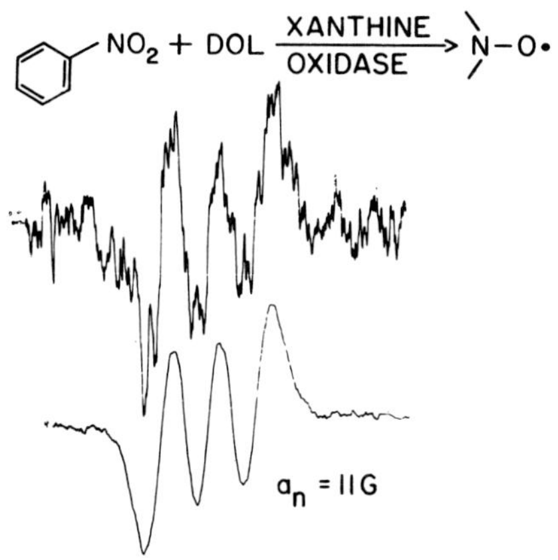

FIGURE 1. Nitroxyl radical spectrum formed from the xanthine oxidase reduction of nitrobenzene in the presence of dioleoyl phosphatidyl choline (DOL) vesicles. The two traces represent different ESR instrument settings.

fractions were collected, the solvent removed in vacuo, and the residue dissolved in chloroform/methanol (2/1) for examination by ESR.

RESULTS

The ESR spectrum obtained when nitrobenzene is reduced by xanthine oxidase in the presence of DOL vesicles shows the characteristic three line spectrum of a nitroxyl free radical (Fig. 1) with spin-spin coupling constant (a_n) of 11G. An identical spectrum is obtained when nitrosobenzene is mixed with DOL solutions. In the case of linoleic acid micelles, in which the lipid contains two double bonds, a similar spectrum is obtained (Fig. 2, a_n=11.5) although additional splitting of the signal is evident. Nitrosobenzene added directly to 18:2 micelles produces a similar spectrum. In each case, the samples examined were obtained by column chromatography of the reaction mixture so that the ESR signal arises from a lipid bound nitroxyl adduct. High pressure liquid chromatography (Partisil 10, 4mm x 25cm, $CH_3OH/CH_3CN/H_2O$ 51/33/14) of the DOL adduct, indicates, however, the presence of at least two compounds in the chromatographic fraction containing the nitroxyl entity. Further characterization of compounds is not complete. The additional splitting of the peaks in the case of 18:2 micelle experiment may indicate the presence of more than one nitroxyl product here as well.

As a rough estimate of the yield of the nitroxyl adduct in the 18:2 experiment, various concentrations of the nitroxyl, 4-oxo-2,2,6,6,-tetramethylpiperidinoxy, were added to linoleic acid solutions and the spectra recorded (Fig. 3) at instrument settings

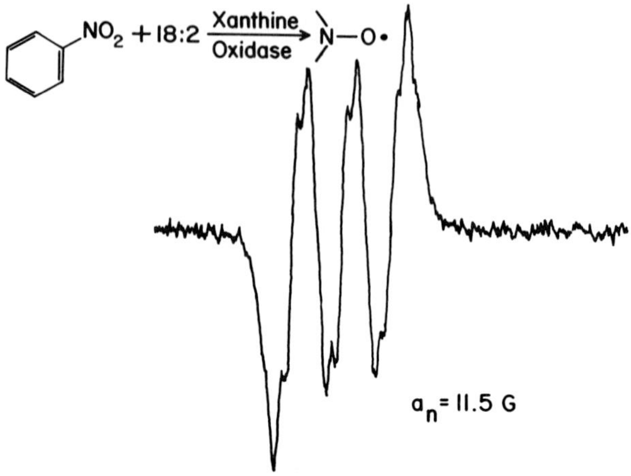

FIGURE 2. Nitroxyl radical ESR spectrum from the xanthine oxidase reduction of nitrobenzene in the presence of linoleic acid (18:2) micelles.

identical to those used for the reductase experiments. The yield of nitroxyl adduct in those experiments is of the order of 10^{-5} M which represents a yield of <1% based on the original concentration of nitrobenzene. In addition to enzymic activation of nitrobenzene by xanthine oxidase, dinitrobenzonitrile can be activated to produce nitroxyl adducts with DOL by both xanthine oxidase and cytochrome c reductase under hypoxic conditions. Extension of

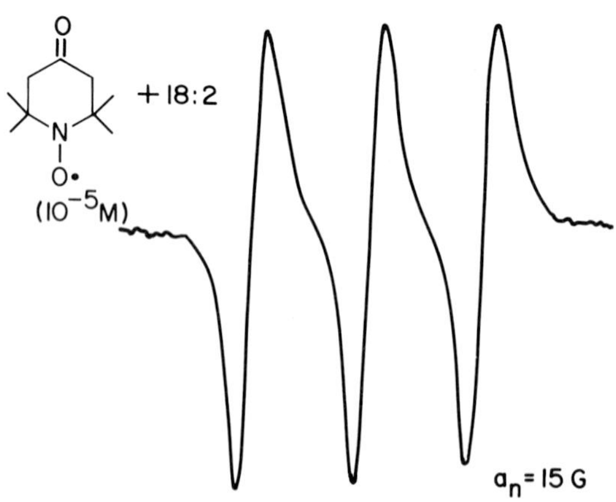

FIGURE 3. ESR spectrum of the nitroxyl, 4-oxo-2,2,6,6-tetramethylpiperidinoxy, in admixture with 18:2 under the conditions used in the nitroreductase experiments.

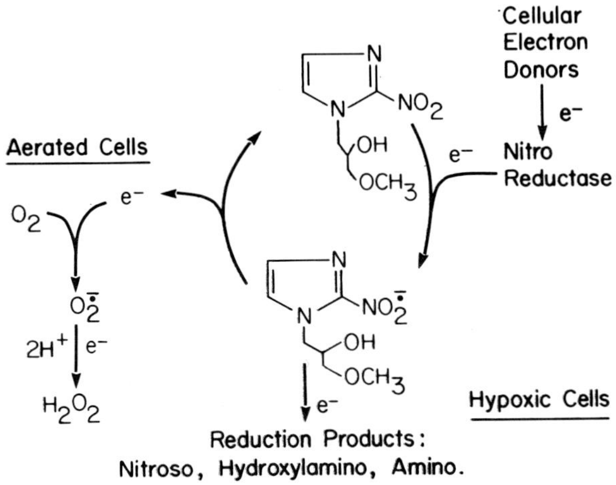

SCHEME 2. Hypothetical electron transfer reactions underlying nitroaromatic toxicity. In aeorbic tissue nitroaromatic radiosensitizers may catalyze $O_2^{\cdot-}$ and H_2O_2 formation. In hypoxic tissue toxic nitroso and hydroxylamines may be formed. In both aerobic and hypoxic tissue nitroaromatic compounds may lead to an altered redox state due to depletion of reducing equivalents in the cells.

these studies to other radiosensitizers is underway.

DISCUSSION

The molecular mechanism of nitroaromatic-induced neuropathies (18-19) is not known. Nitroaromatic toxicity, in general, is linked to reductive pathways in vitro in which the formation of potentially toxic nitroso and hydroxyl amino products is invoked (Scheme 2). In a few cases, hydroxylamines (11) or their decomposition products (20) have been identified but their logical precursor, the nitroso compound, has not been isolated. The trapping of this intermediate by unsaturated lipids demonstrates that arylnitroso compounds are formed as freely diffusible entities—at least in the cases of nitrobenzene and dinitrobenzonitrile. The apparently low yield of nitroxyl radicals (< 1%) in these experiments may be due to the partial release of the nitroso intermediates from the nitroreductases, or poor partitioning of the nitrosobenzenes into the lipid phase, or to the oxidation of the nitroxyl one step farther to the zwitterionic nitrone (Scheme 1). It is conceivable that other nitroaromatic radiosensitizers will produce low yields of nitroxyls for the same reasons in addition to a possibly decreased chemical stability of the nitroso intermediates. While the importance of these reactions in nitroaromatic-induced neuropathies is not established, reactions which might involve the lipid-rich myelin sheath of nerve tissue are obviously of interest. In vitro, the nitroxyl adduct formation occurs only in the absence of oxygen. In vivo, however, it is possible that the binding of nitroaromatic compounds, to otherwise well-oxygenated tissue (21), could occur by similar reductive mechanisms due to compartmentalization within cells or local hypoxia induced by radiosensitizer stimulated oxygen consumption (22). This would be the case in scheme 2 if the radiosensitizer were to catalyze electron transfer to oxygen without itself being consumed.

ACKNOWLEDGMENTS

We thank Mr. G. Bigam of the Spectroscopy Lab, Chemistry Department, University of Alberta for assistance in obtaining the ESR spectra and our colleagues in Radiobiology for helpful discussions. This work was supported in part by a grant from the National Cancer Institute of Canada.

REFERENCES

1. Rudquist, U. and Torssell, K., Acta Chem. Scand., 25,2183 (1971).
2. Whillans, D. W. and Adams, G. E., Int. J. Radiat. Biol., 28, 501 (1975).
3. Raleigh, J. A., Greenstock, C. L. and Kremers, W., Int. J. Radiat. Biol., 23,457 (1973).
4. Weller, J. W. and Hamilton, G. A., Chem. Comm., 1390 (1970).
5. Raleigh, J. A., Chapman, J. D., Borsa, J., Kremers, W. and Reuvers, A. P., Int. J. Radiat. Biol., 23,377 (1973).
6. Raleigh, J. A., Greenstock, C. L., Whitehouse, R. and Kremers, W., Int. J. Radiat. Biol., 24,595 (1973).
7. Raleigh, J. A., Whitehouse, R. and Kremers, W., Radiat. Res., 59,453 (1975).
8. Raleigh, J. A., Kremers, W. and Whitehouse, R., Radiat. Res. 63,53 (1975).
9. Urtasun, R. C., Chapman, J. D. and seven others, Brit. J. Cancer, 37, Suppl. III, 271 (1978).
10. Saunders, M. I., Dische, S., Anderson, P. and Flockhart, I. R., Br. J. Cancer, 37 Suppl. III, 268 (1978).
11. Feller, D. R., Morita, M. and Gillette, J. R., Biochem. Pharmacol. 20,203 (1971).
12. Gillette, J. R., Kamm, J. J. and Sasame, H. A., Mol. Pharmacol., 4,541 (1968).
13. Venulet , J. and Van Etten, R. L., "The Chemistry of the Nitro and Nitroso Groups," Part 2 (H. Feuer, ed.), Interscience, New York, N.Y., 1970, p. 201.

14. Williams, R. T., "Detoxication Mechanisms," Chapman and Hall, London, 1947, p. 130.
15. Sullivan, A. B., J.O.C., 31,2811 (1966).
16. Knight, G. T. and Pepper, B., Tetrahedron, 27,6201 (1971).
17. Floyd, R. A., Soong, L. M., Stuart, M. and Reigh, D. L., Arch. Biochem. Biophys., 185,450 (1978).
18. Bradley, W. G., Karlsson, I. J. and Rassol, C. G., Br. Med. J., 610 (1977).
19. Toole, J. F. and Parrish, M. L., Neurology, 23, 554 (1973).
20. Aufrere, M. B., Hoener, B. and Vore, M., Drug Metabolism and Disposition, 6,403 (1978).
21. Wang, C. Y., Chiu, C. W. and Bryan, G. T., Drug Metabolism and Disposition, 3,89 (1975).
22. Biaglow, J. E., Jacobson, B., Greenstock, C. L. and Raleigh, J., Mol. Pharmacol., 13,269 (1977).

X

ACCELERATED OXYGEN CONSUMPTION BY CATECHOLAMINES IN THE PRESENCE OF AROMATIC NITRO AND NITROSO COMPOUNDS. IMPLICATIONS FOR NEUROTOXICITY OF NITRO COMPOUNDS

Rajagopalan Sridhar, Oklahoma Medical Research Foundation, 825 N.E. 13th Street, Oklahoma City, Oklahoma 73104

The neurotoxicity of the hypoxic cell sensitizers such as misonidazole (Ro-07-0582) (MIS) and to a lesser extent metronidazole (Flagyl) (MET), greatly reduces their usefulness in cancer therapy (1,2). The mechanism of neurotoxicity of MIS is unknown at present. The interactions of catecholamines with nitro and nitroso compounds were studied in view of the possible involvement of catecholamine type neurotransmitters in drug-induced neurotoxicity.

The auto-oxidation rates of catecholamines such as isoproterenol and norepinephrine were enhanced in the presence of several nitro and nitrosoaromatic compounds. Oxygen consumption was measured polarographically using a Clark-type electrode at 30°C. Nitro and nitroso compounds (60 X final concentration in DMSO or ethanol), catecholamines (30 X final concentration in water) and potassium phosphate buffer (0.1 M, pH 7.5) were used. All the nitro compounds tested, except metronidazole (2mM), increased the rate of oxygen consumption by catecholamines (10 mM). Misonidazole (MIS) (2mM) doubled the rate of oxygen consumption by isoproterenol (10 mM) whereas the more strongly electron-affinic compound, 4-nitroquinoline-N-oxide (NQO) was twenty times more effective than MIS. The following order of relative effectiveness was found for the nitro-aromatic compounds (reduction potential in parenthesis): MIS (-0.30) < Nifuroxime (-0.30) < 3,5-dinitrobenzonitrile (-0.16) < 4-nitropyridine-N-oxide (-0.25) < 2,4,6-trinitrobenzenesulfonic acid (-0.20) < 1-(2,4-dinitrophenyl)pyridinium chloride (-0.10) < NQO (-0.15). The generation of hydrogen peroxide during the oxidation was demonstrated by addition of catalase which caused oxygen evolution as monitored by the oxygen electrode whereas development of a pink, red or brown color indicated the formation of quinones. Inclusion of catalase (100 μg/ml; Sigma Chemical Co.) at the beginning of the experiment caused from 0 to 30% inhibition of oxygen uptake depending on the nitro compound tested. This inhibition by catalase was not evident in the case of metronidazole and misonidazole whereas the maximum inhibition was seen in the case of NQO. The addition of superoxide dismutase (100 μg/ml; Sigma Chemical Co.) inhibited oxygen consumption by isoproterenol even in the absence of nitro or nitroso compounds by about 20%. When both catalase (100 μg/ml) and superoxide dismutase (100 μg/ml) were added, the rate of oxygen consumption by

ISBN 0-12-592050-4

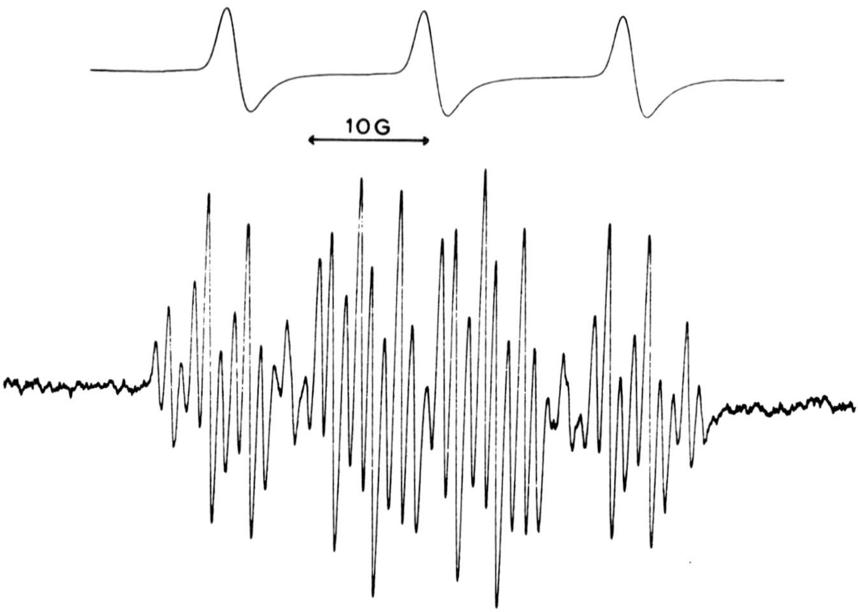

FIGURE 1. Top trace—the first derivative ESR spectrum 2,2,6,6-tetramethylpiperidinol-1-oxyl (TMPN) standard. Bottom trace—the first derivative ESR spectrum of phenylhydronitroxyl radical generated by treatment of nitrosobenzene with isoproterenol. Instrument settings were: modulation amplitude 0.1 G; modulation frequency 100 KH_Z; receiver gain 5 X 10^2; microwave power 25 mW; microwave frequency 9.526 GH and scan rate 3.33 G/min. A similar ESR spectrum was obtained for the hydronitroxyl radical generated by another route, viz., reduction of nitrobenzene with zinc and ammonium chloride followed by aerial oxidation.

isoproterenol - NQO mixture was reduced to 40% of the value obtained under enzyme-free conditions. These results point to the formation of hydrogen peroxide and superoxide anions in these reactions. Nitrosobenzene and 2-nitrosotoluene were more effective than all the nitro compounds tested in stimulating oxygen consumption by isoproterenol. Hydronitroxyl free radicals were generated when nitroso compounds were treated with catecholamines. Thus, treatment of nitrosobenzene with isoproterenol gave phenylhydronitroxyl free radical whose ESR spectrum was recorded (Fig. 1). Free radicals were also generated when nitrosobenzene was treated with sodium ascorbate.

The nitroso or nitroaromatic compound mediated depletion of catecholamines requires particular attention in terms of neurotoxicity. The decrease in heart rate and core temperature (or rectal temperature) observed in mice after administration of misonidazole (3) or metronidazole (4) may reflect depletion of catecholamines such as norepinephrine. Ingestion by humans, of the rat poison, N-3-pyridylmethyl-N'-p-nitrophenylurea causes (in addition to a decrease in blood pressure and rectal temperature) peripheral neuropathy and central nervous system dysfunction (5).

Thus, there is reason to believe that neurotoxicity of nitro compounds may be due to depletion of oxygen, catecholamines and ascorbate in nerve tissue with concomitant generation of toxic species such as hydroxyl, hydronitroxyl and superoxide free radicals as well as nitroso and quinonoid derivatives.

ACKNOWLEDGEMENT

I wish to thank Dr. C. E. Smithen of Roche Products, Ltd., Welwyn, England for the generous supply of misonidazole.

REFERENCES

1. Dische, S., Saunders, M. I. and Flockhart, I. R., Br. J. Cancer, 37, Suppl. III, 318 (1978).
2. Urtasun, R. C., Chapman, J. D., Feldstein, M. L., Band, R. P., Rabin, H. R., Wilson, A. F., Marynowski, B., Starreveld, E. and Shnitka, T., Br. J. Cancer, 37, Suppl. III, 271 (1978).
3. Von Burg, R., Conroy, P. J. and Passalacqua, W., Br. J. Cancer, 40,134 (1979).
4. Haynes, M. J. and Inch, W. R., Int. J. Radiat. Oncol. Biol. Phys. 1,1125 (1976).
5. Le Witt, P. A., New Engl. J. Med., 302,73 (1980).

DISCUSSION

FIELDEN: I would like to ask Dr. Raleigh just how certain he is of the assignment of the ESR Spectra to nitroxyl radicals. I am thinking of some experiments we did in 1971 where a flowing solution of nitrofurantoin was irradiated and then rapidly frozen (within 15 ms). The ESR spectrum of the frozen solution showed a radical with a typical nitrogen three line splitting. However, we felt that we could not unequivocally assign the signal to a nitroxyl radical. Unfortunately, we could not pursue this work.

RALEIGH: I think that is a fair question. But I would ask you what else it could be. The splitting of the lines is of the expected size and the spectra looks very much like 4-oxo-2,2,6,6-tetramethylpiperidinoxy. In addition we have eliminated the possibility that the signal is arising from a small lipid-free molecule by chromatographing the material before ESR.

SRIDHAR: Drs. Floyd, Hampton, Stewart and I have a paper accepted for publication in Applied Spectroscopy about ene-type addition of nitroso compounds to tetramethylethylene and methyloleate. We believe that the assignment of nitroxyl structure to the oxidized adducts is accurate.

BIAGLOW: Just a question for Dr. Sridhar or Dr. Raleigh. One of the side effects of the nitro compounds in human beings is neurotoxicity. Nitro compound reaction with Vitamin C and/or catecholamine may cause this. Do you care to comment as to whether or not the reaction with these drugs is involved in the cytotoxic effects?

SRIDHAR: I think that is a very pertinent question. In the nervous tissue the catecholamines and ascorbate are present in considerable amounts. Nitro and nitroso compounds have the capacity to deplete the catecholamine content as well as the ascorbate content by speeding up the auto-oxidation. The oxidation products of catecholamines are toxic. Dehydroascorbate is not innocuous. Moreover, hypoxia results because of accelerated O_2 consumption. Further, it is known that the combination of ascorbate with misonidazole is more toxic to hypoxic cells than misonidazole alone. Nitroso derivatives, if produced, have the capacity to react with nerve lipids via ene-type addition. I believe that at least some of the reactions I have mentioned lead to neurotoxicity especially in the case of the more reactive nitro compounds. There is indirect evidence for catecholamine depletion by nitro compounds. Von Burgh and Conroy, and Inch and Haynes have shown lowering of heart rate and core temperature due to administration of misonidazole and metronidazole. This might be related to catecholamine depletion. Then there is a commercial rat poison (Vacor) which has a nitroaromatic structure and has been used by humans to commit suicide. These persons developed peripheral neuropathy, hypotension and hypothermia. This points to interaction with the cathecholamine.

COHEN: Has it been definitely established that the aromatic nitrocompounds deplete catecholamines from peripheral nerves of brain, or is that a supposition?

SRIDHAR: There is a paper by Le Witt in New England Journal of Medicine (1980) where there is reference to studies concerning the catecholamine levels in blood. I believe a drop in blood levels of catecholamines has been seen.

CAUGHEY: Can I bring up the role of superoxide once again? I would like to ask Dr. Biaglow or Dr. Raleigh whether or not there is any need of invoking the participation of superoxide in this? Oxygenated species can very easily react with the one-electron donors to form H_2O_2 directly without any formation of superoxide.

For example oxyhemoglobin reacts very rapidly with one-electron donors to form H_2O_2 directly, many orders of magnitude faster than the production of superoxide from oxyhemoglobin. Other oxygenated species are likely to behave similarly.

BIAGLOW: Yes, I would say that what you are saying is very important if you are talking about the aerobic toxicity of potent carcinogens such as four nitroquinoline-l-oxide (4NQO), which is reduced in air to 4 hydroxylamine-l-oxide (4HAQO) unlike many of the other nitro compounds. A possibility of two-electron transfer exists. We know that reduced menadione will react with O_2 to produce H_2O_2 without the intermediate $O_2^{\overline{\cdot}}$ radical species being formed at all.

CAUGHEY: In oxyhemoglobin a one-electron donor contributes one electron, the reduced Fe contributes the other electron and so you effect, as far as dioxygen is concerned, a two-electron reduction. The iron ends up as Fe(III) and the dioxygen as peroxide. So these need only be one-electron producing agents if they are reacting with an oxygen that is bound to something else, even a flavin.

BIAGLOW: I agree, but my mechanistic scheme has been based on measurements by Wardman and by Mason in purified systems of the rate of electron transfer from the nitro radical to O_2.

CAUGHEY: My question really was: has peroxide been directly quantitated in this system, has superoxide been quantitated in this system?

BIAGLOW: Peroxide (H_2O_2) has been quantitated, but $O_2^{\overline{\cdot}}$ (radical) has not.

WARDMAN: In the simple chemical experiment of electron transfer from the nitroaromatic radical-anion to O_2 (for metronidazole), I showed that the absorption spectrum of the product in the ultraviolet was identical to that of $O_2^{\overline{\cdot}}$. Sealy <u>et al</u>. have also used a spin trap to show $O_2^{\overline{\cdot}}$ was produced from the nitrofuran ion radical-anion.

BIAGLOW: With whole cells we are dealing with slightly different conditions. I think one of the more important differences that we have seen is the inhibitory effect of metronidazole on cellular respiration. However, if we add cyanide to cells we stimulate O_2 utilization with this drug. With metronidazole we could never demonstrate the formation of H_2O_2. Notedly, the more electron-affinic compounds produce H_2O_2.

WILLSON: Could we go back to the reaction of ascorbate for a moment? Is this metal catalyzed?

BIAGLOW: We studied the effect of EDTA and DTPA. EDTA inhibited O_2 utilization, but DTPA stimulated the O_2 consumption for the ascorbate nitro compound mixtures.

WILLSON: The reason I asked that question was that we have been doing some work with nitro compounds and sulfhydryls and looking at the way Fe catalyzes these reactions. Cysteamine will reduce some nitroimidazoles in the presence of a small amount of Fe. But interestingly, if you keep the amount of Fe the same and increase the cysteamine concentration you do not increase the rate indefinitely: you go through a maximum and the rate of reaction falls. This has been suggested as a possible mechanism for cysteamine protection against radiation damage: at high concentrations it chelates the metal strongly and stops it from entering into other reactions.

BIAGLOW: Ascorbate or cysteamine can donate one electron to produce the nitro radicals. However, the nitroso compound produced under anaerobic conditions is very reactive with thiols but not with ascorbate. Therefore, with thiols you may have a quenching effect.

FLOYD: I would like to update some information on the nitroso compound lipid adduct. We have now demonstrated that this occurs in vivo when one puts in the proximate carcinogens. So, I think that if indeed one is giving nitro compounds one has to begin to consider the fact that this is a real event that might occur in living systems. The second point is, it seems to be that Biaglow's system is very good for testing the toxicity of superoxide per se since he has a flux of superoxide, an increased flux, using the nitro compounds. Have you tested under conditions where you have inhibited the catalase system for increased killing of cells? It seems to me that this is a very beautiful system where you can begin to actually get some answers to this question.

BIAGLOW: We studied 4-nitro-quinoline-1-oxide cytotoxicity with cells in air. DMSO protected against cytotoxicity but catalase or superoxide dismutase together or alone had no effect on the cytotoxicity of that drug. I know that there have been some recent studies. Koch found no protection of aerobic misonidazole toxicity by catalase. There is one other important biochemical effect in air with nitro compounds: that is the inhibition of DNA synthesis that Dr. Olive has shown. How important this inhibition is for the cytotoxicity of nitro compounds in vitro is not known. But the point that I would like to make is that the nitro compounds are less toxic in air.

WHILLANS: I would like to ask Dr. Raleigh or perhaps Dr. Floyd about the lifetimes or stabilities of these nitroso intermediates. We have been interested in isolating reduction intermediates of misonidazole and in order to take advantage of the high dose rate possibility we have been using the formate/radiation reduction system. Certainly in this environment the nitroso compound is so easily reduced that we never see anything but the four-electron reduction product.

ASMUS: Maybe I could give an answer from the chemist's point of view. I think Dr. Raleigh wondered why you could not detect a nitroso compound whereas it was easy to detect species which were further down in the reduction sequence. Nitroso compounds really behave like oxygen as you said. Oxygen (O_2), we have been shown, reacts with a one-electron reduced nitro radical and so does the nitroso group. Nitrosobenzene, for example, readily reacts with the reduced nitrobenzene radical in an aqueous system with a rate constant of about 10^7. If you produce some nitroso compound this will react with reduced nitro radical and you have no chance to really accumulate the nitroso compound. The intermediate between nitroso and the phenyl hydroxylamine will be formed and even that has a chance to be reduced further in reaction with another reduced nitro radical.

WILLSON: Dr. Asmus, you have demonstrated this with nitrobenzenes, but do you think this would also work with nitroimidazoles?

ASMUS: There is nothing particular about nitrobenzene. It worked the same with substituted nitrobenzenes, so I would think, in principle, it would apply to any kind of nitro compound. The rate constants may differ, of course. This would have to be investigated.

RALEIGH: The point is not the reactivity of the nitroaromatics here, it is the question of whether nitrosoaromatics in a cell are ever released from the enzyme surface to react with biological molecules.

SRIDHAR: I think Dr. Asmus has a good point, and that reaction depends very much on the particular nitroso compound. If you chemically reduce nitrobenzene you can actually isolate the hydroxylamine but it is very difficult to isolate nitrosobenzene from nitrobenzene. Thus it is better to make nitroso compound via the hydroxylamine. I believe you can make the nitroso derivative of misonidazole by reduction. The temperature has to be kept low: the yield of nitroso derivative is low. Azo and azoxy derivatives of misonidazole are, of course, the major products, especially at room temperature.

ASMUS: Again, this is not surprising. If you go up from the phenyl hydroxylamine—there is again a very simple chemical way by adding the ·OH radical—eventually you get to nitrosobenzene. But you never get as far up as to the reduced nitrobenzene radical and that was the one which reduced the nitrosobenzene.

SAGONE: In regard to the activation of these nitroaromatic compounds, the pentose shunt pathway supplying NADPH is considerably distant from nitro reductase enzymes. Will a trans-hydrogenase type of reaction (NADH + HADPH) be involved in the activation? The radiosensitivity would vary tremendously depending on the shunt enzymes and the trans-hydrogenase activities of the cell line. For example, with polymorphonuclear lymphocytes there is no trans-hydrogenase activity.

$$NADPH + NAD^{+} \rightarrow NADH + NADP^{+}$$

Could you comment on that? Also H_2O_2 generated by your system would compete with the NADPH for reducing equivalents. Could you indicate the relative importance of the drug activation and H_2O_2 reduction pathways?

BIAGLOW: That's a very complicated question. I must point out that in air with some of the less electron-affinic compounds such as misonidazole there is no thiol loss, suggesting that there is no problem for the intracellular enzyme systems to reduce H_2O_2. Glycolysis provides NADH and, of course, trans-hydrogenase produces NADPH. The pentose cycle production of NADPH does seem to be adequate. However, problems may arise with low cell density or in the absence of glucose. Increased drug to cell ratio at low cell density may cause a rapid depletion of non-protein thiols.

WILLSON: Misonidazole reduces sulfhydryl levels. Is this temperature dependent? Is it glutathione that is reacting? We do see quite a strong temperature dependence on the rate of reaction of misonidazole with glutathione in the presence of a small amount of Fe.

BIAGLOW: Yes, there is no thiol loss at 0°C with misonidazole. The misonidazole-linked loss of thiols produces an intermediate that is not regenerated back to the free thiol. Presumably a nitroso intermediate has reacted. However, nitro radical destruction of NPSH may also occur. In contrast, when we perturb the redox system of the cell with an oxidant such as diamide, which oxidizes glutathione to GSSG, there is regeneration of the thiol via the pentose cycle.

WILLSON: Well, I think it is quite clear from what we have seen that sensitizers can react in a number of different ways. Dr. Wardman has shown how the one-electron oxidation potential plays a significant role in determining the activity of these sensitizers. Most of them are oxidizing and one, chlorpromazine, is nucleophilic. Most react like O_2 in some ways. But while radical reactions seem involved, surely biochemical reactions may also be important.

X

EXCITED STATE CHEMISTRY OF O_2

Session Chairman, Christopher S. Foote, Los Angeles, California

RECENT EXPERIMENTAL DEVELOPMENTS IN THE SINGLET OXYGEN ENE REACTION WITH OLEFINS

L. M. Stephenson, Hydrocarbon Research Institute and Department of Chemistry, University of Southern California, Los Angeles, California 90007

The interaction of visible light with colored organic molecules, particularly dyes, can lead to excited states which are capable of transferring energy to oxygen to produce singlet oxygen. This simple set of reaction conditions, combined with the high reactivity of the singlet oxygen, gives this system a great deal of importance in both biological and environmental settings, as well as the organic synthesis laboratory (1).

This paper, as well as the four other papers at this symposium series, will be concerned with the ene reaction, by far the most common and most important of the singlet oxygen reactions. See Equation (1).

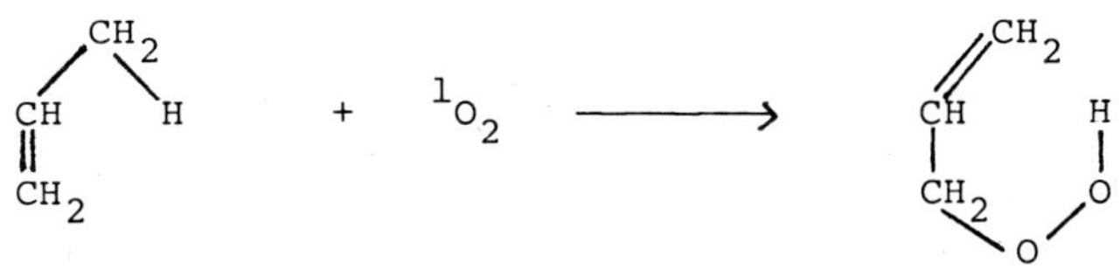

We will concentrate here on mechanistic considerations. This reaction has provided a significant test for reaction mechanism methods over the last twenty years, most importantly, providing the opportunity to test experimental and theoretical tools on a very fast chemical reaction.

At least four distinct mechanisms have been seriously discussed for this reaction. Shown in Fig. 1, these are commonly termed concerted, in which bond making and breaking steps are occurring together; diradical and/or dipolar, in which the C-O bond forms first; and perepoxide, a fully bonded intermediate, presumably detectable under some reaction conditions.

In recent years, dramatically new experimental insight into this reaction has emerged. We wish to review this and provide a comprehensive explanation for the data.

Singlet oxygen is a very electrophilic species. This is shown by the series:

ISBN 0-12-592050-4

propylene < 2-butene < trimethylethylene < tetramethylethylene

essentially no reaction $\Delta H^{\ddagger} \simeq 0$

Paquette (2) has noted a good correlation between olefin highest occupied molecular orbital (HOMO) level and reactivity.

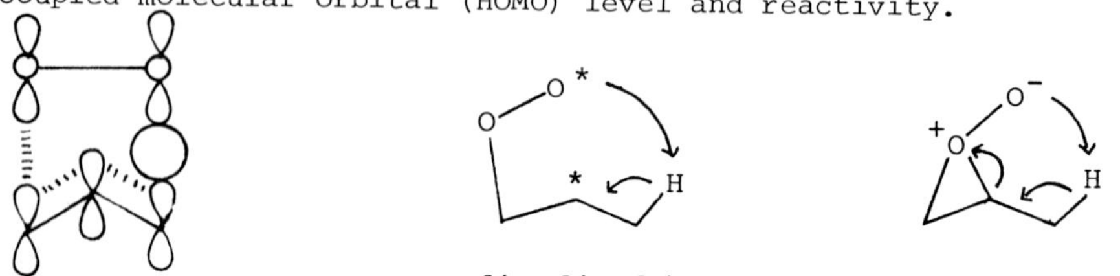

concerted diradical/dipolar perepoxide

FIGURE 1. Proposed mechanisms for singlet oxygen reactions.

Professor Gollnick (3) presents additional data in this volume supporting this idea.

The reactions of singlet oxygen are highly stereospecific. We have tested this aspect of the reaction on two systems (4). The optically active di- and trisubstituted olefins differ greatly in reactivity. The reactions of both are highly stereospecific, however. In Woodward-Hoffman language, both are completely suprafacial (see Equations 2 and 3).

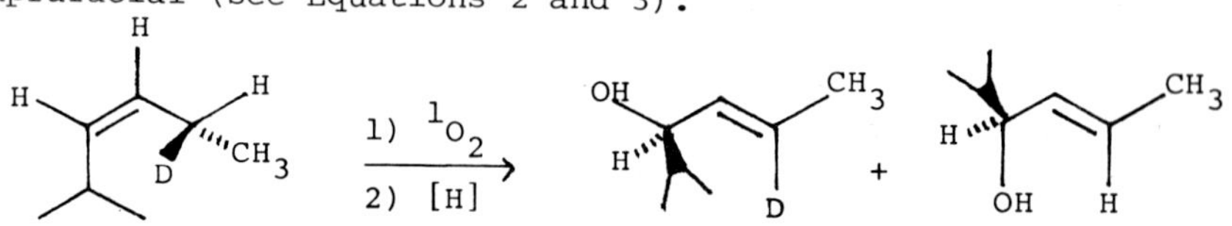

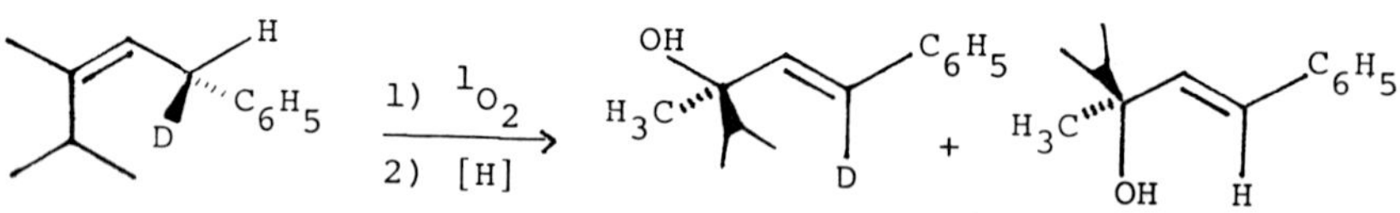

Curiously, these reactions showed no isotope effect, with $k_H/k_D \simeq 1.05$ in each case. Several other workers (5) had found effects in other systems to be larger, $k_H/k_D \simeq 1.4 - 1.8$. This discrepancy remained unresolved until 1979.

Isotope effects of a significant magnitude are only observed when the competing groups have cis relationships (6). This is most obvious for the two compounds shown in Equations (4) and (5). In the first entry, CH and DC bonds have both geminal and

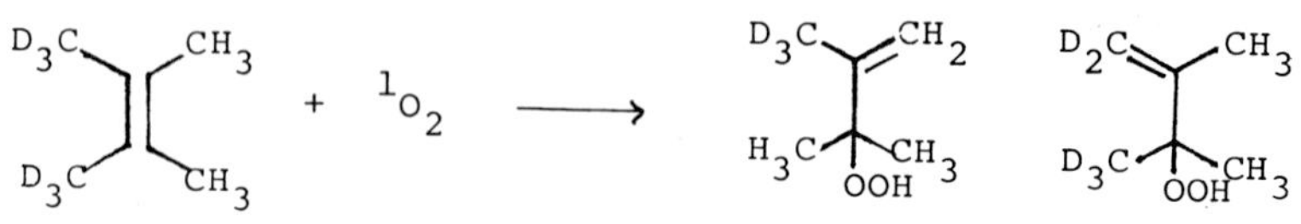

$k_H/k_D = 1.05$

trans relationships, but show no isotope effect, while in the second entry, the cis relationship of CH and DC units allows a competition.

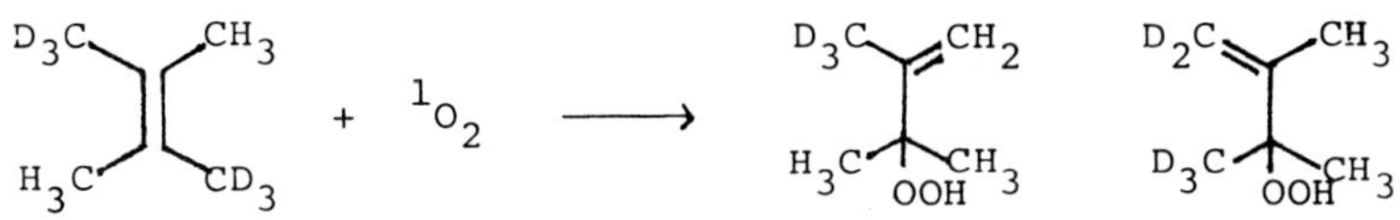

$k_H/k_D = 1.42$

A simple explanation of these latter two data invokes the perepoxide. Considering the two perepoxides which could be formed (in equal yields) from the tetramethylethylene with all six deuterium atoms on one side, it is clear that one structure could remove only H and one could remove only deuterium. Thus, no isotope effect is found (Fig. 2)

Such an intermediate also explains the positive effect, $k_H/k_D \simeq 1.42$, in the other d_6 tetramethylethylene stereoisomer. The stereochemical experiments which show the reaction to be completely suprafacial also support this picture.

Arguing against perepoxide intermediates are recent calculations from Goddard's group (7a) and from the groups of Houk (7b) and Yamaguchi (7c). Below we address an experimental point difficult to reconcile with perepoxides.

Singlet oxygen shows a strong preference for reaction on the crowded side of double bonds (Fig. 3). Data from Conia (8) and Foote (9) (Fig. 4), our work (10) (Fig. 5), and Rautenstrauch and Schulte-Elte (11) (Fig. 6), show that this effect is notable for R = phenyl, alkoxy, and alkyl. Cyclic systems show the same sensitivity. It is not obvious that a perepoxide would prefer to be formed on the more crowded side.

The singlet oxygen ene reaction displays a solvent sensitivity suitable for a reaction with significant charge separation in the transition state. New data from Professor Gollnick's laboratory, and given in his paper can be interpreted in terms of a transition state dipole moment of 5.5 D.

These facts and others have led us to propose (12) that charge transfer interactions dominate the chemistry of this process. In simplest frontier molecular orbital terms, we propose a donation of electron density from the HOMO of the olefin to the lowest occupied level (LUMO) of oxygen. To be comprehensive, this picture must take explicit account of orbitals on allylic hydrogens as well. The importance of interactions with the allylic hydrogens in a complex was first presented by Frimer and Bartlett (13). At this level of complexity the olefin HOMO has the form shown in Fig. 7, illustrated for 2-cis-2-butene. A suitable interaction between olefin HOMO and π^*-like O_2 LUMO then begins to account for all the data presented here (Fig. 8).

FIGURE 2. Possible structure of perepoxide intermediates.

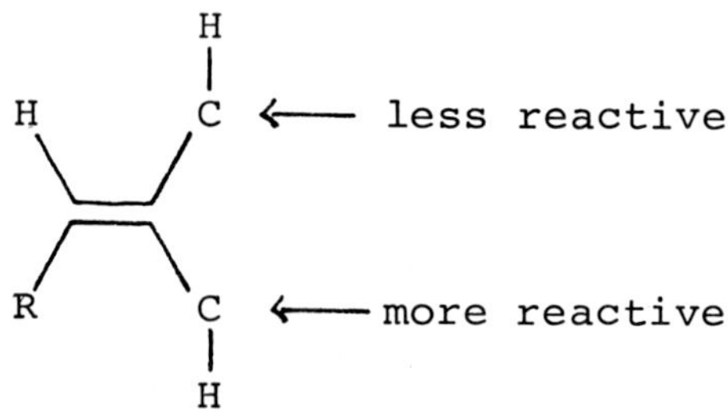

FIGURE 3. Conceivable sites for singlet oxygen reaction.

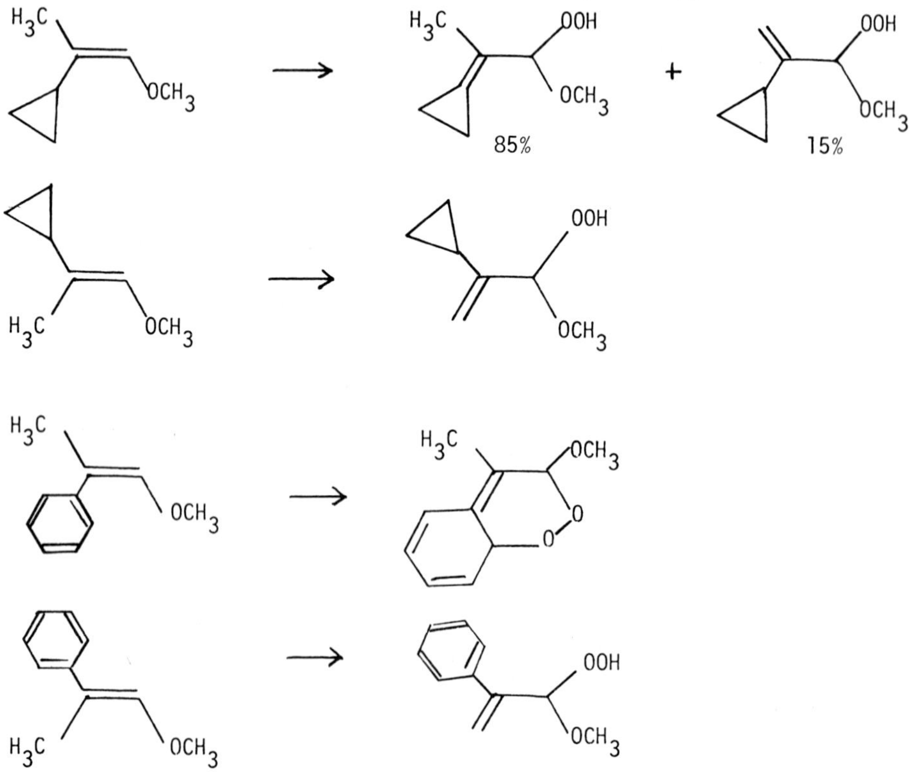

FIGURE 4. Methoxy group directing effects.

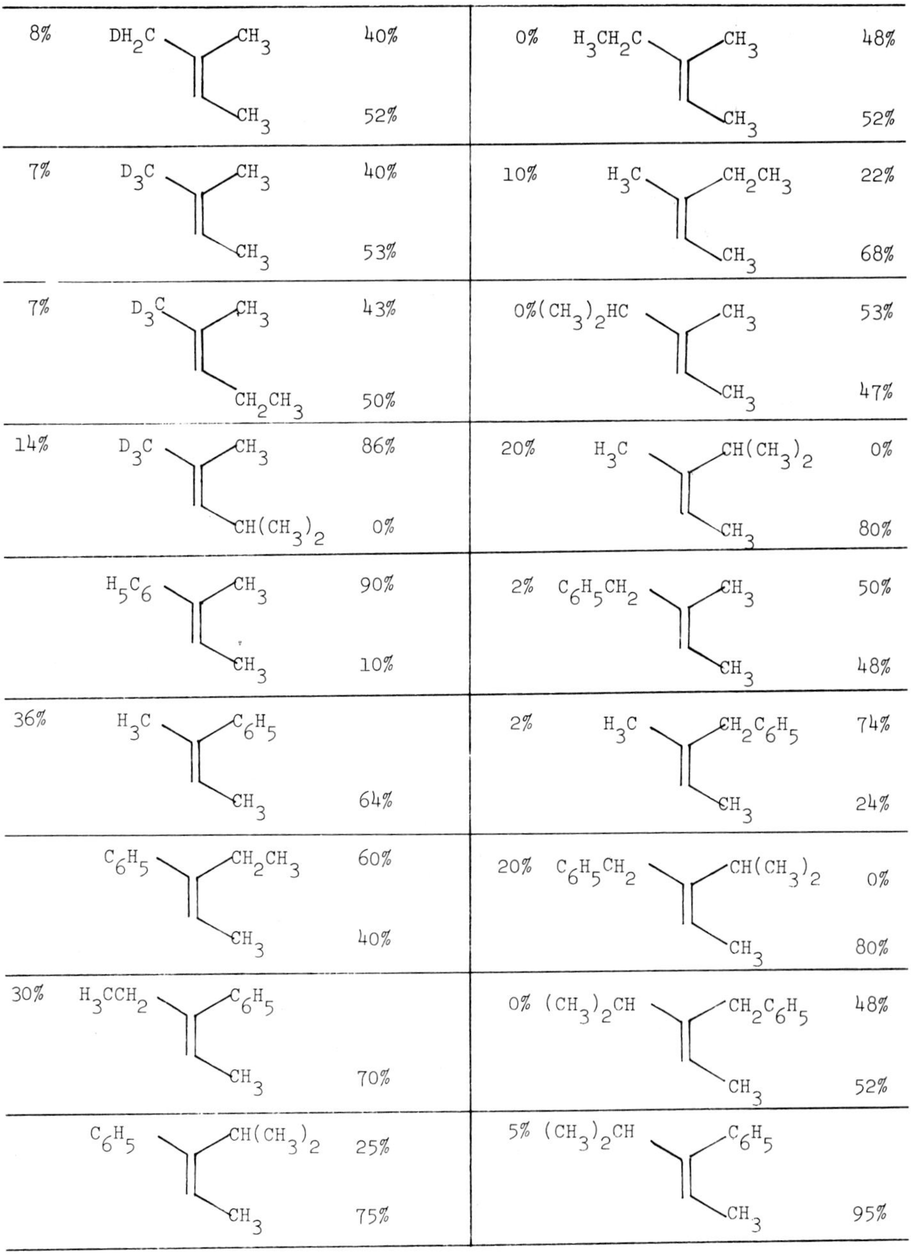

FIGURE 5. Regioselectivity in trisubstituted olefins.

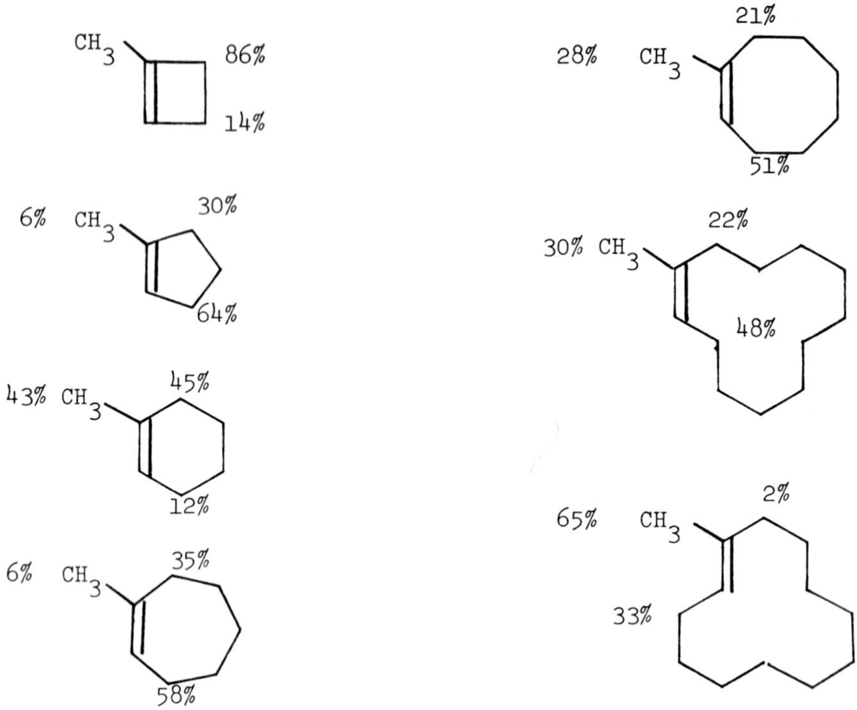

FIGURE 6. Regioselectivity in trisubstituted cycloolefins.

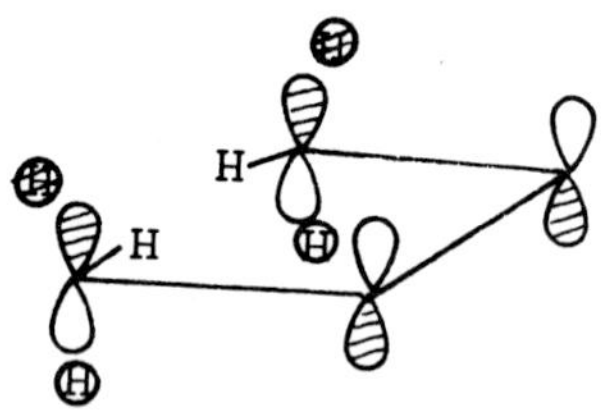

FIGURE 7. Structure of olefin HOMO.

In particular, the regiospecificity favoring crowded side attack, the exclusive cis effect in isotope competitions and the stereospecificity are all accounted for by the orientation provided by this model. Yamaguchi discusses this is greater detail (7c).

Speculations about the motions which follow complex formation are more uncertain at this point. We prefer to move from the initial orienting interactions smoothly in bond making and breaking processes. Except for the calculations of several theorists, we have no substantial basis for eliminating perepoxide

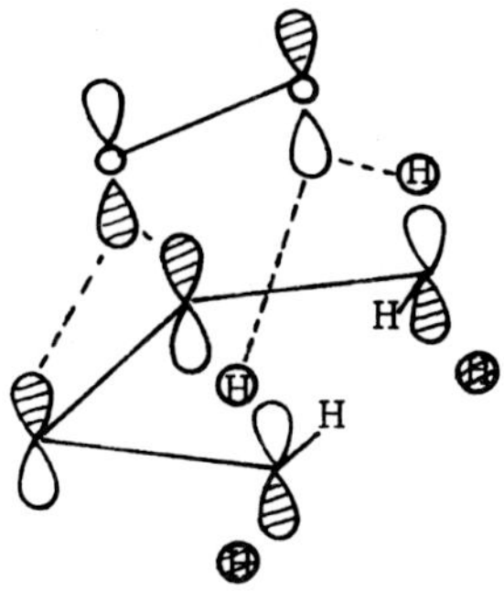

FIGURE 8. Interaction between olefin HOMO and O_2 LUMO.

intermediates, however. The final fine points of distinction will await the results of more sophisticated calculations and experiments before final resolution.

ACKNOWLEDGEMENT

This work was carried forward by four able students: David E. McClure, P. Kevin Sysak, Mary Jo Grdina, and Michael Orfanopoulos. Their experimental and intellectual contributions were significant. This work was supported primarily by the National Science Foundation over the last ten years.

REFERENCES

1. For a review of early mechanistic work, see C. S. Foote, Accts. of Chem. Res., 1,104 (1968). More recent reviews are: Aryeh A. Frimer, Chem. Rev., 79,359 (1979); L. M. Stephenson, M. J. Grdina, M. G. Orfanopoulos, Accts. Chem. Res.,(1980) In press.
2. Paquette, L. A., Liotta, D. C. and Baker, A. D., Tetrahedron Lett., 2681 (1976).
3. Gollnick, K., Hartmann, H. and Paur, H., this volume, page 379.
4. (a) Stephenson, L. M., McClure, D. E. and Sysak, P. K., J. Amer. Chem. Soc., 95,7888 (1973). (b) Orfanopoulos, M. G. and Stephenson, L. M., J. Amer. Chem. Soc., 102,1417 (1980).
5. (a) Kopecky, K. R. and Van de Sande, J. H., Can. J. Chem., 50,4034 (1972).
6. Grdina, M. B., Sr., Orfanopoulos, M. and Stephenson, L. M., J. Amer. Chem. Soc., 101,311 (1979).
7. (a) Harding, L. B. and Goddard, W. A., III, J. Amer. Chem. Soc., 99,4520 (1977); Tetrahedron Lett., 747 (1978). (b) Houk, K., work in press. (c) Yamaguchi, K., Jpn. Chem. Rev., 1,292 (1973); Yamaguchi, K., Fueno, T. and Fukutome, H., Chem. Phys. Lett., 22,466 (1973).

8. Rousseau, G., LePerchec, P. and Conia, J. M., Tetrahedron Lett., 2517 (1977).
9. Lerdal, D. and Foote, C. S., Tetrahedron Lett., 3227 (1978).
10. Orfanopoulos, M. G., Grdina, M. B., Sr., Stephenson, L. M., J. Amer. Chem. Soc., 101,275 (1979).
11. Schulte-Elte, K. H., Muller, B. L. and Rautenstrauch, V., Helv. Chim. Acta, 61,2777 (1978); Schulte-Elte, K. H., Muller, B. L. and Pamingle, H., Helv. Chim. Acta, 62,816 (1979); Schulte-Elte, K. H. and Rautenstrauch, V., J. Amer. Chem. Soc., in press.
12. Stephenson, L. M., Tetrahedron Lett., 1005 (1980).
13. Frimer, A. A., Bartlett, P. D., Boschung, A. F. and Jewett, J. G., J. Amer. Chem. Soc., 99,7977 (1977).

ON THE MECHANISM OF ENE-REACTIONS WITH SINGLET OXYGEN, $^1O_2$ ($^1\Delta_g$): DEPENDENCE OF RATES ON SOLVENT POLARITY AND STRUCTURE

Klaus Gollnick, Herta Hartmann and Hanns Paur, Institut für Organische Chemie der Universität München, Karlstrasse 23, D-8000, München 2, Germany

We have recently reported on the mechanistic details of the ene-reaction with singlet oxygen ($O_2(^1\Delta_g)$; or $^1O_2$) and on the value of this reaction as a synthetic tool in organic chemistry (1). Although there was no rigorous proof available that this reaction occurs in a concerted fashion, there was apparently no need to postulate the intermediacy of a perepoxide, open zwitterion or biradical in a two-step reaction. The overwhelming bulk of experimental results could be explained by assuming a concerted ene-reaction to occur, whereas the assumption of intermediates appeared to be incompatible with a number of experimental results.

Nevertheless, the question of whether the ene-reaction with $^1O_2$ proceeds in a one-step or two-step mechanism is still controversial as is the interpretation of the nature of the intermediate by those authors who favor perepoxides (2-5), 1,4-zwitterionic dipoles (6), or 1,4-biradicals (7) as intermediates in a two-step reaction.

We now wish to report on our results obtained with a series of acyclic and cyclic olefins which should not exert any steric inhibition toward singlet oxygen attack and thus should enable us to study electronic and conformational effects undisturbed by steric hindrance. In order to get some information on the nature of the transition state, the ene-reactions were carried out in a series of solvents with Sensitox^R, a polymer-bound rose bengal, as a photosensitizer for singlet oxygen generation (8). Finally, in order to determine the intermolecular primary kinetic isotope effect and thus to gather information on whether the allylic hydrogen shift to the terminus of the peroxy group occurs in the rate-determining step or else in a separate product-determining step, the competition between 2,3-dimethyl-2-butene and 2,3-dimethyl-2-butene-d_{12} for singlet oxygen was investigated (9).

EXPERIMENTAL RESULTS

Experimental details will be published elsewhere (8,9). Table 1 shows the products and product distributions as they were obtained for tetra-, tri- and di-alkyl substituted acyclic and cyclic olefins in methanol. 2-Methyl-2-butene forms a tertiary hydroperoxide (53%) and, of course, only one secondary hydroperoxide (47%); using 2-trideuteromethyl-cis-2-butene Stephenson (5)

ISBN 0-12-592050-4

TABLE 1. Products and Product Distributions of Sensitox[R]-Photosensitized Oxygenation of Olefins in Methanol at 13°C

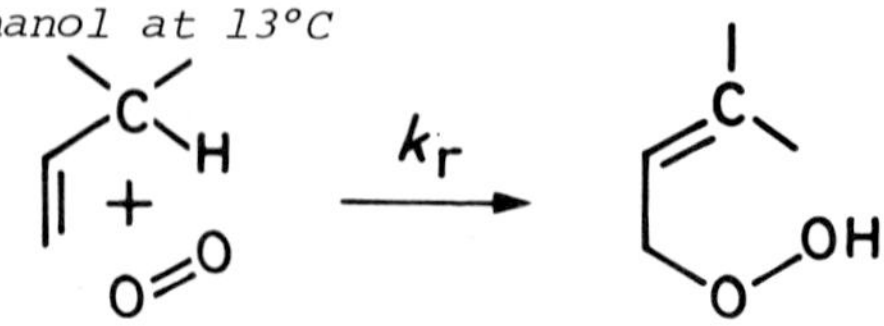

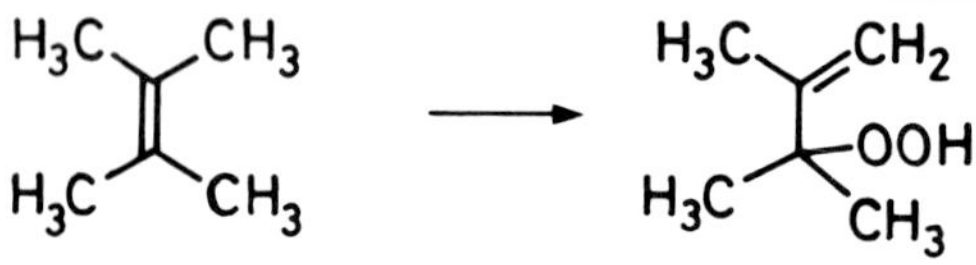

IP = 8.30 eV

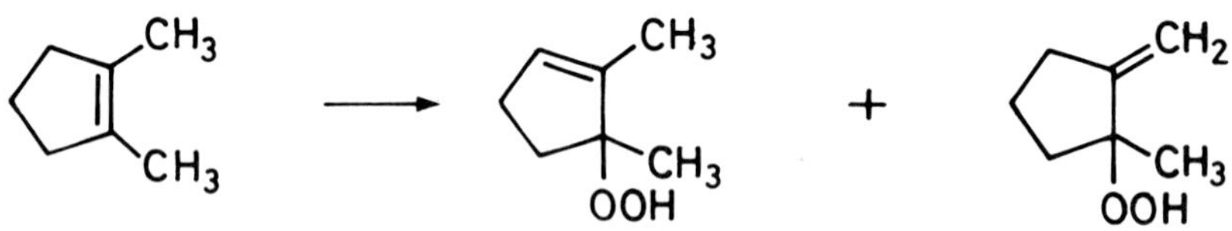

IP = 8.20 eV — 41% + 59%

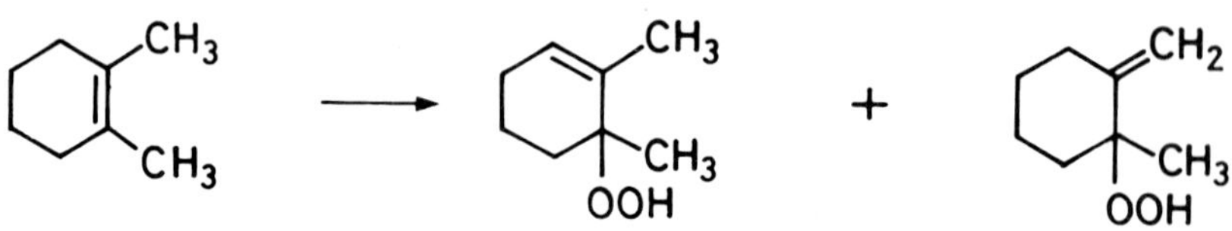

IP = 8.20 eV — 10% + 90%

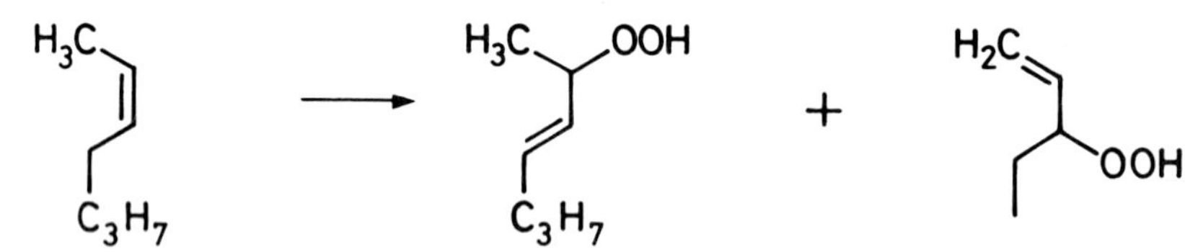

IP = 9.16 eV — 54% + 46%

IP = 9.12 eV

IP = 8.95 eV

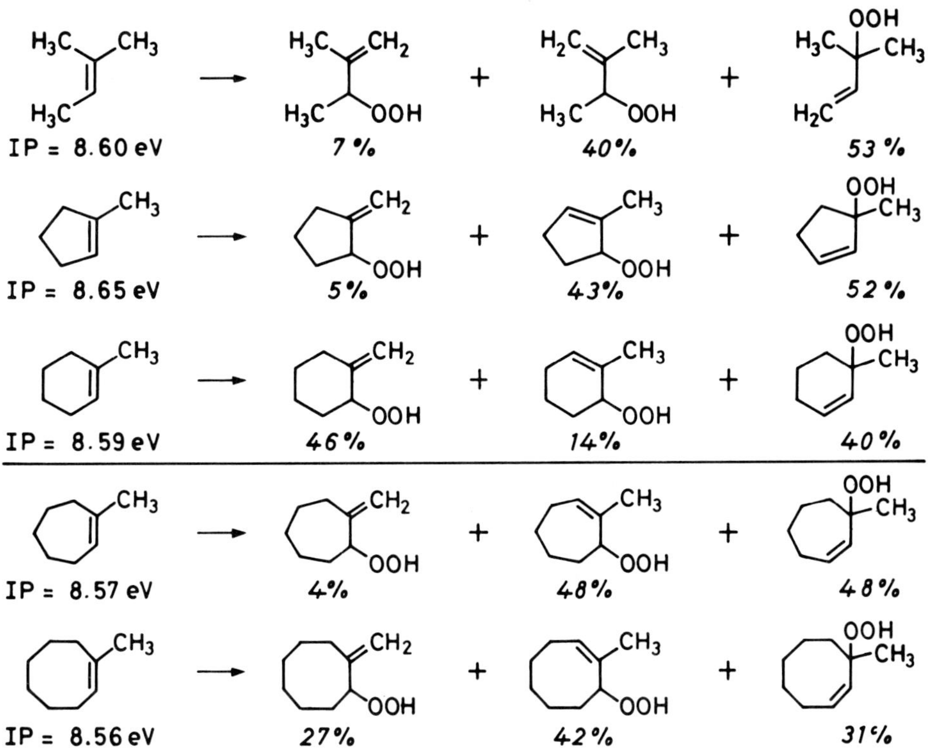

TABLE I. (Continued)

could discriminate between the cis-1-methyl and the trans-2-trideuteromethyl group as hydrogen sources in the ene-reaction with singlet oxygen. Results given for 1-methylcycloheptene and 1-methylcyclooctene were reported by Foote (10). Reactions in acetonitrile, acetone, methylene chloride, pyridine, chloroform, benzene, carbon disulfide, and carbon tetrachloride produced virtually the same results, i.e., no other products than those shown in Table 1 were observed and their percentages do not differ by more than ± 1%.

Figure 1 demonstrates the dependence of the over-all rate, k_r, on the solvent polarity. Plotting log k_r vs $(\varepsilon-1)/(2\varepsilon+1)$, linear correlations are observed as is shown for singlet oxygen reactions with 2,3-dimethyl-2-butene and cyclohexene, the most reactive and the least reactive olefin of the series, respectively. Table 2 contains the over-all rate constants for methanolic solutions, the slopes $\Delta \log k_r/\Delta[(\varepsilon-1)/(2\varepsilon+1)]$, the dipole moments μ (11) and the radii r determined from the mole-volumina assuming the olefins to be spheres. The dipole moments $\mu_{\ddagger}$ of the transition states were calculated by applying equ. (1) (Discussion

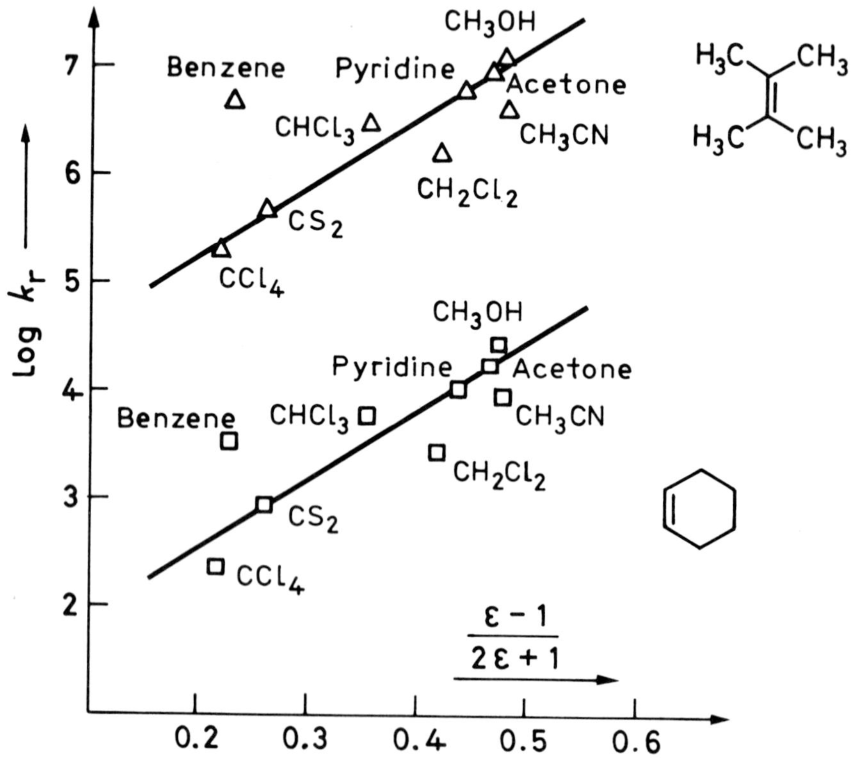

FIGURE 1. Dependence of rate constants k_r of ene-reactions with singlet oxygen on solvent polarity.

TABLE 2. *Dipole Moments $\mu_{\ddagger}$ of the Transition States of Ene-Reactions with Singlet Oxygen*

Olefin	k_r^{MeOH} (L mol^{-1} sec^{-1})	$\dfrac{\Delta \log k_r}{\Delta\left(\dfrac{\varepsilon-1}{2\varepsilon+1}\right)}$	μ(D)	r(Å)	$\mu_{\ddagger}$(D)
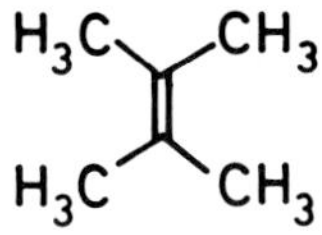	$1.30 \cdot 10^7$	5.75	0	3.61	5.46
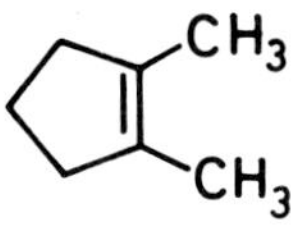	$9.52 \cdot 10^6$	5.64	~0	3.63	5.45
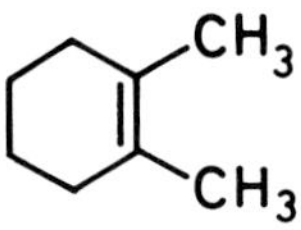	$3.86 \cdot 10^6$	5.95	~0	3.75	5.84
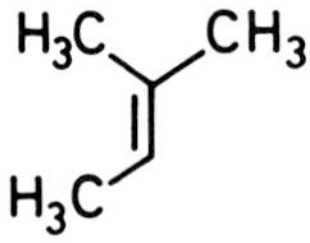	$1.10 \cdot 10^6$	6.18	0.92	3.48	5.40
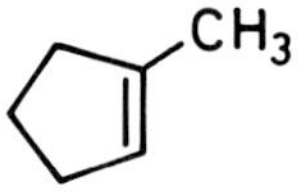	$9.22 \cdot 10^5$	7.19	0.95	3.47	5.92
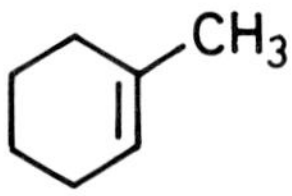	$2.02 \cdot 10^5$	6.47	0.95	3.60	5.88
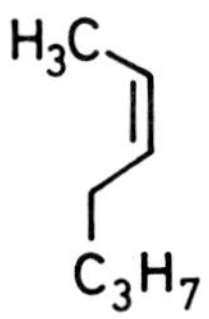	$1.32 \cdot 10^5$	5.50	0.60	3.80	5.70
	$2.21 \cdot 10^5$	7.30	0.93	3.27	5.55
	$2.67 \cdot 10^4$	6.11	0.76	3.42	5.37

Section) with $\mu(^1O_2) = 0$, $r(^1O_2) = 2.23$ A, and $r_{\ddagger} \approx r(\text{olefin}) + (0.23 \pm 0.02)$ A (estimated from the mole-volumina of the transition state which is equal to the sum of the mole-volumina of the reactants minus a double-bond increment).

2,3-Dimethyl-2-butene-d_{12} was prepared from acetone-d_{12} (99.5% D) (9). It contained 0.17% of 2,3-dimethyl-2-butene-d_{11} (^{1}H-NMR: = 1,12 ppm, quintet, J_{H-D} = 2 cps). ^{13}C-NMR: perdeutero compound, 123.3 ppm (olefinic C), 19.4 ppm (C of CD_3-group); perprotio analogue, 123.4 ppm (olefinic C), 20.3 ppm (C of CH_3-group). Photoelectron spectra: lowest IP (perdeutero compound) 8.35 ± 0.03; lowest IP (perprotio analogue): 8.36 ± 0.03 eV. Intermolecular primary kinetic isotope effect, k_H/k_D = 1.11, determined from competition reactions of 1:1-mixtures for singlet oxygen in acetone at 13°C (9).

DISCUSSION

In cases where olefins form more than one allylic hydroperoxide, the product distributions remain completely unaffected by the polarity of the solvents (Table 1). However, the over-all rate constants k_r differ by about two orders of magnitude when solvents of high polarity are compared with those of low polarity (Fig. 1). With the exception of benzene, the logarithms of k_r correlate very well with the solvent polarity, expressed as $(\varepsilon-1)/(2\varepsilon+1)$, and the slopes of these correlations are almost parallel for all the olefins (Table 2). If, however, log k_r is plotted vs. the empirical E_T-solvent parameters (12), less satisfactory linear correlations are observed, with remarkable deviations not only for benzene but especially for methanolic solutions for which the experimental k_r-values are about one order of magnitude lower than they should be according to such correlations. The ene-reactions with singlet oxygen appear to be exceptions to the rule that the logarithms of rate constants correlate much better with empirical solvent parameters than with functions of the solvents' dielectric constant ε. From Fig. 1, however, it seems to be justified to apply the Kirkwood-Laidler-Eyring model (13-15) which considers electrostatic interactions between neutral, but dipolar reactants as the main contributions toward a change of k_r when the reactants are transferred from a medium of dielectric constant equal to unity to a solvent of dielectric constant ε:

(1)
$$\ln k_r = \ln k_o - \frac{N_A}{RT} \cdot \frac{\Sigma - 1}{2\Sigma + 1} \cdot \left\{ \left(\frac{\mu^2}{r^3}\right)_{\text{olefin}} + \left(\frac{\mu^2}{r^3}\right)_{^1O_2} - \frac{\mu_{\ddagger}^2}{r_{\ddagger}^3} \right\}$$

With this equation the dipole moments $\mu_{\ddagger}$ of the transition states could be determined. They are all very similar (Table 2), the mean value being $\mu_{\ddagger}$ = (5.6 ± 0.2) Debye.

From equ. (2) (15)

(2)

$$\Delta S^{\ddagger}_{el} = - \left\{ \left(\frac{\mu^2}{r^3}\right)_{olefin} + \left(\frac{\mu^2}{r^3}\right)_{^1O_2} - \frac{\mu^2_{\ddagger}}{r^3_{\ddagger}} \right\} \cdot \frac{3\Sigma}{(2\Sigma + 1)^2} \cdot \frac{\partial \ln\Sigma}{\partial T}$$

the electrostatic contribution to the entropy of activation is calculated to be -0.6 e.u. for the methanolic solution reaction and -1.2 e.u. for the reaction in carbon tetrachloride, indicating that the electrostatic contribution to $\Delta S^{\ddagger}$ is negligible. This result agrees very well with the recent estimation of $\Delta S^{\ddagger}$ to be about -26 e.u. for methanolic solutions (1) and -23 e.u. for gas phase ene-reactions with singlet oxygen (16). Thus, it is the enthalpy of activation, $\Delta H^{\ddagger}$, which is decreased with increasing solvent polarity. Again, this result agrees very well with the observation that the activation energies for ene-reactions of olefins of Table 1 with $^1O_2$ differ by about 3 kcal/mole if reactions in methanol are compared with those in the gas phase (1,16).

Recently, Harding and Goddard (7) have calculated dipole moments of dioxetanes, 1,4-biradicals, perepoxides and 1,4-zwitterions to be 3.63 D, 2.36 D, 5.59 D, and 10.14 D, respectively (Fig. 2). The experimental value of $\mu_{\ddagger}$ = 5.6 D may therefore be taken as evidence for a perepoxide-like transition state in ene-reactions with singlet oxygen. At least for simple, alkyl-substituted olefins dioxetane-like, biradical-like and zwitterion-like transition states appear to be excluded.

The ene-reactions of olefins with singlet oxygen are associated with activation energies of acout 0.5 to 5.5 kcal/mol (17)

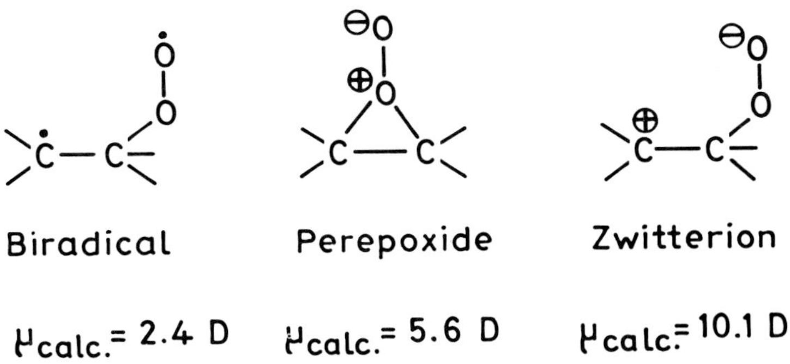

FIGURE 2. Calculated dipole moments of possible intermediates (7).

and activation entropies of about -26 e.u., resulting in free enthalpies of activation between 8 and 13 kcal/mol at room temperature. Since the ene-reactions are exothermic by about 45 kcal/mol (1), a rather early, reactant-like transition state is passed when the reaction proceeds along the reaction coordinate. In this case the PMO theory (18,19) is applicable and may be used to correlate the structure of the reactant olefins with their reactivities in ene-reactions with singlet oxygen.

As Fig. 3 shows, the interaction of the frontier orbitals of propene (as a model olefin) and $^1O_2$ suggests a perepoxide-like transition state. Taking into account that electrophilic singlet oxygen reacts with a nucleophilic olefin, the main interaction of the frontier orbitals is that between HOMO (olefin) and LUMO ($^1O_2$), which leads to the following correlation:

$$\log k_r \sim (E_{HOMO}\ (\text{olefin}) - E_{LUMO}\ (^1O_2))^{-1} \qquad (3)$$

or

$$\log k_r \sim (IP\ (\text{olefin}) - EA\ (^1O_2))^{-1} \qquad (4)$$

with IP (olefins) = lowest ionization potentials of the olefins (1,20), and EA ($^1O_2$) = electron affinity of singlet oxygen,

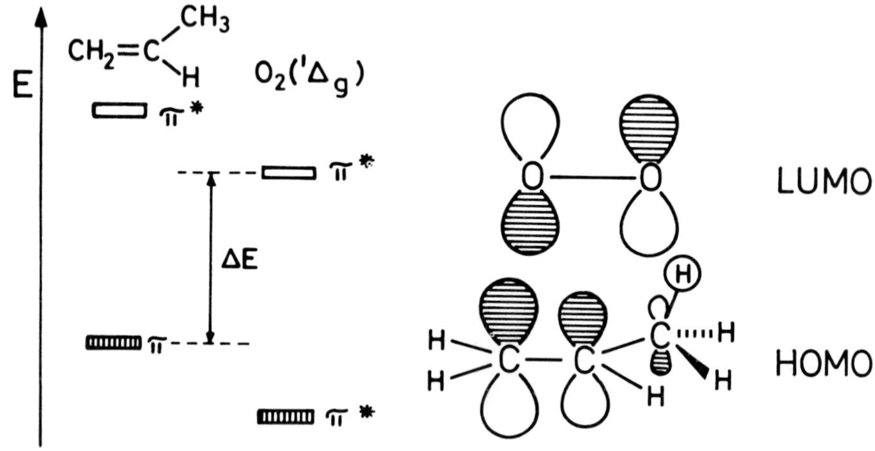

$$\log k_r \sim \frac{1}{E_{HOMO(\text{olefin})} - E_{LUMO}(^1O_2)}$$

$$\log k_r \sim \frac{1}{IP(\text{Olefin}) - EA(^1O_2)}$$

FIGURE 3. Frontier orbital interactions in the ene-reaction of propene with singlet oxygen.

calculated to be 1.8 eV from EA ($^3O_2(^3\Sigma_g)$) = 0.8 eV (21) + excitation energy of singlet oxygen (= 1 eV).

Applying equ. (4) to the over-all rate constants obtained from reactions in methanolic solutions results in linear correlations as shown in Fig. 4. Similar correlations are observed if the rate constants obtained for the other solutions are used. Fig. 4 suggests that there are two different groups of olefins with respect to their reactivity in ene-reactions with singlet oxygen. However, in order to explain reactivity differences, it is more appropriate to correlate the logarithms of the partial rate constants, k_r^{part}, with the reciprocal of the HOMO-LUMO energy differences. The partial rate constants were calculated with the aid of Tables 1 and 2 and are given in Table 3 at the positions of the allylic hydrogens which are used in the respective ene-reaction. For example, with 1,2-dimethylcyclopentene the ene-reaction associated with the use of the endo-cyclic allylic

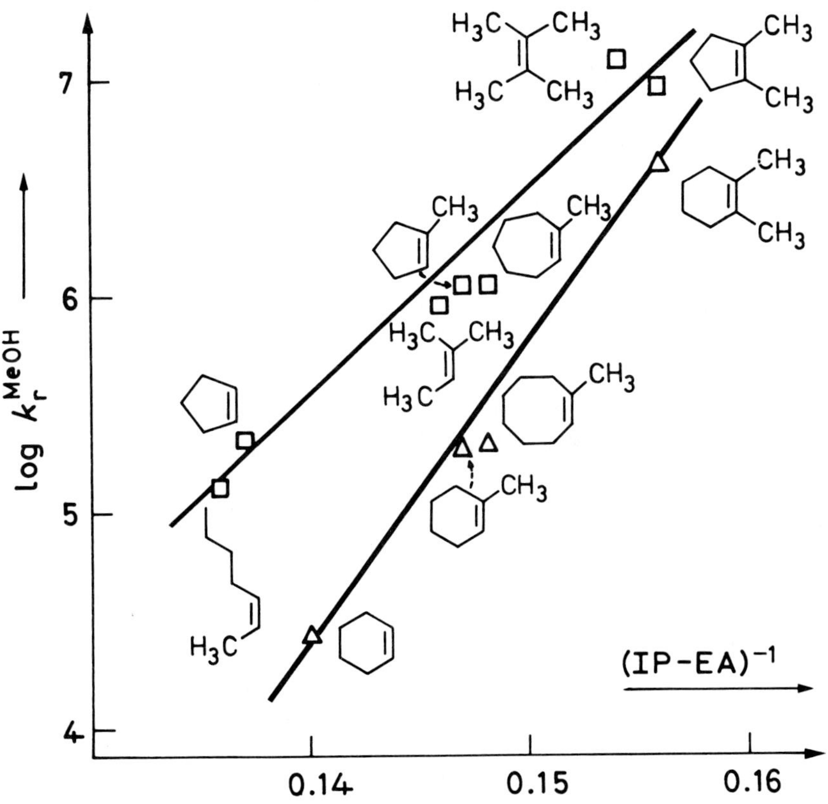

FIGURE 4. Dependence of the overall rate constants k_r of ene-reactions with singlet oxygen on the HOMO energies of the olefins.

TABLE 3. Partial Rate Constants k_r^{part} of the Ene-Reactions with Singlet Oxygen

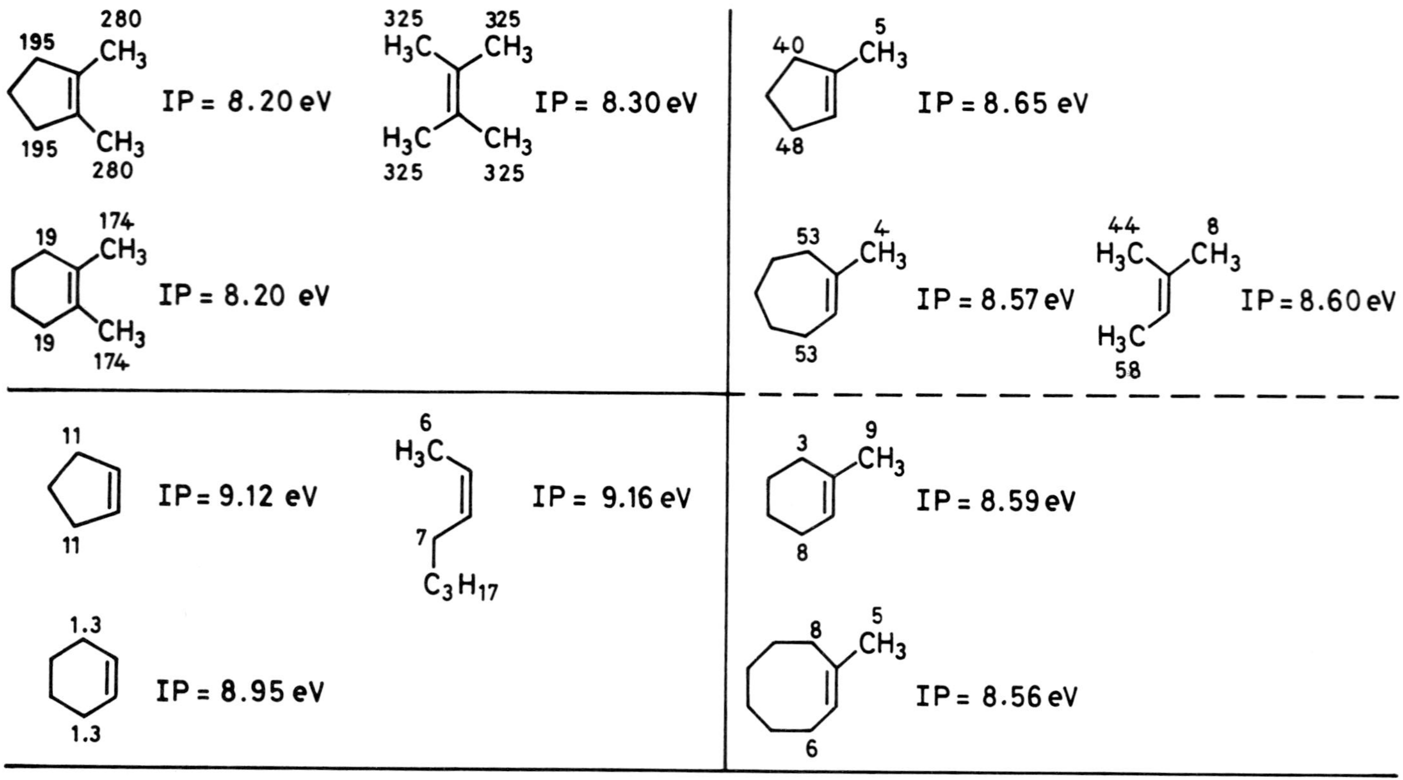

Partial Rate Constants, k_r^{part} (MeOH), in 10^4 L mol^{-1} sec^{-1}-units

hydrogens at one allylic position occurs with $k_r^{part} = 195 \cdot 10^4$ M^{-1} sec^{-1}, the ene-reaction associated with the use of the exocyclic allylic hydrogens of one CH_3 group occurs with $k_r^{part} = 280 \cdot 10^4$ M^{-1} sec^{-1}.

Inspection of Table 3 shows: (1) the allylic hydrogens of the methyl groups of the trisubstituted cyclo-olefins and the endocyclic allylic hydrogens of 1-methylcyclohexene and 1-methylcyclooctene are equivalent; (2) the endocyclic allylic hydrogens of cyclopentenes and cycloheptenes are about ten times as reactive as the endocyclic allylic hydrogens of cyclohexenes and cyclooctenes; (3) if two methyl groups are cis-oriented to each other, the reactivity of the allylic hydrogens is enhanced by a factor of about ten (compare the partial rate constants of 2-methyl-2-butene, compare the partial rate constants of 1-methylcyclohexene with those of 1,2-dimethyl-cyclohexene, of 1-methylcyclopentene with those of 1,2-dimethylcyclopentene, etc.). For acyclic trisubstituted olefins this effect was recently discovered by Schulte-Elte (22,23) and by Stephenson (5).

Plotting log k_r^{part} vs. (IP (olefin) - EA $(^1O_2))^{-1}$, results in two straight correlation lines, almost parallel and about one order of magnitude apart, which emphasize the fact that there are two sorts of allylic hydrogens (Fig. 5).

Before we turn to Figures 6 and 7 which may answer the question why we are dealing with two sorts of allylic hydrogens, we should recall the fact that those allylic hydrogens which can be arranged coplanar (or almost coplanar) with the π-bond are preferentially abstracted during the ene-reactions with singlet oxygen (1,24). In cyclohexenes, the two cis-oriented endocyclic allylic C-H bonds on one face of the molecule cannot be arranged in such a way that the two C-H bonds are disposed in a syn-parallel (or almost syn-parallel) manner, since if one of these C-H bonds is quasi-axial (at C-3, Fig. 6), the other is quasi-equatorial (at C-6, Fig. 6). These two allylic hydrogens do not "support" each other and the reactivity of the quasi-axial allylic hydrogen may be considered as "normal". If there is no second cis-allylic C-H bond present in an olefin as is the case for the methyl groups of the 1-methylcycloalkenes of Table 3, the reactivity of these allylic hydrogens is, of course, "normal." The partial rate constants associated with product formation in which such allylic hydrogens are involved are correlated by the lower curve of Figure 5.

In acyclic olefins, having two cis-oriented CH_3-groups or CH_2R-groups (5,22,23), two cis-allylic C-H bonds may be disposed in a syn-parallel manner to each other (Fig. 7). In cyclopentenes, the two cis-oriented allylic C-H bonds on one face of the molecule are disposed in an almost syn-parallel manner (Fig. 7), and it is obviously this syn-parallel disposition of the two cis-allylic C-H bonds which enhances the reactivity of the otherwise "normal" allylic hydrogens (see upper curve if Fig. 5).

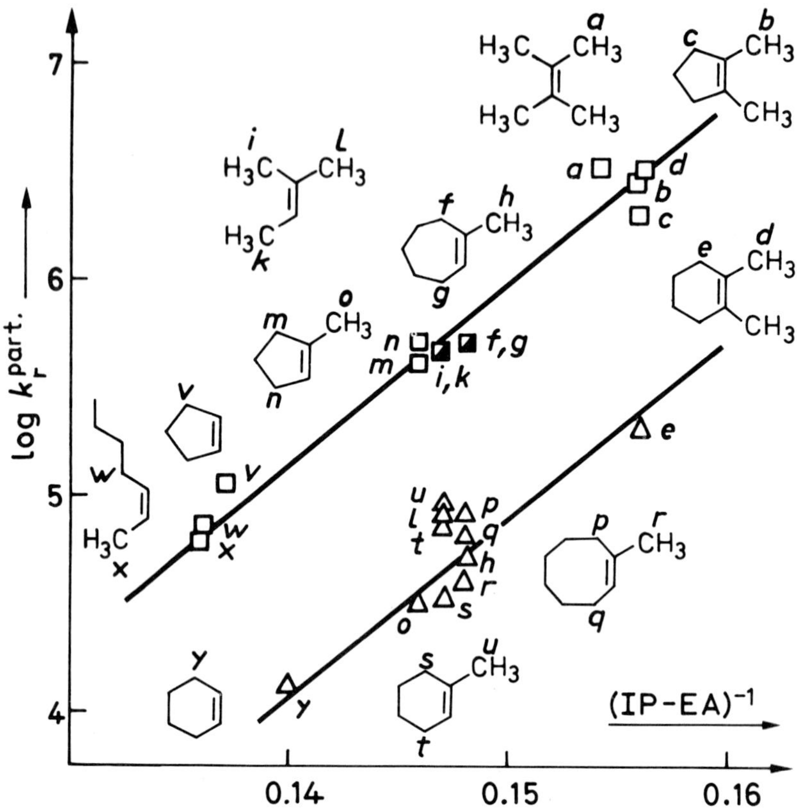

FIGURE 5. Dependence of the partial rate constants k_r^{part} of the ene-reactions with singlet oxygen on the HOMO energies of the olefins.

Fukui's frontier orbital theory (19) is again useful in describing the "co-operative," rate-enhancing effect of the two cis-allylic C-H bonds that can be arranged in a syn-parallel fashion. Figure 8 shows the HOMO-LUMO interactions for the ene-reaction of $^1O_2$ with 2-methyl-2-butene on that side of the molecule which carries two methyl groups. Singlet oxygen approaches the olefin in such a way that a perepoxide-like transition state is passed. In both HOMO-LUMO combinations, the orbitals of the two cis-allylic syn-parallel hydrogens interact with the orbitals of the oxygen atom that is not involved in C-O bond formation. The orbital coefficients, calculated by using the MNDO method (25), indicate that the product ratio due to $^1O_2$ reactions which involve these two methyl groups should be about unity. Obviously, on the other side of the olefin only one allylic hydrogen can participate in such HOMO-LUMO interactions which leads to a transition state of increased energy as compared to a transition

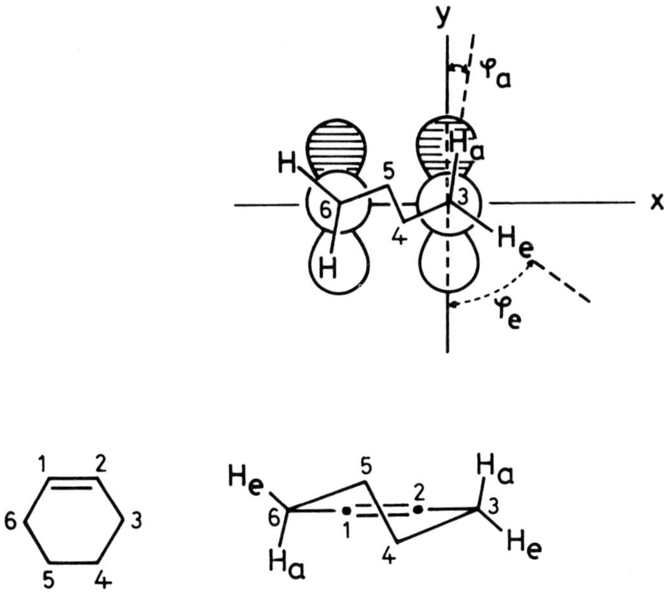

FIGURE 6. Disposition of allylic hydrogens in cyclohexene.

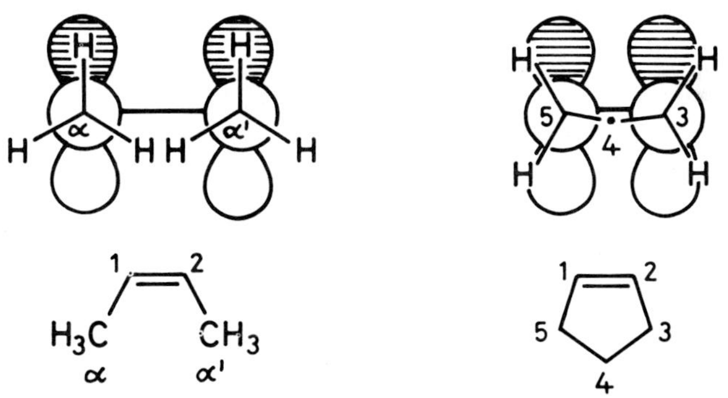

FIGURE 7. Disposition of allylic hydrogens in cis-2-butene and cyclopentene.

state in which two cis-allylic syn-parallel hydrogens take part.

The question now arises how the reaction proceeds further after the perepoxide-like transition state is passed. Does it proceed directly to products (case I, Fig. 9) or is an intermediate (perhaps a perepoxide) involved?

With 2,3-dimethyl-2-butene and 2,3-dimethyl-2-butene-d_{12} we determined $k_H/k_D = 1.11 \pm 0.02$ in acetone at 13°C by following the disappearance of the two compounds in competition reactions

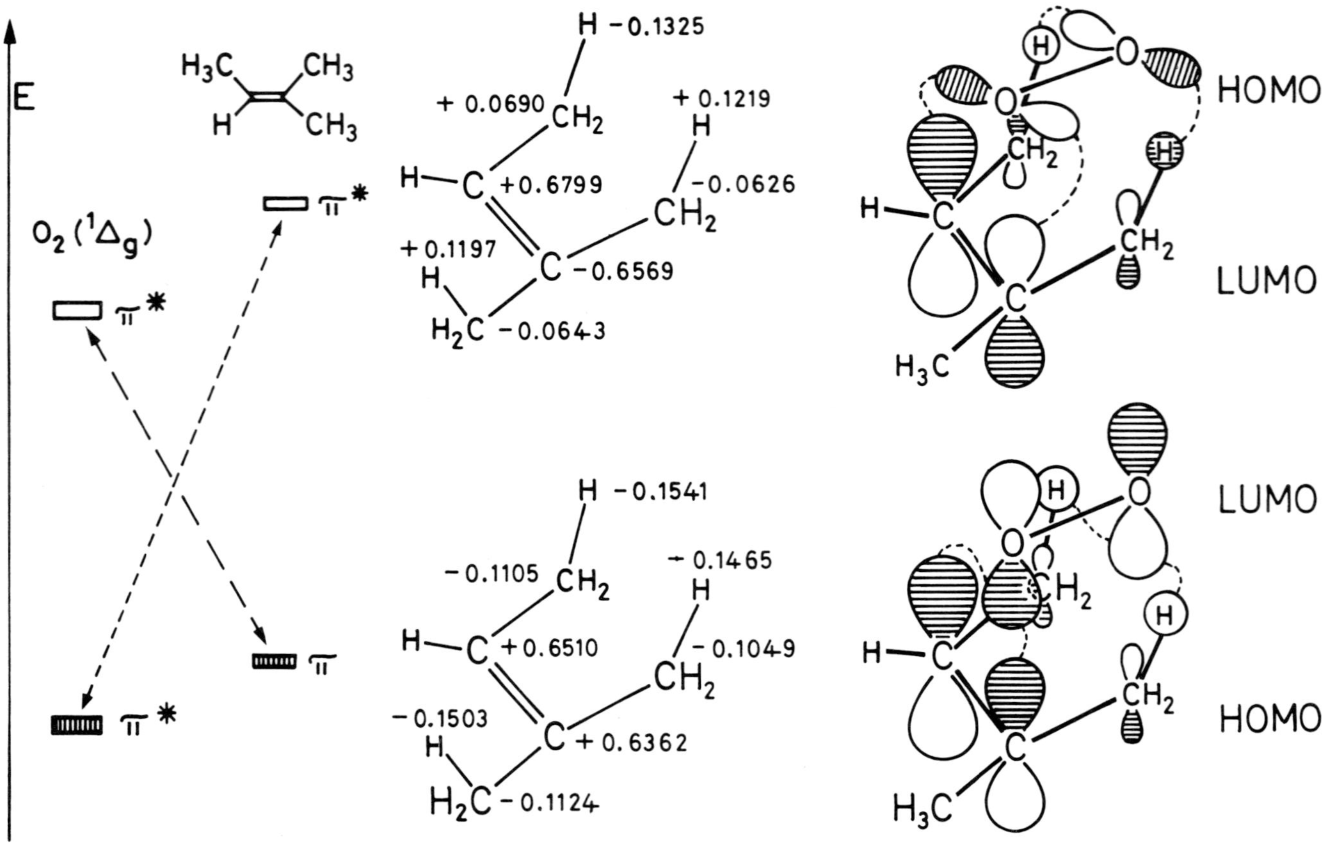

FIGURE 8. Frontier orbital interactions in the ene-reaction of 2-methyl-2-butene with singlet oxygen.

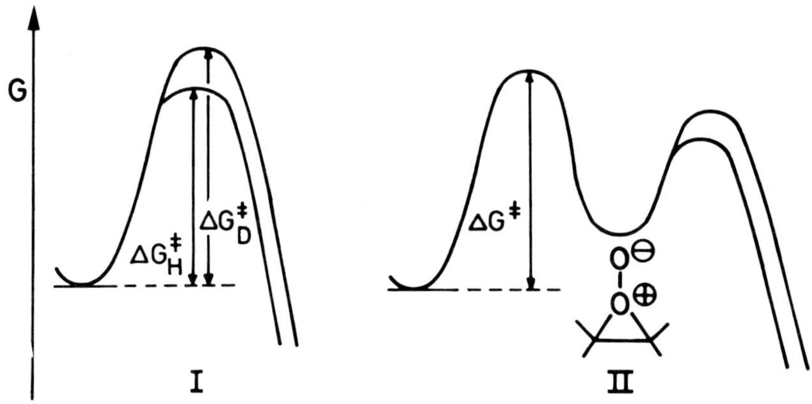

Intramolecular primary kinetic isotope effect:

no destinction between I and II possible

Intramolecular primary kinetic isotope effect:

if $k_H/k_D > 1$: case I = concerted

if $k_H/k_D = 1.00$: case II = two-step

With $(H_3C)_2C = C(CH_3)_2$ / $(D_3C)_2C = C(CD_3)_2$

IP = (8.36 ± 0.03) eV IP = (8.35 ± 0.03) eV

in MeOH: $k_H/k_D = 1.12$

FIGURE 9. Energy profiles for a one-step and a two-step mechanism of ene-reactions with singlet oxygen.

gas chromatographically. Since the ^{13}C-NMR signals of the olefinic C atoms of the two compounds do not differ by more than 0.1 ppm and since the IP's of the two compounds are virtually the same (26), we conclude that the observed effect is due to the intermolecular primary kinetic isotope effect rather than to the secondary kinetic isotope effect.

There are, unfortunately, several mechanisms compatible with this result. If we disregard the possibility that in a two-step mechanism the first step may be reversible, k_H/k_D should be 1.00

if the formation of an intermediate (e.g. a perepoxide, case II, Fig. 9) is rate-determining. Since k_H/k_D is small but clearly greater than unity, a one-step mechanism (case I, Fig. 9) may be indicated, in agreement with the fact that the ene-reaction with singlet oxygen proceeds via a rather early transition state. In this transition state hydrogen abstraction would be involved although only to a small extent. This result would agree very well with the observation of a very small activation energy for the singlet oxygen reaction with 2,3-dimethyl-2-butene (about 0.5 kcal/mol (17).

However, if we allow the first step in a two-step mechanism to be reversible, $k_H/k_D > 1$ may very well measure the energy difference between the second transition states passed during the ene-reactions of singlet oxygen with 2,3-dimethyl-2-butene and its perdeutero analogue, respectively (case II, Fig. 9, with $\Delta G^{\ddagger}$ equal for both reactions but the second transition state being higher in energy for the perdeutero compound compared with the second transition state for the perprotio compound). To the best of our knowledge, there is no experimental evidence for or against a reversible intermediate production available.

For simple alkyl-substituted olefins such as were investigated here, there appears to be no need for the assumption that the ene-reaction with singlet oxygen occurs in a two-step mechanism. We still favor, therefore, a one-step mechanism.

ACKNOWLEDGEMENT

We would like to express our gratitude to the Fonds der Chemischen Industrie, Frankfurt am Main, Germany, for the continuous support of our work.

REFERENCES

1. Gollnick, K. and Kuhn, H. J., "Singlet Oxygen" (H. H. Wasserman and R. W. Murray, eds.), Academic Press, New York, 1979, p. 287.
2. Kearns, D. R., Fenical, W. and Radlick, P., Ann. N.Y. Acad. Sci. 171,34 (1970).
3. Dewar, M. S. J. and Thiel, W., J. Am. Chem. Soc. 97,3978 (1975).
4. Inagaki, S. and Fukui, K., J. Am. Chem. Soc. 97,7480 (1975).
5. Orfanopoulos, M., Grdina, M. B. and Stephenson, L. M., J. Am. Chem. Soc. 101,275 (1979).
6. Jefford, C. W., Tetrahedron Lett. 1979, 985.
7. Harding, L. B. and Goddard, W. A., III, J. Am. Chem. Soc. 102,439 (1980).
8. (a) Gollnick, K. and Hartmann, H., to be published; (b) Hartmann, H., Dissertation, Univ. Munich, Germany, 1979.
9. Gollnick, K. and Paur, H., to be published.

10. Foote, C. S., Pure Appl. Chem. 27,635 (1971).
11. McClellan, A. L., "Tables of Experimental Dipole Moments", W. H. Freeman and Company, San Francisco (1963).
12. Reichardt, C., "Solvent Effects in Organic Chemistry," Verlag Chemie, Weinheim, Germany, 1979.
13. Kirkwood, J. G., J. Chem. Phys. 2,351 (1934).
14. Laidler, K. J. and Eyring, H., Ann. N. Y. Acad. Sci. 39,303 (1940).
15. Frost, A. A. and Pearson, R. G., "Kinetics and Mechanism," J. Wiley and Sons, New York, 1961 (Second Edition).
16. Ashford, R. D. and Ogryzlo, E. A., J. Am. Chem. Soc. 97,3604 (1975).
17. Koch, E., Tetrahedron 24,6295 (1968).
18. Coulson, C. A. and Longuet-Higgins, H. C., Proc. Roy. Soc. A192, 16 (1947).
19. Fukui, K., Acc. Chem. Res. 4,57 (1971).
20. (a) Robin, M. B., "Higher Excited States of Polyatomic Molecules," Vol. 2, Academic Press, New York, 1975; (b) Turner, D. W., Adv. Phys. Org. Chem. 4,31 (1966); (c) Vedeneyev, V. I., Gurvich, L. V., Kondrat'yev, V. N., Medvedev, V. A. and Frankevich, Ye. L., "Bond Energies, Ionization Potentials and Electron Affinities," Arnold, London, 1966; VEB Deutscher Verlag für Grundstoffindustrie, Leipzig, 1971.
21. Pritchard, H. O., Chem. Revs. 52,529 (1953).
22. Schulte-Elte, K. H., Muller, B. L. and Rautenstrauch, V., Helv. Chim. Acta 61,2777 (1978).
23. Schulte-Elte, K. H., Muller, B. L. and Pamingle, H., Helv. Chim. Acta 62,816 (1979).
24. Gollnick, K., Adv. Photochem. 6,1 (1968).
25. Wagner, H. U. and Gollnick, K., unpublished.
26. We are grateful to Dr. P. Potzinger, Institut für Strahlenchemie im Max-Planck-Institut für Kohlenforschung, Mülheim-Ruhr, Germany, for the determination of the ionization potentials from photoelectron spectra.

X

CAN PEREPOXIDES PRODUCE EPOXIDES AND VICE VERSA? RELEVANCE TO THE REACTION OF SINGLET OXYGEN WITH MONO-OLEFINS

C. W. Jefford, W. Knöpfel, L. C. Smith, S. Kohmoto, K. A. Almqvist, M. Jiménez Estrada and G. Barchietto, Department of Organic Chemistry, University of Geneva, 1211 Geneva 4, Switzerland

A subject of enduring interest is the mechanism of the apparently simple reaction of singlet oxygen with mono-olefins. The alternatives should be familiar to most chemists, but nevertheless, by way of introduction, they are now presented in order to define the experimental approach which has been adopted here with the aim of narrowing the field of mechanistic possibilities. Both the formation of dioxetanes and hydroperoxides can be thought of as being the result of a concerted [2+2] cyclo-addition and an ene-type reaction respectively (1). In contrast to the mechanistic economy of these one-step processes, several two-step alternatives have been proposed which have found varying support over the last few years. The first step, the rate-determining step, can give open chain or cyclic zwitterionic intermediates or diradicals. Thereafter, the original allylic hydrogen substituent is abstracted in intramolecular fashion to give the hydroperoxide or when the hydrogen substituent is inaccessible closure occurs to the dioxetane. Although more or less convincing mathematical models have been described (24), an essential difficulty remains in that these intermediates are unknown as discrete species. For example, nitrogen oxides are proper, well-characterized species, but oxygen oxides, with the possible exception of carbonyl oxides (5), are unknown as a class. Furthermore, even if perepoxides were to exist as highly fugitive species, there is no reason to suppose that they would react in the manner suggested. Consequently, we have undertaken and report on two sets of experiments. The first objective was to see if the formation of epoxides is a sure sign of a precursive perepoxide. In other words, can the supernumary oxygen atom be removed in some way, for example by singlet oxygen as kinetic measurements (6) and calculations (3) have indicated. The second set of experiments was based on the primitive idea of allowing selected epoxides to react with ground-state atomic oxygen. Once again, calculations have shown the way and revealed that thermodynamically there is little difference in energy between the heats of formation of epoxides and perepoxides (7). Thus it is reasonable to expect that a perepoxide would form easily by collision of the two species. Depending on the initial epoxide, further reaction might then be expected to give the same products which are obtained by the reaction of singlet oxygen with the corresponding olefin.

ISBN 0-12-592050-4

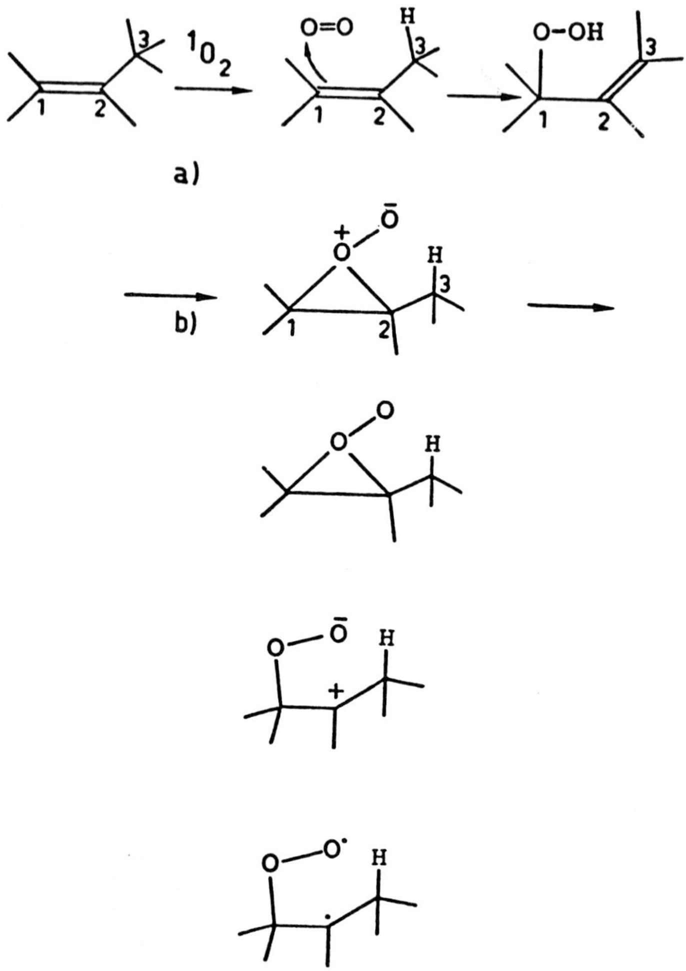

FIGURE 1. Allylic hydroperoxidation of an olefin by singlet oxygen a) ene-type transition state, b) two-step process involving possible intermediates.

Although Staudinger was the first (8) to suggest the perepoxy structure some 68 years ago to account for the course of the autoxidation of ketenes, the experimental finding which provided the first lead towards the real existence of perepoxides was the photooxygenation of adamantylidene-adamantane in solution (9). Sizeable quantities of epoxide are found in addition to the dioxetane, and it has been thought that it owes its origin to a precursive perepoxide which is prevented from undergoing ring enlargement on account of the crowded local environment. In similar fashion, the reaction of singlet oxygen with

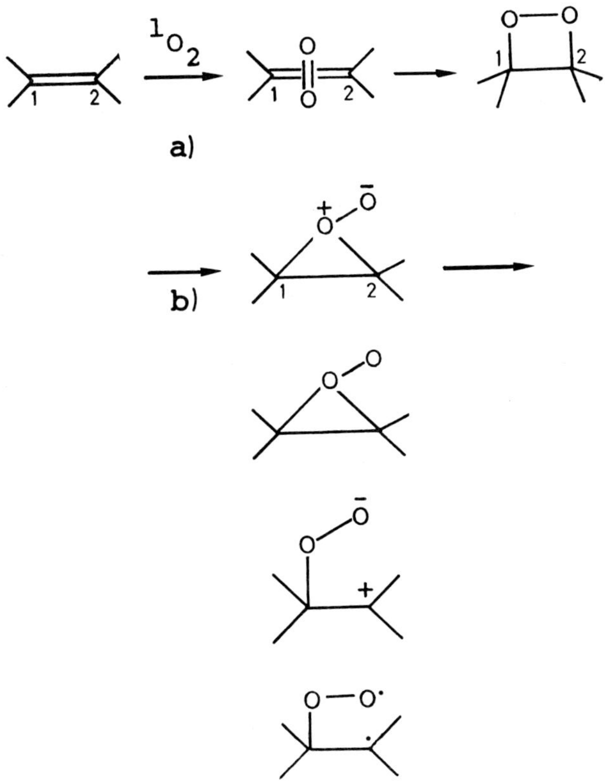

FIGURE 2. Addition of singlet oxygen to an olefin in a) a 2+2 process or b) stepwise involving intermediates.

adamantylidenecamphane gave the conventional dioxetane adduct, but also the Wagner-Meerwein rearranged product (10).

There are two puzzling aspects of these reactions. The first is that the solvent in a minor way is responsible for epoxide formation, although it is not chemically altered itself (11). Secondly, sensitizers like methylene blue give only the dioxetane adduct, whereas anionic sensitizers such as rose bengal and its congeners give substantial amounts of the epoxide. Dewar has calculated that singlet oxygen itself could deoxygenate the intermediate perepoxide to give epoxide and ozone and also that the reaction of excess olefin can also bring about disproportionation to two molecules of epoxide (3).

We have found that by carrying out experiments of the Kautsky-type where the sensitizer is sequestered and immobilized on a resin, that the proportion of epoxide component falls drastically, in fact almost to zero when rose bengal is used. Methylene blue behaves normally, no trace of epoxide being found at all, with only the dioxetane being formed. The small amounts of

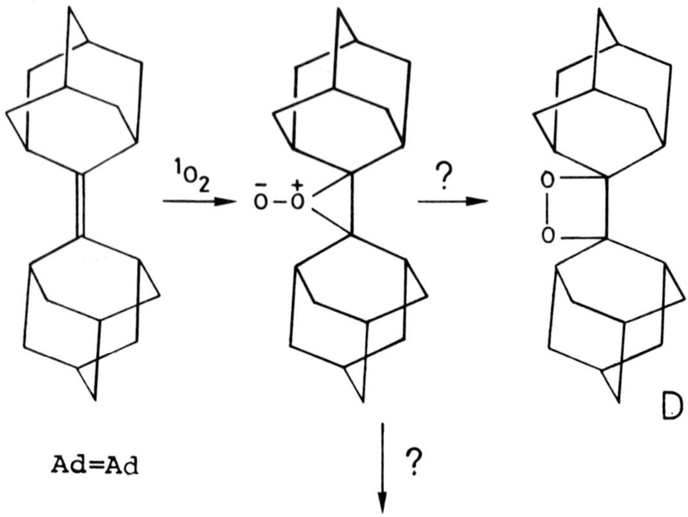

		D	E
TPP	EtOAc	100	0%
TPP	CCl_4	80	20
MB	MeCOMe	95	5
RB	MeCOMe	5	95

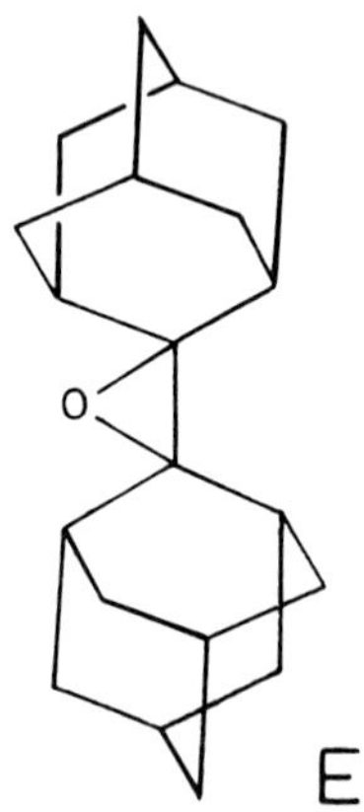

FIGURE 3. Photo-oxygenation of adamantylideneadamantane in solution using tetraphenylporphin (TPP), methylene blue (MB) and rose bengal (RB), by permission American Chemical Society.

epoxide found with rose bengal, some 2 to 4%, probably arise by contact between the two sets of particles. This result is nicely confirmed by using a non-photochemical source of singlet oxygen. Previously we found that the use of triphenylphosphite ozonide gave substantial amounts of epoxide on reaction with adamantylideneadamantane. However, when the endo-peroxide of 1,4-dimethylanthracene was used as the source of singlet oxygen, it was found in all cases that only dioxetane was produced. If methylene blue were admixed, once again dioxetane was the only result. The chemical admixture of rose bengal, erythrosin, eosin

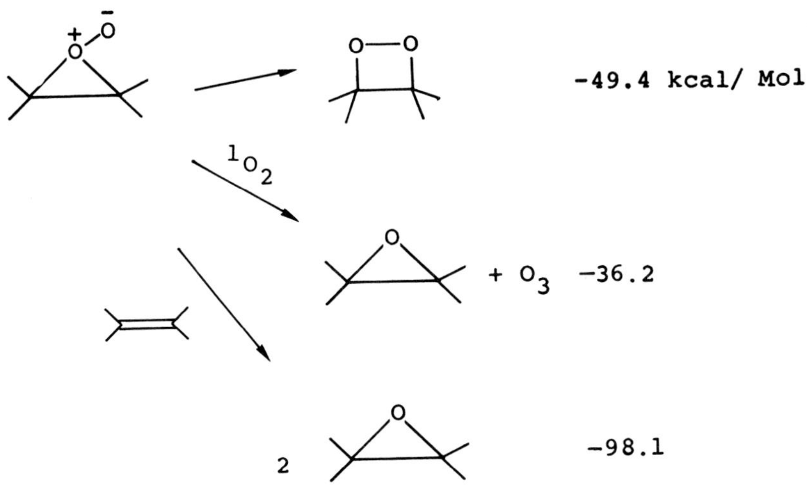

FIGURE 4. Exothermicities calculated for some reactions of a perepoxide (ref. 3), with permission American Chemical Society.

Table 1.

Photo-oxygenation of Adamantylideneadamantane (Ad=Ad) on Solid Supports

Ad=Ad	Support	O_2, hν →	O—O / Ad—Ad (dioxetane)	O / Ad—Ad (epoxide)	Con-version	Rxn time h
0.326 mMol	Florisil	MB	100	0	74%	24
0.161	Silica	IRA 120	100	0	59%	6
0.282	Florisil	RB	98	2	95	30
0.347	Silica	IRA 401	96	4	94	20

IRA: Support, 1:4, 2% Ad=Ad on support

and fluorescein, all gave varying amounts of epoxide in addition to the dioxetane. It now appears that the efficiency of epoxidation probably depends upon the different solubilities in acetone of the members of the fluorane family. The amount of epoxide formed is proportional to the solubility of the sensitizers. Rose bengal is completely soluble in acetone, whereas erythrosin and eosin are partially soluble and fluorescein is almost insoluble.

Although no proof is available, we believe that sensitizers of the anionic type behave as nucleophiles toward singlet oxygen

Table 2.

Chemical Singlet Oxygenation of Ad=Ad in the Dark*

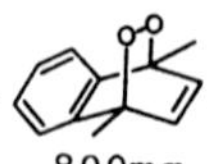

Rxn. #	Rxn. time	Ad=Ad + 70mg	+ (800mg)	+ sens ⟶ acetone	Ad=Ad,	Ad—Ad (O—O)	Ad—Ad (O)
1	45h			--	10	90	0%
2	72			RB, 70mg	16	7	77
3	98			Er, 60	14	27	59
4	98			Eo, 47	12	72	16
5	110			F, 26	18	80	12
6	39			MB, 45	30	70	0

* at room temperature

to give hydroperoxy anionic species, which subsequently break down to hydrogen peroxide under the reaction conditions. This idea is borne out by the fact that adamantylidene-adamantane reacts readily with 30% aqueous hydrogen peroxide in acetone or methanol at room temperature to give exclusively the epoxide. Under the same conditions, styrene and cyclohexene are totally inert. Clearly, in the solid phase hydrogen peroxide and similar peroxidic intermediates, if produced, as they are non-gaseous, cannot react with the olefinic substrate.

It goes without saying that the fact that epoxidation is dependent on the nature of the sensitizer rules out any mechanisms in which epoxide could be formed by singlet oxygen attacking the transient perepoxide.

A second set of experiments consisted of generating atomic oxygen and allowing it to react with epoxides which would be models for the dioxetane cleavage reaction and the hydroperoxidation reaction. First it should be mentioned that atomic oxygen is a long-lived species which is conveniently produced by the irradiation of a mixture of mercury vapor and nitrous oxide by means of a low-pressure mercury lamp (12). Excited mercury atoms then collide with nitrous oxide to yield nitrogen and atomic oxygen.

Few classes of organic compounds have been investigated so far. These consist for the most part of mono-olefins, dienes, various aromatic systems and enol ethers (13). It transpires that atomic oxygen is an electrophile, it epoxidizes olefins and it generally behaves as a biradical. The initial products are "hot" and usually undergo further rearrangement. Addition to double bonds is nonstereospecific and when such a bond is present hydrogen abstraction does not take place. However, in the case of cycloalkanes, hydrogen abstraction favors tertiary over secondary, which in turn is favored over primary hydrogen substituents. It appears that hydrogen abstraction occurs in such a way to generate a radical pair which immediately combines within the ambit of the radical cage. Interestingly enough, apart from the enol ethers in which the double bond is the reactive site, no other compounds containing oxygen have been studied so far.

Our expectation was that an epoxide having nucleophilic properties associated with its lone pair would collide with an atom of oxygen to generate initially a charge-transfer complex which would ultimately rearrange to the zwitterionic perepoxide or alternatively could open to the oxydiradical species. These expectations were completely dashed by the behavior of adamantylidene-adamantane epoxide. This epoxide was conveniently prepared by allowing atomic oxygen to react with the olefin absorbed on fluorisil. However, prolonged exposure of the epoxide to atomic oxygen was without effect, except that Wagner-Meerwein rearrangement eventually occurred to the corresponding ketone. Tests show that this result was due to traces of acid produced by nitrous oxide.

Our second experiment also proved to be disappointing. The 1,2 epoxide of naphthalene was prepared and submitted; however, under the experimental conditions it rapidly rearranged to α-naphthol. α-Naphthol itself on reaction with atomic oxygen gave only tars. A more promising result was found with naphthalene itself. Its oxidation with atomic oxygen was reproducible and gave essentially one product, the trans-*o*-formylcinnamaldehyde. The structure was confirmed by comparison with an authentic sample prepared by potassium periodate oxidation of the *cis*-1,2-dihydro-1,2-diol of naphthalene (14). In both cases the *cis* isomer was obviously initially formed, which underwent ready rearrangement in the presence of traces of acids to the *trans* isomer

This was an unusual result and contrasted with the oxygenation of phenanthrene accomplished by the γ-irradiation of the carbon dioxide (15). In this latter case, the epoxide and corresponding phenol were the chief products, and only traces of the dialdehyde were obtained.

Originally, we were tempted to believe that an epoxide had formed which acquired a second atom of atomic oxygen to give the peroxy species and that rearrangement and cleavage gave the dialdehyde. We are now of the opinion that the first-formed product is the diradical which undergoes cleavage to the carbene-aldehyde species. This would be expected to have a long life-time on account of the stabilization offered by the benzene ring and under

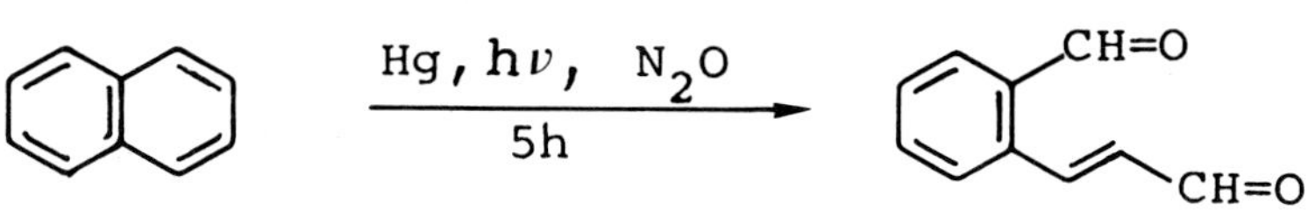

1g
10g florisil

0.25g(21% yield)

FIGURE 5. Oxygenation of naphthalene.

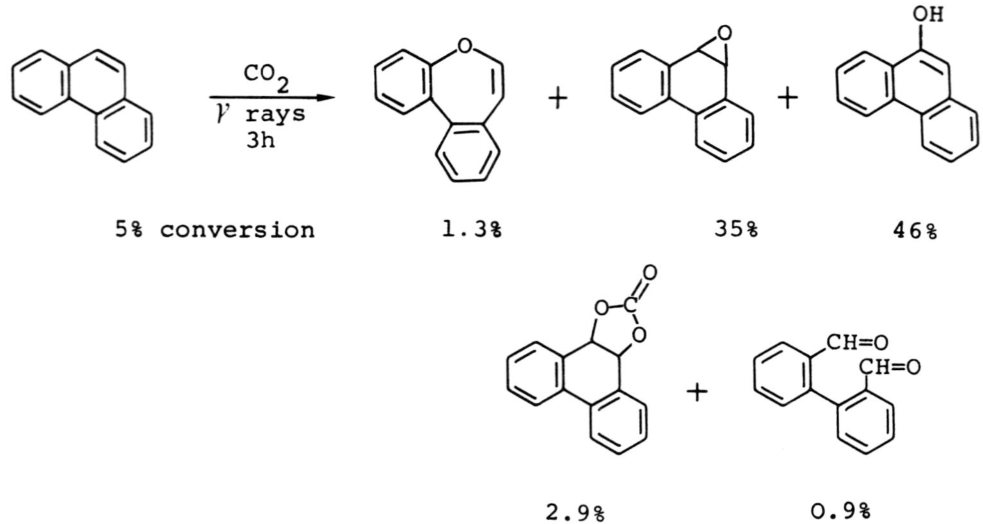

FIGURE 6. Oxygenation of phenanthrene (ref. 15), by permission The Chemical Society of Japan.

the oxidizing conditions, would give the cis dialdehyde which would subsequently rearrange to the observed product.

Our hopes were raised somewhat by the reaction of the epoxide of tetraphenylethylene with atomic oxygen. Although the main product was the rearranged pinacol, small amounts of benzophenone were obtained. It would be tempting to think that this last product derives from the perepoxy species. However, it is difficult to rule out the cleavage of the starting material which would be predisposed to give the ketone and the diphenyl carbene which would ultimately dimerize.

The epoxide of tetramethylethylene proved more tractable. It was injected at 80° into a stream of nitrous oxide and mercury vapor and irradiated. One product was obtained, namely the same hydroperoxide which was formed by the reaction of singlet oxygen with tetramethylethylene. The two products were identical in all respects.

SUMMARY

The aforementioned results show that rose bengal is an epoxidizing agent for those olefins which react readily with hydrogen peroxide. In other words, the hindered olefin, adamanthylidene-adamantane may be a special case. Moreover, sensitizers of the fluorane-type, as they are anions, seem to engender peroxidic species with singlet oxygen in solution.

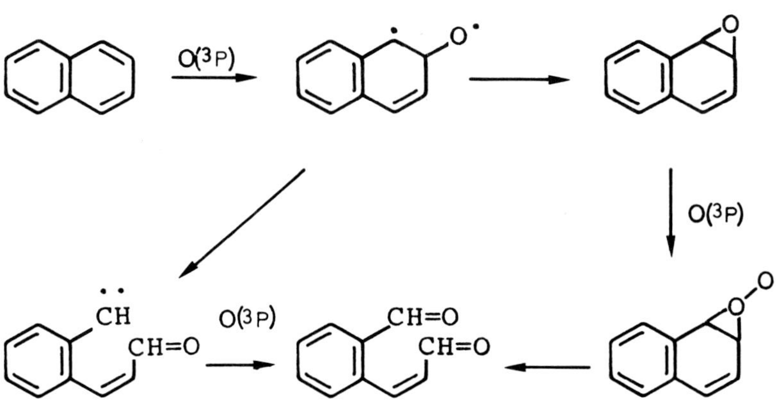

FIGURE 7. Possible mechanisms for oxygenation of naphthalene.

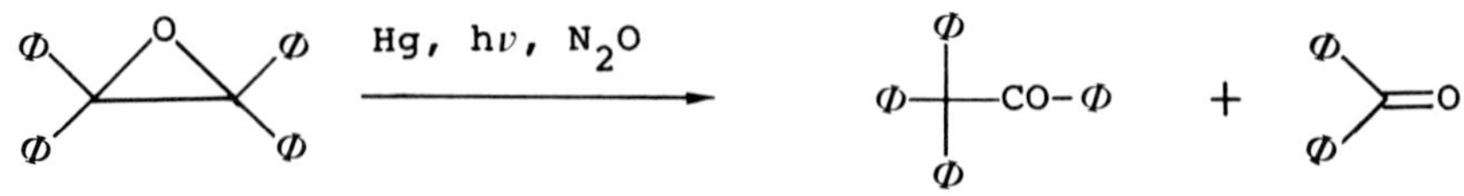

200 mg., 40h. 28% conversion 48 mg. (24%) 8 mg. (4%)
200 mg , 65h. 93% - " - 170 mg. (84%) 20 mg. (9%)

FIGURE 8. Oxygenation of tetraphenylethylene epoxide.

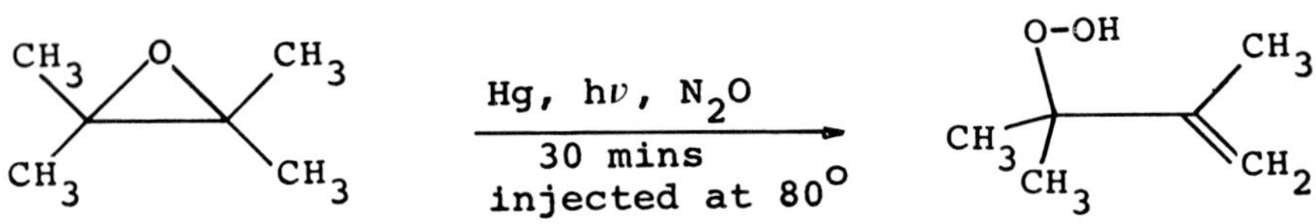

FIGURE 9. Oxygenation of 2,3-dimethylbutene spoxide.

The mechanistic model, namely the reaction of atomic oxygen with epoxides, is far from perfect, but shows promise. Parallels with the dioxetane reaction have yet to be found. However, two reasonably valid models appear to exist for the hydroperoxidation experiment, namely the reaction of epoxides containing what were initially available allylic carbon-hydrogen bonds. Unfortunately, these last results do not permit a distinction to be made between the Goddard diradical or the zwitterionic perepoxide.

REFERENCES

1. Frimer, A. A., Chem. Rev. 79,359 (1979).
2. Yamaguchi, K., Fueno, T., Saito, I. & Matsuura, T., Tetrahedron Letters 3433 (1979).
3. Dewar, M. J. S. and Thiel, W., J. Am. Chem. Soc. 97,3978 (1975); Dewar, M. J. S., Griffin, A. C., Thiel, W., and Turchi, I. J., ibid. 97,4439 (1975).
4. Harding, L. B. and Goddard, W. A., III, J. Am. Chem. Soc. 102,439 (1980).
5. Higley, D. P. and Murry, R. W., J. Am. Chem. Soc. 96,3330 (1974).
6. Bartlett, P. D., Chem. Soc. Rev., 5,149 (1976).
7. Harding, L. B. and Goddard, W. A., III, J. Am. Chem. Soc. 99,4520 (1977).
8. Staudinger, H., "Die Ketene," Verlag von F. Enke, Stuttgart, 1912, pp. 49-56.
9. Schaap, A. P. and Faler, G. R., J. Am. Chem. Soc., 95,3381 (1973).
10. McCapra, F. and Beheshti, I., J. C. S. Chem. Comm. 517 (1977).
11. Jefford, C. W. and Boschung, A. F., Helv. 66,2673 (1977).

12. Cvetanović, R. J., Adv. Photochem. 1,113 (1963).
13. Havel, J. J., J. Am. Chem. Soc. 96,530 (1974); Hori, A., Takamuka, S. and Sakurai, H., J. Org. Chem. 42,2318 (1977).
14. Jeffrey, A. M., Yen, H. J. C., Jerina, D. M., Patel, T. R., Davey, J. F. and Gibson, D. T., Biochem. 14,575 (1975).
15. Goto, S., Hori, A., Takamuka, S. and Sakurai, H., Bull. Chem. Soc. Japan 51,1569 (1978).

A TWO-STAGE CONCERTED (TSC) MECHANISM FOR ENE REACTION BETWEEN SINGLET MOLECULAR OXYGEN AND OLEFINS

Kizashi Yamaguchi, Department of Chemistry, Faculty of Engineering Science, Osaka University, Toyonaka, Osaka 560, Japan

The well-established criterion of Brown (1) is used for diagnosing whether the transition state (TST) for ene reaction of olefins with singlet molecular oxygen ($O_2(^1\Delta_g)$; or $^1O_2$) is cyclic or noncyclic. The ratio, $k_r = k_{exo}(A)/k_{exo}(B)$, of rates of exo addition to norbornene (A) and 7.7-dimethylnorbornene (B) is calculated by using the rate constants measured by Jefford et al. (2) as shown in Table 1, where the k_r-values for other reactions are also listed (1). The kr-value for the ene reaction is considerably smaller than that of epoxidation by peracid, whereas it is larger than the value for the free radical addition (R=·SPh). Therefore the TST for ene reaction should be a "weakly bonded" cyclic form instead of a noncyclic one.

Previously (3) three different oxygen-approach models have been considered for the ene reaction between $^1O_2$ and olefins as shown in Fig. 1. Model (A) corresponds to the so-called "ene" reaction path, where $^1O_2$ synchronously attacks the allyl hydrogen and the terminal C-atom, whereas in model (B) $^1O_2$ simultaneously attacks the allyl hydrogen and the center of the C-C double bond. In model (C) $^1O_2$ attacks only the C-C double bond; this may correspond to the "perepoxide"-like reaction path.

The calculated result (Table 2) based on the Murrel-type perturbation method (4) has predicted that the path corresponding to B is the most preferred in the case of propene plus $^1O_2$ (1 and 2 in Fig. 2) system because of the predominance of the CT interaction, which can be grasped by the HOMO (1)-LUMO (2) interaction scheme (Fig. 2A). This result indicates the importance of simultaneous attack of $^1O_2$ on allyl hydrogen and the C-C double bond in the initial stage of the ene reaction, but it does not imply that the C-O bond formation and allyl hydrogen abstraction should

TABLE 1. Brown's k_r-Values for Reactions

Reactions	k_r	*Reactions*	k_r
Peracid[a)]	*1000*	$Hg(OAc)_2, H_2O$[c)]	*58*
H_2N_2[a)]	*950*	*PhSH (hν)*[c)]	*30*
PhSCl[a)]	*1820*	*HOAC*[c)]	*2.8*
9-BBN[a)]	*480*	*HCl*	*2.2*
Singlet oxygen[b)]	*254*		

a) ref. 1, b) present work, c) ref. 2.

ISBN 0-12-592050-4

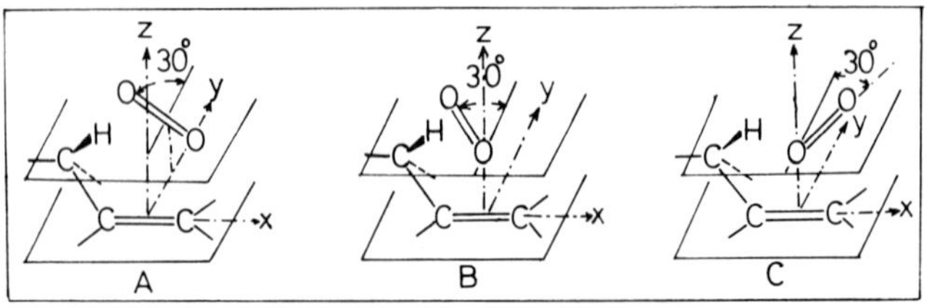

FIGURE 1. Approach models for propene plus $^1O_2$ system.

take place at the same time (one-stage model (Fig. 1A)). The diradical character of $^1O_2$ (5) has disappeared in the hydrogen-transfer stage of the path B because of the allowed orbital interaction (Fig. 2B). Thus the two-stage concerted (TSC) mechanism passing through a oriented π-complex (an exciplex) has been presented for ene reactions of simple olefins (X=CRR'R" in Scheme I) as shown in Fig. 3, where other mechanisms are also illustrated for comparison.

The TSC mechanism is quite compatible with various experimental results for ene reactions of olefins: (1) Brown's k_r-value, (2) isotope effects, (3) stereochemistry, (4) cis-effects, and (5) solvents effects. It is also applicable to enol ethers (X=OR in Scheme I), for which both I_a- and I_b-type reactions take place. However, a zwitterionic mechanism is more reasonable for enamines which do not undergo the I_a-type reaction because of the unsymmetric (2+2) approach (Fig. 2C) (6).

TABLE 2. Interaction Energies for Propene (1) Plus Molecular Oxygen (2) System (kcal/mole)

	$^1\Delta_z$			$^1\Delta_x$	$^3\Sigma$	$^1\Sigma$	
Energy	*A*	*B*	*C*	*B*	*B*	*B*	*Energy terms*
E_Q	-0.27	-0.49	-0.22	-0.49	-0.49	-0.49	E_Q:Coulombbic energy
E_K	0.61	0.96	0.56	1.06	1.00	1.00	E_K:Exchange repulsion
E_I	-0.00	-0.00	-0.00	-0.00	-0.00	-0.00	E_I:Inductive energy
E_D	-0.45	-0.54	-0.31	-0.74	-0.74	-0.74	E_D:Dispersion energy
E_{CT}	-0.65	-1.36	-0.55	-0.15	-0.48	-0.55	E_{CT}:Charge-transfer energy

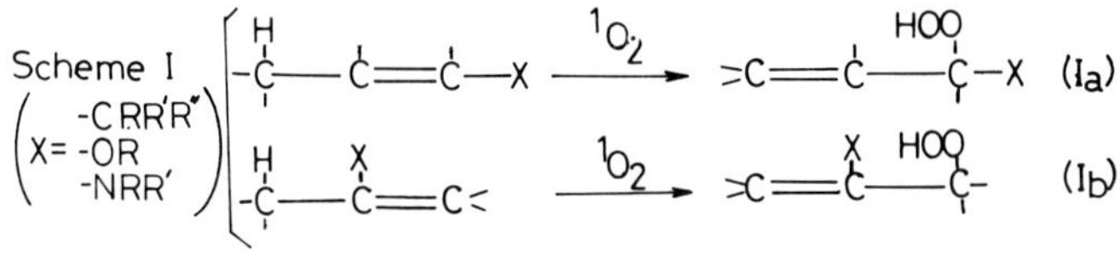

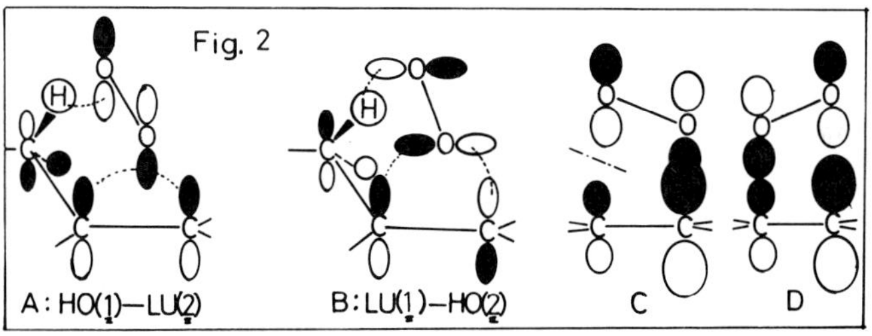

FIGURE 2. Schematic illustration of HOMO-LUMO interactions.

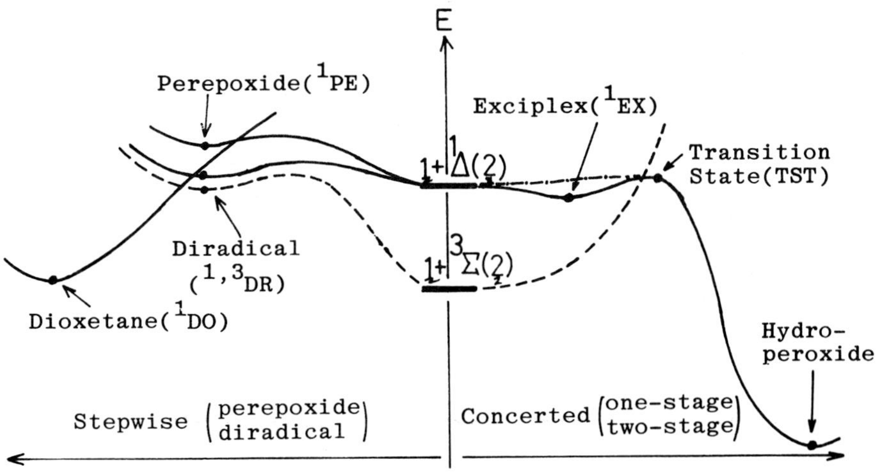

FIGURE 3. Concerted and stepwise machanisms for ene reactions of olefins with $^1O_2$.

REFERENCES

1. (a) Brown, H. C., Kawakami, J. H. and Lui, K. T., J. Am. Chem. Soc., 95,2209 (1973); (b) Brown, H. C. and Lui, K. T., Ibid., 93,7335 (1971).
2. (a) Jefford, C. W., Laffer, M. H. and Boschung, A. F., Ibid., 94,8904 (1972); (b) Jefford, C. W. and Boschung, A. F., Helv. Chim. Acta., 57,2242 (1974).
3. Yamaguchi, K., Jpn. Chem. Rev., 1,292 (1973).
4. Fueno, T., Nagase, S., Tatsumi, K. and Yamaguchi, K., Theoret. Chim. Acta., 26,43 (1972).
5. Yamaguchi, K., Fueno, T. and Fukutome, H., Chem. Phys. Lett., 22,466 (1973).
6. (a) Yamaguchi, K., Fueno, T., Saito, I. and Matsuura, T., Tetrahedron Lett. 3433 (1979); (b) Ibid., 4087 (1980).

DISCUSSION

TRAYLOR: Dr. Gollnick, your ionization potential measurements are accurate only to about ±0.03 E.V.; is not that enough to account for an isotope effect of 1.1 in the tetramethylethylene case?

GOLLNICK: The limit of error on the ionization potential data is too large to exclude a secondary isotope effect. We plan to prepare C_8-complexes to find out whether the C-C double bond's nucleophilicity depends on the deuteration. However, the ^{13}C-NMR-signal and IP-trends given in the experimental part show that the nucleophilicity of the C-C double bond of the deuterated 2,3-dimethyl-2-butene should be higher than that of the perprotio compounds, which means that the secondary isotope effect is expected to be $k_D/k_H > 1$. Thus, the primary kinetic isotope effect (k_H/k_D) should be higher than found (1.11 ± 0.02).

HERRON (TO JEFFORD): Dr. Jefford, I am interested in the O atom reaction you have discussed here. You said that it is possible that it could be abstraction, but it would seem to me that there is no question that it is almost entirely abstraction. The rate constant for the abstraction reaction can be easily calculated knowing the number and types of C-H bonds. So, as a gas kineticist, I would say that the abstraction reaction has got to be going on, and I cannot understand how the products observed are related to oxygen atom attack at the epoxide oxygen atom. That brings up a general problem. There are only a few gas kineticists at this meeting, and I believe we came here to learn something about free radical oxidation in the aqueous phase.

JEFFORD: We all came to find out what everybody else is doing...

HERRON: I have heard considerable discussion of hydroxyl radical chemistry as well as of alkoxy and peroxy radical chemistry, for which there is an extensive body of data which has been obtained in the gas phase. Many of the mechanisms written to account for the observations made in solution do not agree at all with what is known in the gas phase. They should not be so different. The rate constants in solutions differ from the rate constants in the gas phase, but I was not aware that the difference was so great. The rate constants for singlet oxygen reactions are almost identical in the gas phase to those obtained in methanol.

GOLLNICK: One order of magnitude lower.

HERRON: One order of magnitude, but nonetheless the relative rates are probably going to be very similar; and it would seem that one could use the relative rate constants for hydroxyl radical reactions to estimate the rate constant for any hydroxyl radical reaction in solution.

JEFFORD: As atomic oxygen is an electrophile, we expect it to add to the oxygen atom in an epoxide, its most nucleophilic part. Most of our experiments are either done in the gas phase or else the substrate is held on a solid phase and the atomic oxygen flows over it.

HERRON: I do not see how that makes a difference. The relative rate constants for attack at the peroxy atom and at hydrogen atom sites should be the same as they are in the gas phase. So you cannot say it attacks only that peroxy O atom.

SINGH: Your technique, Dr. Jefford, is very elegant, but I see some differences with singlet oxygen. When singlet oxygen reacts with C-C double bonds, the total energy available to it is only about 22 kcal/mole. When you are taking two atoms of oxygen sequentially, the total energy that is available in that reaction with the same double bond is much greater. So the result is likely to be that even though the perepoxide intermediate may form, because of this extra energy available, it could do other things which you would not expect in singlet oxygen chemistry. So from that point of view, to the extent your results are negative, they should not be considered negative from the point of view of perepoxide formation in singlet oxygen chemistry.

JEFFORD: I must say I am a little disappointed. Unfortunately, one of the early experiments we did was to hit the tetramethylethylene epoxide with atomic oxygen and got that marvelous result so we carried on. But all I really see is a parallel with the singlet oxygen reaction. There's no doubt that atomic oxygen gives a lot of energy, perhaps a hundred kcal, to the system. Nonetheless, I think it is a gratifying parallelism. Unfortunately, I am disappointed with adamantylideneadamantane epoxide as it is one of the key experiments. We are investigating, I should say very rapidly, carbonyl compounds. We have done benzil, which gives benzoic acid.

OSMAN: You pointed out correctly that triplet atomic oxygen is an electrophile, but even more electrophilic will be a singlet atomic oxygen. If you can excite the triplet into a singlet, you may be successful in making this perepoxide.

JEFFORD: It's a question of cost. The zinc lamp is a little more expensive than a mercury vapor lamp.

OSMAN (TO GODDARD): Your calculation about the high energy state of the ethylene interacting with singlet oxygen to give perepoxide is not necessarily relevant to the energy of the transition state, or intermediate with allylic hydrogen, because it may have completely different energetics.

GODDARD: No calculations we have done have disproved that there are perepoxide models for the case where you have a number of methyl-like substituents as you have with the ene intermediate. Your point is that our calculations do not, by themselves, disprove the perepoxide model. Our calculations on the perepoxide were for unsubstituted ethylene. The problem with the perepoxide is that once you have made it, you have still got to abstract a hydrogen to obtain the ene product. There are two problems. First, the entropy of activation is way too high to fit with the experiment. And secondly, even given the perepoxide, there is still a barrier to go over to actually abstract the allylic hydrogen. Thus, formally speaking, the perepoxide appears to explain some experiments; however, a real perepoxide is likely to lead to kinetics very different than observed in the ene reaction.

OSMAN: That's exactly my point, because of what you just said about the instability of the intermediate.

GODDARD: The problem with the perepoxide is that when it opens up, it just leads back to the biradical, thus, the perepoxide does not really gain anything mechanistically. The perepoxide might conceivably be sufficiently stabilized that it could be trapped and you might actually be able to observe it. However, as the perepoxide opens up, to complete the ene reaction, it breaks the CO bond and you go right back to the biradical. The electronic structure of the perepoxide is as follows. Start out with an epoxide and consider the two lone pairs, one sticking above the plane and one sticking below the plane. Then form the perepoxide by sticking a singlet oxygen atom onto one of the long pairs, allowing some delocalization of the epoxide lone pair onto the second O atom to form a small dipole moment (5 debye). The dioxygen part of the perepoxide then corresponds to a very high-lying state of O_2 (an excitation energy of 8 or 9 eV). Thus, if you had made a perepoxide, then as soon as you break the CO bond, one of the doubly-occupied orbitals on the outer oxygen moves right on back to the inner oxygen, leading to the biradical intermediate which then abstracts the hydrogen to go to product. If the substituents were to stabilize the perepoxide, formation of the biradical would merely require an additional barrier in order to form the ene product.

OSMAN: Or the concerted transition state.

GODDARD: Yes. The question is what one means by concerted. Many people would say that the biradical pathway in my talk was a concerted reaction, because there was no stable intermediate. That is, the biradical was essentially the saddle point of the reaction, and they would say that my biradical intermediate is merely one point on the concerted pathway.

FRIMER: You showed in one of your slides that when you formed the 1,4-biradical with enol ethers, the radical had the oxygen attacking at the alpha position and the carbon radical ends up beta to the ether group, correct?

GODDARD: What I showed was that the favored intermediate is

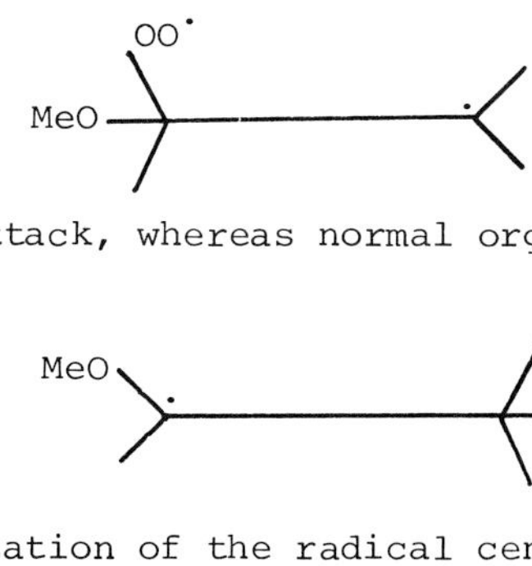

for the dioxygen attack, whereas normal organic intuition would have been

because of stabilization of the radical center by the methoxy. The point I made was that it is lone pair stabilization of the dioxygen by the methoxy that stabilizes attack at the methoxy carbon.

FRIMER: There are two arguments that argue against this. First of all, there is the work that I did with Professor Bartlett on the photooxidation of 4,4-dimethyldihydropyran where you can only get a dioxetane product. The isotope effect for the β-carbon was substantial (k_H/k_D = 0.927) while for the alpha position, there is essentially no isotope effect. That would show that the oxygen is bonding at the beta position rather than at the alpha position. Furthermore, Dr. Jefford brought to my attention last night the work by Havel on the addition of atomic oxygen to enol ethers which also indicates that the oxygen attack is at the beta position of the enol ether.

GODDARD: In the biradical, the ground state is the 3π state and the excited state is the 4π. The 4π state is the one that can close up the form dioxetane without an additional barrier. I believe that most cases involving dioxetane formation generally involve stabilizing the 4π state

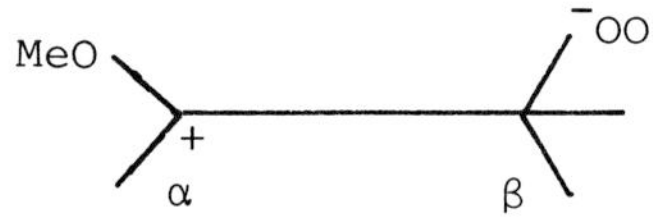

which will always be on the β-carbon, in order for the + charge to be stabilized by the methoxy. But the biradical (3π) state prefers to attack the α-carbon

FRIMER: Would you comment on the work of J. J. Havel and K. H. Chan on the reaction of atomic oxygen with enol ethers which indicates that the biradicals form so that the radical ends up counter to what you suggest?

GODDARD: Was this work with singlet or triplet atomic oxygen?

FRIMER: It was with triplet.

GODDARD: The triplet atomic oxygen reaction need not have anything to do with the singlet dioxygen biradical intermediate.

FRIMER: Well, I doubt there would be a substantial difference in orientational preference between singlet and triplet biradicals. You owe me a calculation.

GODDARD: The argument against the biradical model using the rotational barrier of 1,4 biradicals is not an argument at all. It is well known that the rotational barrier for $-\dot{C}H_2$ radical is around 0.5 kcal (as in your calculation), whereas for $-\dot{C}\lt^{Me}_{H}$, the barrier is 1.5 to about 2 kcal, and for $-\dot{C}\lt^{Me}_{Me}$ the barrier is about 4 kcal. The 1,4 biradical involved in the ene reaction necessarily involves at least one methyl group on the radical carbon and the rotational barrier is large enough to explain the ene reaction using the biradical model. A nice test of the biradical

would be

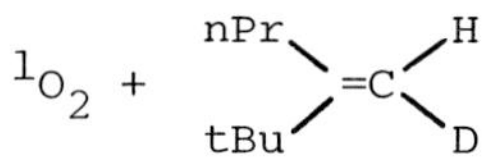

The mechanism suggests that $^1O_2$ could catalyze *cis*-*trans* isomerization.

YAMAGUCHI: There are s-*cis* and s-*trans* isomers for the 1,4-peroxydiradical:

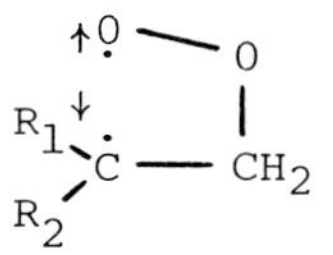

s-*cis*

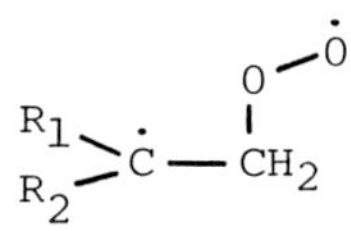

s-*trans*

The latter is more stable than the former in the case of the σπ-diradical state of the propene (**1**) plus O_2 (**2**) system since the exchange interaction between the σπ-diradical electrons is positive (repulsive) in sign. Therefore the rotational barrier of the R_1R_2C-group of the 1,4-diradical is similar to that of the free radical, $R_1R_2\dot{C}$-R_3. The *cis*-*trans* geometrical isomerization reaction should be facile even in the singlet 1,4-diradical, and this must occur before the hydrogen abstraction reaction for which the s-*cis* conformation is essential. Thus, the ene reaction should be nonstereospecific if the singlet 1,4-σπ-diradical is a true reaction intermediate. For instance, the (2+2) cycloaddition between tetrahaloethylene and diene is nonstereospecific, compatible with the singlet 1,4-diradical mechanism. However, stereospecificity has been observed experimentally in the case of ene reactions of olefins with $^1O_2$. Furthermore Brown's k_r-value in Table I of my paper should be similar to that of the PhS· free-radical addition if the 1,4-diradical is a real intermediate. Thus, there is strong evidence for the concerted mechanism for the ene reaction. Further, the present result weakens the arguments used to support the diradical-intermediate mechanisms in related stereo-specific cycloadditions cited in Table I.

FOOTE: I would like to mention to Bill (Goddard) that if he would move his O_2 over a half angstrom to the right, he would be saying exactly the same thing as everyone else this morning.

GODDARD: No, the bonding and electronics are entirely different.

NIKI: I want to ask of Goddard and Yamaguchi what they would expect for the ene type reactions with singlet sigma O_2. Is there anything special in the sigma case, such as the fact that there is extra energy? Now of course, in the gas phase the rate constants for sigma are typically a factor of 100 faster than for the singlet delta state, which is a very slow-reacting species in the gas phase.

GODDARD: First off, my comments are not based on any actual calculations. The structure of $O_2(^1\Sigma_g{}^+)$ is such that the adduct can go through the dioxetane intermediate without an additional barrier. However, this is a very energetic intermediate. In the gas phase, all heck will break loose because there are many possible arrangements. Thus, in general, $^1\Sigma_g{}^+$ dioxygen will be much less selective than the singlet delta state.

YAMAGUCHI: As shown in Table 2 of my paper, the interaction energy (E_{total}) for the $^1\Sigma$ state is smaller than that of the $^1\Delta_z$ component of the Δ_g state, which is responsible for the ene reaction. This implies that an ene-type reaction does not take place even in the olefin-$^1\Sigma(O_2)$ system, as in the case of the olefin plus $^3\Sigma(O_2)$ pair. However, the interaction energy is sufficiently large to quench the $^1\Sigma$ state, compatible with experiments.

STEPHENSON: I think a number of us have been seduced over the years into believing that there were no solvent effects in this reaction because there are typically no solvent effects on product distribution. Now we have heard results today that tell us that there is a significant solvent effect in the reaction. It is faster in more polar solvents than it is in less polar solvents and is slower in the gas phase by a good bit. This suggests to me that the calculations that Bill Goddard and others have done are perhaps just not relevant to the solution phase experiments that are presently being conducted. With the solvent effect results, I think that it is now imcumbent on some of us to do some of these isotope effects and stereochemical experiments in the gas phase, where the results may be significantly different. I think a lot of us did not want to do that for a long time, first of all because they are hard experiments, and secondly because no one thought that they were going to see anything. The fact that there is a real solvent effect now suggests that these experiments are really well worth doing.

BARTLETT: On the subject of the solvent effect, I might call attention to the fact that Frimer's work on the methyldihydropyran system several years ago showed solvents behaving just as Bill Goddard says they should: polar solvents favored dioxetane formation over ene reaction. We were surprised at the time and commented on it. There is one other point in that paper that seems rather pertinent to this discussion: we found γ-tritium isotope effects not only in the ene reaction but in the accompanying dioxetane formation, as if even dioxetane formation had gone through a transition state like a hydrogen bonded perepoxide. The way we put it, I would still put it today, I think: in forming dioxetane the system deviates from the least-motion path and goes through a transition state at least bent toward the perepoxide geometry.

GODDARD: Paul, I wonder if your results might be explained by the following: 1) the ene reaction involves attack at the α-carbon (3π biradical)

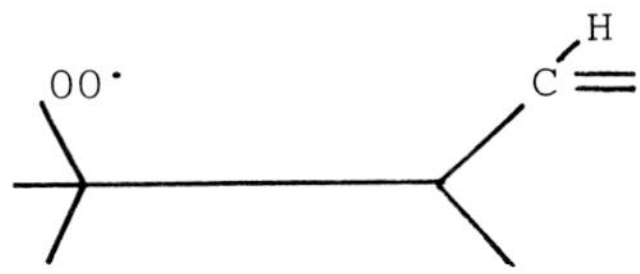

and the isotope effect is due to the γ-H atom.
2) for 1,2-dioxetane formation, the intermediate may be due to attack at the β-carbon (4π zwitterion) and thence show

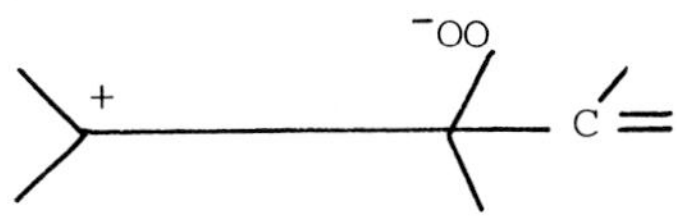

an isotope effect due to the same group as for the ene reaction?

THE SENSITIZED PHOTOOXIDATION OF BIOMOLECULES

Session Chairman, J. D. Spikes, Salt Lake City, Utah

AN OVERVIEW

John D. Spikes and Richard Straight, Department of Biology, University of Utah, Salt Lake City, Utah and Medical Research, Veterans Administration Medical Center, Salt Lake City, Utah

In 1903, Professor H. Von Tappeiner and his students at the Pharmacological Institute in Munich showed that preparations of the enzymes diastase, invertase, and papain were inactivated under aerobic conditions on illumination in the presence of the dye eosin (see 1). This inactivation (termed "photodynamic" inactivation by biologists) was later shown to result from the sensitized photooxidation of certain amino acid residues in the protein molecule. Since then the sensitized photooxidation of many types of biologically-related molecules has been demonstrated, including alcohols, aldehydes, amines, amino acids (in particular cysteine, histidine, methionine, tryptophan, and tyrosine), bilirubin, carbohydrates, a number of drugs, phospholipids, nitrogen heterocyclics, nucleic acids (as well as purines, pyrimidines, nucleosides and nucleotides), unsaturated fats, organic acids, phenols, plant growth regulators, a large number of proteins, steroid hormones, thiols, several vitamins, etc. Also, many kinds of synthetic and naturally occurring organic and organo-metal molecules, as well as some inorganic ions, have been shown to be photodynamic sensitizers. These include acridines, flavins, ketones, phenothiazines, polyenes, porphyrins, quinones, xanthenes, ruthenium bipyridyl, cadmium sulfide, Cu^{2+} and Fe^{3+} ions, etc. (2-4). In general, good photodynamic sensitizers show a high quantum yield of triplet formation, and the sensitized photooxidation of biomolecules is typically mediated by this state of the sensitizer, primarily because of its relatively long lifetime (2-4).

In the usual photodynamic systems with biomolecules in nonreactive solvents, two major reaction pathways are available to the triplet sensitizer. In one (Type I or free radical reactions), the triplet sensitizer reacts directly with the molecule being photooxidized (the "substrate"), typically by a hydrogen or electron abstraction process. The resulting radicals react further by various pathways, often with the formation of an oxidized form

ISBN 0-12-592050-4

of the substrate and the regeneration of ground state sensitizer. In some cases, reactive substrate intermediates (probably radicals) combine to form covalent substrate-substrate photoadducts; for example, protein-protein (5,6) and protein-nucleic acid (7) photoadducts are readily formed. Sensitizer and substrate radicals can also interact with the formation of covalent sensitizer-substrate adducts, as is frequently observed with proteins; the photoreaction is favored by non-covalent pre-binding of the sensitizer with the protein (8). Substantial amounts of superoxide and hydrogen peroxide are produced in some Type I systems. In the other pathway (Type II), triplet sensitizer reacts with ground state oxygen (which is in the triplet state) by an energy tranfer process to give singlet oxygen ($^1\Delta_g$) which then reacts with the substrate. In some cases, small amounts of superoxide can result from electron transfer between triplet sensitizer and ground state oxygen. Thus the sensitized photooxidation of biomolecules may involve free radicals and/or the oxidizing species singlet oxygen superoxide and hydrogen peroxide. It has been reported that hydroxyl radical is produced by illuminated hematoporphyrin (9); to what extent this species might be involved in the sensitized photooxidation of bimolecules is not known.

It is not easy to work out the total array of reactions involved in the sensitized photooxidation of a given biomolecule. A number of mechanistic tests have been developed to estimate the relative involvement of Type I and Type II processes in a photooxidation reaction, but in general, none of them give completely unambiguous results. This topic was discussed in some detail at the Pinawa conference (10); also see (11). One commonly used technique is to compare the photooxidation rates of the substrate in deuterium oxide and in water. If the rate is significantly higher in deuterium oxide, it suggests that the reaction is mediated by singlet oxygen, since that species has a much longer lifetime in that solvent; however, the technique is valid only under certain reaction conditions (3,10,11). Another approach is to examine the effects of inhibitors which are not oxidized by free radical processes, but which quench singlet oxygen with high efficiency. Azide is widely used for this purpose in aqueous systems; unfortunately there are few if any quenchers that are completely specific for singlet oxygen. A promising technique is to compare the structures of the observed photooxidation products with those known to be produced from the substrate by free radical processes and by externally generated singlet oxygen. Catalase and superoxide dismutase have been useful with some reactions in determining the extent of the participation of hydrogen peroxide and superoxide, respectively. Flash photolysis techniques, although not used to nearly the same extent as steady state kinetic approaches, can often provide direct information about the reactivity of the sensitizer triplet with oxygen, substrates and possible inhibitors as well as information on the possible interactions of semioxidized and/or semireduced sensitizer with substrates and inhibitors

(12). Some "singlet oxygen quenchers" also react efficiently with sensitizer triplet; this possibility can be examined with flash techniques. In summary, it is usually necessary to use a multiple-technique approach in order to elucidate the mechansims of a given sensitized photooxidation process.

The mechanism(s) of the photooxidation process in a given photodynamic system is strongly dependent on the chemical-photochemical properties of the sensitizer, the state of the sensitizer (i.e., whether it is free or bound), the chemical nature of the substrate, the chemical nature and dielectric constant of the solvent and the reaction conditions including pH, reactant concentrations (sensitizer, substrate, oxygen, buffer), the temperature, and, possibly, the wavelength of light used (2-4,8). Both Type I and Type II processes can occur simultaneously in some photodynamic systems; in others, the reaction pathway can be switched from Type I to Type II (or vice versa) by changing the reaction conditions.

To illustrate the effects of some of the variables described above, Table 1 presents quantum yields of oxygen uptake during the photooxidation of phenylalanine and some of its derivatives as measured with two sensitizers at two pH values. With eosin as sensitizer, each of the derivatives shows a marked increase in quantum yield as the pH is raised from 7 to 11, while with 3-methyl lumiflavin the change with pH is small or negative. With phenylalanine, the yield is very much greater (over 100-fold at pH 8) with 3-methyl lumiflavin as sensitizer. When an -OH group is put on the phenylalanine ring in the 4-(para-) position to give p-tyrosine, the quantum yield at both pH values is increased 30-fold with eosin as sensitizer, but is changed only slightly with the

TABLE I. Quantum Yields for Oxygen Uptake During the Photooxidation of Phenylalanine and Some of its Derivatives at pH 8 and 11 as Sensitized by Eosin Y and by 3-Methyl Lumiflavin

	*Eosin Y**		*3-Methyl Lumiflavin*	
Substrate	*pH 8*	*pH 11*	*pH 8*	*pH 11*
Phenylalanine	*0.0008*	*0.0046*	*0.10*	*0.076*
p-Tyrosine	*0.024*	*0.145*	*0.13*	*0.18*
4-Methoxyphenylalanine	*0.0013*	*0.0050*	*0.16*	*0.36*
N-Acetyltyrosine	*0.0038*	*0.054*	*0.025*	*0.031*

*The reaction mixtures were 1mM in substrate, 3 μM in sensitizer, 0.125M in sodium phosphate buffer and 0.22mM in oxygen at the start of illumination (517 nm for eosin and 453 nm for the flavin); the temperature was 25°C. *Data from ref. 12.*

flavin. When the -OH group is blocked, as in 4-methoxyphenylalanine, the quantum yield with eosin is reduced to essentially the same low value observed with phenylalanine; in contrast, the yield increases somewhat with 3-methyl lumiflavin. Blocking the amino group of tyrosine, as in N-acetyltyrosine, reduces the quantum yield with both sensitizers at both pH values; this suggests that the free amino group is involved in some way in the sensitized photooxidation of tyrosine (13). In summary, then, these data show the very large effects that can be produced with changes in pH, in the sensitizer, and by different substituent groups on the substrate.

REFERENCES

1. Von Tappeiner, H., Ber. Deut. Chem. Ges., 36,3035 (1903).
2. Spikes, J. D. and Livingston, R., Adv. Radiat. Biol. 3,29 (1969).
3. Foote, C. S., "Free Radicals in Biology," Vol. II (W. A. Pryor, ed.), Academic Press, New York, N.Y.,(1976),p. 85.
4. Spikes, J. D., "Photomedicine: Light in Disease and Therapy" (J. D. Regan, ed.), Plenum Press, New York, N.Y., In press, (1980).
5. Girotti, A. W., Lyman, S., and Deziel, M. R., Photochem. Photobiol. 29,1119 (1979).
6. Dubbelman, T. M. A. R., de Goeij, A. F. P. M. and Van Steveninck, J., Photochem. Photobiol. 28,197 (1978).
7. Helénè, C., "Aging, Carcinogenesis and Radiation Biology" (K. C. Smith, ed.), Plenum Press, New York,(1976), p. 149.
8. Jori, G. and Spikes, J. D., this volume, p. 441.
9. Beuttner, G. R. and Oberley, L. W., Prog. Amer. Soc. Photobiol., 7,178 (1979).
10. Foote, C. S., Photochem. Photobiol. 28,922 (1978).
11. Foote, C. S., "Biochemical and Clinical Aspects of Oxygen" (W. S. Caughey and H. Caughey, eds.), Academic Press, New York, N.Y.,(1978), p. 603.
12. Rizzuto, F. and Spikes, J. D., Photochem. Photobiol. 25,465 (1977).
13. Straight, R. and Spikes, J. D., Photochem. Photobiol. 27,565 (1978).

DISCUSSION

STEVENS: As point of information, I think the triplet state is the origin of singlet oxygen in these dye sensitizers because the singlet-triplet splitting is less than the 8000 cm^{-1} necessary to produce O_2 ($^1\Delta_g$). In aromatic hydrocarbons where the singlet triplet splitting is much greater than 8000 cm^{-1}, O_2 ($^1\Delta_g$) can be produced by quenching of the singlet state.

PHOTOOXIDATION OF BIOLOGICAL MODEL COMPOUNDS

Christopher S. Foote, Department of Chemistry, University of California, Los Angeles, Los Angeles, California 90024

Damage to organisms caused by light in the presence of dyes has been known for many years (1,2). This process is referred to as "photodynamic action," and is now known to cause damage primarily to certain specific chemical targets. This damage causes enzyme deactivation (primarily through destruction of histidine, methionine, tryptophan, tyrosine, and cysteine residues), nucleic acids are damaged (the guanine residues are the primary target), and membranes are damaged (with unsaturated fatty acids and cholesterol being typical targets). Although chemical products of many of the target molecules are not yet known, a few reactions have been well elucidated and characterized. I will concentrate on those substrates for which the primary products have been well characterized. In many cases, of course, the primary products are unstable and very rapidly give secondary products at room temperature. While some of these secondary products have been characterized, I will not discuss them in any detail. In many other cases, the primary products have not yet been characterized and may never be because they are too unstable thermally or because they are themselves rapidly photooxidized.

There are two principal mechanisms of photooxidation (Fig. 1) (2). These have been characterized as "Type I" and "Type II" processes by Gollnick and Schenck (3). The Type I process is characterized by primary interaction of substrate or solvent with an excited state of the sensitizer to give radicals by hydrogen atom or electron transfer. Subsequent reactions of these radicals with oxygen can give oxygenated products. Distinguishing the exact mechanism of the reaction in complex systems is often not at all easy, and some guidelines have been suggested (4). In the Type II process, the excited sensitizer reacts with oxygen to produce singlet molecular oxygen, which reacts in turn with substrates to give the oxygenated products.

TYPE I CHEMISTRY

Only a few examples of this process will be quoted here, since primary products of this process with biological model substrates have been well characterized in only a few instances. Amines undergo dealkylation in many cases by an electron or hydrogen abstraction process; a sequence such as that shown below is probably involved (5). However, a Type II process is sometimes involved (Fig. 2a). Many alcohols react following hydrogen abstraction to give initial radicals (Fig. 2b) which react to give

ISBN 0-12-592050-4

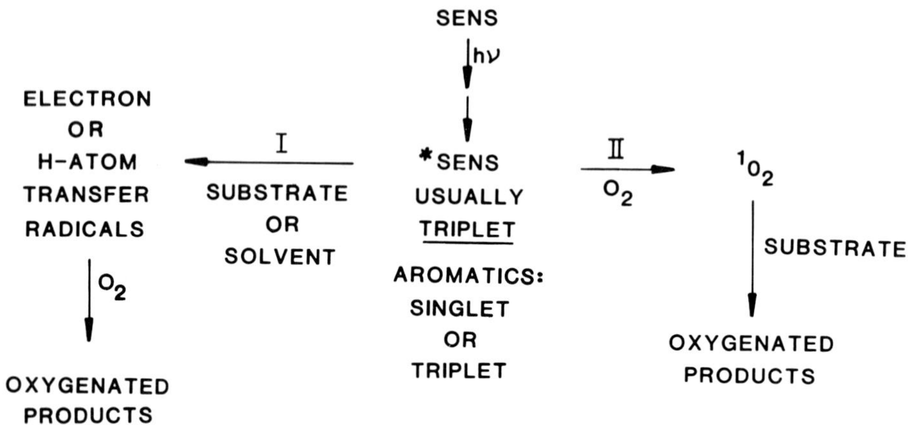

FIGURE 1. Photooxidation Mechanisms.

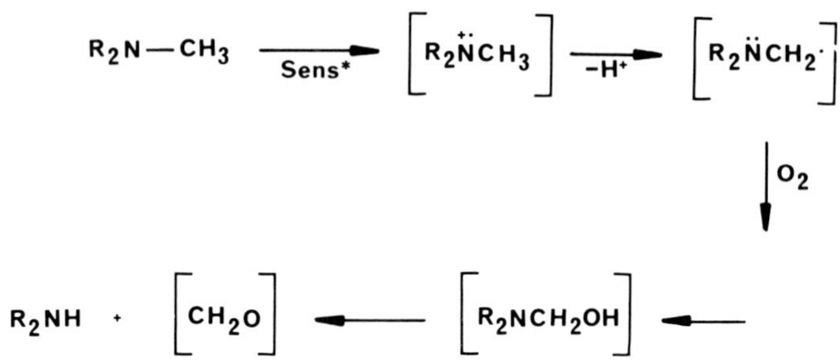

FIGURE 2a. Scheme for dealkylation of amines.

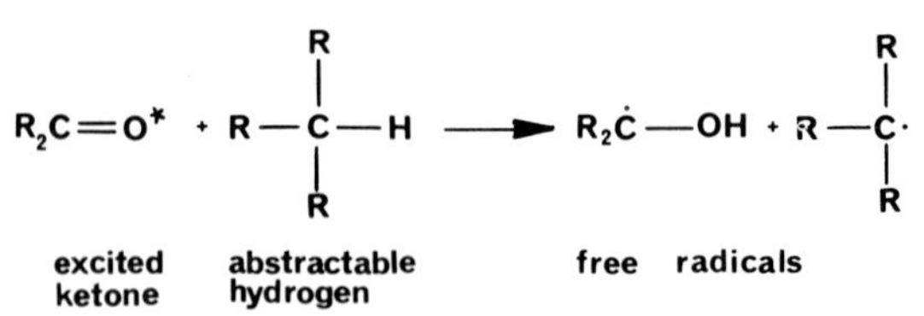

FIGURE 2b. Radical formation by hydrogen abstraction.

α-hydroxyhydroperoxides; these are easily decomposed to give ketones plus H_2O_2, which may be the primary products under some conditions (6).

TYPE II CHEMISTRY

The primary products of the Type II photochemistry have been much better characterized, partly because they have been more extensively studied. There are five principal classes of reaction (2). Class I is a reaction of dienoid compounds (including many heterocycles, aromatic compounds, and related substrates) in which oxygen adds 1,4 to the diene system to give a cyclic endoperoxide. In many cases these endoperoxides are quite stable, but in other cases they do not survive at room temperature (Fig. 3) (2,3,7).

The second class of reaction is analogous to the "ene" reaction, and results in the oxidation of olefins with allylic hydrogens to allylic hydroperoxides in which the double bond has shifted its position. In many cases the hydroperoxides are stable compounds; they can also serve as initiators for autoxidation (Fig. 4) (8).

A third class of reaction involves electron-rich olefins and a few other types of compounds in which a [2 + 2] addition of oxygen to the electron-rich double bond occurs, producing dioxetanes (9). Many dioxetanes are stable at room temperature, but others (particularly those with electron-rich substituents) break down well below room temperature (Fig. 5) (9,10).

FIGURE 3. Endoperoxide formation.

FIGURE 4. Hydroperoxide formation in ene-reaction.

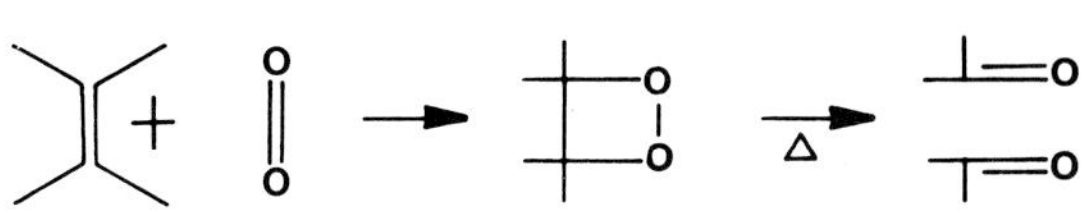

FIGURE 5. Dioxetane formation and fragmentation.

A fourth class of reaction involves oxidation of sulfides to sulfoxides; intermediates are present in this reaction which are capable of being trapped by certain nucleophiles such as sulfides or sulfoxides (Fig. 6) (11).

A fifth class of reaction, which has not yet been studied extensively, is that of phenols; phenols which have two or three alkyl substituents are reactive enough to react readily with singlet oxygen to give hydroxydienones, as shown below (Fig. 7) (12).

The five classes of reaction of singlet oxygen given above provide a basis for the understanding of the reaction of singlet oxygen with biological target molecules. Similar products can be obtained in some cases by other mechanisms such as Type I reaction or free radical oxidation.

The reactions of biological target molecules are summarized below.

METHIONINE (13)

Methionine undergoes photooxygenation, both as the free amino acid and in peptides, on photosensitization by various dyes. The degradation of methionine is responsible for the loss of activity of several important enzymes which are damaged in photodynamic action. The Type II product (singlet oxygen) is the sulfoxide (Fig. 8); in Type I processes, loss of the carboxyl group and eventual formation of methional is the observed course (14).

$$2R_2S \xrightarrow[O_2]{h\nu/Sens} 2R_2\overset{+}{S}-\overset{-}{O}$$

FIGURE 6. Conversion of sulfides to sulfoxides.

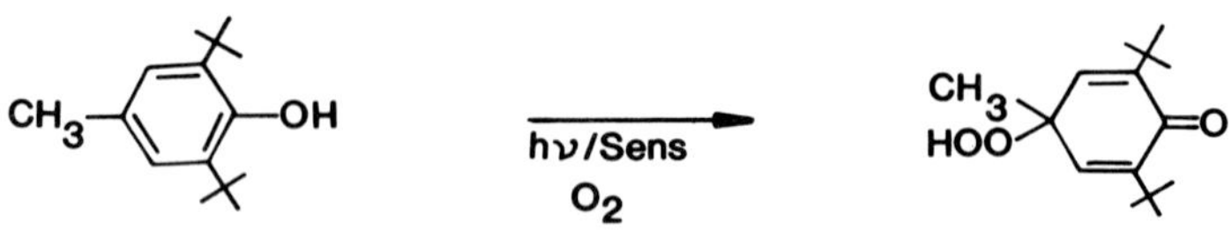

FIGURE 7. Oxidation of phenols.

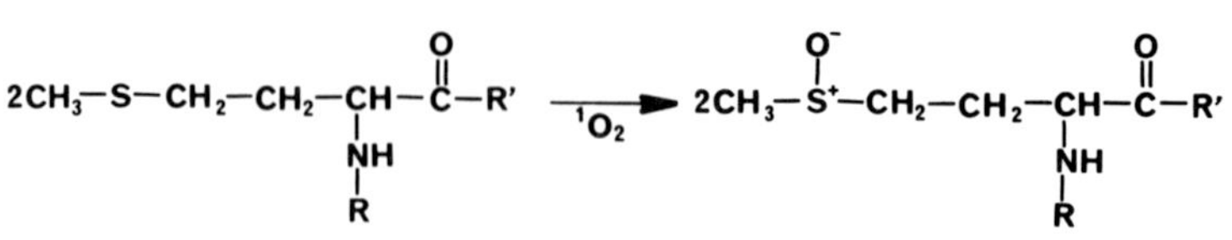 (4)

FIGURE 8. Type II reaction of methionine.

A detailed study of the reactions of methionine has been carried out (13). It was shown that the reactions are dependent on pH. When the pH is low enough that the amino group is protonated (or when the amino group is acylated, as it would be in a peptide), a comparatively simple reaction occurs to give the sulfoxide as the primary detectable product. When the amino group is free (as at high pH when it is not acylated), an intermediate "dehydromethionine" is formed; this intermediate subsequently undergoes slow hydrolysis to give methionine sulfoxide in a surprisingly complex process (Fig. 9).

HISTIDINE

Histidine is one of the more important target amino acids for singlet oxygen. It is very reactive and the first isolable products are usually products of extensive degradation of the heterocyclic ring. Model imidazoles have been studied for many years; one course of reaction which is observed is formation of a hydroperoxide which can undergo ready cleavage as shown below (Fig. 10) (7, 15).

In a study carried out entirely at low temperatures, we were able to show that the primary product of histidine and various related imidazoles is the endoperoxide (16). These endoperoxides are very unstable, particularly when R_1 = H, undergoing decomposition at temperatures as low as -80° . (Fig. 11) In one case at least, it was shown that the primary product of decomposition

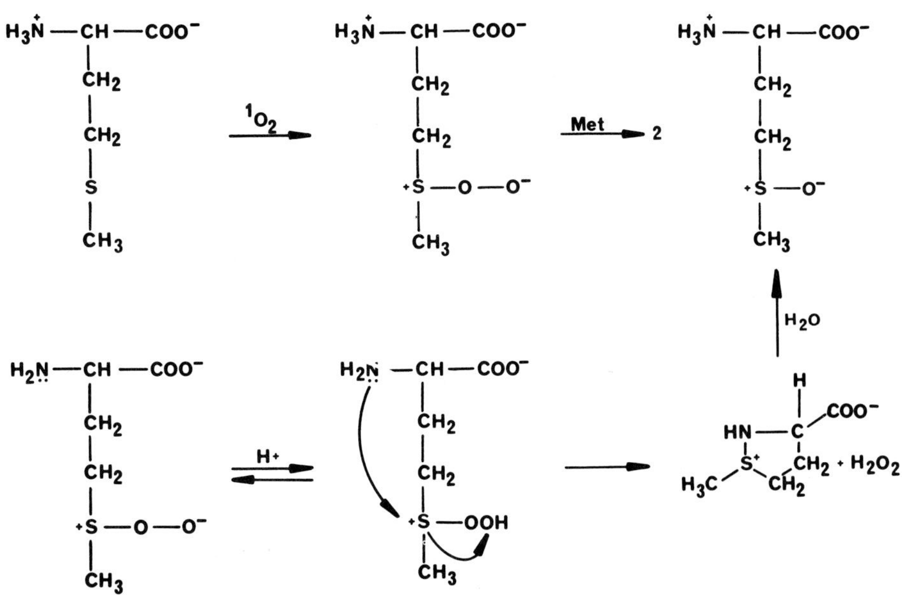

FIGURE 9. Reaction scheme for methionine sulfoxide production.

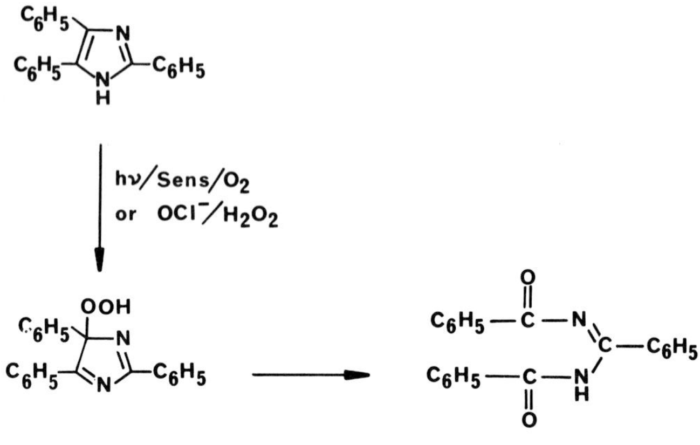

FIGURE 10. *Type II cleavage of model imidazole.*

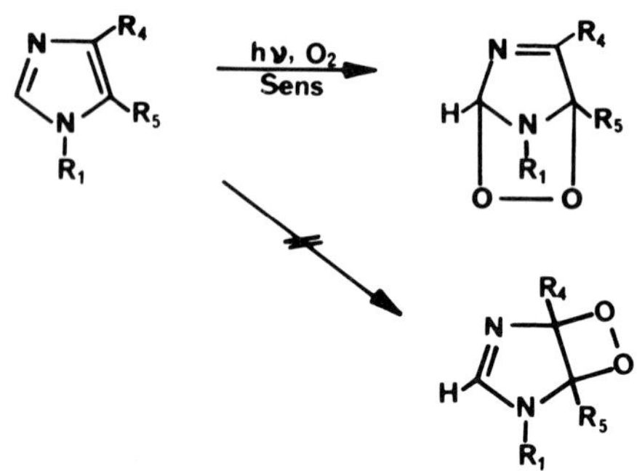

FIGURE 11. *Endoperoxide formation from histidine.*

appears to be the hydroperoxide. On warming, a complex mixture of products is formed which has not been well characterized.

TYROSINE

Tyrosine has not been well studied from the point of view of products. It is not reactive enough to react efficiently with singlet oxygen (12) and is reported to undergo "ring oxidation" under conditions which have not been well defined. Usually, the reaction is a Type I process (2,17).

TRYPTOPHAN

Tryptophan reacts by a mechanism which is not yet certain to involve singlet oxygen; the products are extremely complex in many cases and the course of reaction depends upon many variables (7). At least in one case, an intermediate 1,2-dioxetane has been characterized (18). The primary isolable product appears to be

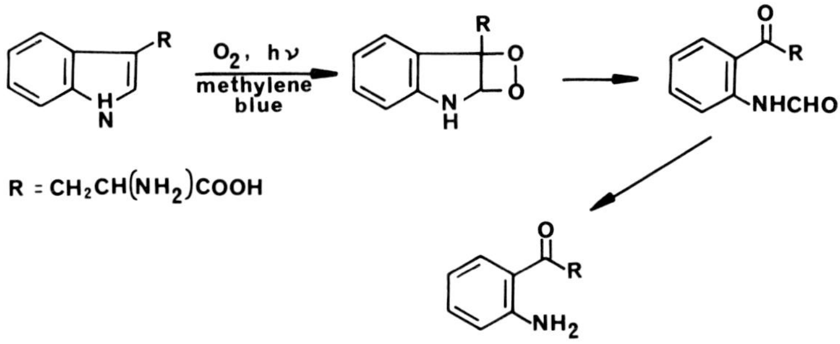

FIGURE 12. Oxidation scheme for tryptophan.

N-formyl kynurenine, which is very readily deformylated to kynurenine (Fig. 12) (19).

CYSTEINE

Cysteine undergoes reactions which are not yet well characterized; it is not certain whether the reaction is with singlet oxygen or not. Under certain conditions, sulfinic and sulfonic acids have been isolated; in other cases disulfide is the isolable product (Fig. 12) (1,2,20).

An important target for photooxidative damage appears to be membranes (1,2,14). Lipid peroxidation is frequently believed to be responsible for loss of membrane stability and biological function. Hydroperoxidation is often an important consequence of this process. Olefins (such as unsaturated fatty acids) can undergo two classes of reaction: after hydrogen abstraction by radicals, they give allylic free radicals which can (in principle, at least) react at either allylic position to give two different allylic hydroperoxides; complex product mixtures are usually found under these conditions, including alcohols, ketones, and epoxides (Fig. 14) (2).

The reaction with singlet oxygen is much simpler; the hydroperoxides which are formed are those in which the double bond is shifted exclusively to the allylic position. When more than one product is possible, often all are formed (as with fatty acids). However, in other cases the product formed is more specific. For example, cholesterol has been used as a probe for singlet oxygen rather frequently (21). In this case, the principal product

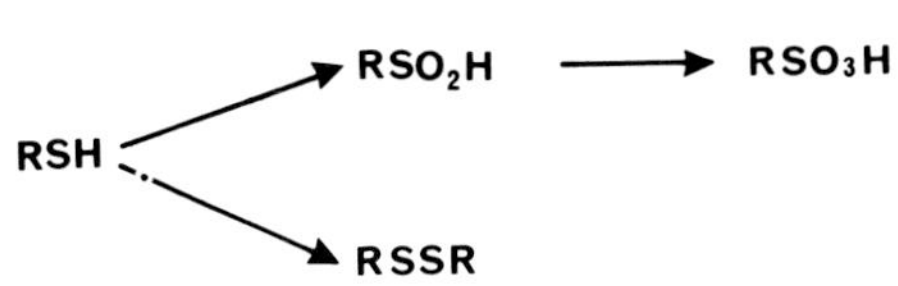

FIGURE 13. Cysteine oxidation products.

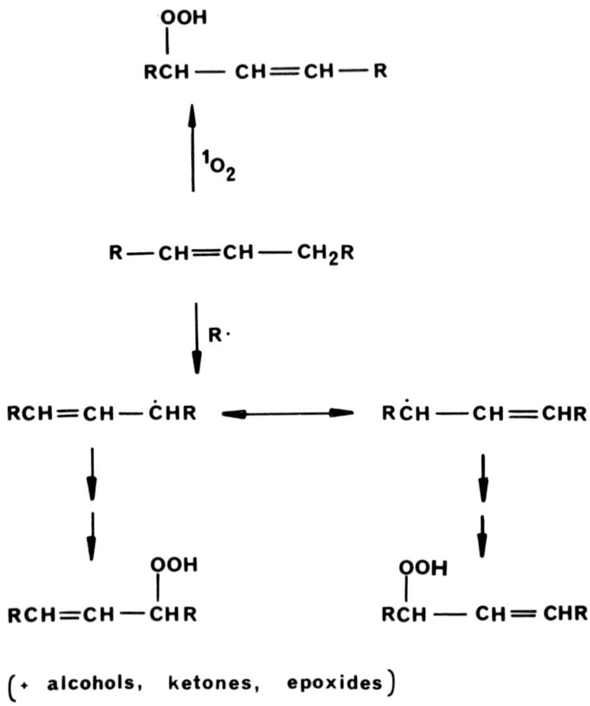

FIGURE 14. Oxidative pathways in unsaturated lipids.

formed is the 5-α hydroperoxide; other peroxides are formed in only a minor amount with singlet oxygen. Under radical conditions, on the other hand, a complex product mixture including the 7-α and β hydroperoxides and many other products is formed. The analysis of this mixture is complicated by the fact that the 5-α product can rearrange to the 7-α and β, although not apparently the reverse. Cholesterol has been suggested to be useful as a marker for the presence of singlet oxygen in biological systems; however, one drawback is its low reactivity, which results in poor sensitivity (Fig. 15).

The reactions of singlet oxygen with nucleic acids have not yet been well characterized (7). Solubility is a problem in many of these cases and primary products are very difficult to determine. Complex product mixtures are usually involved; they can be rationalized as deriving from endoperoxides, dioxetanes, or hydroperoxides in many cases, but the exact course of the reaction is by no means clear. For example the reaction of pyrimidines gives urea, ammonia, CO_2, and other fragments; a possible rationalization is shown below (Fig. 16) (7).

In one case, uracil derivatives have been studied in liquid ammonia at low temperatures; apparently a hydroperoxide is the initial intermediate at low temperature in these cases; on

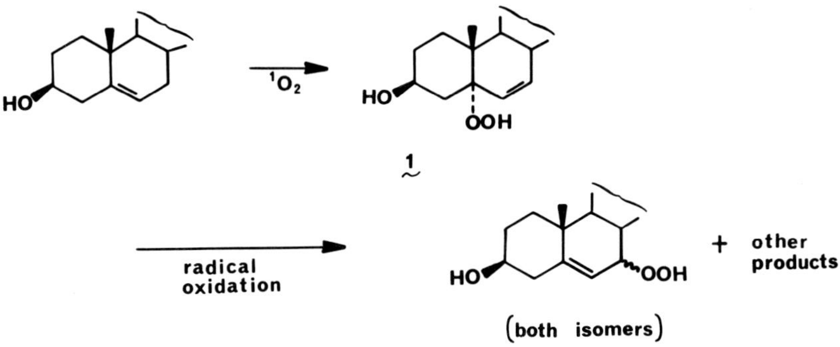

FIGURE 15. Oxidative pathways in cholesterol.

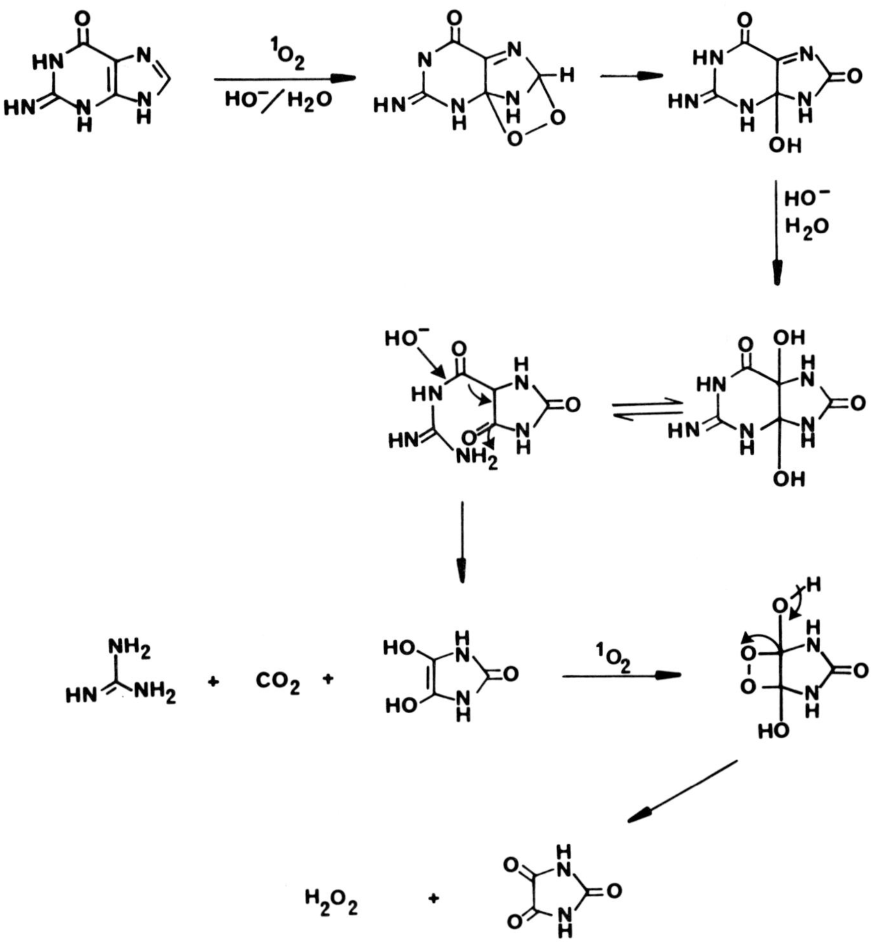

FIGURE 16. Fragmentation of pyrimide by singlet oxygen.

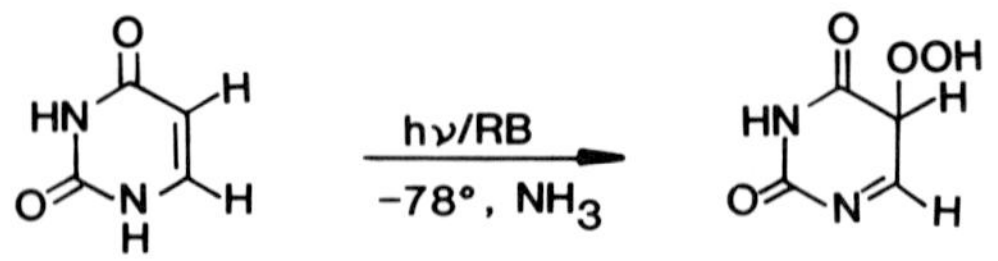

FIGURE 17. Hydroperoxide production from uracil.

warming, this product cleaves (22). However, the biological significance of this work has not been established, since uracil is not normally a significant target for singlet oxygen (Fig. 17) (1,2).

MISCELLANEOUS COMPOUNDS

α-Tocopherol is an important biological antioxidant. It is an excellent free radical chain terminator, and this may be its primary mode of action. However, it is interesting that tocopherol also reacts with singlet oxygen; two reactions have been characterized. One is a quenching reaction in which singlet oxygen is deactivated to the ground state; this reaction is accompanied by a slower irreversible reaction to give product (Fig. 18) (23).

The radio of the rates at which these two processes occur is solvent dependent (4,23). Under most conditions, the quenching reaction is much faster than the reaction to form products. The initial products have been characterized under simple conditions and are exactly analogous to products from substituted phenols (the hydroperoxydienones) (24). Two isomers have recently been shown to exist, which can be separated by chromatography (25). On standing or under certain other conditions, these peroxides break down to give epoxyquinones and quinones, which are the isolated products when care is not taken to preserve the primary products (Fig. 19) (24).

Bilirubin is another interesting biological substrate. This compound has been studied extensively because of the importance of phototherapy to treatment of neonatal jaundice (26). In neonatal jaundice, bilirubin deposits in the skin of the infant. Phototherapy (exposure of the infant to light) causes the disappearance of the bilirubin and seems to prevent the brain damage often associated with the jaundice. It was originally suggested

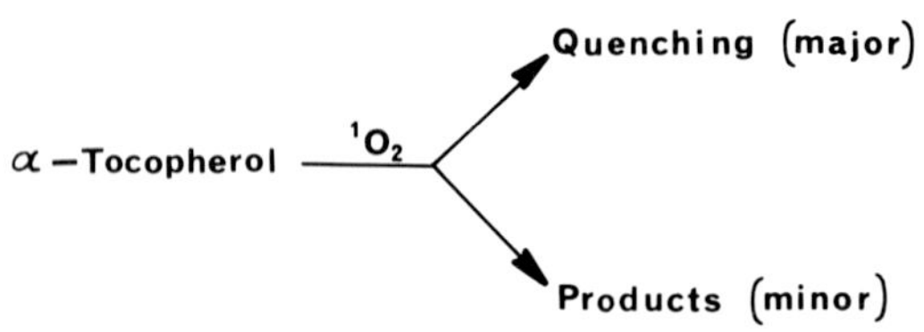

FIGURE 18. Vitamin E interaction with $^1O_2$.

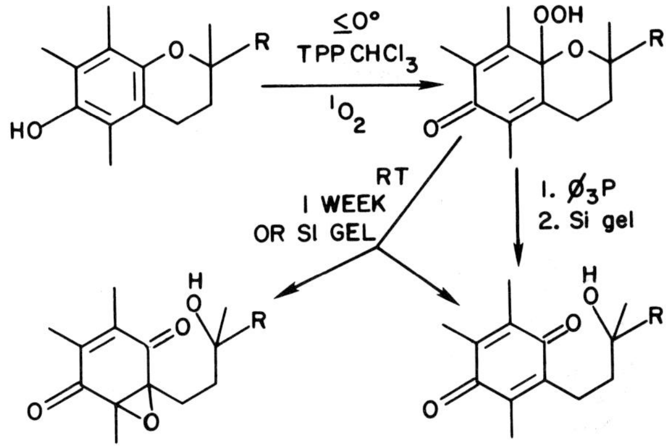

FIGURE 19. Intermediates in α-tocopherol oxidation.

that the disappearance was due to photooxidation, since bilirubin is a weak sensitizer for singlet oxygen and an excellent acceptor (27). Indeed, bilirubin reacts faster than any other known substrate for singlet oxygen. The products of this reaction have been characterized in vitro by Lightner and by Bonnett; they include products of cleavage of the double bond and products which apparently derive from addition of singlet oxygen across dienoid systems (Fig. 20) (28).

However, recent work has suggested that the photooxidation of bilirubin may not be as important to phototherapy as its photoisomerization; bilirubin undergoes ready photolysis to give a photoisomer which is much more water soluble than bilirubin and can be readily excreted; this may be the key reaction for the phototherapy (29).

CAROTENES

Carotenes appear to act as biological protective agents

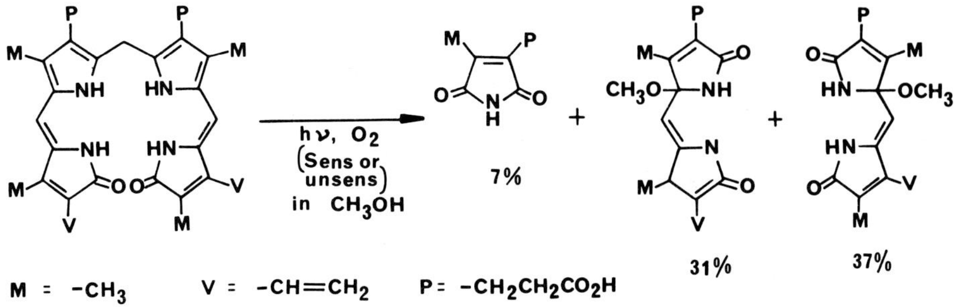

FIGURE 20. Products from bilirubin oxidation.

FIGURE 21. *Energy transfer between $^1O_2$ and β-carotene.*

against singlet oxygen in some cases (30). Most photosynthetic organisms contain carotenoids; in many cases, carotenoidless mutants are known, but light in the presence of oxygen is toxic to them. Also, normal organisms in which carotenoid synthesis has been blocked chemically are subject to killing by light in the presence of oxygen. Thus, the carotenoids have been known for a long time to exert an important protective function in photosynthetic organisms (30). One mechanism for this protective action may be the quenching of singlet oxygen (2,4,31). It has been shown that singlet oxygen is quenched at a diffusion-controlled rate by β-carotene; this process is very efficient, and results in destruction of the carotene in only about one in 10^4 deactivating collisions (Fig. 21).

The dependence of the efficiency of this process on the number of conjugated double bonds in the molecules suggests that a major decrease in the efficiency of quenching of singlet oxygen occurs between 9 and 7 conjugated double bonds; the biological protective activity falls off sharply in this region as well, suggesting that the mechanisms of quenching of singlet oxygen and of protective action against photodynamic damage may be the same (2,4,31). The probable mechanism of the protective reaction is energy transfer from singlet oxygen to the carotene to give a metastable carotenoid triplet which readily reverts to the ground state.

An important application of these phenomena has been found recently in the treatment of erythropoietic protoporphyria (32). Patients with this disease deposit porphyrins in the skin. The patients are very sensitive to light, and the porphyrins appear to act as photosensitizers. A recently developed symptomatic treatment for this disease involves administering β-carotene to the patient; the carotene deposits in the skin, giving a slight yellowish cast. The photosensitivity syndrome is simultaneously relieved (32). It is probable that the mechanism of this protective action also involves quenching of singlet oxygen, but this has not yet been established.

This article has been a brief summary of certain reactions of singlet oxygen with certain important biological target molecules; it is not certain that these reactions are involved in photodynamic action, but it is likely that they are, at least in some cases. These reactions need to be elaborated in more complex systems, a process which is only in its infancy.

ACKNOWLEDGEMENT

The work in our group has been supported by NSF and NIH grants.

REFERENCES

1. Spikes, J. D. and Livingston, R., Adv. Radiat. Biol., 3,29 (1969); Spikes, J. D. and MacKnight, M. L., Ann. N. Y. Acad. Sci., 171,149 (1970); Krinsky, N. I., "The Survival of Vegetative Organisms" (T. G. R. Gray and J. R. Postgate, eds.), Cambridge University Press, New York, 1961, p. 209; Spikes, J. D., Ann. N. Y. Acad. Sci., 244,496 (1975); Spikes, J. D., this volume, p. 421. Grossweiner, L. I., Current Topics in Radiation Research Quarterly, 11,141 (1976).
2. Foote, C. S., Science, 162,963 (1968); Foote, C. S., "Free Radicals in Biology," Vol. II (W. A. Pryor, ed.), Academic Press, New York, (1976), p. 85; Foote, C. S., "Active Oxygen and Medicine" (A. P. Autor, ed.), Raven Press, New York, (1980).
3. Gollnick, K., Advan. Photochem., 6,1 (1968); Gollnick, K. and Schenck, G. O., "1,4-Cycloaddition Reactions" (J. Hamer, ed.), Academic Press, 1967, p. 255; Schenck, G. O. and Koch, E., Z. Elektrochem., 64,170 (1960).
4. Foote, C. S., "Biochemical and Clinical Aspects of Oxygen" W. S. Caughey, ed.), Academic Press, New York, 1979, p. 603; Foote, C. S., "Singlet Oxygen" (H. H. Wasserman and R. W. Murray, eds.), Academic Press, New York, 1979, p. 139.
5. Bartholemew, R. F. and Davidson, R. S., Chem. Commun., 1174 (1970); Cohen, S. G., Stein, N. and Chao, H. M., J. Am. Chem. Soc., 90,521 (1968);
5a. Lindner, J. E., Kuhn, H.-J. and Gollnick, K., Tetrahedron Lett., 1705 (1972); Fisch, M. H., Gramain, J. C. and Olesen, J. A., Chem. Commun., 13 (1970).
6. Schenck, G. O., Becker, H.-D., Schulte-Elte, K.-H. and Krauch, C.-H., Chem. Ber., 96,509 (1963).
7. Wasserman, H. H. and Lipshitz, B. H., "Singlet Oxygen" (H. H. Wasserman and R. W. Murray, eds.), Academic Press, New York, (1979), p. 430.
8. Gollnick, K. and Kuhn, H.-J., "Singlet Oxygen" (H. H. Wasserman and R. W. Murray, eds.), Academic Press, New York, (1979), p. 287.
9. Schaap, A. P. and Zaklika, K. A., "Singlet Oxygen" (H. H. Wasserman and R. W. Murray, eds.), Academic Press, New York, (1979), p. 174.
10. Bartlett, P. D. and Landis, M. E., "Singlet Oxygen" (H. H. Wasserman and R. W. Murray, eds.), Academic Press, New York, (1979), p. 244.

11. Foote, C. S. and Peters, J. W., "XXIIIrd International Congress of Pure and Applied Chemistry, Special Lectures," Vol. 4, Butterworth & Co. Ltd., England, 1971, p. 129; Foote, C. S. and Peters, J., J. Am. Chem. Soc., 93,3795 (1971).
12. Thomas, M. J. and Foote, C. S., Photochem. Photobiol., 27, 683 (1978); see also I. Saito and T. Matsuura, "Singlet Oxygen" (H. H. Wasserman and R. W. Murray, eds.), Academic Press, New York, (1979), p. 511; Matsuura, T., Yoshimura, N., Nishinaga, A. and Saito, I., Tetrahedron, 28,4933 (1972); Pfoertner, K. and Böse, D., Helv. Chim. Acta, 53,1553 (1970); Saito, I., Kato, S. and Matsuura, T., Tetrahedron Lett., 239 (1970).
13. Sysak, P. K., Foote, C. S. and Ching, T.-Y., Photochem. Photobiol., 26,19 (1977).
14. Spikes, J. D. and McKnight, M. L., "Research Progress in Organic, Biological and Medicinal Chemistry," Vol. 3 (U. Gallo and L. Santamaria, eds.), American Elsevier Co., New York, (1972), p. 124; S. Cohen and S. Ojanpera, J. Am. Chem. Soc., 97,5633 (1975).
15. Sonnenberg, J. and White, D. M., J. Am. Chem. Soc., 86,5685 (1964); White, E. H. and Harding, M. J. C., ibid., 86,5686 (1964).
16. Ryang, H.-S. and Foote, C. S., J. Am. Chem. Soc., 101,6683 (1979).
17. Weil, L., Gordon, W. G. and Buchert, A. R., Arch. Biochem. Biophys., 33,90 (1951).
18. Saito, I., Matsuura, T., Nakagawa, M. and Hino, T., Acc. Chem. Res., 10,346 (1977).
19. Savige, E. W., Aust. J. Chem., 24,1285 (1971).
20. Jori, G., Galiazzo, G. and Scoffone, E., Int. J. Protein Res., 1,289 (1969); Weil, L., Arch. Biochem. Biophys., 110, 57 (1965).
21. Kulig, M. J. and Smith, L. L., J. Org. Chem., 38,3639 (1973); Nickon, A. and Bagli, J. F., J. Am. Chem. Soc., 81,6330 (1959); ibid., 83,1498 (1951); Schenck, G. O., Gollnick, K. and Neumüller, O.-A., Justus Liebigs Ann. Chem., 603,46 (1957); Schenck, G. O. and Neumüller, O.-A., ibid., 618,194 (1958); Smith, L. L. and Hill, F. L., J. Chromatogr., 66,101 (1972); Smith, L. L., Teng, J. I., Kulig, M. J. and Hill, F. L., J. Org. Chem., 38,1763 (1973).
22. Vickers, R. S. and Foote, C. S., Boll. Chim. Farm., 109,599 (1970).
23. Fahrenholtz, S. R., Doleiden, F. H., Trozzolo, A. M. and Lamola, A. A., Photochem. Photobiol., 20,505 (1974); Foote, C. S., Ching, T.-Y. and Geller, G. G., ibid., 20,511 (1974); Grams, G. W. and Eskins, K., Biochem., 11,606 (1972); Stevens, B., Small, R. D., Jr. and Perez, S. R., Photochem. Photobiol., 20,515 (1974).
24. Clough, R. L., Yee, B. G. and Foote, C. S., J. Am. Chem. Soc. 101,683 (1979).

25. Yamauchi, R. and Matsushita, S., Agric. Biol. Chem., 43,2151 (1979).
26. "Metabolism in the Newborn" (D. Bergsma, D. Y.-Y. Hsia and C. Jackson, eds.), Williams and Wilkins, Baltimore, MD, 1970; "Phototherapy in the Newborn: An Overview" (G. B. Odell, R. P. Schaffer and A. P. Simopoulos, eds.), National Academy of Sciences, Washington, (1972); Cremer, R. J., Perryman, P. W. and Richards, D. H., Lancet, 1,1094 (1958).
27. Bonnett, R. and Stewart, J. C. M., Biochem. J., 130,895 (1972); McDonagh, A. F., Biochem. Biophys. Res. Commun., 44, 1306 (1971); "Phototherapy in the Newborn: An Overview" (G. B. Odell, R. P. Schaffner and A. P. Simopoulos, eds.), National Academy of Sciences, Washington, (1972), p. 56; Foote, C. S. and Ching, T.-Y., J. Am. Chem. Soc., 97,6209 (1975).
28. Bonnett, R. and Stewart, J. C. M., J. Chem. Soc. Chem. Commun., 596 (1972); Lightner, D. A. and Quistad, G. B., FEBS Lett., 25,94 (1972).
29. McDonagh, A. F. and Ramonas, L. M., Science, 201,829 (1978); McDonagh, A. F., Palma, L. A. and Lightner, D. A., Science, 208-145 (1980).
30. Krinsky, N., "Carotenoids" (O. Isler, ed.), Birkhauser Verlag, Basel, (1971), p. 669; Krinsky, N. I., "Pure & Appl. Chem.", Vol. 51, Pergamon Press Ltd., England, (1979), p. 649.
31. Foote, C. S. and Denny, R. W., J. Am. Chem. Soc., 90,6233 (1968); Foote, C. S., Chang, Y. C. and Denny, R. W., J. Am. Chem. Soc., 92,5219 (1970); Farmilo, A. and Wilkinson, F., Photochem. Photobiol., 18,447 (1973).
32. Deleo, V. A., Poh-Fitzpatrick, M., Mathews-Roth, M. M. and Harber, L. C., Am. J. Med., 60,8 (1976); Mathews-Roth, M. M., Pathak, M. A., Fitzpatrick, T. B., Harber, L. C. and Kass, E. H., N. E. J. Med., 282,1231 (1970).

DISCUSSION

ASMUS: Looking at the data you presented on the sulfur-containing compounds, I noticed that there seem to be great similarities in the chemistry of singlet oxygen and ·OH radicals. I think there is no time and need to go into the details of methionine, but there are lots of similar species, and the interaction of sulfur and nitrogen is just the same when ·OH radical induces oxidation. One also observes the dependence on the solute concentrations and the dependence on the protonation state of amino group and carboxyl group. You also mentioned the formation of disulfide, and sulfonic and sulfinic acids from thiols in alternative routes. Now I would suggest, in analogy to the ·OH radical reactions, that whenever you observe the acids that this goes through disulfide. Disulfides, once they are formed from the oxidation of the thiols, are known to undergo reactions with ·OH radicals to form these acids. So you may have a similar reaction here.

FOOTE: The thiol sulfinates are produced by reaction of singlet oxygen with sulfides and are not very reactive to singlet oxygen and probably would survive. I do not think this reaction has been very well characterized.

STEVENS: On the question of Type I and Type II reactions, I wonder if one can get some information from the stability of the sensitizer? When you photooxidize these amino acids with methylene blue, the methylene blue disappears. It not only acts as a sensitizer but is presumably reacting as well. I wonder whether one can use sensitizer consumption as a criterion of some type I reaction taking place?

FOOTE: In some cases where Type I reactions occur the sensitizer is bleached. However, there are many cases where unstable and very reactive peroxides are formed which are capable of bleaching dyes. For example, the furan endoperoxides react with dyes. The histidine endoperoxides also react directly with the dyes to bleach them.

WINTERLE: I am very interested in the hydroperoxy derivative obtained from Vitamin E that you mentioned. We have been reacting Vitamin E with aklyl peroxy radicals, and we see a very similar product at this point. I am very interested—have you done a controlled hydrolysis of this peroxy ketal?

FOOTE: We did quite a bit of uncontrolled hydrolysis before we were able to hold onto it. This was published in JACS about a year and a half ago. We have not done much other work with it. Matsushita may have; he has a recent paper in a Japanese journal.

WINTERLE: One more question—do you propose that you have free phenoxyl radicals present during this reaction?

FOOTE: That would require a little too long to answer here. We think they are not free, but we do not have much evidence on it. We do have spectroscopic evidence in one case (obtained by Mike Thomas) that phenoxyl radical-like species are an intermediate... but perhaps not free.

PHOTOSENSITIZED OXIDATIONS IN COMPLEX BIOLOGICAL STRUCTURES

Giulio Jori[1] *and John D. Spikes*[2]

[1]*Centro C.N.R. Emocianine, Istituto di Biologia Animale, Università di Padova, 35100 Padova, Italy*
[2]*Department of Biology, University of Utah, Salt Lake City, Utah*

INTRODUCTION

Although some of the biochemical lesions in cells resulting from the photoexcitation of endogenous and exogenous sensitizers in the presence of oxygen are well-characterized, in general neither the detailed chemistry of the photoprocesses nor the mechanisms leading to the injury or death of cells is well understood. This is partly due to the large number of organic and inorganic compounds (including flavins, porphyrins, acridines, phenothiazines, xanthenes, ketones, quinones, polyenes, haloaromatics, Cu^{2+} and Fe^{3+} ions, etc.) which can act as sensitizers for Type I (hydrogen or electron transfer) or Type II (singlet oxygen) photoprocesses in cells (1-3). With the presence of multiple sensitizers, photoreactions involving different cellular targets may occur simultaneously, even where narrow wavelength bands are used for illumination.

The photophysical parameters of a sensitizer, including the singlet-triplet intersystem crossing quantum yield and the triplet lifetime, as well as the lifetimes and reactivities of the possible intermediate species ($^1O_2$, $O_2^{\bar{\cdot}}$, H_2O_2, e_s^-, dye and substrate radicals) are strongly dependent on the properties of the surrounding medium (1,3). Thus, different solvents may increase or decrease the efficiency of a photosensitizer of even shift a photoreaction mechanism from Type I to Type II, or vice-versa. Living cells are intrinsically non-homogeneous and thus provide a great variety of microenvironments for photosensitizers and substrates; for this reason it is difficult to extrapolate information obtained from photochemical studies in homogeneous solutions to the cell (3,4). In particular, most sensitizers are bound noncovalently in different sites in cells thus often altering their photochemical behavior. Further, the cell environment is dynamic; this requires that movements of sensitizers and substrate be taken into account in developing descriptions of the mechanisms of photosensitization phenomena in cells. In spite of these complications, relevant model systems do provide some clues as to how selected factors influence the course of a photoprocess in vivo, as described in the present paper.

ISBN 0-12-592050-4

BIOCHEMISTRY OF IN VIVO PHOTOSENSITIZATION

Ito made an important step in attempting to rationalize the accumulated information on cell photosensitization by the proposal to classify photoprocesses on the basis of sensitizer localization and of accessibility of the substrate and molecular oxygen to triplet state sensitizer (5). He distinguishes three major situations. In the first, sensitizer remains outside of the cell or is located in the cell membrane, as with thiazines and some other cationic dyes (6,7). On illumination, unsaturated lipids and cholesterol in the membrane are readily attacked at the ethylenic linkage. The easy abstraction of H atoms by triplet sensitizers leads to free radical oxidation processes resulting in the ultimate formation of hydroperoxides (8) or in chain scission (9). Alternatively, allylic hydroperoxides, endoperoxides and 1,2-dioxetanes can result from attack by photochemically generated $^1O_2$ (8, 10). Certain amino acid side chains in proteins are photooxidized with approximately equal efficiency. Typically, only electron-rich sites are photosusceptible, including the unprotonated imidazole ring of histidine, the phenolate anion of tyrosine, the indole ring of tryptophan, the thioether group of methionine and the thiol group of cysteine (1,3). Some reaction pathways involve radical-type species and can result in the formation of homo- or hetero-adducts between spatially adjacent protein molecules (11) and between protein and lipid molecules (12). Singlet oxygen does not attack carbohydrates appreciably (13), however ketonic and quinoid photosensitizers promote H-abstraction pathways resulting in the formation of onic acids. The photosensitized cross-linking of carbohydrates with other biomolecules has not been reported (14). These various photochemical changes severely alter membrane function: ion pumps and sugar and amino acid transport systems are inhibited, several enzymic activities are lost (e.g., cytochrome oxidase, acetylcholinesterase, ATPase, glyceraldehyde-3-phosphate dehydrogenase), and cells often lyse because of increased osmotic fragility.

In the second situation, the sensitizer penetrates into the cell and localizes at various cytoplasmic sites, as with certain xanthene dyes, thiopyronine and some porphyrins (1,5). These sensitizers probably bind in complex equilibria with such cytoplasmic components as the membranes of cytoplasmic organelles (mitochondria, lysosomes, etc.), with ribosomes and microsomes, and with cytoplasmic macromolecules (proteins, t-RNA). Both proteins (enzymes) and t-RNA are likely targets of photodynamic attack; the latter is preferentially photooxidized at guanine residues by both Type I and Type II mechanisms in reactions that are not well-characterized (1,3). Other pathways involving energy transfer from triplet sensitizer to nucleic acid bases and the formation of covalent sensitizer-nucleic acid adducts have also been reported (14). Protein-protein and protein-RNA cross-links may also be formed, probably by Type I processes (14). Cytoplasmic enzymes such as aminoacyl-tRNA-synthetases as well as some involved in respiratory sequences are inactivated (1,5). Cell-injury and killing are also mediated by the photoinduced release of lysosomal hydrolytic enzymes into the cytoplasm (15).

The third cellular situation involves penetration of the sensitizer into the nucleus; in this case, DNA components are the principal sites of damage on illumination (5). The same kinds of chemical changes as described for t-RNA occur, although competition between the reaction pathways is influenced both by the mode of binding (stacking, intercalation) and by the photophysical properties of the sensitizer. DNA-protein photoadducts are often formed. Acridine dyes frequently sensitize these nuclear effects with the principal biological consequences of DNA strand breakage and the production of mutations.

In summary, the photodynamic treatment of cells is a very complex phenomenon, and may be further complicated by the partition of endogeneous and exogeneous sensitizers among various cell binding sites (16). It should be mentioned that the presence of relatively high concentrations of free sensitizer in the medium around cells may optically screen complexed sensitizer molecules and thus alter the course of the photoprocess. Thus, in the future, it is of the greatest importance to identify the actual binding sites of sensitizers in cells as well as to determine binding affinity and the saturation concentration of the sensitizer.

PHOTOSENSITIZED OXIDATIONS BY POLYMER-BOUND SENSITIZERS

Cell constituents such as proteins, nucleic acids and polysaccharides form non-covalent complexes with a variety of photosensitizers. In many cases, such binding increases the photochemical efficiency of the sensitizer (17-20). Several examples of this behavior are given in Table I. The data shown have only a qualitative significance because of the somewhat different experimental conditions used in the measurements with free and bound dyes. This effect, of obvious importance in cellular photosensitization, results from the reduced probability of collisional deactivation of bound electronically-excited dye molecules; thus dyes located in the interior of macromolecules often show phosphorescent emission at room temperature (21). Further, the binding of dyes in polymeric matrices often increases the steady-state triplet population, as might be expected from the typical bathochromic shift of the absorption peak of bound dyes and the consequent smaller singlet-triplet splitting (22).

The enhanced photosensitizing efficiency of bound dyes is particularly evident in the case of nucleic acids (23). Illumination of calf thymus DNA in the presence of intercalating and non-intercalating dyes (Table II) shows that, with comparable light absorption, dye binding increases photodamage (24,25). Photochemical efficiency is favored by a tight and rigid environment for

TABLE I. Effect of Ground State-Complexation on the Photosensitizing Efficiency of Selected Dyes

Substrate	*Photosensitizer*	*Measured Effect*	*Increased Efficiency*	*Reference*
Histidine	*RB + 1% PMA vs. Free RB*	*Rate of oxygen consumption*	*1.78*	*17*
Polyadenylic acid	*PolyA-bound Pf vs. Free Pf*	*EPR signal amplitude due to peroxydic radicals*	*1.87*	*18*
Lysozyme Denat. lysozyme	*Protein-bound Rf vs. Free Rf*	*Quantum yield for the photo-sensitized inactivation*	*11.43*	*19*
Native HSA Denat. HSA	*Protein-bound Br vs. Free Br*	*Rate of photooxidation of tryptophyl residues*	*2.27*	*20*

Abbreviations: RB, rose bengal; PMA, polymethacrylic acid; Pf, proflavine, Rf, riboflavin; HSA, human serum albumin; Br, bilirubin.

TABLE II. Effect of the Mode of Dye Binding on the Photosensitized Induction of Free Radicals on Calf-Thymus DNA

Dye	*Mode of binding*	*Absorption maximum (nm)*	*Molar absorptivity*	*Radical Yield (a.u.)*
Proflavine	*Intercalation*	*444*	*33,500*	*1,091*
Chloroquine	"	*329*	*19,700*	*625*
9-Aminoacridine	"	*398*	*10,850*	*325*
Quinacrine	"	*425*	*8,900*	*125*
Ethidium bromide	"	*480*	*5,600*	*162*
Thiopyronine	*Stacking*	*564*	*46,960*	*450*
Pyronine	"	*538*	*13,650*	*175*
Acridine red	"	*548*	*8,320*	*98*

Data from refs. 24 and 25. Experimental conditions: dye concentration, 0.04-0.50 mM; DNA concentration, 1.56 mM; solvent, aqueous 0.1 M NaCl; light source, high-pressure Hg lamp (500 W) with wavelengths below 320 nm removed by a cut-off filter.

the sensitizer as found in photooxidation studies with intercalated proflavine-DNA complexes; supercoiled double-stranded DNA is approximately ten times more photolabile than the corresponding circular single-stranded DNA (25). Analogous results are observed with proteins. For example, the rate of inactivation of lysozyme as sensitized by bound riboflavin, increases with decreasing temperature (this is the reverse of the temperature dependence observed with unbound sensitizers—see ref. 26); the effect was enhanced at -20°C in non-freezing viscous media (19). Analogously, the photoinactivation rate of eosin-complexed lysozyme is greater than that of the uncomplexed enzyme (27).

The ground state complexing of dyes with macrobiomolecules can also influence the mechanism of the photoprocess, usually by enhancing the probability of a direct interaction between the triplet sensitizer and substrate through H or e^- transfer. This is a consequence of the close spatial relationship between the bound dye and the photosensitive functional groups of the substrate. The various possible fates of the light-generated radicals are shown in Fig. 1. The $SO_2^{\cdot}$ species usually react further to form stable photoproducts, while the substrate and dye radicals (S· and D·, respectively) may recombine (2) with the formation of intramolecular crosslinks, covalent substrate-substrate photoadducts (S-S) and covalent substrate-dye photoadducts (S-D). The latter reaction is frequently observed with proteins (19,28-31) as summarized in Table III. In fact, the photogeneration of

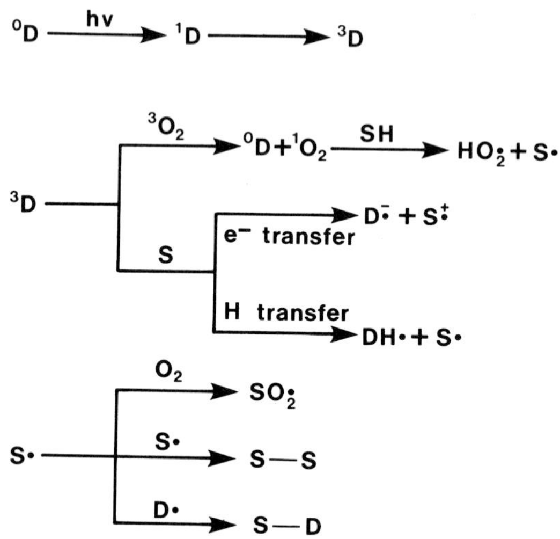

FIGURE 1. Schematic representation of the reaction pathways open to the dye (D) and substrate (S) radicals formed in the early stages of photosensitized processes; S-S is a covalent substrate-substrate adduct and S-D is a covalent substrate-dye adduct.

TABLE III. Photoadduct Formation with Different Protein-Sensitizer Ground State Complexes

Protein	*Sensitizer*	*Estimated yield of photoadduct (%)*	*Reference*
BSA	*Fl, AO, MB*	*5-10*	*28*
HSA	*TCSA*	*5*	*29*
RNase	*Fl*	*5*	*28*
Lysozyme	*Rf*	*n.d.*	*19*
Histone type II	*MOP*	*n.d.*	*30*
BSA, HSA, RNase	*MOP*	*10-15*	*30*
HSA	*Br*	*3-5*	*31*

Abbreviations: BSA, HSA = bovine and human serum albumin; Fl, fluorescein; AO, acridine orange; MB, methylene blue; TCSA, 3,3-, 4',5-tetrachloro-salicylanilide; Rf, riboflavine; MOP, 8-methoxy-psoralen; Br, bilirubin; n.d. not determined.

dye-protein covalent bonds is of such general occurrence that the detection of such products should be a reliable indication that a radical-type photoprocess is operative, even in living systems (31). Dye-protein pre-binding appears to be essential for the production of covalent dye-protein photoadducts. For example, neither 8-methoxypsoralen nor bilirubin bind to denatured albumin, and illumination of the mixtures did not give sensitizer-protein photoadducts (30,31). Further, tetrachlorosalicylanilide, which shows no ground state interaction with free amino acids or gamma-globulin, does not photobind to these substrates (29).

Ito (5) suggested that the prevalence of Type I mechanisms in the sensitized photooxidation of dye-biopolymer complexes was a consequence of poor accessibility of ground state oxygen to the bound sensitizer molecules, since $^1O_2$ production requires energy transfer from triplet sensitizer to oxygen by an overlap interaction (32). However, the photodamage to DNA as sensitized by both acridine orange and proflavine is mediated primarily by $^1O_2$ when the systems are illuminated at the absorption peak of dye molecules stacked on the outside of the DNA (24,25). Also, DNA base radicals photogenerated by e^- transfer from intercalated proflavine are readily scavenged by ground state oxygen; thus other factors may be as important as oxygen-accessibility in controlling the competition between the various photoreaction pathways possible in dye-biopolymer complexes. For example, in the 1:1 eosin-lysozyme complex, the bound eosin sensitizes the photoinactivation of the enzyme largely by a $^1O_2$ mechanism, although the Type I mechanism also occurs (27). Hematoporphyrin and bilirubin form 1:1 complexes of comparable stability with human serum albumin ($k{=}\sim 10^7\ M^{-1}$); fluorescence studies indicate that the porphyrin binding site is readily accessible to the aqueous solvent whereas the bilirubin cleft is deeply buried (20). Both dyes sensitize the photooxidation of the single tryptophyl residue in the human serum albumin molecule by a pure $^1O_2$ mechanism, but the efficiency of the hematoporphyrin-promoted photoprocess is at least eight-fold higher. However bilirubin photobinds to the albumin with a yield approximately 25 times greater than that for hematoporphyrin.

Although only a few systems have been examined, the data suggest that radical-mediated processes are more efficient in low polarity media while singlet oxygen mechanisms tend to predominate in relatively polar environments. An examination of the hematoporphyrin-sensitized photooxidation of tryptophan in homogeneous solutions of different polarity shows that the $^1O_2$ pathway is efficient in polar media (e.g., in water-formamide or water-30% methanol mixtures) whereas a Type I mechanism becomes more important as the dielectic constant of the medium goes below 20, as in low water-high methanol mixtures (33).

Finally, it should be pointed out that other structural and experimental factors can influence the relative participation of Type I and Type II processes in sensitized photoreactions (e.g., stabilization of dye or substrate radical ions by vicinal charges; the relative distance from and reactivity of functional groups in the substrate toward the triplet dye, $^1O_2$, and other reactive photointermediates, etc.). Photosensitization studies involving dyes complexed at known sites on macrobiomolecules of known three-dimensional structure as well as sensitized photochemical

TABLE IV. *Reactivity of $^1O_2$ in Homogeneous Solution and in Aqueous Micellar Dispersions*

Sensitizer	Substrate	Reaction rate constant ($M^{-1}s^{-1}$)	Quantum yield for $^1O_2$	Reference
Pyrene(MeOH)	DPBF(MeOH)	7.8×10^8	0.6	35
Pyrene(SDS)	DPBF(SDS)	2.2×10^{10}	0.9	
Pyrene(DTAC)	DPBF(DTAC)	2.8×10^{10}	1.0	
Chlorophyll(aq)	Imidazole(aq)	2.0×10^7		36
Chlorophyll(Triton)	Imidazole(aq)	2.9×10^7	0.7-0.8	
Methylene blue(EtOH)	DMA(EtOH)	3.8×10^8		37
Methylene blue(aq)	DMA(DTAC)	7.5×10^8		
Eosin(aq)	DMA(DTAC)	9.1×10^8		
ACN(Benzene)	DPBF(Benzene)	9.4×10^8		38
ACN(SDS)	DPBF(SDS)	10.9×10^8		
ACN(CTAB)	DPBF(CTAB)	6.2×10^8		
ACN(Igepal)	DPBF(Igepal)	6.7×10^8		
Methylene blue(aq)	DPBF(SDS)	10.0×10^8		38
Methylene blue(aq)	DPBF(CTAB)	6.2×10^8		
Methylene blue(aq)	DPBF(Igepal)	6.5×10^8		

Abbreviations: MeOH, methanol; EtOH, Ethanol; SDS, sodium dodecyl sulfate; DTAC, dodecyltrimethylammonium chloride; aq, aqueous; Tri, Triton X-100; CTAB, cetyltrimethylammonium bromide; ACN, 2-acetonaphthone; DPBF, 1,3-diphenylisobenzofuran; DMA, 9,10-dimethylanthracene.

experiments with aqueous micellar dispersions (see next section) will, hopefully, shed further light on the questions raised in this section.

SENSITIZED PHOTOOXIDATIONS IN AQUEOUS MICELLAR SYSTEMS

The diffusibility and reactivity of photogenerated oxidizing agents in cells may be affected by abrupt phase changes and variations in medium polarity because of the highly heterogenous and compartmentalized structure. For this reason, aqueous dispersions of micelles above the critical micellar concentration are frequently used as working models of biological assemblies because of the presence of both hydrophobic and hydrophilic regions in these amphipathic surfactant aggregates. In this approach various combinations of sensitizers and substrates solubilized in either the micellar or the aqueous phase have been illuminated as models of the *in vivo* situation.

In 1939 Kautsky (34) showed that photogenerated $^1O_2$ could react with substrates at some distance from its site of generation. More recently it has been found that $^1O_2$ can penetrate freely into the interiors of ionic and neutral micelles and react with acceptors dissolved therein. The results of representative studies (35-38) are listed in Table IV; these tabulated data permit several important generalizations. For example, the observed high quantum yield of $^1O_2$ generation inside micelles and its rapid diffusion across phase boundaries supports the conclusion drawn from experiments on the oxygen-quenching of the fluorescence of micellized aromatic compounds, i.e., that oxygen diffuses through micelles in an essentially unrestricted fashion (39). Supporting this same conclusion, the 2- acetonaphtone triplet is quenched by oxygen with the same efficiency whether the ketone is dissolved in benzene or is incorporated into sodium dodecylsulfate (SDS) micelles (38). In general, the reaction rate of $^1O_2$ is at least as great with micellized substrates as with the same substrates in solution (38,40). Ionic micelles do not alter the lifetime of $^1O_2$ appreciably as demonstrated by the observation that varying the micellar concentration has little effect on the quantum yield of substrate photooxidation (37,38). In contrast, neutral micelles (as prepared with Igepal CO660 or Brij 35) may reduce the lifetime of $^1O_2$ as a result of excitation transfer from the electronic levels of $^1O_2$ to vibrational modes of the terminal hydroxyl groups of the surfactant (38).

Another conclusion is that the photoreaction rate with a given substrate-sensitizer appears to depend on the micellar outer charge. For example, the photooxidation of 1,3-diphenylisobenzofuran (38) and 9,10-dimethylanthracene (41) is most efficient with the substrates incorporated into anionic micelles. With methylene blue and other cationic dyes, this may result from adsorption of the dye on the external shell of the micelles (analogous to the enhanced photosensitizing efficiency of biopolymer-bound dyes described in the preceding section). However, the phenomenon also occurs with neutral dyes (37) and with micellized dyes (Table IV); thus other factors must be involved such as the micellar dimensions and the orientation of the substrate in the micelles (which could determine the accessibility of the substrate to the attacking agent—see ref. 38). The suggestion that the efficiency of the photoprocess is controlled, at least in part, by the orienting capability of the micelle is corroborated by recent studies which show that photoreactions in micelles have a high degree of regiospecificity (42). Finally, the effects of micellar properties on $^1O_2$ penetration rates should be considered; for example, only about one out of ten $^1O_2$-SDS micelle collisions results in penetration (43).

In addition to the $^1O_2$ studies just discussed, a small amount of work has been done on photoinduced electron transfer from micellized donors to acceptors in the aqueous phase. For example, the photoreduction of anthraquinones by zinc tetraphenyl porphyrin (44), of tryptophan by aromatic hydrocarbons (45) and of inorganic cations by phenothiazines (46) and hematoporphyrin (47) have been reported. It might be expected that the micellar surface charge would influence the kinetics and yields of electron transfer processes across the phase boundary. Experiments show, however, that a forward transfer process of the type $(\text{Donor})_{\text{mic}} + (\text{Acceptor})_{\text{aq}} \rightarrow (\text{Donor}^+)_{\text{mic}} + (\text{Acceptor}^-)_{\text{aq}}$ is always favored more in the presence of SDS micelles than in homogeneous solution. This occurs because the negative charge of the micelle markedly decreases the rate of the back reaction. For example, the electron transfer reaction from an aqueous medium to SDS-dissolved substrates may be reduced by a factor of up to 10^4 (48).

Conversely, the positive charge of cationic micelles attracts the electron towards the surface and extends the duration of the encounter, thus enhancing the interaction with substrate (49). The behavior of an electron in a cationic micelle system containing a reactive target molecule may be described by the following sequence of events: (i) The trapping of an electron from the bulk phase by the potential field of the micelle. The encounter rate constant of an electron with cetyltrimethylammonium bromide (CTAB) micelles is estimated to be at least 10^{12} M^{-1} s^{-1}, and is one order of magnitude lower for non-ionic surfactants (49); (ii) Intramicellar electron scavenging by the reactive target molecule located on the micellar surface or in the micellar core; and (iii) Intermicellar electron migration. The solvated electron half-life is 325-400 ns in the presence of 4 mM CTAB while the residence time of the electron at the micellar surface is 5 ns (50); thus each electron may visit a number of micelles before reacting. These same mechanisms may be used in describing the observed efficient electron transfer from CO_2^- to pyrene (49) and from the biphenyl radical to carotene (50) in aqueous CTAB systems; these processes, incidentally, would not be expected to occur in homogeneous solutions. Attempts to extend these studies to biological systems would be timely, although several details

TABLE V. Hematoporphyrin-Sensitized Photooxidation of Indoles in Aqueous Micellar Dispersions

Irradiated System	Substrate location[a]	Rate constant ($M^{-1}s^{-1}$)	Acceleration by D_2O	Azide conc. for 50% inhibition	% Inhibition by 0.5 mM ferricyanide
HP(aq) + TAM(aq)	--	1.74	4.10	5.89 mM	71.2
HP(SDS + TAM(SDS)	Deep in micellar core	1.95	1.62	2.01 mM	1.8
HP(Tri) + TAM(Tri)	Internal micelle regions	2.27	2.72	0.70 mM	7.8
HP(CTAB) + TAM(CTAB)	Stern layer of micelle	1.06	5.56	0.29 mM	42.0
HP(aq) + TRP(aq)	--	1.33	2.01	7.52 mM	43.4
HP(SDS + TRP(aq)	--	1.59	1.66	8.00 mM	68.4
TP(Tri) + TRP(aq)	--	2.14	2.06	3.39 mM	47.2
HP(CTAB) + TRP(CTAB)	Stern layer of micelle	0.97	4.03	0.70 mM	42.3

(a) *Determined on the basis of fluorescence emission parameters and fluorescence quenching by acrylamide and Cs^+ ions according to Eftink and Ghiron (52).*

Abbreviations used: HP, hematoporphyrin; aq, aqueous; TAM, tryptamine; TRP, L-tryptophan; SDS, sodium dodecyl-sulfate; Tri, Triton X-100; CTAB, cetyltrimethylammonium bromide.

Irradiation conditions: 2 mM indoles plus 0.05 mM hematoporphyrin in pH 10 aqueous solution containing 0.1 M Tris; surfactant concentration was 20 mg/ml in all cases. Irradiation wavelength: 404 nm, isolated with an interference filter from the light emitted by a 500 W high-pressure mercury lamp.

of the electron transfer processes remain to be clarified including effects of modulation of the micellar local electrostatic potential by ionic strength and the effects of different counterions.

The above considerations provide a basis for the interpretation of our recent results (47,51) on the hematoporphyrin-sensitized photooxidation of tryptamine and tryptophan in different aqueous micellar systems (see Table V). These indoles are good candidates for a systematic investigation concerning the influence of the medium composition on the various photosensitization mechanisms, since in homogeneous solutions they are typically photooxidized by a mixed Type I/Type II pathway (47). Under the experimental conditions selected for these studies, the porphyrin and uncharged tryptamine partition into the surfactant aggregates, whereas the negatively charged tryptophan localizes in either the aqueous or micellar phase, depending on the outer charge of the micelles. In these systems, the incorporation of the substrate and/or dye into ionic or non-ionic micelles has only a small effect on the photooxidation kinetics. In contrast, striking differences are observed in the quenching efficiency of azide (a $^1O_2$ -scavenger) and ferricyanide (an electron scavenger) for substrates in different locations. An electron transfer process is important in the photooxidation of tryptamine in homogeneous solution. Ferricyanide is also an effective inhibitor when the tryptamine is localized in the Stern layer of cationic micelles, probably as a result of the combined effects of easy accessibility and of catalysis by the positive charge layer. Ferricyanide inhibition drops to a low level when the tryptamine is buried in neutral or anionic micelles; in this latter case, the low accessibility of the target to electron is probably enhanced by electrostatic repulsion. Substrate accessibility is an even more important factor in the interaction between micellized tryptamine and $^1O_2$, since azide inhibition and D_2O acceleration of the reaction are large only when the tryptamine is near the micellar surface. Kinetic data for tryptophan photooxidation, however, suggest that the positive surface charge of CTAB micelles accelerates reactions involving $^1O_2$. Further, an examination of the inhibition of tryptophan photooxidation confirms the ability of SDS micelles to increase the yield of outward electron transfer to an acceptor in the surrounding aqueous medium.

In summary, it should be stressed that both $^1O_2$ and electrons can diffuse from one micelle through an aqueous medium and attack reactive targets dissolved in a different micelle. The efficiency of the $^1O_2$ reaction may depend on both the substrate accessibility and on the external charge of the micelle, whereas micellar charge is of primary importance in the efficiency of electron-mediated photoprocesses.

THE PHOTOSENSITIZED OXIDATION OF CELL MEMBRANES

The concepts developed in the previous sections can be extrapolated to cell membranes, which represent the subcellular structure whose photosensitized oxidation has been studied most extensively (53). An understanding of the photosensitization pathways in such a complex biological assembly requires the identification of the early stages of photosensitizer action, as well as the quantification of factors such as the dependence on sensitizer concentration, duration of dark and light exposure, distribution of the sensitizer in the membrane regions, etc. Two systematic investigations with cell membranes have been carried out which determine the importance of at least some of these experimental parameters: the photosensitized modification of nerve membranes by fluorescein derivatives (54-55) and of leukemia and ascite tumor cells by porphyrins (56-58).

Although the two sets of investigations were carried out with different types of cells and utilized dyes with different chemical and photochemical properties, the results obtained permit some common conclusions. In particular, in both cases, the most effective photosensitizers of membrane damage were relatively hydrophobic and were able to penetrate into the membrane, localizing at or near its inner surface. The following lines of evidence support this statement:

(i) When cells are exposed to a given concentration of dye, the photosensitizing effectiveness does not develop immediately, but rises with a quasi-sigmoidal time course, reflecting the rate of penetration of sensitizer into the membrane. Also, rinsing the cells prior to illumination results in an asymptotic removal of photosensitizing action.

(ii) The action spectrum for the fluorescein-photosensitized sodium channel block in axons is shifted bathochromically with respect to the dye absorption spectrum in aqueous solution indicating that the photoactive species is in a low polarity environment.

(iii) With both porphyrins and fluoresceins, the relative photosensitizing effectiveness does not correspond to the relative sensitizer triplet populations or to quantum yields of $^1O_2$ generation as estimated for aqueous solutions. A better agreement results when these parameters are corrected for the partition coefficient of the dye in octanol/water and for the relative light absorption in the two solvents at equal concentrations (55).

(iv) Highly water-soluble porphyrins, which do not bind with cells, have a very low photoactivity as compared with partially bound porphyrins. The most effective porphyrins can be extracted from aqueous solution by octanol (partition coefficient 2-12) but not by decane; hence they would be expected to localize at a hydrophobic-hydrophilic interface rather than in the highly hydrophobic core of the cell membrane. Studies of cell partitioning between a physiological medium and a polyethylene glycol-palmitate medium show that porphyrin binding blocks hydrophilic surface areas of the cell.

No definite evidence has been obtained on the state of the intramembranous dyes. Probably, both porphyrins and fluoresceins are at least partially complexed at saturable binding sites since washout experiments show that about 60-80% of the sensitizer diffuses out of the cell with the same ease with which it penetrates, while the remaining fraction cannot be removed unless denaturing conditions are used. Photomodification studies, as discussed below, suggest that specific proteins provide strong binding sites for many kinds of photosensitizers. In this case, the high photoactivity of cell-associated dyes would be in agreement with the increased efficiency of photosensitizers which are complexed with macrobiomolecules as described above.

The relatively minor role of external dye molecules in eukaryotic cell photosensitization, which has also been reported for prokaryotic cells (59) and for model systems utilizing liposomes (60), is surprising, at least if a $^1O_2$-pathway is involved. On the basis of oxygen diffusion coefficients and the lifetimes of $^1O_2$ in aqueous and hydrocarbon media, one can estimate (56) an average diffusional range of between 60 and 160 nm for $^1O_2$, a value substantially greater than a typical membrane thickness of 5-8 nm. Therefore, all membrane areas should be suffused by $^1O_2$ independent of its generation site since the ability of oxygen to diffuse freely through different phases was clearly demonstrated by the photosensitization studies using micellar systems as discussed above. Two hypotheses can be proposed for this anomalous behavior: either a non-$^1O_2$ mechanism is mainly responsible for photodamage to cell membranes or the reactions of $^1O_2$ produced in the membrane are somehow more effective both kinetically and biologically than $^1O_2$ generated external to the membrane.

The available information does not allow a definite choice between these two hypotheses. The involvement of $^1O_2$ in several photosensitized processes in vivo is suggested by the observed D_2O-enhancement and azide-inhibition of the reaction rates (5,11, 56,61,62). However, other evidence favors the occurrence of Type I mechanisms. For example, in the protoporphyrin-sensitized photodegradation of red blood cell membranes, steroids and lipids undergo photooxidation at a lower rate than do proteins (61-63), even though the rate constants for the reaction of $^1O_2$ with lipids and with proteins in homogeneous solutions are of the same order of magnitude. Further, protein cross-linking, which appears to be the major mechanism involved in the functional alterations of erythrocyte membranes by photodynamic treatment, also occurs in lipid-free erythrocyte ghosts and is not prevented by the addition of antioxidants which minimize the formation of lipid peroxides (63). The involvement of a limited number of protein types (e.g., spectrin) in the photosensitized cross-linking of red blood cell membranes also suggests that the process involves photoexcited protein-bound dyes interacting preferentially with spatially adjacent photosensitive targets. Such a conclusion is corroborated by the high photosensitizing effectiveness of intramembranous dye molecules as described above, as well as by the observed tendency of proteins with quaternary structure to yield homocross-linked photoadducts (61,63,65). Only small yields of heteroadducts are formed, which stresses the importance of the mutual motion of sensitizers and substrates (i.e., of membrane fluidity) in controlling the pathway and efficiency of the photoreaction. For example, the sensitivity of several cell lines to hematoporphyrin-sensitized photoinactivation is greatest in the middle of the S phase of the cell cycle (65), where membrane fluidity is maximal. Lowering the temperature from 25°C to 4°C decreases the rate of cell inactivation (65) and of protein-protein heterophotoaddition (66). The mean time between collisions of protein molecules in animal cell membranes is about 0.8 ms for a protein of 2 nm radius at 25°C and becomes extremely long (ca. 0.1 s) at 0°C, so that interactions between different proteins are most unlikely (66).

ACKNOWLEDGEMENTS

The preparation of this paper was supported in part by the Consiglio Nazionale delle Ricerche, Italy, and the National Institutes of Health (Grant No. 1 RO1 GM 27921), U.S.A., under the U.S.A.-Italy Cooperative Program in Science.

REFERENCES

1. Spikes, J. D., "The Science of Photobiology" (K. C. Smith, ed.), Plenum Press, New York, N.Y.,(1977),p. 87.
2. Jori, G. and Spikes, J. D., "Photochemical and Photobiological Reviews" (K. C. Smith, ed.), Vol. 3, Plenum Press, New York, N.Y.,(1978),p. 193.
3. Foote, C. S., "Free Radicals in Biology" (W. A. Pryor, ed.), Vol. II, Academic Press, New York, N.Y.,(1976),p. 85.
4. Ranby, B. and Rabek, J. F., "Singlet Oxygen and Its Reactivity with Organic Compounds and Polymers," Wiley Inc., Chichester,(1978).
5. Ito, T., Photochem. Photobiol. 28,493 (1978).
6. Schwartzel, E. H. and Coney, J. J., Can. J. Microbiol. 20, 1015 (1974).
7. Ito, T. and Kobayashi, K., Photochem. Photobiol., 25,399 (1977).
8. Lamola, A. A., Yamane, T. and Trozzolo, A. M., Science 179, 1131 (1973).
9. Rabeck, J. F. and Ranby, B., Photochem. Photobiol. 30,133 (1979).
10. Denney, R. W. and Nickon, A., Org. React. 20,133 (1973).
11. Girotti, A. W., Biochem. Biophys. Res. Commun. 72,1367 (1976).
12. Orlov, S. N., Malkov, Y. A., Rebrov, V. G. and Danilov, V. S., Biofizika 21,276 (1976).

13. Spikes, J. D. and MacKnight, M. L., "Photochemistry of Macromolecules" (R. F. Reinisch, ed.), Plenum Press, New York, N.Y., (1970), p. 67.
14. Hélène, C., "Aging, Carcinogenesis and Radiation Biology" (K. C. Smith, ed.), Plenum Press, New York, N.Y., (1976), p. 149.
15. Allison, A. C., Magnus, I. A. and Young, M. R., Nature 209, 874 (1966).
16. Pooler, J. P. and Valenzeno, D. P., Photochem. Photobiol. 28, 219 (1978).
17. Bellin, J. S. and Yankus, C. A., Arch. Biochem. Biophys. 123, 18 (1968).
18. Cannistraro, S., Calberg-Bacq, C. M. and Van de Vorst, A., Photochem. Photobiol. 23,137 (1976).
19. Silva, E. and Gaule, J., Rad. Environ. Biophys. 14,303 (1977).
20. Jori, G. and Rubaltelli, F. F., unpublished results.
21. Bourbon, J. and Durantè, M., Ann. N.Y. Acad. Sci. 171,169 (1970).
22. Bellin, J. S., Photochem. Photobiol. 8,383 (1968).
23. Lochmann, E.-R. and Micheler, A., "Physical Chemical Properties of Nucleic Acids" (J. Duchesne, ed.), Vol. I, Academic Press, New York, (1973).
24. Piette, J., Calberg-Bacq, C. M., Cannistraro, S. and Van de Vorst, A., Int. J. Radiat. Biol. 34,213 (1978).
25. Piette, J., Calberg-Bacq, C. M. and Van de Vorst, A., Molec. Gen. Genet. 167,95 (1978).
26. Spikes, J. D. and Glad, B. W., Photochem. Photobiol. 3,471 (1964).
27. Baugher, J. F., Grossweiner, L. I. and Lewis, C., J. Chem. Soc. Faraday Trans. II, 1389 (1974).
28. Brandt, J., Fredriksson, M. and Andersson, L.-O., Biochemistry 13,4758 (1974).
29. Kochevar, I. E., Photochem. Photobiol. 30,437 (1979).
30. Yoshikawa, K., Mori, N., Sakakibawa, S., Mizuno, N. and Song, P. S., Photochem. Photobiol. 29,1127 (1979).
31. Rubaltelli, F. F., Jori, G. and Rossi, E., Lancet, Vol. 2, 1295 (1979).
32. Lamola, A. A., Photochem. Photobiol. 8,601 (1968).
33. Jori, G., "Lasers in Photomedicine and Photobiology" (C. A. Sacchi and R. Pratesi, eds.), Springer Series in Optical Science, Berlin, 1980, in press.
34. Kautsky, H., Trans. Faraday Soc. 35,216 (1939).
35. Miyoshi, N. and Tomita, G., Z. Naturforsch. 33b,622 (1978).
36. Kraljić, I., Barboy, N. and Leickman, J. P., Photochem. Photobiol. 30,631 (1979).
37. Usui, Y., Tsukada, M. and Nakamura, H., Bull. Chem. Soc. Japan 51,379 (1978).
38. Lindig, B. A. and Rodgers, M. A. J., J. Phys. Chem. 83,1683 (1979).

39. Wong, M., Thomas, J. K. and Graetzel, M., J. Am. Chem. Soc. 98,2391 (1976).
40. Gorman, A. A. and Rodgers, M. A. J., Chem. Phys. Letters 55, 52 (1978).
41. Bagno, O., Soulignac, J. C. and Joussot-Doubien, J., Photochem. Photobiol. 29,1079 (1979).
42. Turro, N. J., Liu, K. C. and Chow, M. F., Photochem. Photobiol. 26,413 (1977).
43. Gorman, A. A., Lovering, G. and Rodgers, M. A. J., Photochem. Photobiol. 23,399 (1976).
44. Kano, K., Takuma, K., Ikeda, T., Nakajima, D., Tsutsui, Y. and Matsuo, T., Photochem. Photobiol. 27,695 (1978).
45. Putna, L., Reske, G. and Schmidt, H., Photochem. Photobiol. 30,723 (1979).
46. Moroi, Y., Humphrey-Baker, R., Brown, A. M. and Graetzel, M., Abstracts of VII IUPAC Symposium on Photochemistry, Heverlee, Belgium, (1978), p. 251.
47. Sconfienza, C., Van de Vorst, A. and Jori, G., Photochem. Photobiol. 31,351 (1980).
48. Patterson, L. K. and Graetzel, M., J. Phys. Chem. 79,956 (1975).
49. Graetzel, M., Kozak, J. J. and Thomas, J. K., J. Chem. Phys. 62,1632 (1975).
50. Almgren, M. and Thomas, J. K., Photochem. Photobiol. 31,329 (1980).
51. Jori, G., Rossi, E. and Van de Vorst, A., unpublished results.
52. Eftink, M. R. and Ghiron, C. A., J. Phys. Chem. 30,486 (1976).
53. Lamola, A. A., "Research in Photobiology" (A. Castellani, ed.), Plenum Press, New York, N.Y., (1977), p. 53.
54. Pooler, J. P. and Valenzeno, D. P., Photochem. Photobiol. 30, 491 (1979).
55. Pooler, J. P. and Valenzeno, D. P., Photochem. Photobiol. 30, 581 (1979).
56. Jori, G. and Rubaltelli, F. F., unpublished results.
57. Kessel, D., Biochemistry 16,3443 (1977).
58. Kohn, K. and Kessel, D., Biochem. Pharmacol. 28,2465 (1979).
59. Bagchi, B. and Basu, S., Photochem. Photobiol. 29,403 (1979).
60. Suwa, K., Kimura, T. and Schaap, A. P., Biochem. Biophys. Res. Commun. 75,785 (1977).
61. de Goeij, A. F. P. M., Van Straalen, R. J. C.and Van Steveninck, J., Clin. Chim. Acta 71,485 (1976).
62. Girotti, A. W., Biochemistry 18,4403 (1979).
63. Dubbelman, T. M. A. R., de Goeij, A. F. P. M. and Van Steveninck, J., Photochem. Photobiol. 29,197 (1978).
64. Girotti, A. W., Biochemistry 14,3377 (1975).
65. Christensen, T., Moan, J., Wibe, E. and Oftebro, R., Br. J. Cancer 39,64 (1979).
66. Kiehm, D. J. and Ji, T. H., J. Biol. Chem. 252,8524 (1977).

DISCUSSION

RODGERS: I am interested to hear, Giulio, that you are looking at singlet oxygen reactions in micellar systems. As you are obviously aware, I have been involved in this work myself for a number of years—first of all with Gary Lovering and Tony Gorman at the University of Manchester and recently with Barbara Lindig at the University of Texas at Austin, who is in the audience. I have also worked with fluorescent probes in micellar systems and studied the quenching of the fluorescence by various water and lipid-bound quenchers. Over the years (maybe five) I have worked in this field, I have become increasingly more confused about what micelles are doing and what properties they are contributing to the aqueous phase as compared to a bulk homogeneous phase. My comment is aimed at warning against over-interpretation of rate constants that you observe in different kinds of micellar systems. The big problem is that we often do not know exactly where the probe or substrate really is with respect to the bulk phase, the diffuse double layer, or the lipid regions of micelles. If anybody is interested, Barbara Lindig has a poster upstairs in which she has tried to interpret the rate constants for singlet oxygen reactions with DPBF and other substrates in micellar systems. She has made an attempt to bring some order into the confusion.

JORI: Table V of our paper shows some data which might be relevant to some extent to the point you have raised. We developed a model system for photooxidation studies represented by hematoporphyrin in SDS, CTAB, or Triton X-100 micelles and by indole derivatives (tryptophan and tryptamine) dissolved in either the aqueous solution or the micellar phase. In all cases we calculated the first-order rate constant for the photooxidation of the indole derivative. The exact value of the rate constant *per se* may be of limited interest. However, we tried to correlate it with the location of the substrate outside or inside the micelle, taking advantage of the fluorescent properties of tryptophan and tryptamine. In a sense, we tried to quantify the degree of accessibility of the photooxidation targets by fluorescence quenching with acrylamide or with charges species. The trend which results from these data is that the substrate is quite sensitive to attack by both singlet oxygen and by the electron when it is located in the Stern layer of the micelle. However, when it is inside the micelle, the efficiency of its reaction with electrons decreases dramatically. Singlet oxygen can still bring about an efficient photooxidative attack when the substrate is in the micelle although the efficiency is somewhat reduced as compared with that typical of a homogeneous solution.

LINDIG: I just have one short comment about something you mentioned earlier about sensitizing efficiency. That is, for the cases of sensitizers in various systems, you said that the singlet oxygen sensitizing efficiency increased when those materials were bound. My comment concerns an anomalous case, which is methylene blue. We have observed in some recent work with certain micellar systems that its sensitizing efficiency decreases when it is bound to certain types of micelles. The reason for this appears to be explained by some work of a French group a couple of years ago (R. Bonneau, R. Pottier, O. Bagno, J. Jousson-Dubien, Photochem. Photobiol. 21,159 (1975)) in which they showed that decreased oxidation yields as a function of pH were due to protonation of the triplet state. Since the pK_a of the triplet state of methylene blue is very near neutral pH (7.1), then very small changes in its environment can effect its protonation state. In fact, the protonated triplet is a factor of 20 less efficiently quenched by oxygen. In steady state oxidation studies this might lead you to think that you were seeing a difference in reactivity of your substrate when the effect is, in fact, a difference in the sensitizing efficiency of methylene blue.

JORI: Yes, I agree with you on the point that microenvironment has a strong effect; and, of course, the binding of a dye with a micelle represents a different situation from binding with a macromolecule. However, I agree that one must be careful in drawing general conclusions from the data hitherto available.

MacDONOUGH: In the first hematoporphyrin experiment you showed, you correlated singlet oxygen reactivity with solvent polarity. However, porphyrins aggregate in aqueous medium. Is it possible, perhaps even likely, that the phenomena that you observed merely reflected the effect of solvent polarity on the state of aggregation of the porphyrin?

JORI: Yes. Since my talk was restricted in time, I could not go into detail on that subject. You are quite correct. Both effects are involved. Hematoporphyrin is aggregated in aqueous media and is transformed into a monomeric species when the concentration of the organic solvent increases. However, our data suggest that the aggregated hematoporphyrin in a largely aqueous medium sensitizes mainly via singlet oxygen and the electron transfer mechanism becomes more important only when we move to a low polarity environment.

REACTIONS OF TRYPTOPHAN WITH SINGLET OXYGEN, HYDROXYL RADICAL, AND SUPEROXIDE ANION

Ajit Singh, Mary-Jane Bell, Grant W. Koroll, Walter Kremers and Harwant Singh, Medical Biophysics Branch, Whiteshell Nuclear Research Establishment, Atomic Energy of Canada Limited Research Company, Pinawa, Manitoba, Canada ROE 1LO

INTRODUCTION

Photosensitized, photochemical and radiolytic decomposition of tryptophan (trpH) has been widely studied (1-24). However, many aspects of such decompositions of tryptophan and related derivatives remain to be understood.

In photosensitized and photochemical decomposition, an important role of singlet oxygen ($^1O_2$) has either been suggested or implied (e.g. refs. 18 and 23 and refs therein). N-formyl-kynurenine (FK) is a prominent product in these systems, as well as in radiolysis (1,18 and 23). A reaction of superoxide anion ($O_2^{\cdot-}$) with trpH to produce FK has also been suggested (14,15); incentive for this suggestion came from the observed involvement of $O_2^{\cdot-}$ in the conversion of trpH into FK by tryptophan 2,3-dioxygenase (25-27).

In view of the convertibility of singlet oxygen, superoxide anion and hydroxyl radical (·OH) (28) and the importance of these species in biological systems (29-34), we undertook investigation of reactions of trpH with $^1O_2$, ·OH and $O_2^{\cdot-}$. Our results to date are presented here.

EXPERIMENTAL

(i) Materials. DL-tryptophan, DL-serine, DL-kynurenine, DL-alanine, DL-aspartic acid, glycine, and sodium azide were obtained from Sigma Chemical Company (Sigma grade). N-formyl-kynurenine was from Cal. Biochem (B-grade). Potassium thiocyanate, phosphoric acid, ferrous sulfate, sodium formate, sodium phosphate (monobasic), 2-propanol and hydrogen peroxide were obtained from Fisher (reagent grade). Acetonitrile (Fisher) was of HPLC quality. Tertiary-butanol (Aldrich) was redistilled before use. Tetrabutyl ammonium phosphate (PIC "A") was supplied by Waters Associates. A sample of 2-carboxy-3a-hydroxy - 1,2,3,3a,8,8a-hexahydropyrrolo (2,3b) - indole (HPI) was kindly supplied by Dr. S. Zigman.

Glacial acetic acid was obtained from Allied Chemical Canada (Reagent grade). Tritium-labelled tryptophan was obtained from New England Nuclear (lot 1109-155) with a specific activity of 29.4 mCi/mg. All aqueous solutions were made in triply distilled water. Compressed gases were obtained from Union Carbide Canada. Sensitox II was obtained from Hydron Laboratories, Inc.

ISBN 0-12-592050-4

Before use, N_2O was passed through 6N NaOH using sintered glass, two MnO columns designed to remove O_2 and a water pre-bubbler; nitrogen was passed through two MnO columns and saturated with water; oxygen and the other gases were saturated with water to prevent sample evaporation.

(ii) Techniques. Photolysis experiments were carried out using a Xenon arc lamp with appropriate filters such that singlet oxygen was produced by direct light absorption

$$^3O_2 + h\nu \longrightarrow {}^1O_2 \qquad (1)$$

in a high pressure cell (35,36). Solutions were equilibrated under high pressures of O_2 (up to 13.3 MPa), prior to irradiation.

Quantum yields were determined by use of Sensitox II as the photosensitizer, with the photoirradiator system described by Singh and Vadasz (30). Sensitox II absorbs light in the 400-700 nm region to produce $^1O_2$ with a quantum yield of 0.46 in aerated aqueous solutions (37) (this is the average of the values in CH_2Cl_2 and CH_3OH). For uniform $^1O_2$ production, the Sensitox II beads were maintained in suspension by bubbling O_2 through the sample. After irradiation the Sensitox II beads were removed by filtration.

For gamma radiolysis experiments, a Gammacell 220 (Atomic Energy of Canada Limited, Radiochemical Company) with a dose rate of approximately 6.5 krad min^{-1}, was used. The Fricke dosimeter was used for dose determinations (38). Radiolysis experiments were done under air or in the presence of selected gases. For runs with N_2O the trpH solutions were frozen in liquid nitrogen, vacuum degassed and equilibrated with $\sim$100 kPa N_2O before being sealed.

Thin-layer chromatography of irradiated solutions was carried out on silica gel sheets (Bakerflex IB2) using a solvent system composed of n-butanol, acetic acid (glacial), and water (4:1:1). Location of the products was by fluorescence under a UV light source and by ninhydrin spray (39). FK and K were identified as products, with this technique. Experiments were also done using tritium labelled trpH solutions. These solutions were prepared by mixing a 10^{-3} mol dm^{-3} solution of unlabelled trpH with enough labelled trpH to have approximately 5×10^5 cpm in a 15 μl sample. However, in these solutions self-radiolysis by their tritium content was too large to allow satisfactory quantitative data to be obtained. On the whole, the thin-layer chromatography technique was found unsuitable for quantitative work and the data being reported in this paper were obtained with the high pressure liquid chromatography (HPLC) technique.

A HPLC system was developed in order to separate the major oxidation products of trpH. The system isolated FK, K, HPI and trpH, in addition to five unknown products. An ODS column, 4.6 mm internal diameter, 250 mm in length, packed with RSIL-C18 of 5μ particle size, obtained from Alltech Associates, was used. The solvent contained 0.001 mol dm^{-3} NaH_2PO_4 and 1% acetonitrile with the pH adjusted to 4.5 with 10 N NaOH containing 5 x 10^{-3}mol dm^{-3} PIC "A". The solvent was degassed by heating it to 10°C above the column temperature, before use. The column was maintained at 28°C and a flow rate of 0.8 ml/min allowed a sample to be analyzed in 34 minutes. The detector used was a Dupont model 837 variable wavelength detector set at 240 nm in order to maximize product absorbance and minimize interference by tryptophan. Calibrations were done with individual samples of the known products (FK, K, HPI). Some typical HPLC analysis runs are shown in Fig. 1.

All analyses were done immediately (within minutes, unless otherwise noted) after irradiations because the irradiated solutions were found to change their composition (in hours), even on standing in the dark.

Positive test for peroxide formation, on reaction of $^1O_2$ with trpH, was obtained with acidified potassium iodide solution and starch.

RESULTS AND DISCUSSION

A. Photolysis Studies

(1) Reactions with $^1O_2$. Singlet oxygen was produced by direct irradiation with λ>450 nm, in high pressure optical cells (13 MPa) (i) in the liquid phase in aqueous solutions of trpH (10^{-3} mol dm^{-3}) and (ii) in the gaseous phase by irradiation of the volume only above the liquid phase in cells about one-fourth filled with aqueous solutions. In both cases decomposition of trpH was observed with the formation of FK, K, HPI and an unknown product (Table I). These experiments were done to confirm that $^1O_2$ indeed gives these products with trpH. In all the previously described work (2,3,6-13,16-19, 21,23,24) complications due to reactions of other transient species are possible. In our system, presence of sodium azide in the aqueous solutions drastically reduced the decomposition of trpH, as expected.

Photosensitized decomposition of trpH with Sensitox II gave one product in addition to those obtained with $^1O_2$ (Table 1). The quantum yields of the known products are given in Table 1. Comparison of these with the quantum yield for trpH decomposition shows that FK, K and HPI together account for less than half of the trpH decomposed. It may be noted that the ratios of the yields of HPI, K and FK in the two cases (Sensitox II vs $^1O_2$) are also different.

(2) Comparison of quantum yields. Quantum yields for trpH degradation of ∿0.01 have been reported for direct irradiation of trpH in aqueous solutions in the UV region (2,3,6,13), which is considerably lower than our value of 0.05. This suggests that in direct irradiation of trpH, the efficiencies of $^1O_2$ formation, and oxidation via free radical reactions (reactions 4,7) are rather low.

TABLE I. Radiolysis, Photolysis and H_2O_2 Induced Decomposition of Aqueous Solutions of Tryptophan (10^{-3}mol dm^{-3})

	Yields Identified Products				Yields[d] Unidentified Products				
1. Radiolysis (Dose: 50 Krad). Reactant	G^a(-trpH)	G(HPI)	G(FK)	G(K)	X_1	X_2	Y	Z	Q
(i) $O_{2aq}^{\bar{\cdot}}$ + ·OH(Air)	4.0	0.58	0.68	0.18	1.4	1.2	0	0	0
(ii) $O_{2aq}^{\bar{\cdot}}$ (O_2 + formate)	0^b	0	0	0	0	0	0	0	0
(iii) ·OH + H(N_2O)	3.3	0	0.09	0	3.2	3.6	1.0	0	0
(iv) H + ·OH + e_{aq}^- (deaerated)	0^b	0	0	0	0.4	0.4	0	0	0
2. Photolysis Reactant	ϕ(-trpH)	ϕ(HPI)	ϕ(FK)	ϕ(K)					
(i) $^1O_2$ (sensitox II)	0.05	0.015	0.0008	0^c	0	0	0	0.8	1.6
(ii) $^1O_2$ (13 MPa, $\lambda > 450$nm)[d]	-	1.8	0.6	0.2	0	0	0	0.3	0
3. H_2O_2 (10^{-1}mol dm^{-3})[d]									
(i) Analyzed immediately	-	1.4	2.2	0.2	0.3	0.4	0.4	0	1.8
(ii) Analyzed 32 hrs later	-	0.6	4.8	0.2	0.2	0.2	0.3	0	1.4

a. G=number of molecules formed or destroyed by a dose of 100 eV of absorbed energy (40).
b. The detection limit is ± 0.2.
c. In some runs a trace of K was observed.
d. These are O.D. values; all O.D. values are at 240 nm and have been multiplied by 10^3.

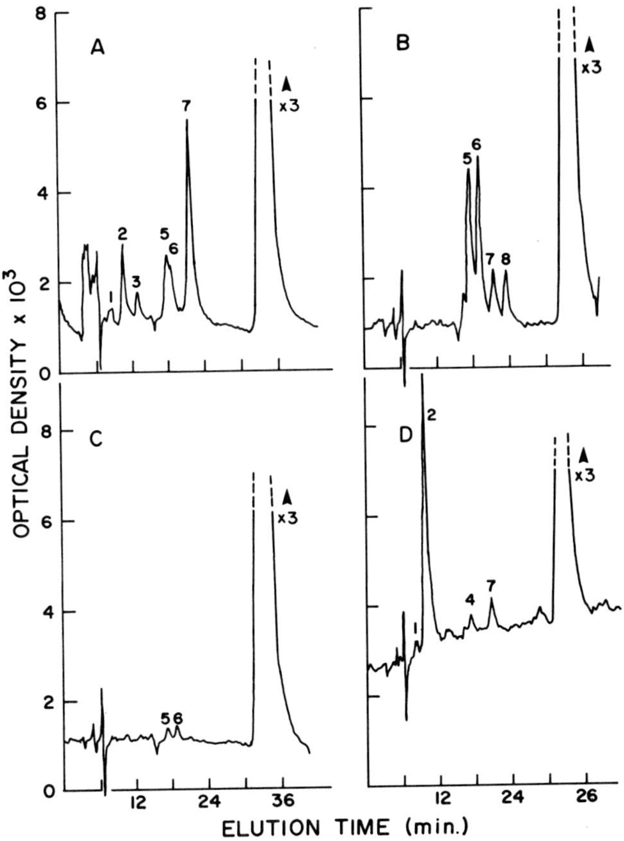

FIGURE 1. Typical HPLC analyses of irradiated aqueous solutions of trpH (10^{-3}mol dm^{-3}): (A) radiolysis, aerated; (B) radiolysis, N_2O added; (C) radiolysis, deaerated; and (D) photolysis, in the presence of Sensitox II and O_2. The product peaks are as follows: 1,Z; 2,HPI; 3,K; 4,Q; 5,X_1; 6,X_2; 7,FK; 8,Y.

$$\text{trpH} \xrightarrow{h\nu} {}^1\text{trpH}^* \tag{2}$$

$$^1\text{trpH}^* \longrightarrow {}^3\text{trpH}^* \tag{3}$$

$$\longrightarrow \text{trp}\cdot + \text{H} \tag{4}$$

$$^3\text{trpH}^* + O_2 \longrightarrow \text{trpH} + {}^1O_2 \tag{5}$$

$$^1O_2 + \text{trpH} \longrightarrow \text{FK, K and HPI + etc.} \tag{6}$$

$$\text{trp} + O_2 \longrightarrow \text{Oxidation Products} \tag{7}$$

The quantum yields reported by Yoshimura and Kato (11) range between 0.006 and ∿0.02; however, as the authors indicate, trpH degradation in those cases might not be via $^1O_2$.

While in most of the studies reported, trpH degradation is greater in the presence of oxygen, Pailthorpe et al (12) report ϕ(-trpH) $\sim$ 0.05 in the presence and absence of O_2 at pH 7. This peculiarity is attributable to the high intensity flash photolysis system used by these authors, in their system, increased radical-radical and radical initiated reactions (following reactions 3 and 4) in the absence of O_2 seem to have balanced the degradation in its presence.

B. Radiolysis Studies

Radiolysis of water produces three highly reactive transients (40)

$$H_2O \rightsquigarrow H + {}^{\cdot}OH + e^-_{aq} \qquad (8)$$

By varying the conditions in aqueous solutions, the relative yields of these transients can be changed. On saturation with N_2O, the yield of ·OH radicals is about doubled (40).

$$e^-_{aq} + N_2O \longrightarrow N_2 + O^{\cdot-} \qquad (9)$$

$$O^{\cdot-} + H_2O \longrightarrow {}^{\cdot}OH + OH^- \qquad (10)$$

In the presence of sodium formate and O_2, the superoxide anion is almost exclusively the reactive transient, due to the following reactions (40).

$$e^-_{aq} + O_2 \longrightarrow O_2^{\cdot-} \qquad (11)$$

$$H + O_2 \longrightarrow HO_2^{\cdot} \rightleftarrows H^+ + O_2^{\cdot-} \qquad (12)$$

$${}^{\cdot}OH + HCOO^- \longrightarrow H_2O + CO_2^{\cdot-} \qquad (13)$$

$$H + HCOO^- \longrightarrow H_2 + CO_2^{\cdot-} \qquad (14)$$

$$CO_2^{\cdot-} + O_2 \longrightarrow CO_2 + O_2^{\cdot-} \qquad (15)$$

We have used these additives to study reactions of ·OH and of $O_2^{\cdot-}$ with trpH.

(1) Reaction of $O_2^{\cdot-}$. As the data in Table I show, there is no detectable decomposition of trpH on radiolysis of its aqueous solution containing sodium formate. This is contrary to the previously published work on the subject (14,15). The following points may assist in understanding the difference in interpretation.

(a) Based on the relative rate constants of ·OH with trpH and with t-BuOH (41), even in 1×10^{-3} mol dm^{-3} solution of trpH, 2% of the ·OH radicals should react with trpH initiating oxidation reactions.

(b) One fate of $O_2^{\cdot-}$ is to produce H_2O_2,

$$2O_2^{\cdot-} \xrightarrow{H^+} H_2O_2 + O_2 \qquad (16)$$

Our work shows that H_2O_2 oxidizes trpH to FK and other products (Table I). Thus any delay in their analyses, after irradiation, could lead to trpH oxidation by H_2O_2 produced in reaction 16.

We should also point out that our result (i.e. no detectable reaction between $O_2^{\cdot-}$ and trpH) is consistent with the published results of Bielski and Shiue (42) who found that k($O_2^{\cdot-}$ + trpH) at pH 7 is only 24 ± 3 L $mol^{-1}s^{-1}$. Our results also have a bearing on the enzymatic conversion of trpH into FK (25-27); we comment on this later in this paper.

(2) Deaerated solutions of trpH. As the results in Table I show, trpH is very stable towards radiolytic degradation, in the absence of oxygen. Since H, ·OH and e_{aq}^- are formed in such solutions, and all of them would react with trpH (4,5,20), it seems that regeneration of trpH from the transients formed predominates e.g. by mechanisms similar to those suggested by Armstrong and Swallow (5).

(3) Oxygenated solutions of trpH. The data in Table I show that in this case the oxidative degradation of trpH is quite high. In addition to FK, K and HPI, two other products are also observed. The ratio of FK, K and HPI is different from that seen in the case of reaction of $^1O_2$ with trpH. The radiolytic oxidation of trpH is attributable to free radical reactions initiated by ·OH.

$$\text{trpH} \xrightarrow{\cdot OH} \text{free radicals} \longrightarrow O_2 \longrightarrow \begin{matrix}\text{Oxidation}\\ \text{Products}\end{matrix} \quad (17)$$

The role of ·OH in reaction 17 has been confirmed by addition of t-BuOH and i-PrOH both of which are ·OH scavengers (41,43). In the presence of these alcohols, the yield of the trpH oxidation products is sharply reduced (unpublished results).

(4) Reaction of ·OH with trpH. In the absence of oxygen, in N_2O saturated solutions, the degradation of trpH is still quite high (Table 1); however, the familiar oxidation products—K and HPI—are absent. A small amount of FK does seem to be formed in this case.

C. Relevant Implications

(1) Monitoring $^1O_2$, $O_2^{\cdot-}$ and ·OH reactions. The data in Table I show that the products and the ratios of similar products from tryptophan are different in the following four cases: (l) $^1O_2$ reactions; (ii) reactions of ·OH; (iii) oxidation reactions initiated by ·OH and (iv) the lack of reaction of $O_2^{\cdot-}$. Thus trpH can be used to monitor these species. Optimization of the conditions of use of trpH for monitoring these species (and search for any other indole derivative which might be even better) is clearly warranted. The advantage of trpH being water soluble can hardly be overemphasized.

(2) Role of $O_2^{\cdot-}$ in enzymatic conversion of trpH into FK. Our result that $O_2^{\cdot-}$ does not detectably react with trpH

necessitates a reassessment of its role in the enzymatic conversion (25-27) of trpH into FK. The following can be surmised from the detailed review on tryptophan 2,3-dioxygenase (26);

(i) the active state of the enzyme has one positive charge less than the inactive form;

(ii) binding of trpH to the enzyme seems to be essential for its conversion to FK;

(iii) superoxides are amongst the compounds that can convert the inactive enzyme to the active one (this requires net transfer of an electron to the enzyme).

We therefore suggest that the role of $O_2^{\overline{\cdot}}$ is to regenerate the enzyme, which then becomes available for further conversion of trpH into FK. Similar reactivation of the indoleamine 2,3-dioxygenase (27) by $O_2^{\overline{\cdot}}$ is also suggested. These suggestions are consistent with a recent paper from Hayaishi's group (44).

ACKNOWLEDGEMENTS

We thank Dr. S. Zigman for supplying us with a sample of HPI. This is AECL no. 6926.

REFERENCES

1. Jayson, G. G., Scholes, G. and Weiss, J., Biochem. J., 57, 386 (1954).
2. Leaver, I. H. and Lennox, F. G., Photochem. Photobiol, 4,491 (1965).
3. Vladimirov, Yu A. and Perrase, N. I., Biophysics, 11,658 (1966).
4. Braams, R., Radiat. Res., 27,319 (1966).
5. Armstrong, R. C. and Swallow, A. J., Radiat. Res., 40,563 (1969).
6. Vladimirov, Yu A., Roshchupkin, D. I. and Fesenko, E. E., Photochem. Photobiol. 11,227 (1970).
7. Spikes, J. D. and MacKnight, M. L., Ann. N.Y. Acad. Sci., 171,149 (1970).
8. Asquith, R. S. and Rivett, D. E., Biochim. Biophys. Acta, 252,111 (1971).
9. Pailthorpe, M. T. and Nicholls, C. H., Photochem. Photobiol., 14,135 (1971).
10. Savige, W. E., Aust. J. Chem., 24,1285 (1971).
11. Yoshimura, A. and Kato, S., Bull. Chem. Soc. Japan, 46,1141 (1973).
12. Pailthorpe, M. T., Bonjour, J. P. and Nicholls, C. H., Photochem. Photobiol., 17,209 (1973).
13. Ghiggino, K. P., Pailthorpe, M. T. and Nicholls, C. H., Symposium on Molecular Rate Processes, Royal Australian Chemical Institute. Warburton, Aug. 1975.
14. Walrant, P., Santus, R. and Grossweiner, L. I., Photochem. Photobiol., 22,63 (1975).

15. Walrant, P., Santus, R., Redpath, J. L. and Lexa, D., C.R. Hebd Seances Acad. Sci., Ser. D., 280,1425 (1975).
16. Churkova, N. I., Kravchenko, N. A., Serebryakov, E. P. and Kaverzneva, E. D., Biochemistry, 41,989 (1976).
17. Saito, I., Takahashi, Y., Imuta, M., Matsugo, S., Kaguchi, H. and Matsuura, T., Heterocycles, 5,53 (1976).
18. Saito, I., Matsuura, T., Nakagawa, M. and Hino, T., Accts. Chem. Res., 10,346 (1977).
19. Nakagawa, M., Watanabe, H., Kodato, S., Okajima, H., Hino, T., Flippen, J. L. and Witkop, B., Proc. Natl. Acad. Sci., 74,4730 (1977).
20. Farragi, M. and Bettelheim, A., Rad. Res., 72,81 (1977).
21. Nakagawa, M. and Hino, T., Heterocycles, 6,1675 (1977).
22. McCormick, J. P. and Thomason, T., J. Am. Chem. Soc., 100, 312 (1978).
23. Sun, M. and Zigman, S., Photochem. Photobiol., 29,893 (1979).
24. Sconfienza, C., Van de Vorst, A., and Jori, G., Photochem. Photobiol., 31,351 (1980).
25. Hirata, F. and Hayaishi, O., J. Biol. Chem., 246,7825 (1971).
26. Fiegelson, F. and Brady, F. O., "Molecular Mechanisms of Oxygen Activation" (O. Hayaishi, ed.), Academic Press, New York, (1976), p. 87.
27. Hirata, F. and Hayaishi, O., J. Biol. Chem., 250,5960 (1975).
28. Singh, A., Photochem. Photobiol., 28,429 (1978).
29. Ito, T., Photochem. Photobiol., 28,493 (1978).
30. Singh, H. and Vadasz, J. A., Photochem. Photobiol., 28,539 (1978).
31. Cohen, G., Photochem. Photobiol., 28,669 (1978).
32. Tauber, A. I. and Babior, B. M., Photochem. Photobiol., 28, 701 (1978).
33. Fridovich, I., Photochem. Photobiol., 28,733 (1978).
34. Petkau, A., Photochem. Photobiol., 28,765 (1978).
35. Keltie, R. J., Koroll, G. W. and Singh, A., Photochem. Photobiol. 28,607 (1978).
36. Koroll, G. W. and Singh, A., Photochem. Photobiol. 28,611 (1978).
37. Schaap, A. P., Thayer, A. L., Blossey, E. C. and Neckers, D. C., J. Am. Chem. Soc., 97,3741 (1975).
38. Sehested, K., "Radiation Dosimetry," (R. J. Berry and N. W. Holm, eds.), Marcel Dekker, Inc., New York, (1970), p. 313.
39. Matheson, A. T., Tigane, E. and Hanes, C. S., Can. J. Biochem. Physiol. 39,141 (1961).
40. Draganić, I. G. and Draganić, Z. D., "The Radiation Chemistry of Water," Academic Press, New York, (1971).
41. Dorfman, L. M. and Adams, G. E., "Reactivity of the Hydroxyl Radical in Aqueous Solutions," NSRDS-NBS46 (1973).
42. Bielski, B. H. J. and Shiue, G. G., "Oxygen Free Radicals and Tissue Damage," CIBA Foundation Symposium no. 63 (New Series), Excerpta Medica, Amsterdam, (1979), p. 43.

43. Anbar, M., Farhataziz, and Ross, A. B., "Selected Specific Rates of Reactions of Transients from Water in Aqueous Solution. II. Hydrogen Atom," NSRDS-NBS51 (1975).
44. Taniguchim T., Sono, M., Hirata, F., Hayaishi, O., Tamura, M., Hayashi, K., Lizuka, T. and Ishimura, Y., J. Biol. Chem. 254,3288 (1979).

DISCUSSION

SANTUS: Is it possible that the information of kynurenine in your experiments is a result of the conditions used for chromatography?

SINGH: We have looked at this possibility over the last two years. All results just described are at neutral pH. We have also done work on the effect of pH on singlet oxygen reaction with tryptophan, and almost in every case we get kynurenine as a product (pH 2-12). So we did look for the possibility of kynurenine arising from hydroylsis of formylkynurenine. On the basis of our results we cannot rule it out; however, it could be a primary product as well.

STRAIGHT: I would like to comment on the cross-linking of proteins by sensitized photooxidations. This is a photodynamic phenomenon that has been observed with several proteins but studied little, at least until very recently. Some of the early work observed protein aggregation, gel formation, and crosslinking with serum proteins (Vodrázka, 1959); albumin and immunoglobulins (Smetana, 1941; Tyler, 1941; Vodrázka, 1961); β-lactoglobulin (Ewil, 1953; Gomyo, 1968); myosin (Venkstern, 1955); and more recently TMV-plant virus protein (Bell and Orlob, 1976), hemoglobin (Girotti, 1979), red cell membrane proteins (Girotti, 1976, 1979 and Van Steveninck et al., 1975, 1980) (also Lepoch et al., 1978). Also, protein-nucleic acid cross-linking recently has been observed to occur between ribosomal RNA and protein by methylene blue sensitized photooxidation (Zook and Fahnestock, 1978).

That such covalent cross-linking occurs is now generally accepted, but the mechanism is unknown as are the conditions for optimal cross-linking. The conditions may vary from protein to protein and sensitized photooxidation cross-linking may be a more general phenomenon than previously realized.

Concerning the mechanism, compared to the photooxidation of amino acids in proteins, we know a great deal about the sensitized photooxidation of amino acids free in solution; although as was pointed out earlier (Foote), there are still some important gaps in our understanding because the mechanisms of photooxidation with some amino acids are very complex. We do know the rate constants for the interaction of histidine, methimine, and tryptophan with singlet oxygen ($^1O_2$). These are nicely summarized at this symposium in the poster presentation by Lindig and Rodgers. I would like to add to that list the rate constant for cysteine-$^1O_2$ interaction which, in collaboration with John Spikes, we have recently determined at pH 7 to be 5.4 x $10^5 M^{-1}s^{-1}$ and at pH 9 to be 3.6 x $10^7 M^{-1}s^{-1}$. The difference is due to the fact that cysteine is primarily in the RSH form at pH 7 and primarily in the RS^- anionic form at pH 9.

Where it has been studied, the conditions that favor singlet oxygen mechanism appear to enhance cross-linking and singlet oxygen inhibitors inhibit cross-linking. Therefore, singlet oxygen seems to be involved in the cross-linking mechanism at least as an initiator but not necessarily at the exclusion of redox initiated mechanisms.

Cross-linking could occur as a dark reaction mediated by relatively stable radical intermediates or by reaction between stable photoproducts as was suggested by Venkstern (1953) to explain the photosensitized oxidative polymerization of myosine. He proposed that a side chain aldehyde (RCHO) product possibly formed by photooxidation of tryptophan could react with primary amine groups (RNH_2) to form a Schiff base (R - C = N - R) cross-link. No direct evidence of this mechanism is available, and the cross-linking I have recently observed with ribonuclease which contains a lot of primary amine groups but no tryptophan suggests at least that other mechanisms may be important.

Also, at least in the case of eosin, the dye sensitizer does not seem to be involved in the protein-protein cross-linking. The reaction is more efficient at lower dye concentrations.

Cross-linking is enhanced in more complex, multiple chain proteins such as IgG (Antivenom) and in non-covalently complexed single chain proteins such as the natural two-chain dimer Crotoxin and the four-chain complex hemoglobin.

Careful kinetic and product studies with model proteins such as ribonuclease, lysozyme, myoglobin, β-lactoglobin, egg albumin, hemoglobin, and IgG should be useful in determining the mechanism of cross-linking.

H. SINGH: I just want to make a comment on one of the comments that you made regarding the cross-linking in ribosomes. I am aware of the work that you just described; I think the doses that are used in that study are very, very high. And also, in relation to the biological activity, whether ribosomes are still active or not is a question that one must consider when looking at the cross-link. One can find cross-links, sure; but I think that, in case you want to relate it to the living system, you have to consider the biological aspect.

STRAIGHT: That is a good point, and I think that same point holds true with the membrane protein cross-linking. Even though it correlates very nicely with the photohemoylsis, there is no direct evidence that the cross-linking is really the major reason for the photohemolysis. There is clearly protein cross-linking, but what effect it has I do not know. I do think it is important, though, that we understand this photocross-linking in more detail. For example, in the case of fluorescein isothiocyanate (FITC) which is used as a tool to study membrane lipid-protein dynamics and you do get photocross-linking of membrane proteins even though the technique is used under the assumption that there is no effect on membrane structure. I think this is now an unwarranted assumption and further reason to understand sensitized photooxidative protein cross-linking better.

SANTUS: I would like to point out that carotenes are well recognized physical quenchers for singlet oxygen or sensitizer triplet states in solutions in vitro. In fact, when interacting with biological structures such as membranes or proteins, the reactions can be entirely different. For instance, the carotenoid astaxanthin (di-hydroxy-β-carotene 4,4-dione) is tightly bound to a protein in some marine invertebrates and is very stable when interacting with the protein. However, this carotene does chemically quench singlet oxygen at a rate approaching the diffusion-controlled limit. In addition, it also reacts with the superoxide anion.

BUETTNER: I would like for you to expand just briefly on what you said about the superoxide ion and the carotene.

SANTUS: Well, what I could say is that, by producing $O_2^{\overline{\cdot}}$, I could see the bleaching of the carotenoid protein which means there is some reaction involving, at the primary step, $O_2^{\overline{\cdot}}$ and leading to the carotenoid bleaching. What happens in between I do not know. Thus, extrapolation of in vitro molecular photobiological mechanisms to living systems must be carefully done after a close examination of the molecular properties of the cellular structures involved in the adaptation of an organism to solar radiations.

SPIKES: Our next discussant is going to ask a question: What are the molecular bases for cellular damage initiated by photosensitized oxidations?

KRINSKY: I have asked this question to emphasize the gap that still exists between an understanding of the types of reactions which can be observed when studying photosensitized oxidations of model compounds, biological molecules, or even model cell systems with the types of effects which occur at the cellular level. In many studies over the last 80 years, the cellular effect used to assess photosensitized damage has been the death of the cell, an event which does not yield useful information with respect to understanding the molecular processes involved (Krinsky, N. I.,

"The Survival of Vegetative Microbes," Symp. Soc. Exp. Biol. 26, 209-239 (T. G. R. Gray and J. R. Postgate, eds.), Cambridge University Press, Cambridge, U.K., 1976). It is important to investigate sub-lethal damage in photosensitized tissues which should include analysis of three portions of the irradiated cell. These include:

1. Membrane damage, including measurements of altered enzyme activity of transport functions (both carrier-mediated and those associated with free diffusion) and of chemical alteration to membrane constituents.

2. Cytoplasmic damage, including enzyme and/or co-enzyme inactivation, depletion, or accumulation of substrates and products, disruption of regulatory controls, etc.

3. Nuclear, or genomic damage, including alterations in nucleic acid and nucleoprotein structure and function, mutagenesis, and failure to replicate genetic material.

α-DIKETONE SENSITIZED PHOTOEPOXIDATION OF PYRIMIDINES

S. Y. Wang, H. S. Ryang, E. P. Burrows, and T. Nagamatsu, Division of Environmental Chemistry, The Johns Hopkins University, C.O.E.H. - U.S.P.H.S. Hospital, Building 6, 3100 Wyman Park Drive, Baltimore, Maryland 21205

Increasing awareness of the importance of epoxides as reactive intermediates in biological systems and their possible mutagenic effects, and the challenge presented by isolation and characterization of the primary photooxidation products of nucleic-acid bases have prompted us to investigate the possibility of preparing and studying the properties of pyrimidine epoxides (1).

Benzil sensitized photooxidation (2) of 1,3-dimethylthymine [1] (O_2, 0°C, CH_2Cl_2, 1 hr with >320 nm light) led to the formation of cis-1,3-dimethylthymine glycol [2a, 60%], and 5-methyl-5-hydroxy-barbituric acid [3] and (2,3;6,5)di-1,3-dimethylthymino-1,4-dioxane [4] as minor products (1). Using biacetyl in

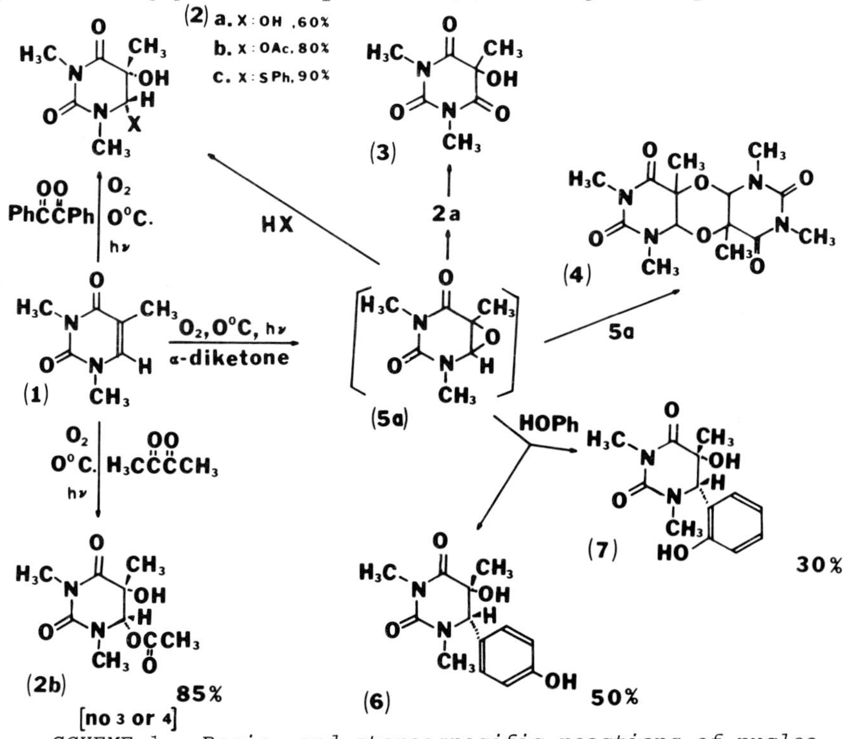

SCHEME 1. Regio- and stereospecific reactions of nucleophiles involving pyrimidine epoxide formed by α-diketone sensitized photooxidation.

ISBN 0-12-592050-4

place of benzil and benzene as solvent, the reaction gave only [2b] in 85% yield; neither [3] nor [4] was detected. Post-irradiation treatment of the former reaction mixture with CH_3COOH also yielded [2b] (80%), but treatment of [3] or [4] with CH_3COOH resulted in no reaction. Similar post-irradiation treatment with thiophenol gave exclusively the cis 6-phenylthio compound [2c] (90%). In contrast, post-irradiation treatment with phenol yielded a 3:2 mixture of isomers later shown by X-ray diffradtion analysis to be the unexpected carbon-carbon bonded adducts [6] and [7] each of cis configuration (3). We interpret these findings in terms of the initial formation of a pyrimidine epoxide [5] which may undergo ring opening by two different mechanistic pathways depending on the nature of the nucleophile. Formation of phenol adducts [6] and [7] may be regarded as electrophilic attack at the O- and p- positions of phenol, but the alternative possibility of a radical coupling mechanism cannot be excluded.

Because 1,3-dimethylthymine epoxide [5a] was too reactive to isolate, more definitive evidence for its formation was sought through studies of the reactions of trans-5-bromo-6-hydroxy-5,6-dihydrothymine derivatives, [8] (4). Treatment of [8a] in THF with water or methanol in the presence of triethylamine yielded mixtures of cis and trans diols [10a] (4:1, respectively) and hydroxyethers [10b] (1:1); the 5-methoxy regiomer was not detected. Treatment of [8b] with pyridine in water gave cis glycol [10c, cis] in a quantitative yield. Also only one isomer [10d] was obtained when [8b] was treated with Et_3N with methanol. In addition, that 100% oxygen migration from C-6 to C-5 occurred during the reactions was proved conclusively by the mass spectrum of cis diol [10e, cis] resulting from utilization of $^{18}OH_2$. Adducts [10f] and [10g] of as yet unestablished stereochemistry have been isolated utilizing the amino acid derivatives N-acetyl cystein methyl ester and N-acetyl tyrosine ethyl ester as nucleophiles under photooxidation conditions as well as from the bromohydrin (3,4).

Recently we were able to isolate and purify thymine epoxide [5b], obtained readily by treatment of trans-5-bromo-6-hydroxy-5,6-dihydrothymine, [8b] with either triethylamine or morpholine in meticulously dried THF at room temperature (5). It is noteworthy and possibly of biological significance that reaction of [5b] with adenosine, guanosine, cytidine, and cytosine (to give, respectively, [10h], [10i], [10j], and [10k] was more rapid than with H_2O or D_2O. In all cases, products consisted of cis and trans isomers with the cis form predominating (60-70%).

The predominant or exclusive formation of cis adducts in the cases of established stereochemistry deserves comment, since S_N2 attack on [5] would lead to trans products. We envision this system as one of those electro-negatively substituted cases where gauche interactions are important and the epoxide intermediate is sufficiently stabilized by zwitterionic contributions [11a↔11b] to allow nucleophilic attack from an energetically favored direction giving cis adducts (6).

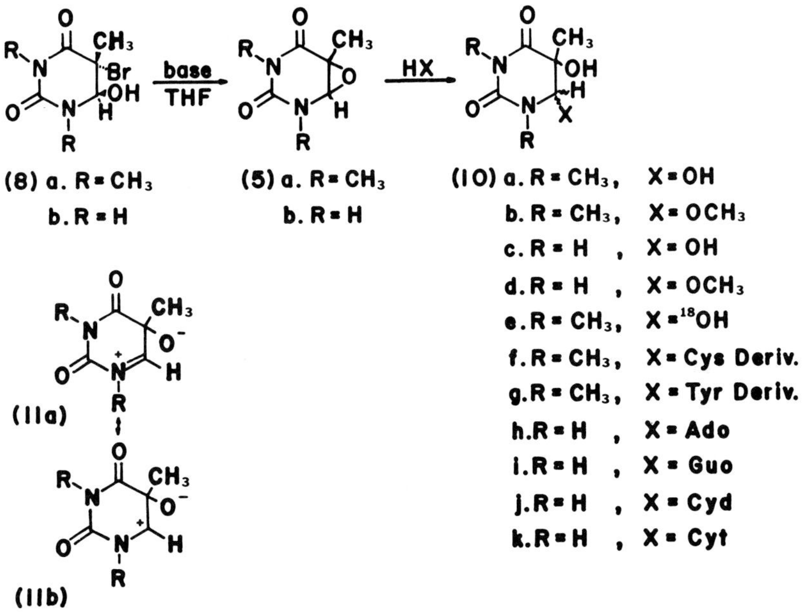

SCHEME 2. Regiospecific reactions of thymine bromohydrins with nucleophiles using bases. Formation of thymine epoxide.

Regarding the photosensitized epoxidation step, a radical mechanism has been proposed and two pathways are possible, i.e. the initial step may involve interaction of the α-diketone with a double bond or with oxygen to yield a diradical (2,7). In methanolic solution, biacetyl sensitized photooxidation of [1] was found to result in a greater than 95% recovery of [1] but the complete oxidation of biacetyl. This finding suggests that the latter alternative is important (1). However, the possibility of a mechanism involving an ionic intermediate cannot be excluded in view of such suggestions concerning the dye-sensitized photooxidation of pyrimidines (8) and indoles (9).

In conclusion, the possibility of attack on pyrimidine epoxides by biomolecules as nucleophiles provides an alternative mechanism for the formation of cross-linkages between nucleic acids and proteins, nucleic acids, membrane constituents, or other biomolecules and thus may be relevant to studies of aging, carcinogenesis, and mutagenesis.

ACKNOWLEDGEMENT

We wish to express our appreciation for the support by grants from the National Institutes of Health (Grants No. 1 P01 ES02300 and 5 R01 GM24238).

REFERENCES

1. Ryang, H. S. and Wang, S. Y., J. Am. Chem. Soc., 100,1302 (1978).
2. Shimizu, N and Bartlett, P. D., J. Am. Chem. Soc., 98,4193 (1976).
3. Burrows, E. P., Ryang, H. S., Wang, S. Y. and Flippen-Anderson, J. L., J. Org. Chem., 44,3736 (1979).
4. Ryang, H. S. and Wang, S. Y., J. Org. Chem. 44,191 (1979).
5. Nagamatsu, T. and Wang, S. Y., unpublished results.
6. Robins, M. J., MacCoss, M., Naik, S. R. and Ramani, G., J. Am. Chem. Soc. 98,7381 (1976).
7. Saltiel, J. and Curtis, H. C., Mol. Photochem. 1,239 (1969).
8. Vickers, R. S. and Foote, C. S., Bull. Chim. Farm. 109,599 (1970).
9. Saito, I., Imuda, M., Takahashi, Y., Matsugo, S. and Matsuura, T., J. Am. Chem. Soc. 99,2005 (1977); Nakagawa, M., Okajima, H. and Hino, J., ibid., 4424 (1977).

PHOTOSENSITIZATION OF BIOMOLECULES IN VIVO

L. I. Grossweiner, Biophysics Laboratory, Physics Department, Illinois Institute of Technology, Chicago, Illinois 60616

INTRODUCTION

The molecular approach to biological photosensitization is based on the assumption that the key processes at the cellular level are initiated by reactions similar to those identified in vitro. The earlier emphasis was on small biomolecules irradiated with visible light or near-UV in the presence of sensitizing dyes and drugs (1,2). More recent findings indicate that the intermediates are the same ones identified in non-biological photosensitization, e.g., excited sensitizer molecules, singlet oxygen (Δ), and superoxide ($O_2^{\bar{\cdot}}$) (3,4). However, the direct extension to living systems presents major problems because the endpoints (e.g., metabolic changes, mutations, cell death) are not easily related to photochemical products. Furthermore, the reaction pathways in vivo may be altered by the microenvironment. The investigation of model systems utilizing high molecular weight biomolecules provides information about the effects of binding and spatial localization on photosensitization mechanisms, particularly active proteins and DNA. Micelles and liposomes have been employed as models for biological membranes. Studies on viruses and cell cultures provide a further extension of complexity where the chemical and biological endpoints may be compared. Representative examples of these investigations are summarized in this paper, emphasizing the dependence of the photosensitization mechanisms on structure. The illustrative figures are taken from current or unpublished work from this laboratory.

THE EFFECT OF SENSITIZER BINDING ON PHOTODYNAMIC ACTION

Binding interactions may alter the relative importance of the Type-1 (dye-substrate) and Type-2 (dye-oxygen) photosensitization pathways compared with low molecular weight substrates. In early work on the photosensitized oxidation of human serum albumin by eosin Y, it was shown that the rate of tryptophan destruction increases with the amount of complexed eosin (5). The decreased flash photolysis yield of triplet eosin in the complex suggests that binding promotes reactive quenching by the protein (6). A similar quenching was observed by laser flash photolysis of the 1:1 lysozyme-eosin complex, in which binding greatly reduced the yield and lifetime of the eosin triplet state (7). However, the effects of sensitizer binding on the photodynamic pathway may be quite complex. The photodynamic inactivation of

ISBN 0-12-592050-4

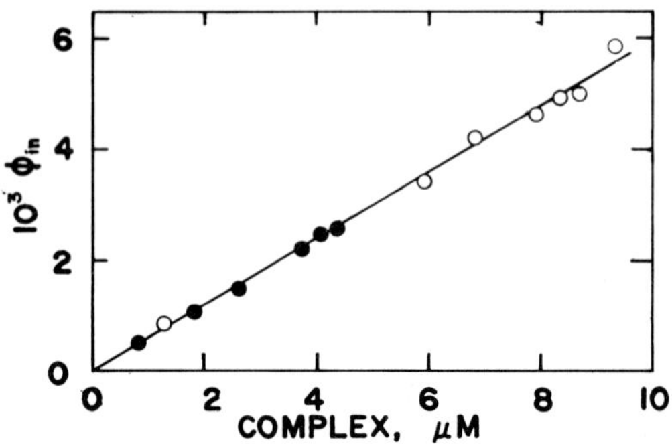

FIGURE 1. Photodynamic inactivation of lysozyme sensitized by eosin Y (air - sat, λ>400 nm). The lysozyme concentration ranged from 6 μM to 300 μM. The eosin concentrations were 9.7 μM (open circles) and 4.2 μM (filled circles) at pH 4.9. The inactivation quantum yield (ϕ_{in}) is a linear function of the concentration of eosin-lysozyme complex. This system showed 5 times higher ϕ_{in} in D_2O and complete protection by 0.03 M azide ion (from ref. 9).

lysozyme is sensitized by many dyes and has been attributed to Δ (8). Typically, the inactivation quantum yield (ϕ_{in}) first increases with increasing enzyme concentration, because of the competition between scavenging of Δ by the enzyme and its decay in the medium, and then decreases at higher enzyme concentrations, because of quenching of the sensitizer triplet state (9-11). However, in the eosin-lysozyme system there was no apparent relationship between ϕ_{in} and the enzyme concentration, although changing the solvent from H_2O to D_2O led to a five times enhancement of ϕ_{in} and there was complete protection by azide ion (9,12). The linear dependence of ϕ_{in} on the amount of the complex in the system indicates that eosin binding strongly sensitizes lysozyme to the inactivating attack of Δ (Fig. 1). Fluorescence measurements have shown that the eosin binding site is located close to two essential tryptophan residues, such that Δ generated by energy transfer from the eosin triplet state can react efficiently with a key lysozyme target (13).

PHOTOSENSITIZATION IN HETEROGENOUS SYSTEMS

The importance of Type-1 reactions can be suppressed by binding of the sensitizer to a soluble polymer. This method has been used to study the photosensitized oxidation of tyrosine and tyramine by methylene blue and thionone, with the dye complexed to the anionic polysaccharide, alginate (14) and the photosensitization of radical formation in polyadenylic acid by complexed proflavine (15). Photodynamic damage to liposomes has been studied with many sensitizers for several endpoints: lysis, marker release, and lipid peroxidation. The participation of Δ in the

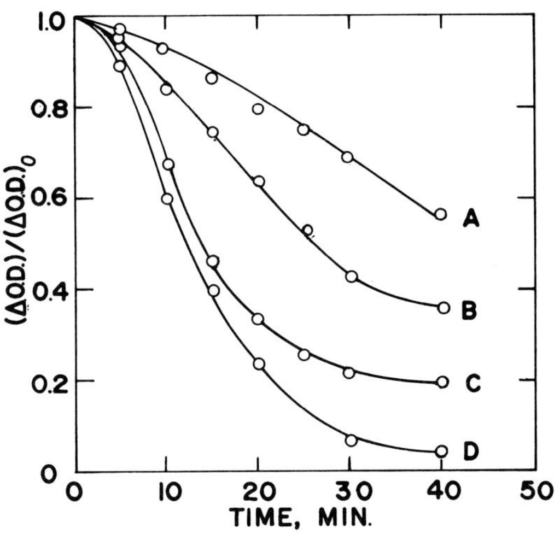

FIGURE 2. Photodynamic lysis of egg lecithin liposomes sensitized by methylene blue (pH 8.5, λ>600 nm): (A) irradiated with oxygen bubbling without dye; (B) irradiated with oxygen bubbling plus dye and 0.04 M azide ion; (C) irradiated with oxygen bubbling plus dye; (D) irradiated with oxygen bubbling plus dye in D_2O.

photosensitized damage to egg lecithin liposomes (frequently incorporating cholesterol and dicetyl phosphate) has been shown for sensitization by toluidine blue (16,17), phenothiazines (18), and methylene blue (19) (Fig. 2). In other experiments the photosensitizer was incorporated in the liposome membrane. The membrane damage was attributed to Δ in the case of all-trans-retinal (20), whereas Type-1 mechanisms were implicated with embedded flavin pigments (21). Little information is available about liposome damage by $O_2^{\overline{\cdot}}$. Current work (Fig. 3) indicates that both $O_2^{\overline{\cdot}}$ and Δ induce lysis of egg lecithin liposomes. The diffusion of Δ from the external medium to acceptors solubilized within micelles has been demonstrated (22,23). In another micellar system, Δ generated by energy transfer from pyrene localized in micelles effectively oxidized an acceptor also localized within micelles (24). These results with liposomes and micelles suggest that Type-2 reactions can be effective with membrane-bound photosensitizers.

VIRUS AND CELL SUSPENSIONS

Although many studies on the photodynamic inactivation of viruses, bacteria and yeast suspensions are reported in the literature (1,25), little information has been available until recently about the primary targets and the reactive intermediates. In a comprehensive investigation on Saccharomyces cerevisiae photosensitized by xanthene, thiazine, and acridine dyes, a positive correlation was found between the inactivation and the

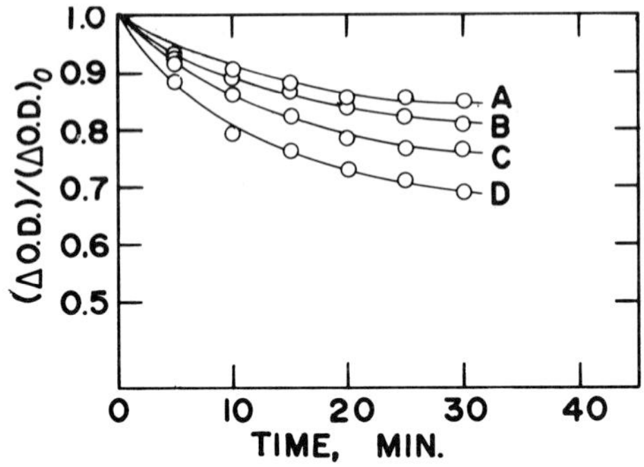

FIGURE 3. Photodymanic lysis of egg lecithin liposomes sensitized by riboflavin (340 nm >λ> 520 nm) in the presence of 0.05 M EDTA: (A) bubbled with oxygen without irradiation or with irradiation in the presence of superoxide dismutase (pH 4 - pH 9); (B), (C), (D) Bubbled with oxygen with irradiation at pH 4.0, 7.0, 9.2, respectively.

participation of Δ, for dyes that penetrate the cell and also for dyes that were judged to remain external to the cell (26). Genetic changes were induced by those dyes that bind to DNA, the major part of which was attributed to Δ. Sensitization by an external dye was shown also in E. coli, where rose bengal was immobilized on polystyrene beads (27). A recent investigation of photosensitized inactivation of T4 phage indicates that diffusion of the dye through the phage capsid is required (28). However, in viruses containing lipids (PM2, Φ6, herpes simplex Type-2) membrane damage involving Δ was implicated in the inactivation mechanism (29). The sensitivity of repair-deficient mutant cells provides a sensitive probe of the relationship between the biochemical lesions and biological endpoints. This technique was employed in an investigation of Proteus mirabilis, in which DNA damage was correlated with the higher sensitivity of the rec^- strain photosensitized by thiopyronine (30). In recent work with E. coli K-12, the lethal effects of furocoumarin DNA monoadducts and interstrand DNA cross-links were investigated with strains sensitive to the different types of lesions (31). The biological significance of the Type-2 mechanism with furocoumarin photosensitizers has not been demonstrated in vivo, although it is known that furocoumarins generate Δ when not complexed to DNA (32-35). In view of the carcinogenic risks identified in the photochemotherapy of psoriasis with near-UV and 8-methoxypsoralen (36), the possible role of Δ in PUVA (psoralen plus UV-A) therapy should be clarified.

LARGE TARGET DIFFUSION KINETICS

The reactions of small agents generated in the external medium (e.g. excited sensitizer molecules, Δ, O_2^-) with large targets may not obey usual biomolecular reaction kinetics (37). In suspensions of liposomes, virus particles, and cells the average separation between individual target units is usually much larger than the diffusion length of the reactive intermediates. Effective reactions are possible only for those intermediates generated within a "reaction volume" surrounding each target unit, the width of which depends on the target radius, the diffusivity of the reactive agent, and its decay lifetime. This situation is evidenced by a linear increase of the reaction quantum yield with target concentration, or equivalently by a constant slope in the conventional, semi-logarithmic survival _vs_ dose relationship. For example, for photodynamic action by Δ in an aqueous suspension of 1 micron diameter cells, this condition obtains at cell concentrations smaller than about 10^{13} ml^{-1} (38). The usual Smoluchowski formula for the bimolecular rate constant of diffusion-limited reactions ($k' = 4\pi RD$) may not be valid for large targets, where R is the target radius and D is the relative diffusion constant of the small reactive agent. The upper limit for reactions of Δ in water is: $R > (D_\Delta \tau_\Delta)^{\frac{1}{2}}$, which is the order of 0.1 micron. The effect of large targets on the effective biomolecular rate constant for reactions of Δ is illustrated in Fig. 4.

DISCUSSION

The analysis of photosensitization _in vitro_ involves a straight-forward application of dilute solution reaction kinetics in terms of the generation and reaction rates of the key intermediates (8). The well-known diagnostic tests for the participation of sensitizer excited states and radicals, Δ, and $O_2^{\cdot -}$ in conjunction with product analysis can lead to a reasonably unambiguous delineation of the overall mechanism. The analysis of mechanism is much more complicated in model biological systems as evidenced by the results summarized above. The binding of the sensitizer to a high molecular weight substrate may make it necessary to consider the parallel reaction pathways for the free and the complexed sensitizer. Binding can have three interrelated effects: (a) modified relative rates of the competing photophysical steps, particularly the intersystem crossing efficiency and the triplet decay lifetime; (b) inhibition of triplet state accessibility to oxygen and other solutes; (c) promotion or inhibition of the reactivity towards targets because of geometrical factors. The overall effect of sensitizer-substrate interactions may be more complex than anticipated. For example, complexing eosin to lysozyme strongly enhances the rate of photosensitized reactions, nevertheless Δ is the dominant inactivating intermediate for the complexed as well as the free dye. The results cited above on

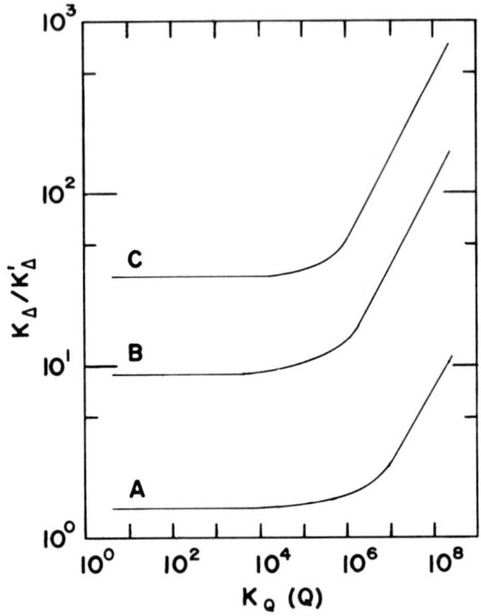

FIGURE 4. Effective bimolecular rate constant (k_Δ) for the reactions of singlet oxygen with spherical targets of radius R in water normalized to the diffusion-limited rate constant (k'_Δ). $k_Q(Q)$ is the pseudo-first order rate constant for quenching of singlet oxygen. The values of R are 0.35 μ (A), 0.6 μ (B) and 2.5 μ (C).

micelles, liposomes and lipid-containing viruses indicate that Δ generation can be significant when the sensitizer is localized in a lipid region. Little information is available about the photosensitized generation of $O_2^{\overline{\cdot}}$ in biological systems, although the dark biochemical generation of $O_2^{\overline{\cdot}}$ is well known (39). Possible sources of $O_2^{\overline{\cdot}}$ in photosensitized reactions are the reduction of oxygen by a dye semiquinone, half-reduced oxidant (e.g. NADH) or a lower oxidation state of a metal ion. Our recent results demonstrating that $O_2^{\overline{\cdot}}$ induces the lysis of egg lecithin liposomes suggests that it may be involved at the cellular level, particularly in view of the increasing evidence that damage to the cell membrane can induce lethality in yeast and bacteria. The diagnostic tests for active oxygen intermediates are likely to be less reliable in vivo than in vitro because of toxicity (e.g. D_2O and N_3^-), segregation of water-insoluble chemical agents in lipid regions, and the effects of metabolism. However, cell studies provide an additional powerful tool for assaying chemical damage to DNA, based on the relative photosensitivity of repair-deficient mutant strains. For example, uvr⁻ and rec⁻ strains of E. coli K-12 were more sensitive to the photodynamic action of acridine orange than the wild type strain and the double mutant, uvr⁻ rec⁻, was more sensitive than either single mutant, indicating that the DNA lesions are repairable by at least two independent pathways. Similar experiments with E. coli K-12 cells exposed to near-UV in the presence of furocoumarins showed that DNA polymerase I is required for the repair of the DNA monoadducts (31).

ACKNOWLEDGEMENTS

The support of the U.S. Department of Energy on Contract No. DE-AS02-76EV02217 and the Department of Health, Education and Welfare on Grant No. GM 20117 is gratefully acknowledged. This is COO-2217-32.

REFERENCES

1. Spikes, J. D., "Photophysiology," Vol. III (A. C. Giese, ed.), Academic Press, New York, 1968, p. 29.
2. Giese, A. C., "Photophysiology," Vol. VI (A. C. Giese, ed.), Academic Press, New York, 1971, p. 77.
3. Spikes, J. D. and Livingston, R., "Advances in Radiation Biology," Vol. 3, Academic Press, New York, 1969, p. 29.
4. Koizumi, S. Kato, Mataga, N., Matsuura, T. and Usui, Y., "Photosensitized Reactions," Kagakudojin Publ Co., Inc., Kyoto, 1978.
5. Grossweiner, L. I. and Kepka, A. G., Photochem. Photobiol., 16, 305(1967).
6. Youtsey, K. J. and Grossweiner, L. I., Photochem. Photobiol., 6,721 (1967).
7. Fisher, G. J., Lewis, C. and Madill, D., Photochem. Photobiol. 24,223 (1976).
8. Grossweiner, L. I., Curr. Topics Radiat. Res. Quart., 11,141 (1976).
9. Kepka, A. G. and Grossweiner, L. I., Photochem. Photobiol., 18,49 (1973).
10. Schmidt, H. and Rosenkranz, P., Z. Naturforsch., 27b,1436 (1972).
11. Schmidt, H. and Rosenkranz, P., Z. Naturforsch., 31c,29 (1976).
12. Lee, E. C. H., M.S. Dissertation, Illinois Institute of Technology (1973).
13. Baugher, J. F., Grossweiner, L. I. and Lewis, C., J. Chem. Soc., Faraday Trans. II, 70,1389 (1974).
14. Seely, G. R. and Hart, R. L., Photochem. Photobiol., 23,1 (1976).
15. Cannistraro, S., Calberg-Bacq, C. M. and Van de Vorst, A., Photochem. Photobiol., 23,137.
16. Anderson, S. M. and Krinsky, N. I., Photochem. Photobiol., 18,403 (1973).
17. Anderson, S. M., Krinsky, N. I., Stone, M. J. and Clagett, D. C., Photochem. Photobiol., 20,65 (1974).

18. Copeland, E. S., Alving, C. R. and Grenan, M. M., Photochem. Photobiol., 24,41 (1976).
19. Muller-Runkel, R., Blais, J. and Grossweiner, L. I., Abstracts, 8th Annual Meeting of American Society for Photobiology, Colorado Springs, Feb. 17-21 (1980); Photochem. Photobiol., 00,000 (1981).
20. Delmelle, M., Photochem. Photobiol., 28,357 (1978).
21. Heelis, P. F., Parsons, B. J., Phillips, G. O., Colombetti, G., Lenci, F. and McKellar, J. F., Photochem. Photobiol., 30, 507 (1979).
22. Gorman, A. A., Lovering, G. and Rodgers, M. A. J., Photochem. Photobiol., 23,399 (1976).
23. Usui, Y., Tsukada, M. and Nakamura, H., Bull. Chem. Soc. Japan, 51,379 (1978).
24. Miyoshi, N. and Tomita, G., Z. Naturforsch., 33b,622 (1978).
25. Spikes, J. D., "The Science of Photobiology" (K. C. Smith, ed.), Plenum Press, New York, 1977, p. 87.
26. Ito, T. and Kobayashi, K., Photochem. Photobiol., 26,581 (1977).
27. Bezman, S. A., Burtis, P. A., Izod, T. P. J. and Thayer, M. A., Photochem. Photobiol., 28,325 (1978).
28. Piette, J., Calberg-Bacq, C. M. and Van de Vorst, A., Int. J. Radiat. Biol., 34,223 (1978).
29. Snipes, W., Keller, G., Woog, J., Vickroy, T., Deering, R. and Keith, A., Photochem. Photobiol., 29,785 (1979).
30. Jacob, J. E., Photochem. Photobiol., 21,445 (1975).
31. Grossweiner, L. I. and Smith, K. C., Photochem. Photobiol., 33,317 (1980).
32. Poppe, W. and Grossweiner, L. I., Photochem. Photobiol., 22, 217 (1975).
33. Poppe, W., Ph.D. Dissertation, Technishen Universitat Hannover (1977).
34. Cannistraro, S. and Van de Vorst, A., Biochem. Biophys. Acta, 476,166 (1977).
35. De Mol, N. J. and Biejersbergen van Henegouwen, G. M. J., Photochem. Photobiol., 30,331 (1979).
36. Stern, R. S., Thibodeau, L. A., Kleinerman, R. A., Parrish, J. A., Fitzpatrick, T. B. and 22 Participating Investigators, N. Engl. J. Med., 300,809 (1979).
37. Grossweiner, L. I., Photochem. Photobiol., 28,97 (1978).
38. Grossweiner, L. I., Photochem. Photobiol., 26,309 (1977).
39. Bors, W., Saran, M., Lengfelder, E., Spottl, R. and Michel, C., Curr. Topics Radiat. Res. Quart., 9,247 (1974).
40. Harm, W., Biochem. Biophys. Res. Commun., 32,350-358 (1968).

DISCUSSION

BORS: I find your results on the reaction of $O_2^{\overline{\cdot}}$ with liposomes quite interesting. Did you do such experiments also with erythrocyte ghosts since Lynch and Fridovich were suggesting that $O_2^{\overline{\cdot}}$ penetrates the erythrocyte membrane?

GROSSWEINER: We have not looked at superoxide with erythrocyte ghosts. Perhaps this is a good opportunity to mention for anyone involved in phototherapy that we have found (and this is unreported) that 3-carbethoxypsoralen which is being taken seriously as an alternative PUVA-sensitizer lyses whole human erythrocytes in the dark.

BUETTNER: In a spin-trapping system, we have observed radical production from the methylene blue system. It appears to have the properties of hydroxyl radical. Dr. Krinsky has made similar observations. We do not know how important this is; but doing various checks such as throwing ethanol (or formate) into the system, we see the radicals that we would expect from hydroxyl radical reacting with those particular species. Could it be that there is some radical component to the methylene blue results that you report here?

GROSSWEINER: That would be one possible explanation for the results. Another possibility is that there is some protein in our egg lecithin, and we may be generating superoxide under alkaline conditions. Jahnke and Frenkel discussed this possibility in their paper in epinephrine photooxidation at the Pinawa conference.

REVERSIBLE O_2 BINDING

Session Chairman, A. Riggs, Austin, Texas

STRUCTURAL ASPECTS OF HEMOGLOBIN LIGAND BINDING

Keith Moffat, Section of Biochemistry, Molecular and Cell Biology, Clark Hall, Cornell University, Ithaca, New York 14853

The first goal in all macromolecular structure determinations, whether of oxygen-binding proteins or of any other macromolecule, is a structural model—a set of atomic coordinates. In the case of oxygen-binding heme proteins, two important questions can then be posed: How different are the functional properties of the protein, such as ligand binding kinetics, from those of model compounds such as free hemes? How different are the structures of the heme or heme-ligand complexes from those of corresponding free hemes? In other words, we wish to know how the presence of the protein—the globin—modifies the structure and reactivity of the heme. Satisfactory answers to these questions then lead to the second goal, the proposal of a molecular mechanism of action, in which a structural basis for all the functional properties of the heme protein is assigned.

This goal can be achieved only if the limitations of both the functional and the structural data which underlie such a mechanism are appreciated. The results of macromolecular crystallography, like those derived from any other technique, are subject to errors both in data collection and in data interpretation. Perhaps because of the lengthy nature of the experiments, protein crystallographic results have almost never been subjected to the true scientific test of reproducibility in another laboratory. In those few cases in which they have (as for example in a comparison of closely similar molecules such as heme proteins subjected to different lattice forces in different crystallographic space groups), the general structural outlines have definitely been confirmed. There can be little doubt that the crystallographically-derived structures provide a close approximation to the average structure of the macromolecule in solution. Nevertheless, numerous small, potentially significant structural differences have also been noted, which may be critical for appraisal of a proposed molecular mechanisms of action.

ISBN 0-12-592050-4

The errors and limitations of macromolecular crystallography may be speedily noted (1). As noted by Frauenfelder, crystallotraphic results are inherently static, not dynamic; they offer a picture averaged over the hours of data collection, and over the 10^{15} or so molecules in a single crystal. At each instant in time, every molecule in the crystal (and a fortiori in solution) has a slightly different conformation, as each group undergoes thermal vibration and perhaps those more profound tertiary structural changes which are still consistent with the integrity of the crystal lattice. The crystallographers then interpret their electron density maps, which really depict the average over a very large conformational ensemble, in terms of a single set of stereochemically acceptable coordinates. This single set of coordinates should not be thought of as representing the single, rigid structure of the macromolecule, but rather a structure, a complicated average at best. As Frauenfelder has remarked, this reservation has sometimes been overlooked, by both enzymologists and crystallographers. The recent studies by the groups of Frauenfelder and Karplus (see for example Refs. 2-4) have led to a renewed appreciation of protein dynamics, and the role of conformational fluctuations and equilibria in enzymatic processes. That such conformational equilibria might be functionally important in hemoglobin was pointed out ten or more years ago (5-7).

Secondly, the lack of resolution inherent in macromolecular structure determination means that atomic resolution is almost never obtained, unlike the situation in small molecule crystallography. This lack of resolution reflects both the tertiary structural fluctuations, and the relatively poor crystal quality of most macromolecular crystals. Thus, atomic coordinates must be inferred from the results of a complicated crystallographic refinement scheme, in which the structural parameters of a trial model are systematically adjusted to provide optimum agreement with observation. Refinement schemes may be carried out in real space (fitting to an electron density map), in reciprocal space (fitting to observed structure factors), and in energy space (minimizing the total energy by means of potential functions which describe the energetic costs of variation in bonded and non-bonded distances and angles), as described in Ref. 1. No agreement has been reached on the best refinement strategy to be adopted, and indeed errors have subsequently been shown to be present in seemingly the most carefully conducted refinement schemes. Particular difficulty has been encountered in deriving satisfactory models for the prosthetic groups in, say, heme proteins or Fe-S proteins. This presents a severe problem, as profound biochemical consequences may flow from a clear demonstration that the prosthetic groups are (or are not) strained with respect to their structure in externally unconstrained model compounds. For example, claims of strain based on preliminary crystallographic results on the prosthetic groups of deoxyhemoglobin (7) and rubredoxin (8) have not been completely substantiated by further crystallographic refinement, or by other measurements. Indeed, based on our own crystallographic results on a hemoglobin reconstituted with a modified, 2,4-dibromodeutero-heme (14), we now believe that even the question of heme orientation in hemoglobin should be reexamined. At least a fraction of the hemes may be inverted from their assumed orientation, by 180° rotation about a line through the α and γ methene carbons. As a further example of concern to this symposium, the current atomic model derived from extensive (though not yet complete) crystallographic refinement of oxyerythrocruorin contains a nearly linear iron-dioxygen complex, in which the iron-dioxygen bond is surprisingly long and the iron is even further out of the plane of the pyrrole nitrogens than in deoxyerythrocruorin (9,10). These details of the model are at variance with the structures of model hemes, and with theoretical considerations. In contrast, an independent crystallographic refinement of oxymyoglobin (11, 12) is yielding a structure for the iron-dioxygen complex which closely approximates that found in model hemes. The reversible oxygen-binding properties of these two, otherwise closely similar proteins are not as divergent as the large apparent differences in structure of the iron-dioxygen complex would suggest. Resolution of this structural discrepancy is essential if valid mechanisms for the structural control of oxygen affinity are to be obtained, and a careful examination of the starting models and refinement procedures used for these two proteins is called for.

These limitations must be borne in mind when considering the crystallographic data on heme proteins, which are extensive. Structures exist for myoglobin, horse and human hemoglobin, erythrocruorin and lamprey hemoglobin, both unliganded and in combination with many ligands: for ferrous hemes, oxygen (for myoglobin and erythrocruorin only, as noted above), carbon monoxide, and nitric oxide; and for ferric hemes, cyanide, azide, thiocyanate, fluoride and imidazole (9-15; see also Footnotes 7-12,21,22 and 24 of Ref. 16). All liganded proteins differ in detail in tertiary structure, and in the case of hemoglobin they also differ in quaternary structure from that of unliganded, deoxyhemoglobin. In the allosteric notation (17), liganded hemoglobins normally exist in the R (relaxed) quaternary structure, which has a high affinity for oxygen; deoxyhemoglobin exists in the T (tense) quaternary structure, which has an unusually low affinity for oxygen, about 0.005 times that of the R state.

The molecular mechanism by which the oxygen affinity of deoxyhemoglobin is lowered, and by which ligand binding is coupled to the change in quaternary structure (and hence to the expression of cooperativity in ligand binding) is under active debate. The latter effect has been described in detail by Baldwin and Chothia (13). In the former, possible structural features have been identified, but there is no agreement as to their relative importance.

They may be classified roughly into three groups: distal processes, proximal processes, and general processes. In the first, Perutz (7,15) noted that the ligand binding sites in deoxyhemoglobin are obstructed, particularly in the β-chains, and that this would tend to lower their affinity. This process, which is based on steric hindrance to ligand binding, has been amplified recently and extended to cover kinetic differences between ligands such as oxygen and carbon monoxide (16,18,19). The second (13,20) notes that the proximal, heme-linked histidine in deoxyhemoglobin (but not in liganded hemoglobin) is asymmetrically disposed with respect to the heme normal, and that it cannot readily move on to the heme normal and closer to the mean heme plane, as is evidently required for combination with ligands (7, 15). This again would tend to lower the affinity, as would such additional proximal processes as hydrogen bonding to this histidine. General processes might be exemplified by the electrostatic model of Warshel (21), where the electric field at the ligand site generated by all the protein charges and dipoles differs in liganded and deoxyhemoglobin, and is such as to oppose the charge separation which is believed to be present in the iron-dioxygen complex $Fe^{\delta+}-O_2^{\delta-}$. That these various processes could lower oxygen affinity has been clearly demonstrated by studies of ligand binding to model compounds in which these structural processes are incorporated by ingenious chemistry (see for example Refs. 22-24); that any of them do contribute to the lower oxygen affinity of deoxyhemoglobin, and to what extent, is still in doubt.

Finally, the affinity of hemoglobin for oxygen may be altered by allosteric effectors such as protons, 2,3-diphosphoglycerate, chloride ions and carbon dioxide. Their major effect is largely to shift the allosteric equilibrium between the R and T structures. The broad outlines of the molecular mechanism by which this is brought about are understood (15), but the details, and in particular those which may arise from changes in tertiary structure superimposed on the change in quaternary structure, are still in doubt.

It is clear that although much remains to be explained, crystallographic results provide the foundation for satisfactory molecular mechanisms for the control of oxygen affinities and rates. Although the details of crystallographic results are subject to some of the limitations noted above, they can be ignored only at your peril.

REFERENCES

1. Blundell, T. L. and Johnson, L. N., "Protein Crystallography," Academic Press, New York (1976).
2. Austin, R. H., Beeson, K. W., Eisenstein, L., Frauenfelder, H. and Gunsalus, I. C., Biochemistry, 14, 5355 (1975).
3. Frauenfelder, H. and Petsko, G. A., Biophysical J., 32, 465 (1980).
4. Karplus, M. and McCammon, J. A., Nature (London), 277, 578 (1979).

5. Moffat, K., J. Mol. Biol., 55,135 (1971).
6. Moffat, K., Simon, S. R. and Konigsberg, W. H., J. Mol. Biol., 58, 89 (1971).
7. Perutz, M. F., Nature (London), 228, 726 (1970).
8. Watenpaugh, K. D., Sieker, L. D., Herriott, J. R. and Jensen, L. H., Acta Cryst., A29, 943 (1973).
9. Weber, E., Steigemann, W., Jones, T. A. and Huber, R., J. Mol. Biol., 120, 327 (1978).
10. Steigemann, W. and Weber, E., "Symposium on Interaction between Iron and Proteins in Oxygen and Electron Transport," Elsevier-North Holland, in press (1980).
11. Phillips, S. E. V., ibid.
12. Phillips, S. E. V., Nature (London), 273, 247 (1978).
13. Baldwin, J. M. and Chothia, C., J. Mol. Biol., 129, 175 (1979).
14. Korszun, Z. R., Bell, J. and Moffat, K., Manuscripts in preparation.
15. Perutz, M. F., Nature (London), 237, 495 (1972).
16. Moffat, K., Deatherage, J. F. and Seybert, D. W., Science, 206, 1035 (1979).
17. Monod, J., Wyman, J. and Changeux, J.-P., J. Mol. Biol., 12, 88 (1965).
18. Szabo, A., Proc. Nat. Acad. Sci. USA, 75, 3108 (1978).
19. Collman, J. P., Acc. Chem. Res., 10, 265 (1977).
20. Gelin, B. R. and Karplus, M., Proc. Nat. Acad. Sci. USA, 74, 801 (1977).
21. Weiss, R. M. and Warshel, A., "Symposium on Interactions between Iron and Proteins in Oxygen and Electron Transport," Elsevier-North Holland, in press (1980).
22. Geibel, J., Cannon, J., Campbell, D. and Traylor, T. G., J. Amer. Chem. Soc., 100, 3575 (1978).
23. Traylor, T. G., Chang, C. K., Geibel, J., Berzinis, A., Mincey, T. and Cannon, J., J. Amer. Chem. Soc., 101, 6716 (1979).
24. White, D. K., Cannon, J. B. and Traylor, T. G., J. Amer. Chem. Soc., 101, 2443 (1979).

OXYGENASES

Session Chairman, D. T. Gibson, Austin, Texas

OXYGENASES—AN INTRODUCTION

D. T. Gibson, Department of Microbiology, The University of Texas at Austin, Austin, Texas 78712

Twenty-five years ago Hayaishi (1) and Mason (2) independently demonstrated that certain enzymes can incorporate molecular oxygen into organic substrates. Hayaishi and his colleagues showed that pyrocatechase (E.C.1.13.1.1.), a microbial enzyme, would oxidize catechol to cis,cis-muconic acid; a reaction in which both atoms of molecular oxygen were incorporated into the ring fission product. In contrast, Mason's observations with a phenolase complex (E.C.1.10.3.1.) showed that one atom of molecular oxygen was incorporated into 3,4-dimethylphenol to form 4,5-dimethylcatechol. Thus, two types of enzymatic oxygen fixation were discovered and the enzymes were ultimately assigned to a class of proteins called oxygenases. Monooxygenases incorporate one atom of oxygen into organic substrates and reduce the other oxygen atom to water whereas dioxygenases incorporate both atoms of molecular oxygen into organic substrates (3).

In the past two decades many different oxygenases have been detected in microorganisms, plants and animals. The ubiquitous distribution of this class of enzymes is a consequence of their importance in cellular metabolism. They play a vital role in the microbial degradation of many natural and man-made compounds, the biosynthesis of certain amino acids, lipids, porphyrins and hormones, and the drug-metabolizing system of mammalian liver. In addition, oxygenases are also intimately involved in the formation of certain reactive electrophiles which have been implicated as the causative agents in chemical carcinogenesis.

Several oxygenases have been obtained in purified form and current studies, including those to be presented in this session, are directed towards elucidating the nature of "active oxygen" and the mechanisms by which it is formed. These, and future, investigations will lead to a greater understanding of the role of oxygen in cellular processes.

ISBN 0-12-592050-4

REFERENCES

1. Hayaishi, O., Katagiri, M. and Rothberg, S , J. Amer. Chem. Soc., 77,5450 (1955).
2. Mason, H. S., Fowlks, W. L. and Peterson, E., J. Amer. Chem. Soc., 77,2914 (1955).
3. Hayaishi, O., Proceedings of the Plenary Sessions, Int. Congr. Biochem., Abstr. 6th, 33,31 (1964).

MECHANISMS OF METAL-CONTAINING MONOOXYGENASES

Volker Ullrich and Hartmut Kuthan, Department of Physiological Chemistry, 665 Homburg-Saar, Federal Republic of Germany

INTRODUCTION

According to Hayaishi (1) the group of enzymes catalyzing the introdution of molecular oxygen into organic compounds are termed "oxygenases." They are subdivided into "dioxygenases" and "monooxygenases" depending on their abilities to insert both oxygen atoms or only one into their substrates. This can be conveniently established by the use of the oxygen isotope $^{18}O_2$. Monooxygenases require a two electron donor (DH_2) as a cofactor according to the sum equation first postulated by Mason (2)

$$RH + O_2 + DH_2 \longrightarrow ROH + H_2O + D$$

This reaction is strongly exergonic, since both the formation of the hydroxyl group and the reduction of an oxygen atom to water have a high negative free energy of formation (3). However, the energy of activation necessary to split a CH bond is quite high, so that the two electron reduction is believed to drive the formation of a reactive intermediate which is called "active oxygen." In comparison with monooxygenases, the dioxygenases act on a group of more readily oxidizable substrates, like phenols or even diphenols, so that a less reactive oxygenating intermediate is required. Therefore dioxygenases generally are independent of an external electron donor and a Lewis acid-activated dioxygen molecule (4) or a metal ion-oxy complex (5) have been proposed as oxygenating intermediates.

From a chemist's point of view a model for the active oxygen species involved in monooxygenases would not only be of great importance as a tool to study the enzyme mechanism but also of practical value if it could be generated catalytically for the oxidation of organic compounds. Despite many efforts towards this goal no satisfying results have been achieved yet.

It is the object of this contribution to outline the mechanistic principles involved in the enzymatic oxygen activation processes and to especially point out how monooxygenases overcome the manifold difficulties and side reactions that the chemist is left with when dealing with oxidations by molecular oxygen.

GENERAL PATHWAYS OF OXYGEN ACTIVATION IN MONOOXYGENASES

Since monooxygenases depend on a two electron reduction of the dioxygen molecule to generate active oxygen, this species ought to be on the redox level of peroxide or atomic oxygen.

ISBN 0-12-592050-4

Indeed, for all monooxygenases investigated so far this seems to be true, like for the flavin-dependent monooxygenases (6) or for the nonheme as well as heme containing enzymes as will be pointed out later. For iron and copper dependent monooxygenases the transfer of the two electrons in subsequent one electron steps must be postulated. The involvement of a redox-active metal ion carries several advantages for the enzymic catalysis, since it offers a high affinity for the dioxygen binding process and a sufficient stabilization of the active oxygen and its intermediates to guarantee the specificity of the oxygen atom transfer to the substrate. The transfer of the reducing equivalents can also be mediated through the metal ion. Based on the known features of oxygen and its reactions in biological systems a schematic presentation of the dioxygen activation process can be proposed:

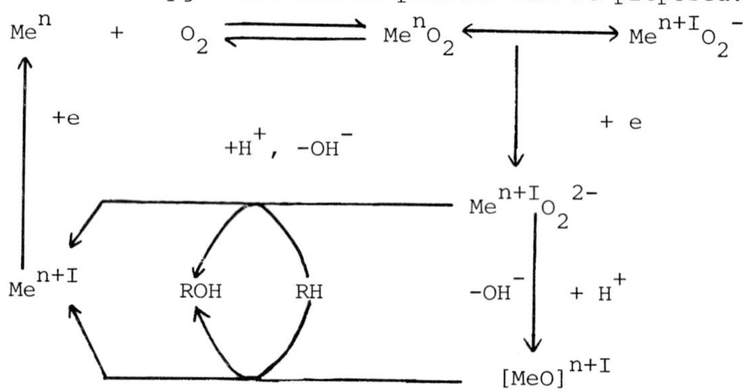

Me = metal center

Because of the electron accepting properties of the dioxygen molecule, the binding of dioxygen generally occurs to the reduced state of the metal ion, i.e. Cu(I) or Fe(II). The resulting oxycomplexes are well-known species characterized spectrally or kinetically, but they tend to be unstable since they may disproportionate in the presence of protons to yield dioxygen and hydrogen peroxide. In aqueous media this would be the most likely fate of an oxycomplex. Hydrogen peroxide also would be unstable in the presence of reduced metal ions and would undergo reduction to OH-radicals. This explains why aqueous model systems for monooxygenases consisting of metal ions, reducing equivalents and dioxygen behave essentially like a Fenton-system which is known to hydroxylate organic molecules by free OH-radicals (7). These radicals can hydroxylate aromatic compounds, but do not form epoxides and attack aliphatic hydrocarbons only through chain reactions which makes them unlike the enzymatic active oxygen and unsuitable for defined chemical hydroxylations (8).

The question then arises how monooxygenases overcome the difficulties leading to this pathway in model systems. Two principal mechanisms exist by which the enzymes can avoid the crucial decomposition of the oxycomplex. One is a very rapid transfer of the second electron before the oxycomplex has time to dissociate. This seems to be verified in some bacterial iron-dependent monooxygenases in which an iron-sulphur center is closely associated with the nonheme iron in charge of dioxygen activation (9). A consequence of a rapid reduction of the oxycomplex should be a lack of inhibition of the reaction by carbon monoxide, since binding of CO will occur only at the ferrous ion and would compete for dioxygen binding. In case of a rapid electron transfer, however, the steady state concentration of the oxycomplex would be too small to allow CO to compete effectively:

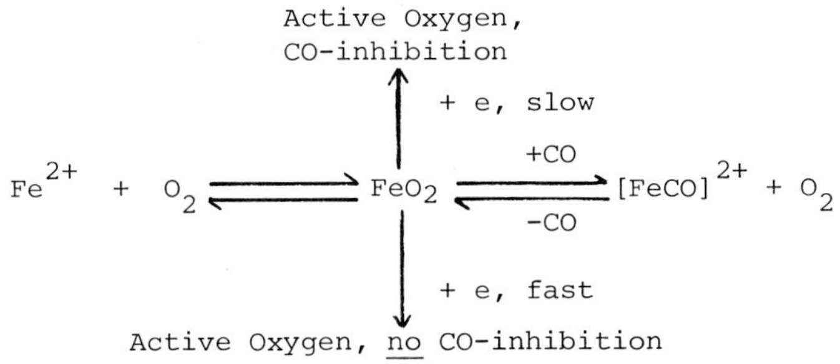

The second possible pathway for avoiding a destruction of the oxycomplex is its stabilization in an appropriate liganding environment. This is the case in cytochrome P450-dependent monooxygenations where the oxycomplex is stable at room temperature for at least one minute (10,11), so that even at slow-rates of its reduction (12) a complete conversion to the active oxygen species is guaranteed. The high steady state concentration of the oxycomplex explains the extreme sensitivity of some of these monooxygenations against CO (13). However, in cases of a more rapid transfer of the second electron to the oxycomplex even cytochrome P450-dependent monooxygenases will become CO-insensitive (14).

The porphyrin skeleton is an ideal ligand for iron enzymes, since the fixation of four ligand positions by the pyrrols inhibits ligand exchange. A second factor adding to the stabilization of a heme-oxycomplex is a lipophilic environment provided by the enzyme which makes protonation as probably the first step in the decomposition of the oxycomplex less likely. Shielding effects and steric factors may be the reason that among hemoproteins large differences in the stability of oxycomplexes are found. Oxymyoglobin and oxyhemoglobin are the most stable ones, whereas peroxidase compound III and oxycytochrome P450 are considerably less stable (15). Little is known about the stabilization of the oxycomplex in copper-containing monooxygenases.

The activation process for the dioxygen molecules starts with the transfer of an electron to the oxycomplex. This was experimentally shown for the oxycytochrome P450 complexes of the steroid-11β-monooxygenase (16) and of the bacterial comphor-5-exo-monooxygenase (17). Theoretically it must be concluded that a peroxide complex is the direct product of this reduction step but the chemical properties and the stability of such a complex are not known. In the first scheme it is pointed out that even splitting of the O-O bond may occur under formation of an "oxenoid" complex, $[FeO]^{3+}$. The evidence for such an active oxygen complex is derived from cytochrome P450 catalyzed reactions (18).

THE REACTION MECHANISM OF CYTOCHROME P450

It was already mentioned that the oxygen activation process at cytochrome P450 involves the formation of an oxycomplex. In the case of cytochrome $P450_{CAM}$ it is sufficiently long-lived to be reduced to the active oxygen complex before it is decomposed. If the second electron is not provided the oxycomplex releases superoxide ion which is detected as hydrogen peroxide (19). Substrate usually is present at the active site since no enzymic reduction occurs in the absence of camphor (20). If chemical reduction is achieved in the absence of camphor the halflife time decreases by a factor of 50 and a strong pH-dependence of the decomposition is observed (21). At a pH-value below 6.6 the autoxidation rate increases drastically indicating that protonation of the oxycomplex leads to a rapid release of $HO_2\cdot$.

With microsomal cytochrome P450 the situation is less clear, because several different forms of cytochrome P450 are present and their reduction seems to occur even without a substrate being bound to the active site. Since microsomes generate hydrogen peroxide at a rate which very often exceeds the rate of substrate monooxygenation we were interested in the mechanism of its formation. Experiments with a reconstituted system had revealed that most of the hydrogen peroxide originated from $O_2^{\overline{\cdot}}$ ions released from oxycytochrome P450 (22,23). A similar study in microsomes was hampered by the fact that contaminating superoxide dismutase did not allow a quantitative determination by the succinylated cytochrome-c test. However, qualitatively it is shown in Fig. 1 that the photochemical action spectra for H_2O_2 and $O_2^{\overline{\cdot}}$ formation are very similar indicating a common pathway.

With a calibration curve standardized by $O_2^{\overline{\cdot}}$ generation from the xanthine oxidase system it was possible to also quantitate the formation of superoxide (Table I). According to these results more than 50% of the oxycomplex formed decompose to $O_2^{\overline{\cdot}}$ radicals. This is a surprisingly low efficiency. The rates of $O_2^{\overline{\cdot}}$ formation were found to remain almost unchanged at lower oxygen pressure being even 90% at an oxygen pressure of 20 torr. This would indicate a high $O_2^{\overline{\cdot}}$ release also under in vivo conditions.

The oxycomplex of cytochrome P450 generates the active oxygen complex by taking up another electron as it was shown for two different monooxygenases (16,17). This should give rise to a peroxide complex; however, such a species does not seem to be very stable in heme enzymes. In peroxidases hydrogen peroxide forms an enzyme substrate complex but there is ample evidence that in this so-called compound I only one oxygen atom is left (24).

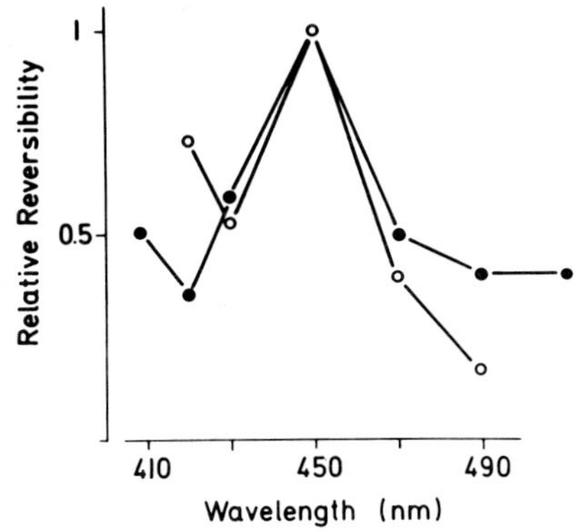

FIGURE 1. Photochemical action spectra of superoxide anion and hydrogen peroxide formation in rat liver microsomes. ($O_2^{\overline{\cdot}}$ = full circles, H_2O_2 = open circles). H_2O_2 assay: The reaction medium (final volume 2.07 ml) containing 50 mM Tris/HCl, pH 7.6, 0.5 mM 7-ethoxy-coumarin, 1 mM NaN_3 and 0.16 mM NADPH was equilibrated with a gas mixture (80% CO and 20% O_2) for 6 min. 50 µl microsomal suspension (22.5 mg/ml) were added and the reaction mixture gassed for another minute. The incubations were carried out for 3 min at 23°C in the dark or in monochromatic light of various wavelength and terminated by transferring 1 ml of the incubation medium to 0.5 ml of 22.5% (w/v) TCA. H_2O_2 was determined as described above. $O_2^{\overline{\cdot}}$ assay: Sample cuvettes containing a suspension of 0.4 mg/ml PB-microsomes, 18.2 µM succinylated cytochrome c, 0.5 mM 7-ethoxycoumarin, 0.2 mM EDTA and 0.2 mM KCN in 50 mM potassium phosphate buffer, pH 7.6 of final volume 0.83 ml at 22°C.

TABLE I. Stoichiometry of the Microsomal Monooxygenase System. Superoxide anions were determined as described (23), benzphetamine-N-dealkylation by formaldehyde release and NADPH-oxidation spectrophotometrically at 340 nm.

Product[2)]	Stoichiometry of the Microsomal[1)] Monooxygenase System nmol · mg protein^{-1} · min^{-1}	
Superoxide, $O_2^{\overline{\cdot}}$	42.8	
Hydrogen peroxide, H_2O_2	27.8	44.1
Benzphetamine N-dealkylation	16.3	
NADPH oxidation		45.4

[1)] Livers from PB-induced rats.
[2)] Typical experiment, means of triplicates.

According to model investigations by Dolphin (25) it is likely that such a species oxidizes the porphyrin to a cation radical leaving the iron as the ferryl ion:

$$Fe^{III} + H_2O_2 \xrightarrow[- H_2O]{} \underset{\text{unstable}}{Fe^{V} = O} \longrightarrow Fe^{IV}OH \cdot$$

If the electron density at the porphyrin ring would be responsible for the release of a hydroxyl ion from the peroxide intermediate then this could be expected to occur also at cytochrome P450 if a weakly acidic group is available at the active site. Support for an $[FeO]^{3+}$ complex as active oxygen in cytochrome P450 comes from our earlier observation that iodosobenzene can substitute for NADPH and O_2 and can cause the same oxidations in the presence of oxidized cytochrome P450 (26). Unlike in peroxidases, however, the fifth ligand of cytochrome P450 is not a histidine but a thiolate ligand according to a number of indications (27,28,29,30). It seems feasible therefore that in cytochrome P450 the thiolate rather than the porphyrin ring would donate an electron to the FeO entity resulting in the following hypothetical structures:

$$S^- \; Fe^{V} = O \longleftrightarrow S^- \; Fe^{IV} - \overline{\underline{O}} \cdot \longleftrightarrow S - Fe^{III} - \overline{\underline{O}} \cdot$$

In favor of radical-like oxenoid species would be the finding of an extraordinary high isotope effect of around 11 (31,32,33) and a partial loss of retention of the configuration at the hydroxylated carbon atom (33). By its charge donation the thiolate ligand seems to make the oxygen atom a suitable leaving group. This would explain the unique role of the heme-thiolate group in monooxygenation reactions.

Final experimental evidence for the oxenoid nature of the active oxygen complex is its fate in the presence of uncouplers. These are defined as quasi-substrates which initiate the oxygen activation process but are unable to accept the oxygen atom because of chemical or sterical reasons. Perfluorinated hydrocarbons are such compounds which cause an NADPH oxidation of microsomes similar to the corresponding hydrogen derivatives but only half of the dioxygen consumption was found (34). The stoichiometry of NADPH oxidation to dioxygen uptake suggested the formation of water from the active oxygen complex and indeed no extra formation of hydrogen peroxide was observed in the presence of azide to inhibit catalase (Table II).

TABLE II. Effect of Uncouplers on H_2O_2 Formation in Liver Microsomes from PB-pretreated Rats. H_2O_2 was determined by the ferrithiocyanate method (19).

		H_2O_2 - Production (nmol x min^{-1} x mg $protein^{-1}$)		
	Compound added	NADPH	NADPH + NADH	NADH
I	Ethanol	8.8		
	Cyclohexane	19.1		
	Perfluorocyclohexane	14.3		
II	Ethanol	14.8		
	n-Hexane	29.8	32.6	3.0
	Perfluoro-n-hexane	22.8	25.2	3.0

UNCOUPLING AS A TOOL TO DIFFERENTIATE BETWEEN ACTIVE OXYGEN SPECIES

The proof of water formation from the cytochrome P450 active oxygen complex would be less convincing if in other monooxygenases hydrogen peroxide would not indeed be the product of an uncoupling process. For flavin dependent bacterial monooxygenases this has been shown (35), as well as for bacterial iron-dependent monooxygenases (36).

This not only demonstrates that non-heme iron containing monooxygenases form a peroxide complex as an alternate possible form of active oxygen, but also confirms the postulated role of the porphyrin ring in donating electrons to the peroxide precursor of cytochrome P450 under formation of the oxenoid complex.

In conclusion one can make use of the uncoupling reaction in monooxygenases to establish the nature of the active oxygen intermediate as summarized in the following scheme

Product in Presence of Uncoupler	Postulated Active Oxygen	Type of Oxygenase
H_2O_2	MeO_2^{2-} or ROOH	Monooxygenase or Dioxygenase
+ 2 e, H_2O	MeO	Monooxygenase

The design of uncouplers is relatively easy since substitution of the target CH bond by a CF bond will be sufficient in specific monooxygenases, whereas less specific enzymes require some further substitutions in the substrate molecules.

The last column in the scheme points to some inherent difficulties in the nomenclature of oxygenases. In principle a peroxide intermediate as active oxygen will allow the transfer of only one oxygen atom to the substrate or both. Examples are known in which both oxygen atoms are transferred to the substrate under consumption of two electrons and one dioxygen molecule (31,37). These enzymes are dioxygenases by the definition of Hayaishi (1), but monooxygenases by their stoichiometry (2). In principle it is possible that a given enzyme catalyzes mono- as well as dioxygenations if a peroxide is the intermediate active oxygen.

ACKNOWLEDGEMENTS

This work was supported by the Deutsche Forschungsgemeinschaft, Sonderforschungsbereich 38, grant Ll.

REFERENCES

1. Hayaishi, O., "Oxygenases," Academic Press, New York, (1962).
2. Mason, H. S., Adv. Enzymol., 19,79 (1957).
3. Jung, C. and Ristau, O., Pharmazie, 33,329 (1978).
4. Lipscomb, J. D., personal communication.
5. Feigelson, P. and Brady, F. O., "Molecular Mechanisms of Oxygen Activation" (O. Hayaishi, ed.), Academic Press, New York-London, (1974), p. 87.
6. Flashner, M. S. and Massay, V., "Molecular Mechanisms of Oxygen Activation" (O. Hayaishi, ed.), Academic Press, New York-London, (1974), p. 245.
7. Loebl, H., Stein, G. and Weiss, J., Chem. Soc., (1950), 2704.
8. Staudinger, Hj. and Ullrich, V., Z. Naturforschg., 19b, 409 (1964).
9. Bernhardt, H. F., personal communication.
10. Gunsalus, I. C., Tyson, C. A., Tsai, R. and Lipscomb, J. B., Chem.-biol. Interactions, 4,75 (1971).
11. Ishimura, Y., Ullrich, V. and Peterson, J. A., Biochem. Biophys. Res. Commun. 42,140 (1971).
12. Debye, P., Hui Bon Hoa, G. and Gunsalus, I. C., Croat. Chem. Acta, 49,309 (1977).
13. Cooper, D. Y., Levin, S., Narashimhulu, S., Rosenthal, O. and Estabrook, R. W., Science, 147,400 (1965).
14. Peterson, J. A., unpublished.
15. Yamazaki, I., "Molecular Mechanisms of Oxygen Activation" (O. Hayaishi, ed.), Academic Press, New York-London, (1974), p. 535.
16. Schleyer, H., Cooper, D. Y., Levin, S. S. and Rosenthal, O., "Biological Hydroxylation Mechanisms" (Boyd, Smellic, eds.), Academic Press, New York, (1972), p. 187.
17. Gunsalus, I. C., Meeks, J. R., Lipscomb, J. D., Debrunner, P. and Münck, E., "Molecular Mechanisms of Oxygen Activation" (O. Hayaishi, ed.), Academic Press, New York-London, (1974), p. 559.

18. Ullrich, V., "Topics of Current Chemistry," Vol. 83,67 (1979).
19. Hildebrandt, A. G. and Roots, J., Arch. Biochem. Biophys., 171,385 (1975).
20. Tasi, R. L., Ph. D. Thesis, Univ. of Illinois, Urbana, 1969.
21. Lipscomb, J. D., Sligar, S. G., Namtvedt, M. J. and Gunsalus, I. C., J. Biol. Chem., 251, 1116 (1976).
22. Estabrook, R. W., Kawana, S., Werringlore, J., Kuthan, H., Tsuji, H., Gram, H. and Ullrich, V., Acta Biol. Med. Germanica, 38,423 (1979).
23. Kuthan, H., Tsuji, H., Graf, H., Ullrich, V., Werringloer, J. and Estabrook, R. W., FEBS-Letters, 91,343 (1978).
24. Schonbaum, G. R. and Lo, S., J. Biol. Chem., 247,353 (1972).
25. Dolphin, D., Formann, A., Borg, D. C., Fajer, J. and Felton, R. H., Proc. Natl. Acad. Sci. USA, 68,614 (1971).
26. Lichtenberger, F., Nastainczyk, W. and Ullrich, V., Biochem. Biophys. Res. Commun., 70,939 (1976).
27. Bayer, E., Hill, H. A. O., Röder, A. and Williams, R. J. P., Chem. Commun., 109 (1969).
28. Stern, J. O. and Peisach, J., J. Biol. Chem., 249,7495 (1974).
29. Ullrich, V., Ruf, H. H. and Wende, P., Croatica Chem. Acta, 49,213 (1977).
30. Philson, S. B., Ph.D. Thesis, Univ. of Illinois, Urbana, 1977.
31. Forster, A. B., Jarman, M., Stevens, J. D., Thomas, P. and Westwood, J. H., Chem. Biol. Interactions, 9,327 (1974).
32. Hjemeland, L. M., Aronow, L. and Trudell, J. R., Biochem. Biophys. Res. Commun., 76,541 (1977).
33. Groves, J. T., McClusky, G. A., White, R. E. and Coon, M. J., Biochem. Biophys. Res. Commun., 76,541 (1977).
34. Staudt, H., Lichtenberger, F. and Ullrich, V., Eur. J. Biochem., 46,99 (1974).
35. White-Stevens, R. H. and Kamin, H., Biochem. Biophys. Res. Commun., 38,882 (1970).
36. Bernhardt, H. F., Erdin, N., Staudinger, Hj. and Ullrich, V., Eur. J. Biochem., 35,126 (1973).
37. Kobayashi, S., Kuno, S., Hada, N., Hayaishi, O., Kozuka, S. and Oae, S., Biochem. Biophys. Res. Commun., 16,556 (1964).

PRELIMINARY EVIDENCE SUPPORTING IRON PEROXIDE INTERMEDIATES AND NUCLEOPHILIC ADDITION IN THE LUNG CYTOCHROME P-450-CATALYZED BIOACTIVATION AND TOXICITY OF 3-SUBSTITUTED FURANS

J. E. Garst[1] *and B. J. Wilson*[2]

[1]*Department of Animal Science, College of Agriculture, University of Illinois, Urbana, Illinois 61801*
[2]*Department of Biochemistry, School of Medicine, Vanderbilt University, Nashville, Tennessee 37232*

McMurtry and Mitchell have demonstrated the toxicity of a number of 2-substituted furans, including the diuretic furosemide (1); however, the most acutely hazardous furans appear to be 3-substituted (2). 3-Methylfuran, arising naturally from vegetation-released isoprene, has been reported as an abundant air pollutant in Washington, D.C. (2). Though less toxic than most 3-substituted furans, the high volatility, unavoidable exposure, and both lung and kidney toxicity of this agent combine to make it a significant health hazard (2). 4-Ipomeanol (1-(3-furyl)-4-hydroxy-1-pentanone) is a 3-substituted furan found in mold-contaminated sweet potatoes (Ipomea batatas) (2). With a lethal dose in cattle of 9 mg/kg, 4-ipomeanol has destroyed many livestock in the South (2). Perilla ketone (3-furylisoamyl ketone), a 4-ipomeanol congener, is the most potent natural 3-substituted furan with a 50% lethal dose in mice (3) and rabbits (4) of 5 and 4 mg/kg, respectively. Unlike 3-methylfuran and 4-ipomeanol, perilla ketone is a lung-selective toxin in both the mouse and rabbit (3,4). Several other potent 3-furyl ketones are found with perilla ketone in the abundant and bovine-toxic mint plant, Perilla frutescens (3).

Several groups have demonstrated that the lung toxicities of 3-methylfuran and 4-ipomeanol are catalyzed by non-ciliated bronchiolar or "Clara cell" cytochrome P-450 (2,5). Specifically, pretreatment with the mixed function oxidase inhibitor piperonyl butoxide substantially reduced the toxicity and radiolabel covalent binding of 3-methylfuran and 4-ipomeanol (2). Wilson has found similar effects of piperonyl butoxide on the toxicity of perilla ketone (6). The discovery that anti cytochrome b_5 antibody markedly affected the metabolic activation and covalent binding of 4-ipomeanol, suggests that second electron transfer to cytochrome P-450 proceeds through cytochrome b_5 (2).

It is postulated that toxin bioactivation occurs in three stages: First, association of the substrate with enzyme, forms the complex ES, which is in equilibrium with the activated transition state ES*. Second, transition state ES* undergoes an irreversible chemical transformation producing an enzyme-bound

ISBN 0-12-592050-4

reactive metabolite RE. Third, the reactive metabolite, R binds to protein macromolecules as BR, leading to the observed effects. Boyd and coworkers demonstrated that radiolabel binding of these furans is rapid, reaching 50% lung-binding within about ten minutes after intravenous administration to rats (2). These furan toxins bind so rapidly that the reactive metabolite fails to escape the tissue of activation, clearly indicating that stage 3 is not likely to be the slow step in the process (2).

This paper will indirectly relate the toxicity of 3-substituted furans to stages one and two, support furan bioactivation through P-450-catalyzed NUCLEOPHILIC ATTACK, and propose dual nucleophilic and electrophilic mechanisms for metabolism by cytochrome P-450. Hansch, and Al-Gailany and coworkers have reported that the apparent Km (spectral Ks) of cytochrome P-450 substrates correlates well with the logarithm of their octanol:water partition coefficient (7,8). This is probably due to interactions with the hydrophobic lipoprotein in active cytochrome P-450 (9,10). If the Km does correlate with the log P of 3-substituted furans, the P-450 metabolite-induced toxicity of these heterocycles should display a triphasic (parabolic) dependence on substrate log P.

Having developed a rapid HPLC method for determination of the octanol:water log P (11), we have correlated the measured log P of electronically similar 3-furyl ketones with their log LD_{50} in 19-26 gram male Swiss mice (Fig. 1). LD_{50} data were obtained by the probit method using fifty or more mice; the standard error is

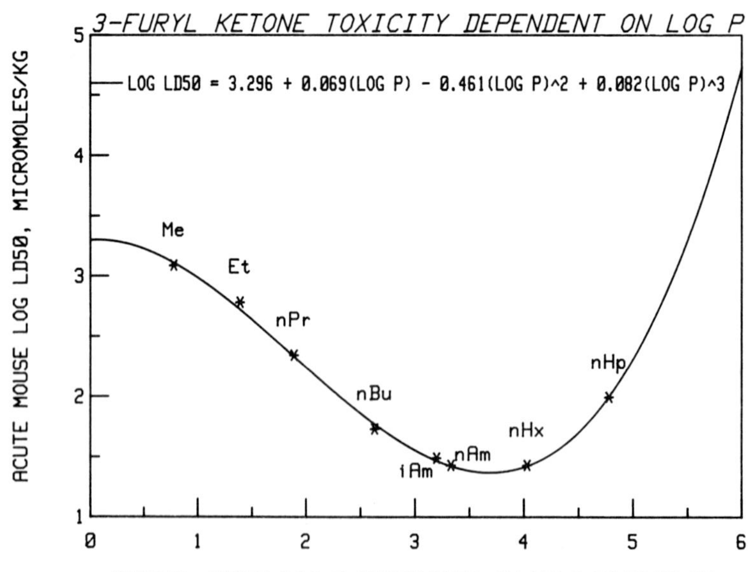

FIGURE 1. Dependence of the logarithm of the 42 hour-acute mouse 50% lethal dose (micromoles/kg) of various 3-furyl ketones on the logarithm of the octanol:water log partition coefficient.

roughly 5% of the LD_{50} value. Compounds were administered in late afternoon as single 50 mcl intraperitoneal injections using dimethylsulfoxide diluent.

Yet another important discovery, probably also based on the relationship between Km and log P, is the apparent correlation between the furan log P and microscopic pathology: all these 3-substituted furans are lung toxic, but those with log P values less than about one are also liver toxins. Those 3-substituted furans with log P values less than about two and more than about one are also kidney toxic, but those with a log P within about one unit of the vertex appear to be lung-selective (Fig. 1). These relationships strongly suggest enzyme-substrate-complex-formation is driven by hydrophobic binding in the catalytic site.

When the Km is sufficient (near the log P vertex), formation of the reactive metabolite from the enzyme substrate complex should depend on the relative velocities of furan bioactivation (e.g. stage 2). Those velocities should depend on the extent that the specific furan further affects the enzymatic energy barrier to reactive metabolite formation. A lower energy transition state barrier favors faster metabolite production and, if that metabolite is toxic, enhanced toxicity. Thus, the cytochrome P-450-mediated toxicities of halogenated methanes $XCCl_3$ have been related to the dissociation energy of C-X (12).

The Hammett sigma para scale qualitatively reflects electronic or energy effects in aromatic ring systems and is based on substituent effects on the acidity of benzoic acid (13). Negative values reflect substituent electron donation to the aromatic ring, while positive values reflect substituent electron withdrawal from the aromatic ring (13). However, the carbon-13 nuclear magnetic resonance (NMR) shift of a single atom reflects this energy factor much more accurately (14). Fig. 2 relates the high correlation between literature Hammett constants and measured C-13 NMR shifts for the #5 furan carbon. Fig. 3 relates the 42 hour-acute mouse log LD_{50} of selected 3-substituted furans to their measured #5 C-13 NMR shifts relative to internal tetramethylsilane. The actual shift-to-atom assignments are based on studies by Gronowitz and coworkers (15). Each compound has a log P sufficient to preclude that factor from influencing furan toxicity. Few such examples are available where mammalian toxicity can be quantitatively predicted by two physical factors. Complete reports detailing the log P and C-13 NMR studies will appear elsewhere.

Since the log P and the carbon-13 NMR shift reflect binding to- and activation by cytochrome P-450, this work can also suggest the mechanism of cytochrome P-450 bioactivation for these agents. In chemistry, highly linear Hammett plots establish the plotted compounds as having the same mechanism of reaction (13). Furthermore, the sign of the slope distinguishes an electrophilic (-) from a nucleophilic (+) reaction. The absolute value of the slope rarely exceeds six and indicates the intensity of charge

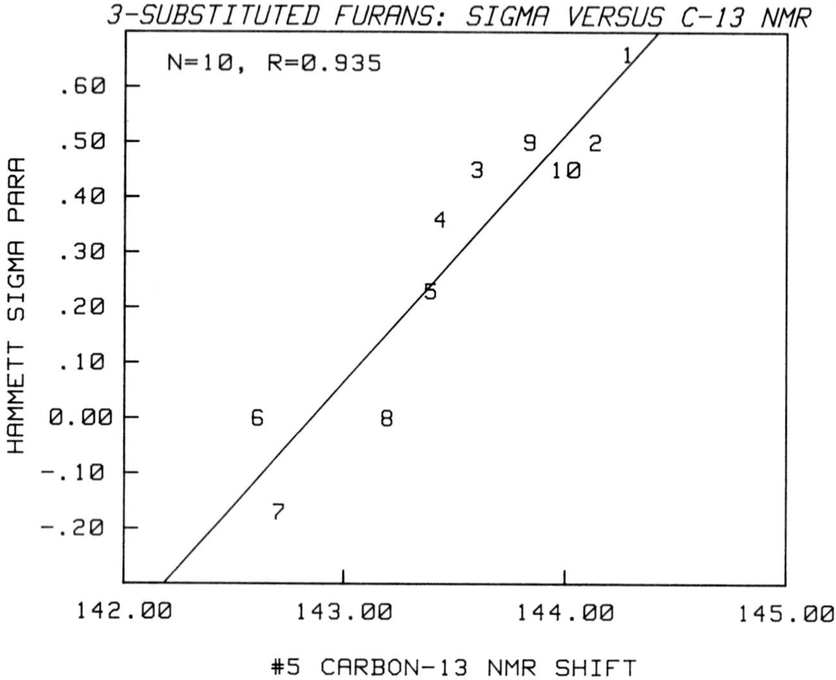

FIGURE 2. Correlation between the #5 furan carbon-13 nuclear magnetic resonance shift in $CDCl_3$ (ppm downfield of tetramethylsilane) and the literature Hammett sigma para for: 1, 3-cyanofuran; 2, 3-furylisoamyl ketone; 3, isobutyl 3-furoate; 4, N-isobutyl 3-furamide; 5, 3-bromofuran; 6, furan; 7, 3-methylfuran; 8, 3-furylmethanol; 9, 3-furylphenyl ketone; and 10, 3-furoic acid.

development in the transition state (13). Typically, Hammett plots are based on the dependence of the logarithm of the ratio of substituent to parent rates on literature Hammett values. However, Hammett constants fail to consider differences between ketones, for example, and this probably accounts for the scatter in Fig. 2. By interpolating corrected Hammett constants using measured #5 C-13 NMR shifts (Fig. 2), this small variation can be smoothed, yet maintain intact both the Hammett sigma para scale and its interpretation.

Using the reciprocal relationship between LD_{50} and toxicity (e.g. velocity) and interpolated Hammett constants, a biological Hammett plot has been constructed (Fig. 4). Two factors are of greatest importance. First, the linear correlation coefficient

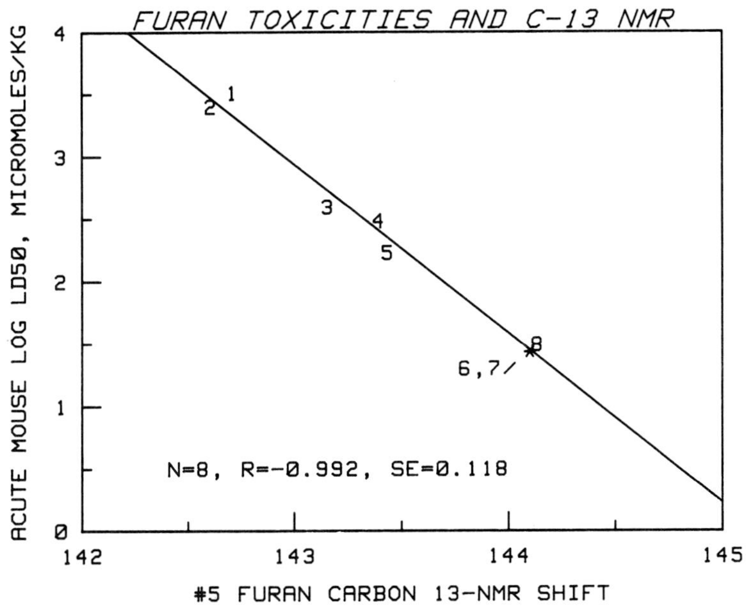

FIGURE 3. Correlation between the #5 furan carbon-13 nuclear magnetic resonance shift in $CDCl_3$ (ppm downfield of tetramethylsilane) and the logarithm of mouse 42 hour 50% lethal dose (micromoles/kg) for: 1, 3-methylfuran; 2, furan; 3, 1-(3-furyl)-4-methyl-1-deutero-pentan-1-ol; 4, 3-bromofuran; 5, N-isobutyl-3-furamide; 6, 3-furylamyl ketone; 7, 3-furylhexyl ketone; and 8, 3-furylisoamyl ketone.

(R=0.992) is high, indicating that the same mechanism is operative among these agents. Secondly, the positive slope of the Hammett plot is near +3, suggesting substantial furan stabilization of negative charge in the reaction transition state and strongly implicating NUCLEOPHILIC ATTACK ON AROMATIC SUBSTRATE. Subsequent studies will compare the slope of the kinetic Hammett plot determined from relative rates of in vitro 3-substituted furan lung metabolism to the slope of the biological Hammett plot in Fig. 4.

Little is understood about the mechanism of cytochrome P-450 operation after formation of the ferrous superoxide intermediate +2FeOO- and before release of water and product (9,10,16,17). Decomposition of +2FeOO- to an electrophilic monooxygen iron oxene has been proposed to account for aliphatic oxygen insertion, formation of isoelectronic carbene adduct, and epoxide products (17). Discovery that P-450 converts oxygen-labelled tert-hydroperoxides into labelled substrate without requiring reductant, further sup-

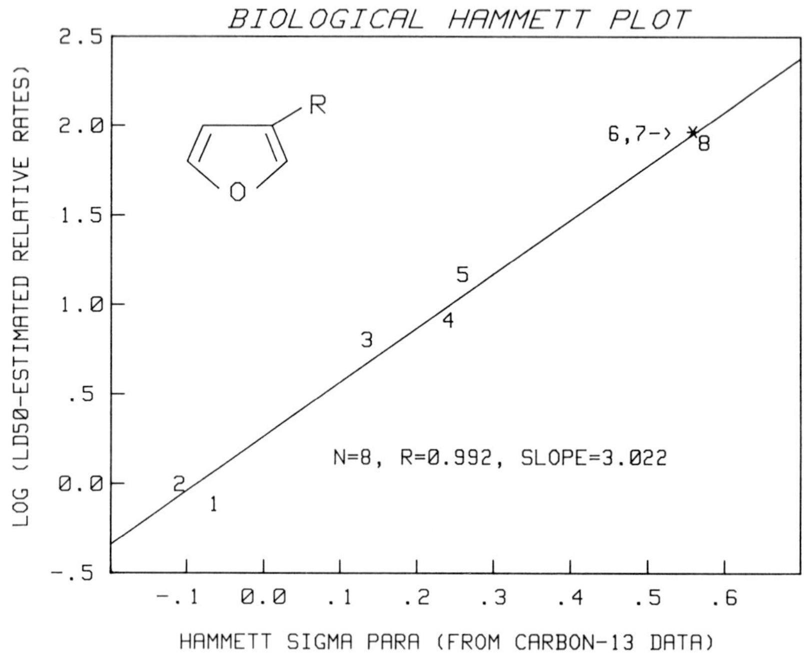

FIGURE 4. Biological Hammett plot [log(k/LD50 of 3-substituted furan divided by k/LD50 for furan)] versus literature Hammett sigma para. Numbers identify structures listed in the legend of Fig. 3 and k is a constant such that k/LD50 defines the velocity of metabolism for that furan.

ported the oxene electrophile (17). However, such a single oxygen electrophile should not afford dioxygenated products, nor display nucleophilicity.

While this biological Hammett plot is certainly not the kinetic Hammett data needed to unequivocally invoke nucleophilic attack on substrate by lung cytochrome P-450, it is suggestive of such a process. In view of this data and evident electrophilic aliphatic oxygen insertion, we propose dual nucleophilic- and electrophilic cycles for substrate metabolism by lung cytochrome P-450 (Fig. 5). Our dual cycles are a modification of that by Estabrook and Werringloer (16). These cycles take into account the data of Ullrich and coworkers (this volume), supporting release

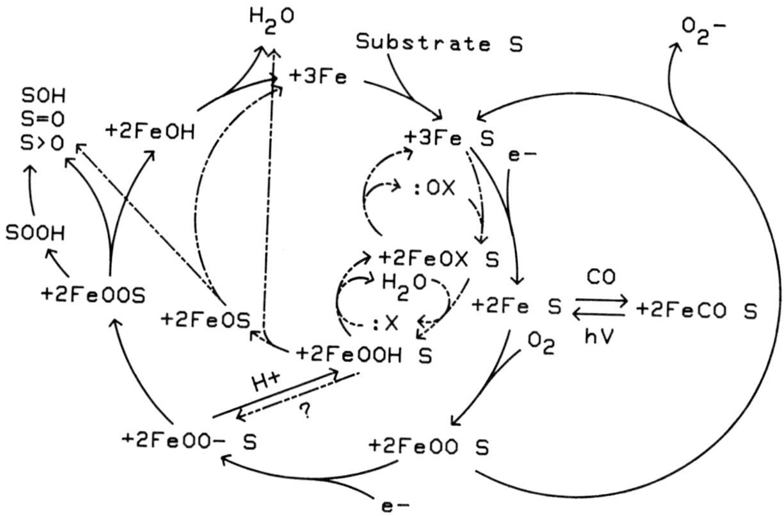

FIGURE 5. Proposed cycles for nucleophilic [▬▬▬] and electrophilic [▪▪▬▬▪▪] substrate metabolism by cytochrome P-450. The substrate S in products +2FeOOS or HOOS could bear a transient charge after nucleophilic addition, depending on the structure of S.

of superoxide anion from the initial, one-electron-reduced ferrous dioxygen species. However, superoxide anion is unlikely to be the species affecting the 3-substituted furans, since second electron addition by cytochrome b5 appears necessary (2). The proposed pathways can account for hydroperoxide formation, proton abstraction, oxygen insertion, and substrate hydroxylation processes typical of cytochrome P-450, as well as nucleophilic attack.

Because of electron withdrawal by iron, ferric hydroperoxide -ferrous hydrosuperoxide (+3Fe -:OOH ↔ +2Fe .OOH) should be acidic, in analogy to the enhanced acidity that fluorine electron withdrawal imparts to pertrifluoracetic acid (estimated pKa=3.7) relative to peracetic acid (pKa=8.2) (18). Thus, at physiological pH, ferric hydroperoxide should exist primarily in the anionic resonance forms, ferric peroxide-ferrous superoxide (+3Fe $O_2^{=}$ ↔ +2FeO_2^-). The oxyanions of many metal hydroperoxides and superoxides are potent nucleophiles (19,20). Sligar and coworkers recently invoked ferrous superoxide nucleophilic attack on lipoic acid to produce a peracyl iron species, possibly accounting for increased hydroxylation rates by cytochrome P-450 (21). Ugo and coworkers have also demonstrated the nucleophilicity of oxygen in both platinum and silver dioxygen complexes (22).

If a nucleophilic cytochrome P-450 process is suggested, the 3-substituted furans should be susceptible to nucleophilic reagents, as is the case (23-25). Such an addition of ferrous super-

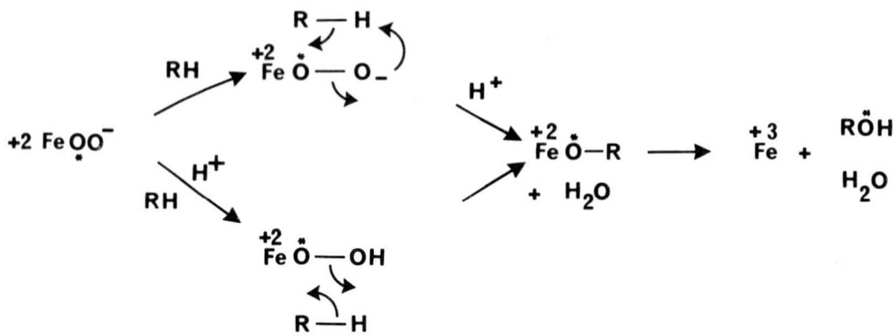

FIGURE 6. Proposed carbon or nitrogen hydroxylation reactions by cytochrome P-450 ferrous superoxide intermediate of acidic or neutral aliphatic agents. The asterisk identifies one oxygen atom.

oxide could produce hydroperoxide products; several more stable hydroperoxides have been reported as cytochrome P-450 metabolites (26). To be considered a likely toxin, they should produce pulmonary effects; hydroperoxides product pulmonary damage in rodents and rabbits, animals most susceptible to the furans (26-28). Although highly reactive hydroperoxides may be difficult to demonstrate, it is encouraging to those efforts that epoxide hydratase inhibitors fail to alter in vitro 4-ipomeanol covalent binding (2).

A nucleophilic ferrous superoxide should also have basic properties, which could account for carbene formation and insertive hydroxylation. The formation of P-450-carbene adducts suggested for various methylenedioxygenzenes (29), could involve intital base-catalyzed proton- (or free radical) abstraction, followed by loss of an internal leaving group in analogy to the formation of dichlorocarbene from chloroform (30). Alternatively, proton transfer from substrate to ferric superoxide will produce ferric hydroperoxide, whose iron-bound, highly electrophilic oxygen could recombine (concertedly?) with the resultant carbanion forming alcohol and water (R=C, Fig. 6). In regard to these two processes, proton abstraction from chloroform is the rate-limiting step in chloroform toxicity. Although some chloroform metabolism via the dichlorocarbene route is evident, most research favors the proton abstraction-hydroxylation pathway, affording trichloromethanol and its decomposition product phosgene (31). Concerted proton abstraction from a protonated amine should similarly afford N-hydroxylated product and water (R=N+, Fig. 6).

Dissociation of hydroperoxy anion from the ferric hydroperoxide complex could produce hydrogen peroxide and, perhaps through hydroxyl radical, this could catalyze a variety of misleading, but clearly P-450-mediated radical-hydroxylation or abstraction reactions (17). Substantial hydrogen peroxide production during P-450-catalyzed amine dealkylation is one such example (16). Since protonated amines would be highly unfavored in a very hydrophobic environment, perhaps catalytic proton transfer from an active-site imidazole to amine, to ferrous superoxide-ferric peroxide could lead to hydrogen peroxide production. By this route less basic amines (e.g. amides) might favor the concerted N-hydroxylation, producing less hydrogen peroxide, while more basic amines might favor proton exchange, hydrogen peroxide production, and resultant N-dealkylation. In any event most basic amines generate hydrogen peroxide, whie the non-basic polyfluorinated aliphatic hydrocarbons generate very little hydrogen peroxide (16).

Protonation by water could also convert ferrous superoxide to ferric hydroperoxide (Fig. 6). Perhaps, the hydrophobic catalytic site (lipoprotein) has evolved to minimize such random ferrous superoxide anion protonation, thus, maximizing nucleophilicity and basicity and minimizing release of hydrogen peroxide. The specificity of oxygen addition is probably dependent on the hydrophobic substrate's orientation in the catalytic site, which is in turn probably dependent on the specific cytochrome P-450 lipoprotein.

The final metabolic product, be it alcohol, epoxide, or ketone, may be determined by the chemistry of the intermediate iron adduct. Anaerobic Lewis base addition of a nucleophilic group, such as the terminal oxygen in alkyl hydroperoxides, iodosobenzene, nitroalkanes, nitrite, and amine oxides or in theory any species :OX (O in Fig. 5 need not be limited to oxygen) to substrate-complexed ferric P-450 should give the substrate-ferrous complexes (+2FeOX, Fig. 5). These complexes (see Fig. 5 and tert-peroxide example in Fig. 7A) should undergo precedented nucleophilic attack at the electron-deficient oxygen [*] adjacent to iron, either by water directly or via enzymatic assistance, or by an enzymatic nucleophile :N (perhaps the very close, heme-amino nitrogen) (32). This reaction displaces tert-alcohol, iodobenzene, nitrosoalkane, nitric oxide, amine oxide, or, in general, the reduced leaving-group :X (Fig. 5) (17,33-35), while producing +2Fe*OOH or enzyme bound +2Fe*ON, either of whose oxygen is susceptible to nucleophilic attack by the electron-rich R-H bond, thus effectively inserting the original oxidant oxygen [*] into substrate (Fig. 6) (17,33,34). This process could be reversible if :X is sufficiently nucleophilic; alternatively, a poor leaving group X may inhibit attack on +2FeOX; either event could affect enzyme operation, and therefore substrate metabolism and toxicity.

A second mode of iron adduct degradation is likely when a carbanion, perhaps arising from initial nucleophilic attack on an aromatic ring or olefin, is adjacent to the hydroperoxide carbon and can displace the leaving group +2FeO- to form epoxides by a well-established nucleophilic process (Fig. 7B) (32,36). In addition, iron-peroxide intermediates can also decompose by alkyl rearrangement producing alcohol and aldehyde or ketone (Fig. 7C (37).

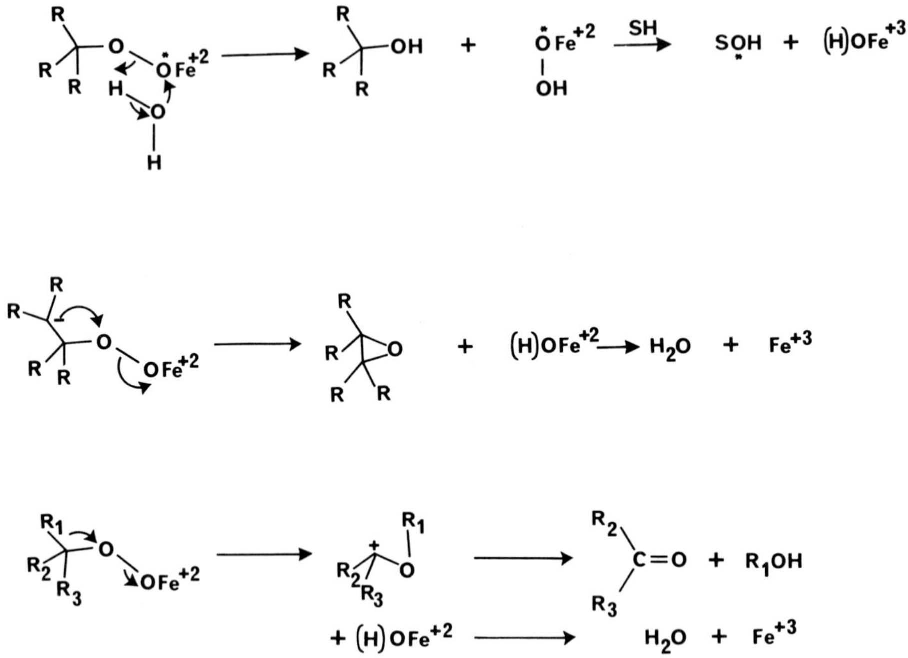

FIGURE 7. Suggested degradation of cytochrome P-450 ferric peroxide intermediates by (A) nucleophilic or perhaps enzymatic hydrolysis, (B) carbanion displacement, and (C) hydrogen or alkyl migration. The asterisk identifies one oxygen atom.

Numerous toxicological observations could be explained by mechanisms invoking nucleophilic attack on substrate with possible hydroperoxide intermediates. One example involves cytochrome P-450 metabolism of vinyl chloride to form both alpha-chloroacetaldehyde and chloroethylene oxide (38). These products can account for the deleterious effects of vinyl chloride, yet neither product can account for the degradation of cytochrome P-450 heme (38). However, the decomposition of iron peroxide- or free hydroperoxide intermediates could account for both metabolites, as well as heme degradation (Fig. 8 (17).

Others have reported that the destructive heme binding by allyl isopropyl acetamide (AIA) and its congeners does not proceed via the epoxide metabolite, although the terminal olefin is essential to heme destruction (39). Michael-like nucleophilic addition of +2FeOO- to the olefin in these agents could lead to cyclobutane ring formation by intermediate carbanion attack on the carbonyl carbon. Alkoxide anion-mediated ring-opening could then displace +2FeO- and simultaneously lead to formation of the epoxide product, as with vinyl chloride. Various conformationally-restricted analogs of AIA which do not cause heme destruction (30), cannot easily form the proposed cyclobutane ring intermediate; furthermore, formation of such an intermediate should not be greatly

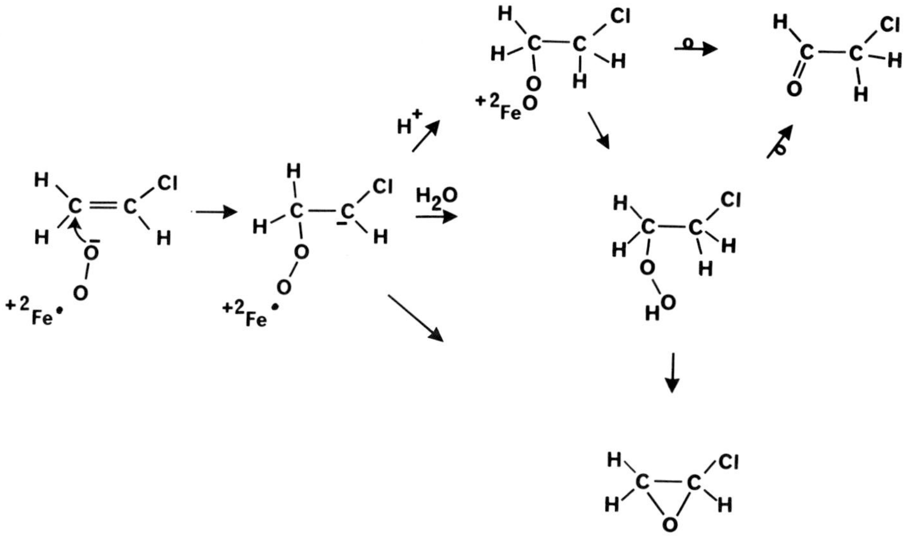

FIGURE 8. Proposed cytochrome P-450 metabolism of vinyl chloride to ferric peroxide and/or hydroperoxide intermediates capable of accounting for both products and heme destruction.

altered by different carbonyl substituents (e.g. ester, ketone, etc.) or even replacement of the carbonyl group by a displaceable methoxy ether, as was discovered (39). Again, hydroperoxides, formed by hydrolysis of suggested iron peroxide intermediates, could account for the reaction product as well as the heme destruction, so typical of these reactive intermediates (17).

Another process might be similarly explained; nucleophilic addition to methyl- and halobenzenes favors para-attack (40), while cyclization of the resultant meta carbanion, would afford the observed 3,4-benzene epoxides (41). Interestingly, already para substituted p-dichlorobenzene binds to tissue (e.g. metabolism) much less readily than either monochloro- or o-dichlorobenzene (42).

Toxicity related to para hydroxylation of N-arylamides might also involve hydroperoxides. Initial N-hydroxylation, followed by their anaerobic -:ONArAc (e.g. :OX) addition to the ferric P-450 and the homolytic cleavage of the N-O linkage, would form +2FeO- and Ar+NAc. The +2FeO- decomposes to water and ferric P-450, which may then recycle, but the now electron-deficient cationic nitrogen forces aromatic ring oxidation. Subsequent nucleophilic addition to the reactive para cation perhaps by oxygen-regenerated P-450 ferrous superoxide, could afford transient secondary hydroperoxides, which might rearrange to imidoquinones in analogy to an established chemical mechanism for conversion of N-hydroxyamines into para-hydroxy arylamines (43). Imidoquinones, possibly from N-hydroxy precursors, have precedent as reactive cytochrome P-450 metabolites of N-acyl-N-arylamines, but the role of precursor hydroperoxides in this process is unclear (44).

The proposed dual mechanisms suggest the unique ability of cytochrome P-450 to alter virtually any substrate entering the catalytic site. Although epoxide metabolites may mediate various toxic and carcinogenic effects, hydroperoxide intermediates can lead to reactive epoxides (32,36), hydroxyl- and alkoxyl radicals (45), as well as carbonyl compounds (37); more importantly, hydroperoxides may alter protective mechanisms (46) or even govern cytochrome P-450 levels. By concentrating on the effects of possible hydroperoxide products we may understand other toxicological processes and speed progress in metabolism, nutrition, aging, and cancer.

ACKNOWLEDGEMENTS

This work was supported by NIEHS grants ES 02193 (JEG) and ES 0569 (BJW) and the University of Illinois College of Agriculture (Hatch 20-323) and Research Board. The authors would like to thank Professors I. C. Gunsalus, V. Ullrich, B. Griffin, T. Burka, and W. Porter for their helpful comments. Portions of this research were presented at the 1979 and 1980 Society for Toxicology meetings in New Orleans, Louisiana and Washington, D.C.

REFERENCES

1. McMurtry, R. J. and Mitchell, J. R., Toxicol. Appl. Pharmacol., 42,285 (1977).
2. Boyd, M. R., "Reviews in Biochemical Toxicology," (E. Hodgson, J. R. Bend and R. M. Philpot, eds.), Vol. 2, Elsevier/North Holland, New York, N.Y., (1980), p. 71.
3. Wilson, B. J., Garst, J. E., Linnabary, R. D. and Channell, R. B., Science, 197,573 (1977).
4. Unpublished results, W. Thurlbeck and J. E. Garst.
5. Serabjit-Singh, C. J., Wolff, C. R., Philpot, R. M. and Plopper, C. G., Science, 207,1469 (1980).
6. Unpublished results, B. J. Wilson.
7. Hansch, C., "Correlation Analysis in Chemistry: Recent Advances" (N. B. Chapman and J. Shorter, eds.), Plenum, New York, N.Y., (1978), p. 397.
8. Al-Gailany, K. A. S., Houston, J. B. and Bridges, J. W., Biochem. Pharmacol., 27,783 (1978).
9. Coon, M. J. and White, R. E., "Dioxygen Binding and Activation by Metal Centers" (T. G. Spiro, ed.), Wiley, New York, N.Y., (1980).
10. White, R. E. and Coon, M. J., Ann. Rev. Biochem., 49, in press.
11. Unpublished results, J. E. Garst.
12. Recknagel, R. O., Glende, E. A., Jr., and Hruszkewycz, A. M., "Free Radicals in Biology" (W. J. Pryor, ed.), Vol. 3, Wiley, New York, N.Y., (1977), p. 97.

13. Exner, O., "Advances in Linear Free Energy Relationships" (N. B. Chapman and J. Shorter, eds.), Plenum, New York, N.Y., 1972, p. 1.
14. Wehrli, F. W. and Wirthlin, T., "Interpretation of Carbon-13 NMR Spectra," Heydon, Philadelphia, PA, (1978), p. 31.
15. Gronowitz, S., Johnson, I. and Hornfeldt, A.-B., Chem. Scripta, 7,211 (1975).
16. Estabrook, R. W. and Werringloer, J., "Drug Metabolism Concepts" (D. M. Jerina, ed.), Am. Chem. Soc., Washington, D.C., (1977), p. 1.
17. O'Brien, P. J., Pharmac. Ther. A., 2,517 (1978).
18. Fieser, L. and Fieser, M., "Reagents for Organic Synthesis," Vol. 1, Wiley, New York, N.Y., (1967), p. 823.
19. Daven, W. C. and Warner, R. J., Tetrahedron Letters, 989 (1977).
20. Fee, J. A. and Valentine, J. S., "Superoxide and Superoxide Dismutases" (A. M. Michelson, J. M. McCord and I. Fridovich, eds.), Academic Press, New York, N.Y., 1977, p. 19.
21. Sligar, S. G., Kennedy, K. A. and Pearson, D. C., Proc. Natl. Acad. Sci., 77,1240 (1980).
22. Ugo, R., Fusi, A., Zanderighi, G. M. and Casella, L., J. Mol. Catalysis, 7,51 (1980).
23. van der Plas, H. C., "Ring Transformations of Heterocycles," Academic Press, New York, N.Y., (1973), p. 190.
24. Takegi, K. and Ueda, T., Chem. Pharm. Bull., 19,1218 (1971).
25. Lutz, R. E. and Reveley, W. G., J. Amer. Chem. Soc., 63,3178 (1941).
26. a. Chen, C. and Tu, M., Biochem. J., 160,805 (1977).
    b. Kehrer, J. P. and Witschi, H., Toxicol. Appl. Pharmacol., 53,333 (1980).
27. Anderson, W. R., Tan, W. C., Tukatori, T. and Privett, O. S., Arch. Path. Lab. Med., 100,154 (1976).
28. Findlay, G. M., Draper, H. H. and Bergan, J. C., Lipids, 5, 970 (1970).
29. Mansuy, D., Battioni, J.-P., Chottard, J.-C. and Ullrich, V., J. Amer. Chem. Soc., 101,3971 (1979).
30. March, J., "Advanced Organic Chemistry: Reactions, Mechanisms, and Structure," 2nd ed., McGraw-Hill, N.Y., (1977), p. 342.
31. Pohl, L. R. in reference 2, vol. 1, (1979), p. 79.
32. Swern, D., "Organic Peroxides, vol. 2, Wiley, New York, N.Y., (1971), p. 60-86, 570.
33. Mansuy, D., Gans, P., Chottard, J.-C. and Bartoli, J. F., Eur. J. Biochem., 76,607 (1977).
34. Sakurai, H., Hermann, G., Ruf, H. H. and Ulrich, V., Biochem. Pharmacol., 29,341 (1980).
35. Powis, G. and Wincentsen, L., Biochem. Pharmacol., 29,347 (1980).
36. House, H. O., "Modern Synthetic Reactions," Benjamin, Menlo Park, CA, (1972), p. 306.

37. Reference 30, p. 1012.
38. Guengerich, F. P. and Watanabe, P. G., "Biochem. Pharmac., 28,589 (1979).
39. Ortiz De Montellano, P. R., Mico, B. A., Yost, G. S. and Correia, M. A., "Enzyme-Activated Irreversible Inhibitors" (N. Seiler, M. J. Jung and J. Koch-Weser, eds.), Elsevier/ North-Holland, New York, N.Y.,(1978),p. 337.
40. Reference 30, p. 591.
41. Lau, S. S., Abrams, G. D. and Zannoni, V., Fourth Intl. Symp. on Microsomes and Drug Oxidations, Ann Arbor, Mi,(July, 1979), Abst # P647.
42. Reid, W. D. and Krishna, G., Exp. Mol. Path., 18,80 (1973).
43. Reference 30, p. 616.
44. Hinson, J. A. in reference 2, vol. 2, 1980, p. 103.
45. Cohen, G. C. and Cederbaum, A. I., Science, 204,66 (1979).
46. Reed, D. J. and Beatty, P. W. in reference 2, vol. 2, 1980, p. 213.

2-METHYL-3-HYDROXYPYRIDINE-5-CARBOXYLIC ACID OXYGENASE: MOLECULAR AND CATALYTIC CHARACTERISTICS

Ganesh M. Kishore and Esmond E. Snell, Departments of Chemistry and Microbiology, The University of Texas at Austin, Austin, Texas 78712

Two related pathways for the bacterial degradation of pyridoxine, outlined in Fig. 1, have been described by our laboratory (1,2). Pathway II is one of the longest chains of inducible metabolic reactions that has been investigated at an enzymatic level. A series of oxidative, hydrolytic, and decarboxylation reactions culminate in the conversion of pyridoxine to 2-methyl-3-hydroxypyridine-5-carboxylic acid (Cpd I), the most oxidized aromatic intermediate in the pathway. A unique oxidative reaction, reaction 7, Fig. 1, catalyzes cleavage of the pyridine nucleus of Cpd I to yield the acyclic product, α(N-acetylaminomethylene)succinate (Cpd A), which is hydrolyzed by a separate enzyme to succinaldehyde, acetate, CO_2 and ammonia. An analogous ring-cleavage reaction also exists in pathway I (reaction C, Fig. 1), but the enzymes catalyzing the two reactions exhibit a high degree of preference for their natural substrates.

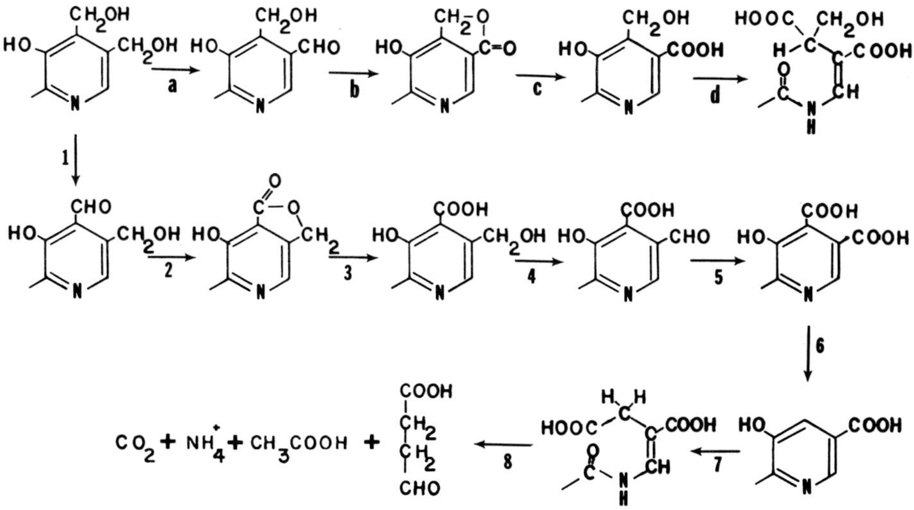

FIGURE 1. Pathways relating intermediates formed during growth of Pseudomonas sp. IA (Pathway I, a through d) or Pseudomonas sp. MA-1 (Pathway, II, 1 through 8) on pyridoxine as sole source of carbon and nitrogen.

ISBN 0-12-592050-4

The enzyme catalyzing reaction 7 (Fig. 1) has been purified to homogeneity from Pseudomonas sp. MA-1 grown on pyridoxine as sole source of carbon and nitrogen (3). The reaction proceeds with the stoichiometry shown in Fig. 2. Experiments with $^{18}O_2$ showed that approximately 1.1 atoms of molecular oxygen are incorporated into the product (3); some incorporation from $H_2{}^{18}O$ was also observed. The reaction is therefore an oxygenase reaction, but differs from oxygenase reactions in which homocyclic aromatic rings are cleaved in the following respects: (a) Reaction 7 is catalyzed by an FAD protein, whereas most enzymes that cleave aromatic rings are metalloproteins. (b) Reaction 7 requires NADH as a cosubstrate; most aromatic dioxygenases do not. (c) Reaction 7 involves both a 1,4-reduction and ring-cleavage by O_2 addition across C(2) - C(3) of Cpd I; only oxygenation with ring cleavage is observed with most dioxygenases. Because of these unique features, we have investigated the mechanism of reaction 7 in greater detail. Some of the results to be presented here have been reported previously (4,5).

PROPERTIES OF Cpd I OXYGENASE

Molecular weight measurements summarized in Table I, show that the native oxygenase is a tetramer composed of subunits of apparently identical size. FAD is released upon denaturation of the protein either by boiling or trichloracetic acid treatment, and from its millimolar extinction of 11.3 at 452 nm, a content of 1.9 moles of FAD per mole (166,000 g) of the native oxygenase was calculated. Native apoenzyme can be conveniently prepared by precipitating the protein with acid-ammonium sulfate, which removes 95-99% of the FAD with a corresponding loss of activity. Addition of FAD to the apoenzyme restores most of the original activity (Table II). FMN, 5-deaza-FAD, and mixtures of FMN and

TABLE I. Molecular Weight Determinations of Cpd I Oxygenase

Method	*Molecular Weight*
Native holoenzyme	
Gel filtration (Sephacryl S-300)	*160,000*
Sedimentation equilibrium	*166,000*
Denatured enzyme	
SDS gel electrophoresis	*42,000*
Native apoenzyme	
Gel filtration	*83,000*

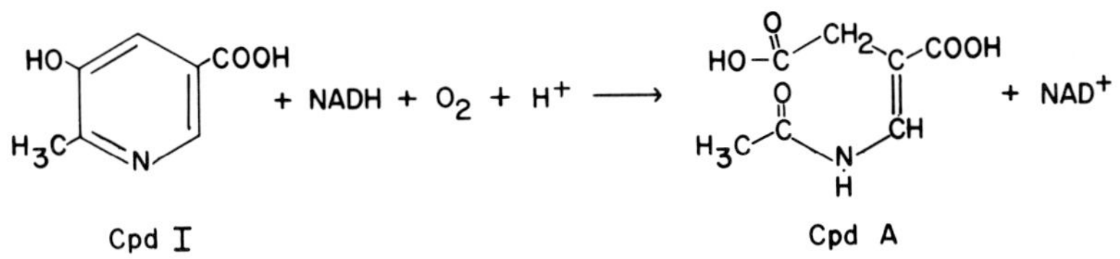

FIGURE 2. Reaction catalyzed by purified Cpd I oxygenase.

adenine nucleotides neither reconstitute the apooxygenase nor inhibit its interaction with FAD. Binding of FAD to the apoprotein is accompanied by extensive spectral and fluorescence changes, and quantitative measurement of these changes also showed that the native enzyme binds only two FADs per tetramer. 1-Deaza-FAD interacts with the apoenzyme as effectively as FAD but the combination is catalytically inert. These results imply that in Cpd I oxygenase, the N(5) of FAD participates in coenzyme binding, while N(1) is essential for the reduction reaction. They distinguish this enzyme from other bacterial external flavooxygenases which form partially competent complexes with 1-deaza-FAD and incompetent complexes with the 5-deaza analogue (6,7).

Spectral titrations show that two moles of Cpd I, i.e., one for each FAD present, bind to the native enzyme, and since both spectral and velocity measurements yield hyperbolic saturation curves characteristic of a single class of binding site, it appears that the two substrate binding sites are equivalent and that the two flavins participate independently in catalysis. Fluorescence measurement with the holoenzyme-Cpd I complex demonstrate a resonance energy transfer from Cpd I (which absorbs at 323 nm) to FAD (which emits at 520 nm). Thus Cpd I is close to (i.e., within the critical distance for energy transfer, <80Å) FAD in the enzyme-substrate complex. X-ray diffraction studies of a crystalline p-hydroxybenzoate hydroxylase·substrate complex (8) also show the substrate to be close to the FAD in that enzyme.

An extensive kinetic study of the oxygenase reaction (4) has established its kinetic mechanism to be that shown in Fig. 3. In agreement with the spectral data just referred to, Cpd I first reacts with the holoenzyme, followed by NADH. Reduction of the complex then occurs, followed by release of NAD^+. Only then can oxygen combine with the reduced complex, and lead to oxygenation and release of Cpd A.

The mechanism of Fig. 3 is similar to that postulated for certain external flavomonoxygenases of bacterial origin (9). Several such enzymes exhibit 4R chirality for NADH or NADPH, and Kaplan and co-workers (10) suggest that this sterochemistry will be a constant feature of such flavoenzymes. A study of the stereochemistry of NADH utilization by Cpd I oxygenase was therefore undertaken.

When (4R)-[4-^{3}H] NADH was used as a substrate, the bulk of the ^{3}H label was labilized and appeared in the aqueous fraction.

TABLE II. Effect of Flavins and Flavin Analogues on the Activity of Cpd I Apooxygenase and its Reactivation by FAD

Addition to Complete System[a,b]	Oxygen Uptake n moles/min
None	1.97
FAD[b]	37.04
FMN	2.23
FAD + FMN	38.43
AMP	1.97
FMN + AMP	2.10
FMN + ADP	2.10
ADP + FAD	34.00
1-d-FAD + FAD	18.4
5-d-FAD + FAD	37.08
1-d-FAD	1.97

[a]The complete system contained 0.2 μmole of Cpd I, 0.2 μmole of NADH and 6 μg of apoenzyme in 1 ml of 0.05 M K-phosphate buffer, pH 7.0.
[b]FAD and its analogues were added as indicated in a concentration of 5 μM.

With (4S)-[4-^{3}H]NADH, however, negligible amounts of ^{3}H appeared in the solvent. Since NADH serves to reduce the enzyme-bound flavin in the Cpd I oxygenase-catalyzed reaction, and since the N(5)-H in reduced FAD exchanges its hydrogen rapidly with water, one would expect the prochiral hydrogen participating in catalysis to appear in the aqueous medium. To establish whether the ^{3}H label in (4S)-[4-^{3}H]NADH was retained in product NAD^+ or incorporated into Cpd A the products of the oxygenase reaction were fractionated over Dowex-1-formate and the radioactivity in each fraction was determined. The results (Table III) show that ^{3}H from (4S)-[4-^{3}H]NADH, was retained by NAD^+, whereas only a small amount of the label from (4R)-[4-^{3}H]NADH remains in the solid residue. Of this latter fraction a slight but detectable amount was incorporated into Cpd A. Control experiments showed that there was

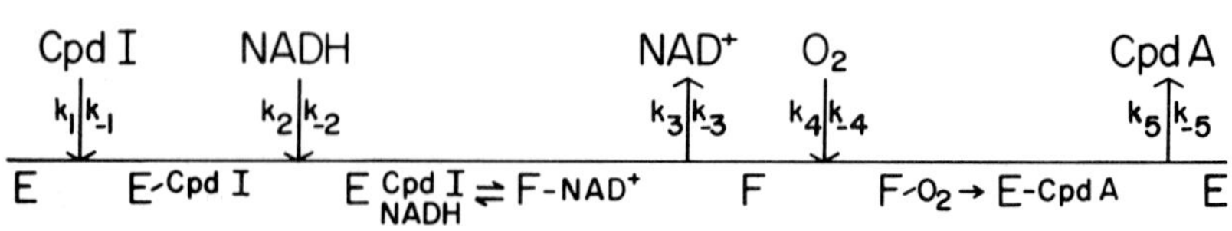

FIGURE 3. Kinetic mechanism of the reaction catalyzed by Cpd-I oxygenase under steady state conditions.

TABLE III. Distribution of ^3H from Stereochemically Labelled NAD^{3}H in Different Products Following Dowex-1-formate Chromatography of the Enzymatic Reaction Mixture

Reaction mixture containing (DPM)	Counts in aqueous fraction (DPM)	Counts in solid residue (DPM)	Counts on NAD$^+$ (DPM)	Counts on Cpd A (DPM)
(4R)-[4-^{3}H]NADH (6.2 x 10^5)	5.9 x 10^5	2.6 x 10^4	2.36 x 10^4	trace
(5S)-[4-^{3}H]NADH (1.03 x 10^5)	3.7 x 10^3	9.9 x 10^4	9.5 x 10^4	N.D.
^{3}HOH + NADH	1.69 x 10^{10}	3.2 x 10^4	N.D.	2.8 x 10^4

N.D. = Not Detectable.

no enzyme-mediated exchange between ^{3}HOH and either Cpd A or Cpd I; however, as shown in Table III, ^{3}H from ^{3}HOH is readily incorporated into Cpd A during its formation from Cpd I. This tritium is associated with C-(4) of Cpd A since upon acid hydrolysis, treatment with 2,4-dinitrophenylhydrazine, and extraction of the hydrazone of succinaldehyde into ethylacetate, the ^{3}H label was associated with the hydrazone. The latter product co-migrated with authentic 2,4-dinitrophenylhydrazone of succinaldehyde in several solvent systems (11).

More quantitative results, shown in Table IV, were obtained by use of NAD^{2}H or ^2H_2O, followed by isolation of Cpd A as its dimethyl ester (3), and mass spectral analysis. There was no specific incorporation of ^{2}H from NAD^{2}H into Cpd A. The high observed incorporation of ^{2}H (or ^{3}H) from D_2O into Cpd A, must therefore be attributed to an exchange between N(5)-H of reduced enzyme-bound FAD and water. This exchange indicates that the N(5)-H of FAD is accessible to the solvent at least transiently during catalysis. This view is also supported by experiments on the effect of free FMN or riboflavin on the stoichiometry of the oxygenase reaction. These flavins, which cannot interact with either the holo- or apo-oxygenase, can cause leakage of electrons from FAD at the active site and thus lead to an abnormally high NADH comsumption (5). This is possible only if the enzyme-bound reduced FAD can interact with free flavins in aqueous solution fairly rapidly.

The above experiments indicate that activation of oxygen for the oxygenase reaction cannot occur by the concerted transfer of hydride ion to Cpd I with further passage of electrons to molecular oxygen, as originally postulated (3). In addition, it is known that the direct reaction of a singlet species such as a reduced pyridine (either NADH or a hypothetical 1,4-reduction

TABLE IV. Extent of Incorporation of ^2H from NAD^2H and ^2H_2O

		Deuterium content of Cpd A	
System	H_2O content of medium	Atom per cent excess	Atom per molecule
(4R)-[4-^2H]NADH* + H_2O	-	0.05	0.006
^2H_2O + NADH	80%	6.27	1

*40 µmoles of >95% ^2H in NADH was used per ml reaction mixture.

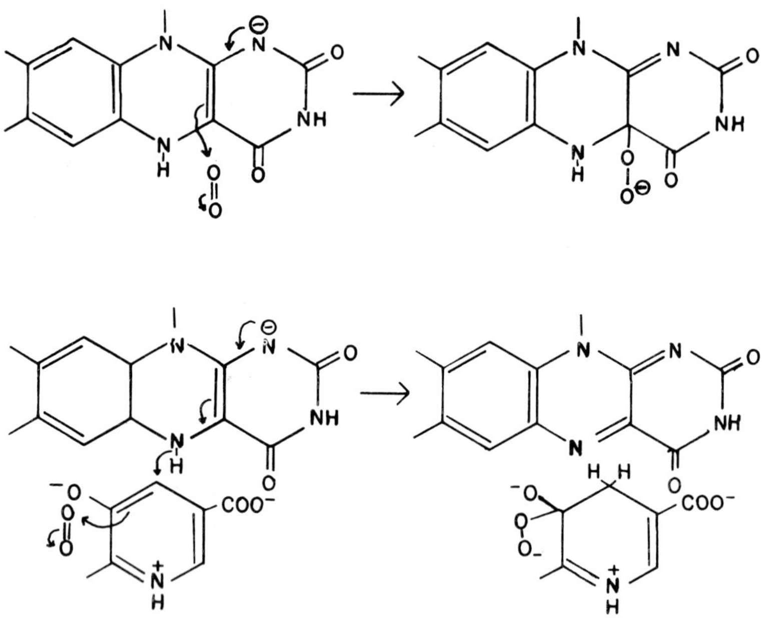

FIGURE 4. Postulated mechanisms for reoxidation of reduced flavin at the active site. In mechanism A, the reduced FAD reacts with O_2 to generate an oxygenated flavin whereas in mechanism B, reoxidation of reduced enzyme is effected by a hydride ion transfer to C(4) of Cpd I and electron transfer.

product of Cpd I) with triplet molecular oxygen is most unlikely because of spin restrictions. Since no metal ions are present in the enzyme (3, unpublished data) activation of oxygen by Cpd I oxygenase must involve its interaction with reduced FAD. Such a postulate, shown in Fig. 4, resembles those in recent reports that a flavin-oxygen adduct is an intermediate in the oxygenation of aromatic substrates by bacterial flavooxygenases (18), and of several substrates by the mammalian microsomal amine oxidase (15). The structure of one such oxygenated flavin (involved in the action of bacterial luciferase) has been established by ^{13}C-NMR studies as the C(4a)-dihydroflavin hydroperoxide (14).

How can oxygenation by a flavin hydroperoxide lead to a coupled reduction of the substrate? Either of two hypothetical mechanisms, presented in Fig. 5, can account for the reduction of the substrate during oxygen insertion and for the observed (3) incorporation of ^{18}O from $H_2{}^{18}O$ into the carboxyacetyl moiety but not into the N-acetyl moiety of Cpd A.

Chemical reactions analogous to the postulated mechanism for the Cpd I oxygenase catalyzed reaction are known. Bristol and Dittmer (16) reported that treatment of 1-methyl- or 1-benzyl-3-carbamoylpyridinium chloride with alkaline hydrogen peroxide resulted in formation of a hydroperoxide (Fig. 6), as the end

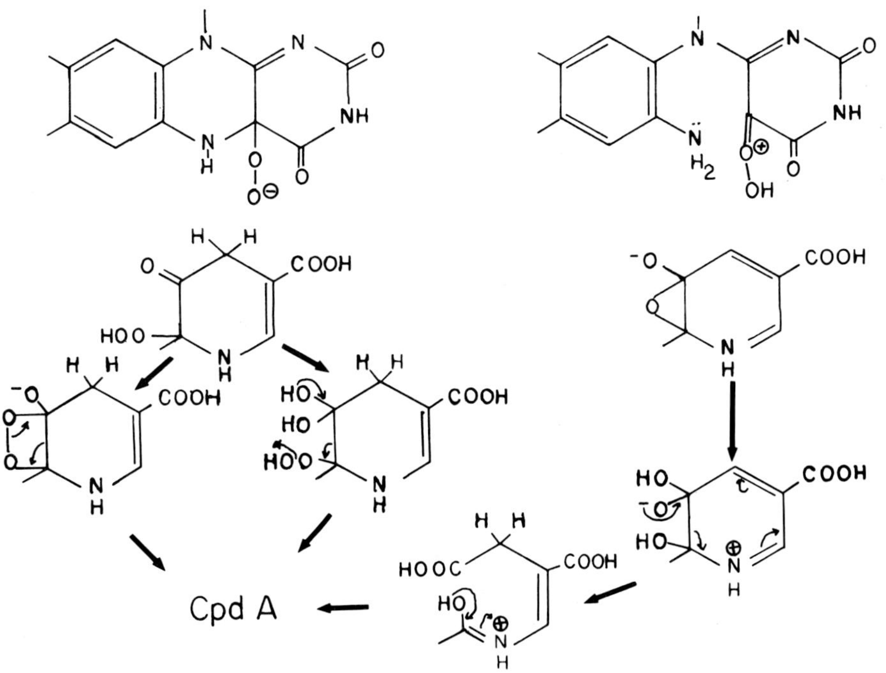

FIGURE 5. Possible mechanisms for the cleavage of Cpd I by Cpd I oxygenase.

TABLE v. *Comparative Properties of Cpd I Oxygenase and Bacterial External Flavomonooxygenases*[a]

Property	Cpd I oxygenase	Bacterial flavin monooxygenase
1) Stereospecificity for NADH or NADPH	4R - Specific	4R - Specific
2) Kinetic mechanism	Cpd I binds first, followed by NADH. After reduction of FAD, O_2 addition occurs.	Same as Cpd I oxygenase
3) Requirement for a phenolic group on the substrate	Absolute	Only exception is imidazoyl acetate monooxygenase, where a heterocyclic nitrogen replaces the phenolic group
4) Newly incorporated oxygen from O_2 is either *o*- or *p*- to the phenolic group on the substrate	Yes. The N-acetyl moiety of Cpd A derives all its oxygen atoms from O_2.	Yes. Hydroxylation is mostly *ortho*. Only in *m*-hydroxybenzoate-5-hydroxylase, hydroxylation is *para*.
5) Requirement for cofactors	Only FAD serves as cofactor. No metal ions are involved.	Same as Cpd I oxygenase
6) Reactivity with free flavins	Rapid	Rapid
7) Inhibition by monovalent ions	Occurs	Occurs
8. Binding of 1-deaza-FAD to the apoenzyme	Occurs; the complex is catalytically incompetent	Occurs; the complex functions as an NADH oxidase

[a]The properties of Cpd I oxygenase are those determined in this laboratory; those tabulated for bacterial flavin monooxygenases are a composite of those reported in literature [for a review, see (9)].

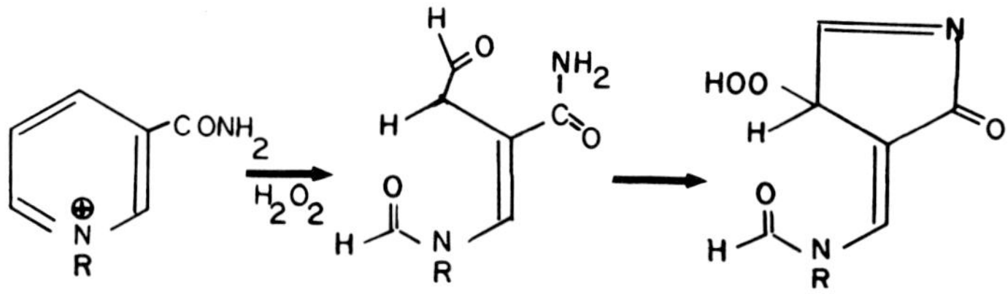

FIGURE 6. Ring cleavage of 1-methyl (or 1-benzyl)-3-carbamoyl pyridinium chloride by alkaline hydrogen peroxide.

product of an extensive rearrangement. These reactions presumably occur with the intermediate formation of an acyclic compound (II), followed by ring closure. Similar reactions may be implicated in the ring transformations and opening of pyridinium salts and pyridinium betaines by hydrogen peroxide (17).

It appears therefore that the unusual dioxygenase-type of ring cleavage catalyzed by Cpd I oxygenase is a consequence of the chemical nature of intermediates formed during the reaction. Although, as shown in Table V, the dioxygenase has many characteristics in common with flavin monooxygenases, it differs from them in that the reducing equivalents of NADH are incorporated into the dioxygenated organic product, rather than being removed by the net elimination of H_2O.

SUMMARY

Intermediates involved in the oxidative pathway by which Pseudomonas sp. MA-1 degrades pyridoxine to acyclic products have been previously characterized in this laboratory. Pseudomonas sp. MA-1 was initially selected for its ability to grow on this vitamin as a sole source of carbon and nitrogen. Degradation proceeds by a series of side chain oxidations and a decarboxylation leading to 2-methyl-3-hydroxypyridine-5-carboxylic acid (Cpd I). This substituted pyridine then undergoes ring-cleavage in an unusual reaction catalyzed by an FAD-oxygenase in the presence of NADH and oxygen, that results in addition of two oxygen and two hydrogen atoms to yield NAD^+ and the acyclic product, α-(N-acetylaminomethylene) succinic acid (Cpd A). Hydrolytic cleavage of Cpd A in a separate process, yields succinaldehyde, acetate, ammonia and carbon dioxide. The homogeneous oxygenase that catalyzes ring-cleavage has a molecular weight of 166,000 and is a tetramer of 43,000-dalton subunits. However, only two molecules of FAD, each of which appear to function independently during catalytic turnover, are present per tetramer. Neither 1-deaza-FAD nor 5-deaza-FAD is catalytically active; the 5-deaza-compound is not bound by the apoenzyme, and the complex of 1-deaza-FAD with the apoenzyme is not reduced by NADH even in the presence of Cpd I.

Under steady state conditions, Cpd I first reacts with the oxygenase; a ternary complex between oxidized enzyme, Cpd I and NADH is then formed. The 4R-hydrogen of NADH is then transferred to FAD, NAD^+ is released, and oxygen binding then occurs. At this stage, the pyridine nucleus of Cpd I is cleaved and Cpd A is released thus regenerating the oxidized enzyme for another catalytic cycle.

Kinetic isotope experiments in the presence of NAD^2H, NAD^3H, and ^2H_2O have shown that hydride-transfer from NADH is only partially rate determining during the overall reaction and that H-addition to the C(4) of Cpd A is not a rate determining step. This H-atom is derived from the solvent and not from NADH. In view of the fact that the enzyme does not contain any metal ions capable of participating in redox catalysis, it is suggested that oxygen activation is accomplished by enzyme-bound reduced flavin and that H-addition to C(4) of Cpd A occurs as a proton instead of as a hydride ion or a hydrogen atom.

ACKNOWLEDGEMENTS

These studies were supported by grants from the Welch Foundation (Grant F-714) and the United States Public Health Service (Grants AM 19898 and AI 13940).

REFERENCES

1. Burg, R. W., Rodwell, V. W. and Snell, E. E., J. Biol. Chem., 235,1164 (1960).
2. Ikawa, M., Rodwell, V. W. and Snell, E. E., J. Biol. Chem., 233,1555 (1958).
3. Sparrow, L. G., Ho, P. P. K., Sundaram, T. K., Zach, D., Nyns, E. J. and Snell, E. E., J. Biol. Chem., 244,2590 (1969).
4. Kishore, G. M. and Snell, E. E., Fed. Proc. 38,319 (1979).
5. Kishore, G. M. and Snell, E. E., Biochem. Biophys. Res. Commun., 87,518 (1979).
6. Entsch, B., Husain, M., Ballou, D. P., Massey, V. and Walsh, C. T., J. Biol. Chem. 255,1420 (1980).
7. Hersh, L. B. and Walsh, C. T., "Methods in Enzymology," Vol. 66, Part E, (D. B. McCormick and L. D. Wright, eds.), Academic Press, New York, N.Y.,(1980), p. 277.
8. Weirenaga, R. K., DeJong, R. J., Kalk, K. H., Hol, W. G. T. and Drenth, J., J. Mol. Biol. 131,55 (1979).

9. Massey, V. and Hemmerich, P., "The Enzymes," Vol. XII, Part B (P. D. Boyer, ed.), Academic Press, New York, N.Y., (1975), p. 191.
10. You, K., Arnold, L. J. and Kaplan, N. O., Arch. Biochem. Biophys. 180,550 (1977).
11. Cavalini, D. and Frontali, N., Biochim. Biophys. Acta 13, 439 (1954).
12. Drysdale, G. R., Spielfel, M. J. and Strittmatter, P., J. Biol. Chem. 236,2333 (1961).
13. Weber, M. M., Kaplan, N. O., San Pietro, A. and Stolzenbach, F. O., J. Biol. Chem. 227,27 (1957).
14. Ghisla, S., Hastings, J. W., Favaudon, V. and Lohste, J.-M., Proc. Natl. Acad. Sci. USA, 75,5860 (1978).
15. Poulsen, L. L. and Zeigler, D. M., J. Biol. Chem. 254,6449 (1979).
16. Bristol, D. W. and Dittmer, D. C., J. Org. Chem. 35,2487 (1970).
17. Katritsky, A. R., Ramsden, C. A., Zakaria, A., Harlow, R. L. and Simonsen, S. H., J. C. S. Chem. Comm. 8,363 (1969).
18. Entsch, B., Ballou, D. P. and Massey, V., J. Biol. Chem., 251,2550 (1976).

GENERAL DISCUSSION

SINGH: Dr. Kishore, in one of the last two slides you showed, you had a reaction with H_2O_2 with ring opening. Was an enzyme present in that reaction?

KISHORE: No. It is a nonenzymatic reaction.

SINGH: Just straight H_2O_2?

KISHORE: Yes.

SINGH: How long does it take?

KISHORE: It takes a couple of hours for completion.

ULLRICH: What was the pH of the reaction?

KISHORE: Alkaline conditions.

MOFFAT: Would you care to comment further on the fact that a tetramer of apparently identical subunits binds only two moles of flavin per mole?

KISHORE: Well, this is not really unique for this system; there has been at least one example of a monooxygenase which contains, for a dimeric unit, a single flavin molecule. There have been examples of several dehydrogenases and oxidases which contain one flavin for every two subunits. One possibility is that there is a common domain between the two subunits which is actually responsible for the finding of the flavins. Although we get a single band on polyacrylamide gel electrophoresis, that does not mean that all the subunits are identical.

WHILLANS: I would be very much interested to see if you modify your substrate, what then are the requirements for binding to the enzyme? Do you find some kind of dead end inhibitors as other people have found for flavoproteins?

KISHORE: We have studied the ability of this oxygenase to bind as well as oxidize several other substrates. The only other substrate which we have found for this enzyme is a 5-pyridoxic acid which is an intermediate in a closely related pathway used for degradation of vitamin B_6 by another bacterium. We are also trying to synthesize some substrates which lack the methyl group. We know that both the hydroxyl and the carboxyl are essential for the reaction and the pyridine nitrogen is probably essential.

PEISACH: I would like to amplify a comment to Dr. Moffat concerning the fact that there are proteins and there are oxygenases which are multisubunit in nature and do not contain a complete complement of prosthetic groups. A specific one that comes to mind is tryptophan dioxygenase, which is quite analogous to hemoglobin insofar as it is a tetrameric hemoprotein that undergoes allosteric control. There are only two hemes per tetramer in that case, even though there are two different types of subunits.

ULLRICH: Dr. Garst, I think that you have picked out a substrate which could be very useful in studies of various oxygenases, because it seems to behave differently when compared to all the other substrates that we have. However, I think you might have difficulties in explaining the hydroxylation of benzene to phenol or cyclohexane to cyclohexanol, by your nucleophilic mechanism.

COHEN: I have a question for both Dr. Ullrich and Dr. Peterson about the measurement of superoxide with succinylated cytochrome-c. I understand how you can measure product formation and H_2O_2 accumulation, but when you use a chemical probe such as this, is there not a danger that by mass action you change the equilibrium and overestimate the amount of superoxide produced?

ULLRICH: No, the hydroxylation reaction is not affected. It is only the liberated $O_2^{\cdot}$ in the system which reacts with succinylated cytochrome-c. So you are not doing anything to the system; you just trap something which would be on the sidelines.

COHEN: I wanted to get some clarification about the direct reduction of the ferric form by an electron followed by superoxide reduction. Normally substrate binds first, but how probable is it that it can, in fact, go the other way?

PETERSON: The case that I mentioned related to the adrenal cortex cytochrome P-450 where there is very tight binding of substrate. The kinds of redox potential changes that I described also occur with bacterial P-450 where, first you have substrate binding prior to reduction and oxygenation. I personally do not think that there is going to be a very significant liberation of $O_2^{\cdot-}$. However, in an artificial system one can take the substrate-free form of the enzyme, chemically reduce it with a much more powerful reducing agent, and then introduce O_2 into the system. Under these conditions Gunsalus has demonstrated essentially quantitative liberation of $O_2^{\cdot-}$. But that is a different type of system and it does not operate physiologically, for those enzymes which are highly specific. The ferric form is not reduced in the absence of the substrate. In microsomal systems, where the substrate binds very less tightly, then the potential of having it dissociate is possible.

FEE: I have a question for Dr. Ullrich. Your microsomes are really operating at an extremely low efficiency, and I wonder if they are in some way damaged during the isolation technique or whether cytoplasmic cofactors may control the uncoupling in some way.

ULLRICH. We have of course looked first at whether or not we are dealing with an artificial system. These microsomal fractions are the best ones we have prepared. We improved the preparation to the point where there is no lipid peroxidation, or any other damaging processes. Under these conditions we are looking at the effects of $O_2^{\cdot-}$. These microsomes were for a long time under hydroxylating conditions and there was no damage. With respect to the cytosolic factors, Orrhenius and coworkers have reported that liver cells produce H_2O_2 by the P-450 system. This observation indicates that this is really a physiological process. We are now performing liver perfusion experiments to see how much H_2O_2 is really formed during such a process. However it seems that *in vivo* all the H_2O_2 is trapped by glutathione peroxidase and that catalase plays probably a minor role. So the next thing we have to do is look for the output of oxidized glutathione in the system. And I think we will find it.

FEE: But the peroxide comes mainly from the lysosomes or microbodies. But in this case you essentially have an unlimited source of both substrates and I would think that such a leak would produce such a large amount of peroxide that it would really be toxic and therefore I suggest perhaps the uncoupling that you observed here in terms of $O_2^{\cdot-}$ production is unnecessarily large.

ULLRICH: First the nomenclature: you said uncoupling, we do not call this uncoupling.

FEE: I am sorry.

ULLRICH: Uncoupling is defined by a decomposition of the active oxygen to water.

FEE: Now I am talking about the superoxide leak. Not the uncoupling, right?

ULLRICH: The question is why $O_2^{\cdot-}$ and H_2O_2 are formed? And as I said there are explanations for that, because, as Peterson said, in the microsomal system you do not have this strict correlation of substrate binding forming the high spin complex and then reduction. It seems to me that there is an equilibrium of high and low spin without substrate being present. The only thing you have to do is to remove this unknown sixth ligand and then you get a five-coordinated complex with the right potential. And it seems that there are some forms of P-450 present which are present only as high spin complexes. These can be reduced and may form $O_2^{\cdot-}$ since no substrate is there which protects the oxy complex as shown for P-450 CAM.

GAS PHASE OXIDATION AND ENVIRONMENTAL IMPLICATIONS

Session Chairman, H. Niki, Dearborn, Michigan

STRATOSPHERIC CHEMISTRY AND THE OZONE PROBLEM

Carleton J. Howard, Aeronomy Laboratory, National Oceanic and Atmospheric Administration, U. S. Department of Commerce, Boulder, Colorado 80303

INTRODUCTION

Oxygen is a unique component in the earth's atmosphere because it is naturally present in two allotropic forms, O_2 and O_3. Everyone recognizes the crucial role that the atmosphere plays as a reservoir for this life sustaining element but it also has a second critical role as a filter for harmful solar ultraviolet radiation. In this function both O_2 and O_3 are important.

The total amount of radiation from the sun measured at the top of the earth's atmosphere is about 1370 W m^{-2} (1). About 1% of this radiation is at wavelengths shorter than 300 nm. Radiation at wavelengths shorter than Lyman alpha (122 nm) is capable of ionizing certain components of the atmosphere and is absorbed at high altitudes, creating the ionosphere. Radiation at wavelengths between 122 and about 200 nm is filtered out by oxygen. Radiation at wavelengths between 200 nm and 300 nm is absorbed by ozone. The altitude dependence of the solar flux for four wavelengths between 200 and 350 nm is given in Fig. 1. These curves are from calculations by Liu (2) for an overhead sun and give the photon flux in units of photons per cm^2 per second for a 1 nm wavelength interval. It can be seen that there is no significant absorption of radiation in this region above about 50 km altitude. Below 50 km the 200 nm radiation penetrates more deeply than that at 250 nm because there is a minimum in the combined O_2 and O_3 absorption coefficients at that wavelength. The maximum in the O_3 absorption coefficient falls at 250 nm so radiation at that wavelength is cut off above 30 km. The small increase in the 350 nm flux above 10 km results from multiple scattering of the uv radiation by air molecules.

One significance of the role of ozone as a uv filter is shown in Fig. 2. This figure shows the wavelength dependence of the sunlight intensity at the earth's surface (photons per cm^2 per second per nm wavelength interval), the ozone absorption cross

ISBN 0-12-592050-4

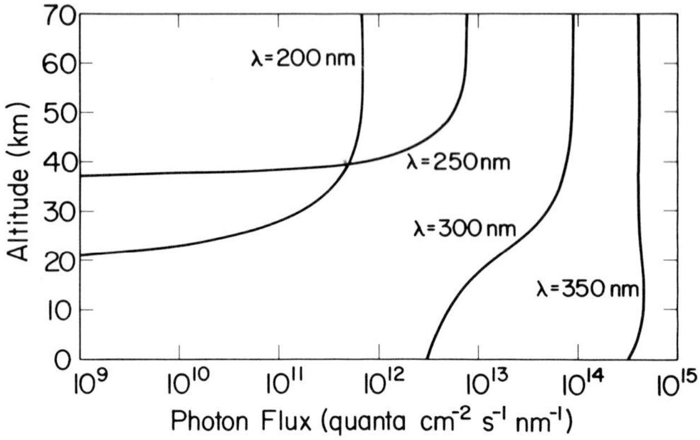

FIGURE 1. Altitude dependence of solar uv flux at four different wavelengths with an overhead sun. From Liu (2).

section (per cm^2) and the relative sensitivity of DNA (per photon) (3). The solar spectrum in this case was calculated for a 25° zenith angle. The relative biological sensitivity scale indicates the susceptibility of DNA to damage by uv radiation, a factor that is at least similar to the spectral sensitivity for skin cancer (4). Increased incidence of skin cancer is one possible direct effect of increased exposure to uv radiation; other effects on

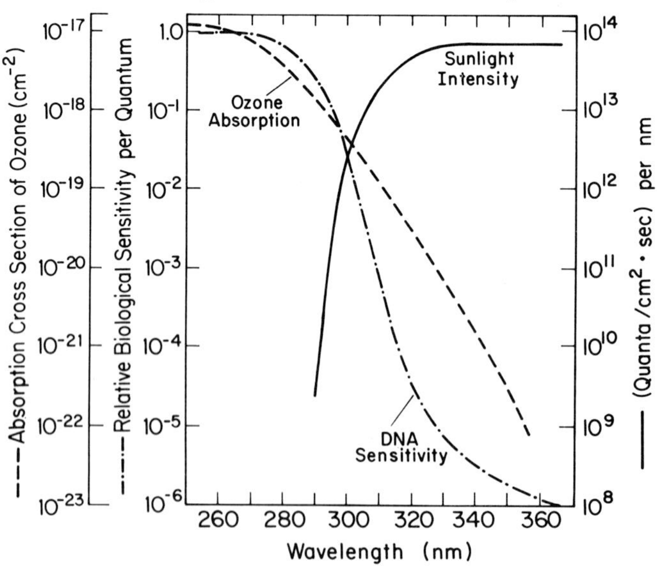

FIGURE 2. Wavelength dependence of the sunlight intensity at the earth's surface (25° zenith angle), of the ozone absorption cross section, and of the relative sensitivity of DNA to damage by uv radiation.

the climate, agricultural productivity, and terrestrial and aquatic ecosystems may also result from modification of the concentration of atmospheric ozone (5).

It has been known that ozone is a component of the atmosphere for about 100 years (6). A description of the stratosphere is given in Fig. 3. In this figure the altitude profiles of temperature, pressure, and ozone concentration are shown. The data represents only typical values because considerable variability exists depending upon factors such as season and latitude. The earth's atmosphere is divided into different regions according to its temperature profile. The lowest layer is called the troposphere and is characterized by decreasing temperature with increasing altitude. The upper boundary of this layer is called the tropopause, which is defined as the altitude at which the temperature stops decreasing. The tropopause typically lies at an altitude between 8 and 16 km. The layer above the tropopause is called the stratosphere and is characterized by an increasing temperature with increasing altitude. This condition creates a relatively stable environment with little vertical mixing of the atmosphere. Therefore, gases injected into the stratosphere have a residence time of several years or longer. The upper boundary of the stratosphere is about 50 km, where the temperature stops increasing. This boundary, the stratopause, defines the beginning of the next layer, the mesosphere.

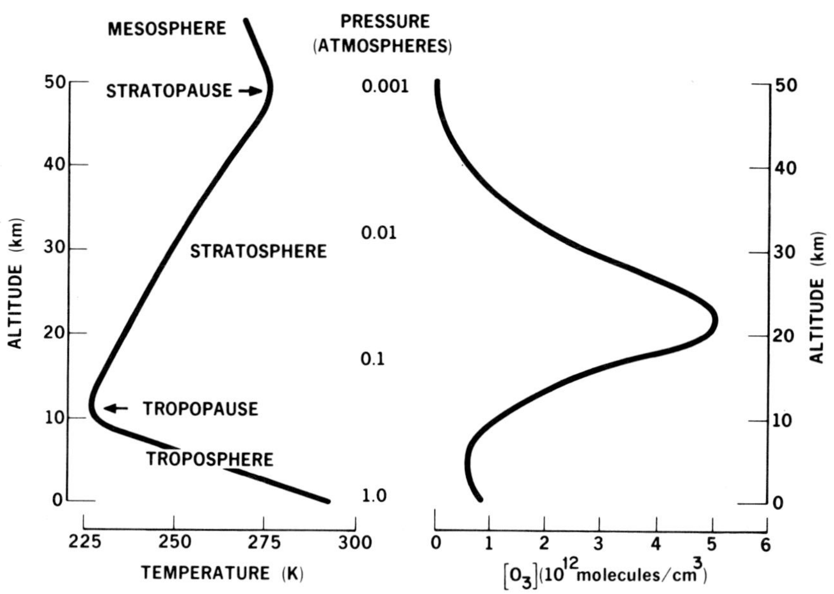

FIGURE 3. *Altitude profiles for temperature, pressure, and ozone concentration in the troposphere and stratosphere.*

The pressure of the atmosphere is given by the barometric law (7) and depends on the temperature and mean molecular weight of the atmosphere. It decreases about one order of magnitude every 16 km (about 10 mi) as shown in the center panel of Fig. 3. The pressure in the stratosphere ranges from about 0.12 atm at the bottom to about 0.001 atm at the top.

The maximum in the ozone concentration profile is located between 22 and 26 km. The concentration at the maximum corresponds to a mole fraction of only about three parts-per-million. If all of the ozone in the atmosphere were compressed to one atmosphere pressure, the resulting layer of gas would be only about 3 mm thick.

OZONE CHEMISTRY

A simple theory to explain the presence of ozone in the atmosphere was given by Chapman in 1930 (8). It consists of the following steps:

$$O_2 + h\nu\ (\lambda < 240\ \text{nm}) \rightarrow 2\ O \qquad (1)$$

$$O + O_2 + M \rightarrow O_3 + M \qquad (2)$$

$$O_3 + h\nu\ (\lambda < 310\ \text{nm}) \rightarrow O_2 + O(^1D) \qquad (3)$$

$$O_3 + O \rightarrow 2\ O_2 \qquad (4)$$

The lack of adequate kinetic data delayed testing Chapman's theory in detail until recently. Most of the energy from the sunlight absorbed in reactions (2) and (3) provides the heat source for the stratosphere.

In 1970 interest in stratospheric photochemistry was stimulated by the suggestion that nitrogen oxides from high-flying supersonic transport (SST) aircraft may affect the natural ozone balance. Crutzen (9) and Johnston (10) independently proposed that the nitrogen oxides generated by the jet engines may act as catalysts for ozone removal via the sequence:

$$NO\cdot + O_3 \rightarrow NO_2^{\cdot} + O_2 \qquad (5)$$

$$NO_2^{\cdot} + O \rightarrow NO\cdot + O_2 \qquad (6)$$

The sum of these reactions is equivalent to reaction (4) in the Chapman mechanism. To make accurate predictions of ozone depletion by NO_x ($NO\cdot + NO_2^{\cdot}$), computer based models of stratospheric photochemistry were developed. The models require data on the transport, composition, and chemistry of the stratosphere. The chemistry of the stratosphere is dominated by free radical reactions, but unfortunately there are very few measurements of the concentrations of these species. A comparison of the concentration measurements of critical trace constituents with model generated profiles gives an important test of the model's validity (11).

It is convenient to analyze the atmospheric reaction schemes in terms of the three types of free radical reactions: (a) initiation, (b) propagation, and (c) termination. For example, the natural initiation reaction for nitrogen oxides in the stratosphere is

$$N_2O + O(^1D) \rightarrow 2\ NO\cdot \qquad (7)$$

N_2O is produced by microbes in the soil and aquatic systems as a part of the natural fixed nitrogen cycle. The N_2O is present in air at a mole fraction of about 0.3 ppm. The $O(^1D)$ atom in reaction (7) is produced in the uv photolysis of ozone as shown in reaction (3). Reactions (5) and (6) are typical of the free radical propagation reactions. The most important termination reaction for the nitrogen oxides is the reaction of $NO_2^{\cdot}$ with $\cdot OH$

$$NO_2^{\cdot} + \cdot OH + M \rightarrow HNO_3 + M. \qquad (8)$$

The nitric acid formed in (8) does not react with ozone and is the relatively inert form in which nitrogen oxides are ultimately rained out of the lower atmosphere.

In 1974 Molina and Rowland (12) traced the chemistry of chlorofluorocarbons, $CFCl_3$ and CF_2Cl_2, which are released into the atmosphere as a result of their use as aerosol propellants, polymeric foam blowing agents, and refrigerants. They found only one significant removal process for these compounds, namely, photolysis by uv radiation in the stratospheric window near 200 nm

$$CF_2Cl_2 + h\nu\ (\lambda < 240\ nm) \rightarrow \cdot CF_2Cl + Cl \qquad (9)$$

They noted that reaction (9) initiates a remarkably efficient chlorine catalyzed destruction of ozone:

$$Cl + O_3 \rightarrow ClO\cdot + O_2 \qquad (10)$$

$$ClO\cdot + O \rightarrow Cl + O_2 \qquad (11)$$

$$\text{net reaction} \quad O_3 + O \rightarrow 2O_2 \qquad (4)$$

As an indication of the potential impact of a small amount of chlorine, Rowland and Molina estimated that a single Cl atom from reaction (9) can remove an average of 10^5 odd oxygen ($O + O_3$) before it is permanently removed from the stratosphere. The most important termination reaction for the chlorine species is

$$Cl + CH_4 \rightarrow HCl + \cdot CH_3. \qquad (12)$$

Methane is present in the lower atmosphere in a mole fraction of about 1.5 ppm.

Another possible threat to stratospheric ozone has been discussed by Crutzen (13) and McElroy et al. (14). They describe a possible change in the natural fixed-nitrogen cycle from the agricultural use of man-made nitrogen fertilizers. The scenario for this disaster is that a fraction of the fertilizer nitrogen is ultimately converted to N_2O and adds to its concentration in the atmosphere. Stratospheric ozone is then perturbed by the reaction sequence (7), (5), and (6).

Although there are uncertainties in the model predictions of ozone perturbations by man-made chemicals, it is important to use the best available data to evaluate their possible impact. Present estimates give relatively small effects from SST flights, and indeed the present fleet is much smaller than it was estimated to be about 10 years ago. Similarly, little effect is presently expected from continued use of nitrogen fertilizers. However, a recent critical evaluation by a special committee of the National Academy of Sciences (15) estimated that continued use of chlorofluoromethanes would lead to an ozone depletion of about 17%. It is expected that the accuracy of evaluations of that type will be improved in the next few years through continued research of the kinetics, photochemistry and composition of the stratosphere. More details and numerous references can be found in reference (15).

REFERENCES

1. "The Solar Output and Its Variation," (O. R. White, ed.), Colorado Associated University Press, Boulder, Colorado (1977), p. 22.
2. Liu, S. C., 1-dimensional model, private communication, (1980).
3. "Environmental Impact of Stratospheric Flight: Biological and Climatic Effects of Aircraft Emissions in the Stratosphere," National Academy of Sciences, Washington, D.C. (1975).
4. "Halocarbons: Environmental Effects of Chlorofluoromethanes Release," National Academy of Sciences, NAS Printing and Publishing Office, Washington, D.C. (1976), p.82.
5. Reference (4) p. 67-80.
6. An excellent brief history of stratospheric ozone is given by: Nicolet, M., Rev. Geophys. Space Phys. 13, 593 (1975).
7. "U. S. Standard Atmosphere, 1976," National Oceanic and Atmospheric Administration, NASA, US Air Force, NOAA-S/T76-1562 U.S. Government Printing Office, Washington, D.C. (1976), p. 6.
8. Chapman, S., Mem. Roy. Meteorol. Soc. 3, 103 (1930).
9. Crutzen, P. J., Quart. J. Roy. Meteorol. Soc. 96, 320 (1970).
10. Johnston, H. S., Science, 173, 517 (1971).
11. Ehhalt, D. H., Rev. Geophys. Space Phys. 16, 217 (1978).

12. Molina, M. J. and Rowland, F. S., Nature, 249, 810 (1974); and Rowland, F. S. and Molina, M. J., Rev. Geophys. Space Phys., 13, 1 (1975).
13. Crutzen, P. J., Ambio, 3, 201 (1974) and Geophys. Res. Lett., 3, 169 (1976).
14. McElroy, M. B., Elkin, J. W., Wofsy, S. C. and Yung, Y. L., Rev. Geophys. Space Phys., 14, 1430 (1976).
15. "Stratospheric Ozone Depletion by Halocarbons: Chemistry and Transport," National Academy of Sciences, Washington, D.C. (1979).

GAS-PHASE CHEMISTRY OF OXYGEN RADICALS

Alan C. Baldwin, Department of Chemical Kinetics, SRI International, 333 Ravenswood Avenue, Menlo Park, California 94025

INTRODUCTION

Oxygen-centered radicals are key intermediates in the gas-phase oxidation of organic compounds, and a detailed knowledge of their reactions is essential to our understanding of the environmental impact of atmospheric pollutants. Ideally, we would like to know the rate and mechanism of all possible elementary reactions. Even if we restrict ourselves to the reactions of oxygen radical species present in the atmosphere under typical atmospheric conditions, we would require an enormous amount of experimental information.

Organic oxy and peroxy radicals, their hydrogen analogues $\cdot OH$ and $HO_2^{\cdot}$, and the various inorganic oxygen radicals can undergo a wide range of reactions including decomposition, isomerization, disproportionation, and combination, as well as bimolecular reactions with many species. We can not hope to have experimental data on all these reactions, and often they can only be studied under conditions of temperature and pressure different from those of the atmosphere. Therefore, it becomes clear that we need a method to make predictions and estimates from the limited experimental data base, to make extrapolations from experimental-to-atmospheric conditions, and to interpret and compare various experiments.

In this paper, the thermodynamic formulation of transition state theory (TST) will be summarized, and its strengths and weaknesses described. Since it is the only realistic, generally applicable, method available, it will then be used to interpret the available experimental data on oxygen-centered radicals. One of the strengths of this approach to TST is that it is particularly easy to consider families of similar compounds in a consistent manner. Thus, the discussion of reactions will be grouped according to the following families: hydroxyl and hydroperoxyl, aliphatic peroxy radicals, aliphatic oxy radicals, aromatic radicals, and biradicals. Although hydroperoxyl radicals behave very similarly in most respects to their hydrocarbon analogue peroxy radicals, these are considered separately because there is far more detailed data available on these species. Hydroxyl is considered separately from organic oxy radicals since its reactions are considerably different.

ISBN 0-12-592050-4

TRANSITION STATE THEORY (TST)

The transition state theory expression for a thermal rate constant (k_T) may be written as:

$$k_T = \left(\frac{kT}{h}\right) \exp(-\Delta G_T{}^{\ddagger}/RT) \tag{1}$$

where k = Boltzmann's constant, h = Planck's constant, and $\Delta G_T{}^{\ddagger}$ is the free energy change between the reactants and the transition state. Since we have no independent way of measuring $\Delta G_T{}^{\ddagger}$, equation (1) cannot be used to predict k_T, and is essentially a definition of $\Delta G_T{}^{\ddagger}$. However, if we represent $\Delta G_T{}^{\ddagger}$ by equation (2),

$$\Delta G_T{}^{\ddagger} = \Delta H_T{}^{\ddagger} - T\Delta S_T{}^{\ddagger} \tag{2}$$

and compare the TST espression with an experimental rate constant represented by the Arrhenius equation (other representation of the experimental temperature dependence can be used, leading to more complex results), we find that the experimental A-factor (A) and activation energy (E) can be identified by equation (3):

$$A = \left(\frac{ekT_m}{h}\right) \exp(\Delta S^{\ddagger}/R); \qquad E = \Delta H^{\ddagger} + RT_m \tag{3}$$

where T_m is the mean experimental temperature.

We can now use the technique of "Thermochemical Kinetics" developed by Benson (1) to estimate $\Delta S^{\ddagger}$ and $\Delta H^{\ddagger}$. This approach is particularly useful in considering groups of reactions where we may 'calibrate' by calculating $\Delta H^{\ddagger}$ and $\Delta S^{\ddagger}$ from measured rate constants, and predict the changes in $\Delta H^{\ddagger}$ and $\Delta S^{\ddagger}$ for unmeasured, but similar, reactions.

At this point, it is useful to catagorize reactions according to the following scheme:

<u>Umimolecular Reactions</u>

Simple Fission: AB → A + B
Complex Fission: Molecule (or radical) → Molecule + Molecule (or radical)
Isomerization: Intramolecular Atom Rearrangement

<u>Biomolecular Reactions</u>

Direct Metathesis: A + BX → AX + B
Addition: A + Molecule → Stable Adduct (reverse of complex fission)
Association: A + B → AB (reverse of simple fission)

It is well known that rate constants for umimolecular reactions and their reverse addition and association reactions are pressure dependent. The TST formulation of the thermal rate constant only applies to the high pressure limit, where the reactions are pressure independent. At lower pressures, these reactions enter the fall-off region, until at sufficiently low pressure the rate constants are linearly dependent on pressure. Given a knowledge of $\Delta H^{\ddagger}$ and $\Delta S^{\ddagger}$, we can use umimolecular reaction rate theories, such as RRKM (2) to predict the pressure dependence with good accuracy. Thus, of the above categories, only direct metathesis reactions should be pressure independent. In certain circumstances, we must modify our simple framework with a single fixed transition state to take account of the failure of this picture to represent the true physical process. For example, in bond (simple) fission or recombination reactions, there is no well-defined maximum along the potential energy surface for the reaction; thus, the concept of a simple fixed transition state is not applicable. Recent work has shown that a transition state that becomes tighter ($\Delta S^{\ddagger}$ decreases) as the temperature increases can well represent experimental data over wide temperature ranges. This non-fixed transition state model enables us to reproduce the small negative temperature dependence observed in many radical combination reactions (1, 3, 4).

Another type of reaction that requires a more detailed analysis is a bimolecular reaction that can proceed through an intermediate bound complex (5). In this case, the general mechanism shown below is applicable:

$$A + B \underset{-1}{\overset{1}{\rightleftarrows}} Y^{*} \overset{2}{\rightarrow} C + D$$

$$Y^{*} \underset{[M]}{\overset{3}{\rightarrow}} Y$$

where the intermediate Y is formed with excess energy (indicated by *) due to the exothermicity of reaction (1); it may then dissociate back to reactants (reaction -1), dissociate to products (reaction 2), or be stabilized by collision with bath gas [M] (reaction 3).

If we analyze the above scheme, we see that the experimental rate constant is given by:

$$k_{expt} \equiv -\frac{1}{[A][B]}\frac{d[A]}{dt} = \frac{k_1(k_2 + k_3[M])}{k_{-1} + k_2 + k_3[M]} \equiv k_1 E(T,M)$$

A number of limiting cases can be identified, corresponding to different values of the efficiency factor, E(T,M).

Case A: $k_3[M] << k_2$

(1) $k_{expt} = K_1k_2$ for $k_{-1} >> k_2$ or

(2) $k_{expt} = k_1$ for $k_{-1} << k_2$

where $K_1 = k_1/k_{-1}$

Case B: $k_3[M] >> k_2$

(1) $k_{expt} = k_1$ for $k_3[M] >> k_{-1}$

(2) $k_{expt} = K_1k_3[M]$ for $k_3[M] << k_{-1}$

In cases A(2) and B(1), the rate-controlling process is reaction (1), and the reaction may be described by TST for this reaction; in some cases, this is a combination reaction, and the difficulties noted above must be considered. In Case A(1) the experimental rate constant is given by a product of three rate constants, each of which may be calculated by TST. The temperature dependence of such a rate constant is very complex. In Case B(2), the experimental rate constant is pressure dependent, as well as having a complex temperature dependence. Note that these are limiting cases; the real experimental result may be a complex (temperature and pressure dependent) intermediate case.

In subsequent sections, a number of different reactions are discussed and analyzed using the TST framework developed.

HYDROXYL AND HYDROPEROXYL RADICALS

The hydroxyl radical (·OH) is a ubiquitous species in the atmosphere, and the fate of most stable species that do not phootlyze is reaction with ·OH. The rates of reaction of ·OH with a large number of hydrocarbons have been measured, and this data has been thoroughly reviewed recently (7). The reactions of ·OH with hydrocarbons are summarized in Table 1. It can be seen that ·OH reactions with alkanes and probably aldehydes are simple direct metathesis reactions proceeding with small activation energies. Barker et al (8) have presented evidence that $O(^3P)$ reacts with formaldehyde by addition. This may indicate that ·OH may also add to aldehydes. The available experimental data are not sufficient to resolve this point.

The reaction of ·OH with alkenes is most easily rationalized by a mechanism involving reversible addition to the double bond, followed by collisional stabilization of the radical adduct as the principal channel. This is Case B in our formation reaction scheme. We would thus expect a transition from Case B(1) to Case B(2) as the pressure is changed and a larger fraction of the adducts are stabilized. Howard (9) has shown that at 300 K, addition is the only pathway for ·OH reaction with ethene. Abstraction must be a minor path for most alkenes at room temperature;

Table 1.

EXPERIMENTAL DATA FOR THE REACTION OF OH WITH HYDROCARBONS

Alkanes

$$\text{e.g., } \cdot OH + CH_3CH_2CH_2CH_3 \rightarrow \cdot CH_2CH_2CH_2CH_3 + H_2O$$

$$\rightarrow CH_3\dot{C}HCH_2CH_3 + H_2O$$

$$\left.\begin{array}{l} A = 1.76 \times 10^{-11}\ cm^3\ molec^{-1}\ s^{-1} \\ E = 1.1\ kcal/mol \end{array}\right\}\ \text{Ref. 5}$$

These Arrhenius parameters are typical of most ·OH reactions with alkanes.

Alkenes

$$\text{e.g., } \cdot OH + CH_3CH{=}CH_2 + M \rightarrow CH_3CH(OH)CH_2^{\bullet} + M$$

$$\rightarrow CH_3\dot{C}HCH_2(OH) + M$$

$$\left.\begin{array}{l} A = 4.1 \times 10^{-12}\ cm^3\ molec^{-1}\ s^{-1} \\ E = -1.08\ kcal/mol \end{array}\right\}\ \text{Ref. 6}$$

Typical Arrhenius parameters:

$$\left.\begin{array}{l} A = 10^{-12} \text{ to } 10^{-11}\ cm^3\ molec^{-1}\ s^{-1} \\ E = -1\ kcal/mol \end{array}\right\}\ \text{Ref. 7}$$

Aldehydes

$$\text{e.g., } \cdot OH + CH_3CHO \rightarrow CH_3CO\cdot + H_2O$$

$$\left.\begin{array}{l} A = 6.87 \times 10^{-12}\ cm^3\ molec^{-1}\ s^{-1} \\ E = -1.01\ kcal/mol \end{array}\right\}\ \text{Ref. 8}$$

however, as the temperature increases, the abstraction channel should become favored as the addition adduct becomes less stable and the abstraction rate increases. Atkinson et al (10) have shown in alkenes with a weak allylic hydrogen, abstraction can occur up to 30% of the total reaction at 300 K.

Both ·OH and $HO_2^{\bullet}$ undergo combination reactions with $NO_2^{\bullet}$ to give nitric and pernitric acid respectively:

$$\cdot OH + NO_2^{\cdot} + M \rightleftarrows HNO_3 + M \quad (4)$$

$$HO_2^{\cdot} + NO_2^{\cdot} + M \rightleftarrows HO_2NO_2 + M \quad (5)$$

These reactions have a complex temperature and pressure dependence, (11) but we have shown that a large amount of data can be reconciled using RRKM theory and our modified temperature-dependent transition state (3). This is illustrated in Fig. 1. Note that data obtained in different pressure regimes characteristic of various experimental techniques can be compared and treated in a consistent manner, and be used to determine $\Delta H^{\ddagger}$ and $\Delta S^{\ddagger}$.

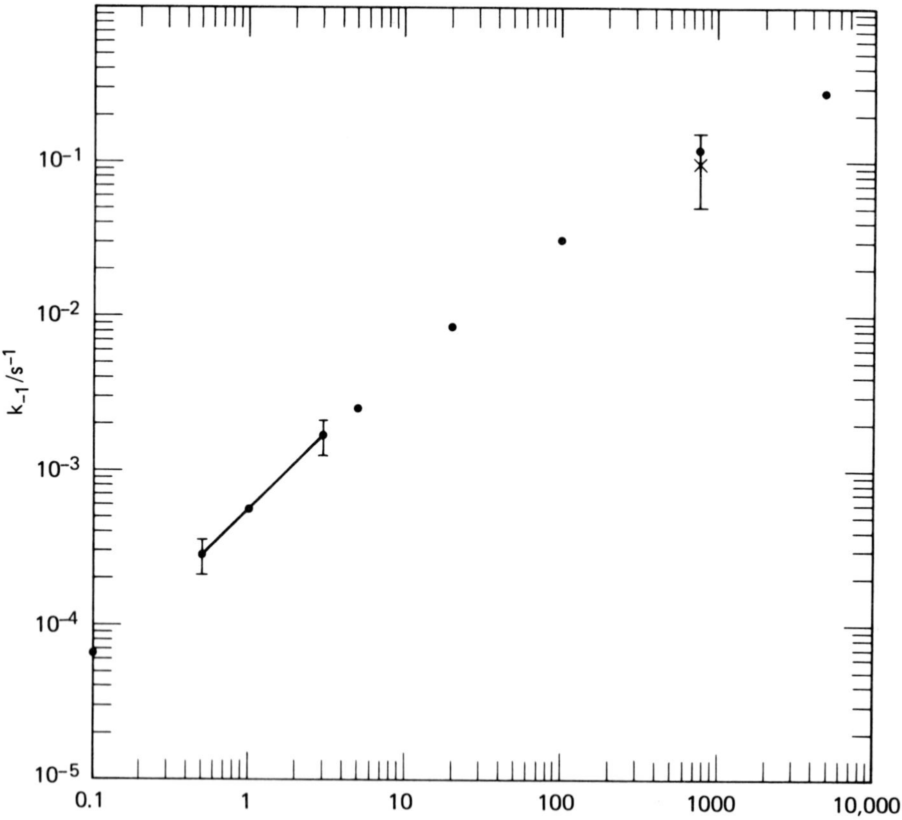

FIGURE 1. Pressure dependence of the reaction of HO_2 with NO_2 at 300 K. • calculated; x ——— experimental

$HO_2^{\cdot}$ also undergoes an important reaction with NO·:

$$HO_2^{\cdot} + NO\cdot \rightarrow \cdot OH + NO_2^{\cdot} \qquad (6)$$

which has a significant effect on the $\cdot OH/HO_2^{\cdot}$ ratio in the atmosphere. This reaction has been extensively studied by Howard (12), who found that from 232 K to 1271 K, the rate constant is given by $k_6 = 3.51 \times 10^{-12} \exp(240/T)$ cm^3 $molec^{-1}$ s^{-1}. If this reaction were a simple direct metathesis reaction, we would not expect to find a negative temperature dependence. If we interpret the reaction in terms of our complex formation reaction scheme, we see that we must be in Case A ($k_3[M] << k_2$). Both Case A(1) and Case A(2) could explain the negative temperature dependence. In Case A(1), k_2/k_{-1}, the ratio of the rate constant for decomposition to products to the rate constant for decomposition back to reactants must have a negative temperature dependence. This ratio depends sensitively on the thermochemistry of the pernitrous acid intermediate, which is not known and cannot be estimated with sufficient precision to decide if this is the correct explanation. Alternatively, in Case A(2), the rate constant k_1 is controlling, and the negative temperature dependence is a consequence of this being a radical-radical combination reaction. Recent measurements (16, 17) of the rate constant for the analagous reaction of methylperoxy radicals (reaction 7) with NO· ; which has approximately the same rate constant at:

$$MeO_2^{\cdot} + NO\cdot \rightarrow MeO\cdot + NO_2^{\cdot} \qquad (7)$$

room temperature as reaction (6) tends to indicate that Case A(2) is the correct explanation. It is much more likely the $k_6 = k_7$ if it is a simple recombination reaction than if it is a balance between two reaction channels. Experiments on the temperature-dependence of reaction (7) should clarify this point.

A number of other reactions of $HO_2^{\cdot}$ are important in the atmosphere; for example:

$$HO_2^{\cdot} + \cdot OH \rightarrow H_2O + O_2 \qquad (8)$$

$$HO_2^{\cdot} + HO_2^{\cdot} \rightarrow H_2O_2 + O_2 \qquad (9)$$

$$HO_2^{\cdot} + O_3 \rightarrow \cdot OH + O_2 + O_2 \qquad (10)$$

$$HO_2^{\cdot} + ClO\cdot \rightarrow HOCl + O_2 \quad \text{or}$$

$$O_3 + HCl \qquad (11)$$

Unfortunately, the measured rate constants for a number of these reactions are extremely difficult to understand and rationalize. In this respect, the kinetics of $HO_2^{\cdot}$ radicals are one of the least well understood sets of reactions of importance in the atmosphere. Reaction (9) has a very complex temperature and pressure dependence and is sensitive to the presence of water vapor (13). Reaction (10) has a very low A-factor (14) and reaction (11) has a complex temperature dependence that indicates multiple reaction channels (15). None of these reactions seem to proceed by a simple direct metathesis, although there is no satisfactory explanation for this fact. Individual reactions can be rationalized to some extent by postulating stable (often cyclic) intermediate complexes, but a consistent picture is at present elusive. Hopefully, the extensive experimental work now in progress will clarify, rather than complicate, the picture.

ALKYL AND ACYL PEROXY RADICALS

As noted previously, alkanes, alkenes, and aldehydes in the atmosphere primarily undergo reaction with ·OH (either abstraction or addition) to give carbon-control radicals which then rapidly add oxygen to form an alkyl or acyl peroxy radical. In the presence of NO· or $NO_2^{\cdot}$ these peroxy radicals react rapidly according to reactions (12) or (13):

$$RO_2^{\cdot} + NO^{\cdot} \rightleftarrows RO^{\cdot} + NO_2^{\cdot} \quad (12)$$

$$RO_2^{\cdot} + NO_2^{\cdot} + M \rightleftarrows RO_2NO_2 + M \quad (13)$$

Experimental results on the reactions of peroxy radicals with NO· and $NO_2^{\cdot}$ are given in Table 2.

Table 2.

EXPERIMENTAL DATA FOR THE REACTIONS OF PEROXY RADICALS WITH NO AND NO_2

Radical	*Reactant*	*k(∼ 300K)* cm^3 $molec^{-1}$ s^{-1}	*Ref.*
$CH_3O_2^{\cdot}$	$NO^{\cdot}$	8.0×10^{-12}	16
$CH_3O_2^{\cdot}$	$NO^{\cdot}$	1.8×10^{-11}	17
$CH_3C(O)O_2^{\cdot}$	$NO^{\cdot}$	2.59×10^{-12}	18
$CH_3C(O)O_2^{\cdot}$	$NO^{\cdot}$	3.24×10^{-12}	19
$CH_3O_2^{\cdot}$	$NO_2^{\cdot}$	1.52×10^{-12}	20
$CH_3C(O)O_2^{\cdot}$	$NO_2^{\cdot}$	1.4×10^{-12}	18
$CH_3C(O)O_2^{\cdot}$	$NO_2^{\cdot}$	1.05×10^{-12}	19

There are no experimental results on the temperature dependence of reaction (12), but as previously discussed, the rate constants are very similar to that for $HO_2^{\cdot}$ + NO· at room temperature. It seems likely that these reactions should have a similar mechanism and therefore a small negative temperature dependence due to the rate-determining step being a radical combination process.

The combination and decomposition reactions (13) and (-13) are analagous to the reaction of $HO_2^{\cdot}$ and $NO_2^{\cdot}$ discussed earlier. For R containing two or more carbon atoms, the reactions should be in the high pressure limit at 1 atmosphere pressure and 300 K. The bond strength in acylperoxy nitrates is ∿27 kcal/mol; if we estimate the bond strength for alkylperoxy nitrates as ∿23 kcal/mol based on HO_2NO_2, the difference in stability could explain why acylperoxy nitrates are formed in the polluted troposhere, whereas alkylperoxy nitrates are not.

ALKOXY RADICALS

As we have seen, the fate of most hydrocarbons in the atmosphere is reaction with ·OH, followed by addition of oxygen to give a peroxy radical, followed by reaction with NO· to give an oxy radical. However, alkoxy radical chemistry is extremely complex and depends sensitively on the size and structure of the radical. In general, alkoxy radicals may undero three reactions:

Decomposition: $$RO\cdot + M \rightarrow R' + R''CHO + M \quad (14)$$

Reaction with Oxygen: $$RO\cdot + O_2 \rightarrow R'CHO + HO_2 \quad (15)$$

Isomerization: $$RO\cdot + M \rightarrow \cdot ROH + M \quad (16)$$

The large number of different hydrocarbons present in the atmosphere and the possibility of ·OH attack at various sites means that there are an enormous number of different alkoxy radicals generated in the atmosphere. In this case, the use of TST enables us to rapidly generalize in a quantitative manner the available experimental data, and to estimate the large number of unmeasured reaction rate constants.

The available data on alkoxy radical decomposition reactions is summarized in Table 3. The details of the application of TST cannot be given here, but the results of a simple correlation of $\Delta S^{\ddagger}$ and $\Delta H^{\ddagger}$ with molecular properties yields the results shown in the estimated column of Table 3. These results may now be simply generalized to include a whole range of radicals (30).

Table 3

EXPERIMENTAL VALUES FOR RO· DECOMPOSITION RATES[a]

Radical	log A	log E	log A_{est}	E_{est}	Ref.
EtO·	13.7	33.1	13.7	22.1	21
i-PrO·	16.1	20.6	14.6	17.4	22,23
s-BuO·	16.4	18.0	14.4	13.9	24
	14.9	15.3		14.2	25
t-BuO·	15.1	16.2	15.2	16.3	26-29

[a]Units: E in kcal/mole; A in s^{-1}

There is little experimental data on the reaction of alkoxy radicals with oxygen to give a carbonyl compound and $HO_2^{\cdot}$. The only rate constant known (31) is for reaction (17):

$$CH_3O\cdot + O_2 \rightarrow CH_2O + HO_2^{\cdot} \qquad (17)$$

However, we may take as a first approxiamtion that $\Delta S^{\ddagger}$ is constant for all reactions of this type (corrected for reaction path degeneracy), and that $\Delta H^{\ddagger}$ may be correlated with the enthalpy of reaction. The results of this calculation are shown in Table 4.

Table 4

ESTIMATES: RO· + O_2 REACTIONS[a]

Reaction	log $(A)_{est}$	I $E_a = 4.0$	II $E_a=10.6 + 0.25 \times (\Delta H_R^o)$	III $E_a=11.5 + 0.29 \times (\Delta H_R^o)$
		k (min^{-1})	k (min^{-1})	k (min^{-1})
$CH_3O\cdot + O_2$	8.5	2.0×10^5	2.0×10^5	2.0×10^5
EtO· + O_2	8.3	1.3×10^5	8.2×10^5	1.3×10^6
n-PrO· + O_2	8.3	1.3×10^5	8.2×10^5	1.3×10^6
i-PrO· + O_2	8.0	6.7×10^4	1.5×10^6	3.7×10^6
n-BuO· + O_2	8.3	1.3×10^5	3.5×10^5	5.8×10^5
s-BuO· + O_2	8.0	6.7×10^4	1.1×10^6	2.2×10^6

[a]Effective first-order rate constants at 300 K in air (2.1×10^5 ppm O_2).

Alkoxy radical cyclic isomerization reactions were first suggested by Carter et al (32) based on product analysis in a smog chamber, and later observed experimentally in our laboratory (33). Unvortunately, quantitative measurements on a radical isomerism to give another radical are extremely difficult and have not been achieved. If we take as an example n-butoxy, reaction (13):

$$CH_3CH_2CH_2CH_2O\cdot + M \rightarrow CH_2CH_2CH_2CH_2OH + M \qquad (18)$$

we can estimate the A-factor simply from other cyclic reactions as $\sim 10^{11}\ s^{-1}$. The activation energy is estimated as the sum of strain energy required to form the ring, and the intrinsic activation energy for abstraction of an H-atom by an alkoxy radical: Table 5 shows the estimated rate constants for several reactions.

Table 5

ESTIMATED RO· ISOMERIZATION REACTION RATES

Reaction[a]	*log A* (s^{-1})	*E(kcal/mole)*	$k(min^{-1})$
$\cdot OCCCC \rightarrow HOCCCC\cdot$	*11.4*	*7.7*	3.7×10^{7}
$C\overset{O\cdot}{C}CC \rightarrow C\overset{OH}{C}CC\cdot$	*11.7*	*13.1*	8.6×10^{3}
$HOCCCCO\cdot \rightarrow HO\dot{C}CCCOH$	*11.2*	*6.5*	1.9×10^{8}
$HO\overset{O\cdot}{C}CCCOH \rightarrow (HO)_2CCC\dot{C}OH$	*11.2*	*6.5*	1.9×10^{8}
$(HO)_2CCC\overset{O\cdot}{C}OH \rightarrow (HO)_2\dot{C}CCC(OH)_2$	*10.9*	*4.6*	2.2×10^{9}
$(HO)_2\overset{O\cdot}{C}CCC(OH)_2 \rightarrow (HO)_3CCC\dot{C}(OH)_2$	*10.9*	*4.6*	2.2×10^{9}
$C\overset{OH}{C}CCO\cdot \rightarrow \cdot C\overset{OH}{C}CCOH$	*11.4*	*7.7*	3.0×10^{8}

[a]*Notation:* $C\overset{OH}{C}CCO\cdot \rightarrow \cdot C\overset{OH}{C}CCOH$ *represents* $CH_3\overset{OH}{C}HCH_2CH_2O\cdot \rightarrow \cdot CH_2\overset{OH}{C}HCH_2CH_2OH$

AROMATIC RADICALS

In the preceding discussion, no mention has been made of aromatic compounds. even though a significant fraction of the hydrocarbons in the atmosphere are aromatic. Unfortunately, this is because very little quantitative information on the reactions of aromatic species, particularly aromatic oxygen radicals, is available. Taking toluene, the most studied compound as an example, the reaction of ·OH with toluene proceeds principally by addition to give an ·OH adduct radical (34), analagous to the alkenes discussed earlier:

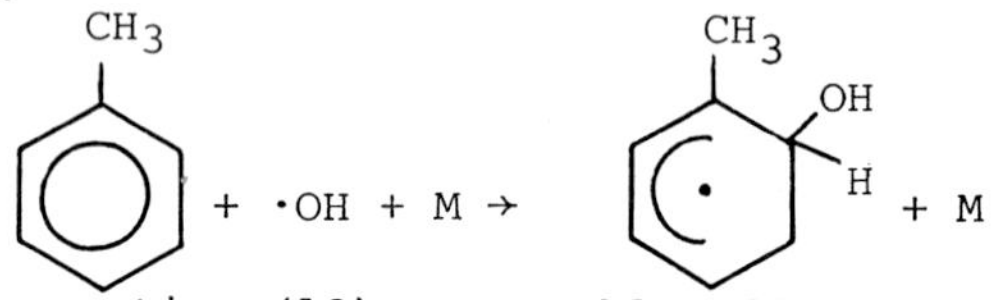

(19)

The product of reaction (19) presumably adds oxygen, reaction (20):

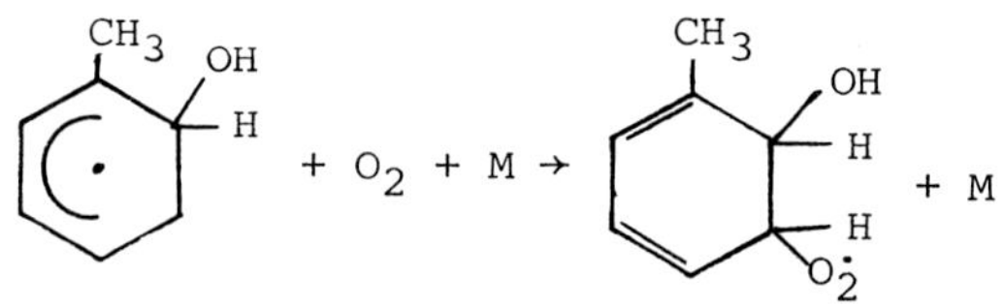

(20)

Atkinson et al (34) have presented arguments based on smog-chamber product analysis that the peroxy radical formed in reaction (20) can undergo cyclization; it is also possible that it undergoes reactions with NO, followed by ring cleavage. At present the lack of thermochemical data on aromatic compounds preclude any detailed analysis. Experimental work in this area is urgently needed.

OZONOLYSIS AND BIRADICALS

The reaction of alkenes with ozone is an important reaction in polluted atmospheres. Herron and Huie (35) found that the reaction of alkenes with O_3 could be explained by the following mechanism shown for ethene:

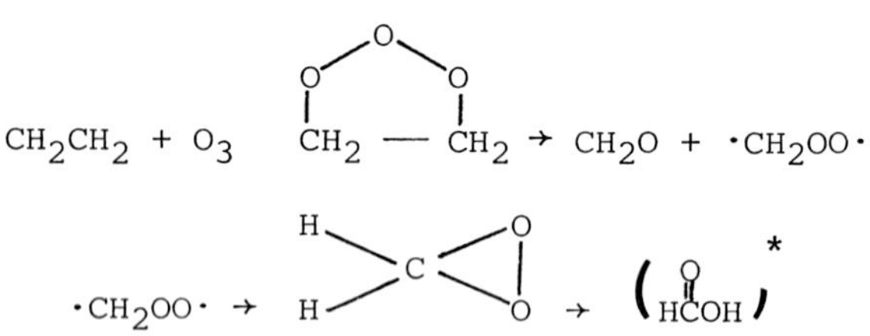

The Criegee intermediate decomposes to give formaldehyde and a biradical intermediate that subsequently rearranges to give formic acid containing a large excess energy (∿ 150 kcal/mol) due to the exothermicity of the reaction. By making reasonable estimates for $\Delta H^{\ddagger}$ and $\Delta S^{\ddagger}$ for the possible decomposition routes of HC(O)OH, we can calculate, using RRKM theory, the relative importance of the various channels for highly excited formic acid. This is shown in Table 6.

Table 6

DECOMPOSITION RATE CONSTANTS FOR FORMIC ACID CONTAINING ∿150 KCAL/MOL

	ΔH kcal mole^{-1}	E_a kcal mole^{-1}	log A s^{-1}	log$k$ k/s^{-1}
$HCOOH \rightarrow H_2 + CO_2$	-3.6	> 50	13	< 11.7
$\rightarrow H_2O + CO$	6.3	> 30	12	< 10.3
$\rightarrow H\overset{O}{\overset{\parallel}{C}}O + H$	106.6	107	14	10.3
$\rightarrow H + \overset{O}{\overset{\parallel}{C}}OH$	92.6	93	14	11.1
$\rightarrow H\overset{O}{\overset{\parallel}{C}} + OH$	108.9	109	16	12.3

SUMMARY AND CONCLUSIONS

Our knowledge of the kinetics of oxygen radical reactions varies greatly depending on the reaction. In recent years, the very accurate rate constants that have been measured for some atmospherically important reactions have forced a re-evaluation of the simple TST approach to chemical kinetics. In many cases, theory lagged experiment and could only by used 'after the fact' to rationalize and explain experiments. With the growing realization that 'complex formation' and non-fixed transition state effects are the rule rather than the exception, we hope that we may be able to predict future reactions as a real test of our knowledge. In other areas of atmospheric chemistry, it is apparent that estimation and correlation, based on transition state theory, have been extremely useful and continue to offer the only practical method for the prediction of the many rate constants that are unknown.

ACKNOWLEDGEMENT

I would like to thank my colleague David M. Golden for his assistance, and the insights he has given me concerning transition state theory and baseball.

REFERENCES

1. Benson, S. W., "Thermochemical Kinetics," 2nd Ed., John Wiley and Sons, Inc., New York (1976).
2. Robinson, P. J. and Holbrook, K. A., "Unimolecular Reactions," John Wiley and Sons, Inc., London (1972).

3. Baldwin, A. C. and Golden, D. M., J. Phys. Chem., 82, 644 (1978).
4. Golden, D. M., J. Phys. Chem., 83, 108 (1979).
5. Perry, R. A., Atkinson, R. and Pitts, Jr., J. N., J. Chem. Phys., 64, 5314 (1976).
6. Atkinson, R. and Pitts, Jr., J. N., J. Chem. Phys., 63, 3591 (1975).
7. Atkinson, R., Darnall, K. R., Lloyd, A. C., Winer, A. M. and Pitts, Jr., J. N., Adv. Photochem., 11, 375 (1979).
8. Chang, J. S. and Barker, J. R., J. Phys. Chem., 83,3059 (1979).
9. Howard, C. J., J. Chem. Phys., 65, 4771 (1976).
10. Atkinson, R., Perry, R. A. and Pitts, Jr., J. N., J. Chem. Phys., 67, 3170 (1977).
11. (a) Smith, G. P. and Golden, D. M., Int. J. Chem. Kinetics, 10, 489 (1978). (b) Baldwin, A. C. and Golden, D. M., J. Phys. Chem., 82, 644 (1978).
12. Howard, C. J., J. Chem. Phys., 71, 2352 (1979); J. Amer. Chem. Soc. In Press.
13. Hamilton, E. J. and Lii, R. R., Int. J. Chem. Kinetics, 9, 875 (1977); DeMore, W. B., J. Phys. Chem., 83, 1113 (1979); Cox, R. A. and Burrows, J. P., J. Phys. Chem., 83, 2560 (1979); Thrush, B. A. and Wilkinson, J. P. T., Chem. Phys. Lett., 66, 441 (1979).
14. Howard, C. J., Private communication. See Reference 4.
15. Stimpfle, R. M., Perry, R. A. and Howard, C. J., J. Chem. Phys., 71, 5183 (1979).
16. Adachi, H. and Basco, N., Chem. Phys. Lett., 63, 490 (1979).
17. Plumb, I. C., Ryan, K. R., Steven, J. R. and Mulcahy, M. F. R., Chem. Phys. Lett., 63, 255 (1979).
18. Adachi, H. and Basco, N., Int. J. Chem. Kinetics, 12, 1 (1980).
19. Cox, R. A. and Roffey, M. J., Environ. Sci. Technol., 11, 8900 (1977).
20. Hendry, D. G. and Kenley, R. A., J. Amer. Chem. Soc., 99, 3197 (1977).
21. Leggett, C. and Thynne, J. C. J., J. Chem. Soc., A, 1188 (1970).
22. Ferguson, J. M. and Phillips, L., J. Chem. Soc., 4416 (1965).
23. Cox, D. L., Livermore, R. A. and Phillips, L., J. Chem. Soc., B, 245 (1966).
24. East, R. L and Phillips, L., J. Chem. Soc., A, 1939 (1967).
25. Batt, L. and McCulloch, R. D., Int. J. Chem. Kinetics, 8, 911 (1976).
26. Cadman, P., Trotman-Dickenson, A. F. and White, A. J., J. Chem. Soc., A, 2296 (1971).
27. Bire, F. W., Danby, C. J. and Hinshelwood, C. M., Proc. R. Soc., London, Ser. A, 239, 154 (1957).
28. Quee, N. J. and Thynne, J. C. J., Trans. Faraday Soc., 63, 2970 (1967).

29. McMillan, G. R., J. Amer. Chem. Soc., 82, 2422 (1960).
30. Baldwin, A. C., Barker, J. R., Golden, D. M. and Hendry, D. G., J. Phys. Chem., 81, 2483 (1977).
31. Barker, J. R., Benson, S. W. and Golden, D. M., Int. J. Chem. Kinetics, 9, 31 (1977).
32. Carter, W. P. L., Darnall, K. R., Lloyd, A. C., Winer, A. M. and Pitts, Jr., J. N., Chem. Phys. Lett., 42, 22 (1976).
33. Baldwin, A. C. and Golden, D. M., Chem. Phys. Lett., 60, 108 (1978).
34. Darnall, K. R., Atkinson, R. and Pitts, Jr., J. N., J. Phys. Chem., 83, 1929 (1979); Atkinson, R., Carter, W. P. L., Darnall, K. R., Winer, A. M. and Pitts, Jr., J. N., Int. J. Chem. Kinetics, In press.
35. Herron, J. T. and Huie, R. E., J. Amer. Chem. Soc., 99, 5430 (1977).

SMOG CHEMISTRY AND URBAN AIRSHED MODELING

Roger Atkinson and Alan C. Lloyd, Environmental Research and Technology, Inc., 2625 Townsgate Road, Suite 360, Westlake Village, California 91361

INTRODUCTION

With the countinuing increase in the geographical areas of the world impacted by air pollution, further demands have been placed on control officials and atmospheric chemists to understand the interaction of pollutants in the atmosphere, and the most effective means for their control (1,2). Early attempts at controlling ozone (O_3) or photochemical oxidant were through the use of a linear rollback of hydrocarbon emissions (3), but it was soon realized that this was a technically unacceptable method of approach since, for instance, it neglected the role of nitrogen oxides emissions (4). Studies were then initiated, using smog chambers under controlled conditions, to obtain data on the ozone forming potential of irradiated NO_x-organic-air mixtures (5-7), and these and later experimental investigations have led to the progressive development of detailed chemical models (8-15).

Unfortunately, the chemistry in these models is much too extensive to be incorporated into urban airshed models, which must incorporate NO_x, SO_x, hydrocarbon and carbon monoxide emissions, meteorology, and transport, and, because of significant computer constraints, it is hence feasible to include only a limited chemistry package (16,17). Consequently, "lumped" chemistry approaches were developed in which certain hydrocarbon classes and species were lumped together and their reactions generalized. Examples of early chemical lumped parameter models were those developed by Hecht, Seinfeld and Dodge (18) for incorporation into the urban airshed model of Reynolds et al (19), and the model of Eschenroeder and Martinez (20).

There have been siginificant improvements in our knowledge of the chemistry of polluted atmospheres during the last five years, and these changes have been incorporated into some of the more recent chemical models developed both for specific hydrocarbons and also for urban airshed models. However, this is obviously a continuing process and there is still a need for a better understanding of the chemistry involved in order to address quantitatively such issues as the control of urban and regional ozone, control of sulfates and nitrates, the role of SO_2 and NO_x emissions in acid rain formation, and the mechanism of formation of organic aerosols and their relationship to visibility reduction. In addition, there is continued speculation about the role

ISBN 0-12-592050-4

of natural hydrocarbon emissions (21-26).

This paper will focus on the current status of our knowledge of atmospheric chemistry and on the areas of our present uncertainties and of major advances. The application of this chemical knowledge to urban airshed models and the type of agreement that can be expected from airshed models and ambient air data will be addressed. We shall discuss the current and future improvements in our understanding of atmospheric chemistry, and important areas for future work will be identified.

ATMOSPHERIC CHEMISTRY

The atmospheric chemistry involving hydrocarbon oxidation in the presence of NO_X and SO_X is conveniently summarized in Fig. 1. This shows the conversion of NO· to $NO_2^{\cdot}$ and the formation of O_3. The role of radical species in converting NO· to $NO_2^{\cdot}$ is shown, these radicals being formed by the attack of ozone or hydroxyl radicals on the various classes of hydrocarbons.

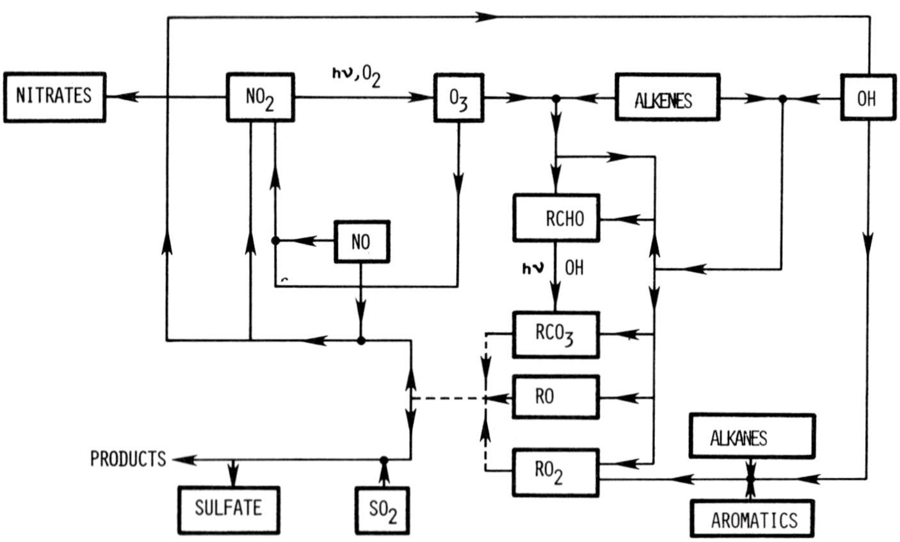

FIGURE 1. Schematic representation of current knowledge of the overall feature of gas-phase photochemical smog formation.

The organics present in ambient air can be categorized into three general classes of hydrocarbons, as shown in Table 1 (27): alkanes, such as n-butane, isobutane, and the pentanes; alkenes (ethene, propene, butenes); and aromatics (benzene, toluene, and xylenes). Oxygenates, such as the aldehydes shown in Fig. 1, are another class. These oxygenates are mostly composed of aldehydes,

TABLE 1

Hydrocarbon Composition of Tulsa Area Air [27]a

Hydrocarbons	Concentration (ppbc)	Hydrocarbons	Concentration (ppbc)	Hydrocarbons	Concentration (ppbc)
Alkanes		Alkenes		Aromatics	
ethane	7.2	ethene	7.3	toluene	14.8
propane	8.9	propene	2.8	ethylbenzene	2.5
isobutane	12.9	isobutene	4.3	p-xylene	2.0
n-butane	48.7	trans-2-butene	4.7	m-xylene	5.4
isopentane	65.9	cis-2-butene+butadiene	0	o-xylene	2.5
n-pentane	40.9	1-pentene	2.2	isopropylbenzene	0
cyclopentane	4.6	2-methyl-1-butene	3.1	n-propylbenzene	0.9
2-methylpentane	19.6	trans-2-pentene	7.3	m+p-ethyltoluene	0
3-methylpentane	13.0	cis-2-pentene	5.6	1,3,5-trimethylbenzene	1.0
n-hexane	14.0	2-methyl-2-butene	6.0	o-ethyltoluene	0
n-heptane	5.8	isoprene	0.3	1,2,4-trimethylbenzene	1.3
nonane	1.3	4-methyl-2-pentene	0.8	m-ethyltoluene	1.3
decane	1.8	trans-2-hexene	1.3		
Σ ALKANES	230.6	Σ ALKENES	45.8	Σ AROMATICS	31.7

a) 10:33-11:03 A.M., Tulsa Post Office, 7-27-1978.

but smaller amounts of ketones, alcohols, ethers, and esters are also present. Of these, only the aldehydes and ketones are photoactive, but the other species can also react to form radicals.

In order to most logically discuss our present knowledge of the gas phase chemistry of polluted atmospheres, it is best done by consideration of each of these organic classes, together with the inorganic reaction mechanisms.

Our knowledge of the gas phase chemistry occurring in photochemical air pollution systems has advanced rapidly in the last five years or so (2,8-15,28-34), and in general we now have, in most cases, a qualitative understanding of the gas phase chemistry involved. From smog chamber irradiations of hydrocarbon-NO_X-air mixtures in the ppm or sub-ppm (part-per-million) concentration regimes, typical time-concentration profiles such as that shown in Fig. 2 (15) are obtained. Hydrocarbon consumption occurs along

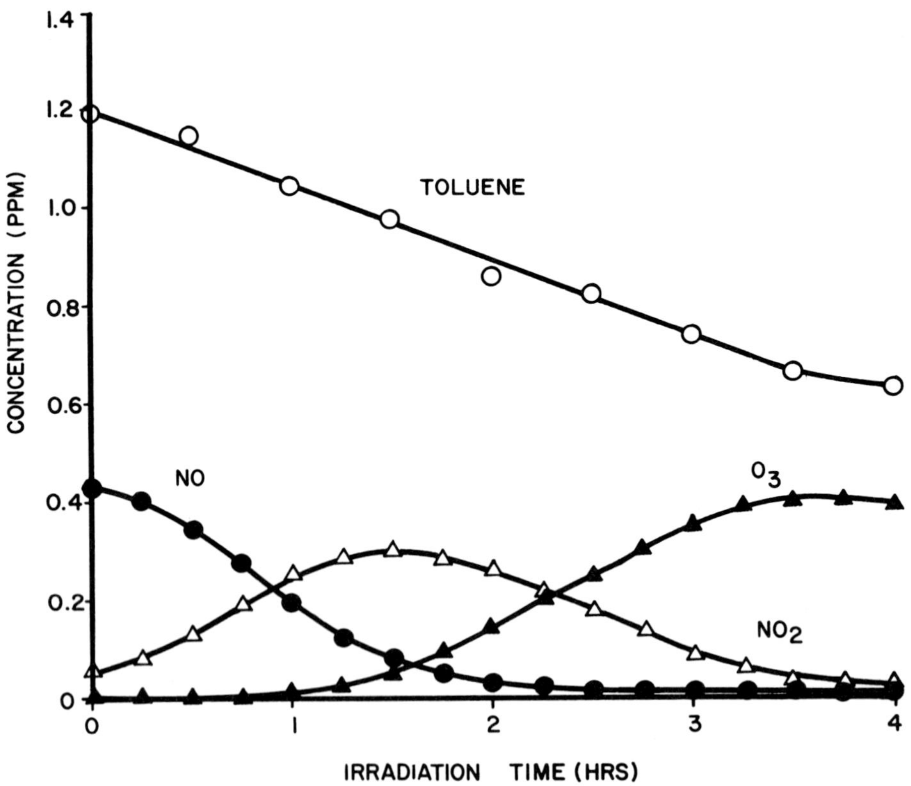

FIGURE 2. Experimental time-concentration profiles for toluene, $NO\cdot$, $NO_2\cdot$ (i.e., NO_X-$NO\cdot$-PAN), and O_3 for an irradiated NO_x-toluene-air mixture (EC-266 [15]).

with NO· to $NO_2^{\cdot}$ conversion, the formation of O_3, intermediate carbonyls, and peroxyacetyl nitrate (PAN). The ultimate goal is to be able, from a set of elementary reactions and their respective rate constants, to computer model these NO_x-photooxidations to within the experimental accuracies, without the need for any adjustable parameters. While chamber dependent effects, such as wall effects manifesting themselves as chamber radical sources, etc., makes this perhaps unrealizable (12), very significant progress has been made in the past decade (8-15).

In the following sections we will discuss briefly the status of our understanding of the gas phase chemistry of the inorganics and of the various classes of organics involved in photochemical air pollution, with attention focusing on the oxy-radicals.

a) Inorganic Reactions

Probably largely because of the impetus provided by the CIAP program regarding depletion of stratospheric O_3 by the supersonic transports, and later by similar considerations in the case of the halocarbons, (30,31) the inorganic reaction mechanisms applicable to the lower troposphere and air pollution chemistry now appear to be reasonably well understood (15,30,31,34); the basic reactions being:

$$NO_2^{\cdot} + h\nu \rightarrow NO\cdot + O(^3P)$$

$$O(^3P) + O_2 + M \rightarrow O_3 + M$$

$$NO\cdot + O_3 \rightarrow NO_2^{\cdot} + O_2$$

followed by the important tropospheric ·OH radical production steps

$$O_3 + h\nu\ (\lambda \lesssim 310\ nm) \rightarrow O('D) + O_2\ ('\Delta_g)$$

$$O('D) + M \rightarrow O(^3P) + M\ (M{=}air)$$

$$O('D) + H_2O \rightarrow 2\ \cdot OH$$

together with the reaction of O_3 with $NO_2^{\cdot}$ leading to NO_3 and N_2O_5 chemistry.

Reaction of hydroxyl radicals with CO:

$$\cdot OH + CO \xrightarrow{O_2} HO_2^{\cdot} + CO_2$$

$$HO_2^{\cdot} + NO\cdot \rightarrow \cdot OH + NO_2^{\cdot}$$

leads to regeneration of ·OH, and this chain oxidation of CO was first postulated around 1969 (35,36). The inorganic chemistry "package" is then rounded out by reactions of ·OH and $HO_2^{\cdot}$ with O_3

and by the termination reactions of $\cdot OH$ with $NO\cdot$ and $NO_2^{\bullet}$ and of $HO_2^{\bullet}$ with $NO_2^{\bullet}$:

$$\cdot OH + NO\cdot \xrightarrow{M} HONO \text{ (nitrous acid)}$$

$$\cdot OH + NO_2^{\bullet} \xrightarrow{M} HNO_3 \text{ (nitric acid)}$$

$$HO_2^{\bullet} + NO_2^{\bullet} \xrightarrow{M} HO_2NO_2 \text{ (pernitric acid)}$$

Because of the efficient regeneration of $\cdot OH + NO\cdot$ from HONO by photolysis and of $HO_2^{\bullet} + NO_2^{\bullet}$ from HO_2NO_2 by thermal decomposition, under ambient lower tropospheric conditions the $\cdot OH + NO_2^{\bullet}$ termination reaction is the most important of these three.

However, we cannot afford to be too complacent, as evidenced by the major "glitches" during the past few years, for example, the discovery of pernitric acid, HO_2NO_2 (37), formed from the reaction of $HO_2^{\bullet}$ with $NO_2^{\bullet}$ (38-40), and its thermal decomposition kinetics (41,42), the determination of the ∿8-40 times faster rate constant for $HO_2^{\bullet} + NO\cdot$ than previously believed (43), and the observation of a pressure dependent rate constant for the reaction of $\cdot OH$ radicals with CO (at least in the presence of O_2) (44-47). It is thus unwise to expect that similar large changes in rate constants and/or elucidation of new reaction pathways will not occur in the future - this is also, of course, particularly true of the organic mechanisms and rate constants discussed in some detail below.

b) Alkanes

It has been established from basic laboratory investigations, supported by smog chamber modeling studies (10-14), that under ambient atmospheric conditions the sole chemical loss process of the alkanes is via reaction with the OH radical:

$$\cdot OH + RCH_2R_1 \rightarrow H_2O + R\dot{C}HR_1$$

followed by, for the simple ($\lesssim C_3$) alkanes:

$$R\dot{C}HR_1 + O_2 \rightarrow RR_1CHOO\cdot$$

$$RR_1CHOO\cdot + NO\cdot \rightarrow RR_1CHO\cdot + NO_2^{\bullet}$$

$$RR_1CHOO\cdot + NO_2^{\bullet} \rightleftarrows RR_1CHO_2NO_2$$

$$RR_1CHO\cdot + O_2 \rightarrow RCOR_1 + HO_2^{\bullet} \quad \text{(reaction with } O_2\text{)}$$

$$RR_1CHO\cdot \rightarrow \left\{ \begin{array}{l} RCHO + R_1^{\bullet} \\ R_1CHO + R\cdot \end{array} \right\} \quad \text{(decomposition)}$$

$$RR_1CHO\cdot + NO_2^{\cdot} \rightarrow RR_1CHONO_2$$

followed by the regeneration of the chain carrier, the $\cdot OH$ radical:

$$HO_2^{\cdot} + NO\cdot \rightarrow \cdot OH + NO_2^{\cdot}$$

The formation of alkylperoxynitrates appears to be of minimal importance due to their short ($\lesssim$ 1 sec at 298°K) lifetime with respect to thermal decomposition back to reactants (12,48, 49). Hence, this scheme leads to the ultimate degradation of the original carbon skeleton via the intermediate formation of carbonyl compounds, these carbonyls then reacting with OH radicals or photodecomposing further. The key intermediate organic radicals in these NO_x photooxidations are then alkyl peroxy (RO_2) and alkoxy (RO) radicals -- hence, their reactions under ambient atmospheric conditions are of prime importance.

Vast advances in our understanding of the reactions of these relatively simple (up to and including C_4) alkylperoxy and alkoxy radicals have taken place during the past few years, especially due to thermochemical estimation techniques by Golden and coworkers (10,29,50), experimental investigations of, for instance, Batt and coworkers (51-53), and smog chamber modeling studies of, for instance, Carter _et al_. (12,54).

While the above reaction scheme is applicable to the smaller alkanes, i.e., methane, ethane and propane, two other processes involving $RO_2\cdot$ and $RO\cdot$ reactions can occur for the larger alkane system:

1) <u>alkyl nitrate formation from $RO_2^{\cdot} + NO\cdot$</u>

This reaction pathway:

$$RO_2^{\cdot} + NO\cdot \rightarrow RONO_2$$

was postulated by Darnell _et al_. (55) from smog chamber product studies of the NO_x photooxidation of n-butane, n-pentane and n-hexane. It was proposed that the reaction of $RO_2^{\cdot}$ radicals with $NO\cdot$ proceeds via

$$RO_2^{\cdot} + NO\cdot \rightarrow RO\cdot + NO_2^{\cdot}$$

$$RO_2^{\cdot} + NO\cdot \rightarrow RO_2NO^*$$

$$RO_2NO^* \rightleftarrows RONO_2^*$$

TABLE 2

Rate Constant Ratios $k_a/(k_a + k_b)$ Derived for the Reaction of RO_2 Radicals with NO

Alkane, RH	$k_a/(k_a + k_b)$
propane	0.04[a]
n-butane	0.083 ± 0.02[b]; 0.086[a]
n-pentane	0.14 ± 0.05[b]; 0.11[c]
n-hexane	0.37 ± 0.08[b]; 0.15 ± 0.03[c]
2,3-dimethylbutane	∿0.1[c]

a Reference 12.
b Reference 55.
c Reference 56.

$$RONO_2{}^* \rightarrow RO\cdot + NO_2^{\bullet} \quad \text{decomposition}$$

$$RONO_2{}^* \overset{M}{\rightarrow} RONO_2 \quad \text{stabilization}$$

with the ratios $k_a/(k_a + k_b)$ derived being given in Table 2; where k_a and k_b are the rate constants for the overall reactions (a) and (b)

$$RO_2^{\bullet} + NO\cdot \rightarrow RONO_2 \qquad (a)$$

$$\rightarrow RO\cdot + NO_2^{\bullet} \quad (b)$$

As expected from unimolecular theory, the stabilization/decomposition ratio k_a/k_b increases with the number of degrees of freedom of the RO radical, and this ratio is (56) less for branched than unbranched alkylperoxy radicals.

Since this reaction pathway forming alkylnitrates from $RO_2^{\bullet}$ + NO· is both a substantial sink for NO_x and a radical termination process, it is obvious that further experimental work is needed to confirm this route and to provide independent quantitative data for the fraction of the overall reaction proceeding via this pathway.

2) Alkoxy radical isomerization

Isomerization of alkoxy radicals, known from cool-flames and hydrocarbon oxidations at elevated temperatures, was first proposed, on the basis of smog chamber product studies, as being of importance in room temperature NO_X-photooxidations by Carter et al. (54), and subsequently by Baldwin et al. (29).

Thermochemical calculations by both groups (29,54,57) have shown that under ambient atmospheric conditions only the 1.4- and 1.5-H shifts of alkoxy radicals (via 5- and 6- membered ring transition states) are of importance (alkyl and alkyl peroxy radicals react with O_2 and NO·, respectively, much faster that the corresponding isomerizations). Thus, the simplest alkoxy radical isomerization is that of the n-butoxy radical:

$$CH_3CH_2CH_2CH_2O\cdot \rightarrow \cdot CH_2CH_2CH_2CH_2OH$$

For $\geq C_4$ alkanes, this then means that the alkoxy radicals formed during their NO_X photooxidations can react via three routes (neglecting combination with $NO_2^{\cdot}$ to form alkyl nitrates): reaction with O_2, unimolecular decomposition or isomerization. For the simple ($\leq C_4$) alkoxy radicals, rate constants for reaction with O_2 and for unimolecular decomposition are becoming available from experimental studies (51-53), and thermochemical estimation procedures are also available, especially for the umimolecular decompositions and isomerizations (29,50-52,57). These latter estimation procedures have been discussed in detail by Golden and coworkers (29,50), by Carter et al. (54,57), and by Batt (51,52), and will not be dealt with in detail here. As an indication of the relative importance of the various alkoxy radical reaction routes under atmospheric conditions, Table 3 summarizes the experimental and/or thermochemically estimated rate data for the simpler alkoxy radicals.

For n-butane, the isomerization pathway only affects ∿15% of the total reaction pathway with the ·OH radical, since the initial radicals are ∿15% n-butyl and ∿85% sec-butyl (12,32, 34); while for the higher alkanes such as n-pentane, of the total number of alkoxy radicals formed, ∿70% can undergo isomerization (32,54), and alkoxy radical isomerization probably dominates for the $\geq C_5$ straight chain alkanes (54). The reactions subsequent to the first isomerization have been a matter of conjecture until fairly recently - since in theory the hydroxy alkoxy radical formed subsequent to the first isomerization could isomerize further, as in Reaction Scheme 1. However, recent experimental data of Niki et al. (59) and of Carter et al. (60) have shown that the α-hydroxy radicals, $RR_1\dot{C}OH$, react with O_2 to form the corresponding carbonyl and $HO_2^{\cdot}$:

$$RR_1\dot{C}OH + O_2 \rightarrow RCOR_1 + HO_2^{\cdot}$$

and so terminate the isomerization sequence after a total of two isomerizations. This is illustrated in Reaction Scheme 2, for the 2-pentoxy radical which forms a δ-hydroxy carbonyl, (competing reactions, i.e., decomposition and reaction with

TABLE 3

Reaction Rates for Selected Alkoxy Radicals at 298°K and 760 Torr of Air

Alkoxy Radical	Reaction Rate, sec^{-1}: Reaction with O_2 [a]	Decomposition [b]	Isomerization [c]
$CH_3O\cdot$	5×10^3	Negligible	No isomerization possible
$CH_3CH_2O\cdot$	1.1×10^4	2×10^{-5}	Negligible
$CH_3CH(O\cdot)CH_3$	1.1×10^4	20	Negligible
$CH_3CH_2CH_2CH_2O\cdot$	1.1×10^4	0.3	$6 \times 10^4 - 6 \times 10^5$
$CH_3CH_2CH(O\cdot)CH_3$	1.1×10^4	5×10^3	120 - 450
$(CH_3)_3CO\cdot$	No Reaction Possible	2×10^3	Negligible

a derived from references 51-53 and 58. $k(CH_3O\cdot + O_2)$ taken to be 1×10^{-15} cm^3 $molecule^{-1}$ sec^{-1} at 298°K[34]; for other alkoxy radicals, the expression given by Batt [52] for ethoxy and isopropoxy radicals ($k = 1.7 \times 10^{-12}$ $e^{-2000/T}$ cm^3 $molecule^{-1}$ sec^{-1}) has been assumed.

b References 29, 51, 52.

c References 29, 57.

O_2, of these alkoxy radicals are expected to be minor; reaction with $NO_2^{\cdot}$ to form $RONO_2$ can also occur, of course).

Reaction Scheme 1.

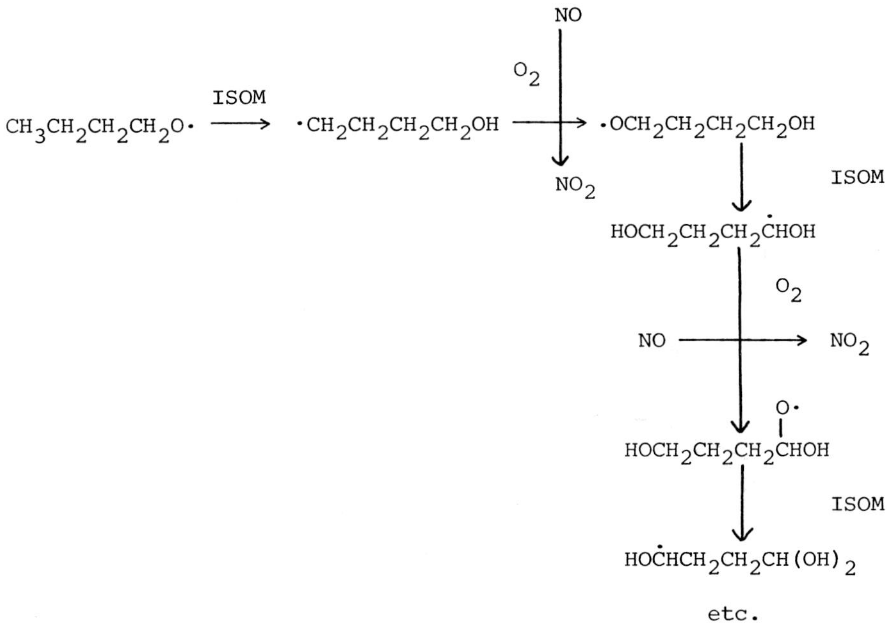

Reaction Scheme 2.

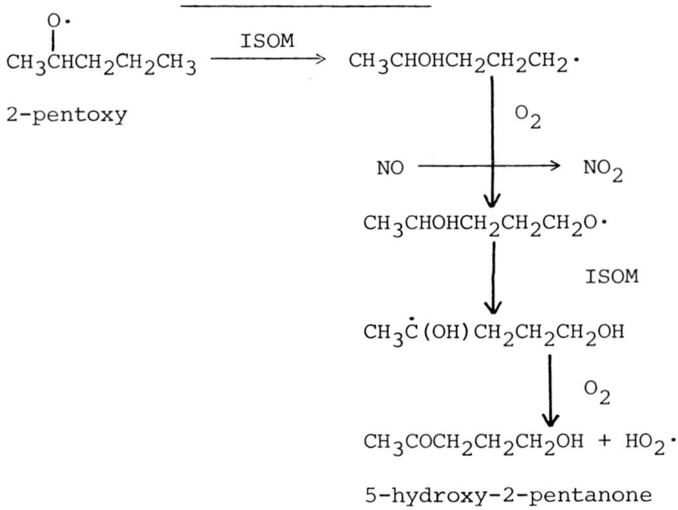

Since these species retain the carbon skeleton, then, at least for the δ-hydroxyketones, further reaction with the ·OH radical will lead to progressively more polyfunctionality and may ultimately lead to aerosol formation.

c) Alkenes

Under ambient atmospheric conditions the alkenes react predominantly with ozone and with ·OH radicals (32). The ·OH radical rate constants are now known reasonably well, to within, for most alkenes, ± 20-25% (32,33,61-66), and the initial reaction pathway for ethene and the methyl-substituted ethenes proceeds essentially totally via ·OH radical addition to the double bond (32,33). However, for alkenes containing longer ($\geq C_2$) side chains, a substantial amount (20 ± 5% for the case of 1-butene (67)) of the overall ·OH radical reaction proceeds via H atom abstraction from the weak allylic C-H bonds (32,66,67):

$$\cdot OH + CH_2{=}CHCH_2CH_3 \rightarrow HOCH_2\dot{C}HCH_2CH_3 + \dot{C}H_2CHOHCH_2CH_3 \quad (a)$$

$$\rightarrow H_2O + CH_2{=}CH\dot{C}HCH_3 \quad (b)$$

with $k_b/(k_a + k_b) = 0.20 \pm 0.05$ at room temperature (67). The major uncertainty, or, more accurately, conflict, is between theoretical calculations (29,50-52) and experimental data (12,59) concerning the fate of the β-hydroxyalkoxy radicals formed from the ·OH-alkene adducts:

$$HOCH_2\dot{C}HCH_3 + O_2 \rightarrow HOCH_2CH(OO\cdot)CH_3$$

$$HOCH_2CH(OO\cdot)CH_3 \xrightarrow{NO\cdot \rightarrow NO_2\cdot} HOCH_2CH(O\cdot)CH_3$$

As with the simple alkoxy radicals, these β-hydroxyalkoxy radicals can react with O_2 or unimolecularly decompose:

$$CH_3CH(O\cdot)CH_2OH + O_2 \rightarrow HO_2\cdot + CH_3COCH_2OH$$

$$CH_3CH(O\cdot)CH_2OH \rightarrow CH_3CHO + \cdot CH_2OH$$

$$\cdot CH_2OH \xrightarrow{O_2} HCHO + HO_2\cdot$$

Theoretical estimates of Baldwin et al. (29), Golden (50), and Batt (51,52) indicate that reaction of O_2, at least for $\leq C_4$ β-hydroxyalkoxy radicals, should predominate (Table 4), while the experimental data, from smog chamber studies (12) and from irradiated HONO-alkene-NO-air mixtures (59), shows conclusively that decomposition of these β-hydroxyalkoxy radicals predominates.

Thus, although experimentally the situation appears clear i.e., β-hydroxyalkoxy radicals decompose rather than react with O_2 under atmospheric conditions, the discrepancy with thermochemical calculations is unsettling, since this thermochemical estimation approach is an invaluable tool in predicting both order of magnitude rate constants and reaction pathways etc. (50). It must be pointed out that this thermochemical approach works very well for the alkoxy and alkylperoxy radical reactions (34), and the β-hydroxyalkoxy radicals do appear to be an exception - the effect of the ·OH substituent may possibly cause more of a perturbation than expected.

Two further reactions of interest concerns:

1) the possibility of isomerization of these β-hydroxyalkoxy radicals (32), the simplest case being the 2-hydroxy-1-butoxy radical derived from internal ·OH radical addition to 1-butene;

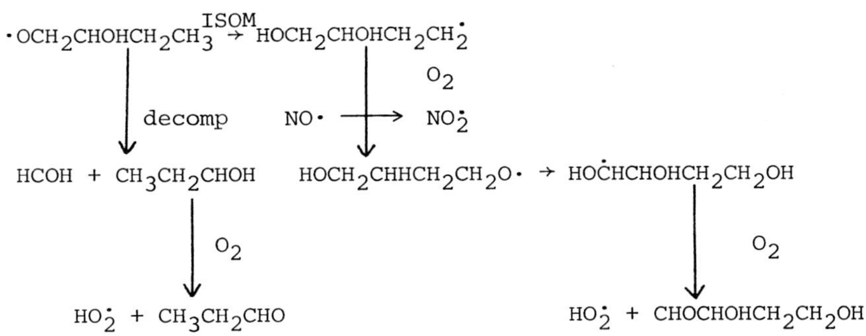

whether these isomerizations in general can compete with the decompositions is not known at present, but appears unlikely (34).

2) the reactions subsequent to H atom abstraction from alkenes. Thus, taking 1-butene as an example, ∿20% of the OH radical reaction yields the allylic radical $CH_2{=}CH\dot{C}HCH_3$ (67), which can presumeably add O_2 to form the peroxy radical

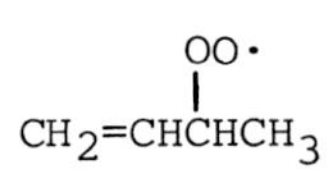

TABLE 4

Estimated Reaction Rates for Selected β-Hydroxy Alkoxy Radicals at 298°K and Atmospheric Pressure

	Reaction Rate sec^{-1}		
Radical	*Reaction with* O_2[a]	*Decomposition*	*Isomerization*[b]
$HOCH_2CH_2O\cdot$	1.1×10^4	<6[c]; 0.2[d]	Negligible
$CH_3CHOHCH_2O\cdot$	1.1×10^4	<2[e]	120-450
$CH_3CH(O\cdot)CH_2OH$	1.1×10^4	$<2 \times 10^3$ [e]	Negligible
$CH_3CH(O\cdot)CHOHCH_3$	1.1×10^4	$\sim 2 \times 10^2$ [c]	120-450

a Assumed to be identical to the reaction rates of O_2 with $C_2H_5O\cdot$ and $(CH_3)_2CHO\cdot$ radicals (52).
b References 29,57.
c Reference 51.
d Reference 50.
e Reference 29.

This radical can, besides reacting with NO· to yield (mainly) the corresponding alkoxy radical, cyclize:

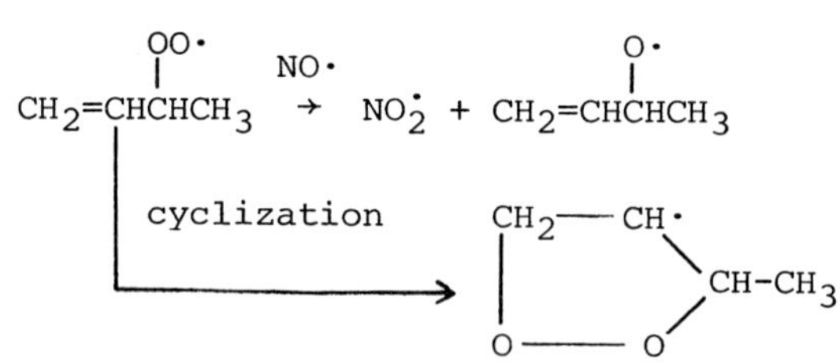

Despite erlier estimates (9,15) that cyclization should dominate under atmospheric conditions, recent experimental data on the NO_x photooxidation of isoprene (68), where methylvinylketone and methacrolein are produced in the early stages, i.e., before significant amounts of O_3 are formed, indicates that, at least for these particular radicals [CH_2=CRC(OO·)R_1CH_2OH, where R, R_1=H and CH_3 or CH_3 and H, formed after reaction of ·OH radicals with isoprene], the reaction with NO· predominates (15).

For the reactions of O_3 with the alkenes, despite well over a decade of investigation, there are still significant areas of uncertainty (33). Thus, while the rate constants for the initial reaction of O_3 with the 1-alkenes seem to be reasonably well known (33), larger uncertainties exist for the internal alkenes - for instance the rate constants at 298°K for trans-2-butene from the two latest studies are 1.76×10^{-15} (69) and 2.60×10^{-15} (70) cm^3 $molecule^{-1}$ sec^{-1}. This topic has been discussed recently (33) and will not be dealt with here.

Recent work by Niki and coworkers (71), Calvert and coworkers (72), and by Herron and Huie (73,74) has shown that the O_3 - alkene reactions proceed mainly by the Criegee route:

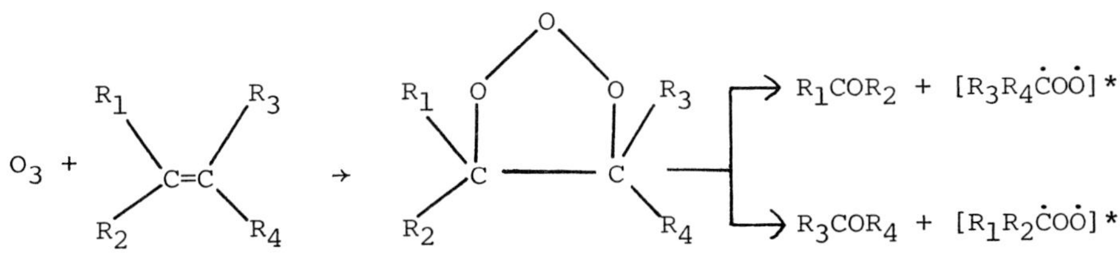

where the symbol [ ]* denotes an initially energy rich species.

The observation of significant yields of secondary ozonide formation (71), including secondary ozonide formation from the ozone-ethene reaction (72), shows that a significant fraction (~20-40%) of the initially energy rich biradicals are thermalized under atmospheric conditions, with rearrangement and/or decomposition accounting for the other reaction pathways of the biradicals. Thus, for example, some of the recently proposed reactions (75) of the $CH_3\dot{C}HOO\cdot$ biradical are shown below:

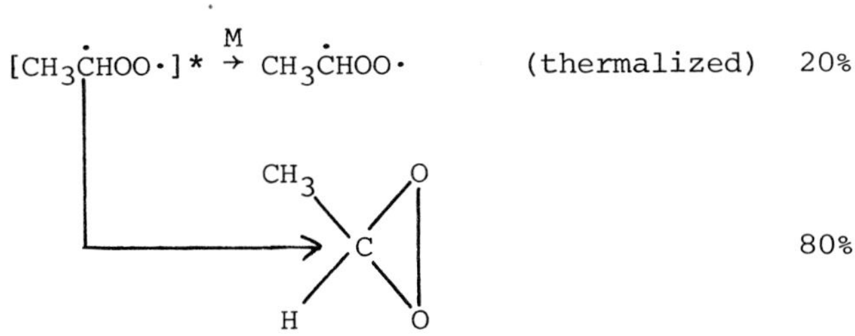

followed by Reaction Scheme 3.

The reactions of the thermalized biradicals ($\cdot CH_2OO\cdot$, $CH_3\dot{C}HOO\cdot$ etc.) need further investigation. They have been shown to react with aldehydes (71,72) (to form secondary ozonides), with SO_2 (71,72,76) (presumeably to form SO_3 and the corresponding aldehyde), and with H_2O (76) as shown in Reaction Scheme 4.

Reaction Scheme 3.

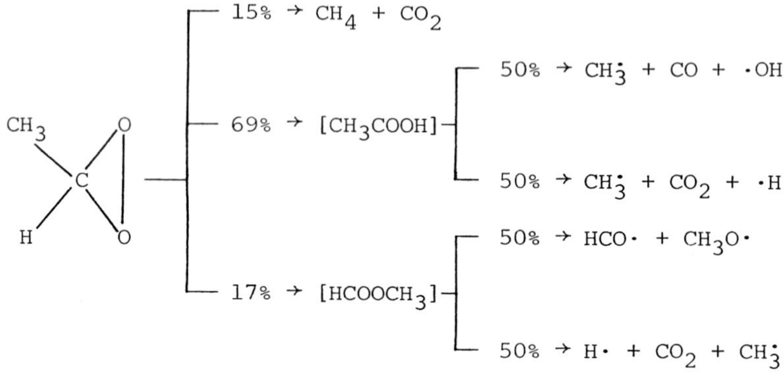

Reaction Scheme 4.

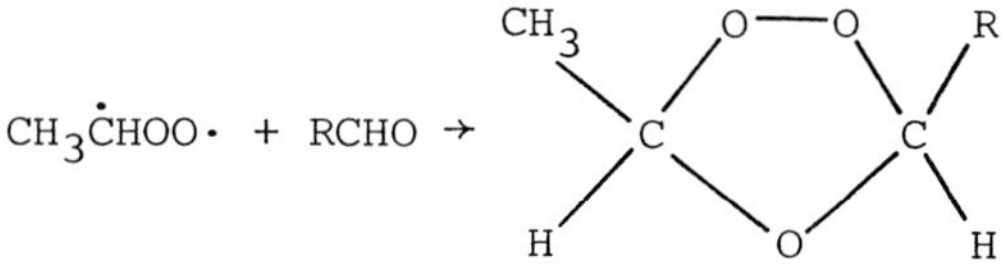

$$CH_3\dot{C}HOO\cdot + SO_2 \rightarrow CH_3CHO(?) + SO_3$$

$$CH_3\dot{C}HOO\cdot + H_2O \rightarrow\rightarrow CH_3COOH + H_2O$$

Further possible important reactions with NO· and $NO_2^{\cdot}$ need to be studied (12,34,75,76):

$$CH_3\dot{C}HOO\cdot + \left\{\begin{matrix} NO\cdot \\ \\ NO_2^{\cdot} \end{matrix}\right\} \rightarrow CH_3CHO + \left\{\begin{matrix} NO_2^{\cdot} \\ \\ NO_3 \end{matrix}\right\}$$

Thus, it is obvious that much work, aimed now more at the finer details of these systems, remains to be carried out. We have a reasonably good qualitative, and in some cases semi-quantitative, understanding of the O_3 - alkene systems in the absence of NO_x, but the finer points regarding the precise splits between different mechanistic pathways, and the relevant rate constant data for these pathways, need to be known (34).

d) Aromatics

Significant advances have been made in recent years as to our understanding of the atmospheric chemistry of the aromatics. It has been known since the early 1970's that the only important chemical loss process of the aromatic hydrocarbons under atmospheric conditions is via reaction with the ·OH radical, and data are now available concerning both the overall ·OH radical rate constants (32,62,77), and the relative amounts of the two reaction pathways (32,77-79); ·OH radical addition to the ring, and H atom abstraction, mainly from the substituent groups:

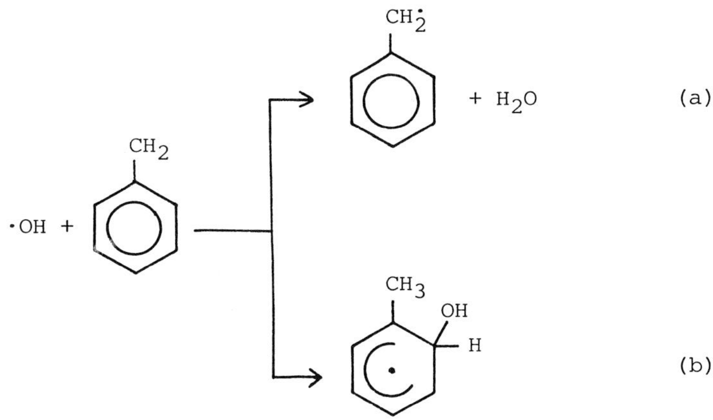

(A)

with, for toluene, $k_a/(k_a + k_b) \simeq 0.15$ (32,77-79).

Under atmospheric conditions the benzyl radical will react to form benzaldehyde and benzyl nitrate - these reaction pathways have been discussed in detail previously (15) and are analogous to the relatively well known alkyl radical reactions.

The uncertainties in the aromatic NO_x-photooxidation mechanisms lie in the fate of the addition adducts (A) - both experimentally and theoretically the situation is unclear, but some data are available which allow a reasonable mechanism to be postulated (15) (it sould be emphasized that much experimental data is needed before this suggested mechanism can be regarded as being confirmed).

The ·OH - aromatic adducts can react with O_2 via two pathways; H atom abstraction to form cresols; or reversible O_2 addition to the adduct (15,32) as in Reaction Scheme 5. At low total pressures and low NO_x concentrations, cresols and benzaldehyde are essentially the sole products (78,79), with ∿85% cresol and ∿15% benzaldehyde (78,79), while at atmospheric pressure, cresol yields of $\lesssim$20% have been derived from smog chamber studies (15, 80), these data then being consistent with a pressure dependent O_2 addition pathway.

Reaction Scheme 5.

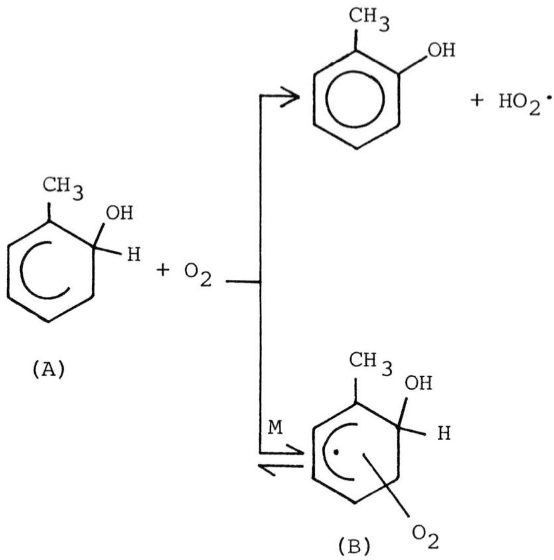

The structures and atmospheric fates of species such as (B) are not known at present; however, certain conclusions can be drawn from recent smog chamber studies. Thus, the prompt formation of biacetyl and other α-dicarbonyls in the NO_X photooxidation of o-xylene (81,82) eliminates the reaction sequence, (and others analogous to it):

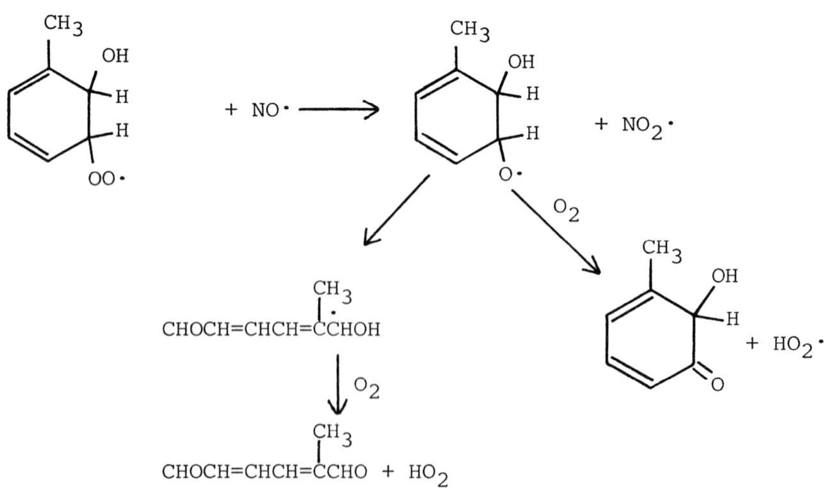

and it appears (15), both to fit the experimental data and from a thermochemical point of view, that the only plausible route is the formation of bicyclic radicals of the form

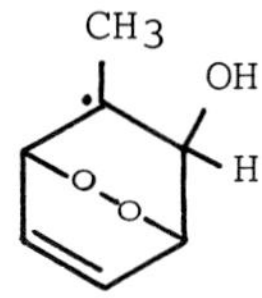

(C)

These bicyclic radicals can then add O_2 and react with NO· similarly to alkyl radicals:

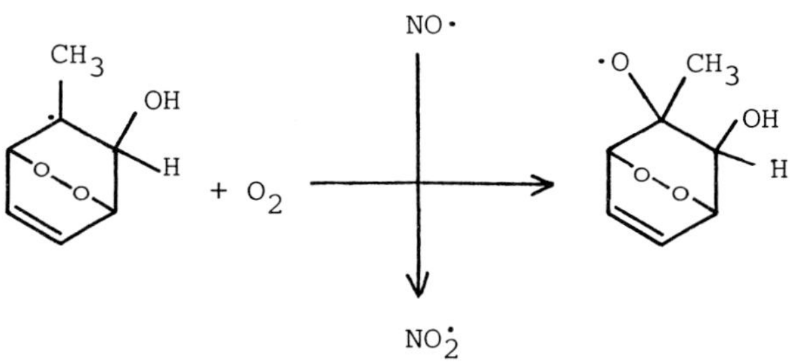

followed by spontaneous decomposition of these bicyclic oxy radicals:

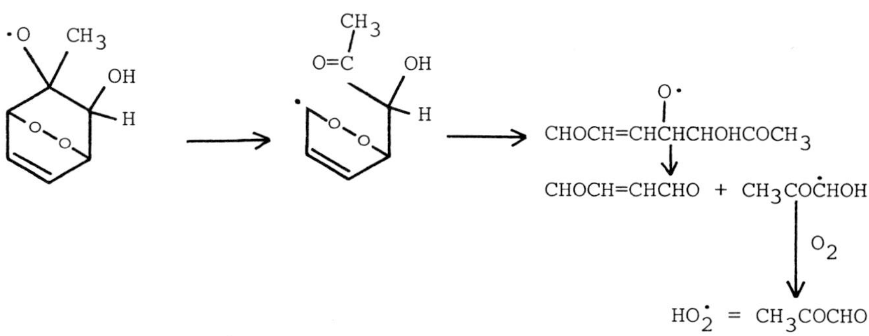

followed by the subsequent reactions (15) of the α-dicarbonyls [glyoxal, methylglyoxal and biacetyl, depending on the parent aromatic hydrocarbon and the precise structure of the bicyclic intermediates (C)], and γ-unsaturated dicarbonyls (CHOCH=CHCHO and CH_3COCH=CHCHO, for example).

Extension of this plausible mechanism to the cresols yields an explanation for their low reactivity, in terms of NO· to $NO_2^{\cdot}$ conversion and O_3-forming potential (15), in that, instead of the formation of the α-dicarbonyls methylglyoxal and biacetyl, which are strong photoinitiators, the cresols should instead lead to the formation of carbonyl acids, such as $CHOCO_2H$ and CH_3COCO_2H, which are expected to be much less photoreactive (15).

In this regard, since glyoxal apparently does not photodissociate rapidly, if at all, in the actinic region to form radicals (but rather photodissociates to yield HCHO + CO (83)) then benzene should also be less reactive, as regards NO· + $NO_2^{\cdot}$ conversion and O_3-forming potential, than may be expected solely on the basis of its ·OH radical rate constant.

Obviously, for the aromatics, though much progress has been made in recent years (15,32,33,79), much more experimental data is needed before we are at the same stage as we presently are for the alkanes or alkenes.

e) Gas Phase SO_2 Oxidation

This topic has been excellently reviewed by Calvert et al. (76), and more recently by Atkinson and Lloyd (34). The major changes since then have come from $HO_2^{\cdot}$ and $RO_2^{\cdot}$ rate constant data which shows that both of these species, i.e., $HO_2^{\cdot}$ and $RO_2^{\cdot}$, (where R=CH_3 and $(CH_3)_3C$) do not react at any significant rates with SO_2 (84-86), and can hence be neglected for atmospheric purposes. Thus, at present the sole homogeneous gas phase reaction pathways depleting SO_2 appear to be via reaction with OH radicals and the Criegee biradicals (34).

AMBIENT AIRSHED MODELS FOR CONTROL STRATEGY EVALUATION AND IMPLEMENTATION

The Clean Air Act and it's amendments have placed significant demands on computer modeling to relate emissions to ambient air quality. For example, modeling is an integral part of the prevention of significant deterioration (PSD) regulations (87) which relate emissions to allowable increments in ambient concentrations of pollutants. At this stage, PSD regulations apply only to SO_2 and total suspended particulates (TSP). In these cases the atmospheric chemistry plays a non-critical role. However, with the proposed PSD regulations for $NO_2^{\cdot}$, ozone and other reactive pollutants (88), additional demands will be placed on models which incorporate the atmospheric reactions of hydrocarbons and NO_X. Currently, such models have been limited in their application to several areas such as Houston, Los Angeles, St. Louis, San Francisco and Denver.

As stated earlier, economics and computing limitations prevent the incorporation of explicit chemical mechanisms into an airshed model. In order to circumvent this problem, at least two approaches have been utilized. These involve "lumping" the hydrocarbons by classes and coming up with a generalized reaction mechanism for these classes (16), or by using the carbon-bond approach (13,14), which partitions the chemical species based on the similarity of their chemical bonding. As an example, the hydrocarbon "lumping" techniques employed by the ERT Environmental Lagrangian Simulator of Transport and Atmospheric Reactions (ELSTAR) (16) and the CalTech Model, developed by McRae and Seinfield (89), lump hydrocarbons in the following classes: alkanes, ethene, other alkenes, aromatic hydrocarbons, formaldehyde, and acetaldehyde.

Following the formulation of such lumped hydrocarbon mechanisms, the mechanisms are then tested out against smog chamber data, preferably using surrogate hydrocarbon mixtures, i.e., a mixture of hydrocarbons representative of the various classes in the ambient atmosphere. While there has been a significant improvement in the level of detail of chemical mechanisms in recent years, the usefulness of these advances has been somewhat negated by the lack of appropriate air quality and emissions data. For example, using the above hydrocarbon lumping schemes requires that both the hydrocarbon emission inventory data and the air quality data have the same resolution as the chemical mechanism. With current technology this is always not the case, particularly for the aldehydes, for which reliable routine measurements are only just becoming possible (90-92). In many places only total hydrocarbon measurements are available for the ambient air, and sometimes the emission inventory, data. These then have to be partitioned into the appropriate classes which inherently involves a great deal of uncertainty. Some experimental programs, such as the Los Angeles Reactive Pollutant Program (LARPP) funded by the Coordinating Research Council (CRC), have obtained detailed hydrocarbon speciated data which has permitted the application of more sophisticated chemical models (93).

For example, ELSTAR has been applied in the Los Angeles Basin for several days for which data are available. Fig. 3 and 4 show, respectively, a trajectory across the Los Angeles Air Basin for July 28, 1974 and the type of agreement between the predicted ozone and $NO_2^{\cdot}$, respectively, with the appropriate air monitoring data (94). Given this level of agreement, one can then vary the emissions inventory to simulate changes in control technologies or control effectiveness to ascertain the equivalent improvement in air quality.

With these more sophisticated models, one can compare the predicted ambient concentrations of some of the more interesting,

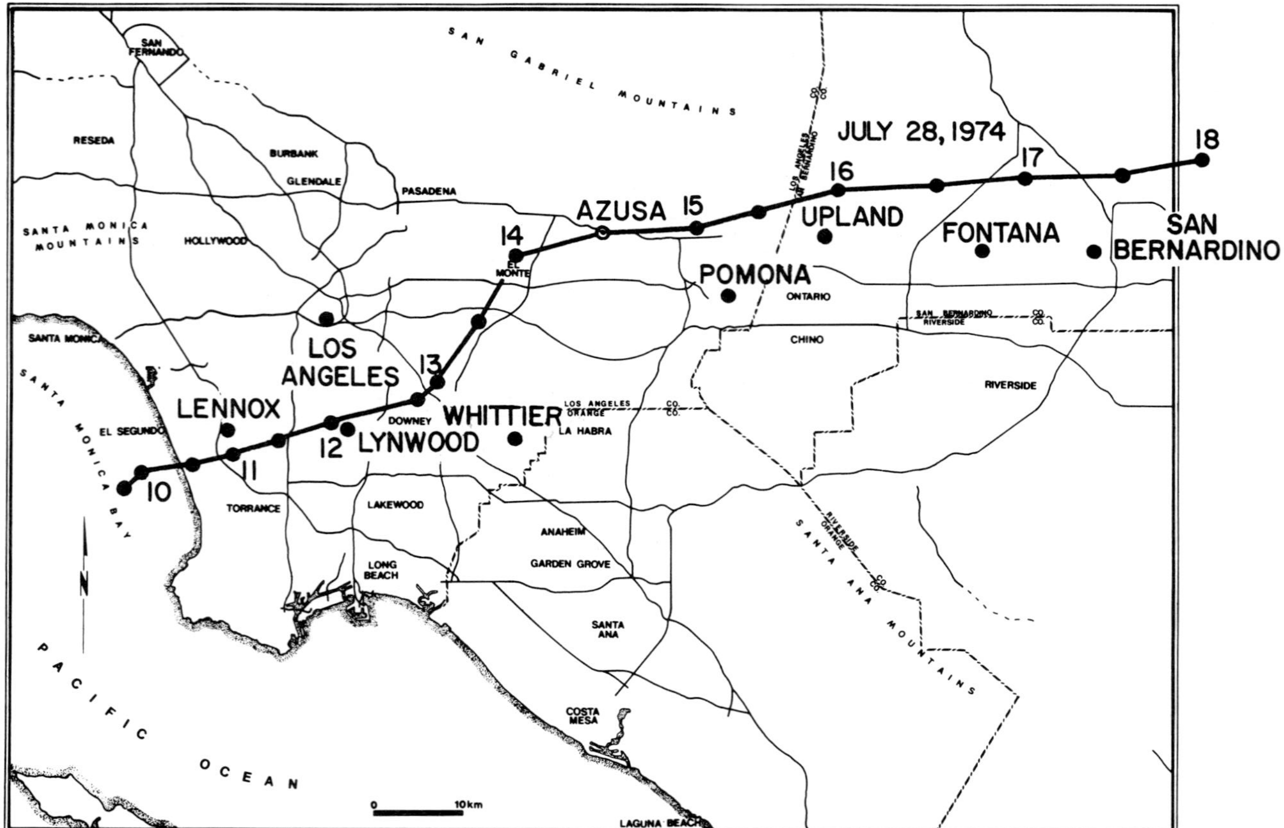

FIGURE 3. Plot of trajectory for June 28, 1974 O_3 episode. From Ref. 94.

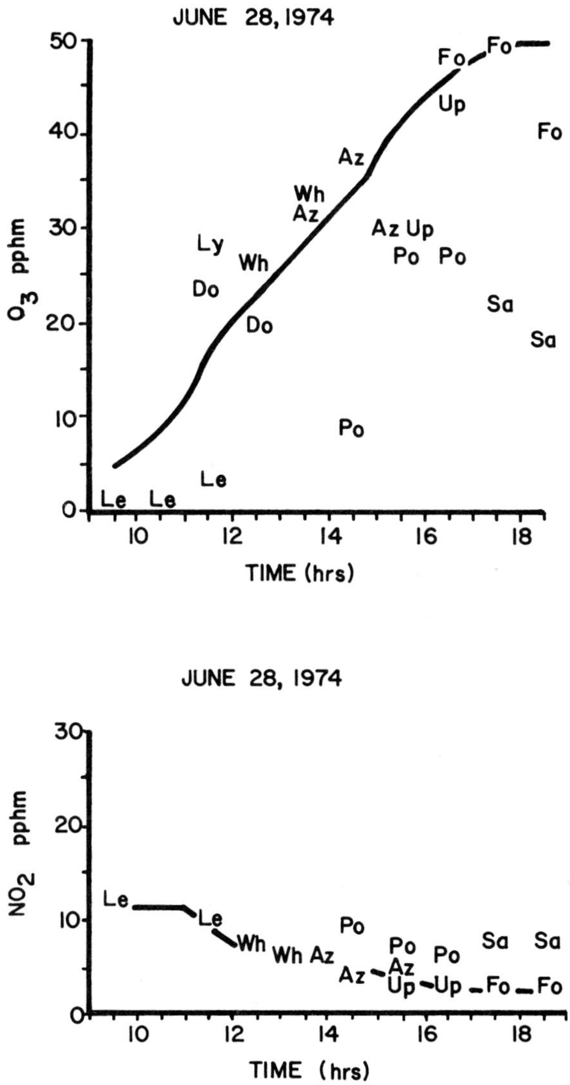

FIGURE 4. Predicted concentrations of O_3 and NO_2 versus time for the June 28, 1974 simulation. Also indicated are the observed hourly average concentrations at stations along the trajectory (Le = Lennox, Ly = Lynwood, Do = Downtown L.A., Wh = Whittier, Az = Azusa, Po = Pomona, Up = Upland, Fo = Fontana, and Sa = San Bernardino). From Ref. 94.

but less well known, species with ambient data. For example, the hydrogen peroxide and peroxyacetyl nitrate concentrations predicted by the model can be compared with those observed in the ambient air, and such comparisons can point the way to deficiencies in the model, the emissions inventory, or in the ambient air measurements.

We are currently in the process of updating the chemical mechanism in our ELSTAR model to incorporate recent information on, among others, ozone-alkene reactions and aromatic hydrocarbon photooxidations under ambient conditions, and with a somewhat lesser degree of lumping than previously used in ELSTAR, i.e., into: propane, lumped $\geq C_4$ alkanes, ethene, 1-alkenes, internal alkenes, benzene, monalkylbenzenes, dialkylbenzenes, HCHO, acetaldehyde, higher aldehydes, and ketones. Fig. 5 and 6 show very recent fits for selected species using this $\sim$80 step mechanism for NO_x-air photooxidations (95) of propene (Fig. 5) and toluene (Fig. 6). It is hoped that an intermediate approach such as this, which is inbetween a "very lumped" mechanism and a multi-hundred reaction explicit mechanism, may provide a better "fidelity" as regards the predictions of time-concentration profiles for a wide range of organic-NO_x photooxidations, as well as being compatible with the more recent detailed ambient atmospheric analyses.

NEEDS FOR FUTURE CHEMICAL MODELING DEVELOPMENT

We have discussed above the current status of our understanding of the important steps of hydrocarbon oxidation in the presence of NO_x, and also SO_2 oxidation, under ambient conditions. Except for those areas identified, substantial strides have been made in our understanding of the gas-phase processes controlling ozone formation under ambient conditions. It is anticipated that other areas which will require attention in the next few years are those identified below:

1) Develop reaction mechanisms for biogenic hydrocarbon species, such as isoprene and the pinenes, so that they may be incorporated into airshed models and their importance in ozone generation or destruction might be established.

2) Gain a better understanding of the fate of NO_x in the atmosphere for both urban and regional scales. This will require not only better understanding of some of the homogeneous mechanisms (for example, nitric acid formation and organic nitrate formation), but also the gas-to-solid or gas-to-liquid reactions involved in nitrate formation. This is becoming increasingly important because of problems associated

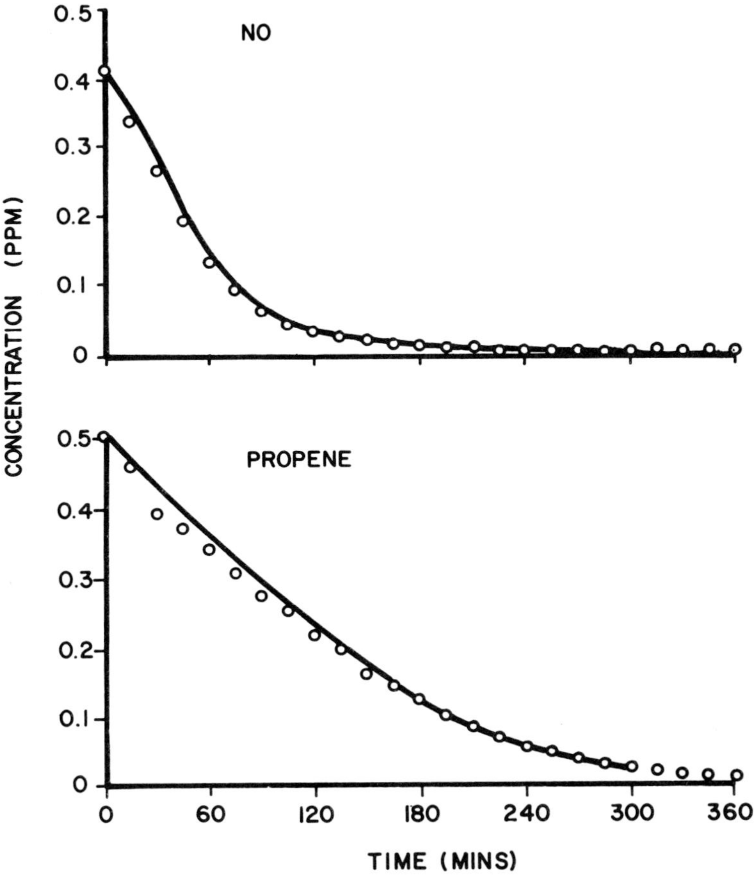

FIGURE 5. Comparison of experimental (O) and calculated (—) time-concentration profiles for selected species in propene-NO_X-air irradiation EC-216. The experimental data are from Ref. 95. (a)

with visibility reduction and acid rain production.

3) Further work is necessary with surrogate hydrocarbon mixtures to better understand the photochemical sulfate production processes and the possible formation of organic sulfur compounds under ambient conditions. Again, this is important because of possible involvement in acid rain and visibility reduction.

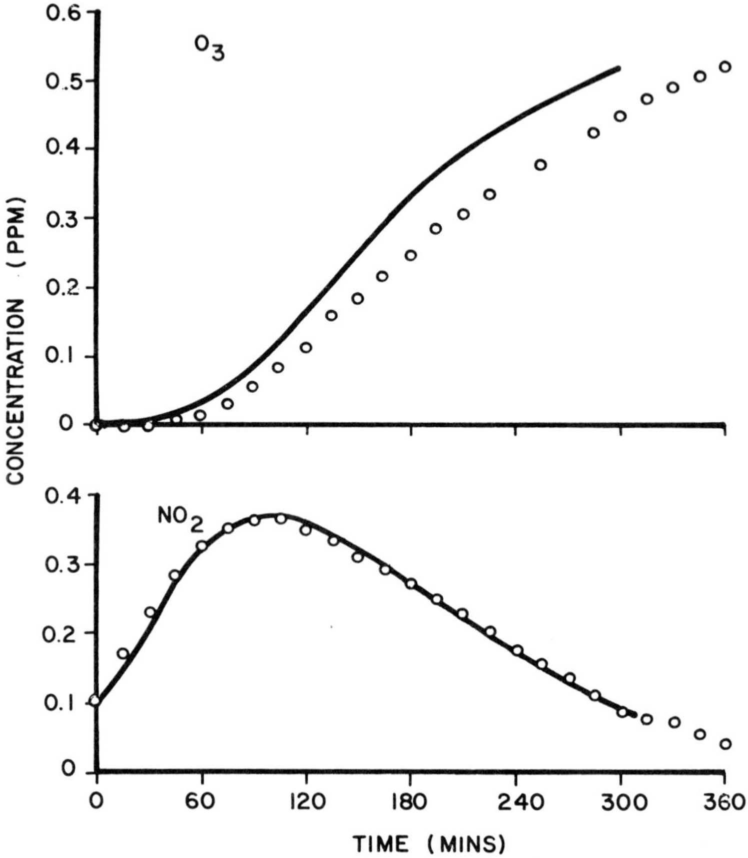

FIGURE 5. (continued) (b)

4) With the increasing attention being given to visibility reduction it is important to be able to understand the gas-to-particle conversion processes producing organic aerosols, little attention having been given to this problem so far. Likewise, it is important to establish the particle loading from these type of processes since they lower the effective UV radiation which in turn drives the photochemical processes for ozone production. It has been well demonstrated that photochemical haze is produced with ozone and that this haze can substantially reduce the UV radiation reaching the surface of the earth (96). This is important for model applications, and emphasizes the necessity for ambient UV radiation measurements for model application and testing studies.

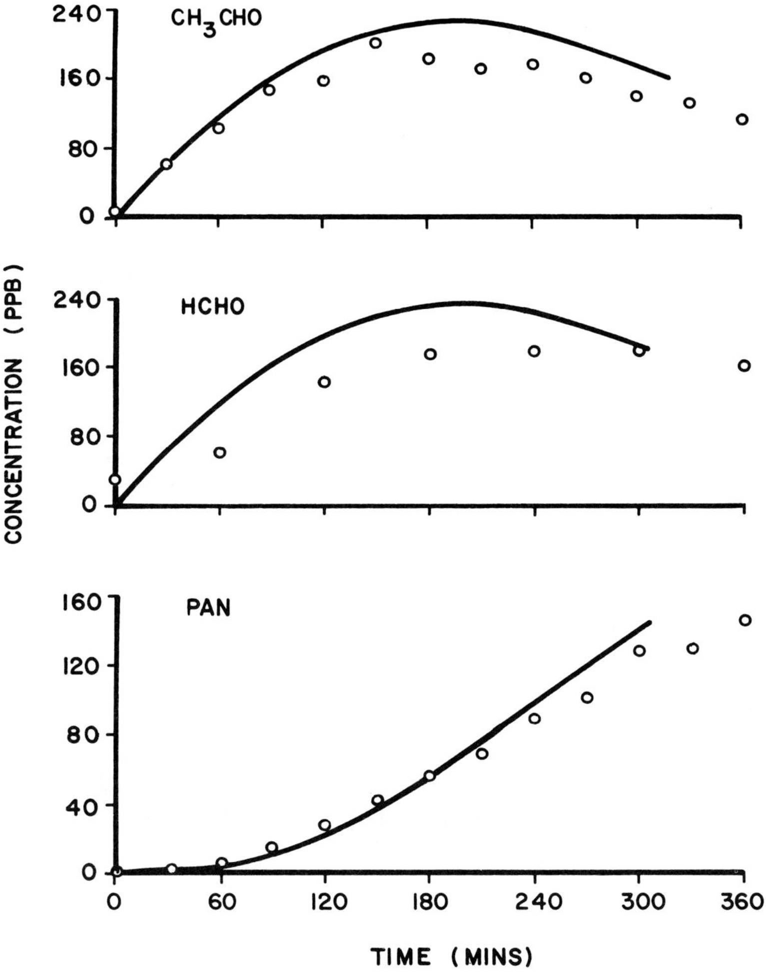

FIGURE 5. (continued) (c)

Finally, with a significant increase in our knowledge of atmospheric chemistry, it is important that similar strides are made in air quality data and in terms of emissions data. It is only when advances are made in all three that the maximum benefit can be obtained from our substantially increased understanding of hydrocarbon oxidation in the presence of NO_X under ambient conditions.

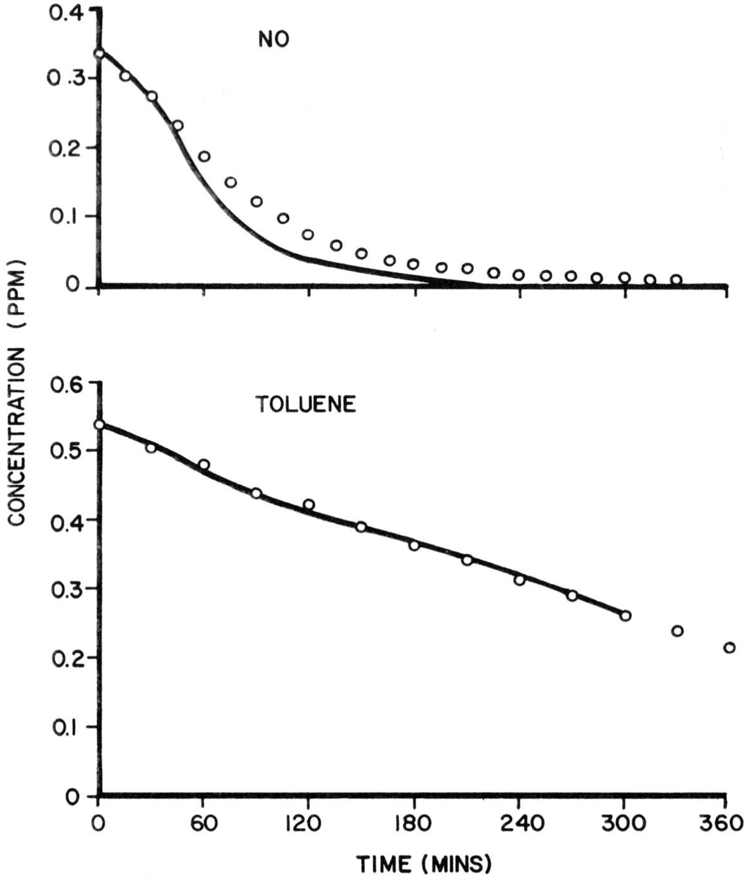

FIGURE 6. Comparison of experimental (O) and calculated (—) time-concentrations profiles for selected species in toluene-NO_X-air irradiation EC-340. The experimental data are from Ref. 95. (a)

ACKNOWLEDGEMENTS

The authors would like to acknowledge the assistance of the United States Environmental Protection Agency (Contract No. 68-02-3280, Project Officer Dr. Marcia Dodge).

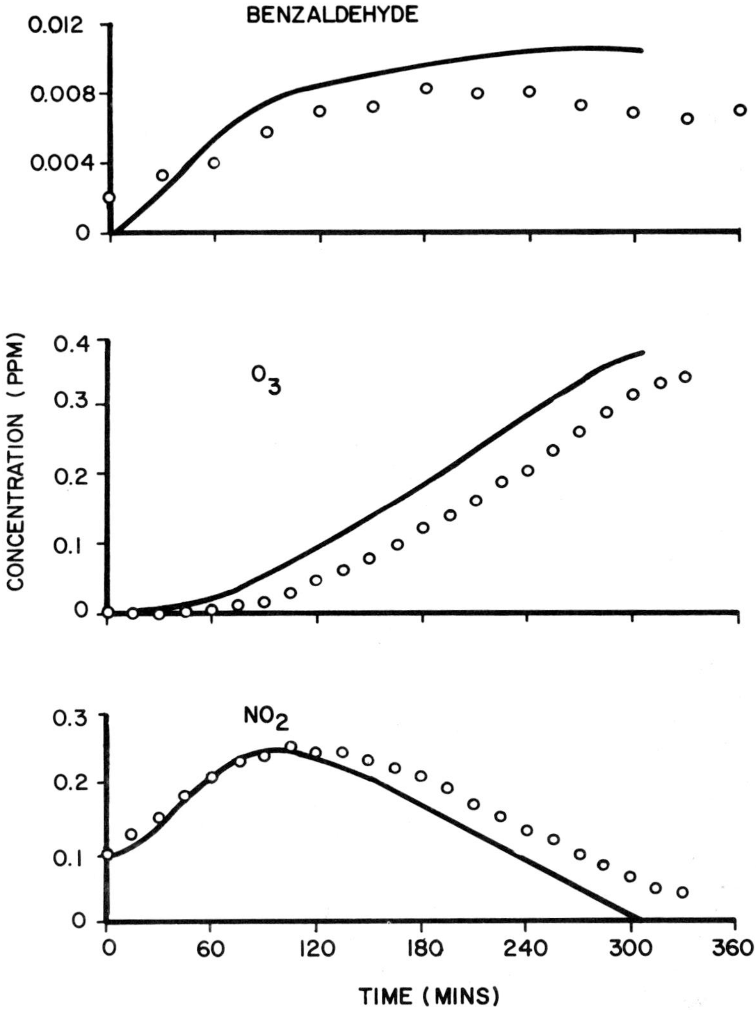

FIGURE 6. (continued) (b)

REFERENCES

1. International Conference on Photochemical Oxidant and its Control, Proceedings, Vol. 1 and 2, EPA-600/3-77-001a, b, January 1977.
2. Finlayson-Pitts, B. J. and Pitts, Jr., J. N., Adv. Environ. Sci. Technol., 7, 75 (1977).

3. Federal Register, 36, 15486, August 14, 1971 (Appendix J).
4. Ozone and Other Photochemical Oxidants, Ch. 5, National Academy of Sciences, Washington, D.C. (1977).
5. Wilson, K. W. and Goyle, G. J., "Investigation of Photochemical Reactivities of Organic Solvents", Final Report, SRI Project PSU-8029, Stanford Research Institute, Irvine, California, Septmember 1970.
6. Glasson, W. A. and Tuesday, C. S., Environ. Sci. Technol. 4, 916 (1970).
7. Laity, J. L., Burstain, I. G. and Appel, B. R., Adv. Chem. Ser., 124, 95 (1973).
8. Niki, H., Daby, E. E. and Weinstock, B., Adv. Chem. Ser., 113, 16 (1972).
9. Demerjian, K. L., Kerr, J. A. and Calvert, J. G., Adv. Environ. Sci. Technol., 4, 1 (1974).
10. Hendry, D. G., Baldwin, A. C., Barker, J. R. and Golden, D. M., "Computer Modeling of Simulated Photochemical Smog", EPA-600/3-78-059, January 1978.
11. Falls, A. H. and Seinfeld, J. H., Environ. Sci. Technol., 12, 1398 (1978).
12. Carter, W. P. L., Lloyd, A. C., Sprung, J. L. and Pitts, Jr., J. N., Int. J. Chem. Kinet., 11, 45 (1979).
13. Whitten, G. Z., Hogo, H., Meldgin, M. J., Killus, J. P. and Bekowies, P. J., "Modeling of Simulated Photochemical Smog with Kinetic Mechanisms", Vol. 1, Interim Report, EPA-600/3-79-001a, January 1979.
14. Whitten, G. Z., Killus J. P. and Hogo, H., "Modeling of Simulated Photochemical Smog with Kinetic Mechanisms", Vol. 1, Final Report, EPA-600/3-80-028a, February 1980.
15. Atkinson, R., Carter, W. P. L., Darnall, K. R., Winer, A. M. and Pitts, Jr., J. N., Int. J. Chem. Kinet.,12, 779 (1980).
16. Lloyd, A. C., Lurmann, F. W., Godden. D. A., Hutchins, J. F., Eschenroeder, A. Q. and Nordsieck, R. A., "Development of the ELSTAR Photochemical Air Quality Simulation Model and its Evaluation Relative to the LARPP Data Base", ETR Document No. P-5287-500, prepared for the Coordinating Research Coucil by Environmental Research and Technology, Inc., Westlake Village, California (1979).
17. Seinfeld, J. H., Reynolds, S. D. and Roth, P. M., Adv. Chem. Ser., 113, 58 (1972).
18. Hecht. T. A., Seinfeld, J. H. and Dodge, M. C., Environ. Sci. Technol., 8, 327 (1974).
19. Reynolds, S. D., Liu, M. K., Hecht, T. A., Roth, P. M. and Seinfeld, J. H., Atm. Environ., 8, 563 (1974).
20. Eschenroeder, A. Q. and Martinez, J. R., Adv. Chem. Ser., 113, 101 (1972).
21. Rasmussen. R. A., J. Air Poll. Contr. Assoc., 22, 537 (1972).
22. Whitby, R. A. and Coffey, P. E., J. Geophys. Res., 82, 5928 (1977).

23. Lonneman. W. A., Seila, R. L. and Bufalini, J. J., Environ. Sci. Technol., 12, 459 (1978).
24. Sandberg, J. S., Basso, M. J. and Okin, B. A., Science, 200, 1051 (1978).
25. Lonneman, W. A. and Bufalini, J. J., Environ. Sci. Technol., 13, 236 (1979).
26. Holdren, M. W., Westberg, H. H. and Zimmerman, P. R., J. Geophys. Res., 84, 5083 (1979).
27. Arnts, R. R. and Meeks, S. A., "Biogenic Hydrocarbon Contribution to the Ambient Air of Selected Areas", EPA-600/3-80-023, January 1980.
28. Finlayson, B. J. and Pitts, Jr., J. N., Science, 192, 111 (1976).
29. Baldwin, A. C., Barker, J. R., Golden, D. M. and Hendry, D. G., J. Phys. Chem., 81, 2483 (1977).
30. "Reaction Rate and Photochemical Data for Atmospheric Chemistry - 1977", (R. F. Hampson, Jr. and D. Garvin, eds.), NBS Special Publication 513, May 1978.
31. "Chemical Kinetics and Photochemical Data for Use in Stratospheric Modeling", Evaluation No. 2, J.P.L. Publication 79-27, April 1979, and revisions thereof.
32. Atkinson. R., Darnall, K. R., Lloyd, A. C., Winer, A. M. and Pitts, Jr., J. N., Adv. Photochem., 11, 375 (1979).
33. "Chemical Kinetic Data Needs for Modeling the Lower Troposphere", (J. T. Herron, R. E. Huie and J. A. Hodgeson, eds.), NBS Special Publication 557, August 1979.
34. Atkinson, R. and Lloyd, A. C., "Evaluation of Kinetic and Mechanistic Data for Photochemical Smog Chamber Modeling", EPA Contract No. 68-02-3280, ERT Document No. P-A040, 1980.
35. Heicklen, J., Westberg, K. and Cohen, N., Report No. 115-69, Center for Air Environment Studies. Pennsylvania State University, University Park, Penn. (1969).
36. Stedman, D. H., Morris, Jr., E. D., Daby, E. E., Niki, H. and Weinstock, B., 160th National Meeting of the American Chemical Society, Chicago, Ill., Sep 14-18, 1970.
37. Niki, H., Maker, P. D., Savage, C. M. and Breitenbach, L. P., Chem. Phys. Lett., 45, 564 (1977).
38. Simonaitis, R. and Heicklen, J., J. Phys. Chem., 80, 1 (1976).
39. Howard, C. J., J. Chem. Phys., 67, 5258 (1977).
40. Cox, R. A. and Patrick, K., Int. J. Chem. Kinet., 11, 635 (1979).
41. Graham. R. A., Winer, A. M. and Pitts, Jr., J. N., Chem. Phys. Lett., 51, 215 (1977).
42. Graham, R. A., Winer, A. M. and Pitts, Jr., J. N., Chem. Phys.,68, 4505 (1978).

43. Howard, C. J. and Evenson, K. M., Geophys. Res. Lett., 4, 437 (1977).
44. Sie, B. K. T., Simonaitis, R. and Heicklen, J., Int. J. Chem. Kinet., 8, 85 (1976).
45. Cox, R. A., Derwent, R. G. and Holt, P. M., J. Chem. Soc., Faraday Trans I, 72, 2031 (1976).
46. Chan, W. H., Uselman, W. M., Calvert J. G. and Shaw, J. H., Chem. Phys. Lett., 45, 240 (1977).
47. Biermann, H. W., Zetzsch, C. and Stuhl, F., Ber. Bunsenges Phys. Chem. 82, 633 (1978).
48. Edney, E. O., Spence, J. W. and Hanst, P. L., J. Air Poll. Contr. Assoc., 29, 741 (1979).
49. Hendry, D. G. and Kenley, R. A., "Nitrogenous Air Pollutants", (D. Grosjean, ed.) Ann Arbor Press, Ann Arbor, Michigan, 1979.
50. Golden, D. M., p. 60 in ref. 33.
51. Batt, L., Int. J. Chem. Kinet., 11, 977 (1979).
52. Batt, L., "Reaction of Alkoxy Radicals Relevant to Atmospheric Chemistry", European Symposium on "Physico-Chemical Behavior of Atmospheric Pollutants", Ispra, Oct. 16-18 1979.
53. Batt, L. and Robinson, G. N., Int. J. Chem. Kinet., 11, 1045, (1979).
54. Carter, W. P. L., Darnall, K. R., Lloyd, A. C., Winer, A. M. and Pitts, Jr., J. N., Chem. Phys. Lett., 42, 22 (1976).
55. Darnall, K. R., Carter, W. P. L., Winer, A. M., Lloyd A. C. and Pitts, Jr., J. N., J. Phys. Chem., 80, 1948 (1976).
56. Carter, W. P. L., Private communication (1979).
57. Pitts, Jr., J. N., Grosjean, D. M., Winer, A. M., Lloyd, A. C. and Doyle, G. J., "Chemical Transformations in Photochemical Smog and Their Application to Air Pollution Control Strategies", Third Annual Progress Report, NSF-RANN Grant, ENV73-02904 A03, October 1, 1975 - December 31, 1976.
58. Barker, J. R., Benson, S. W. and Golden, D. M., Int. J. Chem. Kinet., 9, 31 (1977).
59. Niki, H., Maker, P. D., Savage, C. M. and Breitenbach, L. P., J. Phys. Chem., 82, 135 (1978).
60. Carter, W. P. L., Darnall, K. R., Graham, R. A., Winer, A. M. and Pitts, Jr., J. N., J. Phys. Chem. 83, 2305 (1979).
61. Nip, W. S. and Paraskevopoulos, G., J. Chem. Phys., 71, 2170 (1979).
62. Ravishankara, A. R., Wagner, S., Fischer, S., Smith, G., Schiff, R., Watson, R. T., Tesi, G. and Davis, D. D., Int. J. Chem. Kinet., 10, 783 (1978).
63. Overend, R. and Paraskevopoulos, G., J. Chem. Phys., 67, 674 (1977).
64. Atkinson, R. and Pitts, Jr., J. N., J. Chem. Phys., 63, 3591 (1975), 68, 2992 (1978).
65. Atkinson, R., Perry, R. A. and Pitts, Jr., J. N., J. Chem. Phys. 66, 1197 (1977).

66. Atkinson, R., Perry, R. A. and Pitts, Jr., J. N., J. Chem. Phys., 67, 3170 (1977).
67. Biermann, H. W., Harris, G. W. and Pitts, Jr., J. N., 14th Informal Conf. Photochem., Newport, California, March 30 - April 3 (1980).
68. Arnts, R. R. and Gay, Jr., B. W., "Photochemistry of Some Naturally Emitted Hydrocarbons", EPA-600/3-79-081, September 1979.
69. Huie, R. E. and Herron, J. T., Int. J. Chem. Kinet., Symp. 1, 165 (1975).
70. Japar, S. M., Wu, C. H. and Niki, H., J. Phys. Chem., 78, 2318 (174).
71. Niki, H., Maker, P. D., Savage C. M. and Breitenbach, L. P., Chem. Phys. Lett., 46, 327 (1977).
72. Su, F., Calvert, J. G. and Shaw, J. H., J. Phys. Chem., 84, 239 (1980).
73. Herron, J. T. and Huie, R. E., J. Amer. Chem. Soc., 99, 5430 (1977).
74. Herron, J. T. and Huie, R. E., Int. J. Chem. Kinet., 10, 1019 (1978).
75. Dodge, M. C. and Arnts, R. R., Int. J. Chem. Kinet., 11, 399 (1979).
76. Calvert, J. G., Su, F., Bottenheim, J. W. and Strausz, O. P., Atm. Environ., 12, 197 (178).
77. Perry, R. A., Atkinson, R and Pitts, Jr., J. N., J. Phys. Chem., 81, 296, 1607 (1977).
78. Kenley, R. A., Davenport, J. E. and Hendry, D. G., J. Phys. Chem., 82, 1095 (1978).
79. Hendry, D. G., p. 85, Ref. 33.
80. O'Brien, R. J., Green, P. J., Doty, R. A., Vanderzanden, J. W., Easton, R. R. and Irwin, R. P., "Interaction of Oxides of Nitrogen and Aromatic Hydrocarbons Under Simulated Atmospheric Conditions", in Nitrogenous Air Pollutants, (D. Grosjean, ed.), Ann Arbor Press, Ann Arbor, Michigan, 1979.
81. Darnall, K. R., Atkinson, R. and Pitts, Jr., J. N., J. Phys. Chem., 83, 1943 (1979).
82. Takagi, H., Washida, N., Akimoto, H., Nagasawa, K., Usui, Y. and Okuda, M., J. Phys. Chem., 84, 478 (1980).
83. Calvert, J. G. and Pitts, Jr., J. N., "Photochemistry", Wiley, New York, N. Y. (1966).
84. Graham, R. A., Winer, A. M., Atkinson, R. and Pitts, Jr., J. N., J. Phys. Chem., 83, 1563 (1979).
85. Burrows, J. P., Cliff, D. I., Harris, G. W., Thrush, B. A. and Wilkinson, J. P. T., Proc. Roy. Soc., A368, 463 (1979).
86. Sander, S. P. and Watson, R. T., Chem. Phys. Lett. 77, 473 (1981).
87. Federal Register, 43, 26380, 26388, June 19, 1978 (PSD Regulations).
88. Federal Register, 45, 7800, February 5, 1980 (PSD Regulations).

89. McRae, G. J., Goodin, W. R. and Seinfeld, J. H., "Development of a Second Generation Mathematical Model of Photochemical Air Pollution", California Air Resources Board, Contract A5-046-87 Final Report, 1979.
90. Platt, U., Perner, D. and Patz, H. W., J. Geophys. Res., 84, 6329 (1979).
91. Tuazon, E. C., Winer, A. M., Graham, R. A. and Pitts, Jr., J. N., Adv. Environ. Sci. Technol., 10, 259 (1980).
92. Grosjean, D., Fung, K. and Atkinson, R., "Measurement of Aldehydes in the Air Environment", Paper No. 80-50.4, presented at the 73rd Annual Meeting of the Air Pollution Control Assoc., Montreal, June 22-27, 1980.
93. Calvert, J. G., Environ. Sci. Technol., 10, 256 (1976).
94. Godden, D. A., Lurmann, F. W. and Lloyd, A. C., "Sensitivity Tests, Data Requirements and Accuracy of a Photochemical Dispersion Model - ELSTAR", paper presented at Air Pollution Control Assoc. Meetin, New Orleans, La., November 1979.
95. Pitts, Jr., J. N., Darnall, K., Carter. W. P. L., Winer, A. M. and Atkinson, R., "Mechanisms of Photochemical Reactions in Urban Air", Final Report, EPA-600/3-79-110, November 1979.
96. Stephens, E. R., J. Air Poll. Contr. Assoc., 25, 521 (1975).

POSTER SESSION

Session Chairman, M. A. J. Rodgers, Austin, Texas

PHYSIOLOGICAL FACTORS AFFECTING THE OXYGEN ENHANCEMENT RATIO IN A GAMMA IRRADIATED REPAIR DEFICIENT STRAIN OF ESCHERICHIA COLI

A. H. Ahmed,[1] *S. H. Moss*[2] *and D. J. G. Davies*[1]

[1]*School of Pharmacy and Pharmacology, University of Bath, Bath, Avon, U.K.*
[2]*Department of Radiology, Stanford University Medical School, Stanford, California*

We have previously proposed that the oxygen enhanced inactivation by gamma rays of the excision proficient, recombination deficient strain of Escherichia coli K-12 AB2463 recA is due to more efficient excision of lesions formed under nitrogen (1), and furthermore, that the lesions excised were not pyrimidine dimers. This proposal was based, in part, on the lack of any appreciable oxygen effect in the double mutant E. coli K-12 AB2480 uvrA recA and the independence from the presence of oxygen of the photoreactivable sector. We have now shown that the photoreactivable damage is, depending on the energy of the radiation and the cell concentration, induced largely by ultraviolet Cerenkov radiation and to a lesser extent directly by the ionizing radiation (2). In the course of these studies we have observed that the oxygen enhancement ratio (OER) for the double mutant AB2480 in the absence of photoreactivation is dependent on the pre-irradiation growth conditions of the cells. These findings may correspond to differences in the OER's reported with uvrA recA strains of E. coli by various workers summarized below.

E. coli strain	Radiation	O.E.R.	Ref
AB2480 uvrA recA	1.2 MeV gamma	1.01	(1)
JO307 uvrA recA	220 keV X-rays	<1.0	(3)
AB2480 uvrA recA	25 MeV electrons	1.28	(4)
AB2480 uvrA recA	7 MeV electrons	1.3	(5)
JO307 uvrA recA	1.2 MeV gamma	1.3	(6)
AB2480 uvrA recA	X-rays	1.3	(7)

ISBN 0-12-592050-4

Using stationary phase cells of the strain AB2480 grown in nutrient broth, by varying the amount of aeration during pre-irradiation growth, we have observed OER's after Co_{60} gamma irradiation ranging from 1.3 for standing cultures to 0.8 with vigorously shaken cultures (Fig. 1), the latter reflecting a protection by the presence of oxygen during irradiation. Choosing an intermediate degree of pre-irradiation aeration, we have determined the OER for suspensions of AB2480 in various stages of growth (Fig. 2). There is a drop in OER from 1.25 in log phase (2h cells) to 1.1 in early stationary phase (5h cells) corresponding with a decreased sensitivity in oxic conditions. During late stationary phase (16-24h cells) there is a further decrease in OER to 1.0 corresponding with an increased sensitivity under anoxic conditions. This latter decrease in OER may be accentuated (OER 0.8) at 24h· by more vigorous aeration of the culture (Fig. 1).

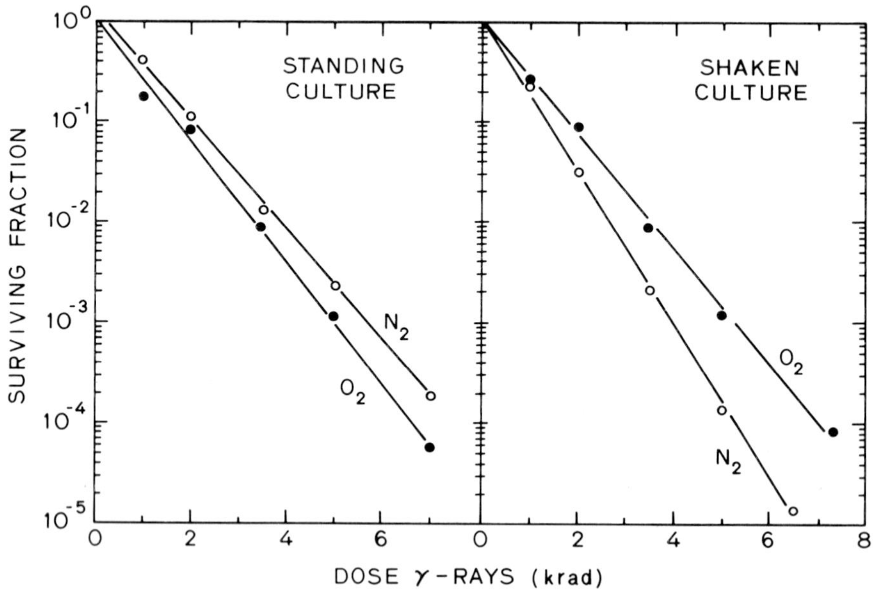

FIGURE 1. Survival curves of E. coli K-12 AB2480 uvrA recA after Co_{60} gamma irradiation under air (●) or nitrogen (o). Stationary phase cells (24h) were used, grown in nutrient broth either with or without aeration by shaking.

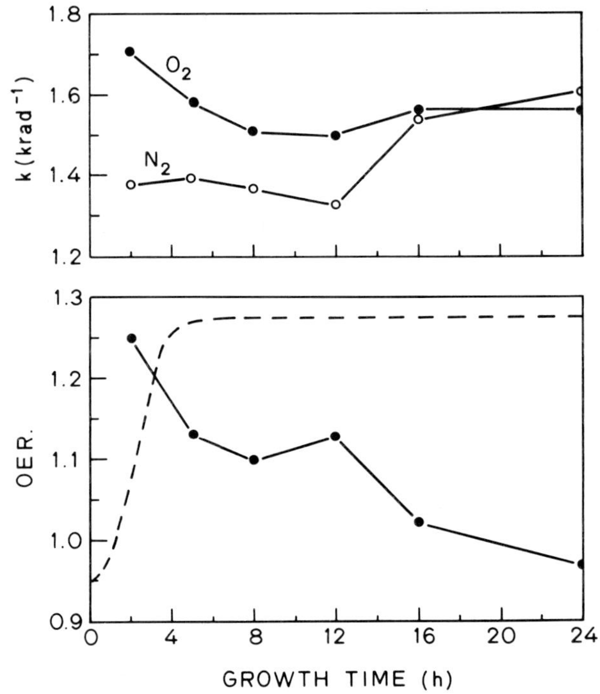

FIGURE 2. Inactivation constants (k) of survival curves of E. coli K-12 AB2480 uvrA recA after irradiation with Co_{60} gamma rays under air (●) or nitrogen (o), and resultant values of OER (●, lower panel) at different stages of growth. The growth cycle, in arbitrary units, is represented by the dashed line. Each point represents the mean of three experiments.

Empirically these results may explain the discrepancies observed in the literature as outlined in the table above; in addition they should be considered in the interpretation of data for any model for OER which makes use of comparisons with repair deficient strains of E. coli. At present we have no mechanistic interpretation of these results. The possibility of changes in concentration of endogenous scavenging compounds, or the role of damage other than DNA damage (e.g., membrane damage) is under consideration.

REFERENCES

1. Vinicombe, D. A., Moss, S. H. and Davies, D. J. G., Int. J. Radiat. Biol., 33,483 (1978).
2. Moss, S. H. and Smith, K. C., Int. J. Radiat. Biol., in press.
3. Brustad, T. and Wold, E., Radiat. Res., 66,215 (1976).
4. Redpath, J. L. and Torotello, M. L., Int. J. Radiat. Biol. 32,505 (1977).
5. Tyrrell, Rex (personal communication).
6. Sapora, O., Fielden, E. M. and Loverock, P. S., Radiat. Res., 69,293 (1977).
7. Rupp, W. D., Zipser, E., von Essen, C., Reno, D., Prosnitz, L. and Howard-Flanders, P., Proceedings of Conference at Carmel on "Time and Dose Relationships in Radiation Biology as Applied to Radiotherapy," Brookhaven National Laboratory Publication 50203 (C-57) (1969), p. 1.

THE REACTIONS OF OXYGEN AND SUPEROXIDE IONS WITH FLAVOSEMIQUINONE RADICALS

Robert F. Anderson, CRC Gray Laboratory, Mount Vernon Hospital, Northwood, Middlesex HA6 2RN, U.K.

INTRODUCTION

Despite many years of investigation the precise mechanism of the oxidation of reduced flavin species remains unknown. The production of superoxide anions, $O_2^{\bar{\cdot}}$ during oxidation has been shown (1) and this species has been suggested to act as an oxidant in subsequent reactions (2). In this paper we report our pulse radiolysis study on the reaction of superoxide ions with flavosemiquinone radicals, $F^{\bar{\cdot}}/FH\cdot$.

MATERIALS AND METHODS

Riboflavin, RF, flavin mononucleotide, FMN and flavine adenine dinucleotide, FAD were obtained from Sigma and sodium formate, p.a. was obtained from Merck. All solutions were prepared in triple distilled water and adjusted to the required pH with phosphate buffers.

Details of the pulse radiolysis equipment used have been published (3).

RESULTS AND DISCUSSION

Superoxide anions were produced by the pulse radiolysis (∿60 Gy in 2 μs) of neutral aerated solutions containing formate anions (0.01 mol dm^{-3}) and a G(radical) value of 6.2 was assumed.

$$H_2O \longrightarrow HO\cdot + H\cdot + e^-_{aq} + H_2O_2 + H_2 \quad [1]$$

$$HO\cdot + HCOO^- \longrightarrow CO_2^{\bar{\cdot}} + H_2O \quad [2]$$

$$H\cdot + HCOO^- \longrightarrow CO_2^{\bar{\cdot}} + H_2 \quad [3]$$

$$e^-_{aq} + O_2 \longrightarrow O_2^{\bar{\cdot}} \quad [4]$$

$$CO_2^{\bar{\cdot}} + O_2 \longrightarrow O_2^{\bar{\cdot}} + CO_2 \quad [5]$$

When flavin (80 μmol dm^{-3}) is also present further electron transfer reactions take place

$$e^-_{aq} + F \longrightarrow F^{\bar{\cdot}} \quad [6]$$

$$CO_2^{\bar{\cdot}} + F \longrightarrow F^{\bar{\cdot}} + CO_2 \quad [7]$$

ISBN 0-12-592050-4

$$FH. \longleftrightarrow F^{\overline{\cdot}} + H^+ \quad pK_a \sim 8.5(4,5,7) \qquad [8]$$

$$F^{\overline{\cdot}} + O_2 \longrightarrow F + O_2^{\overline{\cdot}} \qquad [9]$$

Under the above conditions the only radicals present some 20 μs after the pulse are the FH· and $O_2^{\overline{\cdot}}$ species. The concentration of each species can be deduced by measuring the FH· absorption spectrum in the visible and comparing it with the spectrum obtained in deaerated solution (4,5). It was found that for all three flavins studied 8% of the radical yield produced FH·.

The FH· species monitored at 540 nm decayed much faster than that for its dismutation rate (4,6) and with exponential kinetics. As the FH· species reacts only slowly with oxygen (<4 x $10^4 mol^{-1} dm^3 s^{-1}$(6)) we conclude that the FH· species reacts in a pseudo first order reaction with the excess superoxide anions present.

$$FH\cdot + O_2^{\overline{\cdot}} \longrightarrow FHO_2^- \ (FHO_2H) \qquad [10]$$

Data for all 3 flavins studied is presented in Figure 1. When the dose per pulse is raised ($[O_2^{\overline{\cdot}}]$ increased) the rate of FH· disappearance increases and is consistent with the following kinetic description.

$$\frac{-d[FH\cdot]}{} = \left(\frac{k_9 K_8 [O_2]}{[H^+]} + k_{10}[O_2^{\overline{\cdot}}]\right)[FH\cdot]$$

The rate constant for each flavosemiquinone species reacting with superoxide anions k_{10} is found from the slopes of the plots in Fig. 1 and are presented in Table 1. The second order rate constant as a function of pH for FMN is also presented in Fig. 1 and it is suggested that on protonation superoxide ions react 2 fold faster with flavosemiquinones.

The product of reaction 10 is not the oxidized flavin, but an intermediate (Fig. 2) which shows absorbance bands below 500 nm. The intermediate may well be the same as the proposed intermediate formed on the reaction of reduced flavins with oxygen (8) as the spectrum is similar to the reported partial spectrum seen on the rapid mixing of a reduced flavin with oxygen (9).

TABLE 1. Reaction Rates of Superoxide Anions with Flavosemiquinones

Flavin	$10^{-8}k_{10}(FH\cdot + O_2^{\overline{\cdot}})^a$
RIBOH·	*7.1 ± 0.2*
FMNH·	*3.2 ± 0.2*
FADH·	*2.2 ± 0.2*

a) Units $dm^3 mol^{-1}s^{-1}$

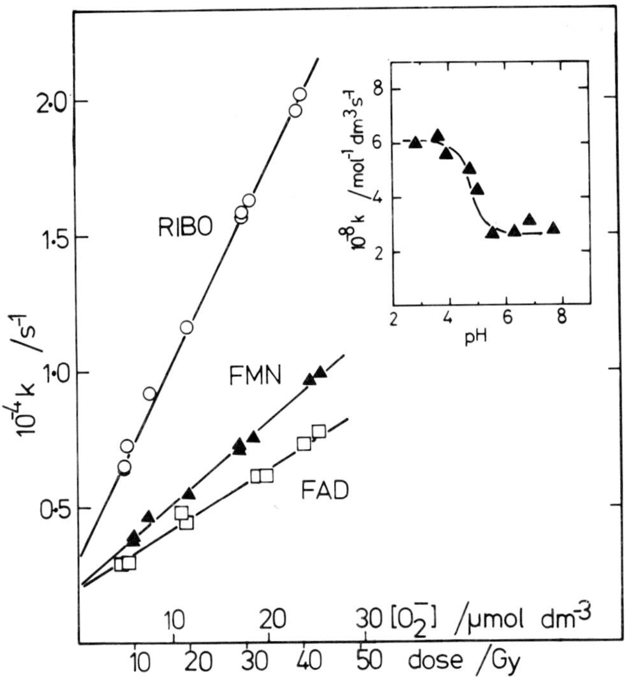

FIGURE 1. First-order rate constants for the disappearance of transients absorbing at 540 nm formed on the pulse radiolysis of neutral aerated solutions containing flavin (80 μmol dm^{-3}) and sodium formate (0.01 mol dm^{-3}). Superoxide anion concentrations were calculated as described in the test. Inset: Calculated second-order rate constants for FMN as a function of pH.

The intermediate slowly decays with exponential kinetics ($k_{11} = 3 \times 10^2\ s^{-1}$) to give the flavin ground state absorption

$$FHO_2 \longrightarrow F + H_2O_2 \qquad [11]$$

The rate of this reaction is independent of [flavin], [O_2] and the dose used.

REFERENCES

1. Ballou, D., Palmer, G. and Massey, V., Biochem. Biophys. Res. Comm., 36, 898 (1969).
2. Massey, V., Palmer, G. and Ballou, D., "Oxidases and Related Redox Systems," (T. E. King, H. S. Mason and M. Morrison,

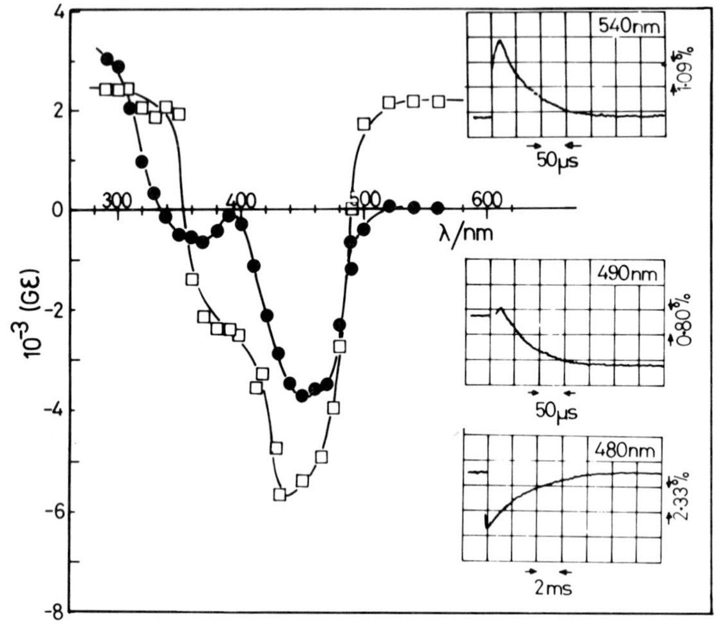

FIGURE 2. Changes in absorption following pulse radiolysis (∿60 Gy) of neutral aerated solutions containing RIBO (80 μmol dm^{-3}) and sodium formate (0.01 mol dm^{-3}).
□ Spectrum measured 20 μs after pulse,
● Spectrum measured 250 μs after pulse.
Inset: Oscilloscope traces of the decay of the transients formed.

eds.), Univ. Park Press, Baltimore (1973), p. 25.
3. Adams, G. E., Boag, J. W. and Michael, B. D., Trans. Faraday Soc., 61, 492, 1674 (1965).
4. Land, E. J. and Swallow, A. J., Biochemistry, 8, 2117 (1969).
5. Anderson, R. F., Ber. Bunsenges. Physik. Chem., 80, 969 (1976).
6. Vaish, S. P. and Tollin, G., Bioenergetics, 2, 61 (1971).
7. Draper, R. D. and Ingraham, L. L., Archiv. Biochem. Biophys., 125, 802 (1968).
8. Gibson, Q. H. and Hastings, J. W., Biochem. J., 68, 368 (1962).
9. Massey, V., Palmer, G. and Ballou, D., "Flavins and Flavoproteins" (Proc. 3rd Int. Symp.) (H. Kamen, ed.), Univ. Park Press, Baltimore, (1971), p. 349.

RING CLEAVAGE AS A MECHANISM OF OXYGEN ACTIVATION BY TETRAHYDROPTERIN MONOOXYGENASES

J. E. Ayling and S. W. Bailey, University of Texas at San Antonio, San Antonio, Texas 78282

Phenylalanine hydroxylase is a mixed function oxygenase, utilizing one atom of molecular oxygen to hydroxylate phenylalanine to tyrosine, the second oxygen atom being reduced to H_2O at the expense of cofactor, tetrahydrobiopterin (6-ℓ-erythro-dihydroxypropyl-tetrahydropterin) (I).

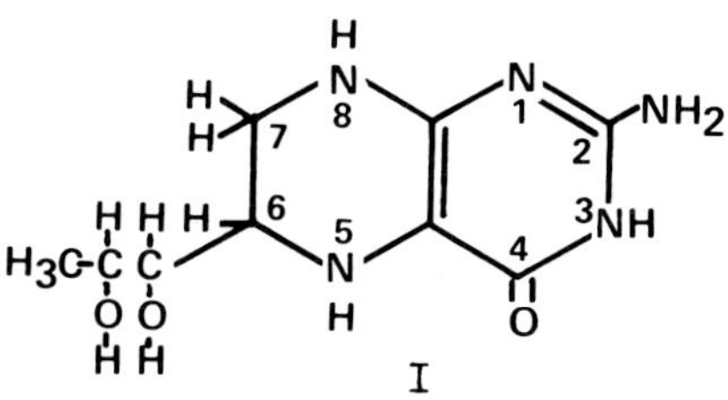

I

The mechanism by which enzyme and cofactor activate molecular oxygen is unknown. The structural similarity of dihydroflavin to tetrahydrobiopterin has prompted speculation that the two classes of mono-oxygenases may function by related intermediates. Of the numerous hypotheses, one proposes that a vinylogous ozone is formed by addition of oxygen to C4a of tetrahydrobiopterin (or dihydroflavin), followed by cleavage of the C4a-N5 bond of the pyrazine ring. The liberated N5, and the carbonyl generated by oxygen donation to substrate, condense, reclosing the ring to form quinoid dihydrobiopterin (1,2). Our recent discovery that highly substituted pyrimidines, which lack a complete pyrazine ring, can serve as cofactors for phenylalanine hydroxylase (3), has provided a means of directly testing this proposal. 2,5,6-Triamino-4-pyrimidinone (TP) (tetrahydropterin minus C6 and C7) and 5-benzylamino-2,6-diamino-4-pyrimidinone (BDP) (6-phenyltetrahydropterin minus C7) possess the same apparent Michaelis constants (0.1 mM and 0.003 mM) and coupling to tyrosine formation as their analogous tetrahydropterins, with maximum velocities 20 times and 60 times slower, respectively. The utilization by phenylalanine hydroxylase of a cofactor lacking a complete pyrazine ring would lead, in the event of a ring cleavage mechanism, to two separate molecules, an oxidized pyrimidine and an amine (Figure 1).

When TP or BDP are used as cofactors for the enzyme catalyzed hydroxylation of phenylalanine, the initial oxidized pyrimidine

ISBN 0-12-592050-4

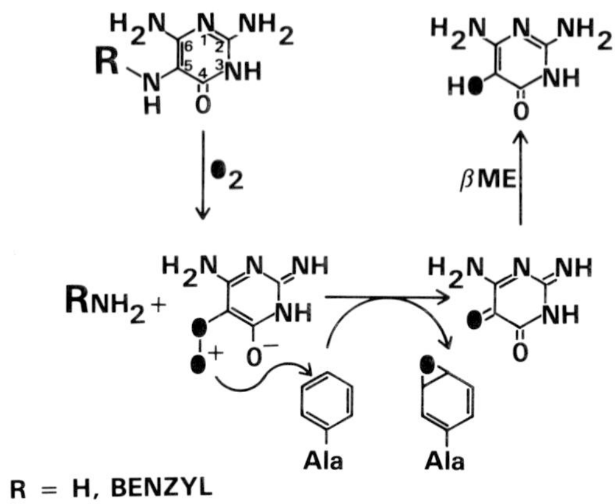

FIGURE 1. Oxygen activation by a ring cleavage mechanism in reaction of phenylalanine hydroxylase with a pyrimidine cofactor. βME = β-mercaptoethanol.

product, which is unstable, is rapidly reduced by mercaptoethanol to a compound which is no longer a cofactor. This compound was identified by its (i) retention on HPLC with a range of different solvent systems, (ii) UV absorbance, (iii) instability in air, (iv) electrochemical activity and (v) mass spectrum, as divicine (2,6-diamino-5-hydroxy-4-pyrimidinone). TP and BDP consumed in a phenylalanine hydroxylase reaction containing mercaptoethanol are both quantitatively converted to divicine, with the production of 0.55 and 0.71 nmole tyrosine pernmole TP or BDP, respectively. The second fragment of BDP, detected by its reaction with fluorescamine, co-chromatographs with benzylamine. The yield of benzylamine is the same whether or not a reducing system is present. Divicine and benzylamine are produced only in the presence of both enzyme and substrate. When oxidized with iodine, or auto-oxidized in air, at neutral pH, the resulting quinoid BDP readily tautomerizes and hydrolyzes to benzaldehyde and TP. Under these conditions no benzylamine is detectable, indicating that cleavage of the 5-amine is enzyme dependent (Table I) (4).

If cleavage of pyrimidine cofactor is a result of formation of the proposed vinylogous ozone, the newly incorporated oxygen at the 5 position of the pyrimidine product would originate from molecular oxygen. Alternatively, if divicine production is due to enzyme catalyzed hydrolysis of the 5-imine of quinoid TP, this oxygen would come from water. To distinguish between these two possibilities, reactions similar to those in Table I containing BDP and mercaptoethanol were run in the presence of $^{18}O_2$. The divicine product was derivatized and analyzed by mass spectrometry. The mass spectrum verified that the oxygen incorporated at the 5-position of the pyrimidine arose from molecular O_2.

TABLE I. Stoichiometry of the Reaction of Phenylalanine Hydroxylase with Pyrimidine Cofactors

Cofactor	Reducing system	Reagent omitted	Products formed, nmole/nmole cofactor		
			Tyrosine	Divicine	Benzylamine
TP	+	-	0.55	1.0	n.a.
	-	-	0.55	n.a.	n.a.
BDP	+	-	0.71	1.0	0.71
	-	-	0.71	n.a.	0.71
	+/-	phenylalanine	0	0/n.a.	0
	+/-	enzyme	0	0/n.a.	0

Reactions were run at 27° in air, and contained 100 µmole Tris-Cl, pH 7.4, 0.25 units phenylalanine hydroxylase, 2,500 units catalase, 1 µmole phenylalanine, 0.05 µmole pyrimidine cofactor and, if a reducing system was included, 100 µmole mercaptoethanol, per 1 ml total volume. n.a. = not assayed.

The incorporation of ^{18}O from molecular oxygen at C-5 of the pyrimidine product of the reaction (analogous to C-4a of the pteridine cofactor) provides evidence for the formation of a cofactor oxygen adduct, and that the position of attack of oxygen is carbon 5 of the pyrimidine cofactor. The concomitant scission of the pyrimidine 5C-N bond (equivalent to C4a-N5 of tetrahydrobiopterin) supports ring cleavage as a mechanism for oxygen activation by the tetrahydrobiopterin dependent aromatic amino acid hydroxylases, and possibly also for the dihydroflavin monooxygenases.

ACKNOWLEDGEMENTS

This investigation was supported by grants CA-20708 and GM-27880 awarded by the NIH, and AX-706 awarded by the Robert A. Welch Foundation.

REFERENCES

1. Hamilton, G. A., "Molecular Mechanisms of Oxygen Activation," (O. Hayaishi, ed.), Academic Press, NY, 1974, p. 405.
2. Entsch, B., Ballou, D. P. and Massey, V., J. Biol. Chem., 251, 2550 (1976).
3. Bailey, S. W. and Ayling, J. E., Biochem. Biophys. Res. Comm. 85, 1614 (1978).
4. Bailey, S. W. and Ayling, J. E., J. Biol. Chem., 255,7774 (1980).

OZONE INJURY AND AGING IN LEAVES: PROTECTION BY EDU

Jesse H. Bennett, Edward H. Lee, Howard E. Heggestad, Ralph A. Olsen and John C. Brown

Plant Stress Laboratory, Plant Physiology Institute, Agricultural Research Center, SEA, USDA, Beltsville, Maryland 20705

During periods of aggravated air pollution known as photochemical smog, ozone (O_3) often builds up to phytotoxic levels in urban atmospheres due to the action of sunlight on automobile exhaust emissions. Ozone is currently rated as the most injurious air pollutant affecting plants in the United States (1). More than 90% of the air pollution damage to vegetation is attributed to photochemical oxidants of which O_3 is the most significant phytotoxic component.

Though O_3 is regarded as a plant-damaging air pollutant when present in elevated concentrations, it has long existed as a natural background component of the atmosphere subject to evolutionary plant adaptation processes. All green plants show some tolerance to O_3, but they vary widely in their susceptibilities. Indeed, every crop species we have studied exhibit varieties and genetic strains that are more sensitive than others. Physiological and toxicological studies have sought to characterize the O_3 stress and determine mechanisms of action. Our findings regarding O_3 phytotoxicity and plant tolerance induced by a new research chemical N-[2-(2-oxo-1-imidazolidinyl)ethyl]-N'-phenylurea (EDU) and are summarized here. Sensitive snap bean, soybean, clover, and watermelon varieties were used as model test plants to develop the experimental data and concepts.

Ozone absorption studies conducted in Physiological Activity and Diagnostic Chambers (PADC) show viable leaf mesophyll quantitatively absorbs O_3 when subjected to subinjurious concentrations (2). Catalytic oxidoreductase enzymes appear to function in the uptake process providing limited tissue protection when saturation concentrations are not exceeded. The rates of absorption are more important than the total dose absorbed by leaves. Minimum doses required to cause any cellular necrosis in sensitive Phaseolus vulgaris L. 'Bush Blue Lake 290' snap bean leaves were above 55 μg O_3/hr/dm^2 leaf area (2). Twice this foliar dose was required if absorbed over a 3-hr period. This suggests that limited scavenging of O_3-generated toxicants in the mesophyll tissues may have occurred. The biological action of O_3 is thought to involve free radicals (3), at least in part. Free radical scavengers can protect against these injurious chemical species. Oxidoreductase enzymes have been implicated in offering protection to cell membranes of aerobic organisms.

ISBN 0-12-592050-4

EDU applied systemically through root application or as a foliar spray effectively transforms O_3-susceptible plants into highly tolerant ones. Up to a 30-fold enhancement in foliar O_3 resistance has been reported (4). Ozone tolerance enhancement in our experiments was correlated with increases in EDU-induced superoxide dismutase (SOD) and catalase activities (5). SOD was elevated in proportion to the O_3-tolerance manifested by 2-4 week old BBL 290 plants given EDU soil applications of 25 to 50 mg EDU/(15-cm dia) pots. Some plants overdosed with 100 mg/pot showed symptoms of EDU toxicity.

In addition to protecting against acute O_3 exposures, EDU retards leaf senescence caused by chronic low-level O_3 treatments (i.e., .10-.20 μg O_3/l, 6-hr/day). Detailed investigations of EDU protection during the vegetative vs fruiting stages of Sugar Baby watermelon plants (*Citrullus vulgaris* Schrad. 'Sugar Baby') showed application at the time of flowering and fruiting gave the most benefit in reducing oxidant damage. Leaves on deflowered plants exhibited greater natural resistance to chronic O_3 stress than plants with melons. These leaves also remained juvenile over a longer period of time.

Time-series experiments using leaf discs floated on aqueous solutions in the dark are standard tests for leaf senescence. Unozonized leaf discs of red clover (*Trifolium pratense* L. "Pennscott') floated on aqueous EDU concentrations ranging between 0 and 500 mg/l EDU showed a progressive retardation of natural senescence in proportion to the EDU test concentration. Optimal concentrations for retarding natural senescence are also optimal for inducing O_3 tolerance in leaves. Biochemical analyses of the leaf tissues indicate EDU arrests senescence by enhancing RNA and protein synthesis in cells (6). We suggest that EDU-induced O_3 tolerance and retardation of aging is mediated through the induction of specific oxy-radical scavenging enzymes and in sustaining RNA and protein synthesis.

REFERENCES

1. Heck, W. W., Taylor, O. C. and Heggestad, H. E., J. Air Pollut. Contr. Assoc., 23, 257 (1973).
2. Bennett, J. H., Heggestad, H. E. and McNulty, I. B., Proc. Fourth Ann. Plant Growth Reg. Working Group Meetings, Hot Springs, Ark., (1977), p. 323.
3. Pryor, W. A., ed., "Free Radicals in Biology," Vol, 1, Academic Press, Inc., New York, N. Y. (1976).
4. Jenner, E. L., Carnahan, J. E. and Wat, E. K. W., Proc. Fifth Ann. Meeting, Plant Growth Regulator Working Group, Blacksburg, Va., (1978), p. 238.
5. Bennett, J. H., Lee, E. H. and Heggestad, H. E., Pl. Physoil. Suppl., 63, 152 (1979).
6. Lee, E. H., Brnnett, J. H. and Heggestad, H. E., Pl. Physoil. Suppl., 64, 62 (1980).

OXYGEN, OXY RADICALS AND SUPEROXIDE DISMUTASE IN CANCER AND CANCER TREATMENT

Garry R. Buettner[1] *and Larry W. Oberley*[2]

[1]*Department of Chemistry, Wabash College, Crawfordsville, Indiana*
[2]*Radiation Research Laboratory, The University of Iowa, Iowa City, Iowa 52242*

Many comparisons have been made recently between superoxide dismutase (SOD) activities in normal and malignant cells (1-19). These studies, of *in vivo* and *in vitro* experimental tumor systems, have included chemically induced, virally induced and spontaneous tumors in rats, mice, and man. In the approximately fifty tumor systems examined to date the cancer cells generally have lower levels of CuZn-SOD than their normal cell counterparts. Exceptions have been found to the lowered CuZn-SOD activity but the Mn-SOD activities of the tumor cells were found, without exception, to be below that found in normal cells. The exciting observation has also been made that the reduction of Mn-SOD in the cell corresponds to the rate of tumor growth, that is, the greater the reduction in Mn-SOD activity the greater a tumor cell's growth rate (17,18). The role of SOD in cancer has recently been reviewed by Oberley and Buettner (1,2).

The presence of Mn-SOD, an inducable enzyme, in the mitochondria would infer the potential for superoxide production in the mitochondria. As yet, no one has demonstrated this in an *in vivo* system. However, Dionisi *et al*. (3) and Bize *et al*. (17,18) have shown that submitochondrial particles from tumor cells can generate superoxide at rates comparable to those isolated from the corresponding normal tissue, thus, the requirement for mitochondrial SOD appears to exist.

In keeping with the preceding observations we have proposed the following model:

> Tumor cells have diminished amounts of Mn-SOD, but at the same time they are capable of producing significant amounts of superoxide resulting in an increase in the numbers of oxygen derived radicals present in the cell. These radicals lead to damage of key subcellular structures, such as the nucleus or the mitochondria, or a shift in the cell's chemistry to avoid and/or compensate for the damage.

Diminished amounts of Mn-SOD activity coupled with superoxide production appears to be a general characteristic of mammalian tumor cells. We believe that this difference between cancerous

ISBN 0-12-592050-4

and normal cells can be exploited in the treatment of cancer. Three different techniques using these properties could be developed.

1. *Increase production of superoxide in the cell.* The tumor cell, with its insufficient SOD activity, would be unable to inactivate the excess superoxide which should then be able to inflict lethal damage.

2. *Inhibition of the CuZn-SOD.* The inhibition of the CuZn-SOD, in a cell already deficient in SOD activity, would expose the cell to the potentially fatal effects of endogenous superoxide.

3. *The addition of pseudo SOD.* The addition of SOD activity to tumor cells may arrest cell division, and thus stop tumor growth. (NOTE: This technique differs from the rationale of most therapies in that the tumor cells would be controlled rather than killed.)

Examples of the first technique are found in two different classes of anti-cancer drugs. Andriamycin and streptonigrin have been shown to stimulate superoxide production (20,21). It has also been shown that drug generated superoxide can cause breakage in circular double stranded DNA. Therefore, it has been suggested that the drug's action is to generate superoxide which could then cause breaks in the DNA of the tumor cell.

The second technique, improving cancer therapy by inhibiting CuZn-SOD, also looks very promising in that normal cells with their full complement of Mn-SOD would be less susceptible than tumor cells with little or no Mn-SOD protection. Experiments using diethyldithiocarbamate as an inhibitor of CuZn-SOD have shown that the toxicity of tumor cells to the anti-cancer drug bleomycin is greatly increased (22).

The third method involves the addition of metal chelates to tumor cells. Some metal coordination compounds such as Cu(II) [3,5-diisopropyl salicylate]$_2$, abbreviated CuDIPS, exhibit SOD activity (23,24). Because of the small size and the lipid solubility of such metal chelates, they are able to penetrate the cell wall thus becoming intracellular superoxide scavengers. Reduction in tumor size, delayed metastases and a significant increase in survival time were observed when mice implanted with Ehrlich carcinoma cells were treated with varying doses of CuDIPS (13,24).

The loss of Mn-SOD activity appears to be a general feature of malignant disease. Thus, the role of superoxide and superoxide dismutase in cancer and cancer treatment should be a promising avenue of research.

ACKNOWLEDGMENTS

This work was supported in part by the Haines Research Fund, The Milheim Foundation for Cancer Research and the Research Corp.

REFERENCES

1. Oberley, L. W. and Buettner, G. R., Cancer Res., 39, 1141 (1979).
2. Oberley, L. W., Oberley, T. D. and Buettner, G. R., J. Med., Hypotheses, 6,249 (1980).
3. Dionisi, D., Galeotti, T., Terranove, T. and Azzi, A., Biochim. Biophys. Acta, 403, 292 (1975).
4. Yamanaka, N. Y. and Deamer, T., Physiol. Chem. Phys., 6, 95 (1974).
5. Sahu, S. K., Oberley, L. W., Stevens, R. H. and Riley, E. F., J. Natl. Cancer Inst., 58, 1125 (1977).
6. Oberley, L. W., Bize, I. B., Leuthauser, S. W. H. C. and Gruber, H. E., J. Natl. Cancer Inst., 61, 375 (1978).
7. Oberley, L. W., Sahu, S. K. and Cheng, F. H. F., Proc. Am. Assoc. Cancer Res., 19, 147 (1978).
8. Yamanaka, N., Ota, K. and Utsumi, H., "Biochemical and Medical Aspects of Active Oxygen," (O. Hayaiski and K. Asada, eds.), Baltimore, MD, (1978), p. 183.
9. Peskin, A. V., Koen, Y. M. and Zbarsky, I. B., FEBS Lett., 78, 41 (1977).
10. Peskin, A. V., Zbarsky, I. B. and Konstantinov, A. H., DOKI. Akad. Nauk. S.S.S.R., 229, 751, (1976).
11. Lankin, U. Z. and Gurevich, S. M., DOKI. Akad. Nauk. S.S.R., 226, 705, (1976).
12. Bozzi, A., Mavelli, I., Finozzi, A. A., Strom, R., Wolf, A. M., Modovi, B. Rotilio, T., Mal. Cell. Biochem., 10, 11 (1976).
13. Oberley, L. W., Leuthauser, S. W. H. C., Buettner, G. R., Sorenson, J. and Oberley, T. D., (A. Autor, ed.), Raven Press, New York, N. Y. In press.
14. Yamanaka, N., Nishida, K. and Ota, K., Physiol. Chem. Phys., 11, 253 (1979).
15. Loven, D. P., Oberley, L. W., Rosseau, F. M. and Stevens, R. H., J. Natl. Cancer Inst., 65,377 (1980).
16. VanBalgooy, J. N. A. and Roberts, E., Comp. Biochem. Physiol., 62B, 263 (1979).
17. Bize, I. B., Oberley, L. W. and Morris, H. P., Cancer Res., 40,3686 (1980).
18. Bize, I., and Oberley, L. W., Abstracts Seventh Biannual Hepatoma Sym., Silver Springs, MD (1979).

19. Galeotti, T., Borrello, S., Seccia, A., Farallo, E., Bartoli, G. M. and Serri, F., Arch. Dermatol. Res., 267, 83 (1980).
20. Gregory, E. M. and Fridovich, I., J. Bacteriol., 114, 1193 (1973).
21. Thayer, W. S., Chem. Biol. Interact., 19, 265 (1977).
22. Lin, P. S., Kwock, L. and Goodchild, N. T., The Lancet, April 7, (1979), p. 777.
23. DeAlvare, L. R., Goda, K. and Kimura, T., Biochem. Biophys. Res. Commun., 69, 687 (1976).
24. Leuthauser, S., Ph.D. thesis, The University of Iowa, (1979).

OXIDATION OF PYRIMIDINE AND PURINE DEOXYRIBONUCLEOSIDES BY REACTIVE OXYGEN SPECIES GENERATED BY THE HYDROLYTIC DECOMPOSITION OF POTASSIUM PERCHROMATE

J. Cadet, A. Balland, L. Voituriez,[1] *B. S. Hahn and S. Y. Wang*[2]

[1]*Laboratoire de Radiobiochimie, Centre d'Etudes Nucléaires de Grenoble 85X, Grenoble, France*

[2]*Division of Environmental Chemistry, The Johns Hopkins University, Baltimore, Maryland 21205*

Hydrolytic decomposition of potassium peroxychromate (K_3CrO_8), which has been shown to be an efficient source of singlet oxygen (1), led to the unexpected non-selective degradation of the fundamental DNA components (2). In the present studies an attempt has been made to determine the nature of the oxygen species involved in the notable degradation of the four pyrimidine and purine 2'-deoxyribonucleosides (dThd, dCyd, dAdo, and dGuo) by treatment with peroxychromate in aqueous solutions. The major decomposition products have been identified as the four cis and trans diastereoisomers of 5,6-dihydroxy-5,6-dihydrothymidine, N(2-deoxy-β-D-erythro-pentofuranosyl) formamide and 5R and 5S optical isomers of N(2-deoxy-β-D-erythro-pentofuranosyl)-5-hydroxy-5-methylhydantoin by comparison of their spectroscopic (^1H and ^{13}C NMR, mass spectrometry, circular dichroism) and tlc or hplc properties with those of authentic samples (3,4). A reasonable mechanism for the formation of these nucleosides is the production of transient thymidine hydroxyhydroperoxides resulting from initial hydroxyl radical addition to the 5,6-pyrimidine bond followed by the fast reaction of molecular oxygen with the resulting organic radical and subsequent reduction step as the following (5,6):

$$\text{T(fragment thymyl)} \xrightarrow{OH^\bullet} {}^\bullet\text{T(OH)} \xrightarrow{O_2} \text{T(OH)}O_2^\bullet \xrightarrow{O_2^{\bullet -},H^+} \text{T(OH)OOH}$$

The release of free thymine (~ 20%) may be accounted for by dismutation between 3' or 5' peroxydic sugar radicals or oxidation of the carbon C(1') after abstraction of the corresponding hydrogen atom by hydroxyl radicals (7). The major role played by the hydroxyl radicals in these reactions is substantiated by the scavenging protective effect of 5% added methanol to the aqueous solution. Under the latter conditions, 2'-deoxyguanosine is partly decomposed whereas the other nucleosides are unaffected. The mechanism of formation of some of the major oxidation products of dGuo, i.e., 2-deoxy-D-erythro-pentose, N(2-deoxy-D-erythro-pentosyl)urea (I), N-(2-deoxy-β-D-erythro-pentofuranosyl) oxamide (II), and (2-deoxy-β-D-erythro-pentofuranosyl)cyanuric acid (III), which are also produced by microwave electric discharges and sensitized photooxidation (8), can be explained by

ISBN 0-12-592050-4

the generation of singlet oxygen O_2 ($^1\Delta_g$)

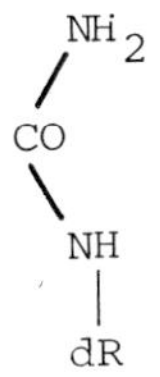

I

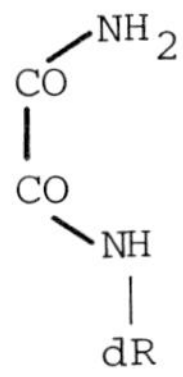

II

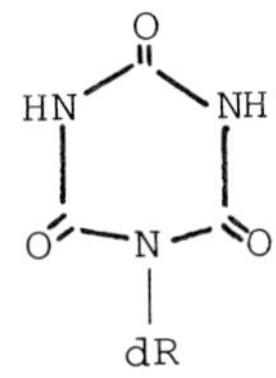

III

Further support for the involvement of O_2 ($^1\Delta_g$) in this reaction is given by the expected increase in the degradation of dGuo in deuterated solvents and at higher pH. Furthermore, a marked protective effect is observed by adding sodium azide, a known scavenger of singlet oxygen. The oxidation of dGuo by O_2 ($^1\Delta_g$) generated by the decomposition of peroxychromate appears to be 10 times less efficient than the degradation of this nucleoside by hydroxyl radicals at neutral pH. It must be emphasized that superoxide anion radicals $O_2^{\cdot -}$ which are also produced in the hydrolysis of this salt (9) are unreactive towards the four 2'-deoxyribonucleosides. In conclusion potassium peroxychromate may be considered as a convenient oxidizing agent for the selective degradation of DNA components either by hydroxyl radicals or by singlet oxygen O_2($^1\Delta_g$).

ACKNOWLEDGEMENT

We wish to express our appreciation for the support by grants from the National Institutes of Health (grants No. 1 PO1 ES02300 and 5 RO1 GM24238).

REFERENCES

1. Peters, J. W., Pitts, J. N., Rosenthal, I., and Fuhr, H., J. Am. Chem. Soc., 94,4348 (1972).
2. Peters, J. W., Bekowies, P. J., Winer, A. M. and Pitts, J. N., J. Am. Chem. Soc., 97,3299 (1975).
3. Cadet, J. and Teoule, R., Bull. Soc. Chim. Fr., 879 (1975).
4. Cadet, J. and Teoule, R., J. Chem. Res. (S) 296 (1977).
5. Scholes, G., "Photochemistry and Photobiology of Nucleic Acids," Vol. 1 (S. Y. Wang, ed.), Academic Press, New York, N.Y., 1976, p. 527.
6. Teoule, R. and Cadet, J., "Molecular Biology, Biochemistry and Biophysics" (J. Hütterman, W. Köhnlein, R. Teoule, and A. J. Bertinchamps, eds.), 27, Springer, Berlin, 1978, p. 171.
7. Dizdaroglu, M., Neuwald, K. and Von Sonntag, C., Z. Naturforsch. 31b,227 (1976).
8. Cadet, J. and Teole, R., Photochem. Photobiol., 28,661 (1978).
9. Paschen, W. and Weser, U., Hoppe-Seyler's Z. Physiol. Chem., 727 (1975).

MICROSOMAL METABOLISM OF HYDROXYL RADICAL SCAVENGING AGENTS

Arthur I. Cederbaum and Gerald Cohen

Departments of Biochemistry and Neurology, Mount Sinai School of Medicine of the City University of New York, New York, New York 10029

The purpose of this communication is to show that in the presence of an NADPH-generating system, liver microsomes can oxidize a variety of hydroxyl radical scavenging agents to appropriate products. The scavengers studied were primary aliphatic alcohols (ethanol and 1-butanol), a secondary alcohol (isopropanol), dimethylsulfoxide (Me_2SO), methional and 2-keto-4-thiomethylbutyrate (KTBA). The microsomes actively oxidized these substrates by a mechanism consistent with a role for the hydroxyl radical, or for a species with the oxidizing power of the hydroxyl radical, e.g., a ferryl radical. Full details of these experiments are described elsewhere (1-7).

Microsomes produced acetaldehyde, butyraldehyde, acetone, methane and ethylene from ethanol, 1-butanol, isopropanol, Me_2SO, methional and KTBA, respectively (Table I). Product formation was both concentration-dependent and time-dependent and was decreased >90% in the absence of NADPH, microsomes, substrate, or when H_2O_2 replaced NADPH. With the exception of ethanol, azide, an inhibitor of microsomal catalase, increased product formation two-to-three-fold (Table I). An explanation is that H_2O_2 is a precursor of $\cdot OH$, and azide prevents the decomposition of H_2O_2 by catalase. Since ethanol may be oxidized by $\cdot OH$ and by

TABLE I. Microsomal Metabolism of Hydroxyl Radical Scavenging Agents in the Absence and Presence of Azide

Scavenger	*Product*	*Rate of Product Formation (nmol/min/mg protein)* -Azide	+Azide
Ethanol	*Acetaldehyde*	*9.85±0.59*	*8.00±0.92*
1-Butanol	*Butyraldehyde*	*3.39±0.59*	*7.98±0.97*
Isopropanol	*Acetone*	*0.98±0.32*	*3.22±0.22*
Me_2SO	*Methane*	*0.03±0.01*	*0.09±0.02*
Methional	*Ethylene*	*0.57±0.09*	*1.45±0.07*
KTBA	*Ethylene*	*0.55±0.09*	*1.25±0.09*

ISBN 0-12-592050-4

TABLE II. Competition Between Hydroxyl Radical Scavenging Agents for Oxidation by Liver Microsomes

		Rate of Reaction (nmol/min/mg protein)	
Substrate	*Competitor*	*-Competitor*	*+Competitor*
10mM Methional	33mM Ethanol	1.39±0.06	0.75±0.07
10mM KTBA	33mM Ethanol	1.42±0.04	1.12±0.06
10mM KTBA	30mM DMPO	3.34±0.13	0.85±0.08
33mM DMSO	33mM Ethanol	0.13±0.02	0.08±0.02
33mM Ethanol	10mM Methional	12.45±0.70	6.44±0.45
33mM Ethanol	10mM KTBA	13.50±0.80	6.55±0.45
33mM Ethanol	33mM Me_2SO	11.65±0.75	4.00±0.25
51mM Ethanol	30mM DMPO	7.23±0.38	2.33±0.26
51mM Ethanol	100mM Mannitol	9.05±1.03	3.34±0.41
51mM Ethanol	67mM Benzoate	8.54±0.90	4.28±0.58
51mM 1-Butanol	94mM Me_2SO	7.24±0.29	3.88±0.32
51mM 1-Butanol	30mM DMPO	4.97±0.48	2.91±0.03
44mM Isopropanol	47mM Me_2SO	2.25±0.21	1.16±0.17
44mM Isopropanol	67mM Mannitol	2.36±0.27	1.33±0.23
44mM Isopropanol	10mM KTBA	2.20±0.18	1.54±0.24

catalase, little effect of azide on ethanol oxidation is observed (1). In other experiments, external addition of H_2O_2 in the presence of NADPH and azide also increased product formation; this latter result is consistent with the interpretation that H_2O_2 serves as a precursor of ·OH. Production of ·OH from H_2O_2 may be via a Fenton-type reaction or an "iron-catalyzed Haber-Weiss reaction" (1,2). Indeed, addition of Fe-EDTA increased product formation (5).

Table II summarizes data which indicate that the oxidation of a ·OH scavenger is decreased in the presence of another ·OH scavenger. When examined in more detail, the kinetics of inhibition were competitive (1,2,4). Urea, which is not an effective ·OH scavenger, did not inhibit product formation (2). The cross competition between the scavengers was specific since the scavengers had no effect on classical drug metabolism pathways such as aminopyrine demethylation or aniline hydroxylation, or on microsomal electron transfer (1,4). Of special interest is the observation that 5,5-dimethyl-1-pyrroline-1-oxide (DMPO) also blocked microsomal metabolism of ethanol, 1-butanol and KTBA (Table II) (6). This free radical spin-trapping agent has been used in ESR studies to detect ·OH produced during microsomal electron transfer (8). Recent experiments (9) indicate that similar results can be obtained with reconstituted systems containing cytochrome P-450 purified from phenobarbital-treated rats.

Control experiments using H_2O_2 in place of NADPH, chelex-100 to remove contaminating iron, and diethylenetriaminepentaacetic acid, indicate that ·OH production is not due to a non-specific interaction between H_2O_2 and "contaminating" iron.

In summary, these studies provide chemical evidence for the production of ·OH by microsomes or by reconstituted microsomal systems during NADPH oxidation. Hydroxyl radicals are utilized for the microsomal metabolism of selective drugs such as alcohols, dimethylsulfoxide, methional and KTBA.

ACKNOWLEDGMENTS

These studies were supported by USPHS Grants AA-03312, AA-04413, AA-03508 and RCDA (AIC) 5K02-00003 from the National Institute on Alcohol Abuse and Alcoholism.

REFERENCES

1. Cederbaum, A. I., Dicker, E. and Cohen, G., Biochemistry, 17, 3058 (1978).
2. Cederbaum, A. I., Dicker, E., Rubin, E. and Cohen, G., Biochemistry, 18, 1187 (1979).
3. Cohen, G. and Cederbaum, A. I., Science, 204, 66 (1979).
4. Cohen, G. and Cederbaum, A. I., Arch. Biochem. Biophys. 199, 438 (1980).
5. Cederbaum, A. I., Dicker, E. and Cohen, G., Biochemistry, 19, 3698 (1980).
6. Cederbaum, A. I. and Cohen, G., Arch. Biochem. Biophys., 204, 397 (1980).
7. Cederbaum, A. I., Qureshi, A. and Messenger, P., Biochem. Pharmacol., In press.
8. Lai, C. S. and Piette, L. H., Arch. Biochem. Biophys., 190, 271 (1978).
9. Cederbaum, A. I., Miwa, G., Cohen, G. and Lu, A. Y. H., Biochem. Biophys. Res. Commun., 91, 747 (1979).

THE INFLUENCE OF SUPEROXIDE ANION RADICAL IN ALTERING THE IN VIVO ADHESIVE INTERACTION BETWEEN GRANULOCYTES AND VENULAR ENDOTHELIUM

R. F. Del Maestro,[1] *H. H. Thaw,*[2] *M. Planker,*[3] *and K.-E. Arfors*[2]

[1]*Laboratory of Experimental Brain Research, Victoria Hospital, Department of Neurosurgery, University of Western Ontario, London, Ontario, Canada*

[2]*Department of Experimental Medicine, Pharmacia AB, Uppsala, Sweden, S-75104*

[3]*Department of Medicine A, University of Dusseldorf, West Germany*

INTRODUCTION

Inflammation is characterized by alterations in the adhesive interaction between polymorphonuclear leukocytes (granulocytes) and vascular endothelium. A variety of changes such as increased granulocyte rolling, decreased granulocyte rolling velocity and the adhesion and emigration of granulocytes have been observed (1, 2). The biochemical basis for this enhanced granulocyte adhesiveness is unknown. Since the activation of granulocytes is associated with the release of large amounts of superoxide anion radical ($O_2^{\bar{\cdot}}$) and the subsequent generation of H_2O_2, hydroxyl radical and other oxidizing species (3), it appeared reasonable to assess the role of these reactive substances on in vivo granulocyte behavior.

MATERIALS AND METHODS

In vivo granulocyte behavior was assessed using the hamster cheek pouch microvascular system (4). Venules (diameter 30-45 μm) were monitored continuously using a T.V. camera with the image displayed on a television monitor and all information stored on a videotape recorder. Granulocyte rolling frequency and granulocyte velocity were measured as described by Atherton and Born (1). Granulocyte adhesion was assessed by counting the number of adherent granulocytes present in a 100 μm venular segment and red cell velocity was measured in the venule under study (5). Free radical generation was induced by the exposure of the cheek pouch to 0.96 mM hypoxanthine and 0.05 units/ml of xanthine oxidase for 1 minute and then recommencing superfusion. Control experiments were performed using denatured enzyme (0.05 units/ml). The role of individual compounds was assessed using 50 μg/ml superoxide dismutase (SOD), 50 μg/ml catalase (CAT) and 10 mM L-methionine in the reservoir fluid.

ISBN 0-12-592050-4

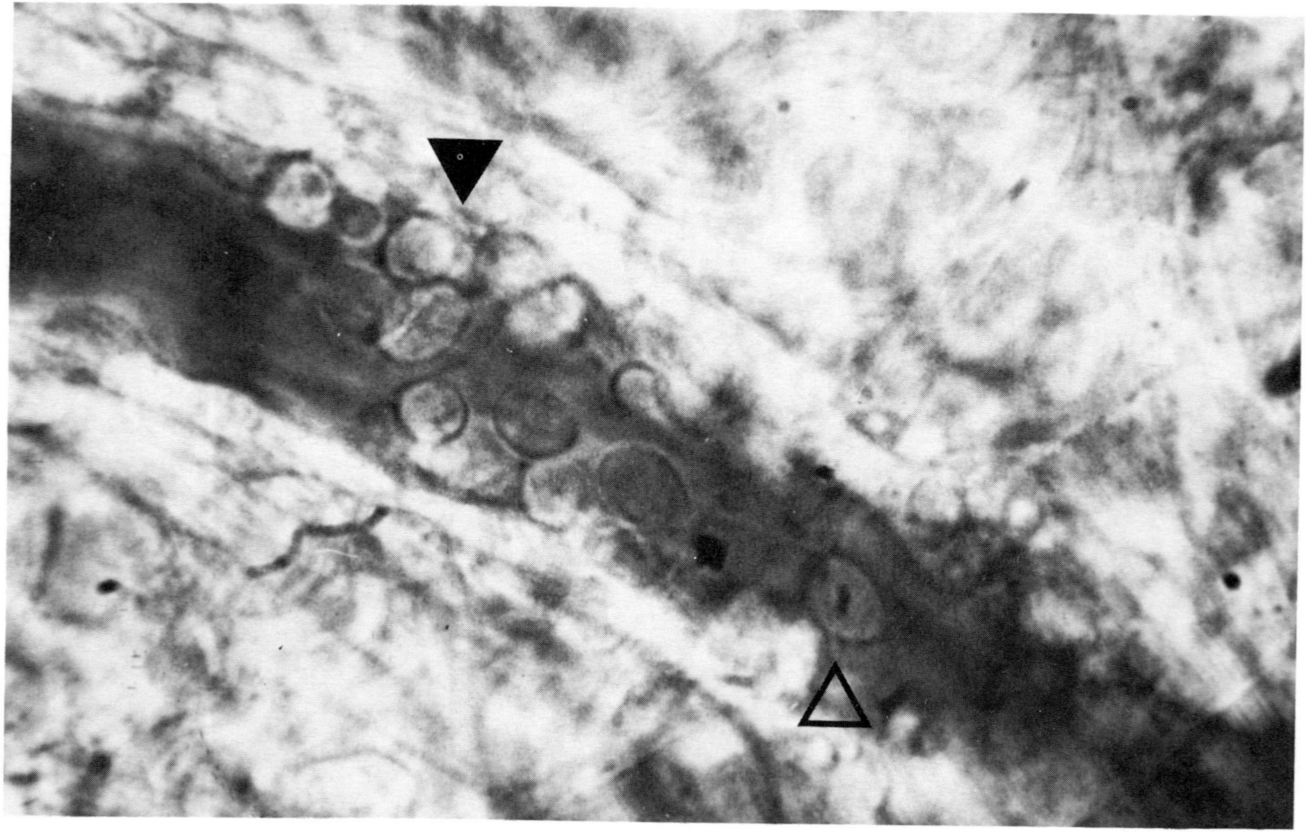

FIGURE 1. In vivo granulocyte behavior observed in a 35 μm diameter venule demonstrating granulocyte adhesion (▲) and granulocyte rolling (△).

RESULTS

The hamster cheek pouch model allowed continuous observation and accurate recording of granulocyte behavior in venules (Fig. 1). No significant differences were noted in the parameters studied in the different groups prior to the addition of enzyme. The replacement of the reservoir fluid with 0.96 mM hypoxanthine and denatured enzyme (0.05 units/ml) resulted in an initial increase in both red cell velocity and granulocyte velocity with a direct association between these two parameters being maintained throughout the experimental period. The application of active enzyme also resulted in an initial increase in red cell velocity dissociated from the observed significant decrease in granulocyte velocity. Marked granulocyte adhesion occurred and granulocyte rolling frequency increased significantly. The incorporation of 50 μg/ml of SOD in the reservoir fluid surrounding the cheek pouch prior to addition of active enzyme completely prevented the alterations in granulocyte behavior seen in the absence of SOD. The use of 50 μg/ml of CAT and 10 mM L-methionine did not modify expected results.

CONCLUSIONS

The enzymatic generation of a flux of $O_2^{\cdot -}$ on the surface of the hamster cheek pouch resulted in an increased granulocyte rolling frequency, increased granulocyte adhesion and a decreased granulocyte velocity. Our results suggest that these changes in granulocyte behavior are dependent on a $O_2^{\cdot -}$ generated chemotaxic product which may be a $O_2^{\cdot -}$ derived lipid hydroperoxide. The usefulness of SOD in experimental models of inflammation (6) may be related to its ability to inhibit the formation of an $O_2^{\cdot -}$ dependent chemotaxic factor generated from the release of $O_2^{\cdot -}$ by activated granulocytes.

ACKNOWLEDGEMENTs

This study was supported in part by the Canadian Medical Research Council.

REFERENCES

1. Atherton, A. and Born, G. V. R., J. Physiol., 222, 447-474 (1972).
2. Marchesi, V. and Florey, H. W., Q. J. Exp. Physiol., 45,343-347 (1960).
3. Babior, B. M., N. Engl. J. Med., 298,659-668 (1978).
4. Svensjo, E., Arfors, K.-E., Arturson, G. and Rutili, G., Upsala J. Med. Sci., 83,71-79 (1978).

5. Arfors, K.-E., Bergquist, D., Intaglietta, M. and Westergren, B., Upsala J. Med. Sci., 80,27-33 (1975).
6. McCord, J. M. and Wong, K., "Oxygen Free Radicals and Tissue Damage," Ciba Sym. 65 Excerpta Medica, Amsterdam (1979), p. 343-351.

HYDROXYCYCLODIENYL PEROXY RADICALS

Miral Dizdaroglu[1] *and Michael G. Simic*[2]

[1]*University of Maryland, Baltimore County, Chemistry Department, Baltimore, Maryland 21228*
[2]*National Bureau of Standards, Washington, D.C. 20234*

Reactions of the OH radical in biological and biochemical systems are extremely complex because of its high reactivity and low selectivity. The rate constants for a variety of substrates have been measured but the resulting final stable products were not determined in most cases. It is well recognized that the reactivity of the OH radical with aromatic molecules is very high (diffusion controlled) but the site of attack is usually more difficult to determine. In the case of simple benzene derivatives the OH radical adds to the benzene ring and gives o-, m- and p-hydroxy products (1).

We have examined the reaction of the OH radical with phenylalanine (Phe) in the presence of oxygen. The initially formed hydroxycyclohexadienyl radical reacts efficiently with oxygen to give a peroxy radical:

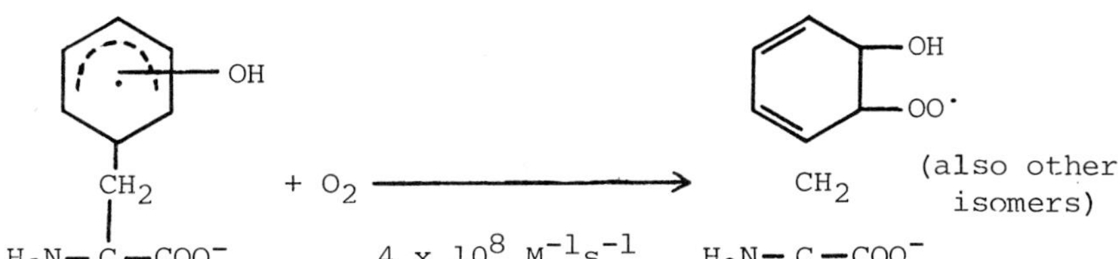

No electron transfer takes place as determined by the benzoquinone (BQ) method (2) since no semiquinone radical was observed.

$$O_2^- + BQ \longrightarrow BQ\cdot^- + O_2$$

$$\text{HO-Phe-OO}\cdot \quad + \quad BQ \quad \not\longrightarrow \quad \text{no reaction}$$

A broad absorption band was observed for HO-Phe-OO· in the 220-350 nm region which disappeared in a mixed order reaction. Hence, it was difficult to determine the mechanism of the HO-Phe-OO· disappearance. The final products were measured under the following conditions: 10^{-3} M Phe, 12.5 mM N_2O, 0.7 mM O_2, dose 75 krads. Under these conditions practically only the HO-Phe-OO· radical is formed and the final products of this peroxy free radical were determined by capillary GC. For this ca. 1 mg of a freeze-dried sample was placed in a Teflon capped vial and trimethylsilylated with 0.4 ml of a mixture of bis(trimethylsilyl) trifluoroacetamide and acetonitrile (1:1) by heating for 30 min

ISBN 0-12-592050-4

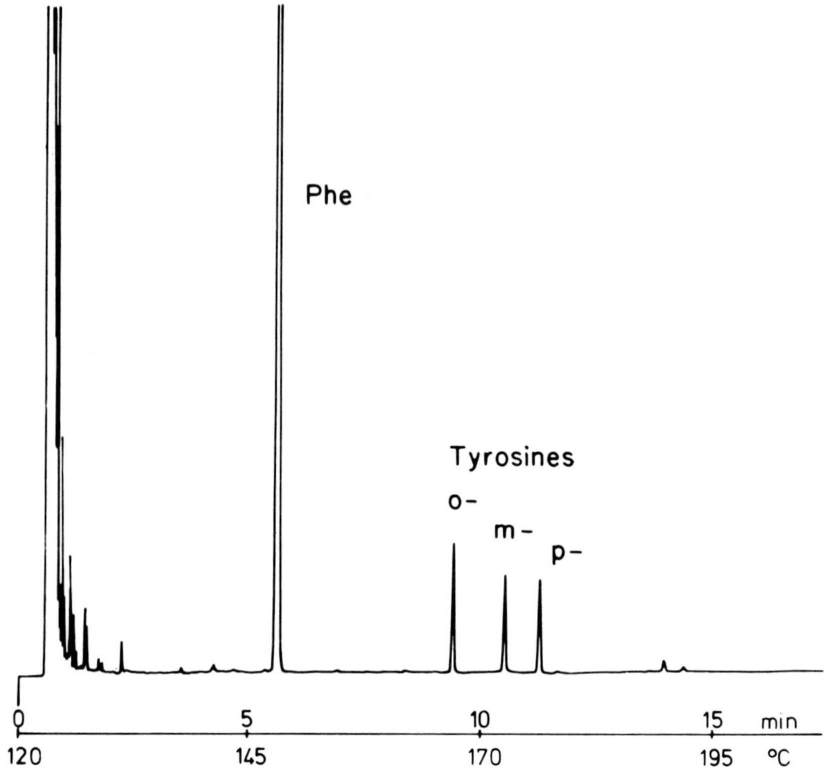

FIGURE 1. Separation of irradiation products of Phenylalanine by capillary GC after trimethylsilylation. Column: fused silica SP-2100, 12 m, 0.2 mm I.D. programmed from 120° to 250° with a rate of 5°/min.

at 120° (3). The trimethylsilylated sample was analyzed by GC using a fused silica capillary column (12 m, 0.2 mm I.D.) coated with SP-2100 (methyl silicone liquid phase). The gas chromatogram is given in Fig. 1. The G-values of o-, m- and p-tyrosine (1.6, 1.1 and 1.1, resp.) were determined by GC using serine as an internal standard.

The formation of o-tyrosine, which is not found in living systems, can be used as a monitor of an OH radical presence in diverse biological and biochemical systems. Hence, Phe provides a highly specific method for detection of OH radicals and we hope that this approach would help in clarification of some physiological reactions which implicate generation and specific reactions of the OH radical in the presence of oxygen.

REFERENCES

1. Volkert, O. and Schulte-Frohlinde, D., Tetrahedron Letts., 2151 (1968).
2. Simic, M. G. and Hayon, E., Biochem. Biophys. Res. Commun., 50,364 (1973).
3. Gehrke, C. W., Nakamoto, H. and Zumwalt, R. W., J. Chromatogr., 45,24 (1969).

ANTIOXIDANT AND RADIOPROTECTIVE PROPERTIES OF LEVAMISOLE

Carroll R. Dobbs, Joseph F. Weiss, K. Sree Kumar[1] *and Michael A. Chirigos*[2]

[1]*Biochemistry Department, Armed Forces Radiobiology Research Institute, Bethesda, Maryland 20014*
[2]*National Cancer Institute, Bethesda, Maryland 20205*

During a survey of the radioprotective properties of drugs that affect the immune response, it was observed that the immunomodulator, levamisole, prolonged survival of irradiated mice of various strains. Levamisole, when administered about 30 min before irradiation, was effective in prolonging the median survival and 30-day survival of male and female BALB/c mice at X-ray doses up to 600 rad (Table 1).

Examination of the structures of levamisole (Fig. 1A) and of an identified in vivo metabolite (1), dl-2-oxo-3-(2-mercaptoethyl-5-phenylimidazolidine or OMPI (Fig. 1B), suggested that the possible conversion of levamisole to a sulfhydryl metabolite might be responsible for any radioprotective or antioxidant effects of this drug. In vitro studies of the antioxidant properties of levamisole (2) showed that at 1.0 and 2.0 mM, levamisole significantly inhibited lipid peroxidation induced in rat liver microsomes, enzymatically (NADPH/ADP-Fe) or nonenzymatically (ascorbate/ADP-Fe). Other studies showed protection or generation of sulfhydryl groups in the incubation systems (3). Lipid peroxidation induced by X irradiation with ADP-Fe as promoter was also inhibited by levamisole as measured spectrophotometrically

TABLE 1
Effect of Levamisole (10 mg/kg ip) on Survival of Irradiated BALB/c Mice

X-RAY RADS	FEMALE: MEDIAN SURVIVAL TIME (days)		FEMALE: 30-DAY SURVIVAL		MALE: MEDIAN SURVIVAL TIME (days)		MALE: 30-DAY SURVIVAL	
	SALINE	LEVAMISOLE	SALINE	LEVAMISOLE	SALINE	LEVAMISOLE	SALINE	LEVAMISOLE
450	–	–	9/10	10/10	–	–	8/9	7/9
500	–	–	5/10	7/10	14.0	–	1/9	6/9
550	12.5	16.5	0/10	4/10	16.0	–	3/10	9/10
600	12.0	14.0	0/10	1/10	15.0	–	2/10	7/10

ISBN 0-12-592050-4

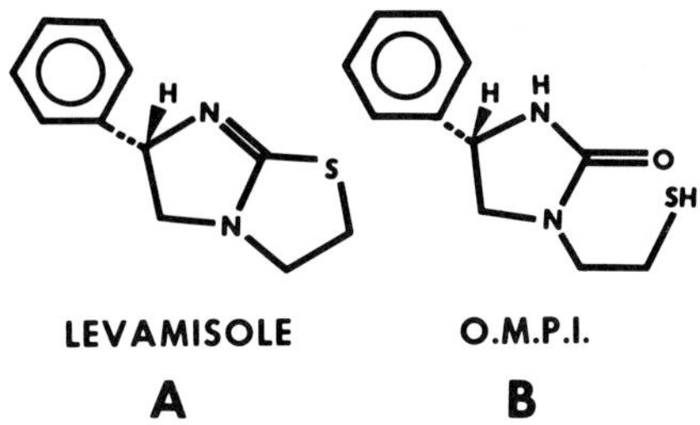

FIGURE 1. Structural formulas of levamisole (A) and its metabolite OMPI (B).

by a reduction in thiobarbituric acid-reactive material (2). A new method of measuring lipid peroxidation—determination of hydrocarbon volatiles—was used also to study levamisole effects in vitro. Quantitative gas chromatographic analysis of n-pentane generated during microsomal incubation showed that lipid peroxidation induced enzymatically was reduced in the presence of levamisole (Fig. 2). Pentane production was inhibited also by the

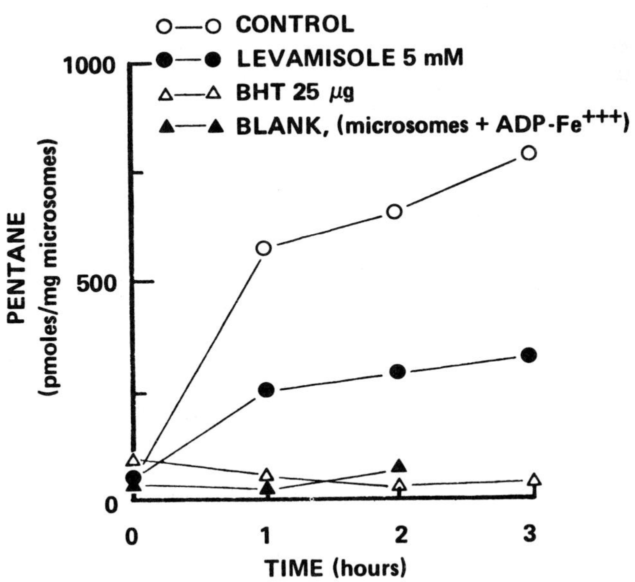

FIGURE 2. Effect of levamisole and butylated hydroxytoluene on microsomal pentane production. Pentane measured in headspace of incubation flasks containing microsomes, NADPH, and ADP-Fe^{+++}.

antioxidant BHT, indicating that the source of pentane was microsomal lipid. Analysis by gas chromatography/mass spectrometry of actively peroxidizing microsomal systems showed that levamisole was being converted to OMPI (3). There was no detectable radiolytic conversion of levamisole to OMPI in an aqueous system (4). Production of OMPI may be responsible for the antioxidant properties of levamisole. However, pure OMPI exhibited a biphasic effect with regard to microsomal lipid peroxidation, acting as a prooxidant below 25 μM and an antioxidant at concentrations higher than 50 μM (3). These studies suggest that OMPI generation and the subsequent effects on cellular oxidative processes are important aspects of the action of levamisole. Other studies have suggested also a relationship between antioxidant properties or effects on sulfhydryl group metabolism and the immunological properties of levamisole and other compounds (5-7). We propose that the effect of synthetic and natural antioxidants on normal oxidative processes in the cell as well as abnormal oxidative stress, such as that resulting from radiation exposure or drug toxicity, may be an important determinant of the immunological properties of these compounds.

REFERENCES

(1) Janssen, P. A. J., Progr. Drug Res., 20, 347 (1976).
(2) Kumar, K. S., Chirigos, M. A. and Weiss, J. F., Int. J. Immunopharmac., 1, 85 (1979).
(3) Kumar, K. S., Dobbs, C. R., Weiss, J. F. and Chirigos, M. A., J. Immunopharmac., 2, 73 (1980).
(4) Dobbs, C. R., Ehlardt, C. E. and May, L., Radiation Effects, 52,91 (1980).
(5) De Brabander, M., Van Belle, H., Aerts, F., Van de Veire, R. and Geuens, G., Int. J. Immunopharmac., 1, 93 (1979).
(6) Harman, D., Heidrick, M. L. and Eddy, D. E., J. Am. Geriat. Soc., 25, 400 (1977).
(7) Hoffeld, J. T., Fed. Proc., 39, 565 (1980).

SOME OXIDATIONS USING SUPEROXIDE AND OTHER RADICAL ANIONS

H. Brian Dunford and Alexander D. Nadezhdin, Department of Chemistry, University of Alberta, Edmonton, Alberta, Canada T6G 2G2

THE HABER-WEISS REACTION

There has been great interest in superoxide anion radical reactions in biology stimulated by the discovery of superoxide dismutase, that is that superoxide is a substrate for erythrocuprein-type enzymes (1). As a result there has been renewed interest in the Haber-Weiss elementary (single step) reaction (2)

$$O_2^{\bar{\cdot}} + H_2O_2 \longrightarrow \cdot OH + OH^- + O_2$$

However this reaction was discounted in thorough kinetic studies (3) and in meticulously performed stoichiometric experiments published in 1947 in which it was demonstrated that the disproportionation reaction

$$2H^+ + 2O_2^{\bar{\cdot}} \longrightarrow H_2O_2 + O_2$$

is unaffected by the presence of excess H_2O_2 (4,5).

We have studied superoxide reactions by using the well established technique of flash photolysis of hydrogen peroxide solutions (6-8)

$$H_2O_2 \xrightarrow{h\nu} 2\cdot OH$$

$$\cdot OH + H_2O_2 \longrightarrow HO_2^{\cdot} + H_2O$$

and the rapid occurrence of the latter reaction is direct proof that the Haber-Weiss elementary reaction cannot be a source of ·OH radicals. Exploration for several elementary reactions which when added together may give an overall Haber-Weiss type reaction is a valuable approach (9) but as a single step process the Haber-Weiss reaction is dead (10).

REACTIONS OF $HO_2^{\cdot}$ WITH ASCORBATE AND NADH

The pH rate profiles for the reactions of superoxide (acting as an oxidizing agent) with ascorbate and NADH have been obtained (8,9). The rate constants are $10^6\ M^{-1}s^{-1}$ and $10^5\ M^{-1}s^{-1}$ for reactions of the hydroperoxyl radical, $HO_2^{\cdot}$, with ascorbate and NADH. The superoxide anion is much less reactive. Thus ascorbate is more reactive than NADH, but ascorbate radicals are more inert than NAD· radicals toward further oxidation (8,9).

ISBN 0-12-592050-4

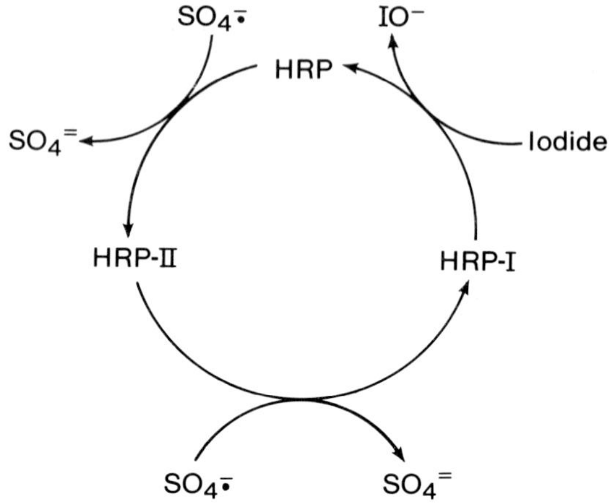

FIGURE 1. A scheme for the oxidation of horse radish peroxidase.

REVERSAL OF THE PEROXIDASE CYCLE

We turn now to the oxidation of an enzyme by radical anions. The photolysis of a potassium persulfate solution either by itself or in the presence of chloride or bromide leads to the production of $SO_4^{\bar{\cdot}}$, $Cl_2^{\bar{\cdot}}$ or $Br_2^{\bar{\cdot}}$. All of these species oxidize horseradish peroxidase compound II to compound I (11). More detailed analysis of the oxidation is possible when native enzyme is oxidized to compound II (12). In this case it is possible to monitor the extent to which attack occurs at various points on the surface of the enzyme as well as at the active site. The ratio of sites on the native enzyme surface oxidized to active site oxidized is 2:1, 8:1 and 5:1 for $SO_4^{\bar{\cdot}}$, $Cl_2^{\bar{\cdot}}$ and $Br_2^{\bar{\cdot}}$ respectively.

The two-electron reduction of compound I directly to native enzyme has been demonstrated using iodide as reducing substrate (13) and this is likely a universal mechanism for enzymatic halide ion oxidation including that by myeloperoxidase where the oxidation of chloride in acid medium is a key component of the phagocytic process (14). Thus the reversal of all steps in the normal peroxidatic cycle has been demonstrated as illustrated in Fig. 1.

REFERENCES

1. McCord, J. M. and Fridovich, I., J. Biol. Chem. 244,6049 (1969).
2. Haber, F. and Weiss, J. J., Proc. Roy. Soc. Ser. A, 147, 332 (1934).
3. Barb, W. G., Baxendale, J. H., George, P. and Hargrave, K. R., Trans. Faraday Soc., 47,591 (1951).

4. George, P., Disc. Faraday Soc. 2,196 (1947).
5. Fee, J. A. and Valentine, J. S., "Superoxide and Superoxide Dismutases," (A. M. Michelson, J. M. McCord and I. Fridovich, eds.), Academic Press, New York, 1977, p. 19.
6. Baxendale, J. H., Radiat. Res. 17,312 (1962).
7. Nadezhdin, A. D. and Dunford, H. B., J. Phys. Chem. 83,1957 (1979).
8. Nadezhdin, A. D. and Dunford, H. B., Can. J. Chem. 57,3017 (1979).
9. Halliwell, B., FEBS Letters, 92,321 (1978).
10. Weinstein, J. and Bielski, B. H. J., J. Am. Chem. Soc. 101, 58 (1979).
11. Nadezhdin, A. D. and Dunford, H. B., Can. J. Biochem. 57, 1080 (1979).
12. Nadezhdin, A. D. and Dunford, H. B., Can. J. Chem. 58,2504 (1980).
13. Roman, R. and Dunford, H. B., Biochemistry, 11,2076 (1972).
14. Klebanoff, S. J. and Clark, R. A., "The Neutrophil: Function and Clinical Disorders," North-Holland, Amsterdam (1978).

WATER RADICALS AND SUPEROXIDE IN THE INTRODUCTION OF NUCLEOTIDE DAMAGE IN γ-IRRADIATED DNA

Ross S. Feldberg and Josephine A. Carew

Department of Biology, Tufts University, Medford, Massachusetts

INTRODUCTION

In this study we report on the use of a damage-specific DNA binding protein from human cells (1) as a probe for nucleotide damage. This protein recognizes nucleotide damage in UV- and γ-irradiated DNA, as well as in DNA deaminated with HNO_2 or $NaHSO_3$ (2). However, the protein does not bind to single-strand breaks, apurinic sites, thymine dimers, single-strand tails or psoralen-induced crosslinks in DNA. Thus, the protein functions as a broad specificity probe for monofunctional base damage, probably by recognizing a minor helix distortion in DNA.

RESULTS

The retention of γ-irradiated T7 DNA on a nitrocellulose filter in the presence of an excess of binding protein was used to follow the introduction of nucleotide damage into the DNA. From dose-response curves for filter binding the percent of DNA retained on the filter after 30 krads of irradiation is shown in Table 1. The Poisson distribution allows calculation of the mean amount of damage introduced per T7 molecule per krad, assuming that a single lesion is sufficient to ensure full retention of the DNA molecule. If this assumption is not true, these data

TABLE 1. Damage introduction by γ-Irradiation

Irradiation condition[a]	*% binding/30 krad*	*lesions/krad/T7*
air	*26.5*	*.0103*
N_2O	*6*	*.0023*
air + I^- (100mM)	*9*	*.0033*
air X NO_3^- (100mM)	*8*	*.0029*

[a]*T7 DNA irradiated at room temperature at 1.3 μg/ml in .01 M Tris-HCl, .001 M EDTA, pH8.*

ISBN 0-12-592050-4

will be an underestimate of the rate of damage introduction. From these data, it is clear that the maximum amount of damage is introduced in air. Under N_2O, in which the aqueous electrons (e^-_{aq}) will be converted to hydroxyl radicals (OH·), very little damage is observed. However, the importance of the OH· to damage introduction is suggested by the inhibition by I^-, a specific OH· scavenger. Inhibition of damage introduction is also seen with the specific e^-_{aq} scavenger, NO_3^-. Since scavenging e^-_{aq} in the presence of oxygen results in a decrease in $O_2^{\cdot-}$ levels, these results suggest that a combination of both OH· and $O_2^{\cdot-}$ are essential for nucleotide damage introduction.

The role of $O_2^{\cdot-}$ was further examined using an enzymatic system to generate it. As shown in Table 2, damage is introduced in a xanthine plus xanthine oxidase system in a reaction dependent both on $O_2^{\cdot-}$ and on H_2O_2. The involvement of OH· is suggested by the inhibition by formate, an OH· scavenger, and a role for iron in OH· generation is suggested by the inhibition by diethylenetriaminepentaacetic acid (DTPA), a specific iron chelator. Finally, the amount of damage introduced was strongly temperature dependent, perhaps indicating a dependence on a partial melting of the DNA.

DISCUSSION

The use of damage-specific DNA binding protein has provided us with a simple and sensitive probe for nucleotide damage in DNA and has allowed us to draw some conclusions about the in vitro introduction of such damage: (1) The hydroxyl radical per se yields little nucleotide damage, as indicated by the result of irradiation under N_2O. (2) Nucleotide damage is introduced both in γ-irradiation and in an $O_2^{\cdot-}$-generating system by a combination of OH· and $O_2^{\cdot-}$ (or some $O_2^{\cdot-}$-derived species).

TABLE 2. Damage Introduction by a Superoxide-Generating System

Components[a]	*% DNA retained*
complete system (55°C)	*36%*
complete system (55°C) − xanthine	*4*
complete system (55°C) + SOD (17 units)	*7*
complete system (55°C) + formate (10mM)	*5*
complete system (55°C) + DTPA (8mM)	*5*
complete system (25°C)	*14*
complete system (25°C) + catalase (1.2 units)	*4*

[a]*Complete system consisted of 0.70 nmoles T7 DNA, 50 nmoles xanthine and 0.005 units xanthine oxidase in a 0.05 M K phosphate buffer, pH 7.8.*

ACKNOWLEDGEMENT

This research was supported by grant CA19419 from the National Cancer Institute, DHEW.

REFERENCES

1. Feldberg, R. S. and Grossman, L., Biochemistry, 15, 2402 (1976).
2. Feldberg, R. S., Nuc. Acids Res., 8, 1133 (1980).

EVIDENCE THAT ALLOXAN AND DIHYDROXYFUMARATE ALTER PANCREATIC ISLET CELL FUNCTION THROUGH THE GENERATION OF OXYGEN FREE RADICALS

L. J. Fischer, Department of Pharmacology, University of Iowa, Iowa City, Iowa 52242

Alloxan has been used to produce diabetes in laboratory animals for over 40 years. During this time the mechanism by which this agent causes destruction of insulin producing cells has remained obscure in spite of intense investigation. Recently, however, evidence has been gathered to implicate the production of hydroxyl free radicals in the mechanism of alloxan cytotoxicity. The initial suggestion that alloxan may damage cells through the production of free radical species of oxygen resulted from biochemical studies using dialuric acid, a reduced form of alloxan (1). These experiments indicated that the autooxidation of dialuric acid produced superoxide anion ($O_2^{\overline{\cdot}}$), hydrogen peroxide (H_2O_2) and hydroxyl free radicals (·OH). Following that report, it was shown that a number of known scavengers of hydroxyl free radicals, when given to mice before alloxan, could protect the animals from diabetes (2). The results of these in vivo experiments were sometimes difficult to interpret because certain of the scavengers (e.g., n-butanol) were shown to protect animals from alloxan by an indirect mechanism and not by scavenging radicals (3).

Recently, this and other laboratories have concentrated on investigating the possible role played by oxygen free radicals in the action of alloxan on isolated pancreatic islets. Experiments with isolated tissue were conducted to help alleviate the complications inherent in the assessment of biochemical mechanisms of toxicity in the intact animal. It is the purpose of this report to briefly review the results of in vitro experiments designed to elucidate the involvement of oxygen free radicals in the action of alloxan.

The exposure of pancreatic islets to alloxan in vitro for short periods (e.g. 5 min) produces detrimental effects that alter the function of islet cells. After exposure to alloxan, islet cells show a deficiency in: glucose stimulated insulin secretion (4), insulin biosynthesis (5), glucose oxidation (6), rubidium uptake and dye exclusion (7). We have utilized glucose stimulated insulin release and its reduction due to alloxan exposure to examine the possible role of oxygen free radicals in the alloxan effect (8). Exposure of rat pancreatic islets to alloxan for 5 min inhibits subsequent glucose stimulated insulin release and this effect is dependent on the initial concentration

ISBN 0-12-592050-4

of alloxan. Using a concentration of 0.15 mg/ml a 60% inhibition was observed and at 0.2 mg/ml alloxan a 90% inhibition occurred. Addition of SOD (1000 U/ml) to islets 10 min prior to alloxan and allowing the enzyme to remain during the 5 min exposure period completely protected the tissue from the effects of alloxan. Heat-inactivated SOD had no protective effect. In addition to SOD, catalase (50 µg/ml) and the iron chelator diethylenetriaminepentaacetic acid (DETAPAC, 1 mM) produced complete protection when used in the same manner. The effectiveness of all of these treatments indicated that ·OH, produced via a scheme which incorporates $O_2^{\bar{\cdot}}$, H_2O_2 and iron, may be involved in alloxan action. Consistent with this, some hydroxyl radical scavengers were found to be effective in preventing the action of alloxan. We found that N,N'-dimethylurea (10 mM), but not thiourea (20 mM) or n-butanol (20 mM) was protective (9). The inability of some radical scavengers to protect from alloxan exposure may be due to their inability to react rapidly enough with ·OH or to be used in high enough concentrations with living tissue.

Results obtained in our laboratory using alloxan and various protectants are entirely consistent with results apparently obtained at the same time by Swedish workers. Grankvist *et al.* (7,10) demonstrated a protective action for SOD, catalase, DETAPAC and some radical scavengers using rubidium accumulation as a test for islet cell function. EDTA, which does not chelate iron in a manner that reduces its ability to allow the production of ·OH from H_2O_2 (11), does not protect from alloxan (10). This further supports the proposal that ·OH generated by a modified Fenton reaction is involved in alloxan action.

If alloxan is causing its insulin-inhibitory action in isolated islets via the generation of ·OH, it should be possible to observe similar actions using other generators of $O_2^{\bar{\cdot}}$ and ·OH. We have recently completed experiments utilizing autooxidizing dihydroxyfumarate (DHF), a known generator of oxygen free radicals (12). In a manner identical to the alloxan studies discussed above, isolated rat pancreatic islets were exposed to DHF for 5 min. A concentration dependent inhibition of subsequent glucose stimulated insulin release resulted (13). SOD, catalase, and DETAPAC, when present before and during the DHF exposure, protected the insulin secreting cells from the insulin inhibitory action caused by DHF treatment. All of these effects were identical to those observed with alloxan exposure and add support to the concept that alloxan acts via the generation of ·OH.

High glucose concentrations can protect cells from alloxan *in vivo* or *in vitro* and this was thought to be due to a competition between glucose and alloxan at (or near) a glucose receptor (5). In experiments with isolated islets, we have found that glucose (5 mg/ml) can protect from the effects of DHF exposure. Thus, DHF and alloxan exposure are alike in regard to protection afforded by glucose and the mechanism of the protection by the sugar may necessarily involve a consideration of oxygen free radicals.

Considering the evidence presented it seems likely that hydroxyl free radical is involved in the deleterious effects of alloxan on the insulin secreting cells. Why alloxan selectively damages the insulin secreting cells and is relatively free from effects on other cells has not been answered. Possible explanations for the selective action of alloxan in light of our knowledge concerning free radical generation include: a) concentration of alloxan or dialuric acid at or near its site of action, b) an increased ability to reduce alloxan to dialuric acid in the insulin cells, c) increased metal catalyzed formation of ·OH from hydrogen peroxide in the insulin cells and d) a relatively limited capability for insulin cells to lower amounts of $O_2^{\bar{\cdot}}$, H_2O_2 or ·OH, by the action of endogenous protective substances. Future research will concentrate on finding which of these possibilities may pertain to the mechanism of alloxan cytotoxicity in the endocrine pancreas.

ACKNOWLEDGMENTS

Support provided by USPHS Grant GM-12675.

REFERENCES

1. Cohen, G. and Heikkila, R. E., J. Biol. Chem., 249, 2447 (1974).
2. Heikkila, R. E., Winston, B. and Cohen, G., Biochem. Pharmacol., 25, 1085 (1976).
3. Schauberger, C. W., Thies, R. L. and Fischer, L. J., J. Pharmacol. Exp. Ther., 201, 450, (1977).
4. Tomita, T., Lacy, P. E., Matchinsky, F. M. and McDaniel, M. L., Diabetes, 23, 517 (1974).
5. Niki, A., Niki, H., Miwa, I. and Lin, B., Diabetes, 25, 574 (1976).
6. Ishibashi, F., Sato, T., Onari, K. and Kawate, R., Endocrinol. Japan., 26, 395 (1979).
7. Grankvist, K., Marklund, L., Sehlin, J. and Täljedal, I. B., Biochem. J., 182, 17 (1979).
8. Fischer, L. J. and Hamburger, S. A., Diabetes, 29, 213 (1980).
9. Fischer, L. J. and Hamburger, S. A., Life Sci., 26, 1405 (1980).
10. Grankvist, K., Marklund, S. and Täljedal, I. B., FEBS Letters, 105, 15 (1979).
11. Buettner, G. R., Oberley, L. W. and Chan-Leuthauser, S., Photochem. Photobiol., 28, 693 (1978).
12. Halliwell, B., Biochm. J., 163, 441 (1978).
13. Fischer, L. J. and Hamburger, S., Endocrinol. In press.

X

PHOTOOXIDATION AND PHOTOREDUCTION REACTIONS SENSITIZED BY HEMATOPORPHYRIN

Albert W. Frenkel,[1] *Leland S. Jahnke,*[2] *and Zbyslaw J. Petryka*[3]

[1]*Department of Botany, University of Minnesota, St. Paul, Minnesota 55108*
[2]*Department of Botany and Plant Pathology, University of New Hampshire, Durham, New Hampshire 03824*
[3]*University of Minnesota, Medical Research Unit, Abbott-Northwestern Hospital, Minneapolis, Minnesota 55407*

Hematoporphyrin sensitizes the photooxidation of reduced nicotinamide adenine dinucleotide phosphate (NADPH) in the presence of molecular oxygen; evidence has been presented that in heavy water (D_2O) this reaction is mediated by singlet oxygen (1). Bodaness and Chan (1) searched for the formation of superoxide in this reaction as indicated by the reduction of ferricytochrome-c (cyt(III)-c) or of nitro blue tetrazolium (NBT), but were unable to find a significant reduction of these electron acceptors. Land and Swallow (2) presented evidence for the production of superoxide in the oxidation of the NAD semiquinone by molecular oxygen in aqueous solutions, and Peters and Rodgers (3) measured the rate constant for the reaction of singlet oxygen ($^1\Delta_g$) with NADH resulting in the formation of the NAD semiquinone and superoxide. Also, Cox <u>et al</u>. (4) had detected by electron spin resonance the formation of a superoxide spin adduct when hematoporphyrin in protic or aprotic solvents was irradiated with white light (with an ultraviolet cut-off filter). Our preparations of hematoporphyrin (Htp), in air saturated aqueous solutions (pH 8.0), excited either with green light (peak emission 530 nm) or with blue-violet light (peak emission 410 nm, u.v. cut-off 335 nm) sensitizes the photooxidation of NADPH or of reduced nicotinamide adenine dinucleotide (NADH); if either cyt(III)-c or NBT are added to this reaction mixture, good rates of reduction of these electron acceptors can be observed during continuous illumination after a lag period of 10 to 30 seconds. Superoxide dismutase (SOD) which inhibits the reduction of cyt(III)-c (5) and of NBT (6) by superoxide is equally effective in inhibiting the rates of reduction of cyt(III)-c or of NBT in the Htp sensitized reactions. The rates of these reductions are dependent on the concentrations of the electron acceptors. In the case of NBT the observed rates can be accounted for quantitatively in terms of the rate constant for its reduction by superoxide and the rate constant for the dismutation of superoxide. Thus it appears that this reduction, which most likely involves superoxide, only indirectly involves

ISBN 0-12-592050-4

sensitized electron transfer, and thus it is truly a dark reaction. On the other hand, the photooxidation of NADH or of NADPH is not inhibited by SOD, but is sensitive to azide. Traces of catalase have no effect on the rates of NBT reduction or NADH photooxidation. In the absence of O_2, the photooxidation of NADH or of NADPH proceeds only in the presence of suitable electron acceptors such as NBT or cyt(III)-c. These anoxic photooxidation/reduction reactions are not inhibited by SOD or by azide and appear to involve a sensitized electron transfer. Previously, we presented evidence that singlet oxygen may be a precursor to superoxide in the methylene blue sensitized photooxidation of epinephrine (7), and that most likely type I and type II reactions are involved in this photooxidation (8,9,10). In the aerobic photooxidation of NADH, sensitized by hematoporphyrin, we have observed only a three to five fold acceleration of this reaction in D_2O. This may indicate again that both type I and type II reactions could well be involved in the Htp sensitized aerobic photooxidation of NADH in aqueous solutions.

REFERENCES

1. Bodaness, R. S. and Chan, P. C., J. Biol. Chem., 252, 8554 (1977).
2. Land, E. J. and Swallow, A. J., Biochim. Biophys. Acta, 234, 34 (1971).
3. Peters, G. and Rodgers, M. A. J., "Oxygen and Oxy-Radicals in Chemistry and Biology"(M. A. J. Rodgers and E. L. Powers, eds.), Academic Press, N. Y. (1981). This volume, p. 705.
4. Cox, G. S. Whitten, D. G. and Giannotti, C., Chem. Phys. Letters, 67, 511 (1979).
5. McCord, J. M. and Fridovich, I., J. Biol. Chem., 243, 5753 (1968).
6. Beauchamp, C. and Fridovich, I., Anal. Biochem., 44, 276 (1971).
7. Jahnke, L. S. and Frenkel, A. W., Photochem, Photobiol., 28, 517 (1978).
8. Foote, C. S., "Singlet Oxygen," (B. Rånby and J. F. Rabek, eds.), Wiley, New York, (1978), p. 135.
9. Grossweiner, L. I. and Kepka, A. G., Photochem. Photobiol., 16, 305 (1972).
10. Kraljić, I., El Mohsni, S. and Arvis, M., Photochem. Photobiol., 27, 531 (1978).

THE PHOTOOXIDATION OF CYCLOPROPENES

Aryeh A. Frimer and Abraham Antebi, Department of Chemistry, Bar-Ilan University, Ramat-Gan, Israel

As a result of our previous studies on the photooxidation of vinylcyclopropanes (1,2) and methylenecyclopropanes (3) we have suggested that $^1\Delta$ is insensitive in the transition state to the relative thermodynamic stability of the products. To pursue this point further we have studied the photosensitized oxidation of cyclopropenes **1**-**3** which bear at least one hydrogen at the 3-position of the ring. The uptake of oxygen proceeded rapidly, but surprisingly the reaction does not involve singlet oxygen; rather, it is a free radical process as the following evidence demonstrates. The reaction is not quenched by Dabco but is dramatically slowed by 2,6-di-t-butylphenol. No reaction occurs in methanol or much below room temperature. Nor were any products observed than $^1\Delta$ was generated by thermal decomposition of triphenylphosphite ozonide. The product distribution is given in the Table. The plethora of products can be rationalized in terms of secondary rearrangements of initially formed hydroperoxide **4** and epoxide **5** as outlined in Scheme 1. All transformations are well precedented processes.

REFERENCES

(1) Frimer, A. A. and Roth, D., J. Org. Chem., 44, 3882 (1979).
(2) Frimer, A. A., Rot, D. and Sprecher, M., Tetrahedron Lett., 1927 (1977).
(3) Frimer, A. A., Farkash, T. and Sprecher, M., J. Org. Chem., 44, 989 (1979).

ISBN 0-12-592050-4

TABLE I.[a] *Product Distribution[b] of the Photooxidation of Cyclopropenes 1-3*
By permission Journal of Organic Chemistry.

Substrate	Products						
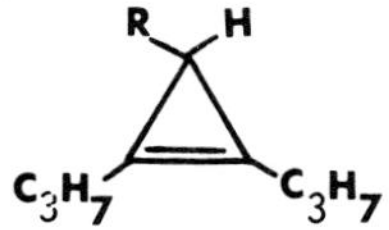	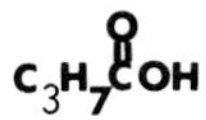	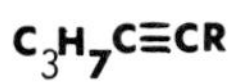	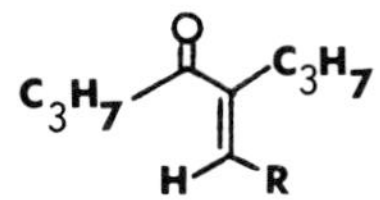	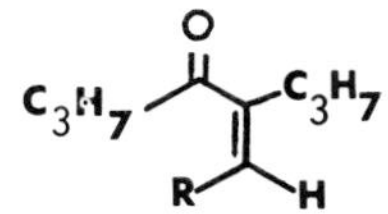	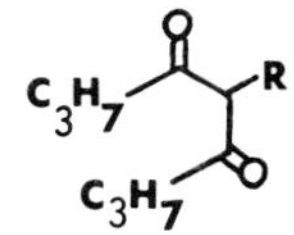	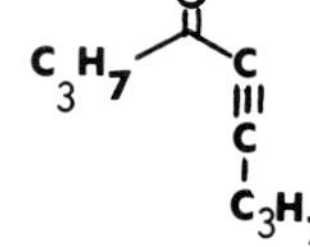	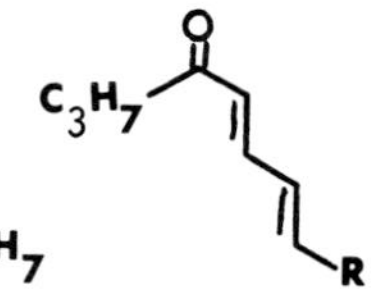
	6	7	8	9	10	11	12
R = $CO_2C_2H_5$ 1	0.43	0.48	0.02	0.07			
R = $COCH_3$ 2	0.49	0.45			0.06		
R = H 3	0.40	[0.40][c]	0.12			0.04	0.04

a) C_3H_7 represents an n-propyl radical; b) mole fraction of total addition products observed; calculated assuming equal gas chromatographic peak area responses; c) product 7c was isolated on a separate GLC column.

SCHEME I. *Proposed Mechanism for the Formation of Products 6-12 By permission Journal of Organic Chemistry.*

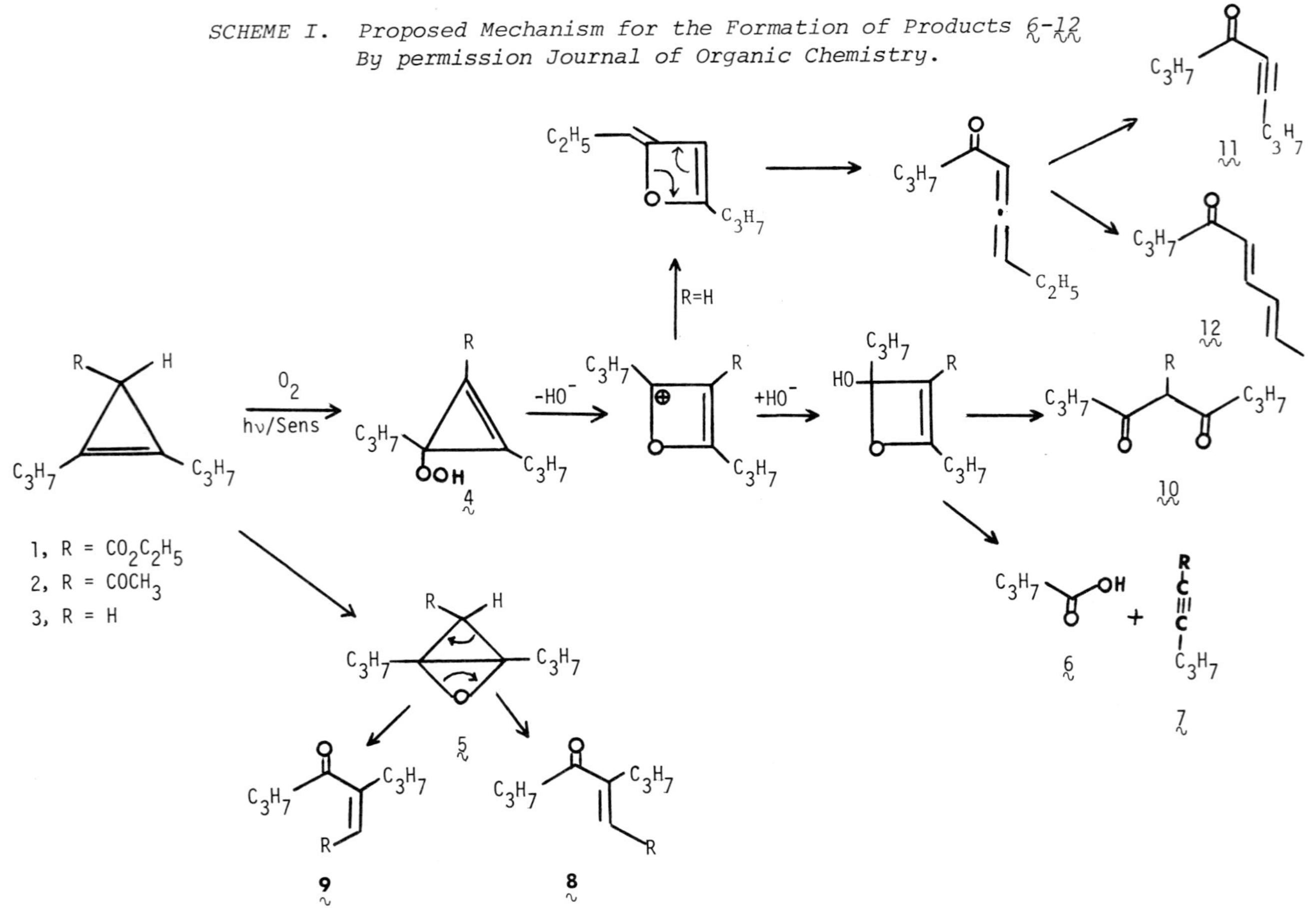

OXASTEROID SYNTHESIS WITH SUPEROXIDE ANION RADICAL
BASE CATALYZED AUTOXIDATION OF ENONES

Aryeh A. Frimer and Pessiah Gilinsky, Department of Chemistry, Bar-Ilan University, Ramat-Gan, Israel

The use of steroidal hormones, such as corticoids, progesterones and various anabolic agents, is well known in modern clinical therapy. However, the useful properties of these drugs are usually accompanied by unwanted side effects. As a result, a variety of heterosteroids have been synthesized in the hope that the undesired properties could be separated from the beneficial effects. Some success (1) has indeed been observed in the case of 2-oxa-3-oxo-Δ^4 steroids [1]. Such cyclic α,β-unsaturated lactones are conveniently synthesized by reducing the corresponding 1-hydroxy-2-oxa-3-oxo-Δ^4 steroids [2] with $NaBH_4$. However, the preparation of these lactols is by no means a trivial matter (1a).

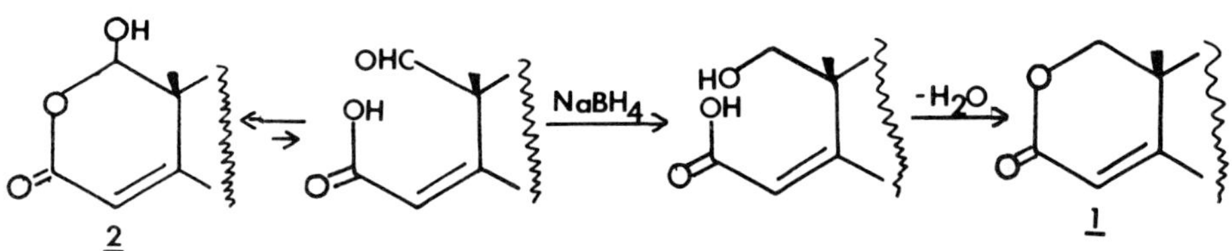

We have developed a solution to this problem as a result of our recent research on the reaction of superoxide anion radical with enones (2). We have reported (2) that 4,4- and 5,5-disubstituted cyclohex-2-en-1-ones [3 and 4] react with KO_2/18-crown-6 in benzene yielding 2-hydroxycyclohexa-2,5-dien-1-ones [5]. Interestingly, if not isolated, these enols are oxidized further to the corresponding lactols [6] in moderate to high yields.

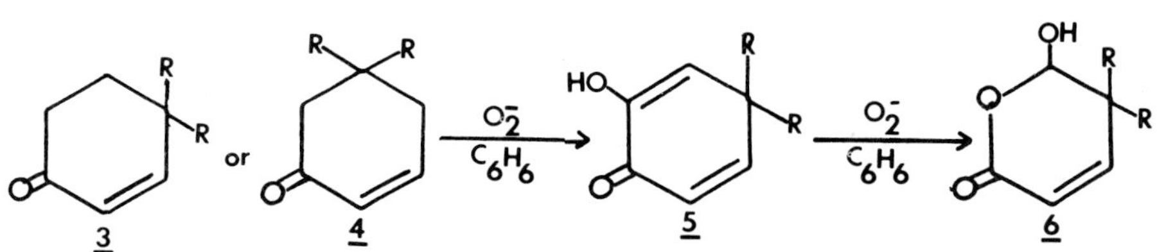

Similarly, 3-cholesten-4-one and testosterone were both converted to the corresponding lactols by reacting the starting enones with O_2^- in benzene for 72 hours in an oxygen atmosphere. The molar ratio of KO_2: 18-crown-6: substrate was 6:3:2. Reduction of

ISBN 0-12-592050-4

the lactols with $NaBH_4$ yielded the desired oxasteroids.

Sawyer (3) has recently argued that from a thermodynamic perspective $O_2^{-\cdot}$ is more likely to interact with labile hydrogens as a strong base (proton removal) than as a radical (hydrogen atom abstraction). Indeed, we have obtained similar results to those just described when KO_2 is replaced by KOH or potassium t-butoxide. The overall conversion is faster and more complete with potassium t-butoxide and slowest with KOH. This is consistent with Sawyer's suggestion (3) that $O_2^{-\cdot}$ is a stronger base in aprotic solvent than HO^-. The results also lead us to believe that the transformation of enone to enol and on to lactol is a series of base catalyzed autoxidative processes. A mechanism for the conversion of enones [3] and [4] to enol [5] has already been described (2). A likely pathway for the oxidation of enol [5] to lactol [6] is outlined below.

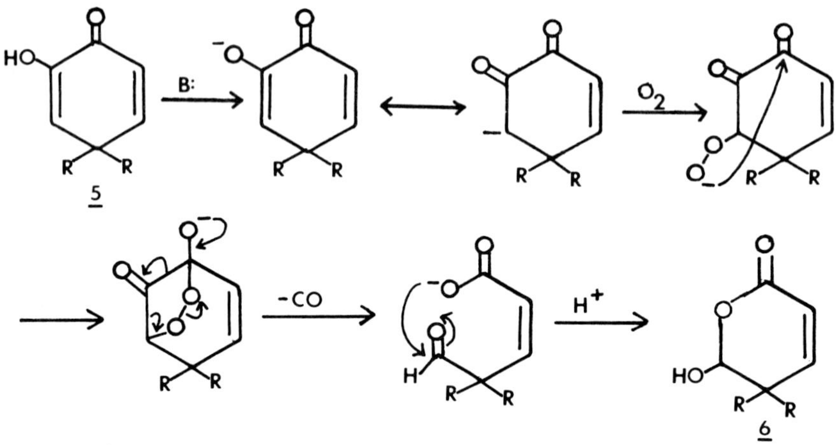

ACKNOWLEDGEMENTS

We would like to acknowledge the partial support of the Israel Commission for Basic Research.

REFERENCES

1. For reviews see a) Pappo, R., Intra-Science Chemistry Reports, 3,105 (1969); b) Singh, H., Kapoor, V. K. and Paul, D., Prog. Med. Chem., 16,35 (1979).
2. Frimer, A. A. and Gilinsky, P., Tetrahedron Lett., 4331 (1979).
3. a) Sawyer, D. T., Gibian, M. J., Morrison, M. M. and Seo, E. T., J. Amer. Chem. Soc., 100,627 (1978); b) Wilshire, J. and Sawyer, D. T., Accts. Chem. Research, 12,105 (1979).

GROWTH-RELATED CHANGES IN TUMOR SUPEROXIDE DISMUTASE CONTENT

T. Galeotti,[1] *G. M. Bartoli,*[1] *S. Santini,*[1] *S. Bartoli,*[1] *G. Neri,*[2] *P. Vernole,*[2] *L. Masotti,*[3] *and C. Zannoni*[4]

[1]*Institute of General Pathology, Catholic University, Via Pineta Sacchetti 644, Roma, Italy*
[2]*Institute of Human Genetics, Catholic University of Rome*
[3]*Institute of Biological Chemistry, University of Parma, Italy*
[4]*Institute of Organic Chemistry, University of Bologna, Italy.*

The decrease of cytosolic and mitochondrial superoxide dismutase (SOD) activity is considered, with a few exceptions, a general feature of experimental tumors (see ref. 1 for review). Data have been provided also on the loss of the copper-zinc enzyme in human tumors (2). This condition may have at least three kinds of implications: i) an increased susceptibility of tumors to γ-therapy; ii) a reduced protection of transformed cells to O_2^--induced lipid peroxidation and iii) an enhancement of the effectiveness of certain chemical carcinogens, such as polycyclic aromatic hydrocarbons, whose metabolic derivatives (e.g. benzpyrene-diols) seem to involve O_2^- radicals in the damage to DNA (3). In the light of these considerations it appeared relevant to measure in the present study the content of cytosolic SOD and the effect of its diminution on intracellular membranes of tumors with different degrees of growth rate and differentiation. Some data will also be presented on the enzyme content of a tumor derived from an actively proliferating tissue, i.e. rat thymus, with the purpose of assessing whether the lowered SOD content is a peculiar property of transformed cells or simply of rapidly dividing cells.

Table I reports the values of SOD content extracted from minimal- (Morris hepatoma 44) and maximal-deviation (Morris 3924A

TABLE 1. Content of cytosolic SOD in rat liver, hepatomas and Ehrlich ascites cells

	SOD (μg/g wet weight)
Rat liver	*218.9 ± 9.8 (6)*
H 44	*54.3 ± 6.9 (4)*
H 3924A	*11.7 ± 1.0 (4)*
H Novikoff	*11.0 (2)*
Ehrlich Lettré	*13.3 (2)*

ISBN 0-12-592050-4

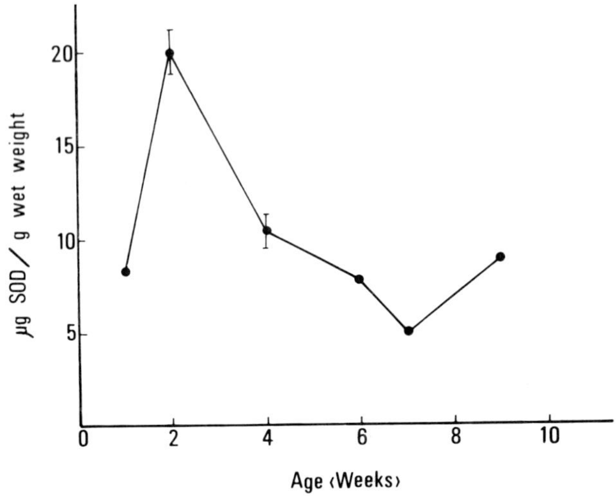

FIGURE 1. Age-related changes of SOD in F344 inbred rat thymus.

and Novikoff hepatomas; Ehrlich Lettré adenocarcinoma) tumors, in comparison to that of rat liver. The enzyme is maximally lowered in the three poorly differentiated and fast-growing tumors, whereas is present at higher values (about 25% of the control) in the hepatoma 44. Fig. 1 shows the behavior of SOD in rat thymus as a function of age. The enzyme reaches its maximum value at about two weeks after birth and decreases gradually to the initial values within 6 to 9 weeks. A thymic lymphoma, induced by i.p. injection of Gross leukemia virus at birth, has a SOD content of 12.8 ± 0.61 (6) μg/g wet weight, that is about 35% lower than the maximum value in normal thymus. These results suggest i) a dependence of SOD content in tumors on growth rate and ii) a specificity of the enzyme decline even for tumors generated from rapidly dividing cell systems. Fig. 2 shows that the rate of malonaldehyde production in microsomal membranes exposed to exogenous lipoperoxidative agents, such as the O_2^- radical generating system xanthine-xanthine oxidase (A) or ascorbate (B), decreases progressively with the increasing growth rate of the tumors. The

TABLE II. Lipid Content and Fatty Acid Composition of Rat Liver and Hepatoma Microsomes

	Rat liver	*H44*	*H3924A*
Lipid-protein	*0.66*	*0.56*	*0.32*
Polyenes/monoenes	*1.5*	*1.0*	*0.6*
Double-bond index	*82*	*60*	*50*

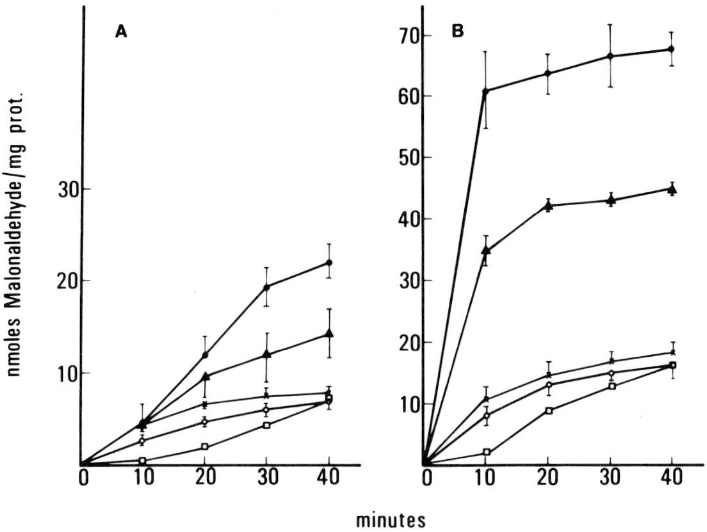

FIGURE 2. Lipid peroxidation induced by O_2^- radicals (A) and ascorbate (B) of microsomes derived from rat liver ( ), Morris hepatoma 44 ( ) and 3924A (o), Novikoff hepatoma ( ) and Ehrlich (X) ascites cells. Means ± SE (3-5 experiments).

decrease in the vulnerability of microsomal membranes and mitochondria (not shown) to peroxidative agents may be ascribed to changes in the intrinsic features of the lipid bilayer which seem to occur in a fashion related to the growth ability of the transformed tissue. Such evidence is given by the changes in the lipid content and fatty acid composition as well as in the structure of the tumor microsomal membranes. As shown in Table II, the lipid to protein ratio and the degree of fatty acid unsaturation decrease gradually from normal to hepatoma 44 and 3924A microsomes. The differences in lipid composition are reflected also by a difference in some physical properties of the bilayer. Table III gives the parameters obtained from ESR spectra of microsomes labeled with 5-nitroxide stearate and 16-nitroxide stearate, which report the order and the freedom of motion of the hydrocarbon chains in the vicinity of the lipid polar head groups and in the lipid-core, respectively. The gradual increase of the order parameters $\langle P_2 \rangle$ and $\langle P_4 \rangle$ is indicative of a more orderly structure of the tumor membranes. Moreover, while the fluidity of the membrane region close to the surface does not seem to be affected, the core of the membrane appears to be less fluid in tumors.

In conclusion it can be proposed that the loss of the protective effect of SOD against O_2^--induced lipid peroxidation in tumors may be a factor responsible for the alterations of the subcellular membrane components. In light of the view that lipid peroxides

TABLE III. ESR Parameters of Doxyl Derivatives in Microsomes from Rat Liver and Morris Hepatomas

	5-NS				16-NS
	$<P_2>$	$<P_{22}>$	$<P_4>$	τ_o(nsec)	τ_o(nsec)
Rat liver	0.60	-0.11	0.10	4.0 ± 0.2	1.4
H 44	0.61	-0.247	0.10	4.0 ± 0.2	1.7
H 3924A	0.67	-0.06	0.20	4.0 ± 0.2	2.1

inhibit cell division (4), the limited amount of peroxides produced by membranes already damaged would result inadequate to control the mitotic activity of tumors in a proportion directly related to the magnitude of SOD content loss.

ACKNOWLEDGEMENTS

This work was supported by CNR grant 79.00410.04 and CT 79.00940.04.

REFERENCES

1. Oberley, L. W. and Buettner, G. R., Cancer Res., 39,1141 (1979).
2. Sykes, J. A., McCormack, F. X. and O'Brien, T. J., Cancer Res., 38,2759 (1978).
3. Lorentzen, R. J. and Tso, P. O. P., Biochemistry, 16,1467 (1977).
4. Player, T. J., Mills, D. J. and Horton, A. A., Biochim. Biophys. Res. Commun., 78,1397 (1977).

MELANIN: A POSSIBLE CELLULAR SUPEROXIDE SCAVENGER

Nigel T. Goodchild, Lester Kwock and Peck-Sun Lin

Department of Therapeutic Radiology, Tufts-New England Medical Center, Boston, Massachusetts

INTRODUCTION

We have demonstrated that inhibition of superoxide dismutase by the copper chelator diethyldithiocarbamate increases the radiosensitivity of human erythrocytes (1) and Chinese hamster cells (2). The data indicate that the scavenging of the superoxide anion radical, $O_2^{\cdot-}$, by superoxide dismutase plays an important role in protecting cells from the adverse effects of ionizing radiation. Thus, due to the nature of melanin which enables it to capture damaging radiation-induced free radicals, thus sparing melanin containing cells from the effects of radiation (3), we wanted to determine: (i) whether melanin could scavenge $O_2^{\cdot-}$ and thus act as a superoxide scavenger; and (ii) whether chlorpromazine, a drug which has been shown to bind to melanin and affect its ability to scavenge free radicals (4), could also inhibit the ability to scavenge $O_2^{\cdot-}$.

RESULTS AND DISCUSSION

Fig. 1 shows the results of a representative experiment with respect to the ability of melanin to scavenge $O_2^{\cdot-}$ which is enzymatically generated by the xanthine-xanthine oxidase system (5).

As shown, complete scavenging of $O_2^{\cdot-}$ generated in this system by melanin can be achieved with concentrations of melanin above 10 μg/ml. These data indicate that melanin can scavenge one free radical known to be generated by ionizing radiation, i.e., $O_2^{\cdot-}$.

Chlorpromazine can form a charge transfer complex with melanin and inhibit the free radical scavenging ability of melanin (4,7). This process has been proposed as one of the possible models for chlorpromazine radiosensitization of melanomas (8). To test whether chlorpromazine could decrease the ability of melanin to scavenge $O_2^{\cdot-}$, the following experiment was performed. Samples of melanin (10 μg/ml) were incubated with various concentrations of chlorpromazine ranging from 0.5 to 5 μg/ml for 24 h at 37°C (maximum chlorpromazine binding occurs at this time). The ability of chlorpromazine-treated melanin to scavenge $O_2^{\cdot-}$ was then compared to similarly incubated melanin controls. As shown in Fig. 2, at all concentrations tested, chlorpromazine can partially inhibit the ability of melanin to scavenge $O_2^{\cdot-}$. At 5 μg/ml of chlorpromazine, approximately 30% inhibition is observed.

ISBN 0-12-592050-4

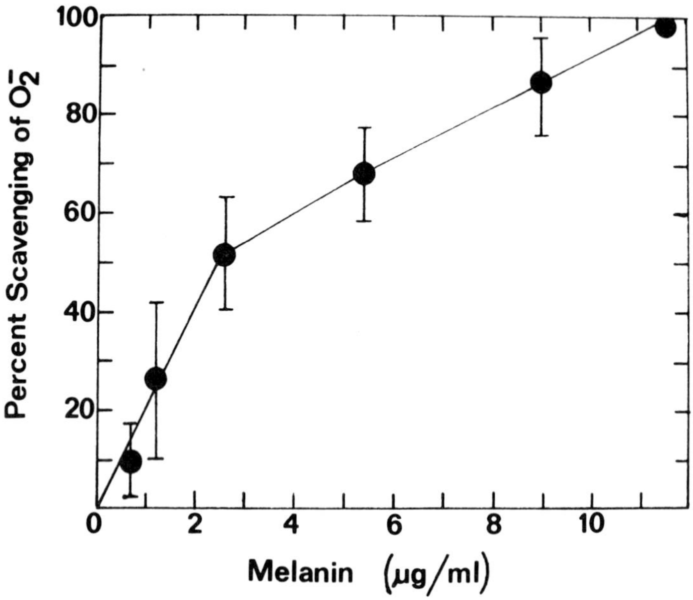

FIGURE 1. $O_2^{\bar{\cdot}}$ scavenging by melanin. $O_2^{\bar{\cdot}}$ was generated using the xanthine-xanthine oxidase system described in ref. 16. This system was modified by the substitution of 0.9% NaCl for the phosphate buffer used in ref. 16. The reaction was adjusted by varying the xanthine oxidase concentration to produce approximately 0.08 O.D. (540 nm) unit change per minute. The quenching of $O_2^{\bar{\cdot}}$ by melanin was determined by addition of 0.4 ml aliquots of various concentrations of melanin giving a final concentration in the range of 0.5-10 μg/ml. An equivalent volume of 0.9% NaCl was added to control samples and all samples were blanked against a melanin-containing control. Melanin samples were prepared by vigorous shaking of melanin in 20 ml of 0.9% saline and then centrifuging at 200 x g for 10 min. and removing the supernatant. The concentration of dissolved melanin was then determined using the method described in ref. 6 and then appropriate dilutions made to give the desired final concentrations. Each data point represents the average of at least nine determinations ± S.D.

It has been shown that 7-hydroxychlorpromazine is the major cellular metabolite of chlorpromazine (9). Furthermore, 7-hydroxychlorpromazine has been shown to be more melanotropic and also a more effective radiosensitizer than chlorpromazine. Thus, the partial efficacy by which chlorpromazine inhibits the $O_2^{\cdot-}$ scavenging ability of melanin in this system may be due to the inability of chlorpromazine to be converted to 7-hydroxychlorpromazine in this system.

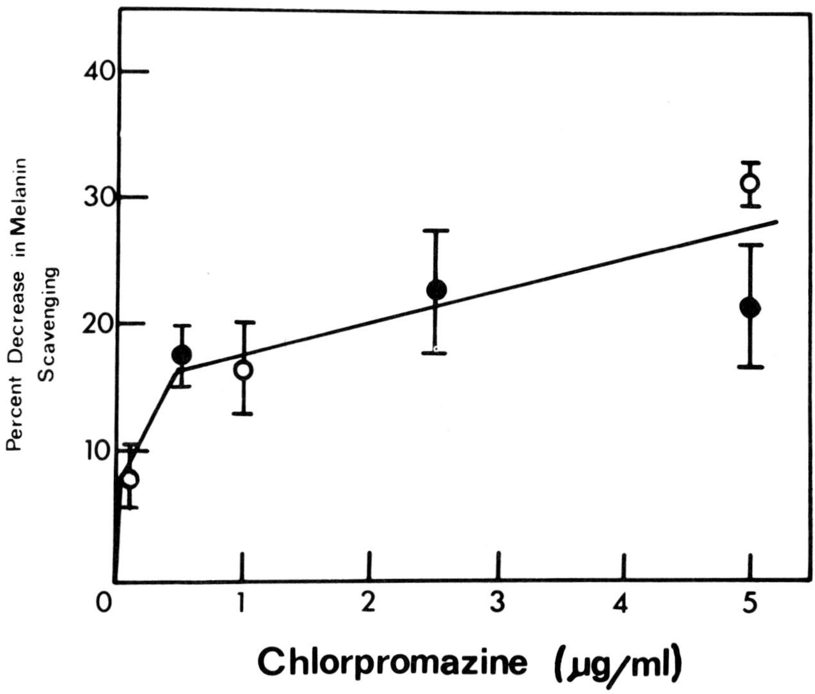

FIGURE 2. Effect of chlorpromazine on the ability of melanin to scavenge $O_2^{\cdot-}$. The open and closed circles represent separate experiments and each point represents the average ± S.D. of at least 8 determinations. In separate experiments, the addition of chlorpromazine did not affect the rate of xanthine-xanthine oxidase generated $O_2^{\cdot-}$.

We have shown that melanin can act as a superoxide dismutase. On the basis of this property, we propose that in melaninized cells melanin can act as a radioprotector in a similar fashion to the observed superoxide dismutase protection afforded to bacteria (1) and mammalian cells (11).

ACKNOWLEDGEMENTS

This work was supported by a grant from the National Cancer Institute, CA-12178.

REFERENCES

1. Stone, D., Lin, P.-S. and Kwock, L., Int. J. Radiat. Biol., 33, 393 (1978).
2. Lin, P.-S., Kwock, L. and Butterfield, C. E., Radiat. Res., 77, 501 (1979).
3. Mason, H. S., Ingram, D. J. E. and Allen, B., Arch. Biochem. Biophys., 86, 225 (1960).
4. Bolt, A. G. and Forrest, I. S., Life Sci., 6, 1285 (1968).
5. McCord, J. M. and Fridovich, I., J. Biol. Chem., 244, 6049 (1969).
6. Wittaker, J. R., Dev. Biol, 8, 99 (1963).
7. Van Woert, M. H., Proc. Soc. Exp. Biol. Med., 129, 165 (1965).
8. Cooper, M. and Mishima, Y., "Biology of Normal and Abnormal Melanocytes," (T. Kawamura, T. B. Fitzpatrick and M. Seiji, eds.), Univ. Park Press, Baltimore (1971), p. 141.
9. Damsker, J. I., Macklis, R. and Brady, L. W., Int. J. Radiat. Oncol. Biol. Phys., 4, 821 (1978).
10. Oberly, L. W., Lindgreen, A. L., Baker, S. A. and Stevens, R. H., Radiat. Res., 68, 320 (1976).
11. Petkau, A., Chelack, W. S. and Pleskach, S. L., Life Sci., 22, 867 (1978).

EFFECT OF SUPEROXIDE ANION ($O_2^{\bar{\cdot}}$) ON THE SURVIVAL OF CHINESE HAMSTER CELLS

N. T. Goodchild, L. Kwock, P. S. Lin and L. Ciborowski

Tufts-New England Medical Center, Department of Therapeutic Radiology, Division of Radiobiology, Boston, Massachusetts 02111

INTRODUCTION

Reports have implicated the superoxide anion ($O_2^{\bar{\cdot}}$) radical as being responsible for part of the observed radiation damage to cells both in vivo and in vitro (1-4). However, ionizing radiation can also cause direct damage which can lead to cell death. In order to determine the extent of $O_2^{\bar{\cdot}}$ involvement in radiation cell killing, we have examined the effect of either extra- or intracellularly generated $O_2^{\bar{\cdot}}$ on the reproductive capacity of Chinese hamster lung cells (DON).

RESULTS AND DISCUSSION

In this study we have utilized the autoxidation of dihydroxyfumaric acid (DHF) (10^{-5} -10^{-3}M) and the xanthine-xanthine oxidase system as our extracellular O_2- generating source (Table 1).

TABLE 1. Effect of O_2^- Generated by Either Autoxidation of DHF or the Xanthine-Xanthine Oxidase System on the Survival of DON Cells

Concentration of DHF (treated for 1h)	*Relative survival*
0	*1.0*
10^{-5}M	*0.92 ± 0.08*
10^{-4}M	*0.89 ± 0.05*
10^{-3}M	*0.63 ± 0.04*
Xanthine-xanthine oxidase treatment (Function of treatment time)	*Relative survival*
0 min	*1.0*
30 min	*0.75 ± 0.09*
60 min	*0.57 ± 0.06*
120 min	*0.64 ± 0.04*

ISBN 0-12-592050-4

TABLE 2. Effect of DDC on the Relative Survival of DON Cells as a Function of Treatment Time and Concentration

Temperature (°C)	Treatment Time (min)	Effect of [DDC] on survival of Don cells (% survival)			
		0	10^{-4}M	10^{-5}M	10^{-6}M
37	10	100± 3.1	66.3±9.5	99.2±15.7	100 ± 1
37	30	100±12.2	32.6±5.4	76.2± 5.1	80.2± 4.1
37	60	100± 7.2	25.2±6.8	76.4± 7.3	100 ± 7.3
37	120	100± 3.5	19.9±7.8	45.7± 2.2	100 ±13.6

To determine the effect of intracellularly generated O_2^-, we have used the superoxide dismutase inhibitor diethyldithiocarbamate (10^{-6} - 10^{-4}M) (Table 2). We have found that either elevating the extracellular or intracellular levels of O_2^- can impair the clonogenic survival of these cells (40-60%).

In addition we found that DON cells incubated in the presence of superoxide dismutase (1 μg/ml) prior to being irradiated at doses < 250 rads protected these cells from the adverse effects of radiation (Table 3). At irradiation levels greater than 250 rads no protection with respect to clonogenic survival was observed even when the levels of SOD were raised to 10 μg/ml. Also, we found that inhibition of intracellular SOD with DDC (10^{-6}-10^{-4}M) radiosensitizes these cells (Table 4). The data suggest that elevation of either extra- or intracellular levels of O_2^- can impair the reproductive capacity of cells and that this radical species is important in the observed cytotoxic effects of radiation.

TABLE 3. Effect of SOD on Survival of Irradiated DON Cells

SOD	0 RAD	50	100	250	500
0	100 ± 4.68	63 ± 9	41 ± 10	25 ± 4	2.3
1 μg/ml	100 ± 1.83	83 ± 12	92 ± 25	47 ± 6	4.4

TABLE 4. Change in Survival of DON Cells after Various Concentrations of DDC or Simultaneous Treatment with DDC and 250 RAD of γ Radiation

		DDC concentration			
RAD	*Control*	$10^{-4}M$	$10^{-5}M$	$10^{-6}M$	$10^{-7}M$
0	159.5± 6 (100)	99.0±7 (62.1)	131.4±15 (82.4)	140.0±18 (87.8)	150.8±15 (94.5)
250	107 ±11 (100)	40.6±6 (37.4)	63.8±11 (59.6)	77.8±10 (72.7)	87.8± 7 (82.1)

ACKNOWLEDGMENTS

Supported by National Cancer Institute Grant CA 26897-01.

REFERENCES

1. Bors, W., Saran, M., Lengfelder, E., Spottl, R. and Michdl, C., Curr. Top. Radiat. Res., 9, 247 (1974).
2. Petkau, A., Chelack, W. S. and Pleskach, S. D., Int. J. Radiat. Biol., 29, 297 (1976).
3. Oberley, L. W., Lingren, A. L., Baker, S. A. and Stevens, R. H., Radiat. Res., 68, 320 (1976).
4. Lin, P. S., Kwock, L., Lui, P. and Hefter, K., Int. J. Radiat. Onc. Biol. Phys., 5, 169 (1979).

SUPEROXIDE DISMUTASE AND ENDOTOXIC SHOCK IN MICE

Brian H. Gray and Sharon Schmidt, Experimental Hematology Department, Armed Forces Radiobiology Research Institute, Bethesda, Maryland 20014

Mice injected with sufficiently high doses of bacterial lipopolysaccharide (endotoxin) undergo physiological stress and death usually within 6 days. Lung stress following endotoxin administration may be due in part to pulmonary granulocyte sequestration followed by release of toxic oxygen radicals (1,2). It has been demonstrated that rats are protected from oxygen toxicity by both previous exposure to 85% oxygen atmospheres and sublethal doses of endotoxin (3,4). Superoxide dismutase (SOD), glutathione peroxidase, and catalase have been linked to this protective effect (5,6). In addition, SOD has been shown to have a protective effect for mice during irradiation (7). This investigation was done to determine the effects of exogenous SOD on mice challenged with endotoxin and to determine endotoxin effects on endogenous SOD, glutathione peroxidase, and catalase in mouse lungs.

B6CBF1 mice, at least 10 weeks old and weighing about 25 g, were housed in a colony where they consumed a standard laboratory diet and chlorinated water ad libitum. Mice were given 0.25-ml intravenous (IV) injections of solutions of SOD (Sigma, bovine blood, type I, 2900 units/mg) prepared by dissolving the enzyme in pyrogen-free sterile saline (Cutter Laboratories). These same mice were injected intraperitoneally (IP) with 0.25 ml of a *Salmonella typhosa* lipopolysaccharide W 0901 solution (Difco Laboratories) made with pyrogen-free sterile saline.

Additional mice were used to prepare lung extracts for enzymatic assays. Mice challenged with 250 μg endotoxin or sterile saline IP were sacrificed, and their lungs were flushed with about 10 ml of saline by injection into the right ventricle after severing the descending aorta. Lungs from 5 animals per group were excised, rinsed in ice-cold 0.05-M sodium phosphate buffer (pH 7.8), minced, rinsed, and homogenized in buffer at a 1:10 (w/v) ratio. Homogenates were centrifuged at 0°-4°C for 15 min at 5000 g max, and the supernatant fluids were assayed for glutathione peroxidase (8) and catalase (9) with the aid of a Varian model 219 spectrophotometer. Aliquots of the supernatant fluids were dialyzed against buffer overnight at about 5°-10°C and assayed for SOD (10). The final assay mixture was stirred continuously, and the reaction rate was calculated 5 to 6 min after epinephrine addition by measuring optical density at 320 nm. One unit of enzyme activity was defined as that quantity that reduced the change in optical density at 320 nm by 50%. Protein was assayed by the method of Bradford (11), using bovine plasma gamma

ISBN 0-12-592050-4

TABLE 1. Thirty-Day Survival of Mice

Endotoxin dose IP	Sterile saline IV	SOD 1 mg IV
500 μg	6/33 (18%)	17/33 (51%)
600 μg	3/21 (14%)	4/14 (29%)
700 μg	3/25 (12%)	7/25 (28%)

globulin as a standard (Bio-Rad Laboratories).

A higher percentage of mice administered IV injections of 1 mg SOD about 1 hour before IP challenge with endotoxin survived than did control mice. The table shows survival of mice challenged with three concentrations of endotoxin. A dose of 1 mg SOD per mouse was chosen because lower concentrations were not as effective in reducing mortality. Also, 4 or 8 mg SOD per mouse enhanced the mortality in endotoxin-challenged animals. There was no mortality in mice injected with sterile saline IP and up to 8 mg SOD IV.

The figure shows percent of control specific activities for glutathione peroxidase, catalase, and SOD after IP injection of 250 ug endotoxin as a function of time. There was a decrease in specific activity of SOD 1 hour after endotoxin injection in mouse lungs. The rapid drop in SOD specific activity following IP injection of 250 ug endotoxin may account for the limited protective effect of exogenous SOD in mice. It is noteworthy that

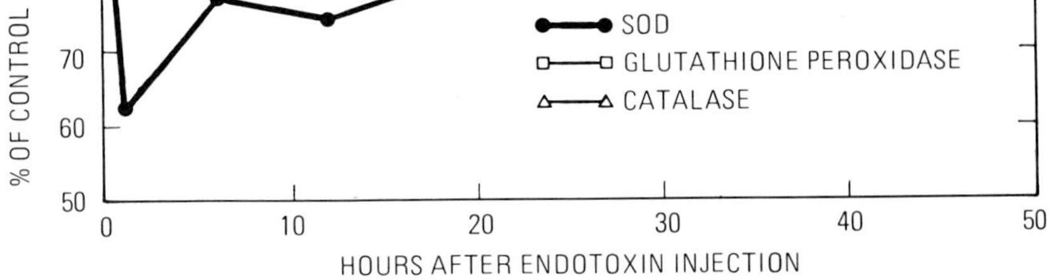

FIGURE 1. Enzyme specific activity expressed as percent of control specific activity for mouse lung supernatant fluids. Five mice were used to prepare superantants at each time point. Value for control specific activity for each enzyme was the average of specific activities determined on each group of controls.

endotoxin effects very with the source of toxin, strain of mouse, and other factors. Also, SOD at sufficiently high levels enhanced death in endotoxin-treated mice. Finally, the radioprotective SOD dose was about 5 times that that yielded protection from moderate endotoxin shock (7).

REFERENCES

1. Stetson, C. A., Jr., J. Exptl. Med., 93, 489 (1951).
2. Sacks, T., Moldow, C. F., Craddock, P. R., Bowers, T. K. and Jacob, H. S., J. Clin. Invest., 61, 1161 (1978).
3. Rosenbaum, R. M., Wittner, M. and Lenger, M., Lab. Invest., 20, 516 (1969).
4. Lee, F., Yam, J. and Roberts, R. J., J. Clin. Invest., 61, 269 (1978).
5. Lee, F. and Roberts, R. J., J. Appl. Physiol.: Respirat., Environ. Exercise Physiol., 47, 577 (1979).
6. Crapo, J. D. and Tierney, D. F., Am. J. Physiol., 226, 1401 (1977).
7. Petkau, A., Photochem. Photobiol., 28, 765 (1978).
8. Paglia, D. E. and Valentine, W. N., J. Lab. Clin. Med., 70, 158 (1967).
9. Beer, R. F., Jr. and Sizer, I. W., J. Biol. Chem., 195, 133 (1952).
10. Sun, M. and Zigman, S., Anal. Biochem., 90, 81 (1978).
11. Bradford, M., Anal. Biochem., 72, 248 (1976).

X

THE REACTION OF NITRONE SPIN TRAPS WITH SINGLET OXYGEN IN AQUEOUS MEDIUM

John R. Harbour, Sandra L. Issler and Michael L. Hair, Xerox Research Centre of Canada, Mississagua, Ontario L5L 1J9

There is a considerable interest in oxygen centered radicals as intermediates in a variety of biological systems. The hydroxyl radical (·OH) and the superoxide radical ($O_2^{\bar{}}$·) are most often implicated as reactive intermediates but the peroxy radicals (RO_2·) are also important. Detection of these radicals directly at room temperature is essentially precluded due to the low level of their steady state concentration. The introduction of the technique of spin trapping (1) with nitrones has offered a means of indirect detection of these oxygen centered radicals. The ·OH and $O_2^{\bar{}}$ adducts (2) and the RO_2· adducts (3) with nitrones are now characterized and have subsequently been utilized in biological systems. When the lifetime of the spin adduct is long compared to the time of measurement, there is a continual buildup of radical concentration. This "integration" characteristic coupled with the fact that the nitroxide spin adducts usually have only six lines of relatively narrow linewidth, leads to a detection system with a sensitivity sufficient for biological systems. In practice, of course, it is necessary to determine the lifetime of the spin adduct in the particular environment and determine the interaction of the spin trap with the system. Despite these difficulties, the tremendous potential of spin trapping to biological systems where oxygen centered as well as other radicals occur, will witness an increased utilization of these traps in wide variety of systems (including non-biological ones). Singlet oxygen ($^1\Delta_g$) is a diamagnetic reactive oxygen intermediate which may also be generated in these systems. It is often difficult to differentiate between reactions caused by $^1\Delta g$ or oxygen centered radicals and spin trapping appears to offer an excellent means of determining the reactive intermediates involved. However, this assumes that nitrones are non-reactive toward $^1\Delta_g$. Ching and Foote (4) have demonstrated that $^1\Delta_g$ will react with certain nitrones in chloroform and that $^1\Delta_g$ can be physically quenched by a nitrone similar to DMPO. We have looked at the reactivity of various nitrone spin traps with $^1\Delta_g$ in aqueous medium and have demonstrated that a chemical reaction does occur. In these experiments, methylene blue (MB) at 1×10^{-4} M was used as a $^1\Delta_g$ generator and an oxygen electrode used to monitor oxygen uptake. It was determined that the spin trap, α-phenyl N-(tertbutyl) nitrone (PBN) when present in the MB aqueous solution, resulted in an

ISBN 0-12-592050-4

oxygen uptake upon illumination. This uptake was linear with PBN concentration up to 1×10^{-2} M. Since $^1\Delta_g$ has a lifetime of 20 μ sec in D_2O vs a lifetime of 2 μ sec in H_2O[5], the replacement of H_2O by D_2O should yield an increased rate of uptake in D_2O. This was confirmed, and a value of uptake D_2O/uptake H_2O ~ 10 was approached. A second means of demonstrating a $^1\Delta_g$ reaction is to include a known quencher of $^1\Delta_g$ into the system. When azide (N_3^-) was added, quenching of the nitrone reaction did in fact occur. The rate constant for the reaction of PBN and $^1\Delta_g$ was determined to be 1.4×10^8 M^{-1} sec^{-1}. Hence the nitrone PBN can chemically react with $^1\Delta_g$ and rapidly deplete O_2 from the system. This reaction occurs with other nitrone traps as well. DMPO had the lowest rate constant of 1.8×10^7 $M^{-1}sec^{-1}$.

It is significant that this reaction can occur. A search for radicals in a system through the use of spin trapping can lead to false conclusions. This $^1\Delta_g$ reaction with nitrones can rapidly deplete or diminish the oxygen concentration and prevent either normal radical reactions which could normally have occurred or the reactions of $^1\Delta_g$ itself. These facts must be considered when employing nitrone spin traps and in subsequent interpretations of the results.

REFERENCES

1. Janzen, E. G., Acc. Chem. Res., 4, 31 (1971).
2. Harbour, J. R., Chow, V. and Bolton, J. R., Can. J. Chem., 52, 3549 (1974).
3. Janzen, E. G. and Liu, I. J., J. Mag. Reson., 9, 510 (1973).
4. Ching, T. -Y. and Foote, C. S., Tetrahedron Lett., 44, 3771 (1975).
5. Kearns, D. R., "Singlet Oxygen", (H. H. Wasserman, R. W. Murray, eds.), Academic Press, New York, (1979), p. 115.

RADIOLYTIC STUDY OF THE ONE-ELECTRON REDUCTION OF A SERIES OF SUBSTITUTED NITRO-IMIDAZOLES IN AQUEOUS SOLUTION

C. R. Hardy, H. Ma and P. O'Neill, Institute of Cancer Research, Department of Physics, Clifton Avenue, Sutton, Surrey, U.K.

Halogenated nitroimidazoles ($ImNO_2$) have been shown (1,2) to have increased radiosensitization efficiencies as compared to those predictable (3) from their one-electron reduction potentials at pH 7 (E^1_7). The properties of the one-electron reduced species of (4-bromo(iodo)-1-methyl-)5-nitroimidazoles and (5-bromo(iodo)-1-methyl)-4-nitroimidazoles in aqueous solution were studied using the techniques of pulse radiolysis and ^{60}Co γ radiolysis to generate these species.

At pH 7 the reaction of $ImNO_2$ with e^-_{aq} and $CO_2^{\bar{\cdot}}$, formed on reaction of $\cdot OH$ and $H\cdot$ with HCO_2^-, result in the production of the one-electron reduced species of the substituted nitroimidazoles ($ImNO_2^{\bar{\cdot}}$) which are stable for >50 ms in the absence of other radicals, e.g. 2-hydroxy-2-methyl-propyl radicals. In the presence of the persistent nitroxyl free radical, norpseudopelletierine-N-oxyl (NPPN), the rate of decay of $ImNO_2^{\bar{\cdot}}$ is first order and dependent upon the concentration of NPPN. The bimolecular rate constants determined for reaction of NPPN with $ImNO_2^{\bar{\cdot}}$ are presented in the Table together with the one-electron reduction potentials determined at pH 7 (E^1_7). The rate constants decrease as the E^1_7 values become more positive; this is consistent with previous observations (4).

From the yield and time dependence of conductivity and optical changes and the dependence of the rate constant upon E^1_7, it is suggested that reaction 1 proceeds

$$ImNO_2^{\bar{\cdot}} + NPPN \rightleftharpoons ImNO_2 + NPPN^- \underset{pK>12}{\overset{+H^+}{\rightleftharpoons}} NPPNH \qquad (1)$$

by an electron transfer process resulting in formation of the hydroxylamine derived fron NPPN.

It was found that the reaction of e^-_{aq} and $CO_2^{\bar{\cdot}}$ with $ImNO_2$ also leads to the production of halide ions (X^-), the yields of which are presented in the Table. The yields of X^- are independent of dose rate. In the presence of NPPN, the yield of X^- is reduced, although when the concentration of NPPN exceeds that of the radiation-induced radicals, NPPN does not completely prevent the production of X^-. The "residual" yields of X^- determined are presented in the Table and are <15% of those determined in the absence of NPPN.

It is proposed that the reaction of e^-_{aq} (or $CO_2^{\bar{\cdot}}$) with the halogenated nitroimidazoles results in formation of either the corresponding radical anion (∿95%) or the nitroimidazoyl radical

ISBN 0-12-592050-4

TABLE I. Yield of Halide on Reduction of $ImNO_2$ in Presence and Absence of NPPN.

	$ImNO_2$	$G(X^-)^{a)}$	$G(X^-)^{b)}_{NPPN}$	$-E^1_7/mV$	$k(ImNO_2 + NPPN)$ $M^{-1}s^{-1}$
4-Br	$5-NO_2$	1.6	0.27	460	1.6×10^7
4-I	$5-NO_2$	1.9	0.10	461	2.1×10^7
5-Br	$4-NO_2$	1.6	0.22	523	4.5×10^7
5-I	$4-NO_2$	2.1	0.18	529	4.8×10^7

a) dose rate 2 krads/min., dose 4-14 krads.
b) $10^{-4}M$ NPPN, $2 \times 10^{-4}M$ $ImNO_2$.

and X^- by electron-dissociative capture.

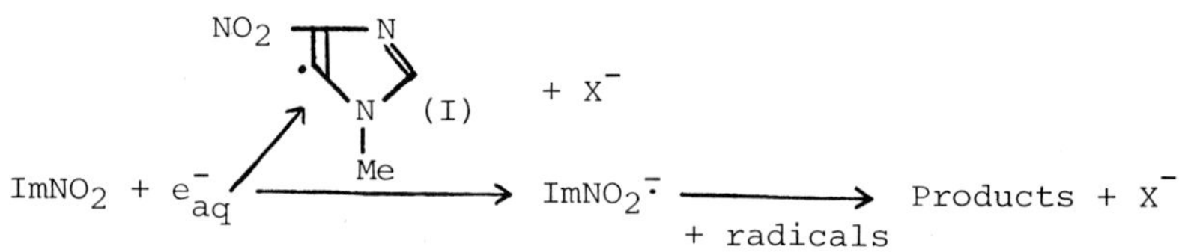

The majority of the halide yield is suggested to result from bimolecular reactions of the radical anion of $ImNO_2$. However, in the presence of NPPN, $ImNO_2^{\cdot-}$ is scavenged (reaction 1) so that production of halide is prevented by this route; the 'residual' yield of X^- arises from electron dissociative capture. (5-Chloro-1-methyl)-4-nitroimidazole on reduction has previously been reported (5) to result in the production of Cl^-.

The question arises as to the importance of carbon-centered radicals, e.g. type I, when investigating the mode of action of radiosensitizers in radiobiology.

ACKNOWLEDGEMENTS

We thank Dr. E. M. Fielden and Prof. G. E. Adams for helpful discussions. This work was funded by the MRC/CRC joint committee and by NCI (USA). Mr. Ma is a visiting fellow from the Institute of Biophysics, Peking, China.

REFERENCES

1. Adams, G. E., Ahmed, I., Fielden, E. M., O'Neill, P. and Stratford, I. J., Cancer Clin. Trials, 3,37 (1980).
2. Stratford, I. J. and Williamson, C., unpublished data.
3. Adams, G. E., Clarke, E. D., Flockhart, I. R., Jacobs, R. S., Sehmi, D. S., Stratford, I. J., Wardman, P., Watts, M. E., Parrick, J., Wallace, R. G. and Smithen, C. E., Int. J. Radiat. Biol., 35,133 (1979).

4. Jenkins, T. C. and O'Neill, P., Stud. Phys. Theor. Chem., 6 (Radiat. Biol. Chem.: Res. Dev.), 21 (1979).
5. Clarke, E. D. and Wardman, P., Int. J. Radiat. Biol.,37,463, (1980).

EFFECT OF OXYGEN ON TRANSFORMING DNA IRRADIATED IN THE PRESENCE OF SULFHYDRYL COMPOUNDS

K. D. Held, H. A. Harrop and B. D. Michael, Cancer Research Campaign Gray Laboratory, Mount Vernon Hospital, Northwood, Middlesex HA6 2RN, England

Recent experiments have demonstrated a relationship between cellular sulfhydryl levels, the sensitization of bacteria by oxygen and electron-affinic compounds, and the kinetics of radiation-induced oxygen-dependent damage (1). In order to expand our understanding of these interactions we have recently begun to extend these studies to a subcellular system involving bacterial transforming DNA, about which there is a considerable amount of information (2,3,4). We hope to answer the questions: 1. Can sulfhydryl-containing compounds chemically repair damage to the DNA by donating a hydrogen atom? 2. If so, does oxygen compete with and block this hydrogen donation? In this system DNA extracted and purified from Bacillus subtilis wild type cells, strain W23, is irradiated under various chemically defined conditions, and damage to the DNA is assessed by determining the ability of the DNA to transform tryptophan-deficient mutant cells, strain 168M, to tryptophan independence. Previous experiments (2) have shown that the radiation sensitivity of transforming DNA in the absence of other additives is less when irradiation is in 100% O_2 than when it is in 100% N_2, i.e., the Oxygen Enhancement Ratio (OER) is about 0.7.

For our initial investigations of the effects of sulfhydryl-containing compounds, dithiothreitol (DTT) was selected because its radiation protective and radiation chemical properties have been investigated (5,6). Typical transforming DNA survival curves are shown in the figure and some selected preliminary data on radiation sensitivity and OER values at various concentrations of DTT given in the table. From these results the following observations can be made. When transforming DNA is irradiated in the presence of micromolar concentrations of DTT, the DNA is protected (∿1.5x at 1 μM) in both N_2 and O_2, and the OER remains about the same as that in the absence of DTT. As the DTT concentration is increased the radiosensitivity of the transforming DNA decreases, but the degree of protection becomes dependent upon whether irradiation is in the presence of O_2 or N_2, i.e., DTT is much more efficient as a radiation protector of transforming DNA in N_2 than in O_2. For example, in 1 mM DTT transforming DNA is protected by a factor of over 600 in N_2, but only by a factor of about 50 in O_2. This results in an OER of about 10.

Sulfhydryl-containing compounds react very rapidly with OH radicals and these results have therefore been analyzed in terms

ISBN 0-12-592050-4

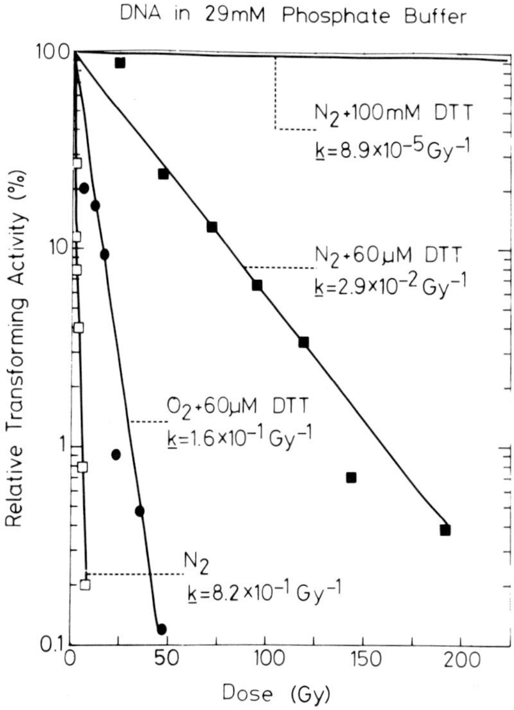

FIGURE. Typical inactivation curves of the biological activity of transforming DNA γ-irradiated in the presence or absence of DTT, under the conditions indicated.

of chemical competition kinetics. A plot of 1/k (the reciprocal of the inactivation constant) versus kC (the bimolecular rate constant for reaction of the additive with ·OH times additive concentration) shows that in O_2 the values obtained in DTT fit

TABLE. Transforming DNA irradiated in the presence of DTT

[DTT],M	k_{N_2} (Gy^{-1})	k_{O_2} (Gy^{-1})	OER
0	7.42×10^{-1}	5.41×10^{-1}	0.73
10^{-6}	5.36×10^{-1}	3.55×10^{-1}	0.67
10^{-5}	3.55×10^{-1}	1.68×10^{-1}	2.1
10^{-4}	1.84×10^{-2}	9.70×10^{-2}	5.3
10^{-3}	1.08×10^{-3}	1.18×10^{-2}	10.9
10^{-2}	2.30×10^{-4}	2.67×10^{-3}	11.6
10^{-1}	8.85×10^{-5}	5.26×10^{-4}	5.8

the same linear expression as do the values previously obtained (4) for irradiation in the presence of ethanol, also an efficient ·OH scavenger. This indicates that in O_2 DTT protects primarily by ·OH scavenging. However, in N_2 the values for irradiation in DTT show a much greater protection than can be accounted for merely by ·OH scavenging. This additional protective effect of DTT in N_2 is consistent with repair of damaged DNA by hydrogen donation from DTT to the damaged site. Such a reaction would be blocked in O_2 by the competing reaction of O_2 with the damaged site.

ACKNOWLEDGEMENT

K. D. H. is supported by a National Science Foundation National Needs Post-doctoral Fellowship No. SPI 7914 4829.

REFERENCES

1. Michael, B. D. and Harrop, H. A., Int. J. Radiat. Oncol. Biol. Phys. (in press).
2. Held, K. D., Synek, R. W., and Powers, E. L., Int. J. Radiat. Biol., 33, 317 (1978).
3. Held, K. D., and Powers, E. L., Int. J. Radiat. Biol., 36, 613 (1979).
4. Held, K. D. and Powers, E. L., Int. J. Radiat. Biol., 36, 665 (1979).
5. Redpath, J. L., Radiat. Res., 54, 364 (1973).
6. Redpath, J. L., Radiat. Res., 55, 109 (1973).

PHOTO-OXIDATION BY RIBOFLAVIN IN MODEL SYSTEMS FOR BOVINE MILK

C. L. Hicks and J. Draper, University of Kentucky, Lexington, Kentucky 40506

Photo-oxidation by immobilized and free riboflavin was observed in model systems utilizing three media; water, 0.05 M phosphate (pH 7.6) and simulated skimmilk ultrafiltrate (SMUF) (1). Comparisons were made on $^1O_2$ producing systems utilizing immobilized rose bengal and methylene blue and a superoxide anion gathering system (2). Traps, quenchers and enzymes used in conjunction with these systems were norbixin (annatto), thioctic acid, 1,4-diazabicyclo [2.2.2] octane (DABCO), 1,3-diphenylisobenzofuran (DPBF), dianisidine, cytochrome c, superoxide dismutase, catalase, methionine, cysteine, various alcohols and potassium iodide.

The water-soluble carotenoid pigment, bixin (3×10^{-8}M), was a $^1O_2$ trap but did not react with $O_2^{\cdot-}$ (Fig. 1a). Reaction rates for bixin (decrease in color at 375 nm) were unaffected when SOD was added to the $^1O_2$ producing systems of immobilized rose bengal and riboflavin. Although bixin was slowly degraded by light, no additional degradation was observed in the $O_2^{\cdot-}$ producing system of xanthine oxidase. Bixin, being water soluble at the concentration used, was more suitable than other traps in determining the effect of $^1O_2$ in aqueous systems.

Production of $^1O_2$ and $O_2^{\cdot-}$ decreased as buffer systems became more complex. Substrates such as methionine, cysteins and dianisidine that reduced riboflavin were effective in inhibiting the production of singlet oxygen. Cysteine was more effective than methionine for inhibiting singlet oxygen production in SMUF (Fig. 1b). When adequate reducing compounds were present little singlet oxygen could be observed in SMUF systems (Fig. 2, line A & B). When reducing agents were present, essentially all changes in absorbance of annatto was due to direct photodecomposition (Fig. 2, line C & D). When reducing agents were not present singlet oxygen production could be observed (Fig. 2, line E & F). Superoxide dismutase enhanced annatto oxidation slightly when dianisidine was used as a substrate (Fig. 2, line C & D). Superoxide dismutase increased the DH· radical concentration (2) which was slightly reactive with annatto. Also SOD increased the rate at which dianisidine (DH_2) (2) disappeared which would shift the pathway from $O_2^{\cdot-}$ to $^1O_2$.

Presumably the sunlight or "activated" flavor (3) resulting from the oxidation of labile amino acids such as methionine to methional and methyl mercaptan in bovine milk is not diminished by the presence of SOD. Although SOD affected the rate of fat oxidation as measured by thiobarbituric acid values (4), no

ISBN 0-12-592050-4

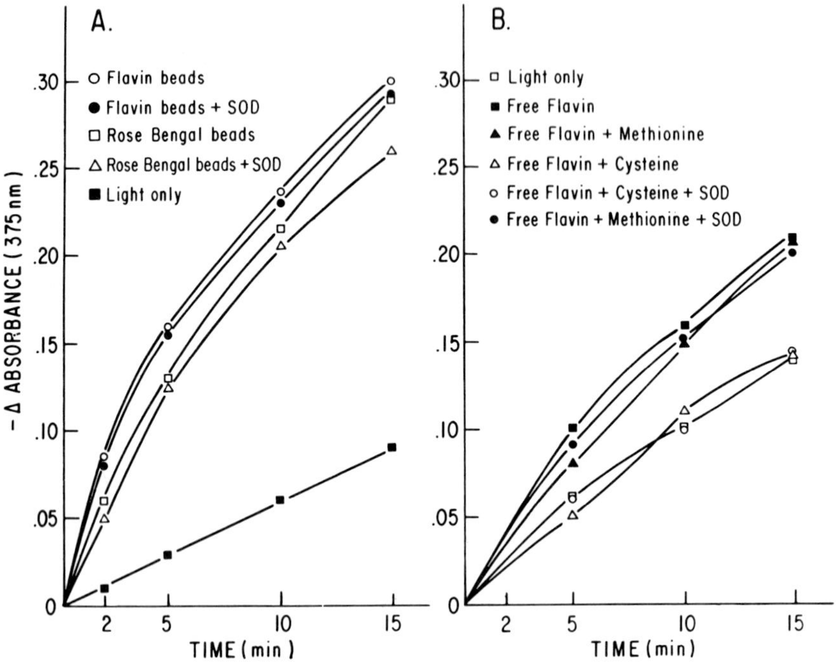

FIGURE 1. Characterization of Norbixin as a $^1O_2$ trap. (a) Light-induced degradation and photooxidation of norbixin with immobilized rose bengal or riboflavin. Reaction mixture included $3 \times 10^{-8}M$ norbixin, 1.35 mls (2 mg volume) immobilized amberlite IRA-410 sensitizer, and 10 μg/ml SOD (where indicated), in H_2O at 25°C while being purged with oxygen. (b) Effect of reducing substrates on free riboflavin catalyzed photooxidation of norbixin. Reaction mixture included $3 \times 10^{-8}M$ norbixin, $2 \times 10^{-6}M$ free riboflavin, 4.0 μg/ml SOD (where indicated), and $2 \times 10^{-3}M$ cysteine or methionine (where indicated) in SMUF (pH 6.6) at 25°C.

change in the rate of methionine utilization could be determined. Therefore, "activated" flavor which occurs when milk is exposed to light for short periods may be an electron transfer mechanism. Presumably when milk is exposed to light for longer durations and an "oxidized" flavor develops (5) the reducing substrates are partially exhausted and $^1O_2$ may be responsible for the "oxidized" flavor, since SOD and catalase should inhibit the Fenton-like reactions (6).

REFERENCES

1. Jenness, R. and Koops, J., Neth. Milk Dairy J., 16,153 (1962).

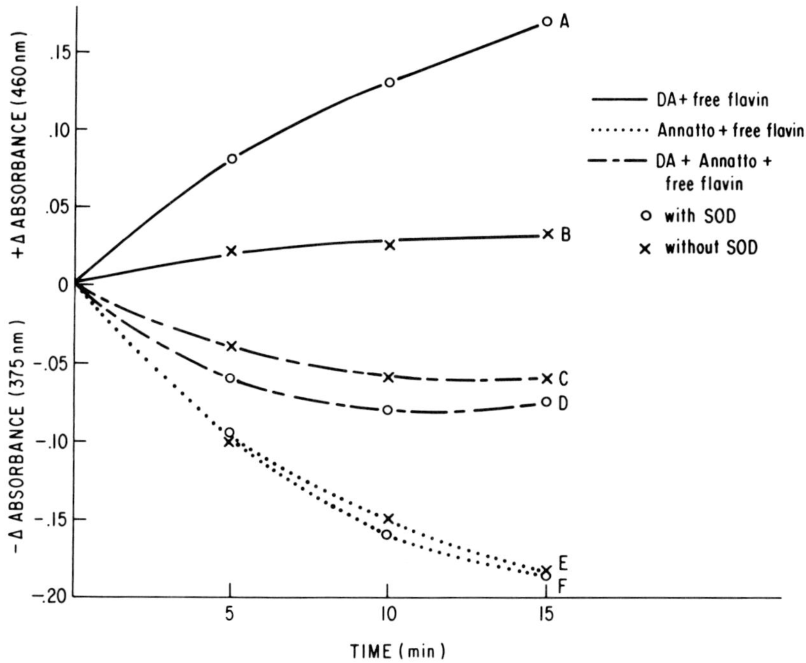

FIGURE 2. Oxidations by riboflavin in skimmilk ultrafiltrate (SMUF). All reaction mixtures contained 4 x 10^{-6}M free riboflavin in SMUF (pH 6.6) at 25°C. Norbixin (Annatto) (3 x 10^{-8}M), dianisidine (DA) (2.0 x 10^{-4}M), and SOD (4.0 μg/ml) were used as indicated.

2. Misra, H. P. and Fridovıch, I., Arch. Biochem. Biophys., 181,308 (1977).
3. Allen, C. and Parks, O. W., J. Dairy Sci., 58,1609 (1975).
4. Holbrook, J. and Hicks, C. L., J. Dairy Sci., 61,1072 (1978).
5. Webb, B. H., Johnson, A. J. and Alford, J. A., "Fundamentals of Dairy Chemistry," 2nd ed., AVI Publ. Co., Inc. (1974), p. 240.
6. Walling, C. and Johnson, R. A., J. Am. Chem. Soc., 97,363 (1975).

OH· RADICAL INDUCED OXIDATION OF SULFUR CONTAINING AMINO ACIDS

K. O. Hiller, M. Göbl, B. Masloch, and K.-D. Asmus

Hahn-Meitner-Institut für Kernforschung Berlin GmbH, Bereich Strahlenchemie D-1000 Berlin 39 Germany

The reaction mechanism of the OH· radical-induced oxidation of S-containing amino acids, in particular methionine, has been investigated in aqueous solutions by radiation chemical methods, mainly pulse radiolysis techniques. The two principal pH-dependent reaction modes of the system, subsequent to initial diffusion-controlled oxidation of the methionine S-atom, have been identified.

In slightly acid, neutral and basic solutions (pH > 3 for methionine) the predominant reaction is a fast intramolecular oxidation of the amino-group by the primary oxidized S-atom. This electron transfer reaction formally yields a nitrogen centered radical cation, which undergoes immediate decarboxylation. Both CO_2 and the resulting α-amino radical have been identified either directly or through subsequent reactions. This oxidation of the amino-group by the oxidized sulfur function requires a free electron pair of the nitrogen. Even in the zwitterionic form (pH = 3-9) the amino-group deprotonation is apparently sufficient to permit reaction—presumably as a result of rapid proton exchange between the amino and carboxyl groups.

Stronger protonation of the amino-group, induced by addition of high concentrations of free protons (pH < 2) or effective proton donors ($H_2PO_4^-$, CH_3COOH etc.), however, depresses the reduction potential of the amino group to such an extent that the oxidation center remains localized at the sulfur atom. The same effect is observed on introduction of electron withdrawing substituents e.g. in N-acetyl methionine. In such cases, the oxidised sulfur atom is temporarily stabilised in a transient radical cation complex,

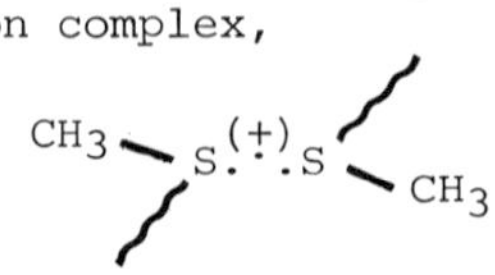

characterized by a 3-electron σ^2/σ^* bond between two S-atoms. The further reactions of this species do not include decarboxylation.

ISBN 0-12-592050-4

SUPEROXIDE DISMUTASE ACTIVITY IN T AND B LYMPHOMA AND ITS POSSIBLE RELATIONSHIP TO RADIATION SENSITIVITY

Rolf D. Issels, Andreas Schalhorn and Eckhardt Thiel, Ludwig-Maximilians-University of Munich, Klinikum Großhadern, Medizinische Klinik III, Marchioninistraße 15, 8000 München 70 West-Germany.

Using polyacrylamide activity stain and isoelectric focusing stain techniques superoxide dismutase activity has been identified in lymphocytes of patients suffering from chronic lymphocytic leukemia (CLL). In lymphocytes of healthy doners Cu, Zn-SOD as well as Mn-SOD could be identified, whereas Mn-SOD is diminished in B- and T-CLL. The increased radiation sensitivity of CLL lymphocytes compared with normal lymphocytes might be related to the diminished amount of Mn-SOD.

The enzyme superoxide dismutase (SOD) ($O_2^{\cdot-}$ - $O_2^{\cdot-}$ oxidoreductase, EC 1.15.1.1.) which catalyzes the dismutation of superoxide free radical anions was first described by McCord and Fridovich (1). Two different forms of SOD in eukaryotic cells are known. One is found in the cytosol and intermembrane space of mitochondria and contains copper and zinc. The other, a manganese containing SOD is found in the matrix of mitochondria (2).

Diminished amounts of Mn-SOD have been found in malignant cells compared to normal cells (3). Using electrophoresis on polyacrylamide gels followed by a specific staining technique for SOD activity in leukemic cells, there was no detectable band of activity due to Mn-SOD (4). The present study shows the results in lymphoma cells of patients suffering from chronic lymphocytic leukemia (CLL) of B and T cell origin. Previous in-vitro studies on CLL lymphocytes showed a percentually augmented rest population of normal lymphocytes after irradiation (5). Radioprotection by SOD in whole-body irradiated mice (6) and high radiation sensitivity of human tissues with a low content of SOD (7) suggest that SOD might be associated with radiation sensitivity.

MATERIALS AND METHODS

Peripheral blood lymphocytes from healthy adult volunteers and from five untreated patients suffering from chronic lymphocytic leukemia (CLL) were isolated by centrifugation in a Ficoll-sodium metrizoate gradient (8). Separated leukemic cell fractions were characterized for T and B cell markers by determination of surface bound immunoglobulins, reaction with anti-T cell sera and rosette formation as described (9). Contaminating monocytes were identified in stained smears cytochemically.

ISBN 0-12-592050-4

The pellet of the different cell fractions was suspended to 25×10^6 cells/ml in doubly distilled water and disrupted by sonification at 0°C. For all experiments the ruptured cell suspension was centrifuged at 27000 g for 30 min at 4°C and the clear supernatant was used in the assay procedures. Protein determination of the cell extracts was done by the Lowry method (10) with bovine serum albumin as the standard. The reduction of nitroblue tetrazolium (NBT) by superoxide resulting in the formation of the blue formazan derivative and its inhibition by SOD was achieved by the method of (11). The gels contained 7.5% acrylamide and electrophoresis was performed at 200 V (2 mA/tube). The gels received equivalent amounts of protein ranging from 50 µg to 150 µg and were stained for SOD activity in the presence or absence of 5 mM cyanide. The cyanide is known to inhibit the Cu,Zn-SOD but does not affect the Mn-SOD, thus allowing the identification of the bands on the gels. Bovine erythrocyte Cu, Zn-SOD prepared according to (12) was used as a marker for Cu,Zn-SOD.

Analytical electrofocusing was carried out on ready-prepared thin-layer polyacrylamide gels containing carrier ampholytes with a range of pH 3.5-9.5 (LKB PAG plates, Bromma, Sweden). Separation of proteins was achieved using a LKB 2103 power supply equipped with a cooling system. Samples containing identical amounts of proteins were layered on the gel. The gels were stained for proteins with Coomassie Brilliant Blue R-250 or for SOD activity (11).

RESULTS AND DISCUSSION

Characterization by immunological methods revealed that the lymphocyte preparation of B- or T-CLL having 95% lymphocytes with B or T markers was quite homogeneous. The remaining cells (5%) were contaminating granulocytes. Fig. 1 demonstrates the SOD activity found in sonicates of B and T leukemic lymphocytes after separation by disc electrophoresis. In contrast to the two bands of activity, which were found in normal lymphocytes (not seen in Fig.) the sonicates of B-CLL and T-CLL gave only one band of activity. Even by overloading the gels with five times the amount of protein there was only the one detectable band of activity in the position of the cyanide sensitive Cu,Zn-SOD. Electrofocusing on thin-layer polyacrylamide gels showed in all sonicates of leukemic lymphocytes one band of activity due to the Cu,Zn-SOD with an isoelectric point of approximately 4.8. In normal lymphocytes there were also two distinct bands of activity in the region of pH 8.0 due to Mn-SOD. A decrease or lack of Mn-SOD was described in human leukemia cells in comparison to mature normal cells (4). Also, a diminished amount of Mn-SOD in the leukemic lymphocyte population of CLL using polyacrylamide activity stain was recently reported (13).

The increased radiation sensitivity of CLL lymphocytes compared with normal lymphocytes is known (14,15). After

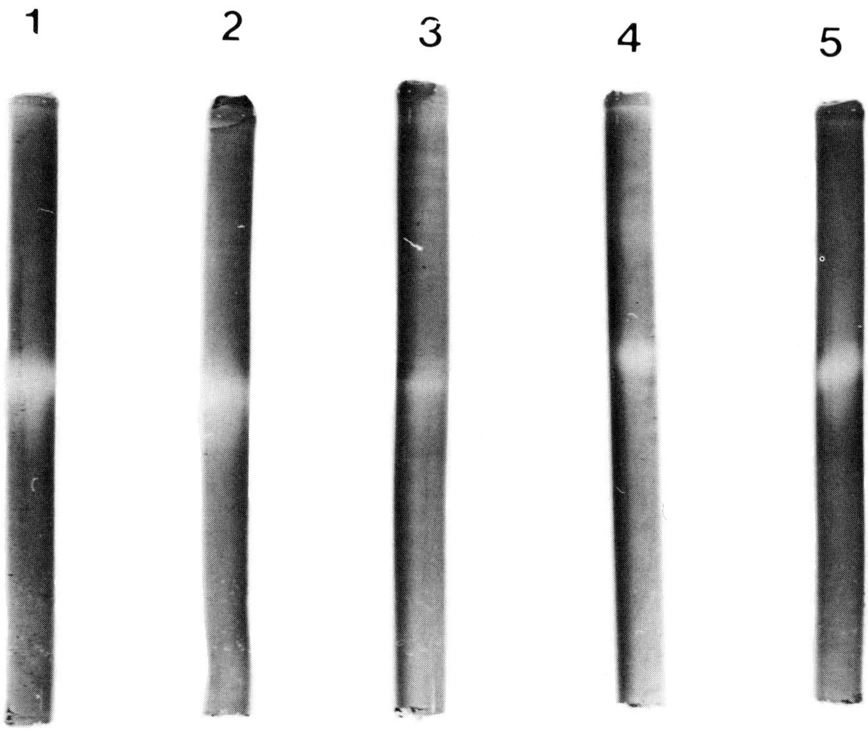

FIGURE 1. Staining for SOD activity after polyacrylamide gel electrophoresis of cell-free extracts of leukemic lymphocytes (CLL) crude soluble extracts of ultrasonically-disrupted lymphocytes were applied to sets of polyacrylamide gels in graded amounts.

B-CLL lymphocytes	*T-CLL lymphocytes*
1. 80 µl	*4. 20 µl*
2. 40 µl	*5. 80 µl*
3. 20 µl	

After electrophoresis the gels were stained for activity by the nitroblue tetrazolium method (11). Discussion of results see text.

extracorporeal irradiation of blood and irradiation of spleen in the treatment of B-CLL we found an increased population of normal lymphocytes with T-cell activity (5). There are some results indicating a possible role of SOD in radio-protection of cells (6,7,16). However, the diminished amount of Mn-SOD in CLL lymphocytes of B and T cell origin might be related to the high radiation sensitivity of these cells.

REFERENCES

1. McCord, J. M. and Fridovich, I., J. Biol. Chem. 243, 5753 (1968).
2. Weisinger, R. and Fridovich, I., J. Biol. Chem. 248, 3582 (1973).
3. Oberlay, L. W. and Buettner, G. R., Cancer 39, 1141 (1979).
4. Yamanaka, N., Ota, K. and Utsumi, T., "Biochemical and Medical Aspects of Active Oxygen" (O. Hayaishi and K. Asada, eds.), University Park Press, Baltimore, 1978, p. 183.
5. Theml, H., Begemann, H., Issels, R., Kaboth, W., Schick, P., Czempiel, H. and Winnewisser, M., Haemat. 8, 43 (1974).
6. Petkau, A., Kelley, K., Chelack, W. S. and Barefoot, C., Biochem. Biophys. Res. Commun. 70, 452 (1976).
7. Sykes, J. A., McCormack, F. X., Jr. and O'Brien, T. J., Canc. Res. 39, 2759 (1978).
8. Boyum, A., Nature, 204, 793 (1964).
9. Thiel, E., Rodt, H., Huhn, D. and Thierfelder, S., Blood 47, 723 (1976).
10. Lowry, O. H., Rosebrough, N. I., Farr, A. L. and Rossdall, R. I., J. Biol. Chem. 193, 265 (1951).
11. Beauchamp, C. and Fridovich, I., Anal. Biochem. 44, 276 (1971).
12. Weser, U., Bunnenberg, E., Cammack, R., Djerassi, C., Flohe, L., Thomas, G. and Voelter, W., Biochem. Biophys. Acta 243, 203 (1971).
13. Issels, R. D., Wilmanns, W. and Weser, U., "The Significance of Superoxide and Superoxidedismutase" (W. H. Bannister and I. V. Bannister, eds.), Elsevier, Amsterdam, 1980, Vol. II, p. 281-285.
14. Johnson, R. E., Canc. Res. 27, 39 (1967).
15. Johnson, R. E., Kagan, A., Gralnick, H. and Fass, L., Cancer 20, 1382 (1967).
16. Marklund, S. L. and Westman, G., "The Significance of Superoxide and Superoxidedismutase" (W. H. Bannister and I. V. Bannister, eds.), Elsevier, Amsterdam, 1980, Vol. II, p. 318-326.

THE PHYSIOLOGICAL ROLE OF THE CHARGE DISTRIBUTION ON SUPEROXIDE DISMUTASE

W. H. Koppenol, Department of Biochemistry and Molecular Biology, Northwestern University, Evanston, Illinois 60201

The second order rate constant for the reaction of the superoxide anion ($O_2^{\cdot -}$) with Cu^{2+}(aq) is approximately three times larger than that for the reaction with bovine copper-zinc superoxide dismutase (SOD). This fact combined with the unreactivity of $O_2^{\cdot -}$ led Fee (1) to suggest that the dismutase activity was a trivial property of the protein. By means of the Von Smoluchowski-Debye equation it can be shown that neither reaction is diffusion controlled and that of every 100 collisions of $O_2^{\cdot -}$ 25 lead to a reaction with Cu^{2+}(aq) and 10 to a reaction with SOD. These numbers differ slightly from an earlier calculation (2) due to the adoption of the value of 8.1×10^9 $M^{-1}s^{-1}$ for the rate constant of Cu^{2+}(aq) with $O_2^{\cdot -}$ (3, 4). Considering (i) that each of the two copper ions in SOD has only one coordination site available, compared to four for every Cu^{2+}(aq), and (ii) that the accessible surface area of the superoxide dismutase molecule, calculated according to Teller's equations (5) to be 11,900 $Å^2$, is 150 times larger than that of a hydrated copper ion (radius estimated to be 2.5 Å), one must conclude that the protein molecule functions most efficiently. To account for this, it was postulated that surface charges guide $O_2^{\cdot -}$ towards the active sites (2). Electrostatic potential field calculations are presented here which indeed show this to be the case.

At pH 7 theoretically all lysines and arginines are protonated and all carboxylic acids are dissociated. Furthermore, the four metal ions each carry a charge +2e (e is the elementary charge), and the bridging imidazolate (His 61) a charge -e. Histidines 41 and 19 have pK's of 10.4 and 6.7, respectively (6,7) and were assigned charges of +1e and +½e. The N-terminus is acetylated and thus uncharged. The net charge of the dimer is then -3e. Potentials were calculated as previously described (8). Results are shown for two parallel cross sections of SOD in Figs. 1 and 2. The potential reaches negative values near the surface of the protein, repelling $O_2^{\cdot -}$, except close to the active sites where the most negative value is only -0.3 kT/e. Neutralization of the charge of arginine 141 (9) increases this electrostatic barrier by 0.4 kT (not shown). Near the copper ions the potential becomes positive and reaches a maximum. The electrostatic potential is expressed in units of kT/e (k is Boltzmann's constant), which facilitates the interpretation of these figures when one considers that every molecule has a kinetic energy of 1½ kT. Due to the assumption of a continuous dielectric medium

ISBN 0-12-592050-4

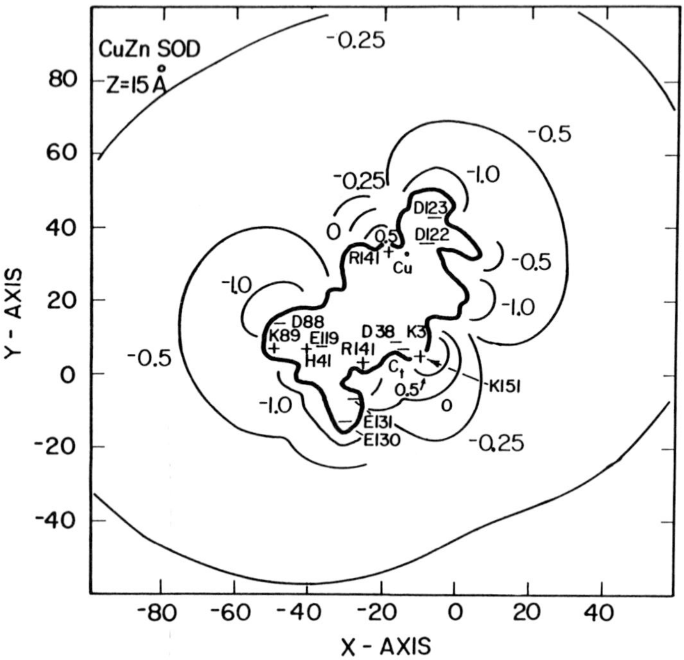

FIGURE 1. *Equipotential lines around bovine copper-zinc superoxide dismutase in a cross section perpendicular to the crystallographic Z-axis, as indicated. Electrostatic potentials are expressed in units of kT/e. The location of positive and negative charges lying between the planes Z + 2 Å and Z - 2 Å are indicated (D = Asp, E = Glu, H = His, K = Lys, R = Arg, M = centre of mass).*

(D = 78.5), the potential inside and near the protein has been underestimated. However, this effect is counteracted by the presence of an inert electrolyte (8). Because of the specific dimeric structure of the protein, the centers of positive and negative charge nearly coincide with the center of mass. This results in a small dipole moment of approximately 80 debye. The net charge and the distribution of the individual charges oppose

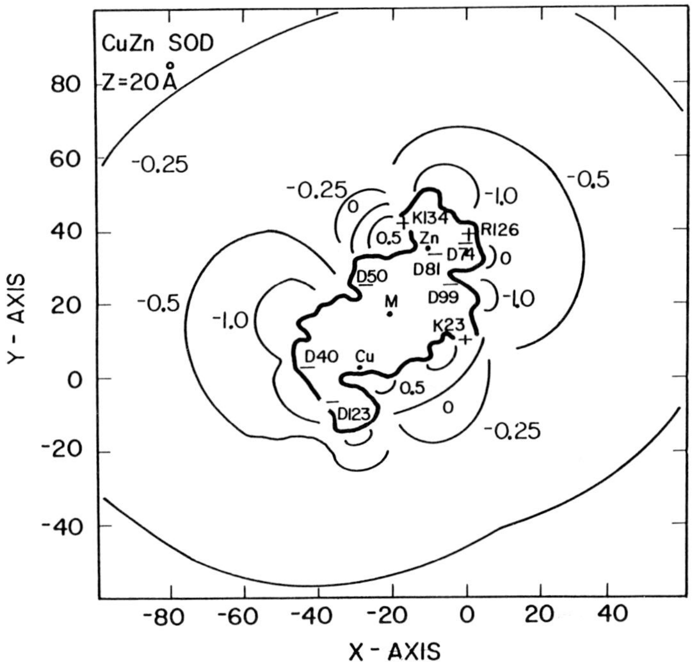

FIGURE 2. See Fig. 1.

binding or a specific orientation of the protein to a negatively charged membrane, which would impair its function as a radical scavenger. It follows that from an electrostatic point of view superoxide dismutase is extremely well suited to react with $O_2^{\bar{\cdot}}$.

If $O_2^{\bar{\cdot}}$ is not reactive, but still effectively scavenged by SOD, it must be the precursor of something more harmful. Shown in Fig. 3 is a scheme which features a Fenton-catalyzed Haber-Weiss reaction and yields MO^{2+} as the reactive species, rather than $\cdot OH$, which is less selective in its reactions. It is concluded that superoxide dismutase reacts with the superoxide anion radical to prevent indirectly the reduction of hydrogen peroxide.

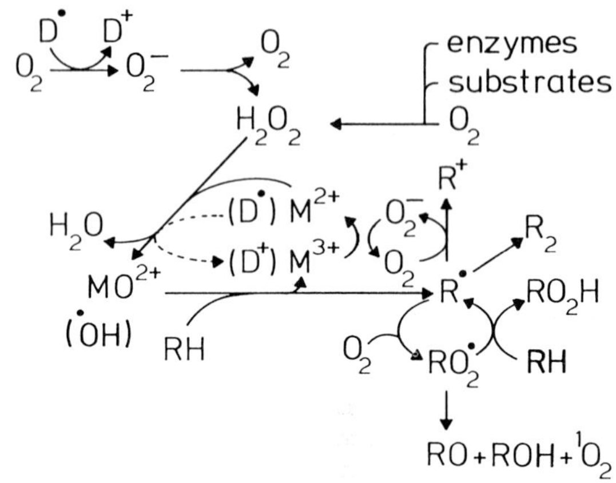

FIGURE 3. Scheme for oxygen toxicity in vivo. D• could represent a semiquinone, a flavin radical, or another organic radical, M^{2+} a Fe(II)ADP complex, RH any biomolecule, R• and R^+ a biomolecule containing $-CH_2-\dot{C}H_2$ and $-CH=CH_2$, respectively.

ACKNOWLEDGEMENTS

Coordinates for these calculations were kindly supplied by Dr. D. C. Richardson (Duke University, Durham, NC). This work is supported by NIH Grant GM 19121.

REFERENCES

1. Fee, J. A., "Oxidases and Related Redox Systems" (T. E. King, H. S. Mason and M. Morrison, eds.), (1981) In press.
2. Koppenol, W. H., "Oxidases and Related Redox Systems" (T. E. King, H. S. Mason and M. Morrison, eds.), (1981) In press.
3. Rabani, J., Klug-Roth, D. and Lilie, J., J. Phys. Chem. 77, 1169 (1973).
4. Butler, J. (1981) unpublished.
5. Teller, D. C., Nature 260,729 (1976).
6. Stoesz, J. D., Malinowski, D. P. and Redfield, A. G., Biochemistry 18,4669 (1975).
7. Cass, A. E. G., Hill, H. A. O., Smith, B. E., Bannister, J. V. and Bannister, W. H., Biochemistry 16,3061 (1977).
8. Koppenol, W. H., Vroonland, C. A. J. and Braams, R., Biochim. Biophys. Acta 503,499 (1978).
9. Malinowski, D. P. and Fridovich, I., Biochemistry 18,5909 (1979).

THE ROLE OF SUPEROXIDE IN MICROSOMAL MONOOXYGENASES

H. Kuthan, H. Graf, H. Tsuji and V. Ullrich, Institute of Physiological Chemistry, The University of The Saarland, D-665 Homburg, Federal Republic of Germany

The NADPH-dependent generation of hydrogen peroxide by liver microsomes, obtained from phenobarbital treated rats, and in a reconstituted cytochrome P-450 system was found to be stimulated about two-fold in the presence of 7-ethoxycoumarin. Similar results were obtained by measuring the NADPH-initiated formation of superoxide anion radicals using microsomal preparations essentially devoid of superoxide dismutase. For these studies a novel method was used to analyze the formation of the superoxide radical. This procedure is based on the unique properties of succinylated cytochrome c which displays only 0.3 to 1% of the NADPH-cytochrome c reductase activity of native cytochrome c towards the purified NADPH-cytochrome P-450 reductase but 30 to 40% of the activity of native cytochrome c in the presence of a superoxide generating system composed of xanthine and xanthine oxidase. Using this technique the following results were obtained: 1. The NADPH-dependent formation of $O_2^{\bar{}}$ by liver microsomes was found to be as sensitive as the formation of H_2O_2 to inhibition by carbon monoxide. 2. The analysis of the photochemical action spectrum revealed a maximal reversibility of the CO-inhibited production of superoxide anions and hydrogen peroxide using monochromatic light of 450 nm wavelength. 3. A ratio of about 2:1 was observed for the rates of the NADPH-dependent formation of superoxide anions and hydrogen peroxide in a reconstituted cytochrome P-450 system. These data indicate strongly that the hydrogen peroxide generated in the presence of NADPH and dioxygen by liver microsomes is the product of dismutation of superoxide anion radicals, arising predominantly from the auto-oxidation of oxy-cytochrome P-450. Minor contribution by auto-oxidation of the reduced NADPH-cytochrome c (P-450) reductase flavoprotein and reduced cytochrome b_5 may be assumed.

ACKNOWLEDGEMENTS

Supported by a grant of the Deutsche Forschungsgemeinschaft, SFB 38, Projekt L_1.

ISBN 0-12-592050-4

FREE RADICALS AND PEA FOLIAGE SENESCENCE: INCREASE OF LIPOXYGENASE AND ESR SIGNALS AND CYTOKININ-INDUCED CHANGES

Y. Y. Leshem, Y. Liftman, S. Grossman and A. A. Frimer

Life Science and Chemistry Departments, Bar-Ilan University, Ramat-Gan, Israel

Recent work in our laboratory (1,2) has suggested that cytokinin (CK), a senescence retarding phytohormone, may act by means of scavenging lipoxygenase-generated free radicals on polyunsaturated fatty acid chains. It also has been suggested that free radicals in turn may be involved in the onset of senescence as shown in fruit (3,4) or aging leaves (5). We endeavored to ascertain whether free radicals do indeed increase with progress of leaf senescence and to elucidate the role of CK in its retardation.

To examine the hypothesis, changes in free radicals were estimated by electron spin resonance (ESR) as related to leaf age, and as influenced by a free radical generating system (6) using xanthine-xanthine oxidase (x-xo) in the absence and presence of CK as a free radical antagonist. Lipoxygenase levels were also determined.

RESULTS AND DISCUSSION

Effects of the x-xo on lipoxygenase activity are consistent with the hypothesis suggesting that the x-xo treatment, which evolves superoxide radicals and its derivatives, markedly increases lipoxygenase levels. In contrast its deleterious effects are reversed by CK. This may be interpreted either as radical quenching or alternatively decrease in free radical production since CK has inhibitory action on xo (7). Results of the ESR experiment (Fig. 1) show that there is an increment of endogenous free radical levels in normally senescing leaves. Young tissue (Fig. 1, upper A) essentially lacks ESR free radical signals while subsequent development of senescence (stage 1-2, Fig. 1, upper B-C) showed increasingly marked signals. The significant change of the original signal by CK treatment (Fig. 1, upper D) indicates definite interaction between the hormone and the nature of the free radicals in the endogenous system. With the progress of ageing, we also observed a triple peaked secondary signal in the lower gaussian scan field (Fig. 1, lower A-D) possessing hyperfine splitting and being accentuated by CK.

The nature of the signals based on g-values suggests two possibilities. a) In the major region (Fig. 1, upper) the onset of ageing (B) has a g-value of 2.0053 which may be an organic

ISBN 0-12-592050-4

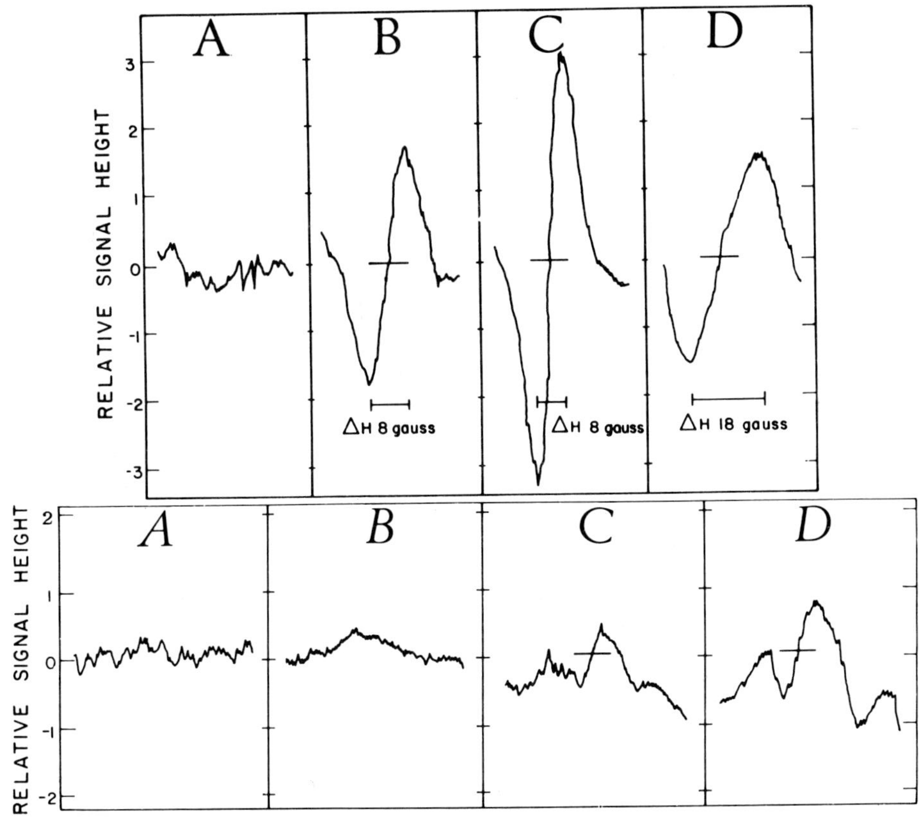

FIGURE 1. ESR profiles of control and CK-treated senescing pea foliage: A = young, B = senescing stage 1, C = senescing stage 2, D = as C + CK. Profiles determined from N_2-frozen tissue strips placed in a flat quartz Varian E-226 cell in a Varian E-line X band spectrophotometer at 9.518 GHz microwave frequency. Upper: The major signal in the 3370-3400 gauss scan range. g-values of B, C and D were 2.0053, 2.0167 and 2.0158, respectively. Lower: The secondary signal in the 3320-3360 gauss scan range. g-values of C and D were 2.0426 and 2.0439, respectively.

free radical (possibly associated with photosystem II). In subsequent steps (C→D) the nature of the radical is changed, suggesting a metallo-protein complex. In general the pattern 1A→D indicates a steady state of reactive free radicals or alternatively a buildup of a stable free radical. The latter is possibly a semi-quinone stabilized by senescence-induced pH changes. b) The secondary signal (Fig. 1, lower) suggests a Cu-protein.

It is tempting to think along the lines of an earlier proposal (8), that an additional function of CK in leaf senescence could be the intracellular complexing of free radicals extracted from reactive sites and binding to some polymer complex. Such stable complexes may serve as "radical dumping grounds" (8). Our data seem to support this hypothesis.

The present experiments indicate that free radicals do increase with ageing, that a free radical generating system increases lipoxygenase activity; this being reversed by CK which also induces definite changes in ESR patterns.

ACKNOWLEDGEMENTS

We would like to thank J. Barnes and K. Tal for their technical help. Thanks are also due to Dr. H. Swartz, Illinois College of Medicine for his kind aid in the interpretation of the ESR signals, and to Drs. Y. Degani of the Weizman Institute, Israel and C. Frenkel of Rutgers University for their helpful discussions.

REFERENCES

1. Grossman, S. and Leshem, Y., Plant Physiol., 59, 411 (1978).
2. Leshem. Y. Y., Grossman, S., Frimer, A. and Ziv, J., "Advances in the Biochemistry and Physiology of Plant Lipids, 3," (L. Appelqvist and C. Liljenberg, eds.), Elsevier/North-Holland Press, Amsterdam (1979), p. 193.
3. Baker, J. E., Lieberman, M. and Kunishi, A. T., Plant Physiol. Suppl., 57, 97 (1976).
4. Brennan, T. and Frenkel, C., Plant Physiol., 59, 411 (1977).
5. Parida, R. K., Kar, M. and Mishra, D., J. Bot., 56, 2937 (1978).
6. McCord, J. M. and Fridovich, I., J. Biol. Chem., 244, 6049 (1969).
7. Henderson, T. R., Skinner, C. G. and Eakin, R. E., Plant Physiol., 37, 552 (1962).
8. Commoner, B., Townsend, J. and Pake, G. E., Nature, 174, 689 (1954).

ANTITUMOR ACTIVITIES OF A COPPER CHELATE WHICH HAS SUPEROXIDE DISMUTASE ACTIVITY AND AN IRON CHELATOR

Susan W. C. Leuthauser,[1] Larry W. Oberley,[1] Terry D. Oberley,[2] John R. J. Sorenson,[3] and Garry R. Buettner[4]

[1]*Radiation Research Laboratory, University of Iowa, Iowa City, Iowa 52242*
[2]*Department of Pathology, University of Wisconsin, Madison, Wisconsin 53706*
[3]*College of Pharmacy, University of Arkansas, Little Rock, Arkansas 72201*
[4]*Department of Chemistry, Wabash College, Crawfordsville, Indiana*

The presence of the manganese containing and the copper and zinc containing forms of superoxide dismutase in mammalian cells have been well documented. However, in many tumor models studied so far, the CuZnSOD was lowered and the MnSOD was reduced to level beyond the limit of detection (1). Superoxide anions were still being generated by tumor mitochondria (2). This creates an increase in the net $O_2^{\overline{\cdot}}$ flux in the tumor cells. This increase may be a contributing factor to the cancerous phenotype of the neoplastic cells. Will replacement of $O_2^{\overline{\cdot}}$ scavenger remove this excess $O_2^{\overline{\cdot}}$ and allow the neoplastic cells to revert to their non-cancerous phenotype? The first part of this study attempts to investigate such possibility.

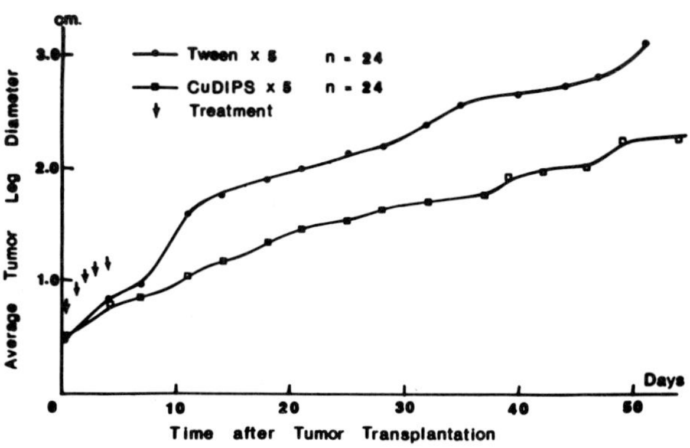

FIGURE 1. Changes in tumor limb diameter after implantation of Ehrlich Carcinoma cells following 5 daily doses of 0.1 ml Tween 80-Saline solution or 0.5 mg of CuDIPS in 0.1 ml Tween 80-Saline, $p < 0.01$.

ISBN 0-12-592050-4

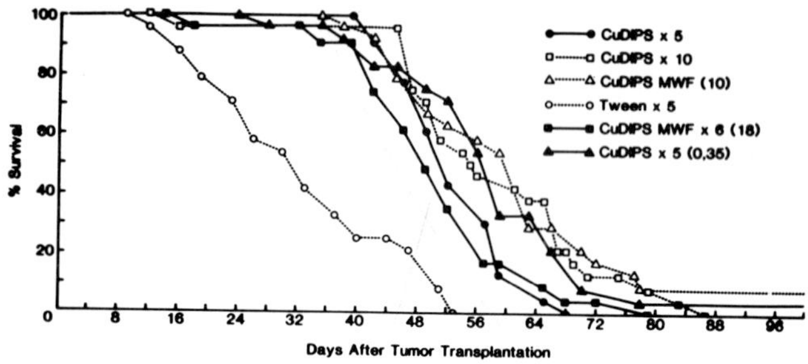

FIGURE 2. Survival of CBA mice implanted with Ehrlich Carcinoma cells after adminstration of various doses of CuDIPS.

Cu(II)[(3,5 diisopropylsalicylate)$_2$] or CuDIPS is an antiinflammatory agent (3) which also had 17,400 ± 1,800 units of SOD activity per mg of the compound. It is lipid soluble with an LD_{50} of 240 ± 33 mg/kg when injected subcutaneously in rats. Histologic studies revealed no CuDIPS toxicity to the mice at the doses that were used. CBA mice were inoculated with 5 x 10^6 Ehrlich carcinoma cells intramuscularly in the right hind limbs, and daily dose of 0.5 mg of CuDIPS in 0.1 ml of 10% Tween 80-Saline solution was injected at the same site for various durations. Fig. 1 shows the difference in tumor growth when 5 daily doses were administered; a reduction of 38% (p=0.01) in the size of the tumor limb diameter resulted. Survival of the CuDIPS treated mice were also increased significantly as indicated in Fig. 2 and Table 1. Autopsies of all the dead animals revealed that massive lung metastasis was the cause of death in all cases. Histologic studies indicated that microscopic nodules of tumor cells were present in the lungs of the control mice as early as 10 days after tumor cells inoculation, but was not observed until after 45 days in CuDIPS treated mice. Similar doses of $CuSO_4$ and DIPS had no beneficial effect on survival of the hosts and growth of the

TABLE I. Summary of Survival Data of CuDIPS Treated Mice Implanted with Ehrlich Tumor Cells

Treatments	Mean Survival	Median Survival	Max. Survival
Tween x 5	33 ± 4* Days	36 Days	53 Days
CuDIPS x 5	54 ± 3	52	68
CuDIPS x 10	58 ± 5	55	87
CuDIPS x 18 MWF x 6	50 ± 4	49	79
CuDIPS x 10 MWF	61 ± 6	62	105**
CuDIPS x 10 0.35 x 5	58 ± 4	59	105**

* Standard Error

** Long Term Survivors

tumor. When 1.2 mg of CuZnSOD isolated from bovine liver was administered daily to tumor mice, no significant effect on the mice was observed. However, morphologically, these tumors became encapsulated instead of having an infiltrative pattern of growth as in the saline treated controls. Native SOD is a high molecular weight compound, and when administered, it probably remains an extracellular $O_2^{\overline{\cdot}}$ scavenger; but CuDIPS has a small molecular size (MW-503) which enables it to penetrate cell membrane. It is likely that the effect of CuDIPS on this tumor model is due to its ability to enter the cell and remove excess $O_2^{\overline{\cdot}}$ intracellularly. Evidence of morphological differentiation of these tumor cells into mammary duct cells was observed in CuDIPS treated mice. CuDIPS has also been shown to inhibit growth and cause differentiation of neuroblastoma cells in our laboratory. Is the SOD activity of CuDIPS the sole cause of its effect on this tumor system? This remains to be investigated in the future.

Superoxide is a reactive species with the generation of ·OH being one of its most important reactions. In an aqueous system, the presence of certain iron-chelator complexes catalyze the formation of ·OH in the presence of H_2O_2 and $O_2^{\overline{\cdot}}$. The release of ·OH has been detected in an iron/picolinic acid/xanthine/xanthine oxidase system by spin trapping techniques (4). In a biological system, picolinic acid withheld iron from the cells. This chelator arrested cell division in cultured rat kidney cells. Such arrest was reversible in the normal cells, but irreversible in the SV40 transformed cells (5). Does this iron-picolinic acid complex enhance the production of ·OH thus causing tumor cell damage? When 5 daily doses of 2.5 mg of picolinic acid were injected into the above tumor model, a reduction of tumor as indicated in Fig. 3 was observed. Histologic studies indicated that massive cell death occurred in the tumors of the picolinic acid treated mice; again, massive lung metastasis was the cause of

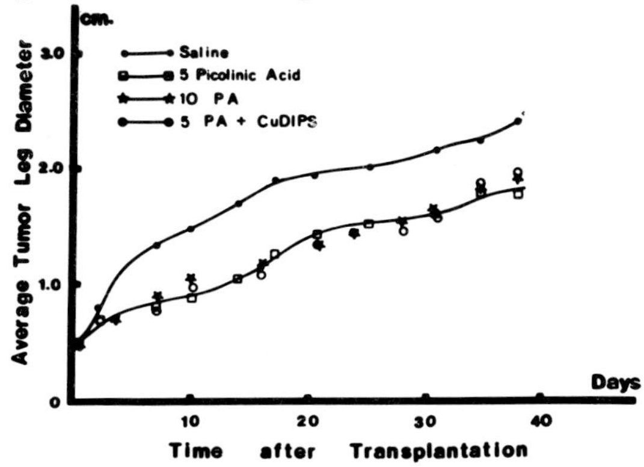

FIGURE 3. Changes in tumor limb diameter after implantation of Ehrlich Carcinoma cells following 5 or 10 daily doses of α-picolinic acid at 2.5 mg per dose or 5 daily doses of α-picolinic acid and 0.5 mg CuDIPS administered simultaneously, $p < 0.01$.

death. Further experiments on possible combination therapy involving CuDIPS and picolinic acid; and the role of picolinic acid in this tumor system are being carried out.

REFERENCES

1. Oberley, L. W. and Buettner, G. R., Cancer Res., 39, 1141 (1979).
2. Bize, I. B., Oberley, L. W. and Morris, H. P., Cancer Res., 40, 3686 (1980).
3. Sorenson, J. R. J., Inflammation 1, 317 (1976).
4. Buettner, G. R. and Oberely, L. W., Biochem. Biophys. Res. Comm. 83, 69 (1978).
5. Fernandez-Pol, J. A. and Johnson, G. S., Cancer Res., 37, 4276 (1977).

DIETHYLDITHIOCARBAMATE BUT NOT PENICILLAMINE AUGMENTS BLEOMYCIN CYTOTOXICITY

P. S. Lin, L. Kwock, G. Misslbeck and K. Hefter

Tufts-New England Medical Center, Department of Therapeutic Radiology, Division of Radiobiology, Boston, Massachusetts 02111

INTRODUCTION

Action of bleomycins on mammalian cells has been extensively studied in recent years. Their effect on DNA is considered to be the major contributing factor to their therapeutic efficacy. The effect of bleomycin on DNA strand scission is inhibited by low concentration of some metal ions, including copper, zinc and cobalt (1) and greatly increased by ferrous (Fe^{+2}) ions (1), the superoxide anion ($O_2^{\bar{\cdot}}$) (2) and hydrogen peroxide (3). The enhancement of bleomycin damage to DNA by these latter agents may be mediated by hydroxyl radicals (4).

Preece et al (5) first suggested using a copper chelator, penicillamine (Pen), to lower the serum copper levels which may reduce cytotoxic effects of bleomycin. We have recently suggested a similar approach using another copper chelator, diethyldithiocarbamate (DDC), which we believe can produce a greater cytotoxicity than penicillamine (6). The anticipated difference is based on the following: (i) DDC is an extremely good copper chelator. (ii) DDC treated cells should have a higher cellular steady-state level of $O_2^{\bar{\cdot}}$ since it is a potent inhibitor of the enzyme responsible for removing this radical species (superoxide dismutase; SOD). The increase in $O_2^{\bar{\cdot}}$ might in turn enhance the bleomycin effect. (iii) Although DDC is a metal chelator, it does not interact with Fe^{+2}. Bleomycin and Fe^{+2} are more effective in degrading DNA than is either species alone. It has been shown that binding of Fe^{+2} to bleomycin generates $\cdot OH$ radicals and that the enhanced effect of bleomycin by $O_2^{\bar{\cdot}}$ is due to its ability to reduce Fe^{+3} to Fe^{+2} (4). Fe^{+3} is ineffective in producing $\cdot OH$ radicals.

RESULTS AND DISCUSSION

Chinese hamster cell (V79) suspensions are prepared by trypsinizing exponentially replicating monolayer cultures. Freshly prepared bleomycin (provided by Dr. W. T. Bardner of Bristol Laboratory, Syracuse, New York) solution is added to the cells' suspension at various final concentrations. After 30 min. treatment at 37°C, the cell suspensions are pelletized and the cells

ISBN 0-12-592050-4

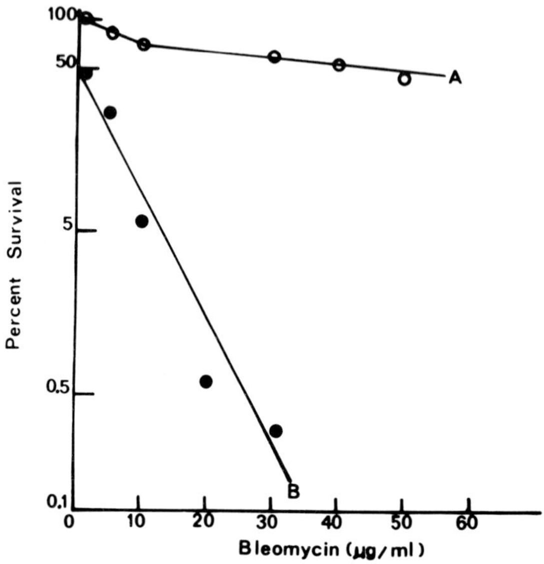

FIGURE 1. The comparison of bleomycin cytotoxicity on V79 cells with or without diethyldithiocarbamate pretreatment as a function of bleomycin dose. A: Bleomycin control; B: DDC treated cell.

washed twice with complete medium. The cells are then plated into plastic petri dishes as in standard cell survival experiments. As shown in Fig. 1 (curve A), the survival response of V79 cells to treatment with increasing doses of bleomycin is composed of two components, a concentration dependent and a concentration independent phase. This upward concavity nature is similar to that reported elsewhere (7,8). Curve B in Fig. 1 is the same bleomycin dose response of V79 cells except that the cells have been pretreated with 10^{-4}M DDC for 30 min. at 37°C. Similar results were obtained in V79 cells treated with 20 μg/ml of bleomycin for different lengths of time. Fig. 2 shows the bleomycin dose response of V79 cells in conjunction with 10^{-4}M penicillamine. These results indicate that DDC augments bleomycin cytotoxicity while penicillamine lacks the ability to alter bleomycin cytotoxicity. Based on the observation illustrated in Fig. 2, it is reasonable to assume that extracellular free Cu^{+2} in the culture medium does not interfere with the effect of bleomycin. Furthermore, if penicillamine is capable of penetrating the cell membrane, then the intracellular free Cu^{+2} is unlikely to be critical to bleomycin cytotoxicity. Therefore, the ability of DDC to inhibit intracellular SOD may indirectly contribute to bleomycin cytotoxicity via the mechanism of $O_2^{\bar{\cdot}}$ (and related products) promoted DNA damage

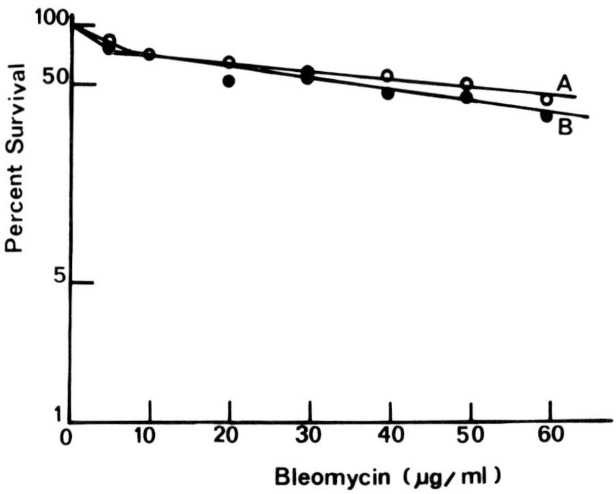

Figure 2. The comparison of bleomycin cytotoxicity on V79 cells with or without added penicillamine as a function of bleomycin dose. A: Bleomycin control; B: Bleomycin and penicilliamine.

(2). This notion is consistent with the hypothesis that localized radicals are produced at the DNA-bleomycin binding site (4, 9) and the possibility that SOD is present in the nuclei of cells (10). However, our current data cannot exclude other possibilities, particularly that DDC may interfere with the bleomycin degradation enzyme. This is now under investigation.

ACKNOWLEDGMENTS

Supported by National Cancer Institute Grant CA26897-01.

REFERENCES

1. Sousville, E. A., Peisach, J and Horwitz, S. B., Biochem., 17, 2740 (1978).
2. Ishida, R. and Takahashi, T., Biochem. Biophys. Res. Commun., 66, 1439 (1975).

3. Nagai, K., Suzaki, H., Tanaka, N. and Umezawa, H., J. Antibiot. Tokyo, 22, 624 (1969).
4. Oberley, L. W. and Buettner, G. R., F.E.B.S. Lett., 97, 47 (1979).
5. Preece, A. W., Light, P. A., Evans, P. A. and Nunn, A. D., Lancet, I, 953 (1977).
6. Lin, P. S., Kwock, L. and Goodchild, N. T., Lancet, I, 777 (1979).
7. Barranco, S. C. and Humphrey, R. M., Cancer Res. 31, 1218 (1972).
8. Takeuchi, T., Takabe, Y., Katsumota, T., Watamable, M. and Umezawa, H., J. Natl. Cancer Inst., 49, 1093 (1972).
9. Takeshita, M., Grollman, A. P., Ohtsubo, E. and Ohtsubo, H., Proc. Natl. Acad. Sci. USA, 75, 5983 (1978).
10. Bartoli, G. M., Faleotti, T. and Azzi, A., Biochim. Biophys. Acta, 497, 622 (1977).

RATE CONSTANTS FOR SINGLET OXYGEN QUENCHING: THE EFFECTS OF AQUEOUS MICELLES

B. A. Lindig and M. A. J. Rodgers

Center for Fast Kinetics Research, University of Texas, Austin, Texas 78712

Rate constants for quenching of singlet oxygen have been measured by the technique of laser flash photolysis using two singlet oxygen probes:

1) 9,10-Anthracenedipropanoic acid (ADPA), which is water-soluble (1),
2) 1,3-Diphenylisobenzofuran (DPBF), which is solubilized in aqueous micelles (2).

Determinations of rate constants were performed for water-soluble quenchers both in D_2O and in D_2O solutions of the anionic surfactant SDS (sodium dodecyl sulfate) or the cationic surfactant CTAB (cetyltrimethylammonium bromide). In neat D_2O, methylene blue was the sensitizer and ADPA was the monitoring compound; in the micellar solutions 2-acetonaphthone was the sensitizer and DPBF was the probe. Measurements were also made for several lipid-soluble quenchers which were contained in the aqueous micelles. Methylene blue and DPBF were the sensitizer and monitor, respectively, in these cases. Table 1 gives a summary of the rate constants obtained by these methods.

For both histidine and methionine, the measured rate constants were independent of the presence of surfactant in the solutions. Comparison of the rate constant measured here for total quenching with literature values (3,4) for the reactive rate constants leads to the conclusion that the quenching is purely reactive (and not physical) for both amino acids in aqueous solution.

All three of the rate constants measured for sodium azide are higher than the reported value of k_q in methanol (2.2×10^8 M^{-1} s^{-1}) (5). The k_q measured in D_2O follows a trend which can be seen from literature values (6,7) of k_q in methanol/water mixtures: namely, that k_q increases with water content in these systems. Evidently k_q is enhanced by the CTAB micelles (relative to D_2O), but decreased by SDS micelles.

The 50% increase in k_q for tryptophan in D_2O over the values measured in SDS and CTAB solution provides evidence that this amino acid is associated with the micelles, resulting in a decrease in its quenching efficiency.

ISBN 0-12-592050-4

TABLE 1. Quenching Rate Constants

Quencher	NEAT D_2O	SDS	CTAB
Sodium azide	5.1×10^8	3.9×10^8	8.2×10^8
Histidine	6.1×10^7	5.9×10^7	5.9×10^7
Tryptophan	6.1×10^7	4.0×10^7	4.2×10^7
Methionine	1.5×10^7	1.8×10^7	1.6×10^7
DABCO	---	1.4×10^6	1.4×10^6
DPBF	---	10.0×10^8	6.7×10^8
PMC	---	4.0×10^8	4.1×10^8
2,3-Dimethylindole	---	7.0×10^8	4.4×10^8

Rate constants given in $M^{-1}s^{-1}$. SDS = sodium dodecyl sulfate, CTAB = cetyl trimethyl ammonium bromide.

Both DPBF and 2,3-dimethylindole, which are micellized singlet oxygen quenchers, yield higher k_q values in SDS than in CTAB micelles. The observation that the DPBF rate constant is higher for such solvents as ethanol, dioxane and methanol than for the less polar solvents benzene, acetone and tetrahydrofuran suggests that the difference in rate constants for SDS and CTAB may be an effect of the solubilization site of the substrate molecule. That is, the DPBF or dimethylindole may be situated nearer the polar interface of the SDS micelle than that of the CTAB micelle, since the latter forms larger aggregates.

While the quenching rate constant for PMC (2,2,5,7,8-pentamethylchromanol) was independent of the type of micelle, it is interesting to compare the k_q obtained with a reported figure for α-tocopherol (an analog) in SDS (8). The rate constant for the tocopherol is some 60% higher, and this may be due to the effect of the hydrocarbon chain (which is absent in PMC) to orient the chroman moiety of the tocopherol near the micelle surface.

When β-carotene was micellized in these systems (by injecting a small volume of a concentrated carotene in dioxane solution) there was <u>no</u> evidence of quenching singlet oxygen by the carotene. Examination of the absorption spectrum of the carotene-SDS micellar solution indicated that the carotene was aggregating, and this evidently results in a lowering of the effective carotene concentration below that required for observable quenching.

ACKNOWLEDGMENTS

The compound ADPA was graciously supplied by A. P. Schaap, Wayne State University. This research was supported by N.I.H. Grant GM-24235. The CFKR is supported by N.I.H. Grant RR-00886 from the Biotechnology Branch of the Division of Research Resources and by the University of Texas at Austin.

REFERENCES

1. Lindig, B. A., Rodgers, M. A. J., Schaap, A. P., J. Am. Chem. Soc., 102, 5590 (1980).
2. Lindig, B. A., Rodgers, M. A. J., J. Phys. Chem., J. Phys. Chem., 83, 1683 (1979).
3. Kraljić, I. and Sharpaty, V. A., Photochem. Photobiol., 28, 583 (1978).
4. Matheson, I. B. C., and Lee, J., Photochem. Photobiol., 29, 879 (1979).
5. Hasty, N., Merkel, P. B., Radlick, P. and Kearns, D. R., Tetrahedron Lett., 49, (1972).
6. Foote, C. S., "Singlet Oxygen," (H. H. Wasserman and R. W. Murray, eds.), Academic Press, New York, N. Y. (1979), p. 139.
7. Gollnick, K., "Singlet Oxygen Reactions with Organic Compounds and Polymers," (B. Rånby and J. B. Rabek, eds.), Wiley, New York, N. Y., (1978), p. 111.
8. Gorman, A. A., and Rodgers, M. A. J., Chem. Phys. Lett., 55, 52 (1978).

SUPEROXIDE DISMUTASE IN THE NORMAL RAT LARGE BOWEL AND IN 1,2-DIMETHYL-HYDRAZINE INDUCED COLON ADENOCARCIMOMA

Dean P. Loven, Larry W. Oberley and Reggie H. Stevens, Radiation Research Laboratory, 14 Medical Laboratories, University of Iowa, Iowa City, Iowa 52242

Previous studies in our laboratory have demonstrated that 1,2-dimethyl-hydrazine induced tumors of the rat large bowel have elevated levels of guanosine 3',5' cyclic monophosphate (cGMP) when compared to normal large bowel (1). Recent work has demonstrated that guanylate cyclase, the enzyme responsible for cGMP formation is activated by free radicals such as superoxide and hydrogen peroxide (2). Superoxide dismutase is responsible for the formation of hydrogen peroxide at the expense of superoxide in cells. We, therefore, measured superoxide dismutase activities to determine if this enzyme has altered in tumor systems.

Tumors were induced by the method of Reddy (3). Frozen samples were assayed by the method of Beauchamp and Fridovich. Nitrotetrazoleum blue (NBT) reduction in the presence of increasing concentrations of protein was measured. One unit of activity was defined as the mg of protein that inhibited NBT reduction by 50%. Mn SOD was measured after the addition of 5 mM cyanide.

Total and Mn SOD was constant down the length of the normal colon. Total activity was unchanged in tissue distant from and adjacent to the tumor, however, Mn SOD was elevated in the normal and depressed in the latter. Total SOD activity in the tumor was greatly elevated, while SOD was depressed (Fig. 1).

The large bowel is a complex tissue composed of various cell types. Mucosal scrapings had total SOD activity that was intermediate between whole tissue and tumor SOD activity. When crypt and villus cells were isolated from the large bowel by the method of Stern (4), the total SOD activity of the crypt was similar to that of the tumor, while Mn SOD was higher than that in the tumor (Fig. 2).

These findings are the first to be reported for a chemically induced tumor. While previous studies have shown depressed levels of SOD to be common, the control tissue has not always been properly defined. This study understores the need for proper control tissues in such studies. Preliminary work on DMH induced tumors of the small bowel indicate a similar pattern of SOD activity exists in the small bowel and small bowel tumor as in the large bowel and large bowel tumor.

ISBN 0-12-592050-4

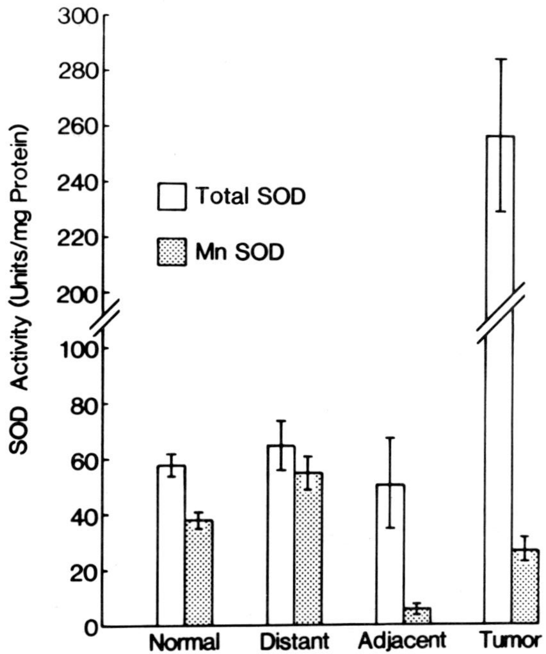

FIGURE 1. Total and Mn SOD activity in various whole tissues of normal rat colon and DMH treated animals. N = 5 animals ± S.E.M.

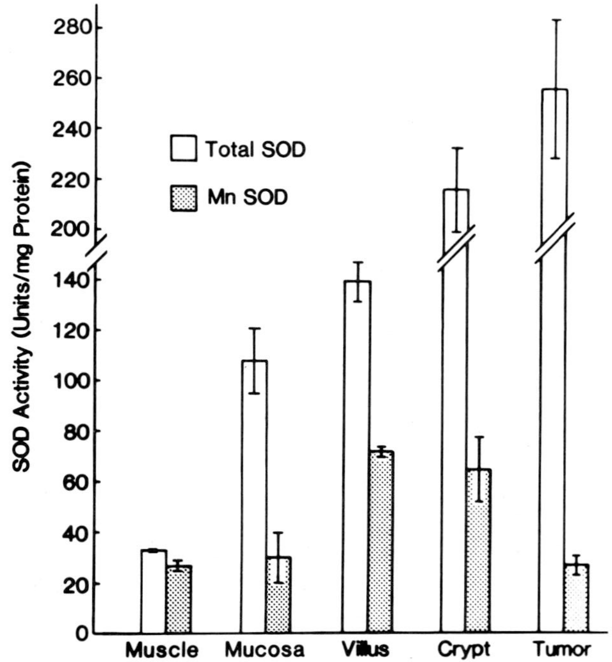

FIGURE 2. Total and Mn SOD activity for various tissue types present in normal rat large bowel. N = 3 animals ± S.E.M.

ACKNOWLEDGEMENTS

Supported by a grant from the Alexander Medical Foundation and CA18327. Data presented in this paper will appear in (5).

REFERENCES

1. Stevens, R. H., Loven, D. P., Osborne, J. W., Prall, J. P. and Lawson, A. J., Cancer Letters, 4,27 (1977).
2. White, A. A. and Crawford, K. K., J. Biol. Chem., 251,7304 (1976).
3. Bandau, S., Reddy, S., Narisawa, T., Wright, P., Vukusich, D., Weisburger, J. H. and Wynder, E. L., Can. Res., 35,287 (1975).
4. Beauchamp, C. and Fridovich, I., Anal. Biochem., 44,276 (1977).
5. Loven, D. P., Oberley, L. W., Rossau, F. M. and Stevens, R. H., J. Natl. Cancer Inst., 65, 377-381 (1980).

CuZn SUPEROXIDE DISMUTASE AND Mn SUPEROXIDE DISMUTASE IN HUMAN TISSUES AND HUMAN MALIGNANT TUMORS

Stefan Marklund[1] *and Gunnar Westman*[2]

[1]*Department of Clinical Chemistry and*
[2]*Department of Oncology, University Hospital of Umeå, Umeå, Sweden*

Extracellularly given superoxide dismutase (SOD) has been shown to protect DNA (1) proteins (2) and cell membranes (3) against ionizing radiation. Myoblasts (4) mycoplasma (5) and bacteria (6,7) are protected by SOD in the medium. Mice given SOD parenterally show increased radioresistance (8). Variations of SOD content in bacteria, induced by growth in different oxygen tensions (6) or by growth in media lacking the prosthetic metals of SOD (7), have indicated the importance of endogenous enzyme in the protection against the oxygen effects. Mammalian cells in culture exposed to diethyl-dithiocarbamate, which inhibits CuZn SOD, show increased radiosensitivity (9,10). Anthracyclin antibiotics and bleomycin appear to exert their toxic action on DNA by way of reactive free radicals derived from oxygen. SOD has been shown in vitro to protect DNA against bleomycin Fe^{2+} complex (11). The anthracyclines, daunomycin and adriamycin apparently easily take part in redox cycles producing superoxide radicals (12) and SOD has been shown to protect DNA (13).

What is the role of endogenous CuZn SOD and Mn SOD in the radiation response of human tumors? Do the enzymes play a role in the response of tumors to radical producing antibiotics?

MATERIALS AND METHODS

Tissues and tumors. Human tissues were obtained within 24 hours after death from the Department of Forensic Medicine from accident or suicide victims without known physical diseases. Macroscopically homogenous pieces of tumor (about 0.5 g) were cut out of operation preparates within an hour after surgery. The tissues and tumors were homogenized and sonicated in 20 volumes of 10 mM potassium phosphate pH 8.0 + 30 mM KCl.

SOD analysis. SOD was determined in terms of its ability to catalyze the disproportionation of $O_2\cdot^-$ in alkaline aqueous solution. The disproportionation was directly studied at A_{250} in a spectrophotometer (14). One unit in the assay is defined as the activity that brings about a decay in $O_2\cdot^-$ at a rate of 0.1 s^{-1} in 3 ml buffer. It corresponds to 8.3 ng human and 4.1 ng bovine CuZn SOD and 65 ng bovine Mn SOD. When the human enzymes are analysed,

ISBN 0-12-592050-4

one unit in the present assay corresponds to 0.024 units CuZn SOD respectively 0.24 units Mn SOD in the xanthine oxidase -Cy C assay (15), which works at physiological conditions. On a physiological activity basis the present method is thus 10 times more sensitive to CuZn SOD, than to Mn SOD.

For protein analysis Coomassie Brilliant Blue G-250 was employed.

RESULTS AND DISCUSSION

The table collects the results of CuZn SOD and Mn SOD analysis in human tissues and tumors. In general there seems to be less CuZn SOD in tumors than in normal tissues. The difference is more pronounced when the results are compared on the basis of wet weight than on the basis of protein content. The "wet weight" results should best describe the in vivo protection activity of the SOD. Mn SOD was detected in all investigated tumors. Again the activities were in general slightly smaller than in normal tissues and again the differences were smaller when the results are compared on the basis of protein content. The method employed for SOD analysis is about 10 times less sensitive for Mn SOD than for CuZn SOD. If that is taken into consideration, it is found that Mn SOD constitutes a large part of the SOD activity of tissues and tumors, in several cases more than half. It is obvious that both enzymes must be considered in an evaluation of the superoxide radical scavenging capacity of human tissues and tumors. It has been reported that the CuZn SOD activities as well as the total SOD activities (16) of various animal tumors are lower than those of normal tissues. In general the present investigation confirms those results, though in some cases larger differences were reported than found here for human tumors. Several investigations have found none or very little Mn SOD (16) in various animal tumors. The idea has been advanced that loss of Mn SOD is intimately related to the cancerous phenotypes (16). Obviously the situation in human tumors is drastically different, here large amounts of Mn SOD are found.

In the investigated tumors there is no obvious relation between radioresistance or response to radical-producing antibiotics and content of Mn SOD and CuZn SOD.

TABLE: Superoxide Dismutase Normal Tissues

	CuZn U/g wet weight	U/mg protein	Mn U/g wet weight	U/mg protein
Liver	60000	480	10000	80
Lung	30000	700	380	9
Renal cortex	20000	300	2200	33
Erythrocytes	19000	55	-	-
Uterus	18000	260	410	6
Brain	16000	550	1000	40
Heart	14000	300	1500	30
Sceletal muscle	12000	200	700	12
Thyroid gland	12000	130	1500	17
Testis	9000	340	340	12
Spleen	9000	110	400	5
Lymphatic node	7000	130	1000	20
Fatty tissue	1000	170	50	11
TUMORS				
Ca mammae				
KD 91	7000	160	270	6.1
IK 97	9200	290	140	4.2
HA 99	6000	200	250	8.1
BH 23	7000	220	480	15
GP 27	5000	99	960	19
SJ 33	15000	290	390	7.5
GB 27	6000	140	120	2.6
TK 40	8800	200	220	4.8
ES 58	5300	120	290	6.6
Fibroadenomatosis				
EJ 34	2300	85	80	3
MK 26	3700	140	160	6.1
Ca ventriculi				
VJ 06	4600	88	640	12
GL 06	7100	170	390	9
Ca recti				
IMN 45	5300	180	450	15
RM 09	5100	130	700	18
SE 12	18900	150	970	12
Ca renis				
IK 32	6100	110	830	14
Ca ves urinaria				
VP 10	10700	230	490	8.8
Ca thyr low diff				
EN 09	4600	94	45	0.9
Malign melanoma				
TF 02	7300	130	200	3.5
Leiomyosarcoma				
KJ 04	18700	425	2200	49

TABLE continued

	CuZn U/g wet weight	U/mg protein	Mn U/g wet weight	U/mg protein
Chondrosarcoma				
SJ 27	3900	290	90	6.6
Sarcoma low diff				
AL 58	9200	320	37	1.0
Benign lipoma				
AN 21	1300	120	67	6.4
Testicular teratocarcinoma				
KH 48	4300	87	72	1.5
Lymphoma				
LS 45	11600	180	840	13
Chron myeloid leukemia				
JL 45 spleen inf	11200	170	450	6.7
AML 08 spleen inf	2000	78	150	5.7
Hodgkins disease				
RP 56	9800	170	880	16
OG 60	9600	111	550	6.4

ACKNOWLEDGMENTS

This study was supported by the Swedish Medical Research Council (04761) and the Lions Research Foundation, Department of Oncology, University of Umeå.

REFERENCES

1. Van Hemmen, J. J. and Meuling, W. J. A., Biochim. Biophys. Acta, 402, 133 (1975).
2. Lavelle, F., Michelson, A. M. and Dimitrijevic, L., Biochem. Biophys. Res. Commun., 55, 350 (1973).
3. Petkau, A. and Chelack, W. S., Biochim. Biophys. Acta, 433, 445 (1976).
4. Michelson, A. M. and Buckingham, M. E., Biochem. Biophys. Res. Commun., 58, 1079 (1974).
5. Petkau, A. and Chelack, W. S., Int. J. Radiat. Biol., 26, 421 (1974).
6. Misra, H. P. and Fridovich, I., Arch. Biochem. Biophys., 176, 577 (1976).
7. Oberley, L. W., Lindgren, A. L., Baker, S. A. and Stevens, R. H., Radiat. Res., 68, 320 (1976).

8. Petkau, A., Chelack, W. S., Pleskach, S. D., Meeker, B. E. and Brady, C. M., Biochem. Biophys. Res. Commun., 65, 886 (1975).
9. Lin, P. S., Kwock, L. and Butterfield, C. E., Radiat. Res., 77, 501 (1979).
10. Westman, G. and Marklund, S., Radiat. Res., 83, 303 (1980).
11. Lown, W. J. and Sim, S. K., Biochem. Biophys. Res. Commun., 77, 1150 (1977).
12. Bachur, N. R., Gordon, S. L. and Gee, M. V., Cancer Res. 38, 1745 (1978).
13. Lown, J. W., Sim, S. K., Majumdar, K. C. and Chang, R. Y., Biochem. Biophys. Res. Commun., 76, 705 (1977).
14. Marklund, S. L., J. Biol. Chem., 251, 7504 (1976).
15. McCord, J. M. and Fridovich, I., J. Biol. Chem., 244, 6049 (1969).
16. Oberley, L. W. and Buettner, G. R., Cancer Res., 30, 1141 (1979).

EXCITED DIMER FROM CHLORPROMAZINE RADICALS
AND SINGLET OXYGEN

Jerzy E. Maruchin, Medical Center of Postgraduate Education, 01-813 Warszawa, Polska

After UV irradiation (303 nm) of aqueous solutions of chlorpromazine (CPZ), free radicals $CPZS^{\cdot+}$ and $CPZ^{\cdot+}$ were obtained (1, 2,3). Under the influence of UV radiation these radicals react with singlet oxygen O_2 ($^1\Delta_g$) and the peroxy radicals $CPZO_2^{\cdot+}$ and $CZPSO_2^{\cdot+}$ appeared. These formed dimers CPZOOCPZ*, CPZSOOSCPZ* in excited states. The cleavage of dimers containing these O-O bonds liberated energy which we measured as luminescence (4,5,6, 3).

To obtain the dimer CPZSOOSCPZ*, an H_2O solution of 1 x 10^{-3}M CPZ at pH 6.01 was placed in a quartz cuvette and irradiated at 22°C with UV light (303 nm) in an Aminco-Bowman spectrophotofluorometer. After 20 sec of irradiation a cloudy green precipitate was formed. This was evaporated under reduced pressure, and after purification (5) a stable dry powder resulted. This was not sensitive to atmospheric oxygen or daylight. Absorption spectra were recorded with a Specord UV-VIS spectrophotometer. The fluorescence intensity was measured with an Aminco microphotometer with HTV 1P-21 photomultiplier. The chemiluminescence was measured by an impulse method using an LKB 1250 Luminometer. The molecular weight of the photoproduct was found with a Varian MAT 771 mass-spectrometer. The concentration of oxygen was determined by a Clark electrode.

The CPZ solution emitted fluorescence around 465 nm. As the concentration of the green photoproduct increased, a new fluorescence between 475 and 485 nm appeared. The isolated green product emitted fluorescence with maximum at 485 nm and had a single excitation maximum at 365 nm. The absorption spectra showed maxima at 245 nm and 741.9 nm. Addition of Li^+ ions into solution increased the intensity of the absorption band and shifted it to 750 nm. It may be assumed that a $CPZS^{\cdot+}$ with a radical center at the S_5 atom was formed, this radical reacting with singlet oxygen to yield the green photoproduct.

Oxidation of the green photoproduct with H_2O_2 or the action of daylight with atmospheric oxygen caused a gradual change of the green color to red, forming another radical with loss of sulfur (absorption at 530 nm).

$$(CPZS^{\cdot+}O)_2 \xrightarrow{\text{oxidation}} (CPZ^{\cdot+}O)_2$$

The CPZ solution irradiated for over 30 min forms red $CPZ^{\cdot+}$ radicals. It may be assumed that $CPZS^{\cdot+}$ radical is the precursor of the red $CPZ^{\cdot+}$ radical. Both radicals with singlet oxygen

ISBN 0-12-592050-4

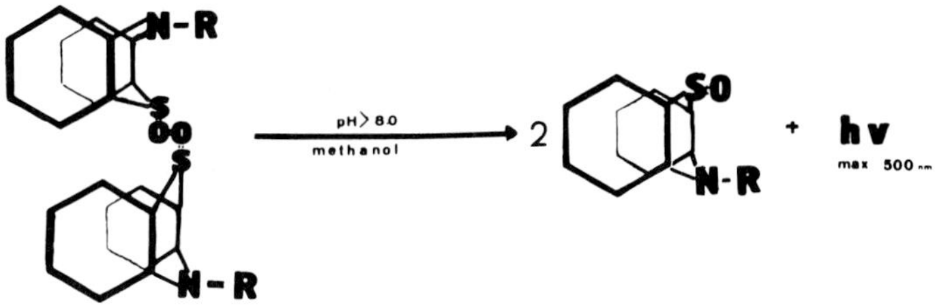

FIGURE 1. The dimeric luminescence emitter from chlorpromazine oxidation (structure of the tricyclic nucleus—"butterfly wing conformation") R-CH_2-CH_2-CH_2-N(CH_3)$_2$.

formed peroxy radicals which had a tendency to dimerize. The molecular weights of the photoproducts were 628 ± 1 d, which approximately confirms the dimeric structure.

The dimers dissolved in methanol at pH 10.0 decomposed with very intense emission of light near 500 nm, forming chlorpromazine sulphoxide (absorbance 225, 240, 275, 302 nm).

The final products of oxidation of chlorpromazine were 2 moles of CPZSO; other compounds were not detected.

The luminescence emitters of the chlorpromazine oxidation (5) are excited dimers of peroxy radicals (CPZOOCPZ and CPZSOOSCPZ), which during decomposition liberate energy in a light emitting process. The cleavage of these O-O bonds of the dimers gives an energy of ca 117.2 kcal/mole. The oxygen bonds have energies from 5.08 eV to 6.48 eV. Excited $^1O_2$ has upper singlet states at 23 ($^1\Delta_g$) and 38 ($^1\Sigma_g$) kcal/mole. These studies showed the existence of luminescence energy at 500 nm.

The free energy can react with certain biological molecules (ATP, FMN, catecholamines, vitamins C and E). The side effects of phenothiazine drugs during long term therapy (skin photosensitivity, melanin deposition, retinal damage, and occular opacity) have been explained by the reaction of "CPZ free radicals." The absorption of the electromagnetic energy liberated from the decomposition of dimers by biological systems may initiate the forenamed side effects.

REFERENCES

1. Lagercrantz, C., Acta Chem. Scand., 15,1545 (1961).
2. Felmeister, A. and Discher, C. A., J. Pharm. Sci., 53,756 (1964).
3. Maruchin, J., "Phenothiazines and Structurally Related Drugs: Basic and Clinical Studies," (Earl Usdiu et al, eds.) Elsevier/North Holland, New York (1980) p. 41-44.
4. Maruchin, J., Abstracts V meeting PTBR, 106, 77 (1979).
5. Maruchin, J., Agressol., 1, 43 (1968).
6. Maruchin, J., Pol. J. Pharmacol. Pharm. 30, 825 (1978).

COAL FLY ASH AND CHEMILUMINESCENCE ASSOCIATED WITH PHAGOCYTOSIS IN BOVINE ALVEOLAR MACROPHAGES

Hara P. Misra, Laboratory for Energy-Related Health Research, University of California, Davis, California 95616

During the combustion of coal in power plant furnaces, small particles of mineral residues are carried with the exhaust gases as fly ash. These fly ash particles consist primarily of fused aluminosilicates together with small quantities, generally less than 5% by weight, of incompletely oxidized carbon (1). High-temperature combustion of coal mobilizes significant quantities of potentially toxic species in stack emissions (2-4) on respirable particles, in part through a vapor condensation mechanism (5-6). It has been shown that the greater mass of volatile elements occurs in thin surface layers (4) thereby potentially increasing biological availability. Since pulmonary alveolar macrophages (PAM) represent the respiratory system's first line of defense against foreign particles, studies were initiated on the effects of fly ash on phagocytosis by PAM.

Allen et al. (7) have reported that polymorphonuclear leukocytes emit flashes of luminescence during phagocytosis of bacteria that can be correlated with metabolic activation of hexose monophosphate shunt, and is also oxygen dependent. Bovine PAM, similar to rabbit PAM and peritoneal macrophage (8), does not elicit chemiluminescence (CL) response after ingestion of zymosan particles. However, in the presence of luminol (5-amino-2,3-dihydro-1,4-phthalazinedione) a CL was observed following phagocytosis of zymosan particles by PAM. CL was measured as the counts per minute (CPM) octained in a tritium channel of a liquid scintillation spectrometer operated in the out-of-coincidence mode.

Fig. 1 demonstrates the CL obtained from 1×10^7 PAM exposed to 4×10^8 heat inactivated, optimized zymosan particles. As shown in this figure, CL rose rapidly from a background level of about 2×10^4 CPM to a maximum of over 16×10^4 CPM after 20 minutes and then gradually decayed to background level over the next several minutes. The magnitude of the response to zymosan increased with zymosan and with macrophage concentration until the particle to cell ratio reached a saturation maximum of 1:1. Fetal calf serum up to 2.8% and lavage fluid up to 2 mg/ml in the reaction mixture had no significant effect. 10 mM Ca^{2+} and 10 mM Mn^{2+} had no detectable effect in this system, whereas 10 mM Mn^{2+} has profound inhibitory effect on chemiluminescence of neutrophils. The response is increased by Mg^{2+} ions and was maximal at 0.7 mM.

As shown in Fig. 1, superoxide dismutase (Sp. Act. 3,000 U/mg) which removes $O_2^{\cdot-}$ anions ($O_2^{\cdot-} + O_2^{\cdot-} + 2H^+ \rightarrow H_2O_2 + O_2$) virtually eliminated CL at 6 µg/ml. Catalase (Sp. Act. 40,000

ISBN 0-12-592050-4

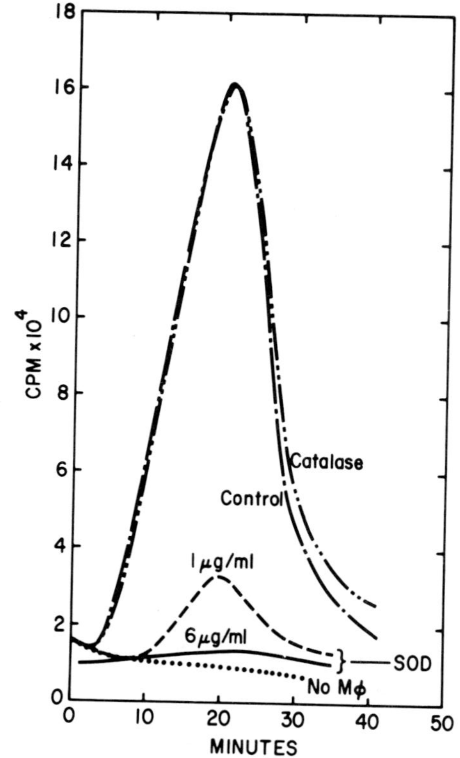

FIGURE 1. Effect of SOD and catalase on phagocytic-associated chemiluminescence.

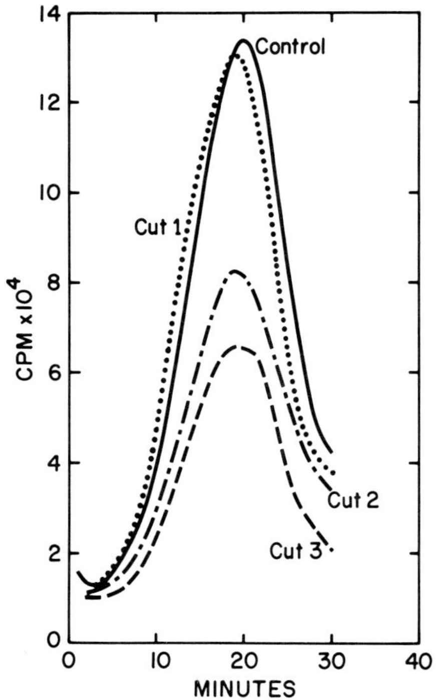

FIGURE 2. Size-dependent of the inhibition of the rate and extent of phagocytic-associated chemiluminescence by stack collected coal fly ash.

U/mg) up to 50 μg/ml had no detectable effect, indicating noninvolvement of H_2O_2 in the CL. Both OH· scavengers, e.g. ethanol and mannitol at 5 mM and singlet O_2 scavenger dimethylfuran at 6 mM inhibited the response by 50%. Since little information is available concerning the reaction of OH· with luminol to produce light, it is possible that both $O_2^{\cdot-}$ and OH· are generated during the phagocytosis by PAM and these radicals might react ($O_2^{\cdot-}$ + OH· → OH^- + $^1O_2$*) to generate singlet O_2 (9) which in turn is responsible for luminol oxidation to generate light.

To evaluate further the potential health hazards of coal combustion products, size-fractionated stack-collected fly ash (collected over a period of several weeks in January and February 1976, using a special sizing system that aerodynamically sized the fly ash in situ during collection) was tested in this system. As shown in Fig. 2, 3 mg/ml of 3μ particles (cut 3) and 6μ particles (cut 2) inhibited the response 71% and 60% respectively, whereas 20μ particles (cut 1) had no significant effect.

Since biological activity depends on chemical state and local concentrations of surface-related compounds and significant enrichment with decreasing particle size was observed for several trace elements (10), the observed inhibition of $O_2^{\cdot-}$ and OH· dependent phagocytosis associated CL in PAM by 3μ and 6μ fly ash particles may have been due to enrichment of certain trace elements or some organic matter yet to be identified.

ACKNOWLEDGEMENT

Supported by U.S. DOE contract number DE-AM03-76SF00472 and Biomedical Research Support grant 79-18.

REFERENCES

1. Fisher, G. L., Prentice, B. A., Silberman, D., Ondov, J. M., Biermann, A. H., Ragaini, R. C. and McFarland, A. R., Environ. Sci. Technol. 12,447 (1978).
2. Bertine, K. K. and Goldberg, E. D., Science, 173,233 (1971).
3. Gordon, G. E., Zoller, W. J. and Gladney, E. S., "Trace Substances in Environmental Health - VII" (D. D. Hemphill, ed.),
4. Linton, R. W., Loh, A., Natusch, F. S., Evans, C. A., Jr. and Williams, P., Science, 191,852 (1976).
5. Natusch, D. F. S. and Wallace, J. R., Science, 186,695 (1974).
6. Kaakinen, J. W., Jorden, R. M., Lawasani, M. H. and West, R. E., Environ. Sci, Technol. 9,862 (1975).
7. Allen, R. C., Stjernholm, R. C. and Steele, R. H., Biochem. Biophys. Res. Commun. 47,679 (1972).
8. Allen, R. C. and Lodge, L. D., Biochem. Biophys. Res. Comm. 69,245 (1976).
9. Arneson, R. M., Arch. Biochem. Biophys. 136,352 (1970).
10. Ondov. J. M., Ragini, R. C., Heft, R. E., Fisher, G. L., Silberman, D. and Prentice, B. A., Natl. Bureau of Standards Special Pub. #464 (1977), p. 567.

A COMPARISON BETWEEN CELL KILLING AND GENETIC DAMAGE IN YEAST IRRADIATED IN THE PRESENCE OF VARIOUS CONCENTRATIONS OF MOLECULAR OXYGEN

R. E. J. Mitchel, P. Unrau, and D. P. Morrison

Radiation Biology Branch, Atomic Energy of Canada Limited, Chalk River, Ontario, KOJ 1JO, Canada

The enhanced lethality of ionizing radiation delivered to cells in the presence of oxygen has been shown to be composed of at least two components in both prokaryotic spores (1) and mammalian systems (2). Since it is possible that cell death could result from either DNA damage or damage to other parts of the cell, for example membranes, we have attempted to estimate the relative contribution of the two types of damage. Oxygen was used as a probe to compare the end point of cell death to an end point which measures only genetic damage. The yeast Saccharomyces cerevisiae (D7.52a,trp-) was suspended in 20 mM phosphate buffer, pH 7.0 at 0°C and exposed to ^{60}Co γ-radiation in the presence of various concentrations of O_2. Genetic damage was measured as trp^+ reversion by gene conversion and showed a linear increase with dose to 10 krad under both oxia and anoxia, conditions which gave 100% survival. In all cases cell suspensions were bubbled with the equilibrating gas prior to and during irradiation.

Survival or gene conversion at intermediate oxygen concentrations was measured after delivery of a single dose of 200 krad at 15 krad/min or 2 krad at 7.5 krad/min respectively. The oxygen response curves obtained for both survival and gene conversion were biphasic with a plateau around 50 μM O_2. Approximately 60% of the total change occurred at O_2 concentrations below the plateau and 40% at higher concentrations. In each case a rapid change began around 10 μM O_2 and the full effect was seen only with oxygen at concentrations higher than that of air. The close similarity in the shapes of the curves suggests that similar events are responsible for oxygen enhancement of killing and genetic damage, and that the major lesion responsible for increased oxic lethality may be DNA damage. Since other work suggests that part of the O_2 enhancement of killing may occur at sites different from the cell nucleus (3,4) we wondered if the biphasic nature of the oxygen response curves may be due to such O_2 reactions. We have attempted to estimate the concentration of O_2 required to successfully compete with naturally occurring cellular components at a site different from, and with a measurement independent of DNA damage. Micrococcus radiodurans possessed an exonuclease and a polysaccharide bound to a mid-wall membrane. Hydroxyl radicals

ISBN 0-12-592050-4

generated by ionizing radiation initiate the release of these molecules from intact living cells into the surrounding medium (5,6). Oxygen protects the wall structure from this type of damage (O_2/N_2 release ratio $\sim$0.4) by interfering with radical transfer to the sensitive site (7). The release of these membrane bound molecules was monitored to estimate the concentration at which O_2 began to compete effectively with naturally occurring molecules for the OH· induced organic radicals. Cell suspensions, equilibrated with bubbling gas at various O_2 concentrations were irradiated to 20 krad at 0°C. The effects of O_2 scavenging in this system were first detectable at about 40 μM O_2, a concentration similar to the plateau region found for gene conversion and survival in yeast. If this wall membrane system in *M. radiodurans* is typical of an organized cellular structure in which organic radicals were formed and subsequently transferred to DNA, then 40 μM O_2 may be a representative O_2 concentration at which oxygen scavenging of these organic radicals could become effective. We propose that reaction of the organic radicals with O_2, to form a peroxy radical, constitutes a shunt around reactions of the organic radical with other naturally occurring organic molecules (which probably include chemical repair types of reactions), and allows a relatively unrestricted transfer path for the peroxy radical center to DNA. This proposal suggests that the second part of the oxygen response curves in yeast could be due to radicals transferred to DNA from other organic molecules, via peroxy type intermediates.

REFERENCES

1. Ewing, D. and Powers, E. L., Science, 194, 1049 (1976).
2. Millar, B. C., Fielden, E. M. and Steele, J. J., Int. J. Radiat. Biol., 36, 1977 (1979).
3. Ito, T. and Ito, A., Radiat. Res., 81, 416 (1980).
4. Shenoy, M. A., Asquith, J. C., Adams, G. E., Michael, B. D. and Watts, M. E., Radiat. Res., 62, 498 (1975).
5. Mitchel, R. E. J., Radiat. Res., 66, 158 (1976).
6. Mitchel, R. E. J., Radiat. Res., 64, 321 (1975).
7. Mitchel, R. E. J., Radiat. Res., 79, 63 (1979).

THE FEASIBILITY OF ONE ELECTRON TRANSFERS FROM MITOCHONDRIAL COMPONENTS TO SINGLET OXYGEN

G. Peters and M. A. J. Rodgers, Center for Fast Kinetics Research, The University of Texas at Austin, Austin, Texas 78712

The excitation of molecular oxygen to produce O_2 ($^1\Delta_g$) (ΔE = 1eV) should result in a concomitant increase in the one electron reduction potential from ∿0.33V (at 1 atm O_2) to ∿0.67V for the excited species (1). The biological ramifications of these arguments are made apparent by considering the one electron redox potential of the various components of the mitochondrial electron transport chain (2) presented in Fig. 1. In principal, mitochondrial components could be susceptible to oxidation by O_2 ($^1\Delta_g$), both in vivo and in vitro, while remaining unaffected by ground state molecular oxygen.

Time resolved laser flash photolysis studies employing a N_2 laser (337 nm) have yielded bimolecular rate constants for O_2 ($^1\Delta_g$) quenching by NADH in aqueous (D_2O) micellar sodium dodecyl sulfate (0.1 mol l^{-1}) systems. The experimental technique described earlier (3) involved generating O_2 ($^1\Delta_g$) via 2-acetonaphthone (10^{-3} mol l^{-1}) sensitizer and monitoring the subsequent removal of DPBF (2.5 x 10^{-5} mol l^{-1}) at varying concentrations of NADH (0-80 x 10^{-5} mol l^{-1}) in air saturated solutions. Under the prevailing reaction conditions O_2 ($^1\Delta_g$) decayed by three competing first order processes:

$$O_2\ (^1\Delta_g) \xrightarrow{k_d} O_2\ (^3\Sigma_g)$$

$$DPBF + O_2\ (^1\Delta_g) \xrightarrow{k_d} -DPBF$$

$$NADH + O_2\ (^1\Delta_g) \xrightarrow{k_d} \text{Products}$$

and $k_{obs} = k_d + k_r[DPBF] + k_q[NADH]$

The resulting kinetic data are presented in Fig. 2 from which a value of k_q = 7.92 x 10^7 l mol^{-1} s^{-1} was extracted. Each point represents a mean of six individual determinations and quoted errors correspond to standard deviations. A similar determination in homogeneous (acetonitrile/methanol 3:1) solution yielded a value of k_q = 7.47 x 10^7 l mol^{-1} s^{-1}.

The NADH quenching mechanism has not been fully elucidated. Single electron transfer from NADH to O_2 ($^1\Delta_g$) would yield NAD· radicals which absorb near 400 nm in aqueous solution (4). Although observations were made at high [NADH] to ensure efficient O_2 ($^1\Delta_g$) capture we were unable to unambiguously demonstrate any absorption due to NAD· radicals. This is probably attributable to the efficiency with which O_2 reacts with NAD· (4) under the

ISBN 0-12-592050-4

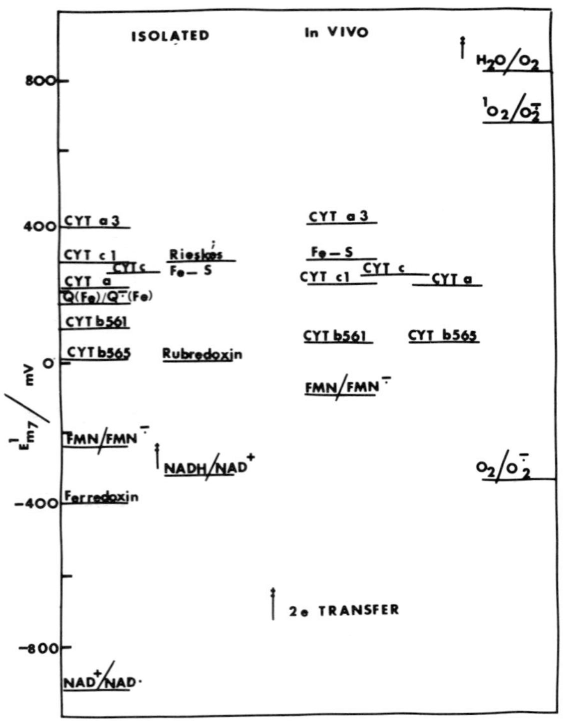

FIGURE 1. One electron redox potentials (vs. NHE) of mitochondrial components at pH 7.

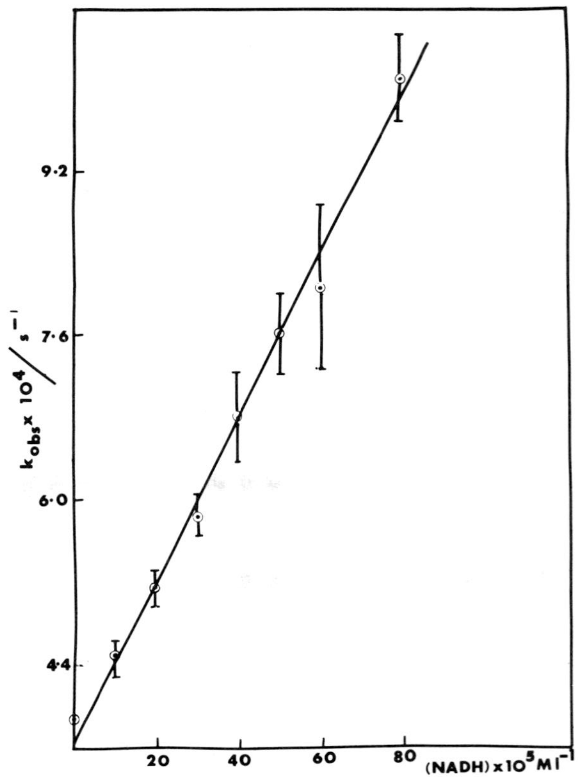

FIGURE 2. Variation of quenching rate constant with [NADH] in an aqueous micellar system.

experimental conditions necessary:

1) $NADH + O_2 (^1\Delta_g) \longrightarrow NAD\cdot + O_2^{\overline{\cdot}}$ $k = 7.92 \times 10^7$ l mol^{-1} s^{-1}

2) $NAD\cdot + O_2 \longrightarrow NAD^+ + O_2^{\overline{\cdot}}$ $k = 2 \times 10^9$ l mol^{-1} s^{-1}

Thus even with [NADH] = 10^{-3} mol l^{-1}, if $[O_2] = 2.7 \times 10^4$ mol l^{-1} then:

Rate of production of NAD· = 7.92×10^4 s^{-1}
Rate of removal of NAD· = 5.4×10^5 s^{-1}

ACKNOWLEDGEMENTS

The CFKR is supported by NIH grant RR-00886 from the Biotechnology Branch of the Division of Research Resources and by the University of Texas at Austin. NIH grant GM 24235 supported this work.

REFERENCES

1. Koppenol, W. H., Nature 262,420 (1976).
2. Erecinska, M., "Tunneling in Biological Systems" (B. Chance et al., eds.), Academic Press (1979), p. 455 and 461.
3. Lindig, B. A. and Rodgers, M. A. J., J. Phys. Chem. 83,1683 (1979).
4. Land, E. J. and Swallow, A. J., Biochim. Biophys. Acta 234, 34 (1971).

BASE RESIDUE RELEASE FROM ^{3}H-THYMINE LABELED DNA IN IRRADIATED *E. coli* UNDER CONDITIONS OF ENZYME INHIBITION

R. C. Richmond[1] *and J. D. Zimbrick*[2]

[1]*Norris Cotton Cancer Center, Dartmouth-Hitchcock Medical Center, Hanover, New Hampshire 03755*
[2]*Office of Health and Environmental Research, Department of Energy, Washington, D.C. 20545*

INTRODUCTION

Ionizing radiation releases damaged and undamaged thymine residues from DNA irradiated in aqueous solution (1,2) and in *E. coli* C cells treated under conditions designed to inhibit enzyme activity (3). The focus of the *in vivo* work reported here is to examine the radiolytic lesions within cellular DNA, and it is for this reason that an attempt is made to exclude enzymatic contribution to *in vivo* product yields from *E. coli* C and *E. coli* K12 cells irradiated in the presence and absence of O_2.

MATERIALS AND METHODS

Methods are similar to previous description (3). Briefly, cells are grown to either stationary or log phase with a constant incorporation of [^{3}H-C_6]-thymine. Cells are irradiated at ice bath temperature under O_2 or N_2 in phosphate buffer with 15 krad ^{60}Co-gamma rays, and then variously treated with 0.1 M EDTA, and lysed with sodium lauryl sulfate (SLS), all at ice bath temperature. This lysate is brought to 37°C and protease digested, dialyzed, and the dialysate is then analyzed by Sephadex G-10 gel filtration for ^{3}H-material released from cellular DNA. Cells studied are *E. coli* C 321 thy^{-} (4), and *E. coli* K12 strains as follows: a) KMBL 1788 *pol*$^{+}$ (5,6); b) KMBL 1787 *polA*$_{1}$ (5,6); and c) SR 111 *recArecB* (7).

RESULTS AND DISCUSSION

A. *E. coli C 321 thy*$^{-}$

The yield of thymine base recovered initially from irradiated oxygenated log phase cells is 3 fold that obtained from cells irradiated in N_2-gassed suspension. Holding these cells at ice bath temperature after irradiation in O_2 for increasing times prior to addition of EDTA results in an increasing recovery of thymine, whereas no such postirradiation increase in recovery of thymine is obtained for cells irradiated in N_2, suggesting

ISBN 0-12-592050-4

postirradiation nuclease activity in the former but not the latter case. Stationary phase cells irradiated in O_2- or N_2-gassed suspensions show no additional development of thymine release during postirradiation holding at ice bath temperature prior to EDTA addition, suggesting that the nuclease activity of log phase cells irradiated in O_2 is absent in stationary phase cells. Physicochemical mechanisms of radiation action likely contribute significantly to the thymine base release assayed for these stationary cells, and for log phase cells irradiated in the absence of O_2.

Eliminating EDTA entirely from the postirradiation treatment of cells results in the near elimination of the thymine base recovered from cells irradiated in N_2. This reversal of radiation-induced thymine base release from anoxic cells depends only on the absence of EDTA, thus suggesting a) that this reversal is a result of some enzymatic process; and b) that in N_2-gassed suspensions thymine is not immediately released from irradiated cellular DNA.

B. E. coli K12

In all three E. coli K12 strains (only stationary phase treated), unirradiated cells generate large and poorly reproducible amounts of oligonucleotide material that amounts to about half the activity of these fractions recovered from irradiated cells. Radiation-induced release of thymine base from pol^+ and $polA_1$ cells is ca. 20 fold and 6 fold, respectively, increased when O_2 is present during irradiation. Thymine release from $polA_1$ is greater than that assayed from pol^+ by ca. 2 fold and 10 fold for cells irradiated in O_2- and N_2-saturated suspensions, respectively.

Deficiency in the 3'→5' exonuclease and polymerase functions of polymerase I coupled with unbalanced 5'→3' exonuclease activity (8) (i.e., comparing pol^+ with $polA_1$) appears to cause increased exonuclease or DNA glycosylase activity in these irradiated cells.

Irradiation of recArecB cells results in a relatively small amount of released thymine base, and this thymine yield is the same both when O_2 is present or absent during irradiation. The amount of thymine base recovered from recArecB cells irradiated in O_2 is 20 fold less than pol^+ cells and 50 fold less than $polA_1$ cells; in N_2 the thymine recovery is the same as pol^+ cells and 8 fold less than $polA_1$ cells. The large reduction in the O_2-dependent thymine base release in recArecB cells suggests that exonuclease V (the recBC gene product) may be involved in exacerbating radiation-induced thymine release from cells irradiated in the presence, but not the absence of O_2. The large amount of thymine base recovered from pol^+ cells and $polA_1$ cells irradiated in O_2-saturated suspension suggests nuclease activity. The thymine base released from all three strains following radiation in N_2 is largely restricted, and it is assumed that the physico-chemical action of radiation likely causes a significant fraction of this yield.

ACKNOWLEDGMENTS

This work supported by NCI Grant CA26045 and NIGMS Grant 18927, DHHS.

REFERENCES

1. Ward, J. F. and Kuo, I., Radiat. Res., 66, 485 (1976).
2. Richmond, R. C. and Simic, M. G., Br. J. Cancer, 37(Suppl. III), 20 (1978).
3. Richmond, R. C. and Zimbrick, J. D., Biochem. Biophys. Res. Comm., 64, 391 (1975).
4. Hewitt, R., Suit, J. C. and Billen, D., J. Bacteriol., 93, 86 (1967).
5. Glickman, B. W., van Sluis, C. A., Heijneker, H. L. and Rorsch, A., Molec. Gen. Genetics, 124, 69 (1973).
6. Heijneker, H. L., Ellens, D. J., Tjeerde, R. H., Glickman, B. W., van Dorp, B. and Pouwels, P. H., Molec. Gen. Genetics, 124, 83 (1973).
7. Kapp, D. S. and Smith, K. C., J. Bacteriol., 103, 49 (1970).
8. Trgovcervic, Z. and Kucan, Z., Nature, 226, 752 (1970).

THE ACTIVE OXYGEN SPECIES IN THE PHENAZINE METHOSULFATE/NADH/O_2 HYDROXYLATING SYSTEM

Helen W. Richter and Walter H. Waddell, Department of Chemistry, Carnegie-Mellon University, Pittsburgh, Pennsylvania

The hydroxylation of benzoic acid in aqueous solution induced by the chemical system consisting of phenazine methosulfate, reduced β-nicotinamide adenine dinucleotide (NADH), and dissolved oxygen has been investigated by a determination of product yields via high pressure liquid chromatography, and by an examination of the effect of the variation of reaction conditions and of selected additives on the product yields. The results of our investigation indicate that in this system the species initiating the hydroxylation reaction is the hydroxyl radical, and that this radical is produced from one-electron reduction of hydrogen peroxide in a Fenton-like reaction. Furthermore, the hydrogen peroxide is produced directly by the transfer of a hydride ion from the fully reduced phenazine methosulfate to molecular oxygen, and not by the disproportionation of superoxide radicals. It is apparent that the sole function of the NADH in the hydroxylation system is the reduction of the phenazine methosulfate.

Generally, it has been assumed for the phenazine methosulfate (PMS^+)/NADH/O_2 system that superoxide radicals are generated when PMS^+, reduced by NADH, reacts with oxygen (1). This chemical system is known to hydroxylate a variety of aromatic compounds (2,3, 4), and it is thought that the reactive chemical species initiating the hydroxylation reaction is the superoxide radical ($O_2^{\cdot-}$) (2,3). Contrary to previous results, Halliwell (4) found that the hydroxylation of ortho- and para- hydroxybenzoic acid to the dihydroxy products by this system was not affected by the addition of superoxide dismutase, and that catalase was an effective inhibitor of the hydroxylation reaction. He concluded that the hydroxylation requires H_2O_2, but that $O_2^{\cdot-}$ is not involved. In view of the utilization of the PMS^+/NADH/O_2 system as a generator of $O_2^{\cdot-}$ in an investigation of the mechanism of action of the enzyme dopamine-β-hydroxylase (EC1.14.17.1) (5), and the frequent use of PMS^+ as an electron carrier in biological studies, we have reexamined the hydroxylation of benzoic acid by the PMS^+/NADH/O_2 system in order to identify the reaction products and to help clarify the reaction mechanism. Benzoic acid was chosen as the substrate since its hydroxylation by this system has been reported previously (2), and since the product yields of its reaction with the hydroxyl radical (·OH) have been carefully determined by Schuler and co-workers (6).

The experimental results indicate that the hydroxyl radical is the chemical species initiating hydroxylation, and that the

ISBN 0-12-592050-4

superoxide radical is not produced to an appreciable extent. These results, coupled with those of Halliwell (4), allow the construction of Scheme I as the proposed reaction mechanism.

Scheme I

$$NADH + PMS^+ \longrightarrow NAD\cdot + PMS\cdot + H^+ \quad (1)$$

$$NAD\cdot + PMS^+ \longrightarrow NAD^+ + PMS\cdot \quad (2)$$

$$2PMS\cdot \underset{-H^+}{\overset{H^+}{\rightleftharpoons}} PMS^+ + PMSH \quad (3)$$

$$PMSH + O_2 \xrightarrow{H^+} PMS^+ + H_2O_2 \quad (4)$$

$$(PMS\cdot + O_2 \xrightarrow{\text{very slow}} PMS^+ + \cdot O_2^-) \quad (5)$$

$$H_2O_2 + PMSH \longrightarrow \cdot OH + OH^- + PMS\cdot + H^+ \quad (6a)$$

$$H_2O_2 + PMS\cdot \longrightarrow \cdot OH + OH^- + PMS^+ \quad (6b)$$

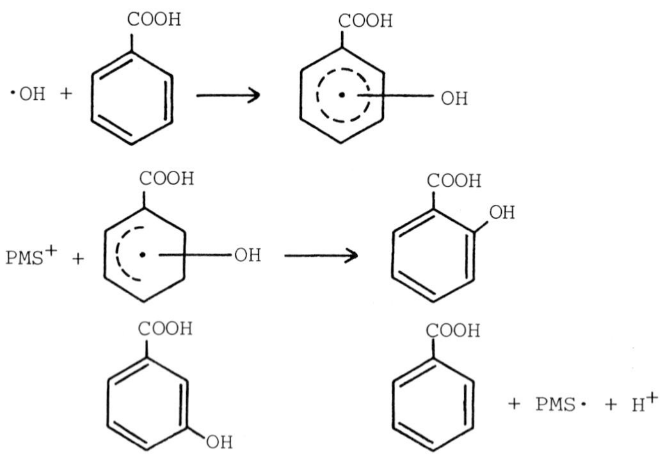

(7)

(8)

It was found that ortho-, meta-, and para- hydroxybenzoic acid are all products of the hydroxylation reaction, contrary to previous reports (2). Furthermore, in the presence of a suitable oxidizing agent, the relative yields of the product isomers is essentially the same as found in radiation chemical studies where reactions 7 and 8 proceed. Inhibition of product formation when formate buffer is used in lieu of citrate buffer, and stimulation of product formation upon addition of Cu^{2+} (in catalytic or substrate concentrations) indicate that ·OH, but not $O_2^{\overline{\cdot}}$, is directly involved in the hydroxylation process and that H_2O_2 derived from $O_2^{\overline{\cdot}}$ is not a precursor of the ·OH. The Fenton-like reactions 6a and 6b are postulated since addition of N,N-bis (2-(bis(carboxymethyl)amino)ethyl) glycine (DTPA or DETAPAC), a reagent which inhibits trace metal catalyzed Fenton-type reactions, had no appreciable effect on product yields. It was demonstrated that the addition of fully reduced PMS^+, i.e., PMSH, to a buffered solution containing H_2O_2 and benzoic acid resulted in the formation of isomeric products even in the <u>absence</u> of dissolved oxygen (product production via reactions 6a-8).

ACKNOWLEDGEMENTS

We wish to acknowledge the support of the Health Research and Services Foundation (Grant U29) and the National Institutes of Health (Grant NS01585).

REFERENCES

1. Nishikimi, M., Appaji Rao, N., and Yagi, K., Biochem. Biophys. Res. Commun., 46,849 (1972).
2. Prema Kumar, R., Ravidranath, S. D., Vaidyanathan, C. S. and Appaji Rao, N., Biochem. Biophys. Res. Commun., 48,1049 (1972).
3. Ravidranath, S. D., Ashok Kumar, A., Prema Kumar, R., Vaidyanathan, C. S. and Appaji Rao, N., Arch. Biochem. Biophys., 165,478 (1974).
4. Halliwell, B., Biochem. J., 167,317 (1977).
5. Liu, T. Z., Shen, J. T. and Ganong, W. F., Proc. Soc. Exper. Biol. Med., 146,37 (1974).
6. Klein, G. W., Bhatia, K., Madhavan, V. and Schuler, R. H., J. Phys. Chem., 79,1767 (1975).

ELECTRON TRANSFER TO OXYGEN FROM REDUCED BIPYRIDINIUM COMPOUNDS AT MICELLAR SURFACES

Michael A. J. Rodgers, Center for Fast Kinetics Research, The University of Texas at Austin, Austin, Texas 78712

Bipyridinium compounds of the type $(R\text{-}NC_5H_5\text{-}C_5H_5N\text{-}R)^{+2}$ where R is an alkyl or substituted alkyl residue are well known as cytotoxins and have excellent herbicidal properties (1). These actions are thought to be related to their ability to act as electron acceptors in biological situations where electron transport from one site to another is crucial to the metabolic process. The free radicals produced by single electron reduction of a bipyridinium species such as $(CH_3\text{-}NC_5H_5\text{-}C_5H_5N\text{-}CH_3)^{+2}$ (paraquat) are stable but do react efficiently with oxygen to undergo electron-transfer production of superoxide

$$V^{\dot{+}} + O_2 \rightarrow V^{++} + O_2^{\dot{-}} \quad [1]$$

where V^{++} and $V^{\dot{+}}$ are the oxidized and reduced forms of paraquat. This reaction has been studied in aqueous and polar organic solutions where (2,3) it was found that the rate constant depends on the difference in reduction potential between V^{++} and O_2 and on the dielectric nature of the solvent medium.

In this work V^{++} ions are adsorbed at the interfacial sites of sodium dodecyl sulfate (SDS) micelles in water dispersion (4) and reduced by hydrated electrons (e^-_{aq}) generated in a pulse-radiolysis experiment.

$$V^{++}_{mic} + e^-_{aq} \rightarrow V^{\dot{+}}_{mic} \quad [2]$$

In the presence of oxygen the reaction

$$V^{\dot{+}}_{mic} + O_2 \rightarrow V^{++}_{mic} + O_2^{\dot{-}} \quad [3]$$

occurs which was followed by observing the decay of the $V^{\dot{+}}$ species at 395 nm by kinetic spectrophotometry. The first order decay constants for [3] for both methyl and benzyl viologens are shown in Table 1. It is noted that the rate constant for [3] is <u>ca</u> seven times slower when the reaction occurs at a micellar surface. The bimolecular rate constant for this reaction was measured in a series of methanol-water binary mixtures at room temperature. As anticipated from reaction rate theory a plot of log k (for MV^{++}) against $1/\varepsilon$ is linear, where ε represents the dielectric constant of the liquid mixture. Interpolation on this plot using the value for methyl viologen in SDS leads to the result that the effective dielectric constant in the region where reaction [3] occurs is near 53.

In another experimental series the temperature dependencies of $k(MV^{\dot{+}} + O_2)$ in water and in 60% ($^V/v$) methanol/water and

ISBN 0-12-592050-4

TABLE I.

System	$10^{-4}k\ /s^{-1}$	$t_{1/2}/\mu s^{-1}$
$10^{-3}M\ MV^{++}$ in H_2O	13.8	5.01
$10^{-3}M\ MV^{++}$ in 0.1M SDS	1.88	36.9
$10^{-3}M\ BV^{++}$ in H_2O	7.57	9.16
$10^{-3}M\ BV^{++}$ in 0.1M SDS	1.07	64.9

k ($BV^{\overset{+}{\cdot}}$ + O_2) in water were determined. Both sets of data show non-linear Arrhenius behavior. This characteristic supports the suggestion that the overall electron transfer reaction proceeds via a complex which equilibrates with reactants:

$$V^{\overset{+}{\cdot}} + O_2 \rightleftharpoons (V...O_2)^{\overset{+}{\cdot}} \longrightarrow V^{++} + O_2^{\overset{-}{\cdot}}$$

ACKNOWLEDGEMENTS

The expert technical assistance of Janet Becker is gratefully acknowledged. The Center for Fast Kinetics Research is supported by NIH grant RR-00886 from the Biotechnology Branch of the Division of Research Resources and by The University of Texas.

REFERENCES

1. Dodge, A. D., Endeavour 30,130 (1971).
2. Farrington, J. A., Ebert, M., and Land, E. J., J. Chem. Soc., Faraday 1, 74,665 (1978).
3. Patterson, L. K., Small, R. D. and Scaiano, J. C., Radiation Res., 72,218 (1977).
4. Rodgers, M. A. J., Foyt, D. C. and Zimek, Z. A., Radiation Res., 75,276 (1978).

CELLULAR RADIATION CHEMISTRY AS A FUNCTION OF RADIATION QUALITY

R. Roots, A. Chatterjee, E. A. Blakely and C. A. Tobias, Division of Biology and Medicine, Lawrence Berkeley Laboratory, Berkeley, California 94720

Various aspects of cellular radiation chemistry have been studied in bacterial and mammalian cells for low linear energy transfer (LET_∞) radiations, including hydroxyl-radical mediated cell inactivation (1-3) and DNA damage (4,5), the effects of the superoxide radical ion (6), and the mechanism(s) of the oxygen effect (2,7,8). A few high LET studies have been done with various biological systems to assess the importance of the radiolytic products of water as a function of radiation quality (9-14). However, most of this work was done with fast neutrons or charged particle beams having highly composite LET values.

We have examined the extent of the hydroxyl radical induced inactivation of mammalian cells in track segment studies for various high-LET radiations, including accelerated particle beams of Li^{3+}, C^{6+}, Ne^{10+}, and A^{18+} of initial energies from 8.5 to 570 MeV/n. Results from comparative experiments with 225 kVp x-rays have been reported elsewhere (15). Cell monolayers, grown in glass vessels, were irradiated in Dulbecco's phosphate buffered saline (PBS) or PBS containing a hydroxyl radical scavenger under aerobic conditions. The decreased cell sensitivity achieved after maximal scavenging of the hydroxyl radical is a measure of cell lethality mediated by this free radical species. In the figure, the radiosensitivities obtained for various radiation qualities have been plotted with respect to track segment mean LET_∞ values. It is seen that the extent of hydroxyl radical dependent lesions is governed by particle velocity and Z number, but LET_∞ is not suitable for correlation. In the case of the low energy particle beams, where fragmentation is minimal, the decrease in the extent of OH-radical induced cell lethality with respect to LET, corresponds to the decrease in $G_{(OH)}$ as a function of LET as depicted in the insert in the figure. The data also indicate that OH-radical mediated damage cannot be completely eliminated even at high-LET values where radical-radical recombination is high.

ACKNOWLEDGEMENT

This work was supported by DOE Contract W-7405-ENG-48.

ISBN 0-12-592050-4

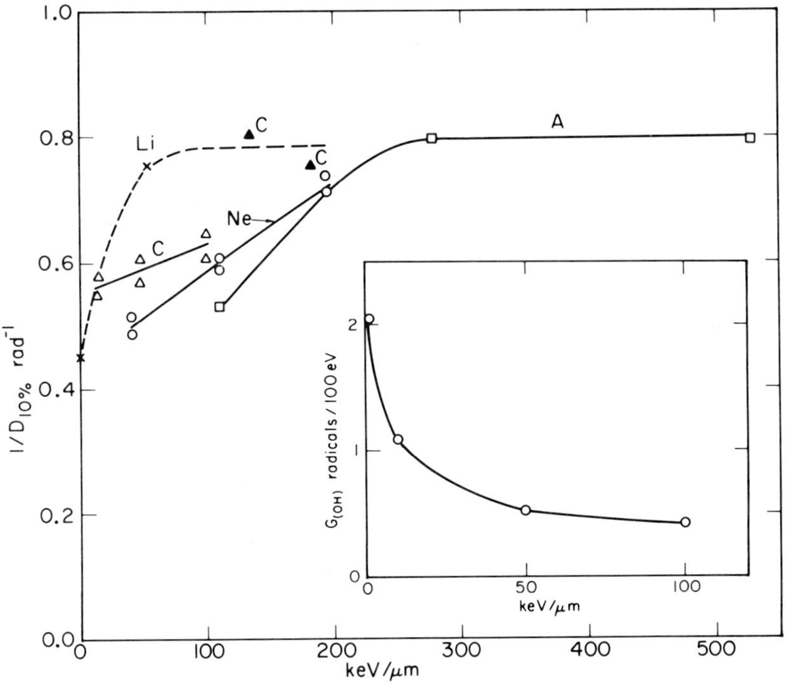

FIGURE 1. Radiosensitivity ($1/D_{10\%}$ rad^{-1}) of Chinese hamster ovary cells protected maximally against OH-induced lethality as a function of track segment mean LET∞ (keV/μm). Accelerated particle beams were used: 11.4 MeV/n Li^{3+} (*), 15.6 MeV/n C^{6+} (▲), 308 MeV/n C^{6+} (Δ), 425 MeV/n Ne^{10+} (o), and 570 MeV/n A^{18+} (□). The insert shows the dependence of $G_{(OH)}$ on the value of LET (neutral pH). Data are taken from a figure published by I. V. Vereschinskii and A. K. Pikaev in Introduction to Radiation Chemistry, Israel Program for Scientific Translations (1964).

REFERENCES

1. Johansen, I. and Howard-Flanders, P., Radiat. Res., 24,184 (1965).
2. Chapman, J. D., Reuvers, A. P., Borsa, J. and Greenstock, C. L., Radiat. Res., 66,291 (1973).
3. Roots, R. and Okada, S., Radiat. Res., 64,306 (1975).
4. Roots, R. and Okada, S., Int. J. Radiat. Biol., 21,329 (1972).
5. Bonura, T. and Smith, K. C., Int. J. Radiat. Biol., 29,293 (1976).

6. Samuni, A. and Czapski, G., Radiat. Res., 76,624 (1978).
7. Roots, R. and Smith, K. C., Int. J. Radiat. Biol., 26,467 (1974).
8. Michael, B. D., Harrop, H. A. and Maughan, R. L., "Proceedings of the Sixth International Congress of Radiation Research" (S. Okada, M. Inamura, T. Terashima and H. Yamaguchi, eds.), Tokyo, 1979, p. 288.
9. Brustad, T., Adv. Biol. Med. Phys., 8,161 (1962).
10. Takeshita, K. and Sawada, S., "Biological Effects of Neutron Irradiation," Vienna, (1974), p. 245.
11. Chapman, J. D., Doern, S. D., Reuvers, A. P., Gillespie, C. J., Chatterjee, A., Blakely, E. A., Smith, K. C. and Tobias, C. A., Radiat. Environ. Biophys., 16,29 (1979).
12. Antoku, S., Int. J. Radiat. Biol., 27,287 (1975).
13. Singh, D. R., Mahajan, J. M. and Krishnan, D., Int. J. Radiat. Biol., 30,585 (1976).
14. Singh, D. R., Mahajan, J. M. and Krishnan, D., Mutat. Res., 37,193 (1976).
15. Roots, R., Blakely, E., Chatterjee, A., Chang, P. and Smith, K., Radiat. Res., submitted, 1980.

EFFECT OF ANTI-INFLAMMATORY AGENTS ON THE HYDROXYL RADICAL (OH·) PRODUCTION OF ZYMOSAN STIMULATED HUMAN GRANULOCYTES (PMNs)

Arthur L. Sagone, Jr., Department of Medicine, The Ohio State University, Columbus, Ohio 43210

Granulocytes play an important role in normal host defense against microorganisms. The ability of granulocytes to kill bacterial organisms is related, in part, to their capacity to generate several reactive oxygen species (ROS) (1-8). These include superoxide ($O_2^{\overline{\cdot}}$), hydrogen peroxide (H_2O_2), hydroxyl radical (OH·), and singlet oxygen (1-8). The production of ROS by phagocytic cells is complex. Superoxide appears to be formed from molecular oxygen in a reaction involving NADPH oxidase (1-8). Subsequently, H_2O_2 may be produced by the dismutation of $O_2^{\overline{\cdot}}$ to H_2O_2 (1-8). The production of OH· may depend, in part, on a Haber-Weiss type of reaction involving the interaction of $O_2^{\overline{\cdot}}$ and H_2O_2 (7,8). In addition, we have recently linked OH· production in PMNs to arachidonic acid metabolism (9). Our observations suggest that OH· is produced during the reduction of endoperoxides in PMNs similar to the production of OH· observed in microsomal systems (10-12). There is now evidence that ROS may also be mediators of the tissue damage occurring in inflammatory diseases (10-12). This has raised the question whether or not one major mechanism of action of anti-inflammatory drugs in vivo is to decrease the release of ROS from phagocytic cells. Several recent studies, therefore, have focused on the capacity of anti-inflammatory drugs to impair the production of superoxide or H_2O_2 by stimulated PMNs in vitro (13-15). Although several drugs impair the production of $O_2^{\overline{\cdot}}$ or H_2O_2, a number particularly non-steroidal anti-inflammatory drugs, do not impair the production of these oxygen metabolites (13-15). However, the direct effect of these drugs on OH· production was not studied (13-15).

We have recently developed a method for the detection of OH· by phagocytic cells using the oxidation of ^{14}C-Carboxyl benzoic acid (8). Using this method, we have studied the effect of several anti-inflammatory drugs on the production of OH· by zymosan stimulated human PMNs in vitro.

MATERIALS AND METHODS

Granulocytes were isolated from the peripheral blood of normal volunteers by dextran sedimentation and Ficoll-Hypaque gradient centrifugation (6,8,9). The production of $^{14}CO_2$ from [^{14}C] carboxyl-benzoic acid by granulocyte suspensions was used to quantitate the generation of OH· production (8,9). The production of

ISBN 0-12-592050-4

$^{14}CO_2$ was measured continuously using the ionization chamber-electrometer method as previously described (6,8,9). Approximately 2×10^7 granulocytes were resuspended in 4.5 ml Dulbecco's phosphate buffered saline (GIBCO) with 50 mg/100 ml of glucose containing [^{14}C]carboxyl benzoic acid (obtained from Amersham Corp. and dissolved in NS) with or without the drug to be studied and placed in a 25 ml triple arm distilling flask. The final concentration of benzoic acid was 0.025 mM. The inlet arm of the metabolic flask was connected to a gas cylinder containing compressed air. The outlet arm of the flask was connected to a 275 ml Cary-Tolbert ionization chamber and a Cary (model 401) vibrating reed electrometer (Varian Instruments). The third arm of the flask was covered with a serum stopper. A duplicate system was used to that $^{14}CO_2$ derived from [^{14}C]benzoic acid could be measured simultaneously from both control and experimental flasks with the drug. The incubation flasks were stirred continuously using a "Magne-4" magnetic stirrer (model 4820-I, Cole-Parmer Instrument Co., Chicago, Ill.) with spin bars in the flasks. After baseline CO_2 production was established, opsonized zymosan particles were added in 0.4 ml of normal saline (4×10^8 particles/ml). CO_2 production was calculated from the peak millivolt reading (see Fig. 1) and expressed as nanomoles of CO_2 produced per hour per 10^9 cells. There was no significant difference in the kinetics of the oxidation of benzoic acid in suspensions with the drug compared to controls. Therefore the peak value achieved after the addition of zymosan was used to compare values between control and experimental flasks. The drug to be studied was dissolved directly into the buffer.

RESULT

Fig. 1 shows the kinetics of OH· production by PMNs during the phagocytosis of zymosan particles. Unstimulated suspensions did not oxidize ^{14}C-Carboxylbenzoic acid. The addition of zymosan induced a prompt increase in the oxidation of this substrate which was maximal between 10 and 20 minutes followed by a gradual decrease.

Table 1 indicates the effect of several anti-inflammatory drugs on the OH· production by zymosan stimulated PMNs. The drugs are divided into three groups. Group I consists of drugs which do not impair the production of $O_2^{\cdot-}$ or H_2O_2 of zymosan stimulated PMNs in the concentrations used. This group included aspirin and indomethacin, phenol, DMSO, and ibuprofen. All of these drugs, in concentrations which can be achieved in vivo, impaired OH· production. Group II drugs consisted of hydrocortisone succinate and 2,3 Dihydroxybenzoic acid. The concentration of the drug which impaired OH· production also impairs the production of other ROS. Therefore, the effect of these drugs on OH· may be secondary to their effects on $O_2^{\cdot-}$ or H_2O_2 production. Phenylbutazone impaired the production of OH· in concentrations of 10^{-6}M to 10^{-4}M and is listed separately since the effects of this drug on the

TABLE 1. *Effect of Anti-Inflammatory Agents on Hydroxyl Radical Generation by Granulocytes Stimulated by Opsonized Zymosan Particles**

	Control	Experimental	% Control
Group I Drugs which do not impair the production of $O_2^{\bar{\cdot}}$ or H_2O_2 in the concentrations used			
Indomethacin			
(3) 10^{-5} M	19.1 ± 3.6	13.0 ± 3.5	68 ± 5.0
(3) 10^{-4} M	16.1 ± 3.9	4.6 ± 0.8	29 ± 2.6
Aspirin			
(3) 10^{-3} M	16.0 ± 4.0	9.0 ± 2.0	57 ± 10.0
Phenol			
(3) 10^{-5} M	18.9 ± 1.8	7.8 ± 1.5	42 ± 4.6
(3) 10^{-4} M	20.4 ± 1.1	2.0 ± 0.2	10 ± 1.8
Ibuprofen			
(3) 10^{-4} M	29.6 ± 8.3	25.0 ± 9.6	82 ± 14.0
(3) 10^{-3} M	38.4 ± 4.9	11.7 ± 3.8	29 ± 5.9
DMSO			
(3) 10^{-3} M	24.0 ± 1.9	18.2 ± 2.7	75 ± 11.8
(1) 10^{-2} M	20.0	7.9	40
Group II Drugs which impair $O_2^{\bar{\cdot}}$ or H_2O_2 in the concentrations used to impair OH· production			
Hydrocortisone Succinate			
(3) 10^{-4} M	13.5 ± 3.0	13.7 ± 2.8	102 ± 2.0
(3) 10^{-3} M	15.4 ± 1.6	5.7 ± 2.7	34 ± 13.0
2,3 Dihydroxybenzoic Acid			
(1) 10^{-3} M	29.2	1.8	6
Group III Effect on $O_2^{\bar{\cdot}}$ and H_2O_2 unclear in concentrations used to impair OH· production			
Phenylbutazone			
(3) 10^{-6} M	26.5 ± 4.7	17.6 ± 2.9	66 ± 2.0
(3) 10^{-5} M	24.4 ± 7.4	5.9 ± 0.9	24.7 ± 5.0
(1) 10^{-4} M	34.9	3.1	9

*Values given indicate the peak rate of $^{14}CO_2$ production from [^{14}C]Carboxyl-benzoic acid in nanomoles per 10^9 PMN per hour from PMN suspensions during the phagocytosis of zymosan particles. The method used is described under Fig. 1. Three paired experiments were done with most concentrations of the drug. The experimental rate (suspension with drugs) compared to the corresponding controls in each experiment is given. For convenience the percentage in the presence of the drug compared to the control is also given in the third column. All experimental values are

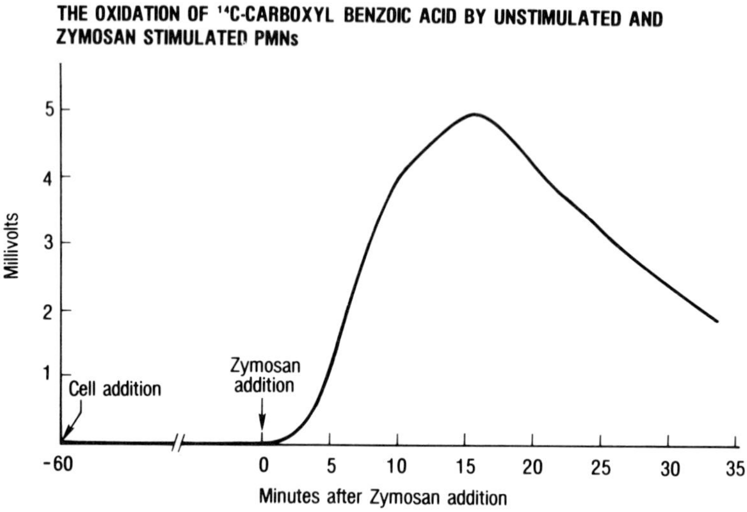

FIGURE 1. This figure demonstrates the oxidation of benzoic acid by granulocyte suspensions (2 X 10^7) both resting and during the phagocytosis of zymosan particles. The Y axis represents the electrical signal in MV generated from $^{14}CO_2$ passing through the ionization chamber and the X axis represents the time of incubation in minutes. As seen in the first part of the curve, unstimulated cells did not oxidize benzoic acid. However, as indicated by the arrow, the addition of opsonized zymosan particles is associated with a prompt oxidation of this substrate.

production of other ROS by PMNs at each concentration is less well established.

DISCUSSION

Our results indicate that several anti-inflammatory drugs have the capacity to impair the release of OH· by zymosan stimulated PMNs. The mechanism by which these drugs impair OH· production, however, varies. DMSO and phenol in concentrations which impair OH· production do not impair the generation of the other reactive oxygen species (16). Thus, these drugs appear to decrease OH· release by scavenging OH· directly. Aspirin and indomethacin in the concentrations used for our experiments are known

TABLE I (Continued)

significantly different from the controls with a $P < 0.05$ except for the 10^{-4} M concentration of hydrocortisone succinate and ibuprofen. None of the drugs caused oxidation of ^{14}C-benzoic acid when incubated with granulocytes in the absence of zymosan particles.

to impair cyclooxygenase activity but do not decrease the production of $O_2^{\cdot -}$ or H_2O_2 (9,13). Thus the inhibitory effects of these drugs appears secondary to their effects on prostaglandin metabolism. However, the possibility that these drugs scavenge OH· directly cannot be excluded and requires further study. Ibuprofen also impaired OH· production. This drug does not impair cyclooxygenase activity (11) and does not impair $O_2^{\cdot -}$ or H_2O_2 in PMNs as determined by the chemiluminescence assay (17). This drug, therefore, most likely scavenges OH· directly, similar to the mechanism postulated in microsomal systems (11).

The impairment of OH· release by hydrocortisone succinate and 2,3 Dihydroxybenzoic acid may be secondary to their effects on $O_2^{\cdot -}$ and H_2O_2 production. The concentration of hydrocortisone succinate which impairs OH· production also impairs $O_2^{\cdot -}$ production (13). Similarly, 2,3 Dihydroxybenzoic acid impairs $O_2^{\cdot -}$ and H_2O_2 production in the concentrations used (19). Since $O_2^{\cdot -}$ and H_2O_2 may be required for optimal OH· production in PMNs via the Haber-Weiss reaction (1-8) the impaired OH· release with these drugs is most likely secondary to decreased $O_2^{\cdot -}$ or H_2O_2 production. However, in view of the similarity of structure between benzoic acid and 2, 3 dihydroxybenzoic acid the possibility that this drug can scavenge OH· directly also requires further study.

Phenylbutazone in the higher concentrations used for our experiments impairs $O_2^{\cdot -}$ and H_2O_2 production (15). Thus the effect of high concentrations of this drug on OH· production may also be secondary to an impairment of the Haber-Weiss reaction. However, since this drug is known to scavenge OH· directly in microsomal systems (12) the possibility that a low concentration of this drug may selectively scavenge OH· without impairing $O_2^{\cdot -}$ or H_2O_2 is currently under investigation.

In summary, our studies indicate that several anti-inflammatory drugs have the capacity to impair OH· release by zymosan stimulated PMNs. Since the concentrations used for our experiments for the most part can be achieved by administration of these agents in vivo, our results suggest that one common mechanism of action of these drugs in vivo may be to impair OH· release from phagocytic cells. The one exception appears to be hydrocortisone succinate, since the concentration required to impair OH· production in vitro would be harder to achieve in vivo. This suggests that the anti-inflammatory effect of steroids may be primarily related to other mechanisms than the impaired generation of ROS. Finally, our results provide further support for the concept that OH· is a mediator of inflammation in vivo.

REFERENCES

1. Babior, B. M., Kipnes, R. S. and Curnutte, J. T., J. Clin. Invest. 52,741 (1974).
2. Babior, B. M., New England J. Med. 298,659 (1978).

3. Johnston, R. B., Jr., Keele, B. B., Jr., Misra, H. P., Lehmeyer, J. E., Webb, L. S., Baehner, R. L. and Rajagopalan, K. V., J. Clin. Invest. 55,1357 (1975).
4. Webb, L. S., Keele, B. B., Jr., and Johnston, R. B., Jr., Infect. Immuno. 9,1051 (1974).
5. Baehner, R. L., Murrman, S. K., Davis, J. and Johnston, R. B., Jr., J. Clin. Invest. 56,571 (1975).
6. Sagone, A. L., Jr., King, G. W. and Metz, E. N., J. Clin. Invest. 57,1352 (1976).
7. Weiss, S. J., Rustagi, P. K. and LoBuglio, A. F., J. Exp. Med. 147,316 (1978).
8. Sagone, A. L., Jr., Decker, M. S., Wells, R., and Democko, C., Biochim. et Biophysica Acta 628,90 (1980).
9. Sagone, A. L., Jr., Wells, R. and Democko, C., Inflammation 4,65 (1980).
10. Egan, R. W., Paxton, J. and Kuehl, F. A., J. Biol. Chem. 251, 7329 (1976).
11. Egan, R. W., Gale, P. H., Kuehl, F. A., J. Biol. Chem. 254, 3295 (1979).
12. Kuehl, F. A., Humes, J. H., Egan, R. W., Ham, E. A., Beveridge, G. C. and Van Arm, C. G., Nature, 265,170 (1977).
13. Lehmeyer, J. E. and Johnston, R. B., Jr., Clin. Immunol. Immunopathol. 9,492 (1978).
14. Oyanaqui, Y., Biochem. Pharmacol. 27,777 (1978).
15. Simchowitz, L., Mehta, J. and Spilberg, I., Arthritis and Rheumatism 22,775 (1979).
16. Repine, J., Eaton, J. W., Anders, M. W., Hoidal, J. R. and Fox, R. B., J. Clin. Invest. 64,1642 (1979).
17. Sagone, A. L., Jr., unpublished observations.
18. Goldstein, I. R., Malmston, C. L., Kaplan, H. B., Jindahl, H., Samuelsson, R. and Weissman, G., Clin. Res. 25,518 (1977).
19. Boxer, L., Allen, J. and Baehner, R., J. Lab. Clin. Med. 92, 730 (1978).

ROLE OF THE HEXOSE MONOPHOSPHATE SHUNT (HMPS) IN CELLULAR PROTECTION AGAINST RADIATION DAMAGE

A. L. Sagone, Jr., W. Roberts, M. Kartha, Departments of Medicine and Radiotherapy, The Ohio State University, Columbus, Ohio

Human blood lymphocytes similar to other tissues have more injury following X-radiation under air than hypoxic conditions (oxygen enhancing effect) (1). We have recently studied the effect of low dose X-radiation on the HMPS of human lymphocytes in order to determine whether or not this pathway provides radiation protection for the cell (1). This pathway has been shown to protect human cells from damage by highly reactive oxygen species in other systems. We found that the HMPS activity was stimulated after aerobic irradiation. However, under hypoxic conditions there was a two-fold higher stimulation of HMPS activity than that observed in aerobically irradiated cells. This unexpected observation, combined with experimental evidence from other laboratories, suggests the following hypothesis: following X-radiation there is a competitive reaction of sulfhydryls and oxygen with hydrogen deletion radicals in macromolecules. Under aerobic conditions the reaction of O_2 is kinetically favored over that of sulfhydryls resulting in irreversible damage explaining the oxygen enhancing effect. These data suggest, therefore, an important role of the HMPS pathway in the protection of hypoxic tissues against irradiation damage.

In order to provide further evidence for a role of the HMPS in the protection of lymphocytes against irradiation damage, we have studied the effect of diamide, in concentrations which stimulate the HMPS, on the radiosensitivity of lymphocytes under air and under hypoxic conditions.

METHODS

Irradiation. Lymphocytes were isolated from defibrinated venous blood of normal healthy volunteers by a modification of the Ficoll-Hypaque technique (1). The mononuclear cells were resuspended in minimum essential medium (GIBCO # 109G) with filtered 20% fetal calf serum and antibiotics (penicillin and streptomycin) so that the final concentration was 3 X 10^6 cell/ml. Aliquots of 1.9 ml were then placed in 25 ml triple armed distilling flasks. The flasks were sealed with serum stoppers and connected to a gas cylinder containing 5% CO_2 in air or 5% CO_2 in nitrogen. The flasks were then gassed for 60 minutes (1). Following this, the side arms were further sealed with parafilm. Diamide (.1 ml of a concentrated solution which had been gassed for 20 minutes with nitrogen) was added to the flasks so that the final concentration

ISBN 0-12-592050-4

was 20 nanomoles/10^6 cell. The cells were irradiated 1 minute after the addition of the diamide via two parallel opposing horizontal beams of 4-MV X-rays from a linear accelerator, using an isocentric set-up. The absorbed doses in rad were calculated to the mid-plane of the irradiation block, where the cell samples were centered. The dose rate of irradiation was approximately 300 rad/minute.

Lymphocyte transformation studies. After the suspensions were irradiated, lymphoblastic transformation to PHA was studied using the microculture system of Parker and Lukes (1,2). The incorporation of thymidine into DNA during the final four hours of the 72 hour culture was used to quantitate DNA synthesis (1,2).

Continuous measurement of HMPS in short term culture. The effect of diamide on the oxidation of ^{14}C-1-glucose of lymphocyte suspensions incubated under air and nitrogen was determined continuously using the ionization chamber electrometer system as previously described (3-5).

RESULTS

The effects of diamide on the oxidation of the ^{14}C-1-glucose of lymphocyte suspension under air and nitrogen atmospheres are indicated in Fig. 1. As indicated, the addition of diamide (20 nanomoles/10^6 cells) resulted in a prompt stimulation of ^{14}C-1-glucose oxidation. The stimulation was similar under aerobic and hypoxic conditions.

The effect of diamide on the radiation sensitivity of human lymphocytes is given in Fig. 2. Incubation of lymphocyte suspensions under hypoxic conditions or with diamide under air and hypoxic conditions did not impair lymphoblastic transformation. Following X-irradiation (600R) under air, LBT is markedly impaired (Fig. 2). The ^{3}H-Thymidine of irradiated suspensions was less than one-third of the controls. Incubation of lymphocytes with diamide did not alter this effect. Under hypoxic conditions lymphocytes had less impairment in LBT following X-irradiation indicating a protective effect. In contrast, suspensions treated with diamide and irradiated under nitrogen had an impairment in LBT which was similar to suspensions irradiated under air.

DISCUSSION

Our results indicate that diamide stimulates the hexose monophosphate shunt activity of human lymphocytes as indicated by the enhanced oxidation of ^{14}C-1-glucose. This reaction is not dependent on oxygen since the rate under hypoxic conditions is similar to that observed for experiments using an air atmosphere. The reaction most likely involves the oxidation of GSH in view of the high rate of reactivity of diamide with this compound (6-9). However, a direct reaction of diamide with NADPH cannot be completely excluded and requires further study.

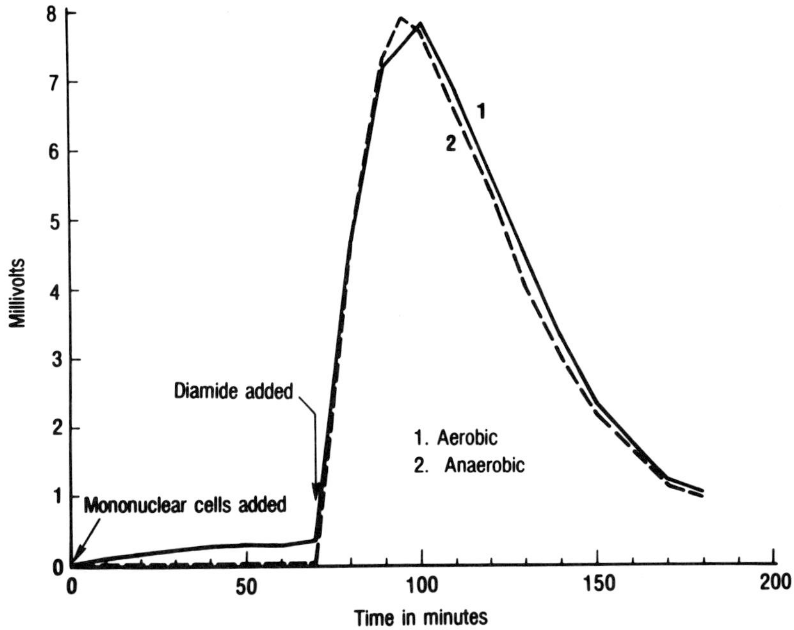

FIGURE 1. Effects of diamide on the oxidation of the ^{14}C-glucose by lymphocyte suspensions under air and nitrogen atmospheres. As indicated by the arrow, the addition of diamide resulted in a prompt increase in the oxidation of ^{14}C-1-glucose which is similar under air and nitrogen. These data indicate that diamide stimulates the HMPS pathway of lymphocytes most likely as a consequence of the oxidation of GSH. The results were derived from a single experiment which is representative of three.

Diamide did not alter the injury induced by X-irradiation in oxygenated suspensions. In contrast, diamide did ablate the protective effect observed when lymphocyte suspensions were irradiated under hypoxic conditions. Thus, diamide induced radiosensitivity under hypoxic conditions.

We have previously suggested that the HMPS pathway may provide protection of hypoxic lymphocytes against radiation damage by repairing hydrogen deletion radicals in macromolecules (1). This mechanism may not work well in oxygenation tissues because the reaction of O_2 with hydrogen deletion radicals is kinetically favored over that of sulfhydryls (10). This model predicts that compounds such as diamide, which reversibly oxidizes GSH and therefore uncouples HMPS activity in the lymphocytes, would enhance X-irradiation damage primarily under hypoxic conditions. This is what we observed in our experiments.

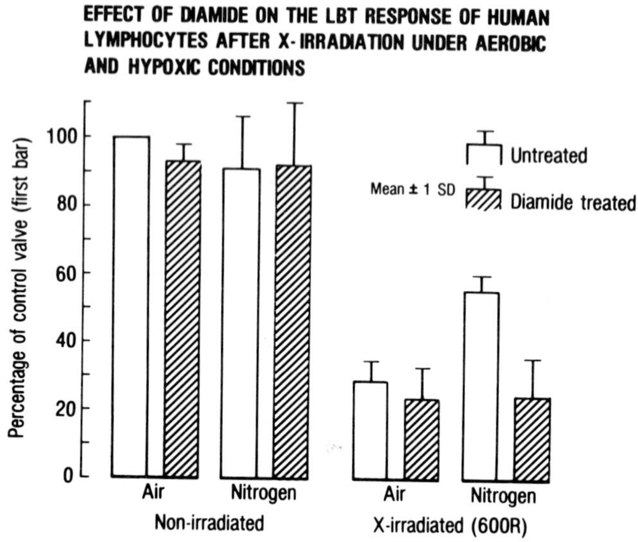

FIGURE 2. Effect of x-irradiation on the lymphoblastic transformation of human lymphocytes. The ^{3}H-Thymidine uptake during the final four hours of the 72 hour culture was used to quantitate DNA synthesis. Bars indicate mean ± SD of 3 experiments. Experimental values are expressed as percentage uptake of the un-irradiated control under air. These control values range from 4-5 X 10^4 CPM/.2 x 10^6 lymphocytes. Pre-incubation of lymphocyte suspensions with diamide under air (2nd Bar) or hypoxic conditions (4th Bar) did not impair LBT. Following X-irradiation (600R) in an air atmosphere, diamide treated suspension (6th Bar) had an impairment in LBT which was similar to the irradiated suspensions without diamide (5th Bar). Suspensions irradiated under hypoxic conditions were protected (7th Bar). However, suspensions incubated with diamide under nitrogen (8th Bar) had an impairment in LBT similar to suspensions incubated under air (5th and 6th Bars).

Therefore our observations provide further evidence that the HMPS is important in the protection of hypoxic tissue against irradiation damage and that the oxygen enhancement effect is the result of an irreversible reaction of oxygen with hydrogen deletion radicals in macromolecules.

REFERENCES

1. Roberts, W., Kartha, M. and Sagone, A. L., Jr., Radiation Research 79,601 (1979).
2. Parker, J. and Lukes, R., Amer. J. Clin. Pathol. 56,174 (1971).
3. Sagone, A. L., Jr., LoBuglio, A. F. and Balcerzak, S. P., Cell Immunol. 14,443 (1974).

4. Sagone, A. L., Jr., Metz, E. N. and Balcerzak, S. P., Biochim. Biophys. Acta 261,1 (1972).
5. Sagone, A. L., Jr., Kamps, S. and Campbell, R., Photochem. and Photobiol. 28,909 (1978).
6. Weiss, S. and Sagone, A. L., Jr., Biochim. et Biophysica Acta 585,620 (1979).
7. Kosower, N. S., Kosower, E. M. and Werthem, B., Biochem. Biophys. Res. Comm. 37,593 (1969).
8. Kosower, E. M., Correa, W., Kinon, B. J. and Kosower, N. S., Biochim. Biophysica Acta 264,39 (1972).
9. Oliver, J. M., Albertini, D. F. and Berlin, R. D., J. Cell. Biol. 71,921 (1976).
10. Howard-Flanders, P., Levin, J. and Theriot, L., Radiation Research 18,593 (1963).

SPECIFIC PHOTOOXIDATION OF HEMOCYANIN ACTIVE SITE PROMOTED BY PHOTOEXCITATION OF THE PROTEIN LIGAND-COPPER CHARGE-TRANSFER TRANSITION

B. Salvato[1], *G. Jori*[2] *and S. Cannistraro*[2]

[1]*Istituto di Biologia Animale, Centro CNR Emocianine, Università di Padova, Italy*
[2]*Instituto di Fisica Università di Perugia, Italy*

Irradiation of Octopus vulgaris hemocyanin with light wavelengths above 320 nm causes the specific photooxidation of a limiting number of eight histidyl residues per protein active site and the irreversible oxidation of Cu^{+} to Cu^{++}. As a consequence, the intensity of the near-UV copper band of the protein decreases according to first-order kinetics with respect to the irradiation time. On the other hand, copper-free hemocyanin was recovered with unchanged spectral properties and amino acid composition even after prolonged irradiation. The rate of the photoprocess is linearly and specifically related with the intensity of absorbed light within the 348 nm-peaking copper-oxygen band. However, the efficiency of histidine photomodification changes as a function of the irradiation time suggesting that the initial destruction of specific targets on the protein molecule affects the accessibility of other residues within the active site to the oxidative attack. On the basis of the above observations, as well as of the ascertained stoichiometry and kinetics of O_2 consumption and Cu^{++} formation, we propose a mechanism for the photoprocess involving:

i) Photoexcitation of the histidine-to-Cu^{++} charge transfer transition with generation of a histidyl radical cation and Cu^{+} and consequent dissociation of $O_2^{\bar{}}$ from its complex with copper;

ii) reaction of the histidyl cation with $O_2^{\bar{}}$ with formation of a histidine monoendoperoxide, which is still able to coordinate copper ions;

iii) binding of a second O_2 molecule to the protein active site with reformation of the charge-transfer complex (protein ligand-Cu^{++}-O_2^{2-}-Cu^{++}-protein ligand); in this way, repetition of the photoprocess according to the above outlined steps may occur leading to complete photooxidation of the eight histidyl ligands;

iv) the oxidative formation of Cu^{++} appears to be a side effect of the photomodification of the metal ligands.

Further support to the proposed mechanism is lent by irradiation experiments performed in the EPR cavity, showing the formation of transient aromatic-type radical signals.

ISBN 0-12-592050-4

ROLE OF CAROTENOPROTEINS IN THE ADAPTATION OF MARINE COPEPODS TO SOLAR LIGHT

R. Santus,[1] *A. Momzikoff*[2] *and C. Salet*[3]

[1]*Laboratoire de Physico-Chimie de l'Adaptation Biologique, Muséum National d'Histoire Naturelle, 43, Rue Cuvier 75231, Paris, Cedex 05, France*
[2]*Institut Océanographique, 195, Rue Saint Jacques, Paris, France*
[3]*Laboratoire de Biophysique, Muséum National d'Histoire Naturelle, 61, Rue Buffon 75005 Paris, France*

INTRODUCTION

Most marine invertebrates and in particular Crustacea contain in relatively large amounts carotenoproteins which, in the case of copepods, are believed to provide the microorganism with a "protective coloration" (1). However it is not known whether the carotenoproteins are acting as an optical filter for the more energetic solar radiations or to protect them against the lethal effect of active oxygen species such as $^1O_2$ or $O_2^{\bar{\cdot}}$ which must be produced in copepods since these microorganisms contain specially good photosensitizers such as flavins and pterins. This study is thus aimed at the determination of the molecular mechanism for the protective effect of the carotenoprotein in connection with the adaptation of the marine copepods Anomalocera pattersoni to solar light.

RESULTS AND DISCUSSION

On Fig. 1 is given the absorption spectrum of a highly purified carotenoprotein extracted from Anomalocera pattersoni. This protein is in fact made of 9 sub-units each containing two components. The intense blue color ($\varepsilon = 1.25 \times 10^5\ M^{-1}\ cm^{-1}$ at 650 nm) is due to a carotene derivative-astaxanthin—acting as a prosthetic group. When unbound, astaxanthin as other carotenes, absorbs at 460 nm. Thus binding to the protein results in a 200 nm bathochromic shift of the astaxanthin spectrum. The protein contains 3 astaxanthins tightly bound to hydrophobic pockets of the sub-units which are responsible for the high stability of the protein (1). The hydrophobic nature of the binding site is confirmed by laser flash spectroscopy by taking advantage of the high quenching efficiency of carotene with respect to the triplet state of aromatic compounds (2). In this instance the FMN triplet was used as a structural probe. Addition of FMN to the carotenoprotein solution more or less reproduces the in vivo conditions since these microorganisms contain relatively large concentrations of flavins (1 riboflavin for 6 astaxanthins). The 353 nm laser

ISBN 0-12-592050-4

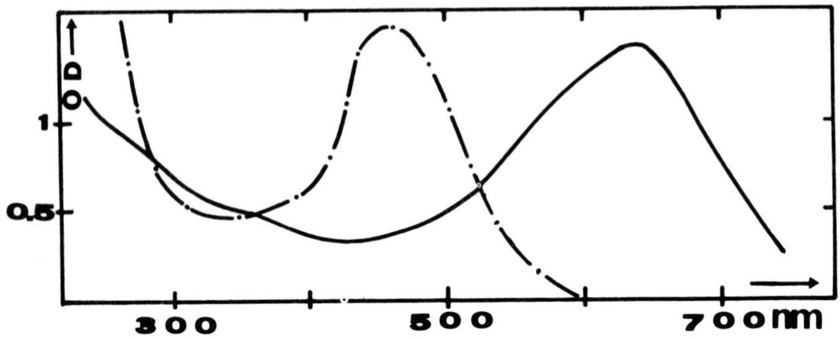

FIGURE 1. Absorption spectrum of Astaxanthin in ethanol (—·—·—) and of the carotenoprotein from Anomalocera pattersoni (———) in pH 7 buffer.

flash spectroscopy shows that no quenching of the FMN triplet state occurs at high protein concentration (∿ 10^{-4} M) which means that the orbital overlap between the carotenoid and the FMN triplet cannot be achieved. The independence of the triplet state lifetime with respect to the protein concentration also means that Trp, Met and His residues which are good FMN triplet quenchers (3) and are present in the protein are not fully accessible to the excited flavin. This observation with the fact that no bleaching of the purified carotenoprotein occurs in O_2 saturated solution on irradiation with λ>320 nm might be interpreted as if the carotenoprotein was acting as a solar screen which is probably one of its functions.

However, addition of various photosensitizers (pterins, flavins and hematoporphyrin) leads upon irradiation to a fast and irreversible bleaching of the carotenoprotein (quantum yield = 0.005: Fig. 2). The $^1O_2$ involvement is easily demonstrated using hematoporphyrin as sensitizer and N_3^- as $^1O_2$ quencher. The $^1O_2$ quenching by carotenes has been shown to mostly involve physical deactivation (2). In the present case, most of the quenching is chemical although we have in other carotenoproteins indication of a good physical quenching ability. N_3^-, which is an excellent $^1O_2$ quencher, induces a strong increase in the bleaching rate of the carotenoprotein (ϕ=0.1) when FMN or pterin are used as sensitizers. This reaction involves $N_3^{\cdot}$ species as deduced from laser flash experiments.

$O_2^{\overline{\cdot}}$ generated (4) in O_2-saturated aqueous solutions of the protein containing FMN and a large excess of a good hydrogen donor (His for instance) is more efficient in promoting the carotenoid bleaching (Fig. 2). The induction period which can be easily seen at the beginning of the irradiation could be due to the $^1O_2$ or $O_2^{\overline{\cdot}}$ reaction with some residues in the protein occurring in competition with the carotenoid bleaching and leading to partial protein denaturation. An alternative explanation would be that

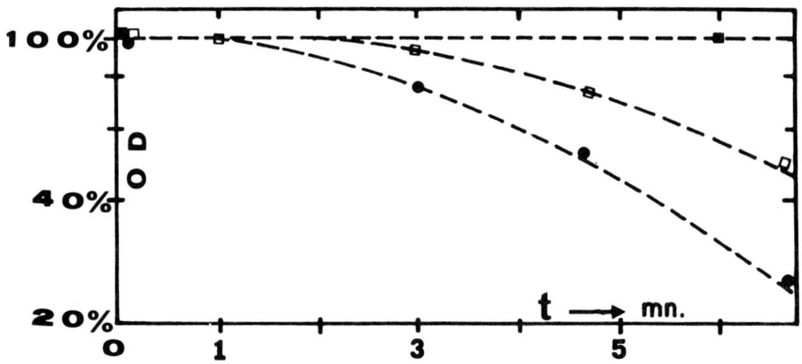

FIGURE 2. OD loss at 650 mn under irradiation of O_2-saturated protein solutions (7 x 10^{-6} M) in presence of flavin mononucleotide (FMN) (o——o), FMN + His (10^{-2} M) (● — ●) and in absence of FMN (■ — ■) or FMN + His-Ar saturated.

most of the $^1O_2$ or $O_2^{\cdot -}$ attack occurs at the astaxanthin level since each sub-unit contains 3 astaxanthins which seem to be crucial for the protein structure. Bleaching of one astaxanthin could weaken the protein structure and speed up the bleaching of the two others.

REFERENCES

1. Zagalsky, P. F. and Heiring, P. J., Comp. Biochem. Physio. 41 B,411 (1972).
2. Bellus, D., "Singlet Oxygen, Reaction with Organic Compounds and Polymers" (B. Randy and J. F. Radek, eds.), John Wiley, New York (1976), pp. 61-109.
3. Hellis, P. F., Parsons, B. F., Phillips, G. O. and McKellar, J. F., Photochem. Photobiol. 28,169 (1978).
4. Michelson, A. M., Biochimie 55,465 (1973).

SPIN TRAPPING OF OXY- AND OTHER RADICALS FROM OXIDATION OF LIPIDS

K. M. Schaich and D. C. Borg

Medical Department, Brookhaven National Laboratory, Upton, New York 11973

The recognition that oxidation of membrane lipids is indeed involved in many pathological processes has stimulated considerable study of the (aut)oxidation free radical chain of lipids and of the relation of various intermediates or products of the chain to specific biological damage. Particular attention has been given to the agents or factors potentially responsible for initiation of lipid oxidation in vivo. Rather than studying free radical intermediates and their effects directly, most investigators have used product reactions and analyses to obtain mechanistic information, understandably, because the techniques are familiar, versatile, and readily available. However, advances in pulse radiolysis and electron paramagnetic resonance (EPR) techniques, have now made possible the generation, detection, and identification of short-lived radical species such as those involved in the initiation and propagation phases of lipid oxidation.

One such EPR technique which has become popular is spin trapping, in which a labile free radical reacts most typically with a nitrone or nitroso compound to form a more stable nitroxide radical. The spectral characteristics of the nitroxide radical give information that may, in the best cases, identify the initial radical (adduct). Spin trapping is physically easy to carry out, frugal with reactants, and especially in conjunction with EPR flow-mixing techniques provides a means of studying free radicals whose steady-state concentrations are otherwise below the detection limits (10^{-8}M) for conventional EPR spectroscopy.

The use of spin traps in biological or other complex reaction systems is not so simple or straightforward as the basic concept implies. Spin traps in general are very reactive compounds, and are particularly susceptible to both oxidative and reductive attack. They are relatively specific in their reactivities with different classes of radicals—not all traps trap radicals equally well, nor does any trap trap all radicals equally well. In addition, spin traps do not differentiate between radicals formed in primary as opposed to secondary or even unrelated reaction sequences.

Because of the difficulties inherent in using spin traps appropriately, identifying spin adducts, and interpreting spin trapping results, our studies with oxidizing lipids have focused first

ISBN 0-12-592050-4

on building a "catalogue" of spin trapping reaction characteristics and identifiable spectra generated in defined or model compounds and reactions, information which can then be used as a basis for interpreting spectra obtained in more complex systems. Reactions have been designed with the fewest reactants and alternative pathways possible, and attention has been given to both competition and reaction kinetics.

Radicals in model systems containing purified unsaturated fatty acids or their hydroperoxides have been generated 1) chemically by electrons or ·OH radicals [titanium (III) or iron (II or III): hydrogen peroxide systems], 2) electrolytically, and 3) by spontaneous autoxidation. Distinct differences were demonstrated in the abilities of the spin traps DMPO (dimethyl pyrrolidine-N-oxide), PBN (phenyl-t-butyl nitrone), and t-NB (2-methyl-2-nitroso-propane) to trap initiating species versus primordial lipid alkyl radicals or secondary oxy-radical species of lipids. Whereas at room temperature DMPO preferentially trapped electrons or oxy-radicals, t-NB preferentially trapped alkyl radicals from the solvent, metal chelator, or wherever else formed in the system. PBN proved to be the slowest reacting but most versatile agent, trapping initiating species, lipid alkyl- and oxy-radicals and other secondary radicals under a variety of reaction conditions. For all three traps, the adduct(s) observed depended on the reaction stage monitored, i.e. the rate of flow or time of addition of spin trap to a static reaction mixture, as well as on the nature and relative concentrations of reactants in a given system. Q-band flow studies utilizing a range of flow rates ($<$ .5 to $\sim$ 3.0 ml/min.) with or without spin traps were quite useful in resolving overlapping spectra, and in addition provided important supplementary information about radical stabilities, competitions, and reaction sequences.

Although this work is far from complete it is already clear that while a single spin trap may be appropriate for studying specific radicals in a simple, defined reaction, the most accurate and complete understanding of radical processes comprising complex reaction sequences, such as those expected in lipid oxidation, requires coordination of non-EPR data with EPR data obtained from several spin traps and from flow experiments, using varying reactant concentrations and reaction conditions.

ACKNOWLEDGEMENTS

This work was supported by the U. S. Department of Energy.

HEAVY ATOM AND SOLVENT EFFECTS ON SPECTRA OF OXYGEN

Ajit Singh and Grant W. Koroll, Medical Biophysics Branch, Whiteshell Nuclear Research Establishment, Atomic Energy of Canada Limited Research Company, Pinawa, Manitoba, Canada

We have studied intermolecular perturbation of the optical transitions of oxygen (1) by compounds containing atoms heavier than oxygen, and by water. Previous work (2) only partially revealed the effects of such perturbations on the basis of comparison of intensity of absorption as a function of oxygen pressure. However, for proper evaluation of these effect, knowledge of the solubility of oxygen in the solvents used is essential. Therefore, we have also determined experimentally, solubilities of oxygen in water, heavy water, Freon-113 and carbon tetrachloride.

For the solubility measurements, high pressure oxygen was collected in a vessel and the decrease in its pressure was recorded when this vessel was connected to another one containing a known volume of the solvent. The solubility of oxygen was then calculated from these data. The measured solubility of O_2 in the four solvents is shown in Fig. 1. Our measured values differ from those derived indirectly by Long and Kearns (2). In the case of water, our values are similar to those reported by Zoss et al. (3) and Frolich et al. (4).

In the cases of SF_6, Freon-113 and CCl_4, the effect of intermolecular perturbations on the optical transitions is rather similar; all the transitions are enhanced though the $^1\Sigma_g^+ \leftarrow {}^1\Sigma_g^-$ is enhanced the least (Table 1). On the other hand, in H_2O and D_2O, only the $^1\Sigma_g^+ \leftarrow {}^1\Sigma_g^-$ transition is observed and the optical density of this transition increases by a factor of ~ 10 in the presence of 1 mol dm^{-3} KI. Similar results are obtained in D_2O.

For solutions of oxygen in all four solvents, evidence of $^1O_2$ formation, by direct light absorption by oxygen, is obtained: in H_2O and D_2O, by reaction of $^1O_2$ with tryptophan (5) and in CCl_4 and Freon-113, by reaction of $^1O_2$ with DPBF (6).

ACKNOWLEDGEMENTS

This is AECL document number 6923.

REFERENCES

1. Khan, A. U. and Kasha, M., J. Am. Chem. Soc., 92,3293 (1970).
2. Long, C. and Kearns, D. R., J. Chem. Phys., 59,5729 (1973).
3. Zoss, L. M., Suciu, S. N. and Sibbitt, W. L., Trans. Am. Soc. Mech. Engrs., 75,69 (1954).

ISBN 0-12-592050-4

TABLE I. *Intermolecular Perturbation of Optical Transitions of Oxygen: Enhancement Factors (as Compared to the Optical Density of O_2 at the Same Concentration (mol dm^{-3}) in Gas Phase)*

Absorption Band (nm)	O_2 + 15% SF_6[1]	O_2 (1 mol dm^{-3} in Freon-113)	O_2 (1 mol dm^{-3} in CCl_4)	O_2 (.12 mol dm^{-3} in H_2O)[2]		O_2 (.12 mol dm^{-3} in D_2O)[2]	
					+1 mol dm^{-3} KI		+1 mol dm^{-3} KI
362	0.20	N	N	N	N	N	N
382	0.14	N	N	N	N	N	N
477	0.16	6	N	N	N	N	N
577	0.17	8	15	N	N	N	N
630	0.24	5	20	N	N	N	N
762	0.04	0.4	0.8	~2[3]	~14[3]	~2[3]	~10[3]
1066	0.23	4	10	N	N	N	N
1266	0.38	9	13	N	N	N	N

[1] O_2 = 11.4 MPa and SF_6 = 2.13 MPa

[2] On the Basic of O_2 solubility, the expected optical density of the bands other than the 762 nm one, is bleow the detection limit for a cell with 5 cm path length (i.e. <5 x 10^{-4}).

[3] The band width of this band is ~ 25 mm as compared to ~ 5 mm for O_2 (gas). Thus, the enhancement based on integrated absorption intensity would be ~ 5 times greater.

[N] could not be measured.

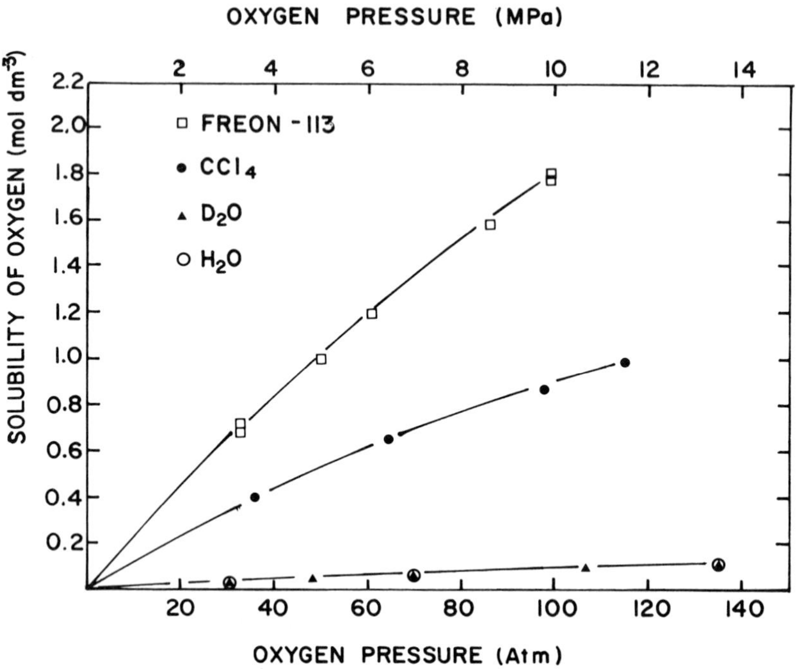

FIGURE 1. Solubility of oxygen in various solvents. 0 H_2O; Δ D_2O; ● CCl_4; ⊐ Freon-113.

4. Frolich, P. K., Tauch, E. J., Hogan, J. J. and Peer, A. A., Ind. Eng. Chem., 23,548 (1931); quoted in "Solubilities" by W. F. Linke, Vol. II, American Chemical Society, Washington, D.C., (1965), p. 1237.
5. Singh, A., Bell, M.-J., Koroll, G. W., Kremers, W. and Singh, H. This volume, p. 461.
6. Singh, A., McIntyre, N. R. and Koroll, G. W., Photochem. Photobiol., 28,595 (1978).

AZIDE RADICAL FORMATION FROM SODIUM AZIDE AND HYDROXYL AND ITS REACTIONS WITH TRYPTOPHAN AND TYROSINE

Ajit Singh, Grant W. Koroll,[1] *and Robert B. Cundall*[2]

[1]*Medical Biophysics Branch, Whiteshell Nuclear Research Establishment, Atomic Energy of Canada Limited Research Company, Pinawa, Manitoba, Canada ROE 1LO*

[2]*Department of Biochemistry, University of Salford, Salford, U.K. M5 4WT*

Sodium azide (or rather, the azide anion, N_{3aq}^-) is a well known singlet oxygen ($^1O_2$) quencher (1-5). However, in many reaction systems, $^1O_2$, superoxide anion ($O_{2aq}^{\bar{\cdot}}$) and hydroxyl radicals ($\cdot OH$) might be present together (6,7). It is important to understand how N_{3aq}^- would react with the latter two species. We have, therefore investigated reactions of N_{3aq}^- with $\cdot OH$ and $O_{2aq}^{\bar{\cdot}}$ and of the azide radical ($N_3^{\cdot}$) with two amino acids, tryptophan (TrpH) and tyrosine (TyrH).

Formation of $N_3^{\cdot}$ in reaction of N_{3aq}^- with $\cdot OH$ has been suggested (8,9).

$$N_{3aq}^- + {}^{\cdot}OH \rightarrow N_3^{\cdot} + OH_{aq}^- \quad [1]$$

followed by its addition (10) to N_{3aq}^-.

$$N_3^{\cdot} + N_{3aq}^- \rightarrow (N_3)_{2aq}^{\bar{\cdot}} \quad [2]$$

These reactions are analogous to those of $^{\cdot}OH$ with the halide and the thiocyanate anions (11,12). An aim of the present work was to compare the reactions of $N_3^{\cdot}$ with those of the halide and the thiocyanate radical anions ($Br_{2aq}^{\bar{\cdot}}$, $(SCN)_{2aq}^{\bar{\cdot}}$, etc.).

Our pulse radiolytic investigations show that the half-life of the $N_3^{\cdot}$ radical formed varies from $\sim$ 30 μs to $\sim$ 5 μs as the concentration of N_{3aq}^- is varied from 10^{-5} to 10^{-1} mol dm^{-3}. This is consistent with reaction [2] for which evidence has been lacking in the past (9). Our work suggests that the reactivity of these two species, $N_3^{\cdot}$ and $(N_3)_{2aq}^{\bar{\cdot}}$, towards the amino acids are different. The decay kinetics of $N_3^{\cdot}$ does not follow simple first or second order kinetics suggesting that several reactions contribute to its decay, e.g. reaction [2] and as follows.

$$N_3^{\cdot} + N_3^{\cdot} \rightarrow N_6 \quad [3]$$

$$N_3^{\cdot} + H \rightarrow HN_3 \quad [4]$$

ISBN 0-12-592050-4

$$N_3^{\cdot} + OH \rightarrow N_3OH \rightarrow \text{Products} \quad [5]$$

There appears to be no reaction between the iso-propyl radical $((CH_3)_2\dot{C}OH)$ and N_{3aq}^- or between $O_{2aq}^{\cdot -}$ and N_{3aq}^-. However, an electron transfer reaction between $(SCN)_{2aq}^{\cdot -}$ and N_{3aq}^- has been observed.

$$(SCN)_{2aq}^{\cdot -} + N_{3aq}^- \rightarrow 2SCN_{aq}^- + N_3^{\cdot} \quad [6]$$

Azide radicals react with tryptophan to give a free radical with the loss of the H atom from its heterocyclic NH.

$$TrpH + N_3^{\cdot} \rightarrow Trp\cdot + HN_3 \quad [7]$$

$$\rightarrow TrpH_{aq}^+ + N_{3aq}^- \quad [8]$$

$$TrpH^+ \rightleftarrows Trp\cdot + H_{aq}^+ \quad [9]$$

The known spectra of Trp· and $TrpH^+$ (11-13) are observed as the pH of the solution is varied. Our results on Trp· formation are consistent with those of Land and Prütz (14). There is evidence of formation of other transient species derived from TrpH as well, e.g. $(N_3\text{-}TrpH)\cdot$.

Similar reactions of $N_3^{\cdot}$ with tyrosine have been observed; in this case the free radical (Tyr·) is formed by the loss of the phenolic H atom (11,12).

ACKNOWLEDGMENTS

This is AECL no. 6917.

REFERENCES

1. Foote, C. S., Fujimoto, T. T. and Chang, Y. C., Tetrahedron Lett., 45 (1972).
2. Göllnick, K., Haisch, D. and Schade, G., J. Am. Chem. Soc., 94, 1747 (1972).
3. Hasty, N., Merkel, P. B., Radlick, P. and Kearns, D. R., Tetrahedron Lett., 49 (1972).
4. Ito, T., Photochem. Photobiol., 28, 493 (1978).
5. Singh, H., and Vadasz, J. A., Photochem. Photobiol., 28, 539 (1978).
6. Tauber, A. I. and Babior, B. M., Photochem. Photobiol., 28, 701 (1978).

7. Singh, A., Photochem. Photobiol., 28, 429 (1978).
8. Kelly, P. and Smith, M., J. Chem. Soc., 1479 and 1487 (1961).
9. Hayon, E. and Simic, M., J. Am. Chem. Soc., 92, 7486 (1970).
10. Kraljić, I., in "Proceedings of the Third Tihany Symposium on Radiation Chemistry," (J. Dobó and P. Hedvig, eds.), Akadémiai Kiadó, Budapest (1972), p. 1405.
11. Adams, G. E., Aldrich, J. E., Bisby, R. H., Cundall, R. B., Redpath, J. L. and Willson, R. L., Radiat. Res., 49, 278 (1972).
12. Posener, M. L., Adams, G. E., Wardman, P. and Cundall, R. B., J. Chem. Soc. Faraday Trans. I., 72, 2231 (1976).
13. Santus, R. and Grossweiner, L. I., Photochem. Photobiol., 15, 101 (1972).
14. Land, E. J. and Prütz, W. A., Int. J. Radiat. Biol., 32, 203 (1977).

EVIDENCE OF HYDROXYL GENERATION DURING ARACHIDONIC ACID METABOLISM BY HUMAN PLATELETS

D. Singh and A. L. Sagone, Jr.

Department of Medicine, Ohio State University, Columbus, Ohio

Reactive oxygen species, probably hydroxyl ($^{\bullet}OH$), have been suggested to be generated during arachidonic acid (AA) metabolism and once released these species can modify the rate and extent of various reactions involved in AA metabolism. We recently studied this phenomenon in washed human platelets. $^{\bullet}OH$ generation was quantitated using ^{14}C-benzoic acid as a specific trap in a continuous ionization chamber system. Resting platelets did not produce any detectable signal, whereas addition of AA resulted in gradual $^{\bullet}OH$ production with peak values detected around 20 minutes. Similar studies conducted under nitrogen or after boiling the platelets almost abolished $^{\bullet}OH$ generation. Indomethacin and aspirin had no significant effect, whereas 5,8,11,14 eicosatetraynoic acid (ETYA) decreased the signal by >90% thus suggesting that $^{\bullet}OH$ is produced primarily through the lipoxygenase pathway. Superoxide dismutase (SOD) and catalase had no effect and, as expected, phenol and mannitol decreased the signal by >50% and 90% respectively. Azide and cyanide reduced the $^{\bullet}OH$ generation by about two thirds. We conclude that $^{\bullet}OH$ is generated during AA metabolism by human platelets. It is primarily produced via lipoxygenase pathway and does require a hemedependent peroxidase. This highly reactive oxidant may play an important role in normal and abnormal hemostasis.

ISBN 0-12-592050-4

PROSTAGLANDIN SYNTHETASE-DEPENDENT SPIN-TRAPPED FREE RADICALS

F. Lee Smith, Robert A. Floyd and Mary P. Carpenter, Oklahoma Medical Research Foundation, Oklahoma City, Oklahoma 73104

Several lines of evidence, including ESR spectra, indicate that free radicals are produced during the biosynthesis of prostaglandins (PG) from arachidonic acid (20:4) (1,2). However, ESR spectra of these radicals have not had sufficient fine structure to allow identification of the radical species. To help solve this dilemma, we have tested the spin-trap, DMPO (5,5-dimethylpyrroline-N-oxide), as a radical scavenger during arachidonate-dependent prostaglandin synthesis by sheep semina vesicle.

Ram semina vesicle microsomes (RSVM) incubated with 20:4 and DMPO produce the ESR spectrum shown in Fig. 1. While the spectral pattern of the spin-trapped radical is similar to that described for superoxide anion radical, the hyperfine splitting constants (A_N = 14.5, A_H^B = 16.2) indicate that the trapped radical is neither a superoxide nor an hydroxyl free radical. Indomethacin and meclofenamic acid, inhibitors of PG synthetase, diminish the amplitude of the ESR signal (Fig. 2). Boiled microsomes incubated with DMPO do not produce a DMPO radical signal, while boiled microsomes in the presence of 20:4 produce an ESR signal characteristic of the DMPO-hydroxyl radical adduct.

The spectrum in Fig. 1 appears to be 20:4-dependent. This signal is not observed when 20:4 is replaced by docosahexaenoic

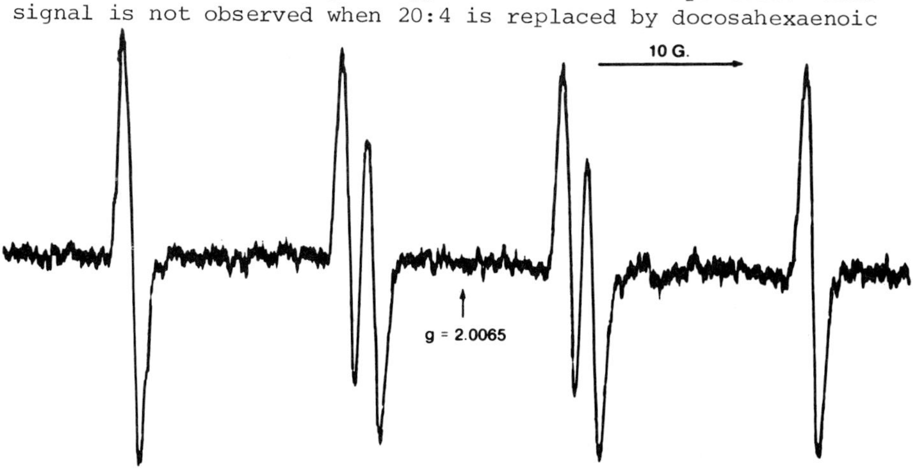

FIGURE 1. ESR signal from 1.2 mg RSVM incubated at 37°C with 820 μM 20:4 and 440 μM DMPO. Scan time = 1 hr., time constant = 3 sec. Modulation amplitude = 0.5 G, gain = 4 x 10^4, scan range = 100 G, modulation frequency = 100 KHz.

ISBN 0-12-592050-4

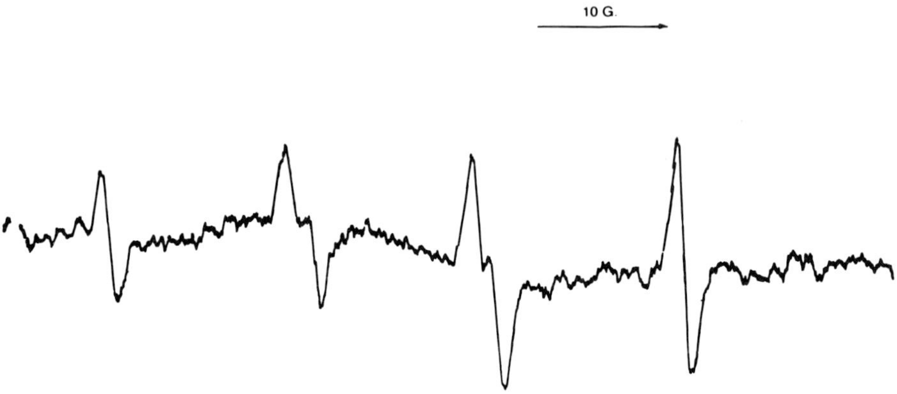

FIGURE 2. ESR signal from 1.2 mg RSVM pre-incubated at 0°C with 1.4 mM indomethacin then incubated at 37°C with 820 μM 20:4 and 440 μM DMPO. Scan time = 16 min., time constant = 3 sec. Modulation amplitude = 2G, gain = 4 x 10^4, scan range = 100 G, modulation frequency = 100 KHz.

acid (22:6), linoleic acid (18:2) or the acetylenic analog of 20:4, ETYA, (eicosatetraynoic acid). A weakened signal results after pre-incubation with ETYA followed by addition of 20:4. Addition of linoleic acid (18:2) in place of 20:4 produces the DMPO-hydroxyl adduct signal. The synthesis of prostaglandins by these microsomes in the presence of 20:4 was analyzed by radioimmunoassay (RIA) (Fig. 3). As expected, addition of PG synthesis inhibitors decreased product formation. Addition of DMPO to the system results not only in a DMPO radical adduct but also affects PG synthesis. Production of PGE_2, the major product of RSVM in the

ADDITION	DMPO	$PGF_{2\alpha}$	PGE_2
		ng. PG./mg./30 min.	
20:4	0	398	2181
20:4 + indomethacin	0	32	128
20:4	+	340	3221

FIGURE 3. Effect of the spin trap, DMPO, on prostaglandin synthesis. $PGF_{2\alpha}$ and PGE_2 were quantified by RIA after extraction into ethyl acetate and partition on silicic acid.

presence of 20:4, is significantly elevated.

The biosynthesis of prostaglandins is believed to proceed through several free radical intermediates. Appropriate use of free radical trapping agents may help identify these radical species and may also provide a means to study regulatory mechanisms that control the prostaglandin synthetic profile.

REFERENCES

1. Nugteren, D. H., Beerthuis, R. K. and Van Dorp, D. A., Recl. Trav. Chim. Pays-Bas, 85, 405 (1966).
2. Egan, R. W., Paxton, J. and Kuehl, F. A., Jr., J. Biol. Chem., 251, 7329 (1976).

THE PHOTOPEROXIDATION OF UNSATURATED ORGANIC MOLECULES XIX. 1,3- DIPHENYLISOBENZOFURAN AS SENSITIZER AND INHIBITOR

B. Stevens, C. N. Christy,[1] and J. A. Ors[2]*

[1]*Department of Chemistry, University of South Florida, Tampa, Florida 33620*
[2]*IBM Watson Research Center, Yorktown Heights, New York 10598*

The photoperoxidation of 1,3-diphenylisobenzofuran (DPBF) in benzene is accompanied by molecular O_2 consumption on a 1:1 stoichiometric basis consistent with formation of the endoperoxide as the primary product. However, the DPBF triplet state appears to react with the endoperoxide with a rate constant of ~$10^7 M^{-1}s^{-1}$ to form unidentified secondary products which could be significant if the endoperoxide/O_2 concentration ratio is allowed to approach 10^2 due to O_2 depletion in the primary photoperoxidation reaction.

The relative rate of DPBF to rubrene consumption is greater when DPBF is initially excited at 365 nm than when rubrene is selectively excited at 546 nm, which is quantitatively consistent with reactive re-encounters of the DPBF-$O_2\,{}^1\Delta g$ molecular pairs formed by $O_2\,{}^3\Sigma\bar{g}$ quenching of DPBF(T_1).

An analysis of the rubrene concentration-time curves in the presence of DPBF as inhibitor is presented to account for the increase in rate of rubrene photoperoxidation as the more reactive inhibitor is consumed.

**To whom correspondence should be addressed.*

ISBN 0-12-592050-4

PROTEIN CROSSLINKING STUDIES IN EOSIN SENSITIZED PHOTOOXIDATION OF SINGLE CHAIN PROTEINS, NON-COVALENTLY LINKED DIMERS (CROTOXIN, MOJAVE TOXIN) AND COVALENTLY LINKED MULTIPLE CHAIN PROTEINS (ANTIVENIN IgG)

R. C. Straight and J. L. Glenn

Venom Research Laboratory, VA Medical Center, Salt Lake City, Utah

Early studies of the sensitized photooxidation of proteins reported a variety of physical-chemical changes suggesting that protein aggregates were produced (1,2). It was suggested that such aggregates were the result of covalent crosslinks (3). More recently, strong evidence has been obtained that covalently crosslinked proteins are produced by dye sensitized photooxidation of erythrocyte membranes (4,5) and hemoglobin (6). Covalent crosslinking has also been reported to occur between ribosomal RNA and proteins by methylene blue-sensitized photooxidation (7). We have studied the eosin-sensitized photooxidation of the single chain proteins (lysozyme, ribonuclease, myoglobin); single chain, non-covalently complexed protein-protein dimers (Crotoxin and Mojave Toxin) from rattlesnake venoms; and covalently linked multiple chain proteins (human IgG and horse IgG antivenin) to determine if crosslinking occurs and what conditions favor crosslinking. Photooxidation reactions were carried out in aqueous solutions containing protein (1.0 - 10.0 mg/ml), eosin (10^{-5}M - 10^{-4}M), oxygen ($2X10^{-4}$M or $1.1X10^{-3}$M), phosphate buffer (0.2M, pH 6-10) and illuminated at 520 nm, Io $\simeq$ $120JM^{-2}s^{-1}$ until the desired μmoles of oxygen were consumed as measured by the *Clark* oxygen electrode. Photooxidized and dark control samples were analyzed by analytical gel-sieving column (1.0 cm X 40 cm) chromatography using Sephacryl S-300 gel (fractionating range 5000-200,000 Daltons) eluted with a mildly denaturing buffer consisting of Tris-HCl (0.1M); NaCl (0.5M) and Urea, 2M. Eluted proteins were detected by absorbance at 280 nm using an ISCO Model UA5, type 6 double cell detector with 0.1 AU full scale sensitivity and recorded as a function of elution time. The test proteins IgG (160,000); egg albumin (45,000); and lysozyme (15,000) were readily resolved by this system with reproducible peak retention times. Retention time of the macromolecule blue dextran was 30±2 min. Samples were also analyzed by analytical polyacrylamide gel electrophoresis in the presence of sodium dodecyl sulfate (SDS-PAGE) and by analytical isoelectric focusing with ampholytes (IEF). The initial sensitized photo-oxidation reaction system of protein 2.5 mg/ml; eosin, $1X10^{-5}$M; oxygen, $2X10^{-4}$M; phosphate buffer, 0.2M, pH 7.5 gave analytical Sephacryl

ISBN 0-12-592050-4

S-300 column peaks retention times for dark control and photooxidized (5.5 μmoles of oxygen consumed) proteins as follows:

Peak Retention Times (Min)

Protein (molecular weight)	Dark Control	Photooxidized		Peak ABS 280 nm
		F	S	F/S
Lysozyme (15,000)	73±2	38	75	.10
Ribonuclease (15,000)	70±2	36(T)	72	<.05
Myoglobin (17,000)	68±3	34(T)	67	<.05
Egg Albumin (45,000)	58±2	31	60	.38
Human IgG (160,000) 35±5(s)	48±2	32	48	.46
Horse Antivenin IgG (160,000) 34±5(s)	46±2	31	44	.40
Crotoxin (22,000) 63	78 85	65	74(T)	83(T) 5.0
Crotoxin A (9,500)	83	--	--	--
Crotoxin B (12,500)	78	--	--	--
Dextran Blue (2,000,000)	30±4	--	--	--

Photooxidation in each case resulted in the appearance of a fast (F) moving peak compared to the dark control sample. The maximum absorbance (280nm) ratios of the fast (F) to slow (S) moving components increased as the complexity of the protein increased. Although retention time is not a linear function of molecular weight for different proteins in this system due to charge and shape interactions the retention times of the fast moving peak of photooxidized proteins were approximately 1/2 the time of slow moving peak. Crotoxin partially dissociates into its two component proteins in the mild buffer used on the column. After photooxidation much less of the two component proteins compared to the crotoxin complex could be dissociated. The same result was found with Mojave toxin, a similar protein-protein dimer. SDS-PAGE analysis of the dark control and photooxidized samples confirmed the analytical gel sieving column results; in each case the dark control protein band (stained with coomassie blue R250) showed a progressive disappearance in the photooxidized sample and an area of diffusely staining material appeared progressing toward bands of higher molecular weight material as photooxidation proceeded. Analytical IEF showed that the photooxidized proteins increased in net negative charge and became more homogeneous in charge with photooxidation. Conditions which appear to favor crosslinking include lower substrate concentration, higher oxygen concentration, higher pH and lower sensitizer concentrations relative to protein concentrations.

The optimum conditions for protein crosslinking by photooxidation have not been established and may vary from protein to protein and the mechanism is not understood, but these results would suggest that photooxidative crosslinking is an important photodynamic phenomenon especially in more complex protein systems.

ACKNOWLEDGMENTS

Supported by Veterans Administration Medical Research Program.

REFERENCES

1. Vodrazka, I. Z., Chem. Listy, 53, 829 (1959).
2. Spikes, J. D. and Straight, R., Ann. Rev. Phys. Chem. 18, 409 (1967).
3. Venkstern, T. V., Biochimija, 18, 97 (1953).
4. Girotti, A. W., Biochemistry, 18, 4403 (1979).
5. Dubbleman, T., De Goeij, A. and Van Steveninck, J., Biochim. Biophys. Acta, 595, 133 (1980).
6. Girotti, A. W., Lyman, S. and Deziel, M., Photochem. Photobiol., 29, 119 (1979).
7. Zook, D. E. and Fahnestock, S. R., Biochim. Biophys. Acta, 517, 400 (1978).

AUTOXIDATION OF 6-HYDROXYDOPAMINE:
INFLUENCE OF IRON AND ASCORBATE

Stephen Gene Sullivan and Arnold Stern, New York University School of Medicine, New York, New York 10016

6-Hydroxydopamine (6HD) autoxidation yields the p-quinone chromophore and H_2O_2 (1). Superoxide ($O_2^{\cdot-}$) is an intermediate of the reaction and functions as a catalyst in the autoxidation (2, 3). Catalase greatly potentiates the action of superoxide dismutase (SOD) in inhibiting 6HD autoxidation (4). In the presence of SOD, oxidation of 6HD by H_2O_2 may become quantitatively more important in the face of decreased availability of $O_2^{\cdot-}$ (4,5).

We have studied 6HD autoxidation at pH 7.4 in the presence or absence of SOD, catalase, ascorbate and iron chelated with EDTA by monitoring changes in absorption at 490 nm, O_2 consumption and H_2O_2 formation. Fe^{+2}-EDTA or Fe^{+3}-EDTA greatly accelerates the rate of 6HD autoxidation and results in consumption of H_2O_2 and bleaching of the chromophore (Fig. 1). Iron functions as a catalyst in 6HD autoxidation making the reaction independent of $O_2^{\cdot-}$ as shown by the lack of effect of SOD in the presence of iron.

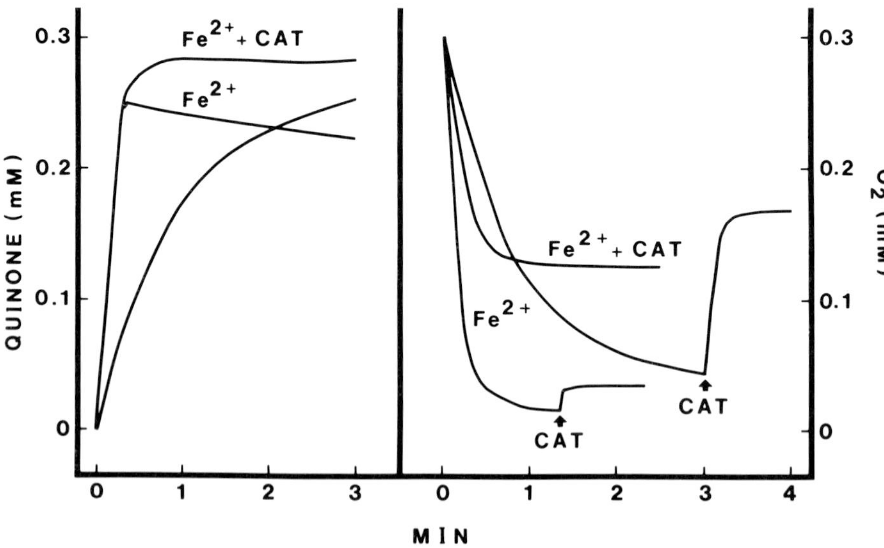

FIGURE 1. Influence of iron on 6HD autoxidation. 6HD autoxidation was followed at pH 7.4 and 25°C. 6HD dissolved in 1 mM KCl, pH 2.0 was injected at zero time. Final concentrations: 0.3 mM 6HD, 0.3 mM EDTA, 100 mM NaH_2PO_4, pH 7.4. Additions where noted: 0.1 mM $FeCl_2$, 10 µg/ml catalase (88,823 U/mg, Calbiochem-Behring).

ISBN 0-12-592050-4

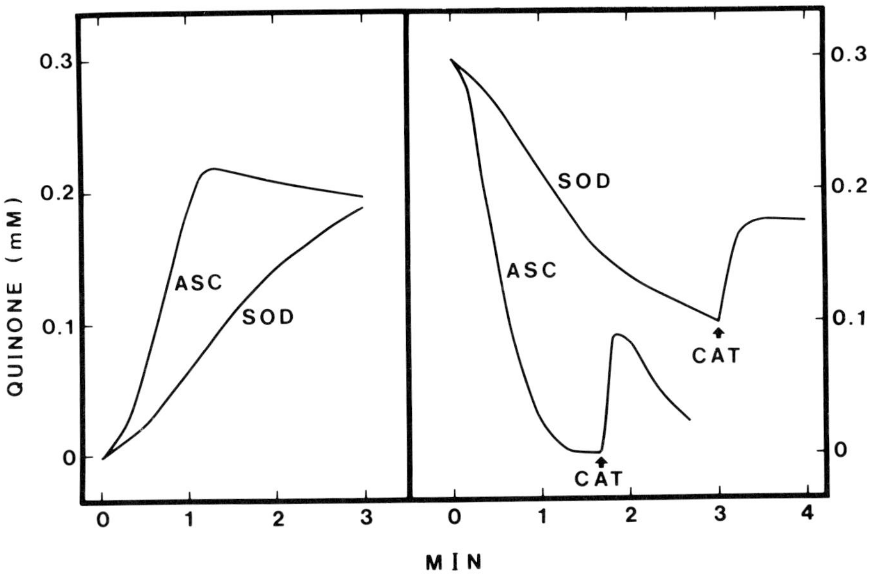

FIGURE 2. Influence of ascorbate and SOD on 6HD autoxidation. Conditions as in legend to Fig. 1. Other additions where noted: 10 mM ascorbate, 0.2 μg/ml SOD (3000 U/mg, Truett Laboratories).

Catalase has a limited inhibitory effect on the rate of 6HD autoxidation. When chelated iron is added, catalase has no effect on the rate of 6HD autoxidation but completely prevents bleaching of the colored product (Fig. 1). This probably occurs by eliminating the reaction of p-quinone with products of the reaction between iron and H_2O_2.

Ascorbate is known to cause cycling between 6HD and its p-quinone resulting in increased O_2 consumption and H_2O_2 formation (6). We observed that ascorbate acts like SOD by causing a lag phase before the linear phase of autoxidation is reached (Fig. 2). This SOD-like effect of ascorbate is probably related to the ability of reduced ascorbate to scavenge $O_2^{\cdot-}$ (7,8). 6HD autoxidation in the presence of ascorbate is independent of $O_2^{\cdot-}$ and H_2O_2 since SOD and catalase have no effect on the reaction. 6HD autoxidation in the combined presence of chelated iron and ascorbate shows the full effects of both additions except that the autoxidation exhibits no lag phase and the iron-catalyzed autoxidation of ascorbate occurs simultaneously with the 6HD reactions.

SOD has no effect on 6HD autoxidation in the presence of chelated iron or ascorbate. $O_2^{\cdot-}$ catalysis of 6HD autoxidation may be of minor importance in biological systems containing metals or reactive reducing substances like ascorbate. Catalase and peroxidative enzymes may protect against 6HD toxicity by preventing the Fenton reaction between iron and H_2O_2.

ACKNOWLEDGEMENT

Supported by NIH grant number 19532.

REFERENCES

1. Graham, D. G., Mol. Pharmacol., 14,633 (1978).
2. Heikkila, R. E. and Cohen, G., Science, 181,456 (1973).
3. Borg, D. C., Schaich, K. M., Elmore, J. J. and Bell, J. A., Photochem. Photobiol., 28,887 (1978).
4. Sullivan, S. G., Sudarsky, L. and Stern, A., "Chemical and Biochemical Aspects of Superoxide and Superoxide Dismutase", (J. V. Bannister and H. A. O. Hill, eds.), Elsevier/North-Holland, New York, (1980) p. 364-371.
5. Liang, Y. O., Wightman, R. M. and Adams, R. N., European J. Pharmacol., 36,455 (1976).
6. Heikkila, R. E. and Cohen, G., Ann. N.Y., Acad. Sci., 258, 221 (1975).
7. Nishikimi, M., Biochem. Biophys. Res. Commun., 63,463 (1975).
8. Nishikimi, M. and Yagi, K., "Biochemical and Medical Aspects of Active Oxygen" (O. Hayaishi and K. Asada, eds.), University Park Press, Baltimore, (1977), pp. 79-87.

"OIL-SOLUBLE" CHLOROPHYLL AS A SENSITIZER FOR PHOTOOXYGENATION

C. Tanielian and L. Golder, Laboratoire de Chimie Organique Appliquée, Institut de Chimie, Université Louis Pasteur, 1, rue Blaise Pascal, 67008, Strasbourg, France

Chlorophyll is one of the most effective sensitizers for photooxygenation of organic substrates (A) (1,2) although Chl was described as a singlet oxygen ($^1O_2$) quencher (3-5). Due to large amounts of time necessary to isolate and purify Chl a and b and the elevated cost of commercially available pure compounds, one is often led to use "oil-soluble" Chl (M) for preparative works.

It was the purpose of this study to show that this very cheap mixture, despite the presence of large amounts of compounds able to act either as diluent or as singlet oxygen quenchers (Q), can be used for kinetics studies, and to determine the rate constants of chemical and physical quenching of $^1O_2$ by M and Chl (in the case of Chl, only the upper limits were estimated).

Oil-soluble chlorophyll M was purchased from Carl Roth KG. The acceptors (A) tetramethylethylene (TME) and 2-methyl-2-pentene (M2P2) were obtained from Fluka. The photooxygenations were performed at room temperature in benzene solutions. The solutions were irradiated (lamp OSRAM HBO 200 W, glass filter MTO J 363a) in a reactor similar to that described by Gollnick et al. (6) and equipped with a device to follow oxygen consumption.

The oil-soluble chlorophyll-sensitized photooxygenation mechanism can be presented by the following scheme

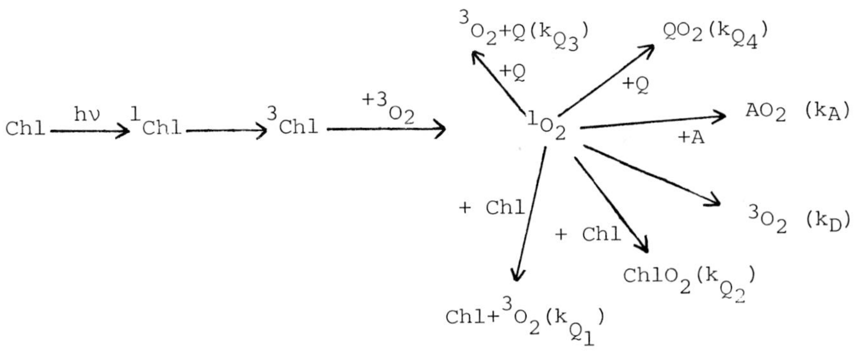

Then, the quantum yield of oxygen consumption is given by

$$\phi_{O_2} = [\gamma_\Delta (k_A[A] + k'_Q[M])] \;/\; [k_A[A] + k_D + k_Q[M]]$$

where γ_Δ is the quantum yield of $^1O_2$ formation, $k'_Q[M] = k_{Q_2}[Chl]$

ISBN 0-12-592050-4

$+ k_{Q4}$ [Q] and $k_Q[M] = (k_{Q1} + k_{Q2})$ [Chl] + $(k_{Q3} + k_{Q4})$ [Q].

When $[\phi O_2]^{-1}$ is plotted against $[A]^{-1}$, straight lines result for each [M]. As $k_A [A] >> k'_Q [M]$ (see later), the ratio of slope to intercept give $\beta = k_D / k_A + k_Q[M] / k_A = \beta_o (1 + k_Q\tau_\Delta[M])$ where $\beta_o = k_D/k_A$ is the "true" β value and $\tau_\Delta = 1/k_D = 25 \times 10^{-6}$ s, the lifetime of $^1O_2$ in benzene (7).

Plots of β vs [M] give straight lines (Fig. 1a and b for M2P2 and TME respectively) for which the intercept gives β_o and the ratio slope/intercept is equal to $k_Q\tau_\Delta$. In the case of M2P2 we find $\beta_o = (4.4 \pm 0.4) \times 10^{-2}$ M,

$k_A = 9.2 \times 10^5 M^{-1}s^{-1}$ $k_Q\tau_\Delta = 1.33 \pm 0.20$ l per g of M

$k_Q = 5.3 \times 10^4$ $l.s^{-1}$ per g of M and for TME $\beta_o = (2.24 \pm 0.08) \times 10^{-3}$ M

$k_A = 1.8 \times 10^7 M^{-1} s^{-1}$, $k_Q\tau_\Delta = 1.35 \pm 0.09$ l per g of M

$k_Q = 5.4 \times 10^4$ $l.s^{-1}$ per g of M. It must be noticed that the determination of $k_Q\tau_\Delta$ value permits to obtain the true β_o value from the measure of β for any [M].

The measure of $\phi' O_2$, the quantum yield of O_2 consumption in the absence of A, permits the determination of k'_Q from the relationship:

$$k'_Q[M] = k_A[A] \Big/ \left[\frac{\phi O_2}{\phi' O_2} \left(1 = \frac{[A]}{\beta}\right) - 1 \right]$$

For [M] = 3 g l^{-1} we find $k'_Q \simeq 1.2 \times 10^3$ and 1.6×10^3 $l.s^{-1}$ per g of M with TME and M2P2 respectively. Thus the chemical quenching of $^1O_2$ by M is very weak with respect to the physical quenching.

From the values of k_Q and k'_Q and taking a value of 5% for the percentage of Chl in M (% Chl was estimated as lying between 5 and 9%), the upper limit for the rate constants k_{Q1} and k_{Q2} of physical and chemical quenching of $^1O_2$ by Chl can be evaluated (no other $^1O_2$ quencher is present in M, i.e. [Q] = 0). We find $k_{Q1} < 10^9$ $M^{-1}s^{-1}$ and $k_{Q2} < 3 \times 10^7$ $M^{-1}s^{-1}$. These values are consistent with those given by Foote et al. (4) ($k_{Q1} + k_{Q2} < 10^9$ $M^{-1}s^{-1}$), Koka and Song (8) (k_{Q2} = (1.2-2) x 10^7 $M^{-1}s^{-1}$) and Krasnovshy (5) ($k_{Q1} + k_{Q2} = 7 \times 10^8$ $M^{-1}s^{-1}$, $k_{Q2} = 4 \times 10^6$ $M^{-1}s^{-1}$).

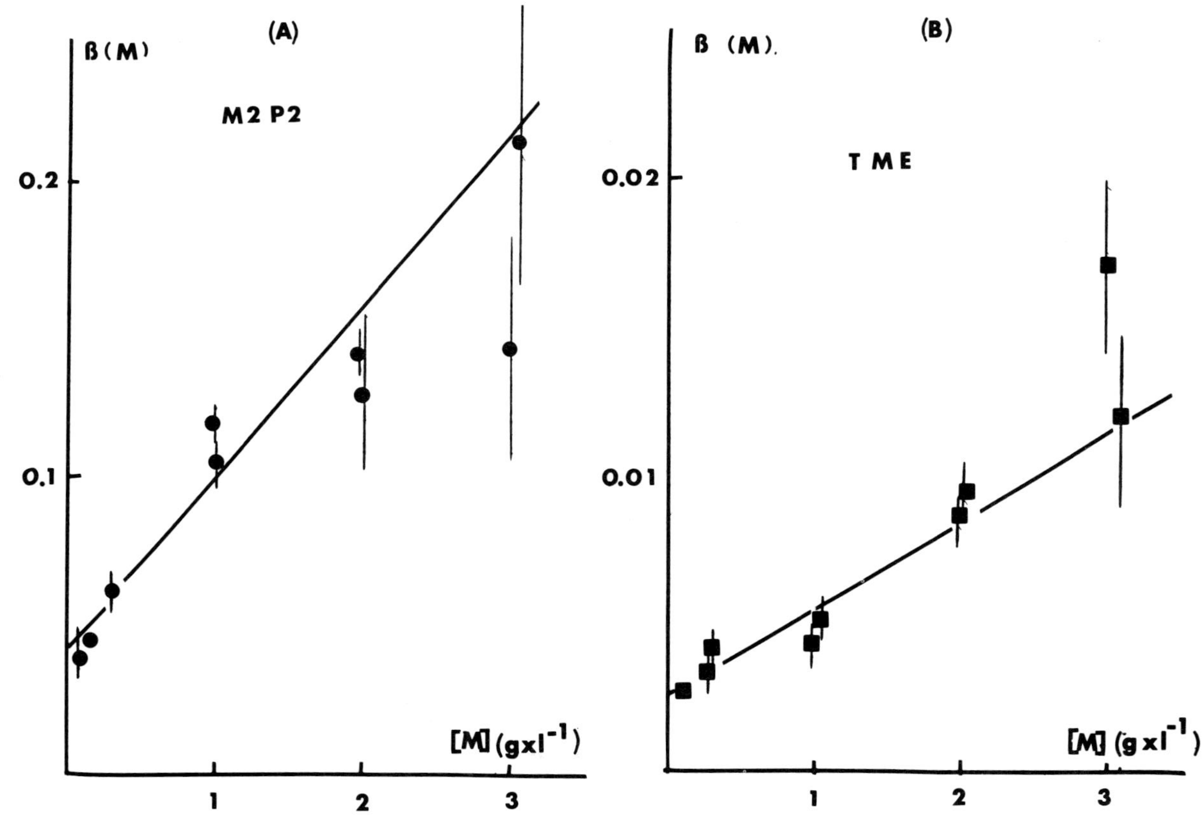

FIGURE 1. Variation of β with oil-soluble chlorophyll concentration: (a) 2-methyl-2-pentene; (b) tetramethylethylene.

REFERENCES

1. Gollnick, K., Adv. Photochem., 6,1 (1968).
2. Foote, C. S. and Denny, R. W., J. Am. Chem. Soc., 93,5168 (1971).
3. Livingstone, R. and Owens, K. E., J. Am. Chem. Soc., 78,3301 (1956).
4. Foote, C. S., Denny, R. W., Weaver, L., Chang, Y. and Peters, J., Ann. N. Y., Acad. Sci., 171,139 (1970).
5. Krasnovsky, A. A., Jr., Photochem. Photobiol., 29,29 (1979).
6. Gollnick, K., Franken, T., Schade, G. and Dörhöfer, Ann. N.Y. Acad. Sci., 171,89 (1970).
7. Wilkinson, F., "Singlet oxygen reactions", (B. Ranby and J. F. Rabek eds.), Wiley, New York, (1978), p. 27.
8. Koka, P. and Song, P. S., Photochem. Photobiol., 28,509 (1978).

ACUTE TOXICITY OF ACTIVE OXYGEN SPECIES ON HUMAN GLIA CELLS

H. H. Thaw,[1] *R. F. Del Maestro,*[2] *E. Arro,*[1] *U. T. Brunk,*[1] *and K-E. Arfors* [1]

[1]*Experimental Medicine, Pharmacia AB, Box 181, S-751 04 Uppsala, Sweden and University of Uppsala, Sweden*
[2]*Laboratory for Experimental Brain Research, Victoria Hospital, University of Western Ontario, London, Canada*

Impeded capacity to handle activated chemical species derived from molecular oxygen has been suggested as an important mechanism in pathologically induced cellular damage (1,2,4). We have investigated the effects of exposing cultures of human glia cells to a flux of active oxygen species generated enzymatically during the aerobic oxidation of hypoxanthine by xanthine oxidase.

$$Enz\text{-}H_2 + 2O_2^{\bar{}} \longrightarrow Enz + 2H^+ + 2O_2^{\bar{}} \qquad (1)$$

$$2H^+ + 2O_2^{\bar{}} \longrightarrow O_2 + H_2O_2 \qquad (2)$$

$$H_2O_2 + O_2^{\bar{}} \longrightarrow {}^{\bullet}OH + OH^- + {}^1O_2 \qquad (3)$$

$$Enz\text{-}H_2 + O_2 \longrightarrow Enz + H_2O_2 \qquad (4)$$

The hypoxanthine-xanthine oxidase enzyme (Enz) model involves both a univalent (reaction 1) and divalent (reaction 4) pathway for oxygen reduction productin superoxide anion radical ($O_2^{\bar{}}$) and hydrogen peroxide (H_2O_2). Dismutation of $O_2^{\bar{}}$ (reaction 3), either spontaneously or as catalyzed by superoxide dismutases (SOD) may then occur, suggesting further interactions (reaction 3) (4) to produce hydroxyl radicals ($^{\bullet}OH$) and possibly singlet oxygen ($^1O_2$). Cells were observed and assessed by time lapse cinematography under light microscopy or electron microscopy, and biochemical studies. Controlled experiments were carried out at 37°C in a blind manner using SOD, cytochrome c^{3+}, catalase, l-methionine, dimethyl sulfoxide, mannitol, ethanol or dexamethasone as biochemical probes. Acute and irreversible morphological changes culminating in cellular death resulted following a one minute exposure of the cells to the hypoxanthine-xanthine oxidase system, whereas no observable changes were seen in appropriate control experiments. Morphological changes (Figs. 1,2) were characterized by: (i) irreversible cellular contraction (Fig. 1), (ii) altered lysosomal activity resulting in the formation of lysosomal aggregates (Fig. 2), (iii) increased plasmalemma ruffling, (iv)

ISBN 0-12-592050-4

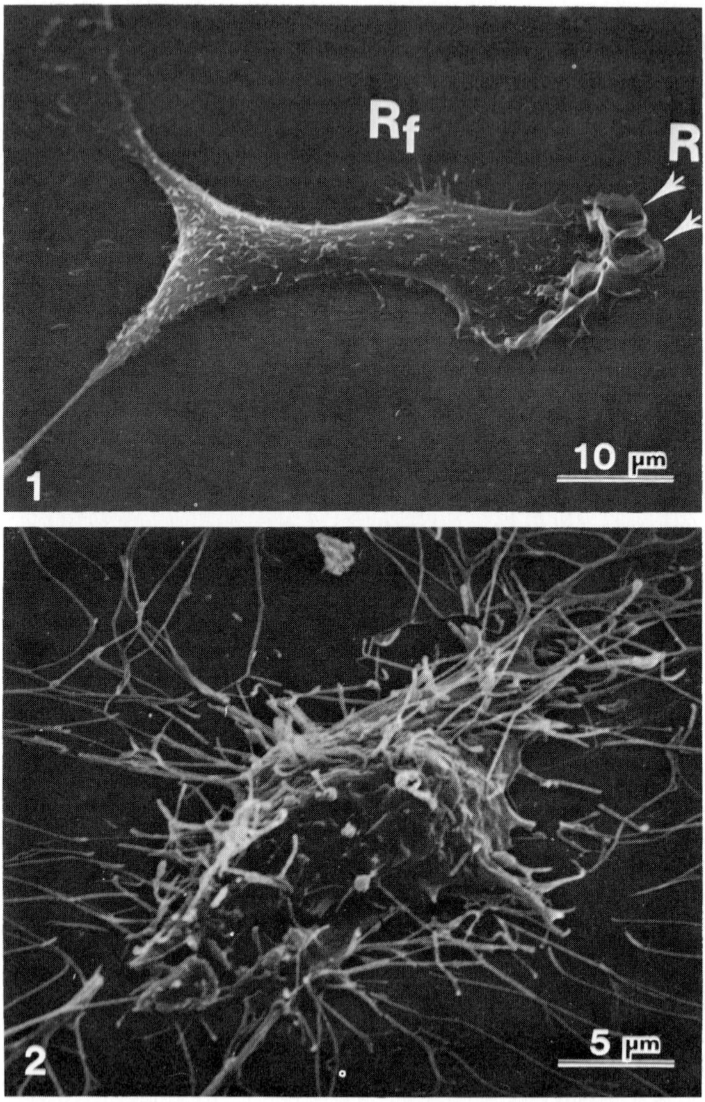

FIGURE 1. Scanning electron micrographs of human glia cells. (1) represents an actively growing cell characterized by a leading lamella with pronounced ruffling (R) activity and a moderate number of microvilli on the dorsal surface. The cell has recently changed direction as is indicated by the retraction fibrils (Rf). Typical appearance of a cell (2) at 120 min following a 1 min exposure to the hypoxanthine-xanthine oxidase system. Cellular contraction, rounding up and the consequent formation of

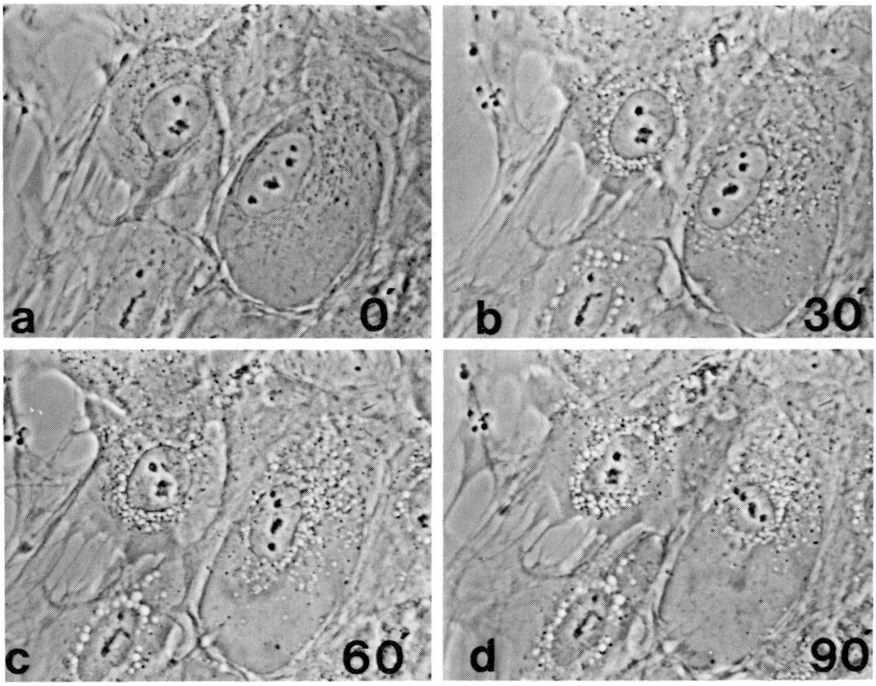

FIGURE 2. Time lapse micrographs of human glia cells at 30 min intervals following a 1 min exposure to the hypoxanthine-xanthine oxidase system. The sequence shows progressive cellular alterations with the formation of large vacuoles often surrounded by secondary lysosomes (small black dots).

increased endocytotic vacuole formation (Fig. 2), and (v) loss of cellular integrity followed by cellular death at 120-180 minutes. Enzymatic assessment of membrane stability provided results that were consistent with the time lapse observations, suggesting that an oxidizing species such as the hydroxyl radical, or a species derived therefrom is the toxic agent.

Our results demonstrate that the exposure of human glia cells 'in culture' to enzymatically generated free radicals derived from oxygen by the hypoxanthine-xanthine oxidase system results in acute and irreversible morphological changes. The cellular damage appears to be mediated through changes in membrane stability and the species responsible is suggested to be $^{\cdot}OH$, or a derivative thereof.

---

numerous retraction fibrils around the cell periphery were representative of the alterations occurring.

REFERENCES

(1) Del Maestro, R. F., Arford, K. -E., "Fourth International Symposium on EC-IC Bypass," Springer-Verlag, New York (in press).

(2) Fridovich, I., "Advances in Neurology," Vol. 96, Raven Press, New York, N.Y., pp. 225.

(3) Haber, F., Weiss, J., Proc. R. Soc. Ser. A., 147, 332 (1934).

(4) Haliwell, B., New Physiologist, 73, 1075 (1974).

A NEW PROCEDURE FOR DETECTING THE PARTICIPATION OF SINGLET OXYGEN IN BIOLOGICAL PEROXIDATIONS

Michael J. Thomas[1] *and William A. Pryor*[2]

[1]*Department of Biochemistry, Bowman Gray School of Medicine, Wake Forest University, Winston-Salem, North Carolina 27109*
[2]*Department of Chemistry, Louisiana State University, Baton Rouge, Louisiana 70803*

Singlet oxygen ($^1O_2$) has been reported to participate in several enzyme catalyzed oxidations. Typically the formation of products by well-characterized $^1O_2$ acceptors that are added to the oxidizing mixture is used as proof of $^1O_2$ participation. However, there is a severe drawback inherent in interpreting the results derived from those molecules frequently used to probe biological oxidations. Most of these acceptors yield only one product with $^1O_2$ and unfortunately this product can be formed by free radical autoxidation. A sensitive probe for $^1O_2$ should: yield at least one product that is easily quantitated, not be formed by autoxidation, and be compatible with the biological system under study. Our interest in polyunsaturated fatty acid (PUFA) oxidation led us to develop a sensitive and easily quantitated test for the participation of $^1O_2$ that uses linoleic acid (a typical PUFA) as the probe molecule.

Dye-sensitized photooxidation of linoleic acid yields four products in approximately equal amounts (1): two conjugated diene hydroperoxides 13-hydroperoxy-cis-9,trans-11-octadecadienoic acid (13-OOH) and 9-hydroperoxy-trans-10,cis-12-octadecadienoic acid (9-OOH) and two nonconjugated diene hydroperoxides 12-hydroperoxy-cis-9,trans-13-octadecadienoic acid (12-OOH) and 10-hydroperoxy-trans-8,cis-12-octadecadienoic acid (10-OOH). In contrast, free radical autoxidation of linoleic acid yields only the conjugated diene hydroperoxides 13-OOH and 9-OOH and their trans,trans analogues (2). Our procedure allows us to readily monitor formation of 10-OOH (as its alcohol 10-OH) in a complex, aqueous mixture. The hydroperoxides are extracted from the acidified (pH 2), aqueous solution, reduced with methanolic $NaBH_4$, converted to their p-bromophenacyl esters (3), and then separated by high pressure liquid chromatography. We are able to detect the formation of nanogram quantities of oxidized linoleate by monitoring the absorption of the p-bromophenacyl moiety at 254 nm. This procedure has been applied to the linoleic acid oxidations induced or catalyzed by xanthine oxidase and soybean lipoxidase.

Xanthine oxidase is a superoxide anion radical ($O_2^{\overline{\cdot}}$) generating enzyme that utilizes various purines and aldehydes as substrates. Superoxide from this source is known to induce the

ISBN 0-12-592050-4

autoxidation of polyunsaturated fatty acids (PUFA), e.g. linoleic acid (4). Several oxygen species, including $^1O_2$ (4), have been proposed to participate in the peroxidation. To test for the participation of $^1O_2$ in this oxidation a 0.05 M phosphate buffer solution (pH 8.1) containing 50 mM acetaldehyde, 1 mM linoleic acid (contaminated with 33 μM linoleate peroxide), 0.1 mM diethylenetriaminepenta-acetic acid, 32.6 μM 12-hydroxystearic acid (as an internal standard), 2.1 M ethanol, and 48 nM xanthine oxidase was sparged with oxygen for 115 min at 25°C. Ultraviolet analysis (235 nm) showed that 0.58 μmoles of conjugated hydroperoxide had been formed. The solution was acidified with 5 N HCl, extracted with 1:1 ether-hexane, and the carboxylic acid residues derivatized. Analysis of the mixture on a 3.9 mm by 30 cm μ-Porasil column using 2-propanol:chloroform:hexane (3/100/897) as the eluant and a flow rate of 4 ml/min showed no 10-OH, indicating that 10-OOH constituted less than 1.2% of the oxidation products, and therefore that less than 5% of the oxidation proceeded via $^1O_2$. Other experiments showed that 10-OOH, prepared by methylene blue sensitized photooxidation of linoleic acid, was not selectively destroyed by xanthine oxidase. Although it is unlikely that any significant fraction of linoleate oxidation proceeds via $^1O_2$, these experiments do not rule out the possibility that significant fraction $O_2^{\cdot-}$ may disproportionate to $^1O_2$ and H_2O_2. However, others (5) have recently reported convincing evidence that $O_2^{\cdot-}$ disproportionation does not produce significant quantities of $^1O_2$.

Soybean lipoxidase will selectively oxidize linoleic acid to 13-OOH. Since this product is also formed with $^1O_2$, several researchers have suggested that the enzyme might produce $^1O_2$. Recent studies using the cholesterol probe have shown that $^1O_2$ is not released into solution during the oxidation of linoleate (6). However, if $^1O_2$ were formed, its concentration at the active site of the enzyme would be higher than in solution. Using linoleic acid as the probe molecule might allow the detection of $^1O_2$ involvement in PUFA oxidation near the enzyme surface. The oxidation was performed at 0°C in oxygen-sparged 0.05 M phosphate buffer (pH 9) containing 0.88 mM linoleic acid and 6 x 10^5 units of lipoxidase. After 12 min the solution was acidified and worked up as above. The hydroxylated material appeared to consist exclusively of 13-OH and no 10-OH or 12-OH could be detected. From this evidence we conclude that free $^1O_2$ does not participate to a significant extent in the peroxidation of linoleic acid by soybean lipoxidase.

REFERENCES

1. Thomas, M. J. and Pryor, W. A., Lipids, accepted for publication (1980).
2. Chan, H. W.-S. and Levett, G., Lipids, 12, 99 (1977).
3. Borch, R. E., Analyt. Chem., 47, 2437 (1975).
4. Thomas, M. J., Mehl, K. S. and Pryor, W. A., Biochem. Biophys. Res. Commun., 83, 927 (1978) and references therein.
5. Foote, C. S., Shook, F. C. and Abakerli, R. B., J. Am. Chem. Soc., 102, 2503 (1980).
6. Teng, J. I. and Smith, L. L., J. Am. Chem. Soc., 95, 4060 (1973).

HYDROPEROXYSTEROLS AS A PROBE FOR THE CYTOCHROME P-450$_{scc}$ HEME POCKET

J. E. van Lier[1] *and C. Larroque*[2]

[1]*University of Sherbrooke Medical Center, Sherbrooke, Quebec, Canada*
[2]*INSERM, U.128, B.P. 5051, Montpellier, France*

INTRODUCTION

The adrenocortical enzyme system responsible for the side-chain cleavage of cholesterol to give pregnenolone consists of cytochrome P-450$_{scc}$ and an electron transport system (1). The sequence of reactions during this bioconversion is believed to proceed through three consecutive in situ oxidations of the C22-H, C20-H and C20-C22 bonds via a ferryl-atomic oxygen complex of P-450$_{scc}$ (2). We have previously studied this reaction by the interaction of the epimeric 20-hydroperoxy derivatives of cholesterol and purified P-450$_{scc}$. In this system these hydroperoxysterols serve both as donor of the reduced oxygen atom to yield the reactive ferryl-oxide and as substrate resulting in a rapid, quantitative and stereospecific hydroxylation of the carbon atom adjacent to the hydroperoxyl group (3). The selectivity of the intramolecular hydroperoxy-glycol rearrangements suggested a restricted conformation of the cholesterol side-chain at the active site of the enzyme. In an attempt to evaluate the relative importance between the forces involved in substrate-binding site and hydroperoxy-ferric heme iron interactions, we have studied the P-450$_{scc}$ mediated decomposition of a number of cholesterol analogs containing hydroperoxy groups at various positions of the sterol side-chain.

MATERIALS AND METHODS

Reference steroids were purchased or prepared by established methods. Bovine andrenocortex mitochondrial cytochrome P-450$_{scc}$ was purified to homogeneity as described by Suhara et al. (4). Heme content was calculated by difference spectroscopy (4). The 25-hydroperoxycholesterol was isolated from autoxidized cholesterol. The synthesis of the remaining hydroperoxysterols was accomplished by peroxidation of the 22- and 24-ketocholesterols; details of the preparations will be presented elsewhere. The hydroperoxysterols were dissolved in methanol (0.5 mM) and stored at -60°C prior to use. Purity was verified by thin layer chromatography (TLC). The interaction between equimolar amounts of P-450$_{scc}$ and hydroperoxysterols was followed by difference

ISBN 0-12-592050-4

spectroscopy and repetitive scanning between 350-500 nm. After no further spectral changes occurred the sample was extracted with methylene chloride and analyzed by TLC or silylated for gas chromatographic (GC) and mass spectral (GC-MS) analysis.

RESULTS AND DISCUSSION

Addition of 5 nmol hydroperoxysterol in 10 µl methanol to an equimolar amount of P-450_{scc} in 2 ml 0.1 M phosphate buffer pH 7.4 at room temperature, lead for all substrates studied to the rapid formation of a single major product. Spectral changes suggested that half of the hydroperoxysterol is transformed within 10 min although some of the products exhibited insufficient induced spectral changes to allow for quantitative analysis. TLC analysis of the methylene chloride extracts obtained after 50 min of incubation indicated the absence of hydroperoxysterols for all substrates studied. The products were identified by comparing their chromatographic and mass spectral properties with authentic material. The 25-hydroperoxycholesterol was converted to the corresponding 25-hydroxycholesterol whereas the 20-hydroperoxy-22-ketosterol gave pregnenolone. The remaining hydroperoxyketocholesterols were converted to their analogues sterol acids via cleavage of the C-C bond between the hydroperoxy and keto groups. Accordingly, the 25-hydroperoxy derivative of 24-ketocholesterol gave cholenic acid, the corresponding 23-hydroperoxysterol gave 23-norcholenic acid and the 23-hydroperoxy derivative of 22-ketocholesterol gave 22,23-bisnorcholenic acid. The rapid transformation of these hydroperoxysterols, regardless of the position of the hydroperoxy substituent, suggests that the interaction between the hydroperoxy groups and the ferric heme iron dominates those interactions involved in sterol attachment to the binding site within the heme pocket. All studies were conducted at room temperature, under which condition the putative ferryl-oxide intermediate eludes detection by spectroscopic means. We are presently evaluating the use of mixed solvent systems and low temperatures for the trapping and visualization of such transient radical intermediates.

REFERENCES

1. Boyd, G. S. and Simpson, E. R., "Function of the Adrenal Cortex", (K. W. McKern, ed.), Academic Press, New York, (1968), p. 49.
2. van Lier, J. E. and Rousseau, J., FEBS Letters, 70, 23 (1976).
3. van Lier, J. E., Rousseau, J. Larroque, C. and Morel, P., "Microsomes, Drug Metabolism and Chemical Carcinogenesis," (M. Coon et al., eds.), Academic Press, New York, N. Y., Vol. I (1980), p. 229.

CELLULAR ACTIVATION OF NITRO COMPOUNDS TO OXYGEN- AND THIOL-REACTIVE INTERMEDIATES

Marie E. Varnes, John E. Biaglow[1] *and Peggy L. Olive*[2]

[1]*Division of Radiation Biology, Case Western Reserve University, Cleveland, Ohio 44106*
[2]*Radiobiology Section, Johns Hopkins Oncology Center, Baltimore, Maryland 21205*

Aerobic reduction. Nitro heterocycles are reduced by dense suspensions of aerobic cells to intermediates that react with oxygen. Superoxide is formed by one-electron transfer from drug radical to oxygen, and peroxide is produced from dismutation of superoxide (1). Peroxide formation can be measured continuously in intact cells using a Clark oxidase electrode (2). A comparison of initial rates of peroxide production in L-929 cells treated with drugs that vary in their half-wave reduction potentials ($E^{1/2}$), rates of cellular reduction and subsequent toxic effects (3) is shown in Table I. Except for niridazole, there is good correlation between $E^{1/2}$ and rates of both oxygen consumption and peroxide production. A plot of $E^{1/2}$ vs log nmol H_2O_2/min (0.1 mM drug) gives a straight line with a slope parallel to the slope of $E^{1/2}$ vs log anaerobic reduction rate (Column 5, Table I). This suggests that the rate-limiting steps for both aerobic and anaerobic drug metabolism are the same. However, we have not as yet quantitated overall R-NO_2-linked electron transfer under aerobic vs anaerobic conditions.

Peroxide production is dependent upon the presence of glucose, which through the pentose cycle provides NADPH for drug reduction. The extent of peroxide formation also varies with addition to the cells of sodium azide, a catalase inhibitor, and NEM, which removes glutathione (GSH) necessary for reduction of peroxide via the GSH-peroxidase-reductase enzyme systems. Our data indicate that GSH peroxidase rather than catalase is more important for peroxide reduction in L-929 cells. For this reason, only 4-NQO, which reacts with intracellular NPSH under aerobic conditions (Column 4, Table I), is able to cause H_2O_2 production in the absence of NEM when used at concentrations as low as 0.1 mM. However, peroxide production in the absence of NEM can occur with the more electron-affinic nitro compounds, such as AF-2 and nitrofurazone, when these are used at concentrations of 1.0 mM and above.

Anaerobic reduction. Under anaerobic conditions nitro heterocyclic compounds are reduced to intermediates that bind to macromolecules (4). We have found that 4-NQO and misonidazole react with NPSH of Ehrlich and V79 cells under hypoxic

ISBN 0-12-592050-4

TABLE I. Peroxide Production, Oxygen Consumption and NPSH Loss in L-929 Cells Treated with Nitro Compounds

Compound[1]	$E^{1/2}$[2]	H_2O_2 formed (mnol/min/10^7 cells)	O_2 lost[3] (mnol/min/10^7 cells)	NPSH remaining % control (aerobic)	Reduction rate relative to NF	NPSH remaining % control (anaerobic)
					1 mM drug	0.5 mM drug
4-NQO	-195	9.10 ± 0 (0.01 mM)		11 ± 2	13.3	0
AF-2	-255	8.80 ± 0.1 (0.05 mM)	10.9 ± 1.0 (0.05 mM)	103 ± 9	1.54	55
Nitrofurazone	-282	6.60 ± 0	14.2 ± 0.2	99 ± 3	1.0	60
Fanft	-310	1.50 ± .04	4.8 ± 0.3	90 ± 4	0.65	78
Misonidazole	-395	0.20 ± .01	1.6 ± 0.3	103 ± 6	0.06	76
Niridazole	-405	1.10 ± .03	2.7 ± 0.5	102 ± 11	0.03	56

[1]*Abbreviations: 4-NQO; 4-nitroquinoline-1-oxide; AF-2; 2,2-(2-furyl)-3-(5-nitro-2-furyl)acrylamide; FANFT; N-(4-5-nitro-2-furyl)-2-thiazolyl)-formamide; NF; nitrofurazone; NEM; N-ethylmaleimide; NPSH; non-protein sulphydryls, principally glutathione.*

[2]*Half-wave reduction potentials were determined as described in (3).*

[3]*For all measurements, cells were incubated at 10^7/ml in 0.9% saline containing 0.05 M phosphate, pH 7.25. Specific conditions for each measurement were as follows: H_2O_2; 1 mM glucose, 0.05 mM sodium azide, 0.2 mM NEM. These conditions gave maximal peroxide production for nitrofurazone. O_2 consumption; 1 mM glucose, 1 mM KCN as inhibitor of both catalase and respiration. NPSH loss; 20 min incubation, 5 mM glucose, 10^{-7}M rotenone to inhibit respiration for aerobic incubations. Control NPSH was 39 ± 4 x 10^{-7} nmol/cell. Relative reduction rate; as described in (3) except 5 mM glucose added. Drug concentrations were 0.1 mM except where indicated.*

conditions (5,6). This reaction, which may partially protect against binding, occurs with all of the compounds listed in Table I. Currently we are investigating the influence of removal of endogenous thiols on drug binding to macromolecules.

ACKNOWLEDGMENTS

This research was supported by Research Grants CA-13747 and CA-24519 awarded by the National Cancer Institute.

REFERENCES

1. Biaglow, J., Varnes, M., Koch, C., and Sridhar, R., "Free Radicals and Cancer," (R. Floyd, ed.), Marcel Dekker, New York, N. Y. (1980). In press.
2. Schroy, C. B., and Biaglow, J., Fed. Proc., 39, 1751 (1980).
3. Olive, P. L., Cancer Res., 39, 4512 (1979).
4. Olive, P. L., "Carcinogenesis, Vol. 4: Nitrofurans," (G. Bryan, ed.), Raven Press, N. Y., (1978), p. 131.
5. Varnes, M. and Biaglow, J., Cancer Res., 39,2960 (1979).
6. Varnes, M., Biaglow, J., Koch, C. and Hall, E., "Radiation Sensitizers", (L. Brady, ed.), Masson Publishing USA, N. Y., (1980) p. 121.

FREE RADICAL-INDUCED TRANSFORMATIONS OF OXYMYOGLOBIN IN AQUEOUS SOLUTION

Kevin D. Whitburn, James J. Shieh, Morton Z. Hoffman,[1] *and Irwin A. Taub*[2]

[1]*Department of Chemistry, Boston University, Boston, Massachusetts*
[2]*Food Engineering Laboratory, U.S. Army Natick Research and Development Command, Natick, Massachusetts 01760*

The radiation-induced changes in the prosthetic group of oxymyoglobin (MbO_2) have been systematically studied in neutral aqueous solution. Compositions of each myoglobin derivative were determined before and after γ-radiolysis by spectral analysis in the visible region, using accurately predetermined molar absorbance data. In order to base mechanistic interpretation on radiation-induced composition changes, a slow thermal post-irradiation process involving H_2O_2 ($t_{1/2}$∿10-20 mins) was extrapolated to zero radiolysis time.

In N_2O-saturated solution,

$$MbO_2 + OH/H \begin{cases} \rightarrow metMb, \; G \sim 3.5 \\ \rightarrow Mb, \; G \sim 2.8 \end{cases}$$

At high doses metMb/Mb radiolysis reactions dominate. In Ar-purged solutions, after making allowance for OH/H-induced transformations, we have

$$MbO_2 + e^-_{aq} \begin{cases} \rightarrow metMb, \; G \sim 2.3 \\ \rightarrow Mb, \; G \sim 0.5 \end{cases}$$

The presence of 0.3 mM O_2 in N_2O solutions of MbO_2 severely depresses the OH-induced conversion efficiency. In O_2-saturated 0.02M HCO_2^- solutions, O_2^- induces negligible MbO_2 conversion.

From known reactivity trends with amino-acid/peptide entities (1,2), the sites of radical reactivity over the metalloprotein, other than the buried prosthetic group, can be reasonably deduced. For OH/H the predominant radicals formed are various peptide H-abstraction and ring addition species. The observed $MbO_2 \longrightarrow$ metMb conversion and an indirect $\{H_2O_2\}$ analysis are consistent with intramolecular electron-transfer from the globin radicals to the oxyferroheme, with redox participation of the bound O_2. The mechanism, presented in Scheme 1, is compatible with features of the cyt(II)-c/e^-_{aq} (3) and cyt(III)-c/OH (4) systems. Those protein radicals unable to initiate one-equivalent transfer may be scavenged by the labile O_2 of the substrate, or the primary product, to generate Mb or Mb-like material. Added oxygen scavenges most of the globin radicals.

ISBN 0-12-592050-4

SCHEME 1

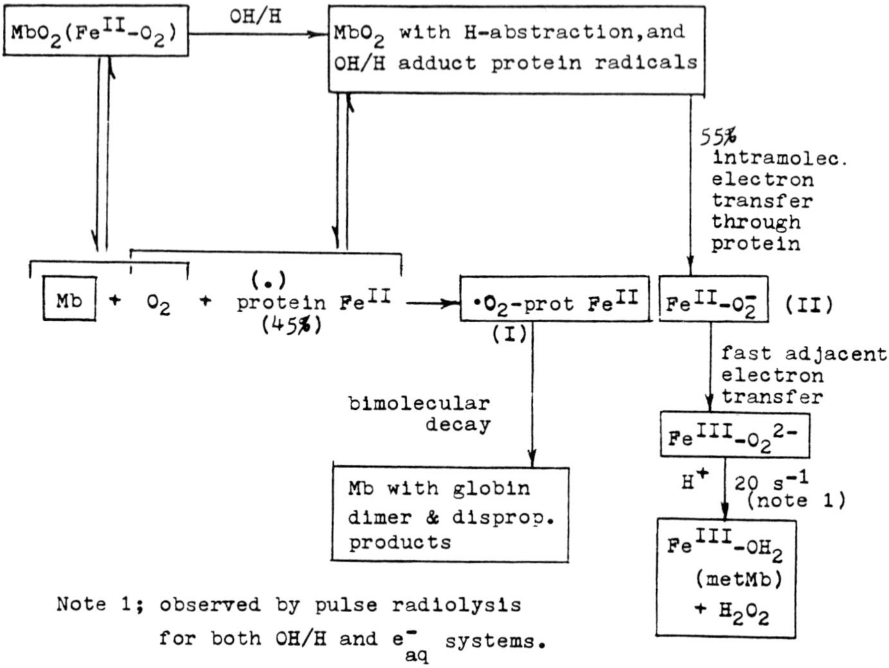

Protonated (2) e^-_{aq}-adducts to peptide carbonyls and ring residues are efficient initiators of MbO_2 redox conversion to metMb; they are inefficient scavengers of the labile O_2, however. Apart from the nature of the initial dispersed reactivity and the different internal transfer efficiency, the MbO_2/e^-_{aq} system shares steps common to the OH/H mechanism.

Given the radiolysis conditions, dose, and time of analysis after irradiation, the overall radiation-induced transformations of MbO_2 can be estimated from G-value and thermal rate data.

REFERENCES

1. Simic, M. G., J. Agric. Food Chem., 26,6 (1978).
2. Taub, I. A., Halliday, J. W. and Sevilla, M. D., Adv. in Chem. Ser., 180, Chapt. 6 (1979).
3. Butler, J., DeKok, J., DeWeille, J. R., Koppenol, W. H. and Braams, R., Biochem. Biophys. Acta, 459,207 (1977).
4. Shafferman, A. and Stein, G., Biochem. Biophys. Acta, 416, 287 (1975).

CONFIGURATION INTERACTION ANALYSIS OF A LUMINESCENT DIOXETANE REACTION PATHWAY

Martin D. Williams[1] *and J. Lawrence Fox*[2]

[1]*Department of Biochemistry and Biophysics, University of Pennsylvania School of Medicine, Philadelphia, Pennsylvania*
[2]*Department of Zoology, University of Texas at Austin, Austin, Texas 78712*

This study employed the stereospecificity of the firefly luminescence reaction and other dioxetane reactions to determine which of the energetically allowed trajectories is also geometrically allowed since the L-isomer of luciferin does not react with oxygen after binding to the luciferase (1). Calculations were performed by successive application of Hubbard two-center electron repulsions to the Hückel integrals which were augmented by an empirically derived electronegativity parameter and an exchange parameter to provide the ground state oxygen triplet (2, 3). These results were utilized in a configuration interaction calculation to produce state correlation diagrams with both Hückel and Möbius routes

In concerted dioxetane synthesis (Fig. 1a) the additions such as triplet oxygen to triplet ethylene might compete with other processes in the route to products by collisional (4), optical, resonance or enzymic excitation of the ethylenic bond. The absence of a barrier to a singlet oxygen reaction with ground state ethylene on the Möbius route suggests a possible role for singlet oxygen (5-7). Moreover, the crossing of the ground state reaction with this trajectory, after a very small barrier is overcome, indicates possible competition between these routes to ground state dioxetane. The ground state route, however, does not distinguish between the Hückel and Möbius additions and so can not be the reaction leading to dioxetane (or zwitterionic hydroperoxide) formation in the firefly luminescence reaction.

The peroxirane synthesis reaction (Fig. 1b) can not be the discriminator between luminescent and non-luminescent routes because the Hückel and Möbius paths coincide in this simplistic model. The peroxirane rearrangement to dioxetane (Fig. 1c) seems rather unlikely on the basis of the existence of barriers to rearrangement by way of the six possible Hückel and Möbius singlet and triplet low-lying trajectories. Reactions involving rearrangement of peroxiranes must proceed through zwitterion formation (Fig. 1d).

The dioxetane zwitterion formation reaction (Fig. 1e) has a unique state correlation diagram. The ground state (thermal) reaction has an allowed trajectory that does not correlate with the

ISBN 0-12-592050-4

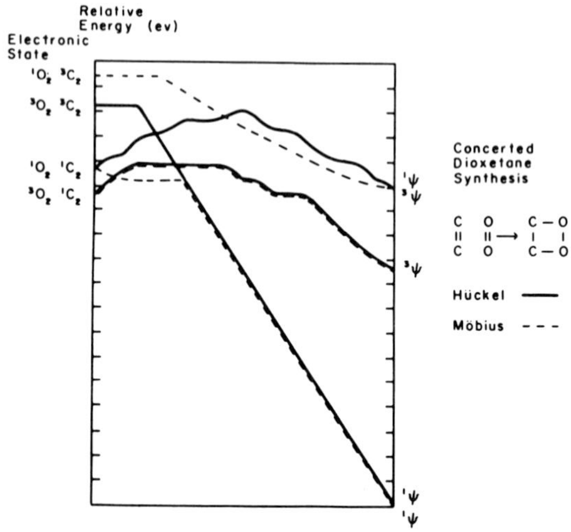

a

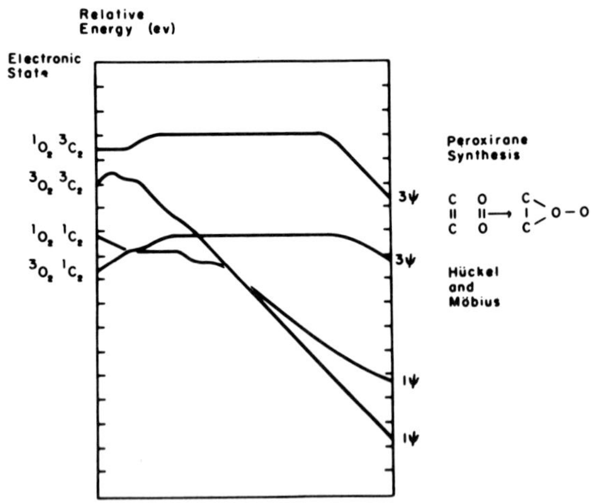

b

FIGURE 1. State correlation diagrams for proposed mechanisms showing Hückel and Möbius pathways. (a and b)

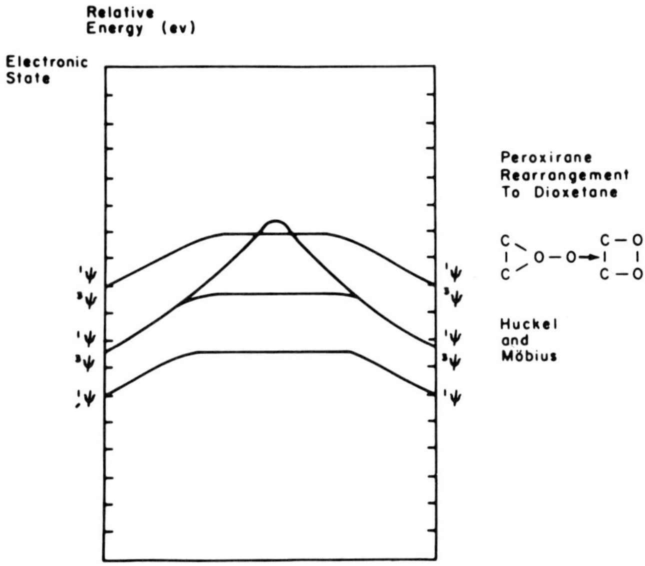

c

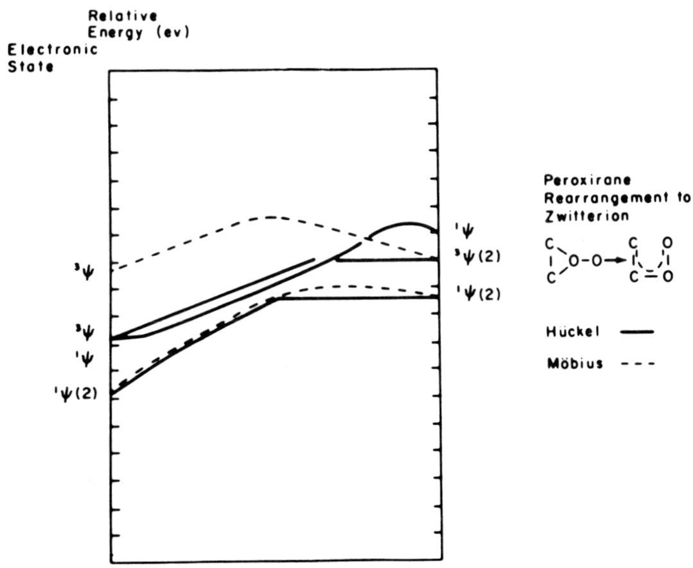

d

FIGURE 1. (continued) (c and d)

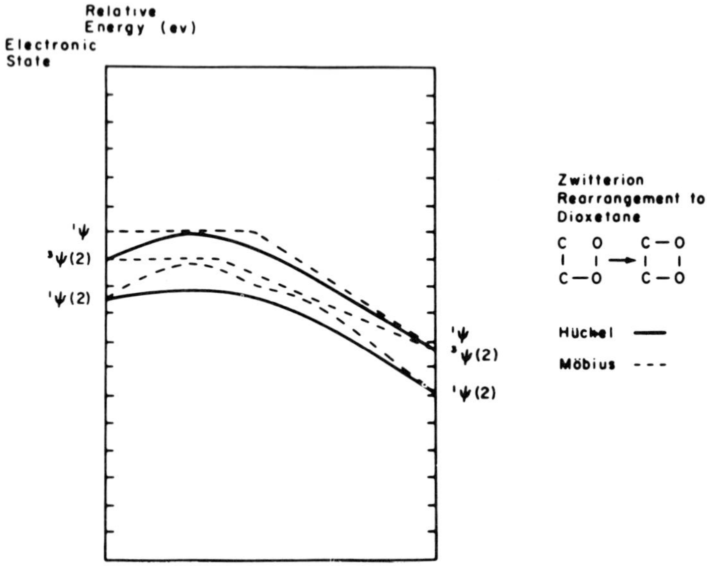

f

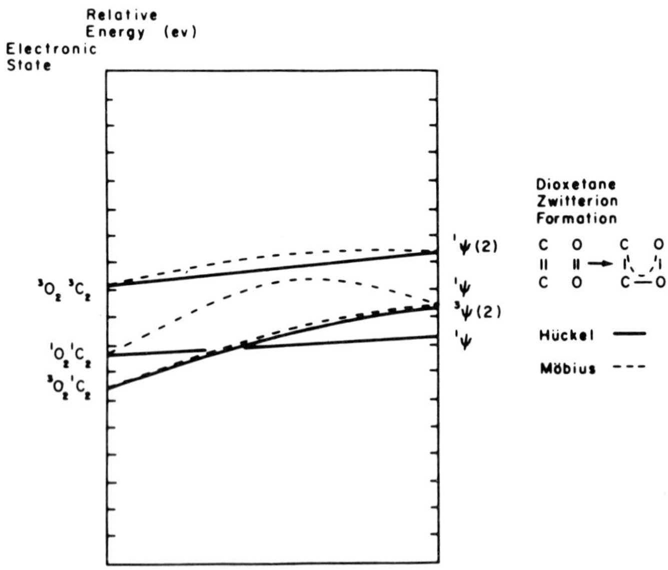

e

FIGURE 1. (continued) (e and f)

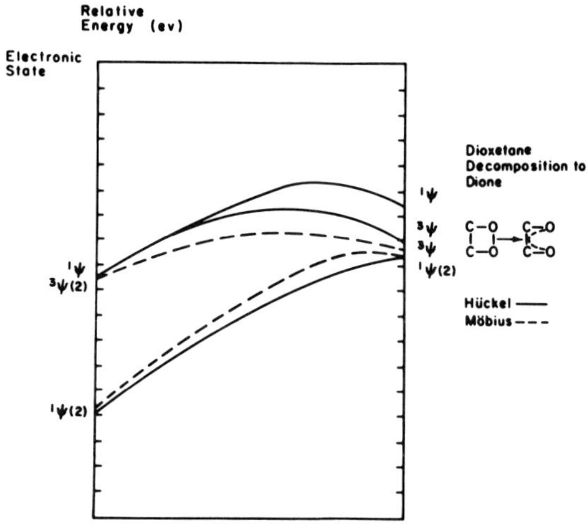

g

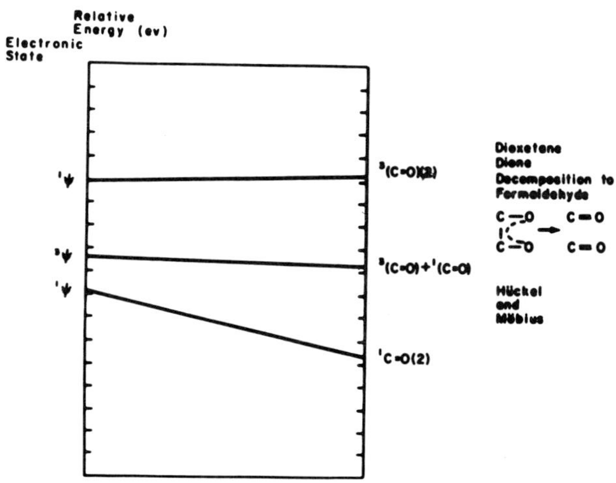

h

FIGURE 1. (continued) (g and h)

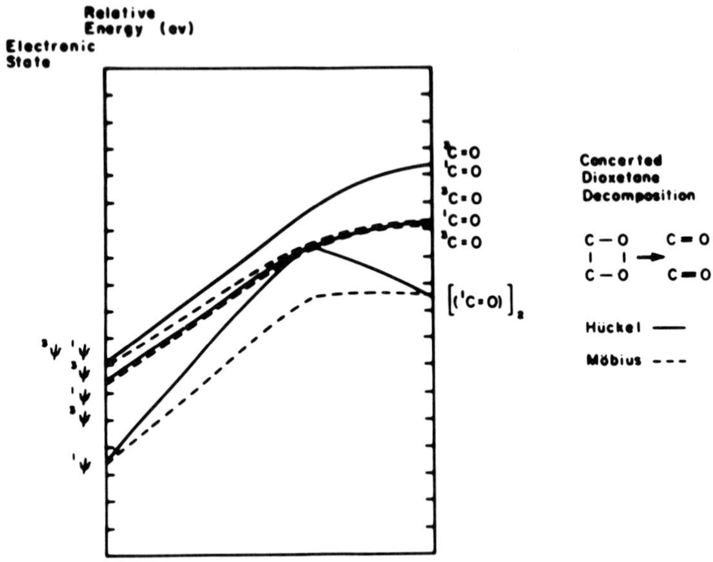

FIGURE 1. (continued) (i)

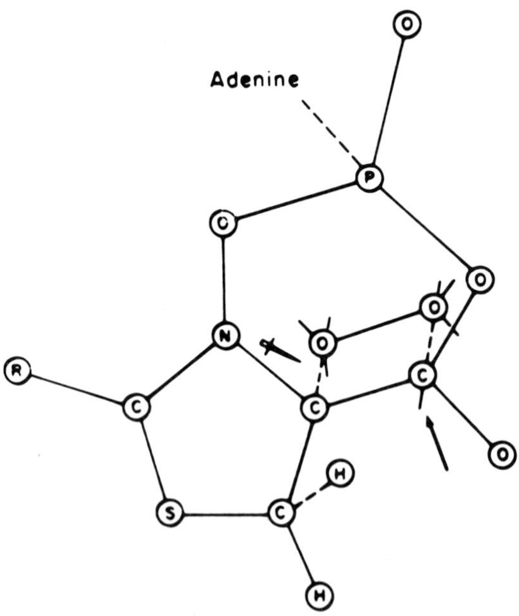

FIGURE 2. Partial model of the keto form of luciferin phosphate showing the orbital (dagger) which determines the molecule's stereospecificity and the lobe (arrow) for the reaction stereospecificity.

ground state of the product zwitterion. Triplet oxygen and singlet ethylene may add to yield the zwitterion in the excited (triplet) state. The reaction initiated by singlet oxygen and singlet ethylene has a trajectory, with practically no barrier, which correlates with the product ground state. This reaction could discriminate between Hückel and Möbius additions. The zwitterion rearrangement to dioxetane reaction (Fig. 1f) could discriminate between the D- and L- forms by either the ground state singlet trajectory or the first excited state triplet route.

The decomposition of dioxetane to the dione (Fig. 1g) has four trajectories: two each of Hückel, Möbius, singlet and triplet that correlate with dione states which lie very close to or at the ground state. If the dioxetane is formed in the ground state, the thermal reaction is allowed but would be favored by the conditions stabilizing the dione product. If the dione is formed in the excited (triplet) state, its decomposition might be the source of the triplet decay observed in dioxetane decomposition. However, luminescent decomposition of dioxetane is presently thought to be a thermolytic reaction. The excited state trajectory for dione decomposition (Fig. 1h) is not degenerate with the ground state of the dione intermediate. Thus, either the reaction does not pass through this intermediate and dioxetane is formed with degenerate ground and excited states with a substantial portion of molecules reacting via the excited state route, or there is an inefficient crossing leading to the dione. In this case, a substantial portion of the molecules reacting by the ground state remain on it despite the action of an intersecting trajectory leading to the ground state dione.

The concerted dioxetane dissociation reaction (Fig. 1i) would seem to be the most likely dissociation if the reactant dioxetane is formed in an excited state. Also the actual barrier height to the ground state Hückel route is very small if crossing occurs between it and the first excited state trajectory. This crossing to a triplet trajectory is favored statistically since it leads to a metastable state. The high energy appearance of the products is an artifact of the calculation in this case.

The oxygen atoms attached covalently to the phosphorus atom of luciferin may now interact with the ring nitrogen atom (Fig. 2). This results in formation of a six-membered ring and is, therefore, a very stable and highly probable configuration in the course of this reaction. This configuration may provide stereospecificity for this reaction. In D-luciferin the presence of this six-membered ring permits interaction of the p orbitals of both oxygen atoms with those of the two carbon atoms involved in dioxetane formation. In the L-form, however, the trigonal bipyramid (sp^2) configuration of the orbitals on the carbonyl carbon puts an orbital with phase opposite to that of the respective orbital of D-luciferin in the reactive position. The reaction with one form (D- or L-) would thus be Möbius in phase relations, the other Hückel.

Results of these calculations and examination of molecular models suggest that luciferyl-dioxetane or dioxetane formation probably proceeds by concerted addition of ground state (triplet) oxygen to an ethylenic bond raised to its first excited state by collision, optical feedback, enzyme activity or resonance interaction. Luciferyl-dioxetane dissociation probably proceeds via a ground state concerted route with crossing to the first excited triplet. In the firefly luciferin reaction the dioxetane intermediate may be stabilized by a six-membered ring completed by the association of a phosphate oxygen atom and the luciferyl nitrogen atom.

ACKNOWLEDGEMENTS

We wish to thank Drs. T. L. Welsher and F. A. Matsen for discussion and access to their computer program. We also wish to thank the N.I.H. for fellowship support to M.D.W. and research support to J.L.F.

REFERENCES

(1) McElroy, W. D., Seliger, H. H. and White, H. H., Photochem. and Photobiol., 10, 153 (1969).
(2) Matsen, F. A., Int. J. Quantum Chem., 10, 511 (1976).
(3) Seitz, W. A., Welsher, T. L., Yurke, B. and Matsen, F. A., J. Amer. Chem. Soc., 100, 4679 (1978).
(4) Evans, D. F., J. Chem. Soc., 1735 (1960).
(5) Kearns, D. R., Chem. Rev., 71, 395 (1971).
(6) Cilento, G., J. Theoret. Biol., 55, 471 (1973).
(7) Hastings, J. W. and Wilson, T., Photochem. Photobiol., 23, 461 (1976).

REACTIONS OF VITAMIN E WITH ALKYLPEROXY RADICALS IN HOMOGENEOUS SOLUTION AND LIPOSOMES

John S. Winterle and Theodore Mill, SRI International, Menlo Park, California 94025

We have determined the stoichiometry for interaction of alkylperoxy radicals and vitamin E (E) by measuring the loss of E under conditions where the absolute rates of peroxy radical production are known, using a well-characterized peroxy radical source. Liposomes prepared from di-lauroyl-phosphatidyl choline (DLPC) were used to provide a reaction matrix for E in pH 7 buffer. Hexane was used as a homogeneous solvent.

Peroxy radicals were produced by thermolysing azobis(2-n-butyl-carboxypropane) (ABCP) in the above aerated media (10^{-2} M in hexane, 10^{-4} M in DLPC) at 50.0 ± .1°C. ABCP and E are sufficiently lipid soluble to be quantitatively sequestered in the lipid bilayers of DLPC. ABCP decays according to first order kinetics to produce carbon radicals which are quickly peroxidized at ambient oxygen pressures:

$$\text{R-N=N-R} \xrightarrow{k_d} 2\underline{a}\text{R}\cdot + (1-\underline{a})\text{R-R} + \text{N}_2 \qquad (1)$$

$$2\underline{a}\text{R}\cdot + 2\underline{a}\text{O}_2 \xrightarrow{\text{fast}} 2\underline{a}\text{RO}_2\cdot \qquad (2)$$

Absolute numbers of radicals produced were calculated from the equation:

$$n\text{RO}_2\cdot = 2\underline{a}(1-e^{-k_d t})[\text{ABCP}]_o \qquad (3)$$

where t is the reaction time at 50.0°C and $[\text{ABCP}]_o$ is the starting concentration of ABCP. Both $\underline{a}$, the cage escape fraction, and k_d, the first order decomposition rate constant of ABCP are solvent specific and must be established independently (1). Vitamin E loss was followed by reverse phase hplc.

In hexane 1.6 ± 0.1 peroxy radicals are consumed for every E destroyed. E is somewhat less efficient than hindered phenols which terminate 2.0 peroxy radicals per molecule under similar reaction conditions (1,2). One product was detected by hplc at 235 nm; appearance of this material was linearly related to the loss of E and it exhibited a uv maximum at 236 nm and a shoulder at 290 nm. Neither tocopherylquinone (4), spirodimer (5), nor 8-hydroxy tocopherone (3) had hplc retention volumes corresponding to the product, although the latter compound has a similar uv spectrum. That the unknown product is an 8a-substituted tocopherone was suggested by its hydrolytic instability, reacting with acid in pH 4.0 (5:2 CH_3CN-0.1 M acetate) buffer with a half-life of 35 minutes to yield α-tocopheryl quinone.

With this information and the similarity of product formation between BHT and E in hexane (1), we tentatively assign this

ISBN 0-12-592050-4

product as 8a-(2'-(2'-n-butylcarboxy)-propylperoxy)-tocopherone, compound I, a peroxy ketal formed by

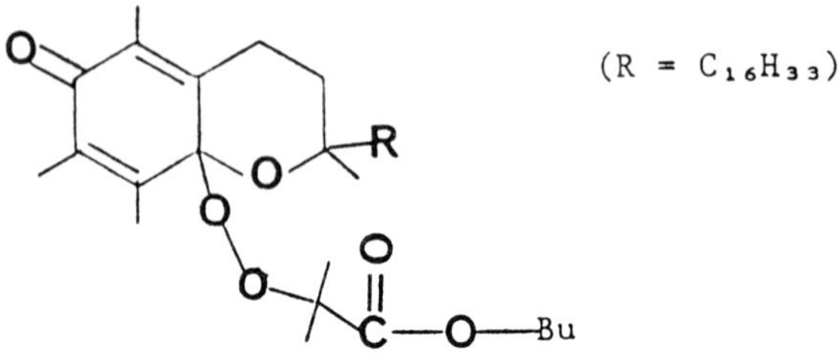

I

the radical coupling of α-chromanoxy and 2-(2-n-butylcarboxy)-propyl peroxy radicals. This structure, if verified, represents a new class of radical products from E, although Clough et al. (6) have identified 8a-hydroperoxy-tocopherone as a photooxygenation product of α-tocopherol. These compounds eventually give tocopheroyquinone from 1 electron radical oxidations.

ABCP (1.6×10^{-4} M bulk conc.) and E ($3\text{-}9 \times 10^{-5}$ M bulk conc.) were reacted in 1% w/v DLPC bilayers suspended in pH 7 imidazole buffer. In the absence of ABCP, E was found to be unstable with time, decaying by a slow first order process ($t_{1/2} = 86 \pm 10$ hr). After correcting for this reaction, the stoichiometry for termination of peroxy radicals by E is 1.3 ± 0.3 peroxy radicals consumed for every E disappearing. Thus E is also less effective than phenolic antioxidants like BHT in scavenging peroxy radicals in lipid bilayers.

REFERENCES

1. Winterle, J. S. and Mill, T., J. Am. Chem. Soc., 102, 6336 (1980).
2. Hammond, G. S., Sen, J. N. and Boozer, C. E., J. Am. Chem. Soc., 77,3244 (1955).
3. Durckheimer, W. and Cohen, L. A., J. Am. Chem. Soc., 86,4388 (1964).
4. Dietzel, J. W. and Empte, W., Hoppe-Seylers Zeit. Physiol. Chemie, 257,173 (1939).
5. Schudel, P., Mayer, H., Metzger, I., Ruegg, R. and Isler, O., Helv. Chim. Acta, 46,636 (1963).
6. Clough, R. L., Yee, B. G. and Foote, C. S., J. Am. Chem. Soc. 101,683 (1979).

CHARACTERISTICS OF DNA MODIFICATION WITH IONIZING RADIATION IN THE PRESENCE OF OXYGEN

Osamu Yamamoto, Research Institute for Nuclear Medicine and Biology, Hiroshima University, Kasumi 1-2-3, Hiroshima 734, Japan

ABNORMAL BASE FOUND IN BASE ANALYSIS FOR THE AGED LIVER DNA AND THE GAMMA-IRRADIATED DNA

Livers of young and adult rats and mice were homogenized in a tissue grinder and then the nuclei were separated by ultracentrifugation. DNA was extracted from the nuclei by the modified Marmur's method (1), and the purified DNA was hydrolyzed with 6 N HCl. The resulting bases were separated by Sephadex G-10 column chromatography while measuring OD_{260}. An abnormal base peak other than those of normal four DNA bases was observed in the adult animal liver DNA analyses, but a very small peak was seen in young ones (Fig. 1). Here a question arose why the area of the abnormal base peak was much larger than the reduced area of the normal base peaks. There should be production of an abnormal base which has a higher UV absorbancy than those of normal bases.

A similar pattern was obtained from the base analysis of calf thymus DNA irradiated with ^{60}Co gamma-rays in O_2-saturated aqueous solution but not in N_2-saturated aqueous solution (1 mg/ml). The area of the abnormal base peak increased with increase in radiation dose (Fig. 2). Therefore, the abnormal base should be produced during oxidation of normal bases.

HIGH UV ABSORBANCY PRODUCTS FROM NUCLEOBASES BY GAMMA-IRRADIATION IN O_2

10^{-3} M aqueous solutions of thymine, cytosine, and adenine containing ^{14}C-radioactivity were irradiated with gamma-rays in O_2 and in N_2, respectively. The solution was eluted better with the use of a starch column (2,3) than with a Sephadex G-10 column in separability. A modified base peak corresponding to the abnormal base peak described in the foregoing was clearly observed on the chromatogram shown with OD_{260} in the O_2-saturated case, but it could not be detected on the chromatogram shown with radioactivity. Fig. 3 shows an example of thymine irradiation. Fig. 4 shows UV absorption spectra of thymine-, cytosine-, and adenine-high UV absorbancy products. OD_{260} of the products are about 10^2 times higher than those of the original bases. This result resolves the question presented in the preceding section.

ISBN 0-12-592050-4

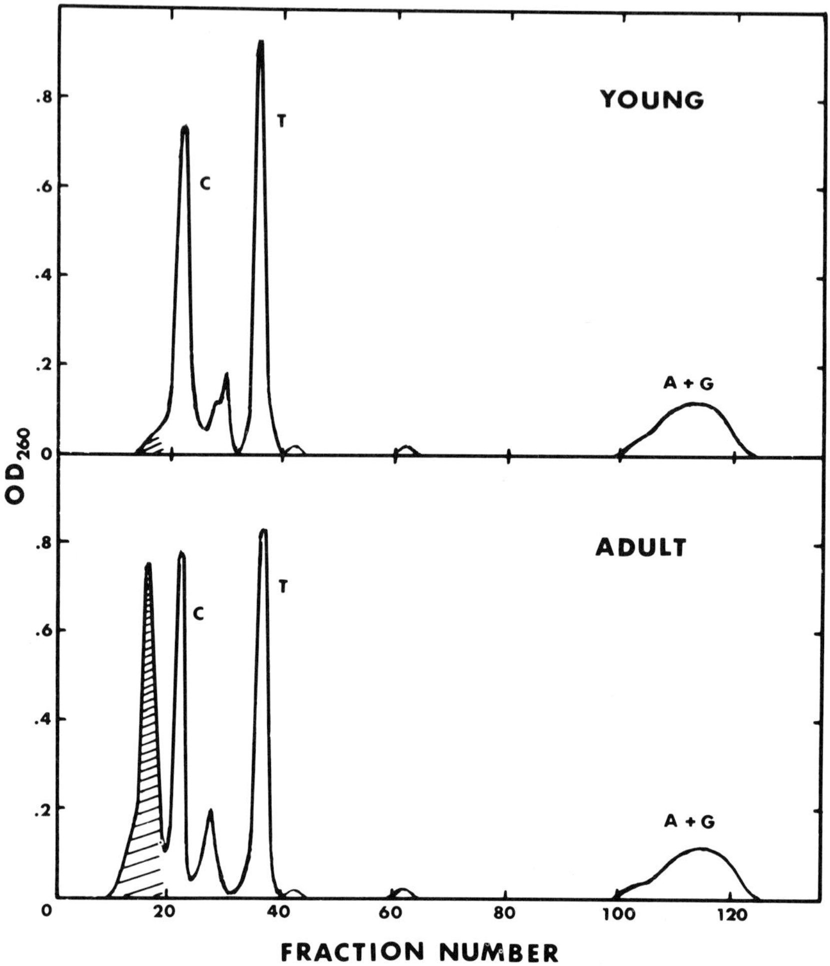

FIGURE 1. Elution pattern of the bases obtained after acid hydrolysis of liver DNA of 3-week and 31-week old rats on Sephadex G-10 column (dimension: 10 mm x 800 mm). The column was eluted with water and the fractions of 2 ml were collected. Shaded peak indicates modified bases.

DNA OXIDATION HYPOTHESIS FOR AGING, MUTAGENESIS, AND CARCINOGENESIS

The appearance of the abnormal base peak in the analysis of the animal liver DNA could be concluded to be due to base oxidation in DNA molecule. In living cells, the essential metabolism is oxidation and there are many kinds of oxygen activating enzymes such as oxidases, dioxygenases and monooxygenases (4). As long as oxygen and oxygen activating enzymes are present in the cells,

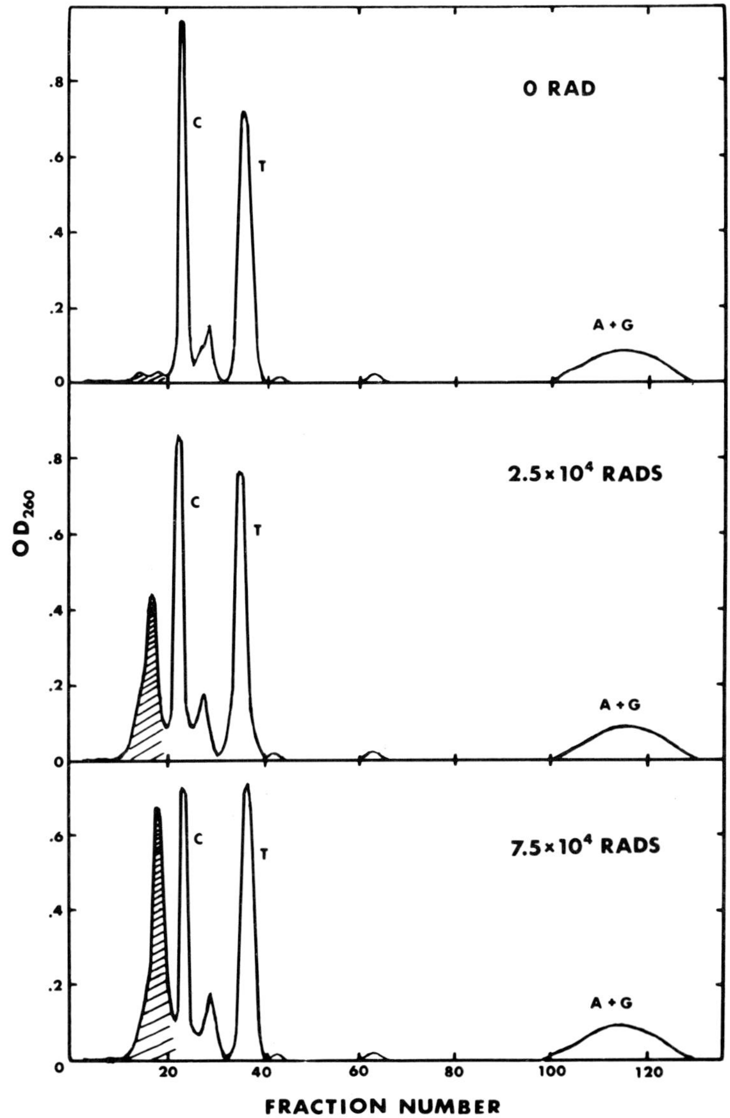

FIGURE 2. Elution pattern of the bases obtained after acid hydrolysis of gamma-irradiated calf thymus DNA (1 mg/ml) on Sephadex G-10 column (dimensions: 10 mm x 800 mm). The column was eluted with water and the fractions of 2 ml were collected. Shaded peak indicates modified bases.

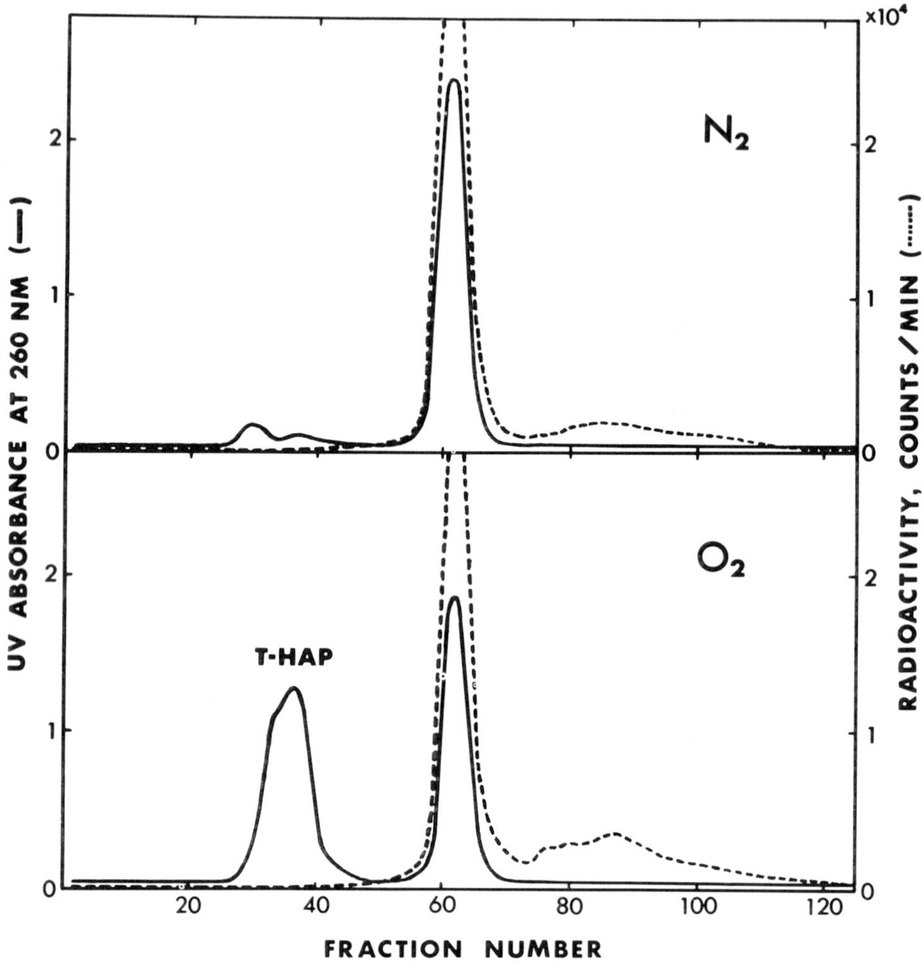

FIGURE 3. Elution patern of gamma-irradiated thymine (10^{-3} M in water) on a starch column (Dimension: 10 mm x 1000 mm). The column was eluted with n-propanol-0.5 N HCl (2 : 1, v/v) and the fractions 2 ml were collected. T-HAP: high absorbancy product derived from thymine.

oxidation of DNA bases is highly possible. Indeed, most of the carcinogens are oxidized to hydroxide form (5-7). Various repair systems for damaged DNA also exist in cells (8). The observation of the abnormal base in the aged animal liver DNA suggests that the base-oxidation damage in DNA is not always reparable. Recently, excision of carcinogen-bound base in DNA was studied by time, but excision was not complete (9). On the basis of these points, DNA oxidation hypothesis for aging, mutagenesis, and carcinogenesis was constructed as follows:

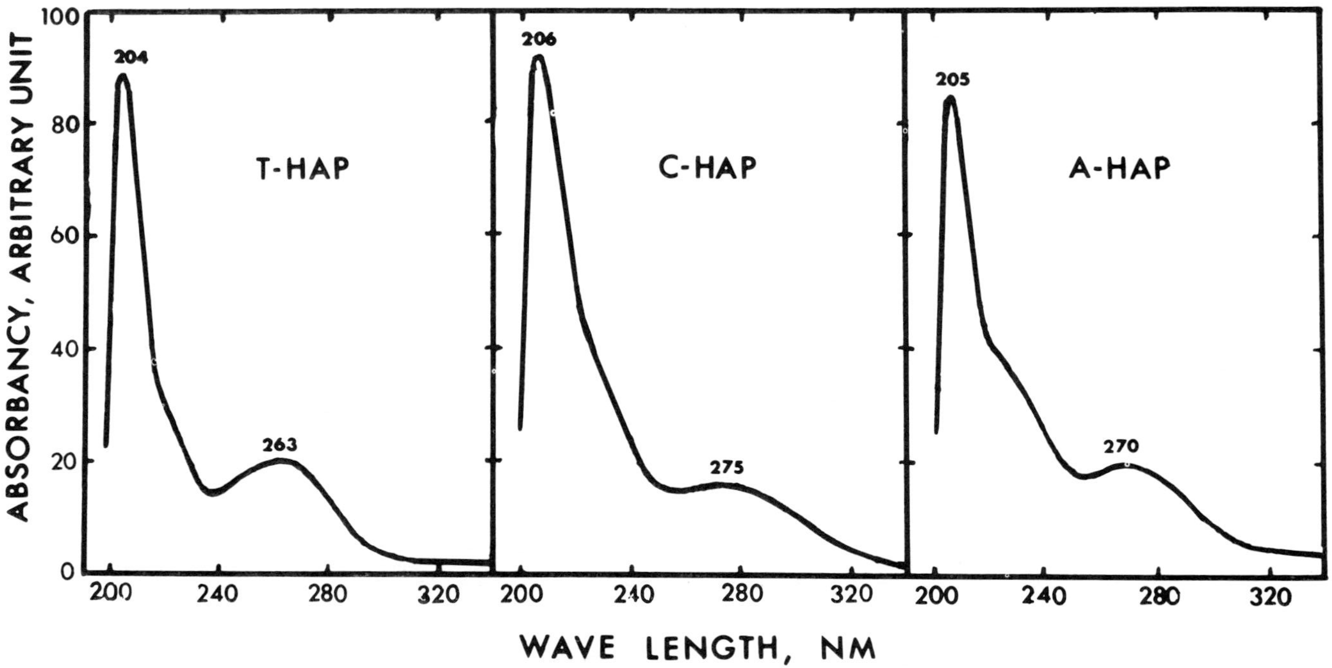

FIGURE 4. UV absorption spectra of the high absorbancy products derived from thymine, cytosine, and adenine (T-HAP, C-HAP, and A-HAP).

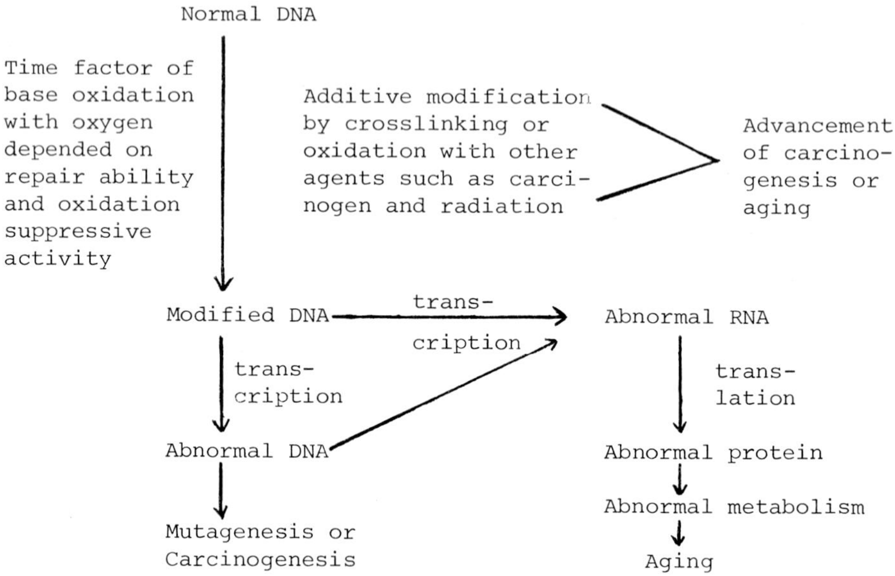

REFERENCES

1. Marmur, J. A., J. Mol. Biol., 3,208 (1961).
2. Daly, M. M. and Mirsky, A. E., J. Biol. Chem., 179,981 (1949).
3. Daly, M. M., Allfrey, V. G. and Mirsky, A. E., J. Gen. Physiol., 33,497 (1950).
4. Karlson, P., "Introduction to Modern Biochemistry" (translated by C. H. Doering), Academic Press, New York, 1975, pp. 224-228.
5. Ts'o, P. and DiPaolo, J., "Chemical Carcinogenesis," Marcel Dekker, New York,(1974).
6. Busch, H., "The Molecular Biology of Cancer," Academic Press, New York,(1974).
7. Magee, P. M., Takayama, S., Sugiura, T. and Matsushima, T., "Fundamental in Cancer Prevention," University of Tokyo Press, Tokyo,(1976).
8. Hanawalt, P. C. and Setlow, R. B., "Molecular Mechanisms for Repair of DNA," Part A and Part B, Plenum, New York,(1975).
9. Shinohara, K. and Cerutti, P. A., Proc. Natl. Acad. Sci. U.S., 74,979 (1977).

OXIDATION OF CO TO CO_2 CATALYZED BY HEART CYTOCHROME c OXIDASE: AN OXYGENASE MECHANISM FOR INTRACELLULAR DETOXIFICATION OF CO

Lawrence J. Young and Winslow S. Caughey, Department of Biochemistry, Colorado State University, Fort Collins, Colorado 80523

More than 90% of the oxygen utilized by oxygen requiring organisms is used to provide energy by being reduced to water at cytochrome c oxidase, a mitochondrial enzyme. Here we report studies which show the enzyme is not only an oxidase but is also a carbon monoxide oxygenase. The conversion of $2CO + O_2$ to $2CO_2$ proceeds enzymatically by a mechanism that is similar to that for the widely studied reaction catalyzed by metals, e.g. platinum. Physiologically, this oxygenase function is a means of detoxifying carbon monoxide within mitochondria.

PREPARATION AND PROPERTIES OF THE OXIDASE/OXYGENASE

The enzyme was obtained in high purity from bovine heart muscle by a method described earlier (1). This preparation contains no other spectrally detectable cytochromes, exhibits high activity, and does not require the addition of detergent to the buffer for solubilization, even up to a 2 mM concentration. Each molecule (mol. wt. ∿200,000) contains two A-type hemes, two coppers with ligands that remain unclear, and about 20% phospholipid, with the remainder accounted for as protein in seven subunits (2). The enzyme is readily interconverted among five redox states. As isolated it is in the fully oxidized state with $2Fe^{3+}$ and $2Cu^{2+}$ metal redox centers; we designate this redox state as CcO(IV). The addition of four electrons yields the fully reduced species CcO(O) with $2Fe^{2+}$ and $2Cu^{+}$. Intermediate states are CcO(I), CcO(II), and CcO(III).

CONVERSION OF CO TO CO_2

Experimental conditions are typically 0.5 ml of a solution of the oxidase 0.8 mM in heme A in 5 mM buffer of pH 7.4 exposed to 86 ml CO and 0.3 ml air at 30°. The reaction is carried out in a cylindrical infrared cell of pathlength 10 cm and volume ca 90 ml. CO_2 production is detected by measuring changes in infrared spectra for the gas phase. Bands characteristic of $^{12}C^{16}O_2$ developed from $^{12}C^{16}O$ and of $^{13}C^{16}O_2$ from $^{13}C^{16}O$. The catalytic production of CO_2 requires molecular O_2.

ISBN 0-12-592050-4

STOICHIOMETRY

Quantitative infrared analysis has shown all CO_2 to come from CO and, under conditions of high CO/O_2 ratios, two CO_2 are formed for one O_2. The enzyme is acting catalytically since many moles of CO and O_2 react per mole of enzyme.

OXYGEN ISOTOPE INCORPORATION INTO CO_2

Use of O_2 with >90% ^{18}O yields $^{12}C^{16}O^{18}O$ in both infrared and mass spectra. The extent of detection of label is limited by the loss of ^{18}O by exchange with ^{16}O of water. These data also support an oxygenase function for the enzyme.

ACTIVATION OF THE OXYGENASE

The fully oxidized resting enzyme as isolated from heart muscle has no detectable CO oxygenase activity. At least partial reduction is necessary for activation. Reduction with cytochrome $\underline{c}^{2+}$, even to the fully reduced species, yields an active CO oxygenase. Also, if the resting enzyme is allowed to stand at room temperature exposed to excess CO, there is a slow formation of partially reduced carbonyl species, first CcO(III)·CO then CcO(II)·CO detected in visible-Soret and infrared spectra. Present evidence supports the conclusion that CcO(II) is much more active than CcO(III).

Two reduced metals appear necessary for an active oxygenase: one metal for CO binding, the other for O_2 binding.

MECHANISMS

The general overall scheme presented in Scheme 1 and as equations 1 through 6 is supported by the above data. A two step process is indicated in which first one CO is oxidized to CO_2, which leaves, and then a second CO reacts with the enzyme to form CO_2. A stoichiometry of one O_2 used per two CO_2 molecules formed was found. Direct oxygen atom transfer is supported by the ^{18}O incorporation experiment. An intermediate with a single oxene-like atom must remain following the first CO oxidation step. Loss of the oxygen atom in the second CO oxidation step leaves the enzyme as unliganded CcO(II) ready for the start of another cycle.

That two metals bind ligands is supported by the apparent requirement for the addition of at least two electrons to CcO(IV) for activation. These two metals that bind ligands must be close to each other, close enough to permit direct oxygen atom transfer from the dioxygen binding site to the bound CO. The most reasonable metals for this are Fe^{2+} and Cu^{+}. The Fe of cytochrome a_3 and the distal Cu are in close proximity (3). We conclude, on the basis of present evidence, that O_2 binds to heme A and is activated for reaction with a CO weakly bound to the immediately

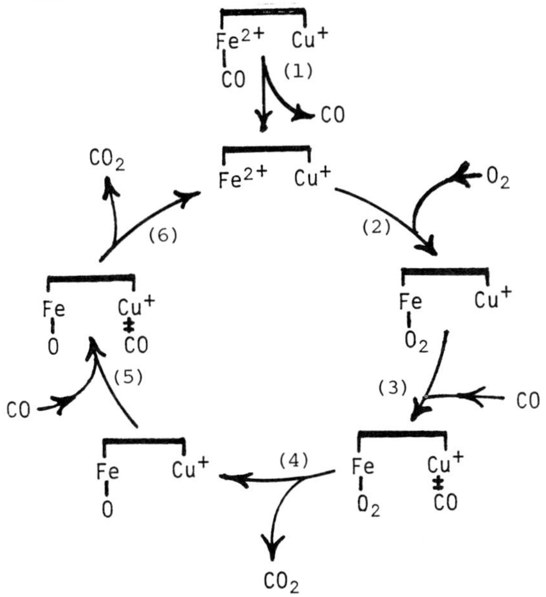

SCHEME 1

(1) $CcO(II)\cdot CO \longrightarrow CcO(II) + CO$

(2) $CcO(II) + O_2 \longrightarrow CcO(II)\cdot O_2$

(3) $CcO(II)\cdot O_2 + CO \longrightarrow CcO(II)\cdot O_2\cdot CO$

(4) $CcO(II)\cdot O_2\cdot CO \longrightarrow CcO(II)\cdot O + CO_2$

(5) $CcO(II)\cdot O + CO \longrightarrow CcO(II)\cdot O\cdot CO$

(6) $CcO(II)\cdot O\cdot CO \longrightarrow CcO(II) + CO_2$

adjacent distal Cu^+. Slowness of the CO to the CO_2 reaction is consistent with the need for replacement of Fe bound CO by O_2 and with no reaction of O_2 with the Cu^+ of CcO(II)·CO where CO is at Fe^{2+}. Respiratory inhibitors other than CO, namely CN^- and N_3^-, inhibit the CO to CO_2 reaction and are shown by infrared spectra to bind readily to CcO(II), presumably at iron.

ANALOGY TO THE LANGMUIR-HINSHELWOOD MECHANISM

The $2CO + O_2 \rightarrow 2CO_2$ reaction on Pt or other metals is a most thoroughly understood catalytic reaction. Consensus supports the Langmuir-Hinshelwood mechanism (4) wherein O_2 binds dissociatively to give chemisorbed oxygen atoms and CO binds only weakly, but must bind. The reaction occurs from contact of the relatively mobile CO with a stationary O atom. In analogy, the CO oxygenase function may similarly involve O_2 binding in a dissociative manner to give chemisorbed oxygen atoms with CO binding only weakly. In turn, the oxygenase, with only a two metal center seemingly possible, demonstrates the lack of a requirement for a multiple metal atom surface.

ANALOGY TO THE CYTOCHROME c OXIDASE MECHANISM

In the $O_2 \rightarrow H_2O$ reaction, if dioxygen binds to cytochrome a_3 dissociatively to give in effect chemisorbed oxygen atoms, which the $CO \rightarrow CO_2$ reaction suggests happens, then the oxidase can be considered to provide a means for breaking the O-O bond without going through a peroxo intermediate (1-3). Each oxene-like oxygen atom may therefore be reduced to the level of oxide (water)—first one O is reduced to leave an intermediate species with a single oxygen atom which is subsequently reduced.

PHYSIOLOGICAL SIGNIFICANCE

CO oxygenase activity is by no means a property unique to the isolated enzyme. Mitochondria and Keilin-Hartree particles also carry out the $2CO + O_2$ reaction. In the 1930s physiologists found evidence for conversion of CO to CO_2 in the whole animal (5). Detoxification of intracellular CO by this reaction may therefore represent an important means of protecting against CO inhibition of mitochondrial oxidative phosphorylation.

ACKNOWLEDGEMENT

This work was supported by USPHS Grant HL-15980.

REFERENCES

1a. Volpe, J. A. and Caughey, W. S., Biochem. Biophys. Res. Commun. 61,502 (1974).
1b. Yoshikawa, S., Choc, M. G., O'Toole, M. C. and Caughey, W. S., J. Biol. Chem. 252,5498 (1977).
2. Caughey, W. S., Wallace, W. J., Volpe, J. A. and Yoshikawa, S., "The Enzymes," Vol. 13 (P. D. Boyer, ed.), Academic Press, Inc., New York,(1976), p. 299.
3. Caughey, W. S., Choc, M. G. and Houtchens, R. A., "Biochemical and Clinical Aspects of Oxygen" (W. S. Caughey, ed.), Academic Press, Inc., New York, 1979, p. 1.
4. Engel, T. and Ertl, G., "Advances in Catalysis," Vol. 28, Academic Press Inc., New York, 1979, p. 1.
5. Fenn, W. O. and Cobb, D. M., Am. J. Physiol., 102,393 (1932).

OXYGEN DEPENDENCE OF ULTRAVIOLET INDUCED CELLULAR DAMAGE IN THE PRIMATE CORNEA

J. A. Zuclich[1] *and W. E. Kurtin*[2]

[1]*Technology Incorporated, Life Sciences Division, San Antonio, Texas 78216*
[2]*Chemistry Department, Trinity University, San Antonio, Texas*

Ultraviolet radiation in excess of ambient levels can result in photokeratitis, a corneal inflammation due to disruption of the epithelial cells. The action spectrum for the cellular disruption covers both the far- and near-UV wavelength ranges, although different mechanisms are operative for the two cases. The near-UV action spectrum for corneal damage in the primate eye (1) is shown in Fig. 1 where it is compared to the action spectrum for UV inactivation of bacterial cells.

Experiments in this laboratory have shown that for UV induced corneal damage, a reciprocity relationship exists between intensity and pulsewidth so that the same total energy is always required to reach the endpoint of an observable corneal clouding (2). This demonstrated that a single-photon photochemical damage mechanism was operative over the range of pulsewidths examined (10^{-8} sec to 10^{4} sec). Further experiments demonstrated an oxygen dependence for the near-UV induced corneal effect (3). Threshold data for the primate eye under oxygen, nitrogen and ambient atmospheres are seen in Table 1. This effect is in contrast to cellular damage induced by far-UV wavelengths which is independent of oxygen concentration.

A final set of experiments examined the cumulative effects (in the cornea) of very long or repeated UV exposures (4). The data plotted in Fig. 2 reveal that the corneal "repair" or recovery proceeds by a simple exponential process with a repair rate, 1/K, of ∿46 hr. With such a slow repair rate, cumulative effects from day to day are likely in certain work environments, and the

TABLE I. Oxygen Dependence of Corneal Epithelial Thresholds

Atmosphere	*Threshold Dose* (J/cm^2)	*95% Confidence Limits* (J/cm^2)
Air (*Control Eyes*)	*82*	*58 - 104*
O_2	*66*	*56 - 77*
N_2	*133*	*116 - 155*

ISBN 0-12-592050-4

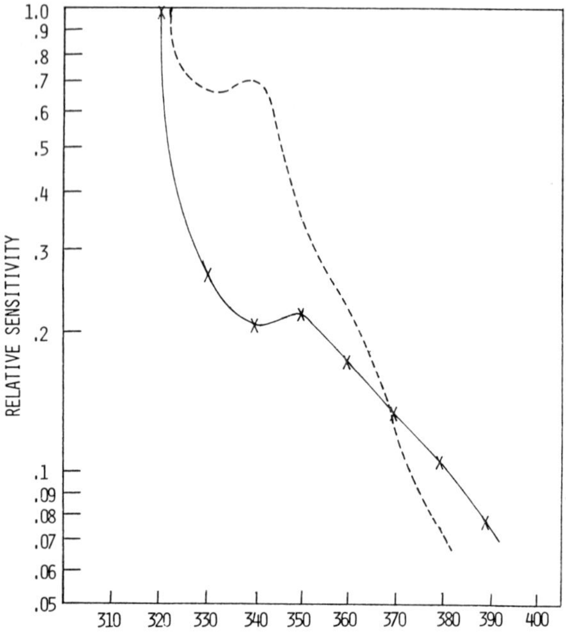

FIGURE 1. Relative sensitivities of primate corneal epithelium and S. typhimurium to absorbed near-UV radiation.

exponential repair model is useful in evaluating the potential hazards associated with repeated UV exposures.

ACKNOWLEDGEMENT

Supported by Contract F33615-77-C-0615, USAF School of Aerospace Medicine. Work carried out at the Laser Effects Branch, USAF School of Aerospace Medicine, Brooks AFB, Texas.

REFERENCES

1. Kurtin, W. E. and Zuclich, J. A., Photochem. Photobiol., 27, 329 (1978).
2. Zuclich, J. A. and Connolly, J. S., Invest. Ophthalmol., 15, 760 (1976).
3. Zuclich, J. A. and Kurtin, W. E., Photochem. Photobiol., 25, 133 (1977).

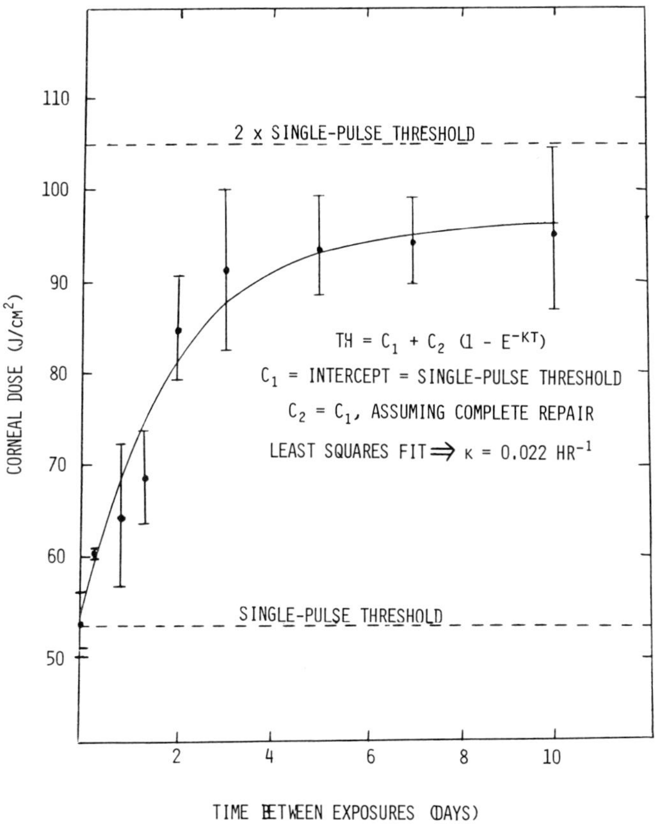

FIGURE 2. Repair of corneal epithelial tissue following subthreshold near-UV rxposures. The total threshold doses for two exposures are plotted as a function of time between the two exposures.

4. Zuclich, J. A., Health Phys., 383,833 (1980).
5. MacKay, T., Eisenstark, A., Webb, R. B. and Brown, M. S., Photochem. Photobiol., 24,337 (1976).

INDEX

B

C

D

E

F

G

H

O

P

R

S

X

Z